Lecture Notes in Computer Science 5575

Arnt-Børre Salberg
Jon Yngve Hardeberg Robert Jenssen (Eds.)

Image Analysis

16th Scandinavian Conference, SCIA 2009
Oslo, Norway, June 15-18, 2009
Proceedings

Springer

Volume Editors

Arnt-Børre Salberg
Norwegian Computing Center
Post Ofice Box 114 Blindern
0314 Oslo, Norway
E-mail: arnt-borre.salberg@nr.no

Jon Yngve Hardeberg
Gjøvik University College
Faculty of Computer Science and Media Technology
Post Office Box 191
2802 Gjøvik, Norway
E-mail: jon.hardeberg@hig.no

Robert Jenssen
University of Tromsø
Department of Physics and Technology
9037 Tromsø, Norway
E-mail: robert.jenssen@uit.no

Library of Congress Control Number: Applied for

CR Subject Classification (1998): I.4, I.5, I.3

LNCS Sublibrary: SL 6 – Image Processing, Computer Vision, Pattern Recognition, and Graphics

ISSN 0302-9743
ISBN 978-3-642-02229-6 Springer Berlin Heidelberg New York

springer.com

Typesetting: Camera-ready by author, data conversion by Scientific Publishing Services, Chennai, India
Printed on acid-free paper SPIN: 12689033 06/3180 5 4 3 2 1 0

Preface

This volume contains the papers presented at the Scandinavian Conference on Image Analysis, SCIA 2009, which was held at the Radisson SAS Scandinavian Hotel, Oslo, Norway, June 15–18.

SCIA 2009 was the 16th in the biennial series of conferences, which has been organized in turn by the Scandinavian countries Sweden, Finland, Denmark and Norway since 1980. The event itself has always attracted participants and author contributions from outside the Scandinavian countries, making it an international conference.

The conference included a full day of tutorials and five keynote talks provided by world-renowned experts. The program covered high-quality scientific contributions within image analysis, human and action analysis, pattern and object recognition, color imaging and quality, medical and biomedical applications, face and head analysis, computer vision, and multispectral color analysis. The papers were carefully selected based on at least two reviews. Among 154 submissions 79 were accepted, leading to an acceptance rate of 51%. Since SCIA was arranged as a single-track event, 30 papers were presented in the oral sessions and 49 papers were presented in the poster sessions. A separate session on multispectral color science was organized in cooperation with the 11th Symposium of Multispectral Color Science (MCS 2009). Since 2009 was proclaimed the "International Year of Astronomy" by the United Nations General Assembly, the conference also contained a session on the topic "Image and Pattern Analysis in Astronomy and Astrophysics."

SCIA has a reputation of having a friendly environment, in addition to high-quality scientific contributions. We focused on maintaining this reputation, by designing a technical and social program that we hope the participants found interesting and inspiring for new research ideas and network extensions.

We thank the authors for submitting their valuable work to SCIA. This is of course of prime importance for the success of the event. However, the organization of a conference also depends critically on a number of volunteers. We are sincerely grateful for the excellent work done by the reviewers and the Program Committee, which ensured that SCIA maintained its reputation of high quality. We thank the keynote and tutorial speakers for their enlightening lectures. And finally, we thank the local Organizing Committee and all the other volunteers that helped us in organizing SCIA 2009.

We hope that all participants had a joyful stay in Oslo, and that SCIA 2009 met its expectations.

June 2009

Arnt-Børre Salberg
Jon Yngve Hardeberg
Robert Jenssen

Organization

SCIA 2009 Oslo June 15-18

SCIA 2009 was organized by NOBIM - The Norwegian Society for Image Processing and Pattern Recognition.

NOBIM Norsk forening for bildebehandling og mønstergjenkjenning

Executive Committee

Conference Chair	Kristin Klepsvik Filtvedt (Kongsberg Defence and Aerospace, Norway)
Program Chairs	Arnt-Børre Salberg (Norwegian Computing Center, Norway)
	Robert Jenssen (University of Tromsø, Norway)
	Jon Yngve Hardeberg (Gjøvik University College, Norway)

Program Committee

Arnt-Børre Salberg (Chair)	Norwegian Computing Center, Norway
Magnus Borga	Linköping University, Sweden
Janne Heikkilä	University of Oulu, Finland
Bjarne Kjær Ersbøll	Technical University of Denmark, Denmark
Robert Jenssen	University of Tromsø, Norway
Kjersti Engan	University of Stavanger, Norway
Anne H.S. Solberg	University of Oslo, Norway
Jon Yngve Hardeberg (Chair MCS 2009 Session)	Gjøvik University College, Norway

Invited Speakers

Rama Chellappa	University of Maryland, USA
Samuel Kaski	Helsinki University of Technology, Finland
Peter Sturm	INRIA Rhône-Alpes, France
Sabine Süsstrunk	Ecole Polytechnique Fédérale de Lausanne, Switzerland
Peter Gallagher	Trinity College Dublin, Ireland

Tutorials

Jan Flusser	The Institute of Information Theory and Automation, Czech Republic
Robert P.W. Duin	Delft University of Technology, The Netherlands

Reviewers

Sven Ole Aase
Fritz Albregtsen
Jostein Amlien
François Anton
Ulf Assarsson
Ivar Austvoll
Adrien Bartoli
Ewert Bengtsson
Asbjørn Berge
Tor Berger
Markus Billeter
Magnus Borga
Camilla Brekke
Marleen de Bruijne
Florent Brunet
Trygve Eftestøl
Line Eikvil
Torbjørn Eltoft
Kjersti Engan
Bjarne Kjær Ersbøll
Ivar Farup
Preben Fihl
Morten Fjeld
Roger Fjørtoft
Pierre Georgel
Ole-Christoffer Granmo
Thor Ole Gulsrud
Trym Haavardsholm
Lars Kai Hansen
Alf Harbitz
Jon Yngve Hardeberg
Markku Hauta-Kasari
Janne Heikkilä
Anders Heyden
Erik Hjelmås
Ragnar Bang Huseby
Francisco Imai
Are C. Jensen
Robert Jenssen
Heikki Kälviäinen
Tom Kavli
Sune Keller
Markus Koskela
Norbert Krüger
Volker Krüger
Jorma Laaksonen
Siri Øyen Larsen
Reiner Lenz
Dawei Liu
Claus Madsen
Filip Malmberg
Brian Mayoh
Thomas Moeslund
Kamal Nasrollahi
Khalid Niazi
Jan H. Nilsen

Ingela Nyström
Ola Olsson
Hans Christian Palm
Jussi Parkkinen
Julien Peyras
Rasmus Paulsen
Kim Pedersen
Tapani Raiko
Juha Röning
Arnt-Børre Salberg
Anne H. S. Solberg
Tapio Seppnen
Erik Sintorn
Ida-Maria Sintorn
Mats Sjöberg
Karl Skretting
Lennart Svensson
Örjan Smedby
Stian Solbø
Jon Sporring
Stina Svensson
Jens T. Thielemann
Øivind Due Trier
Norimichi Tsumura
Ville Viitaniemi
Niclas Wadströmer
Zhirong Yang
Anis Yazidi
Tor Arne Øigård

Sponsoring Institutions

The Research Council of Norway

NOBIM Norsk forening for bildebehandling og mønstergjenkjenning

IAPR

NR Norsk Regnesentral NORWEGIAN COMPUTING CENTER

SINTEF

KONGSBERG

Table of Contents

Human Motion and Action Analysis

Object and Pattern Recognition

Color Imaging and Quality

Multispectral Color Science

Medical and Biomedical Applications

Image and Pattern Analysis in Astrophysics and Astronomy

Face Recognition and Tracking

Computer Vision

Poster Session 1

Poster Session 2

Instant Action Recognition

Thomas Mauthner, Peter M. Roth, and Horst Bischof

Institute for Computer Graphics and Vision
Graz University of Technology
Inffeldgasse 16/II, 8010 Graz, Austria
{mauthner,pmroth,bischof}@icg.tugraz.at

Abstract. In this paper, we present an efficient system for action recognition from very short sequences. For action recognition typically appearance and/or motion information of an action is analyzed using a large number of frames. This is a limitation if very fast actions (e.g., in sport analysis) have to be analyzed. To overcome this limitation, we propose a method that uses a single-frame representation for actions based on appearance and motion information. In particular, we estimate Histograms of Oriented Gradients (HOGs) for the current frame as well as for the corresponding dense flow field. The thus obtained descriptors are efficiently represented by the coefficients of a Non-negative Matrix Factorization (NMF). Actions are classified using an one-vs-all Support Vector Machine. Since the flow can be estimated from two frames, in the evaluation stage only two consecutive frames are required for the action analysis. Both, the optical flow as well as the HOGs, can be computed very efficiently. In the experiments, we compare the proposed approach to state-of-the-art methods and show that it yields competitive results. In addition, we demonstrate action recognition for real-world beach-volleyball sequences.

1 Introduction

Recently, human action recognition has shown to be beneficial for a wide range of applications including scene understanding, visual surveillance, human computer interaction, video retrieval or sports analysis. Hence, there has been a growing interest in developing and improving methods for this rather hard task (see Section 2). In fact, a huge variety of actions at different time scales have to be handled – starting from waving with one hand for a few seconds to complex processes like unloading a lorry. Thus, the definition of an action is highly task dependent and for different actions different methods might be useful.

The objective of this work is to support the analysis of sports videos. Therefore, principle actions represent short time player activities such as running, kicking, jumping, playing, or receiving a ball. Due to the high dynamics in sport actions, we are looking for an action recognition method that can be applied to a minimal number of frames. Optimally, the recognition should be possible using only two frames. Thus, to incorporate the maximum information available per frame we want to use appearance and motion information. The benefit of this representation is motivated and illustrated in Figure 1. In particular, we apply Histograms of Oriented Gradients (HOG) [1] to describe the

A.-B. Salberg, J.Y. Hardeberg, and R. Jenssen (Eds.): SCIA 2009, LNCS 5575, pp. 1–10, 2009.

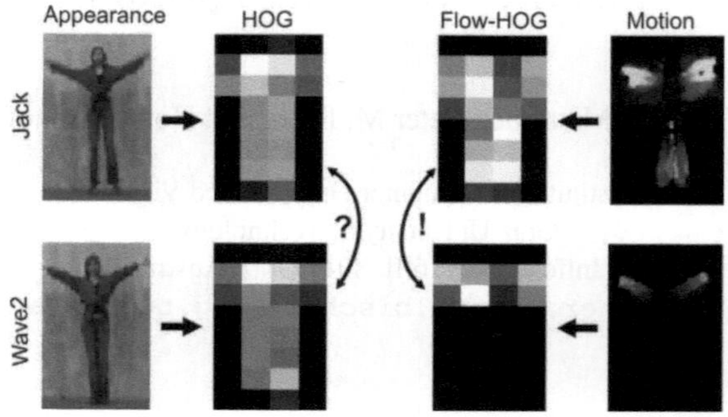

Fig. 1. Overview of the proposed ideas for single frame classification: By using only appearance-based information ambiguities complicate human action recognition (left). By including motion information (optical flow), additional crucial information can be acquired to avoid these confusions (right). Here, the optical flow is visualized using hue to indicate the direction and intensity for the magnitude; the HOG cells are visualized by their accumulated magnitudes.

appearance of a single-frame action. But as can be seen from Figure 1(a) different actions that share one specific mode can not be distinguished if only appearance-based information is available. In contrast, as shown in Figure 1(b), even if the appearance is very similar, additionally analyzing the corresponding motion information can help to discriminate between two actions; and vice versa. In particular, for that purpose we compute a dense optical-flow field, such that for frame t the appearance and the flow information is computed from frame $t-1$ and frame t only. Then the optical flow is represented similarly to the appearance features by (signed) orientation histograms.

Since the thus obtained HOG descriptors for both, appearance and motion, can be described by a small number of additive modes, similar to [2,3], we apply Non-negative Matrix Factorization (NMF) [4] to estimate a robust and compact representation. Finally, the motion and the appearance features (i.e., their NMF coefficients) are concatenated to one vector and linear one-vs-all SVMs are applied to learn a discriminative model. To compare our method with state-of-the-art approaches, we evaluated it on a standard action recognition database. In addition, we show results on beach-volleyball videos, where we use very different data for training and testing to emphasize the applicability of our method.

The remainder of this paper is organized as follows. Section 2 gives an overview of related work and explains the differences to the proposed approach. In Section 3 our new action recognition system is introduced in detail. Experimental results for a typical benchmark dataset and a challenging real-world task are shown in Section 4. Finally, conclusion and outlook are given in Section 5.

2 Related Work

In the past, many researchers have tackled the problem of human action recognition. Especially for recognizing actions performed by a single person various methods exist that yield very good classification results. Many classification methods are based on the analysis of a temporal window around a specific frame. Bobick and Davis [5] used motion history images to describe an action by accumulating human silhouettes over time. Blank et al. [6] created 3-dimensional space-time shapes to describe actions. Weinland and Boyer [7] used a set of discriminative static key-pose exemplars without any spatial order. Thurau and Hlaváč [2] used pose-primitives based on HOGs and represented actions as histograms of such pose-primitives. Even though these approaches show that shape or silhouettes over time are well discriminating features for action recognition, the use of temporal windows or even of a whole sequence implies that actions are recognized with a specific delay.

Having the spatio-temporal information, the use of optical flow is an obvious extension. Efros et al. [8] introduced a motion descriptor based on spatio-temporal optical flow measurements. An interest point detector in spatio-temporal domain based on the idea of Harris point detector was proposed by Laptev and Lindeberg [9]. They described the detected volumes with several methods such as histograms of gradients or optical flow as well as PCA projections. Dollár et al. [10] proposed an interest point detector searching in space-time volumes for regions with sudden or periodic changes. In addition, optical flow was used as a descriptor for the 3D region of interest. Niebles et al. [11] used a constellation model of bag-of-features containing spatial and spatio-temporal [10] interest points. Moreover, single-frame classification methods were proposed. For instance, Mikolajczyk and Uemura [12] trained a vocabulary forest on feature points and their associated motion vectors.

Recent results in the cognitive sciences have led to biologically inspired vision systems for action recognition. Jhuang et al. [13] proposed an approach using a hierarchy of spatio-temporal features with increasing complexity. Input data is processed by units sensitive to motion-directions and the responses are pooled locally and fed into a higher level. But only recognition results for whole sequences have been reported, where the required computational effort is approximately 2 minutes for a sequence consisting of 50 frames. Inspired by [13] a more sophisticated (and thus more efficient approach) was proposed by Schindler and van Gool [14]. They additionally use appearance information, but both, appearance and motion, are processed in similar pipelines using scale and orientation filters. In both pipelines the filter responses are max-pooled and compared to templates. The final action classification is done by using multiple one-vs-all SVMs.

The approaches most similar to our work are [2] and [14]. Similar to [2] we use HOG descriptors and NMF to represent the appearance. But in contrast to [2], we do not not need to model the background, which makes our approach more general. Instead, similar to [14], we incorporate motion information to increase the robustness and apply one-vs-all SVMs for classification. But in contrast to [14], in our approach the computation of feature vectors is less complex and thus more efficient. Due to a GPU-based flow estimation and an efficient data structure for HOGs our system is very efficient and runs in real-time. Moreover, since we can estimate the motion information using a pair of subsequent frames, we require only two frames to analyze an action.

3 Instant Action Recognition System

In this section, we introduce our action recognition system, which is schematically illustrated in Figure 2. In particular, we combine appearance and motion information to

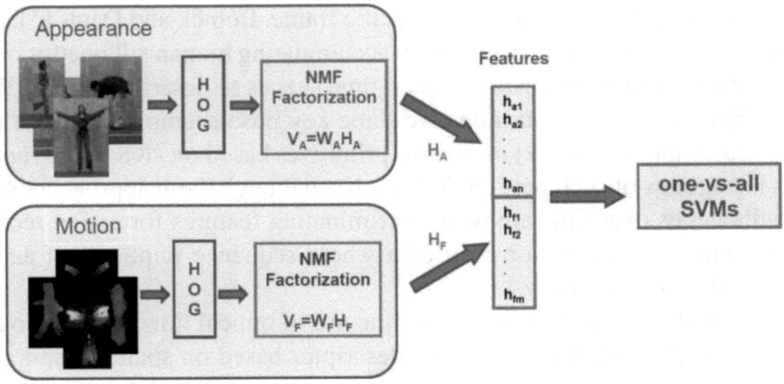

Fig. 2. Overview of the proposed approach: Two representations for appearance and flow are estimated in parallel. Both are described by HOGs and represented by NMF coefficients, which are concatenated to a single feature vector. These vectors are then learned using one-vs-all SVMs.

enable a frame-wise action analysis. To represent the appearance, we use histograms of oriented gradients (HOGs) [1]. HOG descriptors are locally normalized gradient histograms, which have shown their capability for human detection and can also be estimated efficiently by using integral histograms [15]. To estimate the motion information, a dense optical flow field is computed between consecutive frames using an efficient GPU-based implementation [16]. The optical flow information can also be described using orientation histograms without dismissing the information about the gradient direction. Following the ideas presented in [2] and [17], we reduce the dimensionality of the extracted histograms by applying sub-space methods. As stated in [3,2] articulated poses, as they appear during human actions, can be well described using NMF basis vectors. We extend this ideas by building a set of NMF basis vectors for appearance and the optical flow in parallel. Hence the human action is described in every frame by NMF coefficient vectors for appearance and flow, respectively. The final classification on per-frame basis is realized by using multiple SVMs trained on the concatenations of the appearance and flow coefficient vectors of the training samples.

3.1 Appearance Features

Given an image $\mathbf{I_t} \in \mathbb{R}^{\mathbf{m}\times\mathbf{n}}$ at time step t. To compute the gradient components $g_x(x,y)$ and $g_y(x,y)$, for every position (x,y) the image is filtered by 1-dimensional masks $[-1,0,1]$ in x and y direction [1]. The magnitude $m(x,y)$ and the signed orientation $\Theta_S(x,y)$ are computed by

$$m(x,y) = \sqrt{g_x(x,y)^2 + g_y(x,y)^2} \tag{1}$$

$$\Theta_S(x,y) = \tan^{-1}\left(g_y(x,y)/g_x(x,y)\right) \,. \tag{2}$$

To make the orientation insensitive to the order of intensity changes, only unsigned orientations Θ_U are used for appearance:

$$\Theta_U(x,y) = \begin{cases} \Theta_S(x,y) + \pi & \theta_S(x,y) < 0 \\ \Theta_S(x,y) & \text{otherwise} \end{cases} \quad . \tag{3}$$

To create the HOG descriptor, the patch is divided into non-overlapping 10×10 cells. For each cell, the orientations are quantized into 9 bins and weighted by their magnitude. Groups of 2×2 cells are combined in so called overlapping blocks and the histogram of each cell is normalized using the L2-norm of the block. The final descriptor is built by concatenation of all normalized blocks. The parameters for cell-size, block-size, and the number of bins may be different in literature.

3.2 Motion Features

In addition to appearance we use optical flow. Thus, for frame t the appearance features are computed from frame t, and the flow features are extracted from frames t and $t-1$. In particular, to estimate the dense optical flow field, we apply the method proposed in [16], which is publicly available: *OFLib*[1]. In fact, the GPU-based implementation allows a real-time computation of motion features.

Given $\mathbf{I_t}, \mathbf{I_{t-1}} \in \mathbb{R}^{\mathbf{m \times n}}$, the optical flow describes the shift from frame $t-1$ to t with the disparity $\mathbf{D_t} \in \mathbb{R}^{\mathbf{m \times n}}$, where $d_x(x,y)$ and $d_y(x,y)$ denote the disparity components in x and y direction at location (x,y). Similar to the appearance features, orientation and magnitude are computed and represented with HOG descriptors. In contrast to appearance, we use signed orientation Θ_S to capture different motion directions for same poses. The orientation is quantized into 8 bins only, while we keep the same cell/block combination as described above.

3.3 NMF

If the underlying data can be described by distinctive local information (such as the HOGs of appearance and flow) the representation is typically very sparse, which allows to efficiently represent the data by Non-negative Matrix Factorization (NMF) [4]. In contrast to other sub-space methods, NMF does not allow negative entries, neither in the basis nor in the encoding. Formally, NMF can be described as follows. Given a non-negative matrix (i.e., a matrix containing vectorized images) $\mathbf{V} \in \mathbf{R}^{m \times n}$, the goal of NMF is to find non-negative factors $\mathbf{W} \in \mathbf{R}^{n \times r}$ and $\mathbf{H} \in \mathbf{R}^{r \times m}$ that approximate the original data:

$$\mathbf{V} \approx \mathbf{WH} \,. \tag{4}$$

Since there is no closed-form solution, both matrices, **W** and **H**, have to be estimated in an iterative way. Therefore, we consider the optimization problem

$$\begin{aligned} &\min ||\mathbf{V} - \mathbf{WH}||^2 \\ &\text{s.t. } \mathbf{W}, \mathbf{H} > 0 \,, \end{aligned} \tag{5}$$

[1] http://gpu4vision.icg.tugraz.at/

where $||.||^2$ denotes the squared Euclidean Distance. The optimization problem (5) can be iteratively solved by the following update rules:

$$\mathbf{H}_{a,j} \leftarrow \mathbf{H}_{a,j} \frac{\left[\mathbf{W}^T\mathbf{V}\right]_{a,j}}{\left[\mathbf{W}^T\mathbf{W}\mathbf{H}\right]_{a,j}} \quad and \quad \mathbf{W}_{i,a} \leftarrow \mathbf{W}_{i,a} \frac{\left[\mathbf{V}\mathbf{H}^T\right]_{i,a}}{\left[\mathbf{W}\mathbf{H}\mathbf{H}^T\right]_{i,a}}, \tag{6}$$

where $[.]$ denote that the multiplications and divisions are performed element by element.

3.4 Classification via SVM

For the final classification the NMF-coefficients obtained for appearance and motion are concatenated to a final feature vector. As we will show in Section 4, less than 100 basis vectors are sufficient for our tasks. Therefore, compared to [14] the dimension of the feature vector is rather small, which drastically reduces the computational costs. Finally, a linear one-vs-all SVM is trained for each action class using *LIBSVM*[2]. In particular, no weighting of appearance or motion cue was performed. Thus, the only tuning parameter is the number of basis vectors for each cue.

4 Experimental Results

To show the benefits of the proposed approach, we split the experiments into two main parts. First, we evaluated our approach on a publicly available benchmark dataset (i.e., *Weizmann Human Action Dataset* [6]). Second, we demonstrate the method for a real-world application (i.e., action recognition for beach-volleyball).

4.1 Weizmann Human Action Dataset

The *Weizmann Human Action Dataset* [6] is a publicly available[3] dataset, that contains 90 low resolution videos (180 × 144) of nine subjects performing ten different actions: running, jumping in place, jumping forward, bending, waving with one hand, jumping jack, jumping sideways, jumping on one leg, walking, and waving with two hands. Illustrative examples for each of these actions are shown in Figure 3. Similar to, e.g., [2,14] all experiments on this dataset were carried out using a leave-one-out strategy (i.e., we used 8 individuals for training and evaluated the learned model for the missing one.

Fig. 3. Examples from the Weizmann human action dataset

[2] http://www.csie.ntu.edu.tw/ cjlin/libsvm/

[3] http://www.wisdom.weizmann.ac.il/~vision/SpaceTimeActions.html

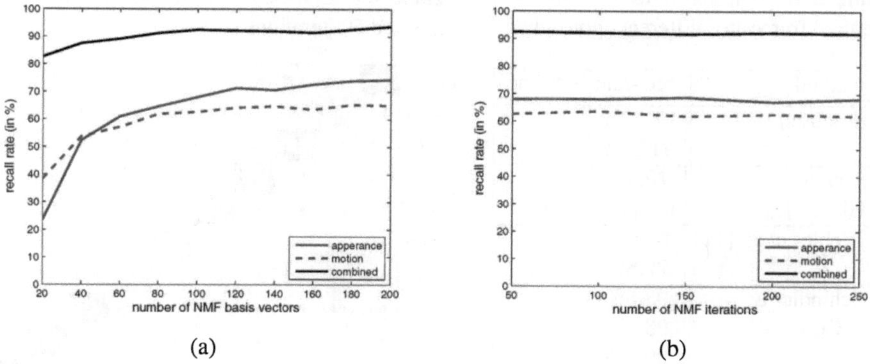

Fig. 4. Importance of NMF parameters for action recognition performance: recognition rate depending (a) on the number of basis vectors using 100 iterations and (b) on the number of NMF iterations for 200 basis vectors

Figure 4 shows the benefits of the proposed approach. It can be seen that neither the appearance-based nor the motion-based representation solve the task satisfactorily. But if both representations are combined, we get a significant improvement of the recognition performance! To analyzed the importance of the NMF parameters used for estimating the feature vectors that are learned by SVMs, we ran the leave-one-out experiments varying the NMF parameters, i.e., the number of basis vectors and the number of iterations. The number of basis vectors was varied in the range from 20 to 200 and the number of iterations from 50 to 250. The other parameter was kept fixed, respectively. It can be seen from Figure 4(a) that increasing the number of basis vectors to a level of 80-100 steadily increases the recognition performance, but that further increasing this parameter has no significant effect. Thus using 80-100 basis vectors is sufficient for our task. In contrast, it can be seen from Figure 4(b) that the number of iterations has no big influence on the performance. In fact, a representation that was estimated using 50 iterations yields the same results as one that was estimated using 250 iterations!

In the following, we present the results for the leave-one-out experiment for each action in Table 1. Due to the results discussed above, we show the results obtained by using 80 NMF coefficients obtained by 50 iterations. It can be seen that with exception of "run" and "skip", which on a short frame basis are very similar in both, appearance and motion, the recognition rate is always near 90% or higher (see confusion matrix in Table 3).

Estimating the overall recognition rate we get a correct classification rate of 91.28%. In fact, this average is highly influenced by the results on the "run" and "skip" dataset. Without these classes, the overall performance would be significantly higher than 90%. By averaging the recognition results in a temporal window (i.e., we used a window

Table 1. Recognition rate for the leave-one-out experiment for the different actions

action	bend	run	side	wave2	wave1	skip	walk	pjump	jump	jack
rec.-rate	95.79	78.03	99.73	96.74	95.67	75.56	94.20	95.48	88.50	93.10

Table 2. Recognition rates and number of required frames for different approaches

method	rec.-rate	# frames
proposed	91.28%	2
	94.25%	6
Thurau & Hlaváč [2]	70.4%	1
	94.40%	all
Niebles et al. [11]	55.0%	1
	72.8%	all
Schindler & v. Gool [14]	93.5%	2
	96.6%	3
	99.6%	10
Blank et al. [6]	99.6%	all
Jhuang et al. [13]	98.9%	all
Ali et al. [18]	89.7	all

Table 3. Confusion matrix for 80 basis vectors and 50 iterations

	bend	run	side	wave2	wave1	walk	skip	pjump	jump	jack
bend	0.96	0.00	0.00	0.00	0.00	0.00	0.00	0.00	0.01	0.02
run	0.00	0.78	0.01	0.00	0.00	0.04	0.17	0.00	0.00	0.00
side	0.00	0.00	1.00	0.00	0.00	0.00	0.00	0.00	0.00	0.00
wave2	0.00	0.00	0.00	0.97	0.03	0.00	0.00	0.00	0.00	0.00
wave1	0.01	0.00	0.00	0.03	0.96	0.00	0.00	0.00	0.00	0.00
walk	0.00	0.03	0.02	0.00	0.00	0.94	0.00	0.00	0.00	0.00
skip	0.00	0.18	0.00	0.00	0.00	0.01	0.76	0.00	0.05	0.00
pjump	0.01	0.00	0.02	0.00	0.00	0.00	0.00	0.95	0.00	0.02
jump	0.00	0.00	0.00	0.00	0.00	0.08	0.02	0.00	0.88	0.00
jack	0.01	0.00	0.01	0.00	0.00	0.00	0.00	0.05	0.00	0.93

size of 6 frames) we can boost the recognition results to 94.25%. This improvement is mainly reached by incorporating more temporal information. Further extending the temporal window size has not shown additional significant improvements. In the following, we compare this result with state-of-the-art methods considering the reported recognition rate and the number of frames that were used to calculate the response. The results are summarized in Table 2.

It can be seen that most of the reported approaches that use longer sequences to analyze the actions clearly outperform the proposed approach. But among those methods using only one or two frames our results are competitive.

4.2 Beach-Volleyball

In this experiment we show that the proposed approach can be applied in practice to analyze events in beach-volleyball. For that purpose, we generated indoor training sequences showing different actions including *digging*, *running*, *overhead passing*, and *running sideways*. Illustrative frames used for training are shown in Figure 5. From these sequences we learned the different actions as described in Section 3.

The thus obtained models are then applied for action analysis in outdoor beach-volleyball sequences. Please note the considerable difference between the training and the testing scenes. From the analyzed patch the required features (appearance NMF-HOGs and flow NMF-HOGs) are extracted and tested if they are consistent with one

Fig. 5. Volleyball – training set: (a) *digging*, (b) *run*, (c) *overhead passing*, and (d) *run sideway*

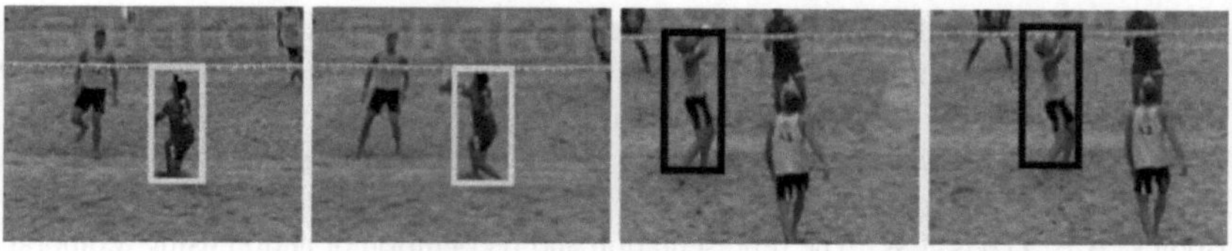

Fig. 6. Volleyball – test set: (left) *action digging* (yellow bounding box) and (right) action *overhead passing* (blue bounding box) are detected correctly

of the previously learned SVM models. Illustrative examples are depicted in Figure 6, where both tested actions, *digging* (yellow bounding box in (a)) and *overhead passing* (blue bounding box in (b)) are detected correctly in the shown sequences!

5 Conclusion

We presented an efficient action recognition system based on a single-frame representation combining appearance-based and motion-based (optical flow) description of the data. Since in the evaluation stage only two consecutive frames are required (for estimating the flow), the methods can also be applied for very short sequences. In particular, we propose to use HOG descriptors for both, appearance and motion. The thus obtained feature vectors are represented by NMF coefficients and are concatenated to learn action models using SVMs. Since we apply a GPU-based implementation for optical flow and an efficient estimation of the HOGs, the method is highly applicable for tasks where quick and short actions (e.g., in sports analysis) have to be analyzed. The experiments showed that even using this short-time analysis competitive results can be obtained on a standard benchmark dataset. In addition, we demonstrated that the proposed method can be applied for a real-world task such as action detection in volleyball. Future work will mainly concern the training stage by considering a more sophisticated learning method (e.g., an weighted SVM) and improving the NMF implementation. In fact extensions such as sparsity constraints or convex formulation (e.g.,[19,20]) have shown to be beneficial in practice.

Acknowledgment

This work was supported be the Austrian Science Fund (FWF P18600), by the FFG project AUTOVISTA (813395) under the FIT-IT programme, and by the Austrian Joint Research Project Cognitive Vision under projects S9103-N04 and S9104-N04.

References

1. Dalal, N., Triggs, B.: Histograms of oriented gradients for human detection. In: Proc. IEEE Conf. on Computer Vision and Pattern Recognition (2005)
2. Thurau, C., Hlaváč, V.: Pose primitive based human action recognition in videos or still images. In: Proc. IEEE Conf. on Computer Vision and Pattern Recognition (2008)

3. Agarwal, A., Triggs, B.: A local basis representation for estimating human pose from cluttered images. In: Narayanan, P.J., Nayar, S.K., Shum, H.-Y. (eds.) ACCV 2006. LNCS, vol. 3851, pp. 50–59. Springer, Heidelberg (2006)
4. Lee, D.D., Seung, H.S.: Learning the parts of objects by non-negative matrix factorization. Nature 401, 788–791 (1999)
5. Bobick, A.F., Davis, J.W.: The representation and recognition of action using temporal templates. IEEE Trans. on Pattern Analysis and Machine Intelligence 23(3), 257–267 (2001)
6. Blank, M., Gorelick, L., Shechtman, E., Irani, M., Basri, R.: Actions as space-time shapes. In: Proc. IEEE Intern. Conf. on Computer Vision, pp. 1395–1402 (2005)
7. Weinland, D., Boyer, E.: Action recognition using exemplar-based embedding. In: Proc. IEEE Conf. on Computer Vision and Pattern Recognition (2008)
8. Efros, A.A., Berg, A.C., Mori, G., Malik, J.: Recognizing action at a distance. In: Proc. European Conf. on Computer Vision (2003)
9. Laptev, I., Lindeberg, T.: Local descriptors for spatio-temporal recognition. In: Proc. IEEE Intern. Conf. on Computer Vision (2003)
10. Dollár, P., Rabaud, V., Cottrell, G., Belongie, S.: Behavior recognition via sparse spatio-temporal features. In: Proc. IEEE Workshop on PETS, pp. 65–72 (2005)
11. Niebles, J.C., Fei-Fei, L.: A hierarchical model of shape and appearance for human action classification. In: Proc. IEEE Conf. on Computer Vision and Pattern Recognition (2007)
12. Mikolajczyk, K., Uemura, H.: Action recognition with motion-appearance vocabulary forest. In: Proc. IEEE Conf. on Computer Vision and Pattern Recognition (2008)
13. Jhuang, H., Serre, T., Wolf, L., Poggio, T.: A biologically inspired system for action recognition. In: Proc. IEEE Intern. Conf. on Computer Vision (2007)
14. Schindler, K., van Gool, L.: Action snippets: How many frames does human action recognition require? In: Proc. IEEE Conf. on Computer Vision and Pattern Recognition (2008)
15. Porikli, F.: Integral histogram: A fast way to extract histograms in cartesian spaces. In: Proc. IEEE Conf. on Computer Vision and Pattern Recognition, vol. 1, pp. 829–836 (2005)
16. Zach, C., Pock, T., Bischof, H.: A duality based approach for realtime tv-l1 optical flow. In: Hamprecht, F.A., Schnörr, C., Jähne, B. (eds.) DAGM 2007. LNCS, vol. 4713, pp. 214–223. Springer, Heidelberg (2007)
17. Lu, W.L., Little, J.J.: Tracking and recognizing actions at a distance. In: CVBASE, Workshop at ECCV (2006)
18. Ali, S., Basharat, A., Shah, M.: Chaotic invariants for human action recognition. In: Proc. IEEE Intern. Conf. on Computer Vision (2007)
19. Hoyer, P.O.: Non-negative matrix factorization with sparseness constraints. Journal of Machine Learning Research 5, 1457–1469 (2004)
20. Heiler, M., Schnörr, C.: Learning non-negative sparse image codes by convex programming. In: Proc. IEEE Intern. Conf. on Computer Vision, vol. II, pp. 1667–1674 (2005)

Using Hierarchical Models for 3D Human Body-Part Tracking

Leonid Raskin, Michael Rudzsky, and Ehud Rivlin

Computer Science Department, Technion,
Technion City, Haifa, Israel, 32000
{raskinl,rudzsky,ehudr}@cs.technion.ac.il

Abstract. Human body pose estimation and tracking is a challenging task mainly because of the high dimensionality of the human body model. In this paper we introduce a Hierarchical Annealing Particle Filter (H-APF) algorithm for 3D articulated human body-part tracking. The method exploits Hierarchical Human Body Model (HHBM) in order to perform accurate body pose estimation. The method applies nonlinear dimensionality reduction combined with the dynamic motion model and the hierarchical body model. The dynamic motion model allows to make a better pose prediction, while the hierarchical model of the human body expresses conditional dependencies between the body parts and also allows us to capture properties of separate parts. The improved annealing approach is used for the propagation between different body models and sequential frames. The algorithm was checked on HumanEvaI and HumanEvaII datasets, as well as on other videos and proved to be effective and robust and was shown to be capable of performing an accurate and robust tracking. The comparison to other methods and the error calculations are provided.

1 Introduction

Human body pose estimation and tracking is a challenging task for several reasons. The large variety of poses and high dimensionality of the human 3D model complicates the examination of the entire subject and makes it harder to detect each body part separately. However, the poses can be presented in a low dimensional space using the dimensionality reduction techniques, such as Gaussian Process Latent Model (GPLVM) [1], locally linear embedding (LLE) [2], etc. The human motions can be described as curves in this space. This space can be obtained by learning different motion types [3]. However, such a reduction allows to detect poses similar to those, that were used for the learning process. In this paper we introduce a Hierarchical Annealing Particle Filter (H-APF) tracker, which exploits Hierarchical Human Body Model (HHBM) in order to perform accurate body part estimation. In this approach we apply a nonlinear dimensionality reduction using the Hierarchical Gaussian Process Latent Model (HGPLVM) [1] and the annealing particle filter [4]. Hierarchical model of the human body expresses conditional dependencies between the body parts, but

A.-B. Salberg, J.Y. Hardeberg, and R. Jenssen (Eds.): SCIA 2009, LNCS 5575, pp. 11–20, 2009.

also allows us to capture properties of separate parts. Human body model state consists of two independent parts: one containing information about 3D location and orientation of the body and the other describing the articulation of the body. The articulation is presented as hierarchy of body parts. Each node in the hierarchy represent a set of body parts called partial pose. The method uses previously observed poses from different motion types to generate mapping functions from the low dimensional latent spaces to the data spaces, that correspond to the partial poses. The tracking algorithm consists of two stages. Firstly, the particles are generated in the latent space and are transformed to the data space using the learned mapping functions. Secondly, rotation and translation parameters are added to obtain valid poses. The likelihood function is calculated in order to evaluate how well these poses match the image. The resulting tracker estimates the locations in the latent spaces that represents poses with the highest likelihood. We show that our tracking algorithm is robust and provides good results even for the low frame rate videos. An additional advantage of the tracking algorithm is the ability to recover after temporal loss of the target.

2 Related Works

One of the commonly used technique for estimation the statistics of a random variable is the importance sampling. The estimation is based on samples of this random variable generated from a distribution, called the proposal distribution, which is easy to sample from. However, the approximation of this distribution for high dimensional spaces is a very computationally inefficient and hard task. Often a weighting function can be constructed according to the likelihood function, as it is in the CONDENSATION algorithm of Isard and Blake [5], which provides a good approximation of the proposal distribution and also is relatively easy to calculate. This method uses multiple predictions, obtained by drawing samples of pose and location prior and then propagating them using the dynamic model, which are refined by comparing them with the local image data, calculating the likelihood [5]. The prior is typically quite diffused (because motion can be fast) but the likelihood function may be very peaky, containing multiple local maxima which are hard to account for in detail [6]. In such cases the algorithm usually detects several local maxima instead of choosing the global one. Annealed particle filter [4] or local searches are the ways to attack this difficulty. The main idea is to use a set of weighting functions instead of using a single one. While a single weighting function may contain several local maxima, the weighting functions in the set should be smoothed versions of it, and therefore contain a single maximum point, which can be detected using the regular annealed particle filter. The alternative method is to apply a strong model of dynamics [7]. The drawback of the annealed particle filter tracker is that the high dimensionality of the state space requires generation of a large amount of particles. In addition, the distribution variances, learned for the particle generation, are motion specific. This practically means that the tracker is applicable for the motion, that is used for the training. Finally, the APF is not robust and suffers from the lack of ability to detect a correct pose, once a target is lost (i.e. the body pose wrongly estimated).

In order to improve the trackers robustness, ability to recover from temporal target loss and in order to improve the computational effectiveness many researchers apply dimensionality reduction algorithm on the configuration space. There are several possible strategies for reducing the dimensionality. Firstly it is possible to restrict the range of movement of the subject [8]. But, due to the restricting assumptions, the resulting trackers are not capable of tracking general human poses. Another approach is to learn low-dimensional latent variable models [9]. However, methods like Isomap [10] and locally linear embedding (LLE) [2] do not provide a mapping between the latent space and the data space, and, therefore Urtasun et al. [11] proposed to use a form of probabilistic dimensionality reduction by GPDM [12,13] to formulate the tracking as a nonlinear least-squares optimization problem. Andriluka et al. [14] use HGPLVM [1] to model prior on possible articulations and temporal coherency within a *walking* cycle. Raskin et al. [15] introduced Gaussian Process Annealed Particle Filter (GPAPF). According to this method, a set of poses is used in order to create a low dimensional latent space. This latent space is generated using Gaussian Process Dynamic Model (GPDM) for a nonlinear dimensionality reduction of the space of previously observed poses from different motion types, such as *walking, running, punching and kicking*. While for many actions it is intuitive that a motion can be represented in a low dimensional manifold, this is not the case for a set of different motions. Taking the *walking* motion as an example. One can notice that for this motion type the locations of the ankles are highly correlated with the location of the other body parts. Therefore, it seems natural to be able to represent the poses from this action in a low dimensional space. However, when several different actions are involved, the possibility of a dimensionality reduction, especially a usage of 2D and 3D spaces, is less intuitive.

This paper is organized as follows. Section 3 describes the tracking algorithm. Section 4 presents the experimental results for both tracking of different data sets and motion types. Finally, section 5 provides the conclusion and suggests the possible directions for the future research.

3 Hierarchical Annealing Particle Filter

The drawback of GPAPF algorithm is that a latent space is not capable of describing all possible poses. The space reduction must capture any dependencies between the poses of the different body parts. For example, if there is any connection between the parameters that describe the pose of the left hand and those, describing the right hand, then we can easily reduce the dimensionality of these parameters. However, if a person will perform a new movement, which differ from the learned ones, then the new poses will be represented less accurately by the latent space. Therefore, we suggest using a hierarchical model for the tracking. Instead of learning a single latent space that describes the whole body pose we use HGPLVM [1] to learn a hierarchy of the latent spaces. This approach allows us to exploit the dependencies between the poses of different body parts while accurately estimating of the pose of each part separately.

The commonly used human body model Γ consists of 2 statistically independent parts $\Gamma = \{\Lambda, \Omega\}$. The first part $\Lambda \subseteq \mathbb{R}^6$ describes the body 3D location: the rotation and the translation. The second part $\Omega \subseteq \mathbb{R}^{25}$ describes the actual pose, which is represented by the angles between different body parts (see. [16] for more details about the human body model). Suppose the hierarchy consists of H layers, where the highest layer (layer 1) represents the full body pose and the lowest layer (layer H) represents the separate body parts. Each hierarchy layer h consists of L_h latent spaces. Each node l in hierarchy layer h represents a partial body pose $\Omega_{h,l}$. Specifically, the root node describes the whole body pose; the nodes in the next hierarchy layer describe the pose of the legs, arms and the upper body (including the head); finally, the nodes in the last hierarchy layer describe each body part separately. Let us define $\Re(\Omega_{h,l})$ as the set of the coordinates of Ω that are used in $\Omega_{h,l}$, where $\Omega_{h,l}$ is a subset of some $\Omega_{h-1,k}$ in the higher layer of the hierarchy. Such k is denoted as $\tilde{l}$. For each $\Omega_{h,l}$ the algorithm constructs a latent spaces $\Theta_{h,l}$ and the mapping function $\wp^{(h,l)} : \Theta_{h,l} \mapsto \Omega_{h,l}$ that maps this latent space to the partial pose space $\Omega_{h,l}$. Let us also define $\theta_{h,l}$ as the latent coordinate in the l-th latent space in the h-th hierarchy layer and $\omega_{h,l}$ is the partial data vector that corresponds to $\theta_{h,l}$. Consequently, applying the definition of $\wp^{(h,l)}$ we have that $\omega_{h,l} = \wp^{(h,l)}(\theta_{h,l})$. In addition for $\forall i$ we define $\Im(i)$ to be a pair $< h, l >$, where h is the lowest hierarchy layer and l is the latent space in this layer, such that $i \in \Re(\Omega_{h,l})$. In other words, $\Im(i)$ represent the lowest latent space in the hierarchy for which the i-th coordinate of Ω has been used in $\Omega_{h,l}$. Finally, $\lambda_{h,l,n}$, $\omega_{h,l,n}$ and $\theta_{h,l,n}$ are the location, pose vector and latent coordinates on the frame n and hierarchy layer h on the latent space l.

Now we present a Hierarchical Annealing Particle Filter (H-APF). A H-APF run is performed at each frame using image observations y_n. Following the notations used in [17] for the frame n and hierarchy layer h on the latent space l the state of the tracker is represented by a set of weighted particles $S^{\pi}_{h,l,n} = \{(s^{(0)}_{h,l,n}, \pi^{(0)}_{h,l,n}), ..., (s^{(N)}_{h,l,n}, \pi^{(N)}_{h,l,n})\}$. The un-weighted set of particles is denoted as $S_{h,l,n} = \{s^{(0)}_{h,l,n}, ..., s^{(N)}_{h,l,n}\}$. The state that is used contains translation, rotation values, latent coordinates and the full data space vectors: $s^{(i)}_{h,l,n} = \{\lambda^{(i)}_{h,l,n}; \theta^{(i)}_{h,l,n}; \omega^{(i)}_{h,l,n}\}$. The tracking algorithm consists of 2 stages. The first stage is the generation of new particles using the latent space. In the second stage the corresponding mapping function is applied that transforms latent coordinates to the data space. After the transformation, the translation and rotation parameters are added and the 31-dimensional vectors are constructed. These vectors represent a valid pose, which are projected to the cameras in order to estimate the likelihood.

Each H-APF run has the following stages:

Step 1. For every frame hierarchical annealing algorithm run is started at layer $h = 1$. Each latent space in each layer is initialized by a set of un-weighted particles $S_{h,l,n}$.

$$S_{1,1,n} = \left\{ \lambda^{(i)}_{1,1,n}; \theta^{(i)}_{1,1,n}; \omega^{(i)}_{1,1,n} \right\}_{i=1}^{N_p} \tag{1}$$

Step 2. Calculate the weights of each particle:

$$\begin{aligned}
&\pi^{(i)}_{h,l,n} \propto w^m\left(y_n, s_{h,l,n}\right) = \\
&k \frac{w^m\left(y_n, \lambda^{(i)}_{h,l,n}, \omega^{(i)}_{h,l,n}\right) p\left(\lambda^{(i)}_{h,l,n}, \theta^{(i)}_{h,l,n} | \lambda^{(i)}_{h,l,n}, \theta^{(i)}_{h,\bar{l},n}\right)}{q\left(\lambda^{(i)}_{h,l,n}, \theta^{(i)}_{h,l,n} | \lambda^{(i)}_{h,l,n}, \theta^{(i)}_{h,\bar{l},n}, y_n\right)} = \\
&k \frac{w^m\left(y_n, \Gamma^{(i)}_{h,l,n}\right) p\left(\lambda^{(i)}_{h,l,n}, \theta^{(i)}_{h,l,n} | \lambda^{(i)}_{h,l,n}, \theta^{(i)}_{h,\bar{l},n}\right)}{q\left(\lambda^{(i)}_{h,l,n}, \theta^{(i)}_{h,l,n} | \lambda^{(i)}_{h,l,n}, \theta^{(i)}_{h,\bar{l},n}, y_n\right)}
\end{aligned} \tag{2}$$

where $w^m\left(y_n, \Gamma\right)$ is the weighting function suggested by Deutscher and Reid [17] and k is a normalization factor so that $\sum_{i=1}^{N_p} \pi^{(i)}_n = 1$. The weighted set, that is constructed, will be used to draw particles for the next layer.

Step 3. N particles are drawn randomly with replacements and with a probability equal to their weight $\pi^{(i)}_{h,l,n}$. For every latent space l in the hierarchy level $h+1$ the particle $s^{(j)}_{h+1,l,n}$ is produces using the jth chosen particle $s^{(j)}_{h,\hat{l},n}$ ($\hat{l}$ is the index of the parent node in the hierarchy tree):

$$\lambda^{(j)}_{h+1,l,n} = \lambda^{(j)}_{h,\hat{l},n} + B_{\lambda_{h+1}} \tag{3}$$

$$\theta^{(j)}_{h+1,l,n} = \phi(\theta^{(j)}_{h,\hat{l},n}) + B_{\theta_{h,\hat{l}}} \tag{4}$$

In order to construct a full pose vector $\omega^{(j)}_{h+1,l,n}$ is initialized with the $\omega^{(j)}_{h,\hat{l},n}$

$$\omega^{(j)}_{h+1,l,n} = \omega^{(j)}_{h,\hat{l},n} \tag{5}$$

and then updated on the coordinates defined by $\Omega_{h+1,l}$ using the new $\theta^{(j)}_{h+1,l,n}$

$$(\omega^{(j)}_{h+1,l,n})|_{\Omega_{h+1,l}} = \wp^{h+1,l}\left(\theta^{(j)}_{h+1,l,n}\right) \tag{6}$$

(The notation $a|_B$ stands for the coordinates of vector $a \in A$ defined by the subspace $B \subseteq A$.) The idea is to use a pose that was estimated using the higher hierarchy layer, with small variations in the coordinates described by the $\Omega_{h+1,l}$ subspace.

Finally, the new particle for the latent space l in the hierarchy level $h+1$ is:

$$s^{(j)}_{h+1,l,n} = \{\lambda^{(j)}_{h+1,l,n}; \omega^{(j)}_{h+1,l,n}; \theta^{(j)}_{h+1,l,n}\} \tag{7}$$

The B_{λ_h} and $B_{\theta_{h,l}}$ are multivariate Gaussian random variables with covariances and Σ_{λ_h} and $\Sigma_{\theta_{h,l}}$ correspondingly and mean 0.

Step 4. The sets $S_{h+1,l,n}$ have now been produced which can be used to initialize the layer *h+1*. The process is repeated until we arrive to the H-th layer.

Step 5. The jth chosen particle $s^{(j)}_{H,l,n}$ in every latent space l in the lowest hierarchy level and their ancestors (the particles in the higher layers that used to produce $s^{(j)}_{H,l,n}$ are used to produce $s^{(j)}_{1,1,n+1}$ un-weighted particle set for the next observation:

$$\begin{aligned} &\lambda^{(j)}_{1,1,n+1} = \tfrac{1}{L_H}\textstyle\sum_{l=1}^{L_H} \lambda^{(j)}_{H,l,n} \\ &\forall i\ \omega^{(j)}(i) = \widehat{\omega}^{(j)}_{\Im(i),n} \\ &\theta_{1,1,n+1} = \wp^{-1}_{1,1}\left(\omega^{(j)}_{1,1,n+1}\right) \end{aligned} \tag{8}$$

Here $\widehat{\omega}^{(j)}_{h,k,n}$ denotes an ancestor of $\omega^{(j)}_{H,l,n}$ in h-th layer of the hierarchy.

Step 6. The optimal configuration can be calculated using the following method:

$$\begin{aligned} &\lambda^{(opt)}_n = \tfrac{1}{L_H}\textstyle\sum_{l=1}^{L_H}\sum_{j=1}^{N} \lambda^{(j)}_{H,l,n}\pi^{(j)}_{h,l,n} \\ &\forall i\ \omega^{(j)}(i) = \widehat{\omega}^{(j)}_{\Im(i),n} \\ &\omega^{(opt)}_n = \textstyle\sum_{j=1}^{N} \omega^{(j)}\pi^{(j)} \end{aligned} \tag{9}$$

where, similar to stage 2, $\pi^{(j)} = w^m\left(y_n, \langle\lambda^{(opt)}_n, \omega^{(j)}\rangle\right)$ is the normalized weighting function so that $\sum_{i=1}^{N_p} \pi^{(i)} = 1$.

4 Results

We have tested H-APF tracker using the HumanEvaI and HumanEvaII datasets [18]. The sequences contain different activities, such as *walking*, *boxing*, *jogging* etc., which were captured by several synchronized and mutually calibrated cameras. The sequences were captured using the MoCap system that provides the correct 3D locations of the body joints, such as shoulders and knees. This information is used for evaluation of the results and comparison to other tracking

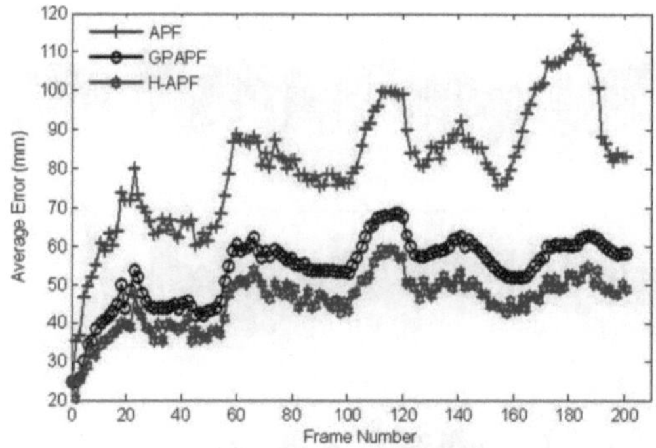

Fig. 1. The errors of the APF tracker (green crosses), GPAPF tracker (blue circles) and H-APF tracker (red stars) for a walking sequence captured at 15 fps

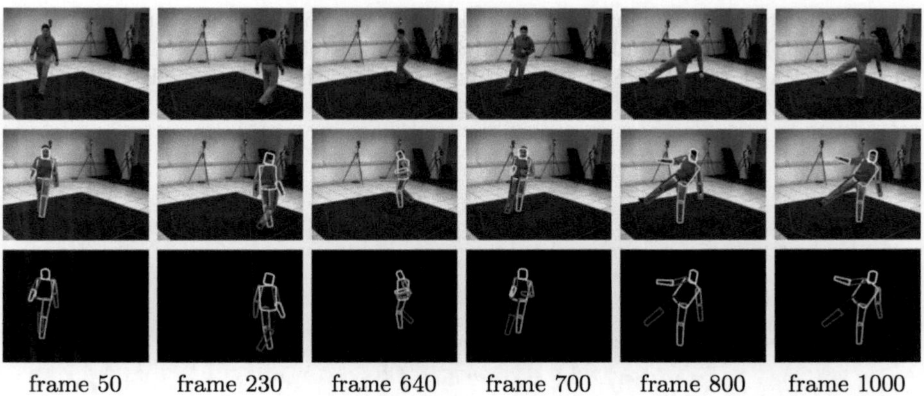

Fig. 2. Tracking results of H-APF tracker. Sample frames from the combo1 sequence from HumanEvalII(S2) dataset.

algorithms. The error is calculated, based on comparison of the tracker's output to the ground truth, using average distance in millimeters between 3-D joint locations [16].

The first sequence that we have used contain a person, walking in a circle. The video was captured at 60 fps frame rate. We have compared the results produced by APF, GPAPF and H-APF trackers. For each algorithm we have used 5 layers, with 100 particles in each. Fig. 1 shows the error graphs, produced by APF (green crosses), the GPAPF (blue circles) and the H-APF (red stars) trackers. We have also tried to compare our results to the results of CONDENSATION algorithm. However, the results of that algorithm were either very poor or very large number of particles needed to be used, which made this algorithm computationally not effective. Therefore we do not provide the results of this comparison.

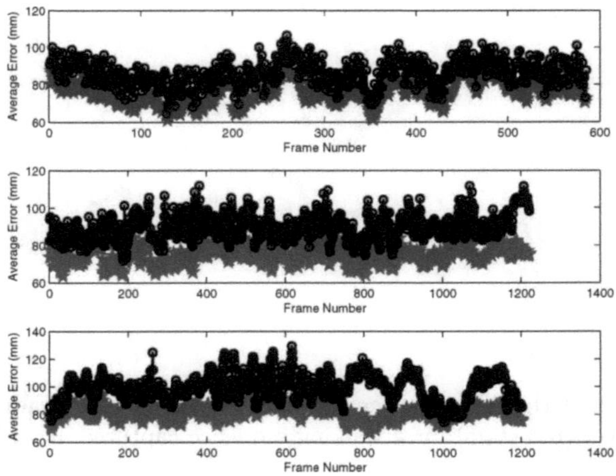

Fig. 3. The errors for HumanEvalI(S1, walking1, frames 6-590)(top), HumanEvalII(S2, frames 1-1202)(middle) and HumanEvalII(S4, frames 2-1258)(bottom). The errors produced by GPAPF tracker are marked by blue circles and the error of the H-APF tracker are marked by red stars.

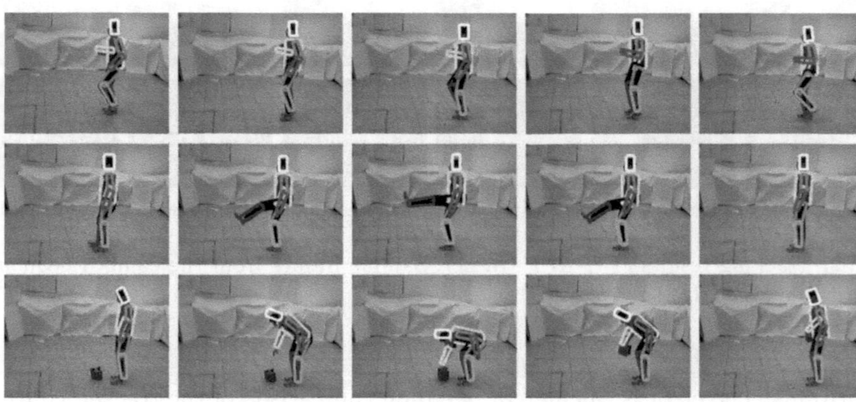

Fig. 4. Tracking results of H-APF tracker. Sample frames from the running, kicking and lifting an object sequences.

Next we trained HGPLVM with several different motion types. We used this latent space in order to track the body parts on the videos from the HumanEvaI and HumanEvaII datasets. Fig. 2 shows the result of the tracking of the HumanEvaII(S2) dataset, which combines 3 different behaviors: walking, jogging and balancing and Fig. 3 presents the errors for HumanEvaI(S1, walking1, frames 6-590)(top), HumanEvaII(S2, frames 1-1202)(middle) and HumanEvaII(S4, frames 2-1258)(bottom). Finally, Fig. 4 shows the results from the *running, kicking and lifting an object* sequences.

5 Conclusion and Future Work

In this paper we have introduced an approach that uses HGPLVM to improve the ability of the annealed particle filter tracker to track the object even in a high dimensional space. The usage of hierarchy allows better detect body part position and thus perform more accurate tracking.

An interesting problem is to perform tracking of the interactions between multiple actors. The main problem is constructing a latent space. While a single persons poses can be described using a low dimensional space it may not be the case for multiple people. The other problem here is that in this case there is high possibility of occlusion. Furthermore, while for a single person each body part can be seen from at least one camera that is not the case for the crowded scenes.

References

1. Lawrence, N.D., Moore, A.J.: Hierarchical gaussian process latent variable models. In: Proc. International Conference on Machine Learning (ICML) (2007)
2. Roweis, S.T., Saul, L.K.: Nonlinear dimensionality reduction by locally linear embedding. Science 290, 2323–2326 (2000)
3. Elgammal, A.M., Lee, C.: Inferring 3D body pose from silhouettes using activity mani-fold learning. In: Proc. Computer Vision and Pattern Recognition (CVPR), vol. 2, pp. 681–688 (2004)
4. Deutscher, J., Blake, A., Reid, I.: Articulated body motion capture by annealed particle filtering. In: Proc. Computer Vision and Pattern Recognition (CVPR), pp. 2126–2133 (2000)
5. Isard, M., Blake, A.: Condensation - conditional density propagation for visual tracking. International Journal of Computer Vision (IJCV) 29(1), 5–28 (1998)
6. Sidenbladh, H., Black, M.J., Fleet, D.: Stochastic tracking of 3D human figures using 2D image motion. In: Vernon, D. (ed.) ECCV 2000. LNCS, vol. 1843, pp. 702–718. Springer, Heidelberg (2000)
7. Mikolajczyk, K., Schmid, K., Zisserman, A.: Human detection based on a probabilistic assembly of robust part detectors. In: Pajdla, T., Matas, J. (eds.) ECCV 2004. LNCS, vol. 3021, pp. 69–82. Springer, Heidelberg (2004)
8. Rohr, K.: Human movement analysis based on explicit motion models. Motion-Based Recognition 8, 171–198 (1997)
9. Wang, Q., Xu, G., Ai, H.: Learning object intrinsic structure for robust visual tracking. In: Proc. Computer Vision and Pattern Recognition (CVPR), vol. 2, pp. 227–233 (2003)
10. Tenenbaum, J.B., de Silva, V., Langford, J.C.: A global geometric framework for nonlinear dimensionality reduction. Science 290, 2319–2323 (2000)
11. Urtasun, R., Fleet, D.J., Fua, P.: 3D people tracking with gaussian process dynamical models. In: Proc. Computer Vision and Pattern Recognition (CVPR), vol. 1, pp. 238–245 (2006)
12. Lawrence, N.D.: Gaussian process latent variable models for visualization of high dimensional data. In: Advances in Neural Information Processing Systems (NIPS), vol. 16, pp. 329–336 (2004)
13. Wang, J., Fleet, D.J., Hetzmann, A.: Gaussian process dynamical models. In: Information Processing Systems (NIPS), pp. 1441–1448 (2005)

14. Andriluka, M., Roth, S., Schiele, B.: People-tracking-by-detection and people-detection-by-tracking. In: Proc. Computer Vision and Pattern Recognition (CVPR), vol. 1, pp. 1–8 (2008)
15. Raskin, L., Rudzsky, M., Rivlin, E.: Dimensionality reduction for articulated body tracking. In: Proc. The True Vision Capture, Transmission and Display of 3D Video (3DTV) (2007)
16. Balan, A., Sigal, L., Black, M.: A quantitative evaluation of video-based 3D person tracking. In: IEEE Workshop on Visual Surveillance and Performance Evaluation of Tracking and Surveillance (VS-PETS), pp. 349–356 (2005)
17. Deutscher, J., Reid, I.: Articulated body motion capture by stochastic search. International Journal of Computer Vision (IJCV) 61(2), 185–205 (2004)
18. Sigal, L., Black, M.J.: Measure locally, reason globally: Occlusion-sensitive articulated pose estimation. In: Proc. Computer Vision and Pattern Recognition (CVPR), vol. 2, pp. 2041–2048 (2006)

Analyzing Gait Using a Time-of-Flight Camera

Rasmus R. Jensen, Rasmus R. Paulsen, and Rasmus Larsen

Informatics and Mathematical Modelling, Technical University of Denmark
Richard Petersens Plads, Building 321, DK-2800 Kgs. Lyngby, Denmark
{raje,rrp,rl}@imm.dtu.dk
www.imm.dtu.dk

Abstract. An algorithm is created, which performs human gait analysis using spatial data and amplitude images from a *Time-of-flight* camera. For each frame in a sequence the camera supplies cartesian coordinates in space for every pixel. By using an articulated model the subject pose is estimated in the depth map in each frame. The pose estimation is based on likelihood, contrast in the amplitude image, smoothness and a shape prior used to solve a Markov random field. Based on the pose estimates, and the prior that movement is locally smooth, a sequential model is created, and a gait analysis is done on this model. The output data are: Speed, Cadence (steps per minute), Step length, Stride length (stride being two consecutive steps also known as a gait cycle), and Range of motion (angles of joints). The created system produces good output data of the described output parameters and requires no user interaction.

Keywords: Time-of-flight camera, Markov random fields, gait analysis, computer vision.

1 Introduction

Recognizing and analyzing human movement in computer vision can be used for different purposes such as biomechanics, biometrics and motion capture. In biomechanics it helps us understand how the human body functions, and if something is not right it can be used to correct this.

Top athletes have used high speed cameras to analyze their movement either to improve on technique or to help recover from an injury. Using several high speed cameras, bluescreens and marker suits an advanced model of movement can be created, which can then be analyzed. This optimal setup is however complex and expensive, a luxury which is not widely available. Several approaches aim to simplify tracking of movement.

Using several cameras but without bluescreens nor markers [11] creates a visual hull in space from silhouettes by solving a spacial Markov random field using graph cuts and then fitting a model to this hull.

Based on a large database [9] is able to find a pose estimate in sublinear time relative to the database size. This algorithm uses subsets of features to find the nearest match in parameter space.

A.-B. Salberg, J.Y. Hardeberg, and R. Jenssen (Eds.): SCIA 2009, LNCS 5575, pp. 21–30, 2009.

An earlier study uses the *Time-of-flight* (*TOF*) camera to estimate pose using key feature points in combination with a an articulated model to solve problems with ambiguous feature detection, self penetration and joint constraints [13].

To minimize expenses and time spent on multi camera setups, bluescreens, markersuits, initializing algorithms, annotating etc. this article aims to deliver a simple alternative that analyzes gait.

In this paper we propose an adaptation of the *Posecut* algorithm for fitting articulated human models to grayscale image sequences by Torr et al. [5] to fitting such models to *TOF* depth camera image sequences. In particular, we will investigate the use of this *TOF* data adapted *Posecut* algorithm to quantitative gait analysis.

Using this approach with no restrictions on neither background nor clothing a system is presented that can deliver a gait analysis with a simple setup and no user interaction. The project object is to broaden the range of patients benefiting from an algorithmic gait analysis.

2 Introduction to the Algorithm Finding the Pose

This section will give a brief overview of the algorithm used to solve the problem of finding the pose of the subject. To do a gait analysis the pose has to be estimated in a sequence of frames. This is done using the adapted *Posecut* algorithm on the depth and amplitude stream provided by a *TOF* camera [2] (Fig. 1 shows a depth map with amplitude coloring). The algorithm uses 4 terms to define an energy minimization problem and find the pose of the subject as well as segmenting between subject and background:

Likelihood Term: This term is based on statistics of the background. It is based on a probability function of a given pixel being labeled background.

Smoothness Prior: This is a prior based on the general assumption that data is smooth. Neighbouring pixels are expected to have the same label with higher probability than having different labels.

Contrast Term: Neighbouring pixels with different labels are expected to have values in the amplitude map that differs from one another. If the values are very similar but the labels different, this is penalized by this term.

Shape Prior: Trying to find the pose of a human, a human shape is used as a prior.

2.1 Random Fields

A frame in the sequence is considered to be a random field. A random field consists of a set of discrete random variables $\{X_1, X_2, \ldots, X_n\}$ defined on the index set I. In this set each variable X_i takes a value x_i from the label set $L = \{L_1, L_2, \ldots, L_k\}$ presenting all possible labels. All values of x_i, $\forall i \in I$ are represented by the vector $\mathbf{x}$ which is the configuration of the random field and takes values from the label set L^n. In the following the labeling is a binary problem, where $L = \{\text{subject}, \text{background}\}$.

Fig. 1. Depth image with amplitude coloring of the scene. The image is rotated to emphasize the spatial properties.

A neighbourhood system to X_i is defined as $N = \{N_i | i \in I\}$ for which it holds that $i \notin N_i$ and $i \in N_j \Leftrightarrow j \in N_i$. A random field is said to be a Markov field, if it satisfies the positivity property:

$$P(\mathbf{x}) > 0 \qquad \forall \mathbf{x} \in L^n \tag{1}$$

And the Markovian Property:

$$P(x_i | \{x_j : j \in I - \{i\}\}) = P(x_i | \{x_j : j \in N_i\}) \tag{2}$$

Or in other words any configuration of $\mathbf{x}$ has higher probability than 0 and the probability of x_i given the index set $I - \{i\}$ is the same as the probability given the neighbourhood of i.

2.2 The Likelihood Function

The likelihood energy is based on the negative log likelihood and for the background distribution defined as:

$$\Phi(\mathbf{D} | x_i = \text{background}) = -\log p(\mathbf{D} | x_i) \tag{3}$$

Using the Gibbs measure without the normalization constant this energy becomes:

$$\Phi(\mathbf{D} | x_i = \text{background}) = \frac{(\mathbf{D} - \mu_{\text{background,i}})^2}{\sigma^2_{\text{background,i}}} \tag{4}$$

With no distribution defined for pixels belonging to the subject, the subject likelihood function is set to the mean of the background likelihood function. To estimate a stable background a variety of methods are available. A well known method, models each pixel as a mixture of Gaussians and is also able to update these estimates on the fly [10]. In our method a simpler approach proved sufficient. The background estimation is done by computing the median value at each pixel over a number of frames.

2.3 The Smoothness Prior

This term states that generally neighbours have the same label with higher probability, or in other words that data are not totally random. The generalized Potts model where $j \in N_i$ is given by:

$$\psi(x_i, x_j) = \begin{cases} K_{ij} & x_i \neq x_j \\ 0 & x_i = x_j \end{cases} \tag{5}$$

This term penalizes neighbours having different labels. In the case of segmenting between background and subject, the problem is binary and referred to as the Ising model [4]. The parameter K_{ij} determines the smoothness in the resulting labeling.

2.4 The Contrast Term

In some areas such as where the feet touches the ground, the subject and background differs very little in distance. Therefore a contrast term is added, which uses the amplitude image (grayscale) provided by the *TOF* camera. It is expected that two adjacent pixels with the same label have similar intensities, which implies that adjacent pixels with different labels have different intensities. By decreasing the cost of neighbouring pixels with different labels exponentially with an increase in difference in intensity, this term favours neighbouring pixels with similar intensities to have the same label. This function is defined as:

$$\gamma(i,j) = \lambda \exp\left(\frac{-g^2(i,j)}{2\sigma^2_{background,i}}\right) \tag{6}$$

Where $g^2(i,j)$ is the gradient in the amplitude map and approximated using convolution with gradient filters. The parameter λ controls the cost of the contrast term, and the contribution to the energy minimization problem becomes:

$$\Phi(\mathbf{D}|x_i, x_j) = \begin{cases} \gamma(i,j) & x_i \neq x_j \\ 0 & x_i = x_j \end{cases} \tag{7}$$

2.5 The Shape Prior

To ensure that the segmentation is human like and wanting to estimate a human pose, a human shape model consisting of ellipses is used as a prior. The model is based on measures from a large Bulgarian population study [8], and the model is simplified such that it has no arms, and the only restriction to the model is that it cannot overstretch the knee joints. The hip joint is simplified such that the hip is connected in one point as studies shows that a 2D model can produce good results in gait analysis [3]. Pixels near the shape model in a frame are more likely to be labeled subject, while pixels far from the shape are more likely to be background.

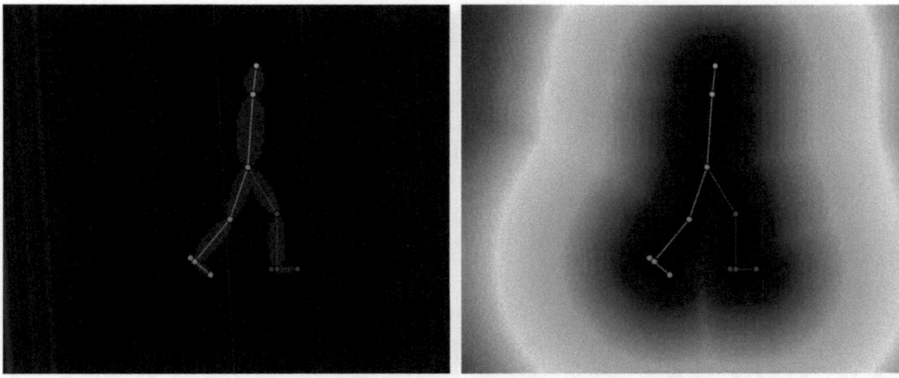

(a) Rasterized model (b) Distance map

Fig. 2. Raster model and the corresponding distance map

The cost function for the shape prior is defined as:

$$\Phi(x_i|\mathbf{\Theta}) = -\log(p(x_i|\mathbf{\Theta})) \tag{8}$$

Where $\mathbf{\Theta}$ contains the pose parameters of the shape model being position, height and joint angles. The probability $p(x_i|\mathbf{\Theta})$ of labeling subject or background is defined as follows:

$$p(x_i = \text{subject}|\mathbf{\Theta}) = 1 - p(x_i = \text{background}|\mathbf{\Theta}) = \frac{1}{1 + \exp(\mu * (\text{dist}(i, \mathbf{\Theta}) - d_r))} \tag{9}$$

The function $\text{dist}(i, \mathbf{\Theta})$ is the distance from pixel i to the shape defined by $\mathbf{\Theta}$, d_r is the width of the shape, and μ is the magnitude of the penalty given to points outside the shape. To calculate the distance for all pixels to the model, the shape model is rasterized and the distance found using the *Signed Euclidian Distance Transform* (*SEDT*) [12]. Figure 2 shows the rasterized model and the distances calculated using the *SEDT*.

2.6 Energy Minimization

Combining the four energy terms a cost function for the pose and segmentation becomes:

$$\Psi(\mathbf{x}, \mathbf{\Theta}) = \sum_{i \in V} \left(\Phi(\mathbf{D}|x_i) + \Phi(x_i|\mathbf{\Theta}) + \sum_{j \in N_i} (\psi(x_i, x_j) + \Phi(\mathbf{D}|x_i, x_j)) \right) \tag{10}$$

This Markov random field is solved using *Graph Cuts* [6], and the pose is optimized in each frame using the pose from the previous frame as initialization.

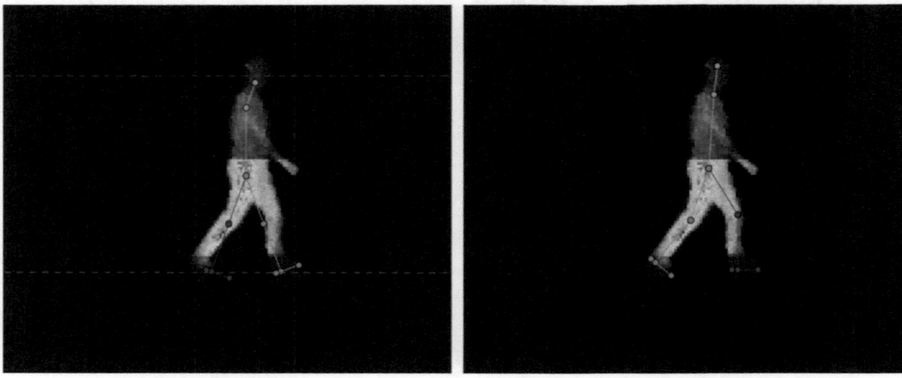

(a) Initial guess (b) Optimized pose

Fig. 3. Initialization of the algorithm

2.7 Initialization

To find an initial frame and a pose, the frame that differs the most from the background is chosen based on the background log likelihood function. As a rough guess on where the subject is in this frame, the log likelihood is summed first along the rows and then along the columns. These two sum vectors are used to guess the first and last rows and columns that contains the subject (Fig 3(a)). From the initial guess the pose is optimized according to the energy problem by searching locally. Figure 3(b) shows the optimized pose. Notice that the legs change place during the optimization. This is done based on the depth image such that the closest leg is also closest in the depth image (green is the right side in the model) and solves an ambiguity problem in silhouettes.

The pose in the remaining frames is found using the previous frame as an initial guess and then optimizing on this. This generally works very well, but problems sometimes arise when the legs pass each other as feet or knees of one leg tend to get stuck on the wrong side of the other leg. This entanglement is avoided by not allowing crossed legs as an initial guess and instead using straight legs close together.

3 Analyzing the Gait

From the markerless tracking a sequential model is created. To ensure local smoothness in the movement before the analysis is carried out a little postprocessing is done.

3.1 Post Processing

The movement of the model is expected to be locally smooth, and the influence of a few outliers is minimized by using a local median filter on the sequences of

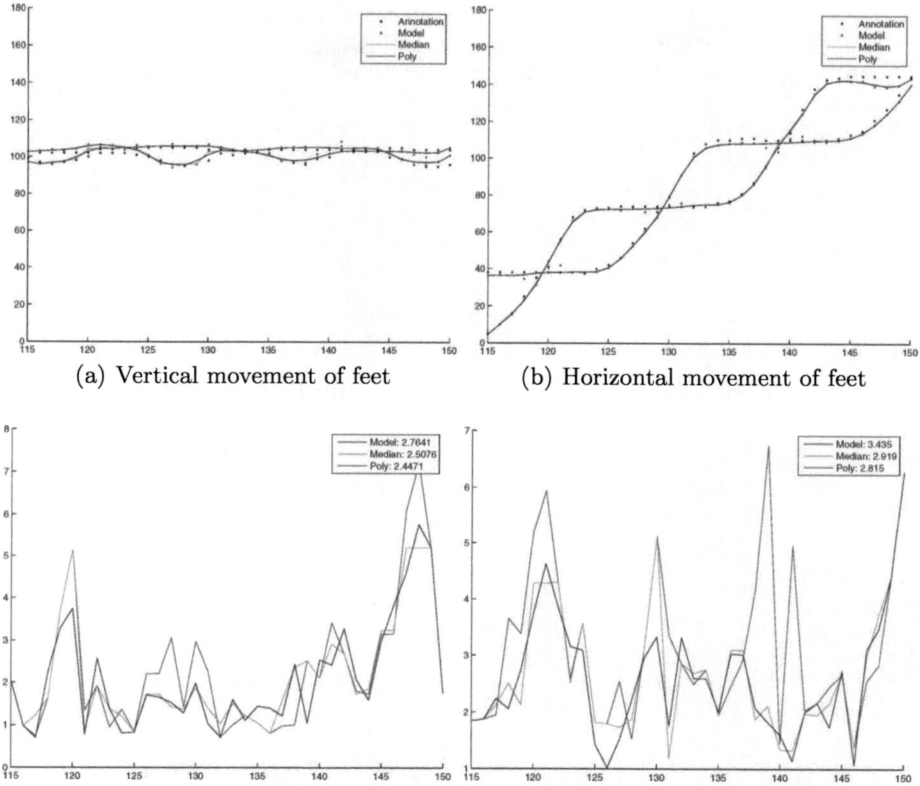

(a) Vertical movement of feet (b) Horizontal movement of feet

(c) Error of right foot (d) Error of left foot

Fig. 4. (a) shows the vertical movement of the feet for annotated points, points from the pose estimate, and for curve fittings (image notation is used, where rows are increased downwards). (b) shows the points for the horizontal movement. (c) shows the pixelwise error for the right foot for each frame and the standard deviation for each fitting. (d) shows the same but for the left foot.

point and then locally fitting polynomials to the filtered points. As a measure of ground truth the foot joints of the subject has been annotated in the sequence to give a standard deviation in pixels of the foot joint movement. Figure 4 shows the movement of the feet compared to the annotated points and the resulting error. The figure shows that the curve fitting of the points gives an improvement on the accuracy of the model, resulting in a standard deviation of only a few pixels. If the depth detection used to decide which leg is left and which is right fails in a frame, comparing the body points to the fitted curve can be used to detect and correct the incorrect left right detection.

3.2 Output Parameters

With the pose estimated in every frame the gait can now be analyzed. To find the steps during gait, the frames where the distance between the feet has a

(a) Left step length

(b) Right step length

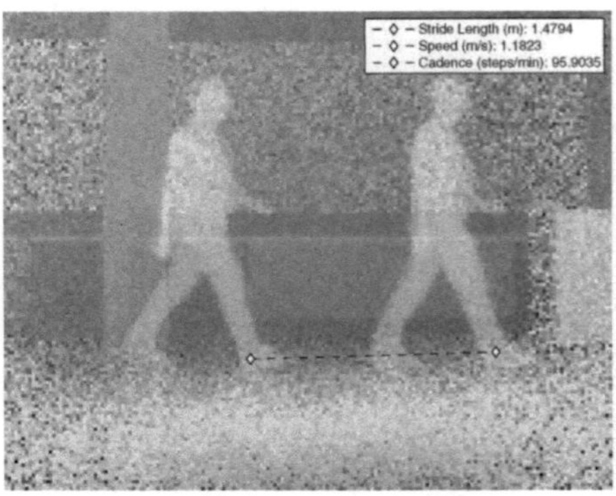

(c) Stride length, speed and cadence

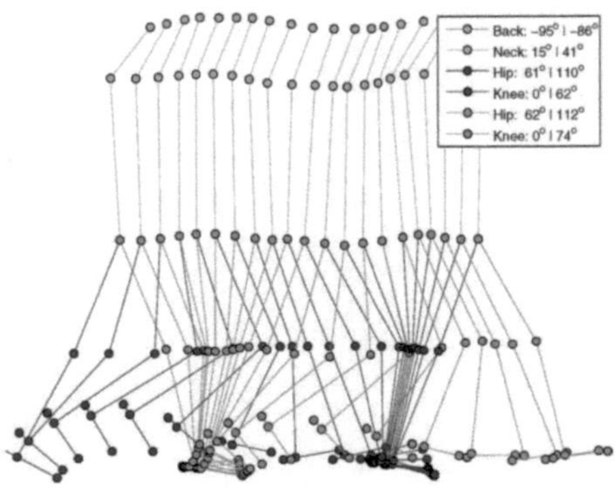

(d) Range of motion

Fig. 5. Analysis output

local maximum are used. Combining this with information about which foot is leading, the foot that is taking a step can be found. From the provided Cartesian coordinates in space and a timestamp for each frame the step length (Fig. 5(a) and 5(b)), stride length, speed and cadence (Fig. 5(c)) are found. The found parameters are close to the average found in a small group of subjects aging 17 to 31 [7], even though based only on very few steps and therefore expected to have some variance, this is an indication of correctness. The range of motion is found as the clockwise angle from the x-axis in positive direction for the inner limbs (femurs and torso) and the clockwise change compared to the inner limbs for the outer joints (ankles and head). Figure 5(d) shows the angles and the model pose throughout the sequence.

4 Conclusion

A system is created that autonomously produces a simple gait analysis. Because a depth map is used to perform the tracking rather than an intensity map, there are no requirements to the background nor to the subject clothing. No reference system is needed as the camera provides a such. Compared to manual annotation in each frame the error is very little. For further analysis on gait the system could easily be adapted to work on a subject walking on a treadmill. The adaption would be that there is no longer a general movement in space (it is the treadmill conveyor belt moving) hence speed and stride lengths should be calculated using step lengths. With the treadmill adaption, averages could be found of the different outputs as well as standard deviations.

Currently the system uses a 2-dimensional model and to optimize precision in the joint angles the subject should move in an angle perpendicular to the camera. While the distances calculated depends little on the angle of movement the joint angles have a higher dependency. This dependency could be minimized using a 3-dimensional model. It does however still seem reasonable that the best results would come from movement perpendicular to the camera, whether using a 3-dimensional model or not.

The camera used is the SwissRanger^{TM} SR3000 [2] at a framerate of about 18 Fps, which is on the low end in tracking movement. A better precision could be obtained with a higher framerate. This would not augment processing time greatly, due to the fact that movement from one frame to the next will be relatively shorter, bearing in mind that the pose from the previous frame is used as an initialization for the next.

Acknowledgements

This work was in part financed by the ARTTS [1] project (Action Recognition and Tracking based on Time-of-Flight Sensors) which is funded by the European Commission (contract no. IST-34107) within the Information Society Technologies (IST) priority of the 6th framework Programme. This publication reflects only the views of the authors, and the Commission cannot be held responsible for any use of the information contained herein.

References

1. Artts (2009), `http://www.artts.eu`
2. Mesa (2009), `http://www.mesa-imaging.ch`
3. Alkjaer, E.B., Simonsen, T., Dygre-Poulsen, P.: Comparison of inverse dynamics calculated by two- and three-dimensional models during walking. In: 2001 Gait and Posture, pp. 73–77 (2001)
4. Besag, J.: On the statistical analysis of dirty pictures. Journal of the Royal Statistical Society. Series B (Methodological) 48(3), 259–302 (1986)
5. Bray, M., Kohli, P., Torr, P.H.S.: Posecut: simultaneous segmentation and 3D pose estimation of humans using dynamic graph-cuts. In: Leonardis, A., Bischof, H., Pinz, A. (eds.) ECCV 2006. LNCS, vol. 3952, pp. 642–655. Springer, Heidelberg (2006)
6. Kolmogorov, V., Zabin, R.: What energy functions can be minimized via graph cuts? IEEE Transactions on Pattern Analysis and Machine Intelligence 26(2), 147–159 (2004)

7. Latt, M.D., Menz, H.B., Fung, V.S., Lord, S.R.: Walking speed, cadence and step length are selected to optimize the stability of head and pelvis accelerations. Experimental Brain Research 184(2), 201–209 (2008)
8. Nikolova, G.S., Toshev, Y.E.: Estimation of male and female body segment parameters of the bulgarian population using a 16-segmental mathematical model. Journal of Biomechanics 40(16), 3700–3707 (2007)
9. Shakhnarovich, G., Viola, P., Darrell, T.: Fast pose estimation with parameter-sensitive hashing. In: Proceedings Ninth IEEE International Conference on Computer Vision, vol. 2, pp. 750–757 (2003)
10. Stauffer, C., Grimson, W.E.L.: Adaptive background mixture models for real-time tracking. In: Proceedings. 1999 IEEE Computer Society Conference on Computer Vision and Pattern Recognition (Cat. No PR00149), vol. 2, pp. 246–252 (1999)
11. Wan, C., Yuan, B., Miao, Z.: Markerless human body motion capture using Markov random field and dynamic graph cuts. Visual Computer 24(5), 373–380 (2008)
12. Ye, Q.-Z.: The signed Euclidean distance transform and its applications. In: 1988 Proceedings of 9th International Conference on Pattern Recognition, vol. 1, pp. 495–499 (1988)
13. Zhu, Y., Dariush, B., Fujimura, K.: Controlled human pose estimation from depth image streams. In: 2008 IEEE Computer Society Conference on Computer Vision and Pattern Recognition Workshops (CVPR Workshops), pp. 1–8 (2008)

Primitive Based Action Representation and Recognition

Sanmohan and Volker Krüger

Computer Vision and Machine Intelligence Lab,
Copenhagen Institute of Technology, 2750 Ballerup, Denmark
{san,vok}@cvmi.aau.dk

Abstract. There has been a recent interest in segmenting action sequences into meaningful parts (action primitives) and to model actions on a higher level based on these action primitives. Unlike previous works where action primitives are defined a-priori and search is made for them later, we present a sequential and statistical learning algorithm for automatic detection of the action primitives and the action grammar based on these primitives. We model a set of actions using a single HMM whose structure is learned incrementally as we observe new types. Actions are modeled with sufficient number of Gaussians which would become the states of an HMM for an action. For different actions we find the states that are common in the actions which are then treated as an action primitive.

1 Introduction

Similar to phonemes being the building blocks of human language there is biological evidence that human action execution and understanding is also based on a set primitives [2]. But the notion of primitives for action does not only appear in neuro-biological papers. Also in the vision community, many authors have discussed that it makes sense to define a hierarchy of different action complexities such as *movements*, *activities* and *actions* [3]. In terms of Bobick's notations, *movements* are action primitive, out of which activities and actions are composed.

Many authors use this kind of hierarchy as observed in the review by Moeslund et al [9]. One way to use such a hierarchy is to define a set of action primitives in connection with a stochastic grammar that uses the primitives as its alphabet. There are many advantages of using primitives: (1) The use of primitives and grammars is often more intuitive for the human which simplifies verification of the learning results by an expert (2)*Parsing* primitives for recognition instead of using the signal directly leads to a better robustness under noise [10][14] (3) AI provides powerful techniques for higher level processing such as planning and plan recognition based on primitives and parsing. In some cases, it is reasonable to define the set of primitives and grammars by hand. In other cases, however, one would wish to compute the primitives and the stochastic grammar automatically based on a set of training observations. Examples for this can be found in surveillance, robotics, and DNA sequencing.

A.-B. Salberg, J.Y. Hardeberg, and R. Jenssen (Eds.): SCIA 2009, LNCS 5575, pp. 31–40, 2009.

In this paper, we present an HMM-based approach to learn primitives and the corresponding stochastic grammar based on a set of training observations. Our approach is able to learn on-line and is able to refine the representation when newly incoming data supports it. We test our approach on a typical surveillance scenario similar to [12] and on the data used in [14] for human arm movements.

A number of authors represent action in a hierarchical manner. Staffer and Grimson [12] compute for a surveillance scenario a set of action primitives based on co-occurrences of observations. This work is used to motivate the surveillance setup of one of our experiments.

In [11] Robertson and Reid present a full surveillance system that allows high-level behavior recognition based on simple actions. Their system seems to require human interaction in the definition of the primitive actions such as *walking, running, standing, dithering* and the qualitative positions (*nearside-pavement, road, driveway, etc*). This is, what we would like to automate.

In [4] actions are recognized by computing the cost through states an action pass through. The states are found by k-means clustering on the prototype curve that best fits sample points according to a least square criterion. Hong et al [8] built a Finite State Machine for recognition by building individual FSM s for each gesture. Fod et al. [5] uses a segmentation approach using zero velocity crossing. Primitives are then found by clustering in the projected space using PCA.

The idea of segmenting actions into atomic parts and then modeling the temporal order using Stochastic Context Free Grammar is found in [7]. In [6], signs of first and second derivatives are used to segment action sequences. These works require the storage of all training data if one wishes to modify the model to accommodate a new action. Our approach eliminate this requirement and thus make it suitable for imitation learning.

Our idea of merging of several HMMs to get a more complex and general model is found in [13]. We propose a merging strategy for continuous HMMs. New models can be introduced and merged online.

1.1 Problem Statement

We define two sets of primitives. One set contains parts that are unique to one *type* of action and another set that contains parts that are common to more than one *type* of action. Two sequences are of the same *type* if they do not differ significantly, e.g., two different walking paths. Hence we attempt to segment sequences into parts that are not shared and parts that are common across sequences types. Then each sequence will be a combination of these segments. We also want to generate rules that govern the interaction among the primitives. Keeping this in mind we state our objectives as:

1. Let $\mathcal{L} = \{X_1, X_2, \cdots, X_m\}$ be a set of data sequences where each X_i is of the form $x_1^i x_2^i \cdots, x_{T_i}^i$ and $x_j^i \in \mathbb{R}^n$. Let these observations be generated from a finite set of sources (or states) $\mathcal{S} = \{s_1, s_2, \cdots s_r\}$. Let $S_i = s_1^i s_2^i \cdots, s_{T_i}^i$ be the state sequence associated with X_i. Find a partition $\mathcal{S}'$ of the set of states

$\mathcal{S}$ where $\mathcal{S}' = \mathcal{A} \cup \mathcal{B}$ such that $\mathcal{A} = \{a_1, a_2, \cdots, a_k\}$ and $\mathcal{B} = \{b_1, b_2, \cdots, b_l\}$ are sets of state subsequences of X_i's and each of the a_i's appear in more than one state sequence and each of the b_j's appear in exactly one of the state sequence. The set $\mathcal{A}$ corresponds to common actions and the set $\mathcal{B}$ correspond to unique parts.

2. Generate a grammar with elements of $\mathcal{S}'$ as symbols which will generate primitive sequences that match with the data sequences.

2 Modeling the Observation Sequences

We take the first sequence of observations X_1 with data points $x_1^1 x_2^1 \cdots x_{T_1}^1$ and generate a few more spurious sequences of the same type by adding Gaussian noise to it. Then we choose (μ_i^1, σ_i^1), $i = 1, 2, ...k^1$ so that parts of the data sequence are from $\mathcal{N}(\mu_i^1, \Sigma_i^1)$ in that order. The value of k^1 is such that $\mathcal{N}(\mu_i^1, \Sigma_i^1)$, $i = 1, 2, ...k^1$ will cover the whole data. This value is not chosen before hand and varies with the variation and length of the data.

The next step is to make an HMM $\lambda_1 = (A^1, B^1, \pi^1)$ with k^1 states. We let A^1 to be a left-right transition matrix and $B_j^1(x) = \mathcal{N}(x, \mu_j^1, \Sigma_j^1)$. All the states at this stage get a label 1 to indicate that they are part of sequence type 1. This model will now be modified recursively.

Now we will modify this model by adding new states to it or by modifying the current output probabilities of states so that the modified model λ_M will be able to generate new types of data with high probability. Let $n - 1$ be the number of types of data sequences we have seen so far. Let X_c be the next data sequence to be processed. Calculate $P(X_c|\lambda_M)$ where λ_M is the current model at hand. A low value for $P(X_c|\lambda_M)$ indicates that the current model is not good enough to model the data sequences of type X_c and hence we make a new HMM λ_c for X_c as described in the beginning and the states are labeled n. The newly constructed HMM λ_c will be merged to λ_M so that the updated λ_M will be able to generate data sequences of type X_c.

Suppose we want to merge λ_c into λ_M so that $P(X_k|\lambda_M)$ is high if $P(X_k|\lambda_c)$ is high. Let $C_c = \{s_{c1}, s_{c2}, \cdots, s_{ck}\}$ and $C_M = \{s_{M1}, s_{M2}, \cdots, s_{Ml}\}$ be the set of states of λ_c and λ_M respectively. Then the state set of the modified λ_M will be $C_M \cup D_1$ where $D_1 \subseteq C_c$. Each of the states s_{ci} in λ_c affects λ_M in one of the following ways:

1. If $d(s_{ci}, s_{Mj}) < \theta$, for some $p \in \{1, 2, \cdots l\}$, then s_{ci} and s_{Mj} will be merged into a single state. Here d is a distance measure and θ is a threshold value. The output probability distribution associated with s_{Mj} is modified to be a combination of the existing distribution and ${b^k}_{s_{ci}}(x)$. Thus ${b^M}_{Mj}(x)$ is a mixture of Gaussians. We append n to the label of the state s_{Mj}. All transitions to s_{ci} are redirected to s_{Mj} and all transitions from s_{ci} will now be from s_{Mj}. The basic idea behind merging is that we do not need two different states which describe the same part of the data.
2. If $d(s_{ci}, s_{Mj}) > \theta$, $\forall j$, a new state is added to λ_M. i.e. $s_{ci} \in D_1$. Let s_{ci} be the rth state to be added from λ_c. Then, s_{ci} will become the $(Ml+r)^{th}$ state

of λ_M. The output probability distribution associated with this new state in λ_M will be the same as it was in λ_c. Hence $b^M{}_{Ml+r}(x) = \mathcal{N}(x, \mu_{s_{c_i}}, \Sigma_{s_{c_i}})$. Initial and transition probabilities of λ_M are adjusted to accommodate this new state. The newly added state will keep its label n.

We use Kullback-Leibler Divergence to calculate the distance between states. The K-L divergence from $\mathcal{N}(x, \mu_0, \Sigma_0)$ to $\mathcal{N}(x, \mu_1, \Sigma_1)$ has a closed form solution given by :

$$D_{KL}(Q||P) = \frac{1}{2}\left(\log\frac{|\Sigma_1|}{|\Sigma_0|} + \mathrm{tr}(\Sigma_1^{-1}\Sigma_0) + (\mu_1 - \mu_0)^T\Sigma_1^{-1}(\mu_1 - \mu_0) - n\right) \quad (1)$$

Here n is the dimension of the space spanned by the random variable x.

2.1 Finding Primitives

When all sequences have been processed, we apply Viterbi algorithm on the final merged model λ_M, and find the hidden states associated with each of the sequences. Let $P_1, P_2, \cdots P_r$ be different Viterbi paths at this stage. Since we want the common states that are contiguous across state sequences, it is similar to finding the longest common substring(LCS) problem. We take all paths with non-empty intersection and find the largest common substring a_k for them. Then a_k is added to $\mathcal{A}$ and is replaced with an empty string in all the occurrences of a_k in $P_i,\ i = 1, 2, \cdots r$.

We continue to look for largest common substings until we get an empty string as the common substring for any two paths. Thus we end up with new paths $P_1', P_2', \cdots P_r'$ where each P_i' consists of one or more segments with empty string as the separator.These remaining segments in each P_i' are unique to P_i. Each of them are also primitives and form the members of the set $\mathcal{B}$. Our objective was to find these two sets $\mathcal{A}$ and $\mathcal{B}$ as was stated in Sec. 1.1.

3 Generating the Grammar for Primitives

Let $\mathcal{S}' = \{c_1, c_2, \cdots c_p\}$ be the set of primitives available to us. We wish to generate rules of the form $P(c_i \to c_j)$ which will give the likelihood of occurrence of the primitive c_j followed by primitive c_i. We do this by constructing a directed graph G which encodes the relations between the primitives. Using G we will derive a formal grammar for the elements in $\mathcal{S}'$.

Let n be the number of types of data that we have processed. Then each of the states in our final HMM λ_M will have labels from a subset of $\{1, 2, \cdots, n\}$, see Fig.1. By way of definition each of the states that belong to a primitive c_i will have the same label set l^{c_i}. Let $\mathcal{L} = \{l_1, l_2 \cdots, l_p\}$ $p \geq n$ be the set of different type of labels received by the primitives. Let $G = (V, E)$ be a directed graph where $V = \mathcal{S}'$ and $e_{ij} = (c_i, c_j) \in E$ if there is a path $P_k = \cdots c_i c_j \cdots$ for

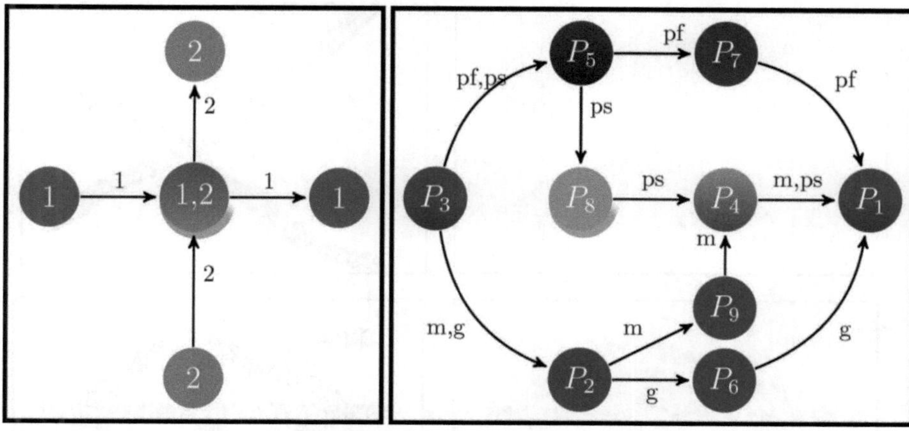

Fig. 1. The figure on the left shows the directed graph for finding the grammar for the simulated data explined in experiments section. Right figure: The temporal order for primitives of hand gesture data. Node number corresponds to different primitives. Multi-colored nodes belong to more than one action. All actions start with P_3 and end with $P1$. Here g=grasp, m=move object, pf=push forward and ps=push sideways.

some k. We have given the directed graph constructed for out test data in Fig. 1.

We proceed to derive a precise Stochastic Context Free Grammar (SCFG) from the directed graph G we have constructed. Let $N = \mathcal{S}'$ be the set of terminals. To each vertex c_i with an outgoing edge with label $l^{e_{ij}}$, associate a corresponding non-terminal $A_{c_i}^{l^{e_{ij}}}$. Let $\mathcal{N} = S \cup \{A_{c_i}^{l^{e_{ij}}}\}$ be the set of all non-terminals where S is the start symbol. For each primitive c_i that occurs at the start of a sequence and connecting to c_j define the rule $S \longrightarrow c_i A_{c_j}^{l^{c_i}}$. To each of the internal nodes c_j with an incoming edge e_{ij} connecting from c_i and an outgoing edge e_{jk} connecting to c_k define the rule $A_{c_i}^{l^{c_i} \cap l^{c_j}} \longrightarrow c_j A_{c_k}^{l^{c_j} \cap l^{c_k}}$. For each leaf node c_j with an incoming edge e_{ij} connecting from c_i and no outgoing edge define the rule $A_{c_j}^{l^{c_i} \cap l^{c_j}} \longrightarrow \epsilon$. The symbol ϵ denotes an empty string. We assign equal probabilities to each of the expansions of a nonterminal symbol except for the expansion to an empty string which occurs with probability 1. Thus $P(A_{c_i}^{l_{ij}} \longrightarrow c_j A_{c_j}^{l_{jk}}) = \frac{1}{|c_i^{(o)}|}$ if $|c_i^{(o)}| > 0$ and $P(A_{c_i}^{l^{e_{ij}}} \longrightarrow \epsilon) = 1$ otherwise.. where $|c_i^{(o)}|$ represents the number of outgoing edges from c_i and $l_{mn} = l^{c_m} \cap l^{c_n}$. Let $\mathcal{R}$ be the collection of all rules given above. For each $r \in \mathcal{R}$ associate a probability $P(r)$ as given in the construction of rules. Then $(\mathcal{N}, \mathcal{S}', S, \mathcal{R}, P(.))$ is the stochastic grammar that models our primitives.

One might wonder why the HMM λ_M is not enough to describe the grammatical structure of the observations and why the SCFG is necessary. The HMM λ_M would have been sufficient for a single observation type. However for several observation types as in final λ_M, regular grammars, as modeled by HMMs are usually too limited to model the different observation types so that different observation types can be confused.

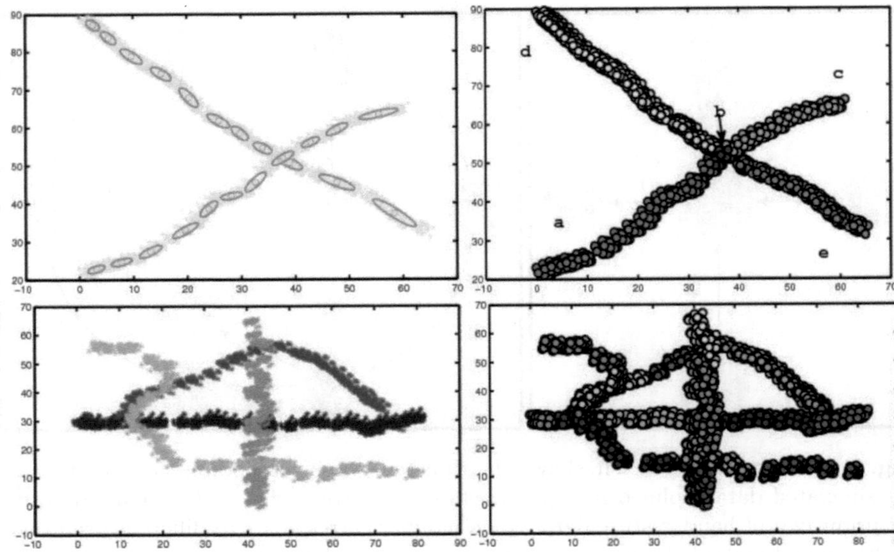

Fig. 2. The top left figure shows the simulated 2d data sequences. The ellipses represent the Gaussians. The top right figure shows the finally detected primitives with different colors. Primitive b is a common primitive and belongs to set $\mathcal{A}$, primitives a,c,d,e belong to set $\mathcal{B}$. The bottom left figure shows trajectories from tracking data. Each type is colored differently. Only a part of the whole data is shown. The bottom right figure shows the detected primitives. Each primitive is colored differently.

4 Experiments

We have run three experiments: In the first experiment we generate a simple data set with very simple cross-shaped paths. The second experiment is motivated by the surveillance scenario of Stauffer and Grimson [12] and shows a complex set of paths as found outside our building. The third experiment is motivated by the work of Vincente and Kragic [14] on the recognition of human arm movements.

4.1 Testing on Simulated Data

We illustrate the result of testing our method on a set of two sequences generated with mouse clicks. The original data set for testing is shown in Fig. 2 at top left . We have two paths which intersect in the middle. If we were to remove the intersecting points we will get four segments. We extracted these segments with the above mentioned procedure. When the model merging took place, the overlapping states in the middle were merged into one. The result is shown in Fig. 2 at top right. The primitives that we get are colored. As one can see in Fig. 2, primitive b is a common primitive and belongs to our set A, primitives a,c,d,e belong to our set B.

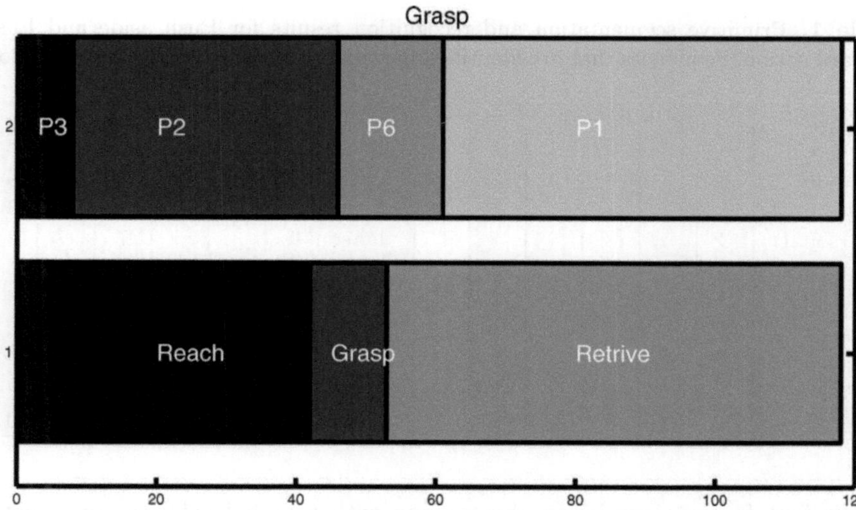

Fig. 3. Comparing automatic segmentation with manually segmented primitives for one grasp sequence. Using the above diagram with the right figure in Fig.1, we can infer that P_3 and P_2 together constitute *approach* primitive, P_6 refers to *grasp* primitive and P_1 corresponds to *remove* primitive.

4.2 2D-Trajectory Data

The second experiment was done on a surveillance-type data inspired by [12]. The paths represent typical walking paths outside of our building. In this data there are four different types of trajectories with heavy overlap, see Fig. 2 bottom left. We can also observe that the data is quite noisy.

The result of primitive segmentation is shown in Fig. 2 on the bottom right. Different primitives are colored differently and we have named the primitives with different letters. As one can see, our approach results in primitives that coincide roughly with our intuition. Furthermore, our approach is very robust even with such noisy observations and lot of overlaps.

Hand Gesture Data. Finally, we have tested our approach on the dataset provided by Vincente and Kragic [14]. In this data set, several volunteers performed a set of simple arm movements such as *reach for object*, *grasp object*, *push object*,*move object* , and *rotate object*. Each action is performed in 12 different conditions: two different heights, two different locations on the table, and having the demonstrator stand in three different locations (0,30, 60 degrees). Furthermore all actions are demonstrated by 10 different people. The movements are measured using magnetic sensors placed on: chest, back of hand, thumb, and index finger. In [14], the segmentation was done manually and their experiments showed that the recognition performance of human arm actions is increased when one uses action primitives. Using their dataset, our approach is able to provide the primitives and the grammar automatically. We consider the 3-d trajectories

Table 1. Primitive segmentation and recognition results for Push aside and Push Forward action. Sequences that are identified incorrectly are marked with yellow color.

Person	Push Aside				
Person 1	3	2	9	4	1
Person 2	3	5	8	4	1
Person 3	3	5	8	4	1
Person 4	3	5	8	4	1
Person 5	3	5	8	4	1
Person 6	3	5	8	4	1
Person 7	3	5	8	4	1
Person 8	3	5	8	4	1
Person 9	3	2	9	4	1
Person 10	3	2	9	4	1

Person	Push Forward				
Person 1	3	5	7		1
Person 2	3	5	7		1
Person 3	3	5	7		1
Person 4	3	5	7		1
Person 5	3	5	7		1
Person 6	3	5	8	4	1
Person 7	3	5	7		1
Person 8	3	5	7		1
Person 9	3	5	8	4	1
Person 10	3	5	8	4	1

for the first four actions listed above along with a scaled velocity component. Since each of these sequences started and ended at the same position, we expect the primitives that represent the starting and end positions of actions will be the same across all the actions.

By applying the techniques described in Sec.2 to the hand gesture data, we ended up with 9 primitives. The temporal order of primitives for actions for different actions are shown in Fig.1. We also compare our segmentation with the segmentation in [14]. We plot the result of converting a grasp action sequence into a sequence of extracted primitives along with ground truth data in Fig.3. We can infer from the figures Fig.1 and Fig.3 that P_3 and P_2 together constitute *approach* primitive, P_6 refers to *grasp* primitive and P_6 corresponds to *remove* primitive. Similar comparison could be made with other actions also.

Using these primitives, an SCFG was built as described in Sec.3. This grammar is used as an input to the Natural Language Toolkit (NLTK, `http://nltk.sourceforge.net`) which is used to parse the sequence of primitives.

Table 2. Primitive segmentation and recognition results for Move Object and Grasp actions. Sequences that are identified incorrectly are marked with yellow color.

Person	Move				
Person 1	3	2	9	4	1
Person 2	3	5	8	4	1
Person 3	3	2	9	4	1
Person 4	3	2	9	4	1
Person 5	3	2	9	4	1
Person 6	3	5	8	4	1
Person 7	3	2	9	4	1
Person 8	3	2	9	4	1
Person 9	3	2	9	4	1
Person 10	3	2	9	4	1

Person	Grasp				
Person 1	3	2	6		
Person 2	3	2	6		1
Person 3	3	5	7	6	1
Person 4		2	6		1
Person 5	3	2	6		1
Person 6	3	2	6		1
Person 7	3	2	9	4	1
Person 8	3	2	6		1
Person 9	3	2	6	7	1
Person 10	3	2	6		1

Results of primitive segmentation for *push sideways, push forward, move,* and *grasp* actions are shown in the tables 1 and 2. The numbers given in the tables represent the primitive numbers shown in Fig. 1. The sequences that are identified correctly are marked with Aqua color and the sequences that are not classified correctly are marked with yellow color. We can see that all the correctly identified sequences start and end with the same primitive as expected. In Tab.2, Person 1 and Person 4 are marked with a lighter color to indicate that they differ in end and start primitive respectively from the correct primitive sequence. This might be due to the variation in the starting and end position in the sequence. We could still see that the primitive sequence is correct for them.

5 Conclusions

We have presented and tested an approach for automatically computing a set of primitives and the corresponding stochastic context free grammar from a set of training observations. Our stochastic regular grammar is closely related to the usual HMMs. One important difference between common HMMs and a stochastic grammar with primitives is that with usual HMMs, each trajectory (action, arm movement, etc.) has its own, distinct HMM. This means that the set of HMMs for the given trajectories are not able to reveal any commonalities between them. In case of our arm movements, this means that one is not able to deduce that some actions share the grasp movement part. Using the primitives and the grammar, this is different. Here, common primitives are shared across the different actions which results into a somewhat symbolic representation of the actions. Indeed, using the primitives, we become able to do the recognition in the space of the primitives or symbols, rather than in the signal space directly, as it would be the case when using distinct HMMs. Using this symbolic representation would even allow to use AI techniques for, e.g., planning or plan recognition. Another important aspect of our approach is that we can modify our model to include a new action without requiring the storage of previous actions for it.

Our work is segmenting an action into smaller meaningful segments and hence different from [1] where the authors aim at segmenting actions like walk and run from each other. Many authors point at the huge task of learning parameters and the size of training data for an HMM when the number of states are increasing. But in our method, transition, initial and observation probabilities for all states are assigned during our merging phase and hence the use of EM algorithm is not required. Thus our method is scalable to the number of states.

It is interesting to note that stochastic grammars are closely related to Belief networks where the hierarchical structure coincides with the production rules of the grammar. We will further investigate this relation ship in future work.

In future work, we will also evaluate the performance of normal and abnormal path detection using our primitives and grammars.

References

1. Barbič, J., Safonova, A., Pan, J.-Y., Faloutsos, C., Hodgins, J.K., Pollard, N.S.: Segmenting motion capture data into distinct behaviors. In: GI 2004: Proceedings of Graphics Interface 2004, School of Computer Science, University of Waterloo, Waterloo, Ontario, Canada, pp. 185–194. Canadian Human-Computer Communications Society (2004)
2. Bizzi, E., Giszter, S.F., Loeb, E., Mussa-Ivaldi, F.A., Saltiel, P.: Modular organization of motor behavior in the frog's spinal cord. Trends Neurosci. 18(10), 442–446 (1995)
3. Bobick, A.: Movement, Activity, and Action: The Role of Knowledge in the Perception of Motion. Philosophical Trans. Royal Soc. London 352, 1257–1265 (1997)
4. Bobick, A.F., Wilson, A.D.: A state-based approach to the representation and recognition of gesture. IEEE Transactions on Pattern Analysis and Machine Intelligence 19(12), 1325–1337 (1997)
5. Fod, A., Matarić, M.J., Jenkins, O.C.: Automated derivation of primitives for movement classification. Autonomous Robots 12(1), 39–54 (2002)
6. Guerra-Filho, G., Aloimonos, Y.: A sensory-motor language for human activity understanding. In: 2006 6th IEEE-RAS International Conference on Humanoid Robots, December 4-6, 2006, pp. 69–75 (2006)
7. Fermüller, C., Guerra-Filho, G., Aloimonos, Y.: Discovering a language for human activity. In: AAAI 2005 Fall Symposium on Anticipatory Cognitive Embodied Systems, Washington, DC, pp. 70–77 (2005)
8. Hong, P., Turk, M., Huang, T.: Gesture modeling and recognition using finite state machines (2000)
9. Moeslund, T., Hilton, A., Krueger, V.: A survey of advances in vision-based human motion capture and analysis. Computer Vision and Image Understanding 104(2-3), 90–127 (2006)
10. Rabiner, L.R., Juang, B.H.: Fundamentals of Speech Recognition. Prentice Hall, Englewood Cliffs (1993)
11. Robertson, N., Reid, I.: Behaviour Understanding in Video: A Combined Method. In: Internatinal Conference on Computer Vision, Beijing, China, October 15-21 (2005)
12. Stauffer, C., Grimson, W.E.L.: Learning Patterns of Activity Using Real-Time Tracking. IEEE Transactions on Pattern Analysis and Machine Intelligence 22(8), 747–757 (2000)
13. Stolcke, A., Omohundro, S.M.: Best-first model merging for hidden Markov model induction. Technical Report TR-94-003, 1947 Center Street, Berkeley, CA (1994)
14. Vicente, I.S., Kyrki, V., Kragic, D.: Action recognition and understanding through motor primitives. Advanced Robotics 21, 1687–1707 (2007)

Recognition of Protruding Objects in Highly Structured Surroundings by Structural Inference

Vincent F. van Ravesteijn[1], Frans M. Vos[1,2], and Lucas J. van Vliet[1]

[1] Quantitative Imaging Group, Faculty of Applied Sciences, Delft University of Technology, The Netherlands
V.F.vanRavesteijn@tudelft.nl

[2] Department of Radiology, Academic Medical Center, Amsterdam, The Netherlands

Abstract. Recognition of objects in highly structured surroundings is a challenging task, because the appearance of target objects changes due to fluctuations in their surroundings. This makes the problem highly context dependent. Due to the lack of knowledge about the target class, we also encounter a difficulty delimiting the non-target class. Hence, objects can neither be recognized by their similarity to prototypes of the target class, nor by their similarity to the non-target class. We solve this problem by introducing a transformation that will eliminate the objects from the structured surroundings. Now, the dissimilarity between an object and its surrounding (non-target class) is inferred from the difference between the local image before and after transformation. This forms the basis of the detection and classification of polyps in computed tomography colonography. 95% of the polyps are detected at the expense of four false positives per scan.

1 Introduction

For classification tasks that can be solved by an expert, there exists a set of features for which the classes are separable. If we encounter class overlap, not enough features are obtained or the features are not chosen well enough. This conveys the viewpoint that a feature vector representation directly reduces the object representation [1]. In the field of imaging, the objects are represented by their grey (or color) values in the image. This sampling is already a reduced representation of the real world object and one has to ascertain that the acquired digital image still holds sufficient information to complete the classification task successfully. If so, all information is still retained and the problem reduces to a search for an object representation that will reveal the class separability.

Using all pixels (or voxels) as features would give a feature set for which there is no class overlap. However, this feature set usually forms a very high dimensional feature space and the problem would be sensitive to the curse of dimensionality. Considering a classification problem in which the objects are regions of interest $\mathcal{V}$ with size N from an image with dimensionality D, the dimensionality of the feature space Ω would then be N^D, i.e. the number of pixels

A.-B. Salberg, J.Y. Hardeberg, and R. Jenssen (Eds.): SCIA 2009, LNCS 5575, pp. 41–50, 2009.

in $\mathcal{V}$. This high dimensionality poses problems for statistical pattern recognition approaches. To avoid these problems, principal component analysis (PCA) could for example be used to reduce the dimensionality of the data without having the user to design a feature vector representation of the object. Although PCA is designed to reduce the dimensionality while keeping as most information as possible, the mapping unavoidably reduces the object representation.

The use of statistical approaches completely neglects that images often contain structured data. One can think of images that are very similar (images that are close in the feature space spanned by all pixel values), but might contain significantly different structures. Classification of such structured data receives a lot of attention and is motivated by the idea that humans interpret images by perception of structure rather than by perception of all individual pixel values. An approach for the representation of structure of objects is to represent the objects by their dissimilarities to other objects [2]. When a dissimilarity measure is defined (for example the 'cost' of deforming an object into another object), the object can be classified based on the dissimilarities of the object to a set (or sets) of prototypes representing the classes.

Classification based on dissimilarities demands prototypes of both classes, but this demand can not always be fulfilled. For example, the detection of target objects in highly structured surroundings poses two problems. First, there is a fundamental problem describing the class of non-targets. Even if there is detailed knowledge about the target objects, the class of non-targets (or outliers) is merely defined as all other objects. Second, if the surroundings of the target objects is highly structured, the number of non-target prototypes is very large and they all differ each in their own way, i.e. they are scattered all over the feature space. The selection of a finite set of prototypes that sufficiently represents the non-target class is almost impossible and one might have to rely on one-class classification.

The objective of this paper is to establish a link between image processing and dissimilarity based pattern recognition. On the one hand, we show that the previous work [3] can be seen as an application of structual inference which is used in featureless pattern recognition [1]. On the other hand, we extend the featureless pattern recognition to pattern recognition in the absence of prototypes. The role of prototypes is replaced by a single context-dependent prototype that is derived from the image itself by a specific transformation for the application at hand. The approach will be applied in the context of automated polyp detection.

2 Automated Polyp Detection

The application that we present in this paper is automated polyp detection in computed tomography (CT) colonography (CTC). Adenomatous polyps are important precursors to cancer and early removal of such polyps can reduce the incidence of colorectal cancer significantly [4,5]. Polyps manifest themselves as protrusions from the colon wall and are therefore visible in CT. CTC is a minimal-invasive technique for the detection of polyps and, therefore, CTC is considered a promising candidate for large-scale screening for adenomatous polyps. Computer aided detection (CAD) of polyps is being investigated to assist the radiologists. A typical CAD system consists of two consecutive steps: candidate detection to detect suspicious locations on the colon wall, and classification to classify the candidates as either a polyp or a false detection.

By nature the colon is highly structured; it is curved, bended and folded. This makes that the appearance of a polyp is highly dependent on its surrounding. Moreover, a polyp can even be (partly) occluded by fecal remains in the colon.

2.1 Candidate Detection

Candidate detection is based on a curvature-driven surface evolution [3,6]. Due to the tube-like shape of the colon, the second principal curvature κ_2 of the colon surface is smaller than or close to zero everywhere (the normal vector points into the colon), except on protruding locations. Polyps can thus be characterized by a positive second principal curvature. The surface evolution reduces the protrusion iteratively by solving a non-linear partial differential equation (PDE):

$$\frac{\partial I}{\partial t} = \begin{cases} -\kappa_2 |\nabla I| & (\kappa_2 > 0) \\ 0 & (\kappa_2 \leq 0) \end{cases} \tag{1}$$

where I is the three-dimensional image and $|\nabla I|$ the gradient magnitude of the image.

Iterative application of (1) will remove all protruding elements (i.e. locations where $\kappa_2 > 0$) from the image and estimates the appearance of the colon surface as if the protrusion (polyp) was never there. This is visualized in Fig. 1 and Fig. 2. Fig. 1(a) shows the original image with a polyp situated on a fold. The grey values are iteratively adjusted by (1) . The deformed image (or the solution of the PDE) is shown in Fig. 1(b). The surrounding is almost unchanged, whereas the polyp has completely disappeared. The change in intensity between the two images is shown in Fig. 1(c). Locations where the intensity change is larger than 100 HU (Hounsfield units) yield the polyp candidates and their segmentation (Fig. 1(d)). Fig. 2 also shows isosurface renderings at different time-steps.

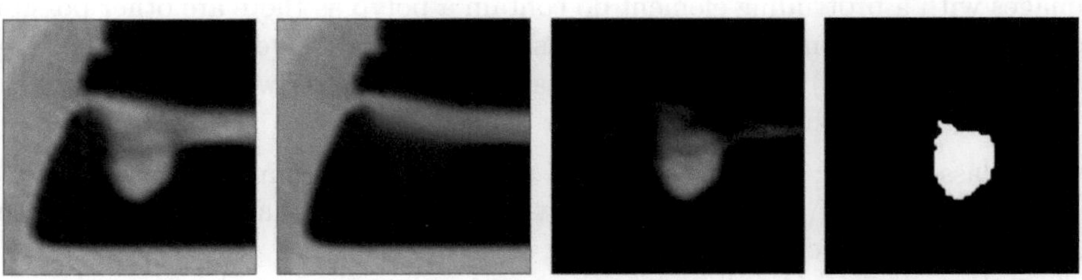

(a) Original (b) Solution (c) Intensity change (d) Segmentation

Fig. 1. (a) The original CT image (grey is tissue, black is air inside the colon). (b) The result after deformation. The polyp is smoothed away and only the surrounding is retained. (c) The difference image between (a) and (b). (d) The segmentation of the polyp obtained by thresholding the intensity change image.

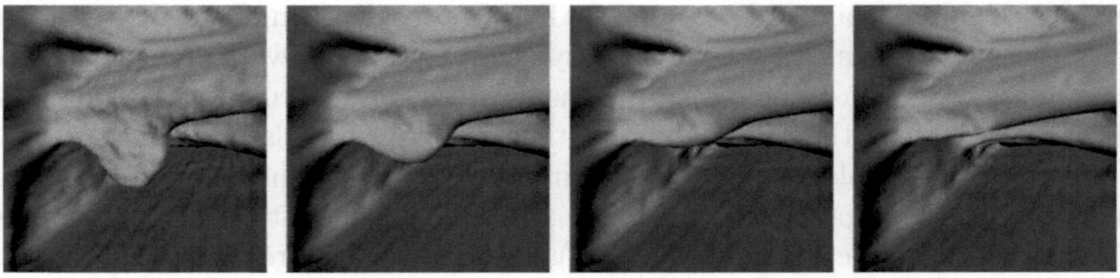

(a) Original (b) 20 Iterations (c) 50 Iterations (d) Result

Fig. 2. Isosurface renderings (-750 HU) of a polyp and its surrounding. (a) Before deformation. (b–c) After 20 and 50 iterations. (d) The estimated colon surface without the polyp.

2.2 Related Work

Konukoglu et al. [7] have proposed a related, but different approach. Their method is also based on a curvature-based surface evolution, but instead of removing protruding structures, they proposed to enhance polyp-like structures and to deform them into spherical objects. The deformation is guided by

$$\frac{\partial I}{\partial t} = \left(1 - \frac{H}{H_0}\right)|\nabla I| \tag{2}$$

with H the mean curvature and H_0 the curvature of the sphere towards the candidate is deformed.

3 Structural Inference for Object Recognition

The candidate detection step, described in the previous section, divides the feature space Ω of all possible images into two parts. The first part consists of all images that are not affected by the PDE. It is assumed that these images do not show any polyps and these are said to form the surrounding class $\Omega_\circ$. The other part consists of all images that are deformed by iteratively solving the PDE. These images thus contain a certain protruding element. However, not all images with a protruding element do contain a polyp as there are other possible causes of protrusions like fecal remains, the ileocecal valve (between the large and small intestine) and natural fluctuations of the colon wall.

To summarize, three classes are now defined:

1. a class $\Omega_\circ \subset \Omega$; all images without a polyp: the surrounding class,
2. a class $\Omega_f \subset \Omega \backslash \Omega_\circ$; all images showing a protrusion that is not a polyp: the false detection class, and
3. a class $\Omega_t \subset \Omega \backslash \Omega_\circ$; all images showing a polyp: the true detection class.

Successful classification of new images now requires a meaningful representation of the classes and a measure to quantify the dissimilarity between an image and a certain class. Therefore, Section 3.1 will describe how the dissimilarities can be defined for objects of which the appearance is highly context-dependent, and Section 3.2 will discuss how the classes can be represented.

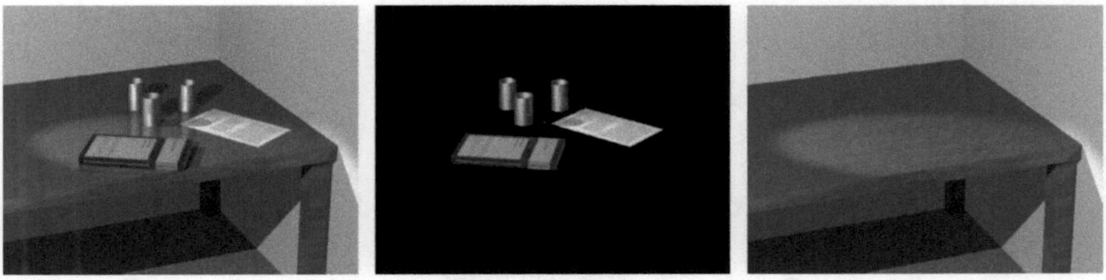

(a) (b) (c)

Fig. 3. (a) Objects in their surroundings. (b) Objects without their surroundings. All information about the objects is retained, so the objects can still be classified correctly. (c) The estimated surrounding without the objects.

3.1 Dissimilarity Measure

To introduce the terminology and notation, let us start with a simple example of dissimilarities between objects. Fig. 3(a) shows various objects on a table. Two images, say $\mathbf{x}_i$ and $\mathbf{x}_j$, represent for instance an image of the table with a cup and an image of the table with the book. The dissimilarity between these images is hard to define, but the dissimilarity between either one of these images and the image of an empty table is much easier. This dissimilarity may be derived from the image of the specific object itself (Fig. 3(b)).

When we denote the image of an empty table as $\mathbf{p}_\circ$, this first example can be schematically illustrated as in Fig. 4(a). The dissimilarities of the two images to the prototype $\mathbf{p}_\circ$ are called $d_{i\circ}$ and $d_{j\circ}$. If these dissimilarities are simply defined as the Euclidean distance between the circles in the image, the triangle-inequality holds.

However, if the dissimilarities are defined as the spatial distance between the objects (in 3D-space), all objects in Fig. 3(a) have zero distance to the table, but the distance between any two objects (other than the table) is larger than zero. This shows a situation in which the dissimilarity measure violates the triangle-inequality and the measure becomes non-metric [8]. This is schematically illustrated in Fig. 4(b). The prototype $\mathbf{p}_\circ$ is no longer a single point, but is transformed into a blob $\Omega_\circ$ representing all objects with zero distance to the table. Note that all circles have zero Euclidean distance to $\Omega_\circ$.

The image of the empty table can also be seen as the background or surrounding of all the individual objects, which shows that all objects have exactly the same surrounding. When considering the problem of object detection in highly structured surroundings this obviously no longer holds. We first state that, as in the first example given above, the dissimilarity of an object to its surrounding can be defined by the object itself. Secondly, although the surroundings may differ significantly from each other, it is known that none of the surroundings contain an object of interest (a polyp). Thus, as in the second example, the distances between all surroundings can be made zero and we obtain the same blob representation for $\Omega_\circ$, i.e. the surrounding class. The distance of an object

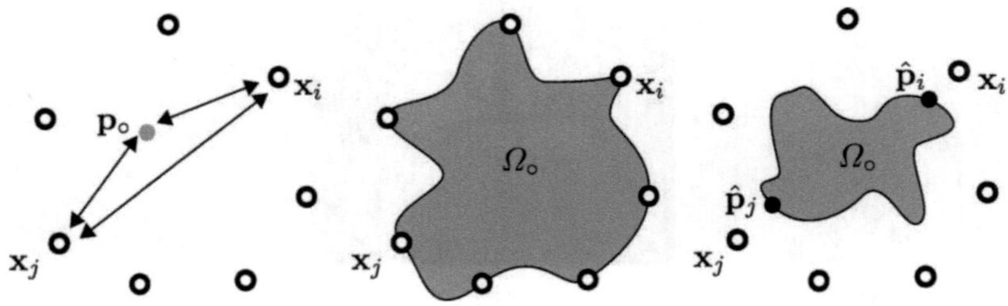

$d_{i\circ} > 0,\ d_{j\circ} > 0$
$d_{ij} \leq d_{i\circ} + d_{j\circ}$
(a)

$d_{i\circ} = 0,\ d_{j\circ} = 0$
$d_{ij} \nleq d_{i\circ} + d_{j\circ} = 0$
(b)

$d_{i\circ} > 0,\ d_{j\circ} > 0$
$d_{ij} \nleq d_{i\circ} + d_{j\circ}$
(c)

Fig. 4. (a) Feature space of two images of objects having the same surrounding, which means that the image of the surrounding (the table in Fig. 3(a)) reduces to a single point $\mathbf{p}_\circ$. (b) When considering spatial distances between the objects, the surrounding image $\mathbf{p}_\circ$ transforms into a blob $\Omega_\circ$ and all distances between objects within $\Omega_\circ$ are zero. (c) When the surroundings of each object are different but have zero distance to each other, the feature space is a combination of (a) and (b).

to the surrounding class can now be defined as a minimization of the distance between the image of the object over all images $\mathbf{p}_k$ from the set of surroundings $\Omega_\circ$

$$d_{i\circ} \triangleq d(\mathbf{x}_i, \Omega_\circ) = \min_k d(\mathbf{x}_i, \mathbf{p}_k) \text{ with } \mathbf{p}_k \in \Omega_\circ.$$

In short, this problem is a combination of the two examples and this leads to the feature space shown in Fig. 4(c). Both images $\mathbf{x}_i$ and $\mathbf{x}_j$ have a related image (prototype), respectively $\hat{\mathbf{p}}_i$ and $\hat{\mathbf{p}}_j$, to which the dissimilarity is the smallest. Again, the triangle inequality does no longer hold: two images that look very different may both be very close to the surrounding class. On the other hand, two objects that are very similar do have similar dissimilarity to the surrounding class. This means that the compactness hypothesis still holds in the space spanned by the dissimilarities. Moreover, the dissimilarity of an object to its surrounding still contains all information for successful classification of the object, which may easily be seen by looking at Fig. 3(b).

3.2 Class Representation

The prototypes $\hat{\mathbf{p}}_i$ and $\hat{\mathbf{p}}_j$ thus represent the surrounding class, but are not available a priori. We know that they must be part of the boundary of $\Omega_\circ$ and that the boundary of $\Omega_\circ$ is the set of objects that divides the feature space of images with protrusions and those without protrusions. Consequently, for each object we can derive its related prototype of the surrounding class by iteratively solving the PDE in (1). That is, $\Omega_s \triangleq \delta\Omega_\circ \cap (\delta\Omega_t \cup \delta\Omega_f)$ are all solutions of (1) and the dissimilarity of an object to its surroundings is the 'cost' of the deformation

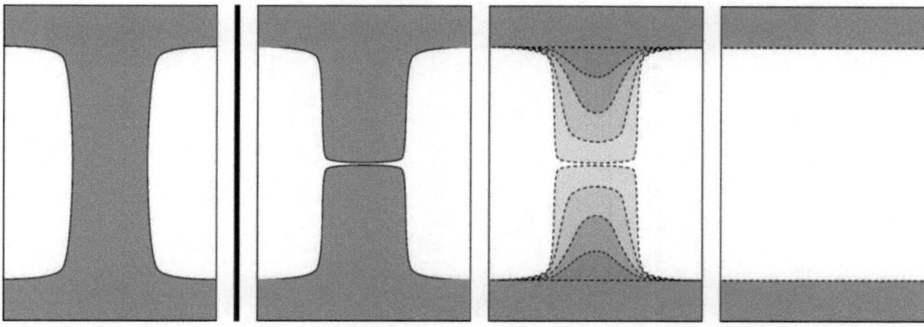

(a) $\mathbf{x}_1 \in \Omega_\circ$ (b) $\mathbf{x}_2$ (c) Deformation (d) $\hat{\mathbf{p}}_2 \in \Omega_s$

Fig. 5. (a–b) Two similar images having different structure lead to different responses to deformation by the PDE in (1). The object $\mathbf{x}_1$ is a solution itself, whereas $\mathbf{x}_2$ will be deformed into $\hat{\mathbf{p}}_2$. A number of structures that might occur during the deformation process are shown in (c).

guided by (1). Furthermore, the prototypes of the surroundings class can now be sampled almost infinitely, i.e. a prototype can be derived when it is needed.

A few characteristics of our approach to object detection are illustrated in Fig. 5. At first glance, objects $\mathbf{x}_1$ and $\mathbf{x}_2$, respectively shown in Figs. 5(a) and (b), seem to be similar (i.e. close together in the feature space spanned by all pixel values), but the structures present in these images differ significantly. This difference in structure is revealed when the images are being transformed by the PDE (1). Object $\mathbf{x}_1$ does not have any protruding elements and can thus be considered as an element of $\Omega_\circ$, whereas object $\mathbf{x}_2$ exhibits two large protrusions: one pointing down from the top, the other pointing up from the bottom. Fig. 5(c) shows several intermediate steps in the deformation of this object and Fig. 5(d) shows the final solution. This illustrates that by defining a suitable deformation, a specific structure can be measured in an image. Using the deformation defined by the PDE in (1), all intermediate images are also valid images with protrusions with decreasing protrudedness. Furthermore, all intermediate objects shown in Fig. 5(c) have the same solution. Thus, different objects can have the same solution and relate to the same prototype.

Let us propose to use a morphological closing operation as the deformation, then one might conclude that images $\mathbf{x}_1$ and $\mathbf{x}_2$ are very similar. In that case we might conclude that image $\mathbf{x}_2$ does not really have the structure of two large polyps, as we concluded before, but might have the same structure as in $\mathbf{x}_1$ altered by an imaging artifact. Using different deformations can thus lead to a better understanding of the local structure. In that case, one could represent each class by a deformation instead of a set of prototypes [1]. Especially for problems involving objects in highly structured surroundings, it might be advantageous to define different deformations in order to infer from structure.

An example of an alternative deformation was already given by the PDE in (2). This deformation creates a new prototype of the polyp class given an image and the 'cost' of deformation could thus be used in classification. Combining

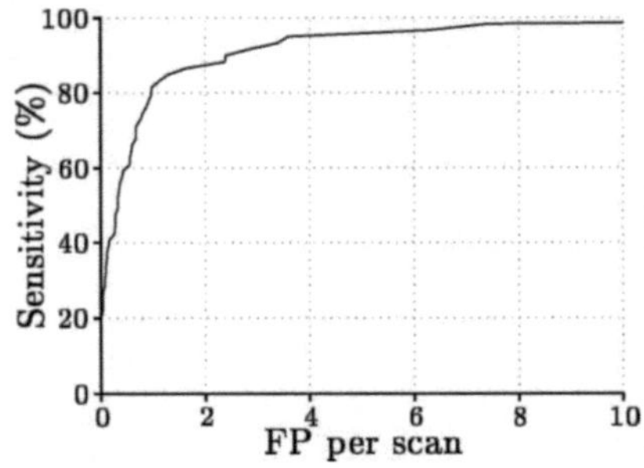

Fig. 6. FROC curve for the detection of polyps $\geq$ 6 mm

both methods thus gives for each object a dissimilarity to both classes. However, this deformation was proposed as a preprocessing step for current CAD systems. By doing so, the dissimilarity was not explicitly used in the candidate detection or classification step.

4 Classification

We now have a very well sampled class of the healthy (normal) images, which do not contain any protrusions. Any deviation from this class indicates unhealthy protrusions. This can be considered as a typical one-class classification problem in which the dissimilarity between the object $\mathbf{x}$ and the prototype $\mathbf{p}$ indicates the probability of belonging to the polyp class. The last step in the design of the polyp detection system is to define a dissimilarity measure that quantifies the introduced deformation, such that it can be used to successfully distinguish the non-polyps from the polyps. As said before, the difference image still contains all information, and thus there is still no class overlap.

Until now, features are computed from this difference image to quantify the 'cost' of deformation. Three features are used for classification: the length of the two principal axes (perpendicular to the polyp axis) of the segmentation of the candidate, and the maximum intensity change. A linear logistic classifier is used for classification. Classification based on the three features obtained from the difference image leads to results comparable to other studies [9,10,11]. Fig. 6 shows a free-response receiver operating characteristics (FROC) curve of the CAD system for 59 polyps larger than 6 mm (smaller polyps are clinically irrelevant) annotated in 86 patients (172 scans). Results of the current polyp detection systems are also presented elsewhere [3,6,12].

5 Conclusion

We have presented an automated polyp detection system based on structural inference. By transforming the image using a structure-driven partial differential equation, knowledge is inferred from the structure in the data. Although no prototypes are available a priori, a prototype of the 'healthy' surrounding class can be obtained for each candidate object. The dissimilarity with the healthy class is obtained by means of a difference image between the image before and after the transformation. This dissimilarity is used for classification of the object as either a polyp or as healthy tissue. Subsequent classification is based on three features derived from the difference image. The current implementation basically acts like a one-class classification system: the system measures the dissimilarity to a well sampled class of volumes showing only normal (healthy) tissue. The class is well sampled in the sense that for each candidate object we can derive a healthy counterpart, which acts as a prototype.

Images that are very similar might not always have the same structure. In the case of structured data, it is this structure that is most important. It was shown that the transformation guided by the PDE in (1) is capable of retrieving structure from data. Furthermore, if two objects are very similar, but situated in a different surrounding, the images might look very different. However, after iteratively solving the PDE, the resulting difference images of the two objects are also similar. The feature space spanned by the dissimilarities thus complies with the compactness hypothesis. However, when a polyp is situated, for example, between two folds, the real structure might not always be retrieved. In such situations no distinction between Figs. 5(a) and (b) can be made due to e.g. the partial volume effect or Gaussian filtering prior to curvature and derivative computations. Prior knowledge about the structure of the colon and the folds in the colon might help in these cases.

Until now, only information is used about the dissimilarity to the 'healthy' class. The work of Konukoglu et al. [7] offers the possibility of deriving a prototype for the polyp class given a candidate object just as we derived prototypes for the non-polyp class. A promising solution might be a combination of both techniques; each candidate object is then characterized by its dissimilarity to a non-polyp prototype and by its dissimilarity to a polyp prototype. Both prototypes are created on-the-fly and are situated in the same surrounding as the candidate. In fact, two classes have been defined and each class is characterized by its own deformation.

In the future, the patient preparation is further reduced to improve patient compliance. This will lead to data with increased amount of fecal remains in the colon and this will complicate both the task of automated polyp detection as well as electronic cleansing of the colon [13,14]. The presented approach to infer from structure can also contribute to the image processing of such data, especially if the structure within the colon becomes increasingly complicated.

References

1. Duin, R.P.W., Pekalska, E.: Structural inference of sensor-based measurements. In: Yeung, D.-Y., Kwok, J.T., Fred, A., Roli, F., de Ridder, D. (eds.) SSPR 2006 and SPR 2006. LNCS, vol. 4109, pp. 41–55. Springer, Heidelberg (2006)

2. Pekalska, E., Duin, R.P.W.: The Dissimilarity Representation for Pattern Recognition, Foundations and Applications. World Scientific, Singapore (2005)
3. van Wijk, C., van Ravesteijn, V.F., Vos, F.M., Truyen, R., de Vries, A.H., Stoker, J., van Vliet, L.J.: Detection of protrusions in curved folded surfaces applied to automated polyp detection in CT colonography. In: Larsen, R., Nielsen, M., Sporring, J. (eds.) MICCAI 2006. LNCS, vol. 4191, pp. 471–478. Springer, Heidelberg (2006)
4. Ferrucci, J.T.: Colon cancer screening with virtual colonoscopy: Promise, polyps, politics. American Journal of Roentgenology 177, 975–988 (2001)
5. Winawer, S., Fletcher, R., Rex, D., Bond, J., Burt, R., Ferrucci, J., Ganiats, T., Levin, T., Woolf, S., Johnson, D., Kirk, L., Litin, S., Simmang, C.: Colorectal cancer screening and surveillance: Clinical guidelines and rationale – update based on new evidence. Gastroenterology 124, 544–560 (2003)
6. van Wijk, C., van Ravesteijn, V.F., Vos, F.M., van Vliet, L.J.: Detection and segmentation of protruding regions on folded iso-surfaces for the detection of colonic polyps (submitted)
7. Konukoglu, E., Acar, B., Paik, D.S., Beaulieu, C.F., Rosenberg, J., Napel, S.: Polyp enhancing level set evolution of colon wall: Method and pilot study. IEEE Trans. Med. Imag. 26(12), 1649–1656 (2007)
8. Pekalska, E., Duin, R.P.W.: Learning with general proximity measures. In: Proc. PRIS 2006, pp. IS15–IS24 (2006)
9. Summers, R.M., Yao, J., Pickhardt, P.J., Franaszek, M., Bitter, I., Brickman, D., Krishna, V., Choi, J.R.: Computed tomographic virtual colonoscopy computer-aided polyp detection in a screening population. Gastroenterology 129, 1832–1844 (2005)
10. Summers, R.M., Handwerker, L.R., Pickhardt, P.J., van Uitert, R.L., Deshpande, K.K., Yeshwant, S., Yao, J., Franaszek, M.: Performance of a previously validated CT colonography computer-aided detection system in a new patient population. AJR 191, 169–174 (2008)
11. Näppi, J., Yoshida, H.: Fully automated three-dimensional detection of polyps in fecal-tagging CT colonography. Acad. Radiol. 14, 287–300 (2007)
12. van Ravesteijn, V.F., van Wijk, C., Truyen, R., Peters, J.F., Vos, F.M., van Vliet, L.J.: Computer aided detection of polyps in CT colonography: An application of logistic regression in medical imaging (submitted)
13. Serlie, I.W.O., Vos, F.M., Truyen, R., Post, F.H., van Vliet, L.J.: Classifying CT image data into material fractions by a scale and rotation invariant edge model. IEEE Trans. Image Process. 16(12), 2891–2904 (2007)
14. Serlie, I.W.O., de Vries, A.H., Vos, F.M., Nio, Y., Truyen, R., Stoker, J., van Vliet, L.J.: Lesion conspicuity and efficiency of CT colonography with electronic cleansing based on a three-material transition model. AJR 191(5), 1493–1502 (2008)

A Binarization Algorithm Based on Shade-Planes for Road Marking Recognition

Tomohisa Suzuki[1], Naoaki Kodaira[1], Hiroyuki Mizutani[1], Hiroaki Nakai[2], and Yasuo Shinohara[2]

[1] Toshiba Solutions Corporation
[2] Toshiba Corporation

Abstract. A binarization algorithm tolerant to both gradual change of intensity caused by shade and the discontinuous changes caused by shadows is described in this paper. This algorithm is based on "shade-planes", in which intensity changes gradually and no edges are included. These shade-planes are produced by selecting a "principal-intensity" in each small block by a quasi-optimization algorithm. One shade-plane is then selected as the background to eliminate the gradual change in the input image. Consequently, the image, with its gradual change removed, is binarized by a conventional global thresholding algorithm. The binarized image is provided to a road marking recognition system, for which influence of shade and shadows is inevitable in the sunlight.

1 Introduction

The recent evolution of car electronics such as low power microprocessors and in-vehicle cameras has enabled us to develop various kinds of on-board computer vision systems [1] [2]. A road marking recognition system is one of such systems. GPS navigation devices can be aided by the road marking recognition system to improve their positioning accuracy. It is also possible to give the driver some advice and cautions according to the road markings.

However, influence of shade and shadows, inevitable in the sunlight, is problematic to such a recognition system in general. The road marking recognition system described in this paper is built with a binarization algorithm that performs well even if the input image is affected by uneven illumination caused by shade and shadows.

To cope with the uneven illumination, several dynamic thresholding techniques were proposed. Niblack proposed a binarization algorithm, in which a dynamic threshold $t(x, y)$ is determined by the mean value $m(x, y)$ and the standard-deviation $\sigma(x, y)$ of pixel values in the neighborhood as follows [4].

$$t(x, y) = m(x, y) + k\sigma(x, y) \qquad (1)$$

where (x, y) is the coordinate of the pixel to be binarized, and k is a predetermined constant. This algorithm is based on the assumption that some of the neighboring pixels belong to the foreground. The word "Foreground" means

A.-B. Salberg, J.Y. Hardeberg, and R. Jenssen (Eds.): SCIA 2009, LNCS 5575, pp. 51–60, 2009.

characters printed on a paper, for example. However, this assumption does not hold in the case of a road surface where spaces are wider than the neighborhood.

To determine appropriate thresholds in such spaces, some binarization algorithms were proposed [5] [6]. In those algorithms, an adaptive threshold surface is determined by the pixels on the edges extracted from the image. Although those algorithms are tolerant to the gradual change of illumination on the road surface, edges irrelevant to the road markings still confound those algorithms.

One of the approaches for solving this problem is to remove the shadows from the image prior to the binarization. In several preceding researches, this shadow removal was realized by using color information. It was assumed in those methods that changes of color are seen on material edges [7] [8]. Despite fair performance for natural sceneries in which various colors tend to be seen, those algorithms does not perform well if the brightness is solely different and no different colors are seen.

Since many road markings tend to appear almost monochrome, we have concluded that the binarization algorithm for the road marking recognition has to tolerate influence of shade and shadows without depending on color information. To fulfill this requirement, we propose a binarization algorithm based on shade-planes. These planes are smooth maps of intensities, and these maps do not have edges which may appear, for an example, on material edges of the road surface or on borders between shadows and sunlit regions. In this method, the gradual change of intensity caused by shade is isolated from the discontinuous change of intensity. An estimated map of background intensity is found in these shade-planes. The input image is then modified to eliminate the gradual change of intensity using the estimated background intensity. Consequently, a commonly used global thresholding algorithm is applied to the modified image.

This binarized image is processed by segmentation, feature extraction and classification which are based on algorithms employed in conventional OCR systems. These conventional algorithms become feasible due to reduction of artifacts caused by shade and shadows with the proposed binarization algorithm.

The recognition result by this system is usable in various applications including GPS navigation devices. For instance, the navigation device can verify whether the vehicle is travelling in the appropriate lane.

In the case shown in Fig.1, the car is travelling in the left lane, in which all vehicles must travel straight through the intersection, despite the correct route heading right. The navigation device detects this contradiction by verifying the road markings which indicate the direction the car is heading for, so that it can suggest the driver to move to the right lane in this case.

It is also possible to calibrate coordinates of the vehicle gained by a GPS navigation device using other coordinates which are calculated from relative position of a recognized road marking and its position on the map.

As a similar example, Ohta et al. [3] proposed a road marking recognition algorithm to give drivers some warnings and advisories. Additionally, Charbonnier et al. [2] developed a system that recognizes road markings and repaints them.

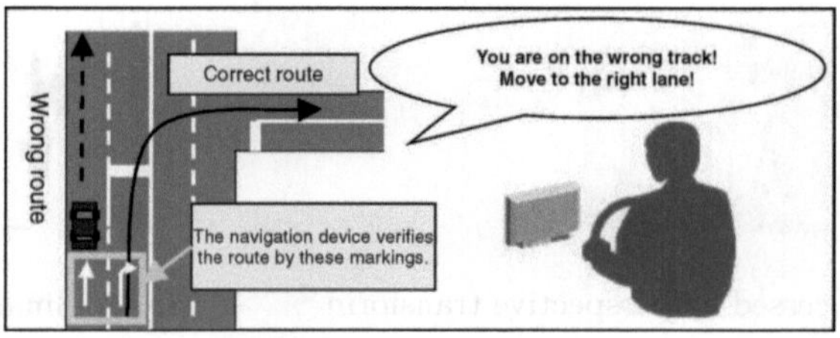

Fig. 1. Lane change suggested by verifying road markings

This paper is organized as follows. The outline of the proposed recognition system is described in Sect.2. Influence of shade and shadows on the images taken by the camera and the binarization result is described in Sect.3. The proposed binarization algorithm is explained in Sect.4. The experimental result of the binarization and the recognition system are shown in Sect.5, and finally, we summarize with some conclusions in Sect.6.

2 Outline of Overall Road Marking Recognition System

The recognition procedure in the proposed system is performed by the following steps: perspective transformation [9], binarization which is the main objective in this paper, lane detection, segmentation, pattern matching and post processing.

As shown in Fig.2, the camera is placed on the rear of the car and directed obliquely to the ground as shown in Fig.3 Since the image taken by a camera in an oblique angle is distorted perspectively, perspective transformation is performed for the image as seen in Fig.4, to produce an image without distortion.

The transformed image is then binarized by the proposed algorithm to extract the patterns of the markings. (See Fig.5) We describe the detail of this algorithm later in Sect.4.

The next step is to extract the lines drawn along the lane on the both sides, in which the road markings are to be recognized. These lines are detected by edges along the road as in the system previously proposed [10].

The road markings, which are shown in Fig.6, are recognized by this system. The segmentation of these symbols is performed by locating their bounding rectangles. Each edge of the bounding rectangles is determined by the horizontal and vertical projection of foreground pixels between the lines detected above.

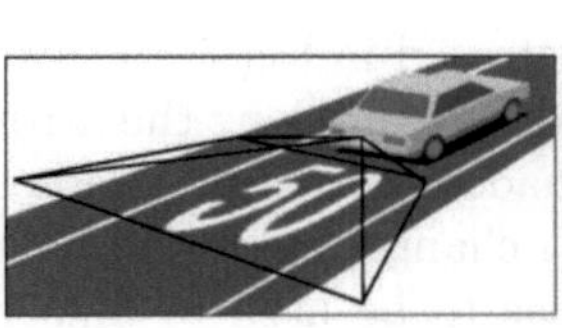

Fig. 2. Angle of the camera

Fig. 3. Image taken by the camera

Fig. 4. Image processed by perspective transform

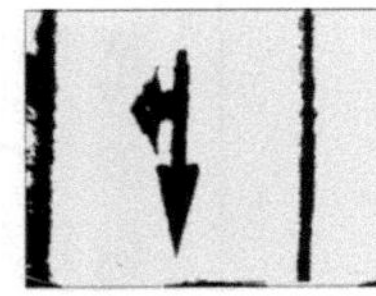

Fig. 5. Binarized image

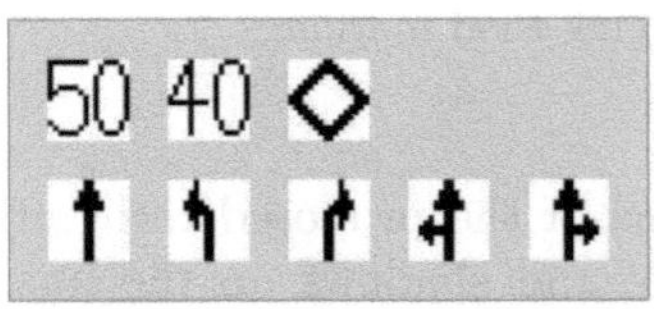

Fig. 6. Recognized road markings

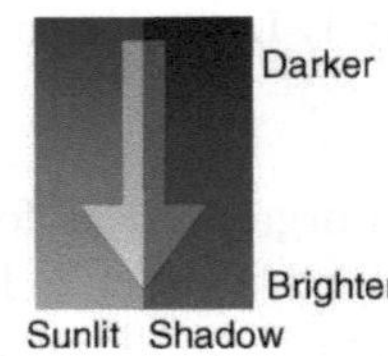

Fig. 7. Road marking with shade and a shadow

The segmented symbols are then recognized by the subspace method [11]. The recognition results are corrected by following post-processes:

- The recognition result for each movie frame is replaced by the most frequently detected marking in neighboring frames. This is done to reduce accidental misclassification of the symbol.
- Some parameters (size, similarity and other measurements) are checked to prevent false detections.
- Consistent results in successive frames are aggregated to one marking.

3 The Influence of Shade, Shadows and Markings on Images

In the example shown in Fig.4, we can see the tendency that the upper right part of the image is brighter than the lower left corner. In addition, the areas covered by the shadows casted by objects beside the road are darker than the rest. As seen in this example, the binarization algorithm applied to this image is to be tolerant to both the gradual changes of intensity caused by shade and the discontinuous change of intensity caused by shadows on the road surface.

For example, these changes of intensity are illustrated in Fig.7. In this example, the gradual change of intensity caused by shade is seen along the arrow, and the discontinuous change of intensity caused by shadow is seen perpendicular to the arrow. From these changes, the discontinuous change on edges of the road marking, the outline of the arrow in this case, has to be used to binarize the image without influence of shade and shadows.

4 The Proposed Binarization Algorithm

In this section, the proposed binarization algorithm is presented.

4.1 Pre-processing Based on the Background Map

In the proposed algorithm, the gradual change of intensity in the input image is eliminated from the input image prior to the binarization by a global thresholding method – Otsu's method [12]. This pre-processing is illustrated by Fig.8 and is performed by producing a modified image (Fig.8(c)) from the input image (Fig.8(a)) and a map of background intensity (Fig.8(b)) with the following equation. This pre-processing flattens the background to make a global thresholding method applicable.

$$g(x,y) = \frac{f(x,y)}{l(x,y)} \tag{2}$$

In this pre-processing, a map of the background intensity called "background map" is estimated by the method described in the following section.

4.2 Estimation of a Background Map by Shade-Planes

In this section, the method for estimating a background map is described.

4.2.1 Detection of Principal-Intensities

An intensity histogram calculated in a small block shown as "small block" in Fig.9 usually consists of peaks at several intensities corresponding to the regions marked with symbols A-D in this figure. We call these intensities "principal-intensity".

The input image is partitioned into small blocks as a PxQ matrix in this algorithm, and the principal-intensities are detected in these blocks. Fig.10 is an example of detected principal-intensities. In this figure, the image is divided into 8x8 blocks. Each block is divided into sub-blocks painted by a principal-intensity. The area of each sub-block indicates the number of the pixels that have the same intensity in the block. As a result, each of the detected principal-intensities corresponds to a white marking, grey road surface or black shadows.

In each block, one of the principal-intensities is expected to be the intensity in the background map at the same position. The principal-intensity corresponding to the background is required to be included in most of the blocks in the proposed

(a) Input image $f(x,y)$ (b) Background map $l(x,y)$ (c) Modified image $g(x,y)$

Fig. 8. A pre-processing is applied to input image

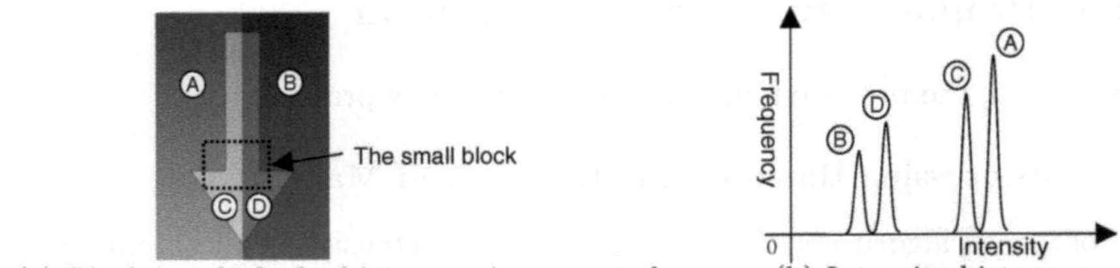

(a) Block in which the histogram is computed (b) Intensity histogram

Fig. 9. Peaks in a histogram for a small block

method. Though, gray sub-blocks corresponding to the background are missing in some blocks at the lower-right corner of the Fig.10.

To compensate the absence of principal-intensities, the histogram averaged in the 5x3 neighbor blocks are calculated instead. Fig.11 shows the result by this modified scheme. As a result, the grey sub-blocks can be observed in all blocks.

4.2.2 The Shade-Planes

In this method, the maps of principal-intensities are called "shade-plane", and a bundle of the plural shade-planes is called a "shade-plane group". Each shade-plane is produced by selecting the principal-intensities for each block as shown in Fig.12. In this example, black sub-blocks among the detected principal-intensities correspond to the road surface in shadows, the grey sub-blocks correspond to the sunlit road surface and the white sub-blocks correspond to markings. The principal-intensities corresponding to the sunlit road surface are selected in the shade-plane #1 and those corresponding to road marking are selected in shade-plane #2.

Principal-intensities in each shade-plane are selected to minimize the following criterion E. This criterion is designed, so that the shade-plane represents gradual change of intensities.

$$E = \sum_{s=1}^{Q} \sum_{r=1}^{P-1} \{L(r+1, s) - L(r, s)\}^2 + \sum_{s=1}^{Q-1} \sum_{r=1}^{P} \{L(r, s+1) - L(r, s)\}^2 \quad (3)$$

where $L(r, s)$ stands for the principal-intensity selected in the block (r, s).

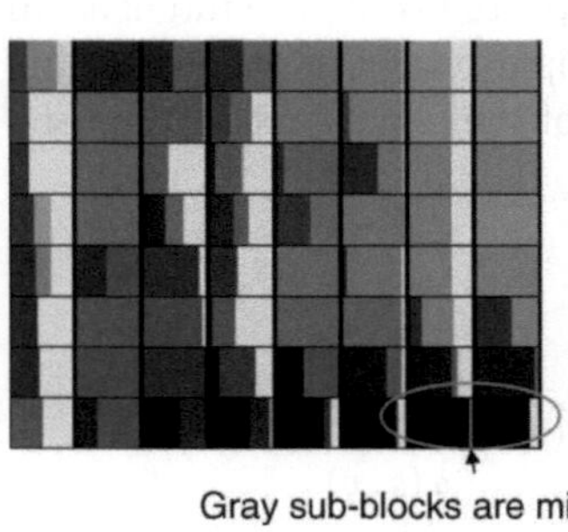

Fig. 10. Results of peak detection

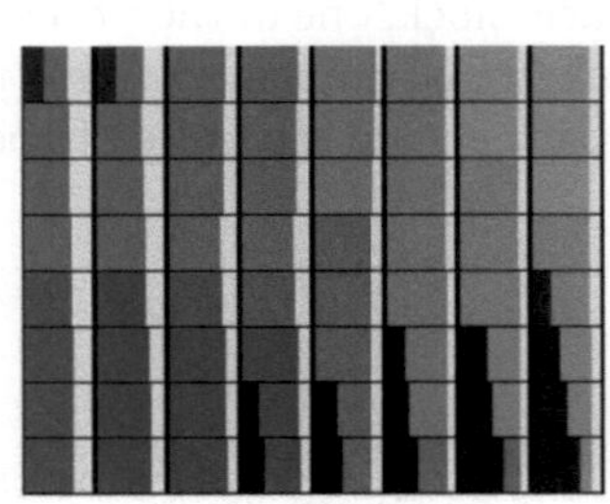

Fig. 11. Results with averaged histograms

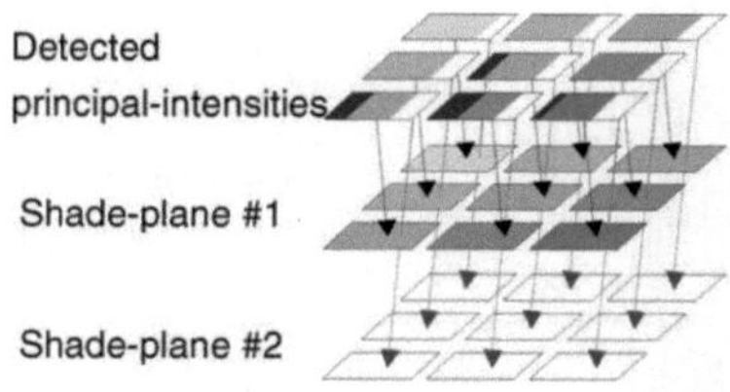

Fig. 12. Shade-planes

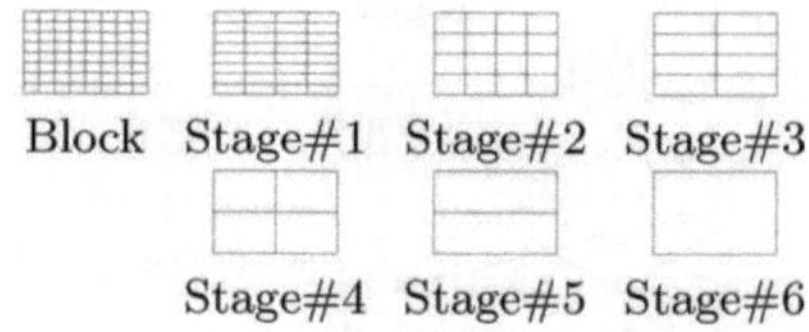

Fig. 13. Merger of areas

The number of the possible combinations of the detected principal-intensities is extremely large. Therefore, a quasi-optimization algorithm with the criterion E is introduced to resolve this problem.

During the optimization process, miniature versions of a shade-plane called a "sub-plane" are created. The sub-planes in the same location form a group called "sub-plane group". The sub-plane groups cover the whole image without overlap altogether. Pairs of adjoining sub-plane groups are merged to larger sub-plane groups step by step, and they finally form the shade-plane group, which is as large as the image. Each step of this process is expressed by "Stage#n" in the following explanation.

Fig.13 shows the merging process of sub-plane groups in these stages. "Blocks" in the Fig.13 indicates the matrix of blocks, and "Stage#n" indicates the matrix of sub-plane groups in each stage. In the stage#1, each pair of horizontally adjoining blocks is merged to form a sub-plane group. In the stage#2, each pair of vertically adjoining sub-plane groups is merged to form a larger sub-plane group. This process is repeated recursively in the same manner. Finally, "Stage#6" shows the shade-plane group.

The creation process of a sub-plane group in stage#1 is shown in Fig.14. In this figure, pairs of principal-intensities from a pair of blocks are combined to create candidates of sub-planes. Consequently, the criterion E is evaluated for each created candidate, and a new sub-plane group is formed by selecting the two sub-planes with the least value of criterion E.

For the stage#2, Fig.15 shows creation of a larger sub-plane group from a pair of sub-plane groups previously created in stage#1. Contrarily to the stage#1, the candidates of the new sub-plane group are created from sub-plane groups instead of principal-intensities.

4.2.3 Selection of the Shade-Planes

A shade plane is selected from the shade-plane group produced by the algorithm described in Sect.4.2.2 as the background map $l(r,s)$. This selection is performed by the following procedure.

1. Eliminate shade-planes similar to another if a pair of shade-planes shares half or more of the principal-intensities.
2. Sort the shade-planes in descending order of the intensity.

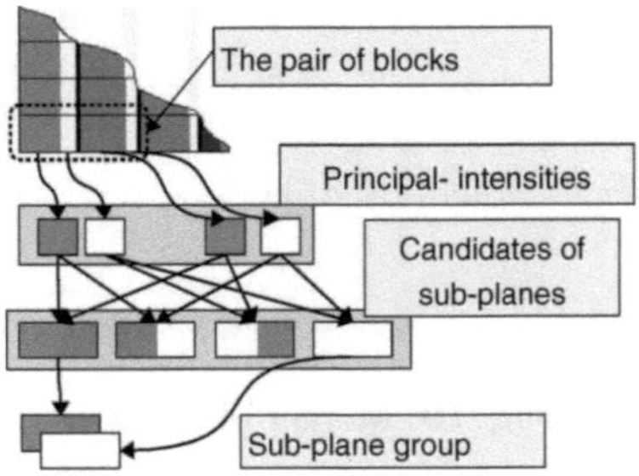

Fig. 14. Sub-planes created in stage#1

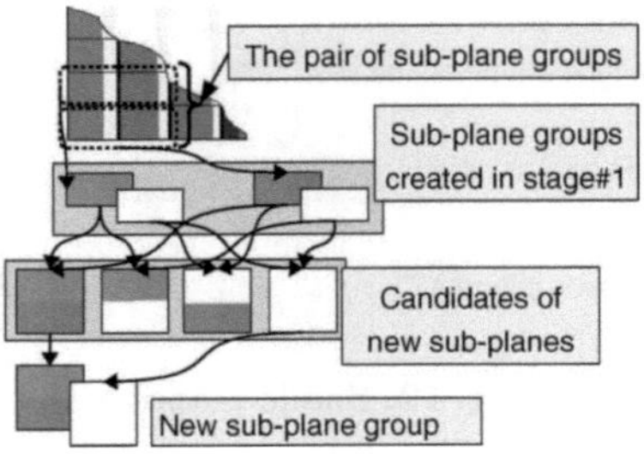

Fig. 15. Sub-planes created in stage#2

3. Select the shade-plane that is closest to the average of shade-planes produced in the preceding K frames. The similarity of shade-planes is computed as Euclidian distance.

5 The Experimental Results

Fig.16 and Fig.17 show the results of the proposed binarization algorithm. In each of the figures, the image (a) is the input, the image (b) is the background map, and the image (c) is the binarization result. As a comparison, the result by Niblack's method is shown in the image (d). Additionally, the image (e) shows the shade-planes produced by the proposed algorithm.

In Fig.16(e), change of intensity corresponding to the marking is seen in "Plane#1" and change of intensity corresponding to road surface is seen in "Plane#2". "Plane#3" and "Plane#4" are useless in this case. These changes of intensity corresponding to the marking and road surface are also seen in Fig.17(e) in "Plane#2" and "Plane#1" respectively.

Contrarily, in the Fig.16(d) and Fig.17(d), the conventional method [4] did not work well.

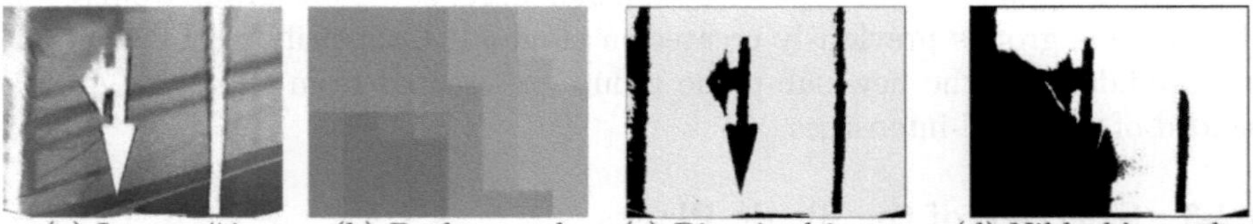

(a) Image #1 (b) Background (c) Binarized image (d) Niblack's method

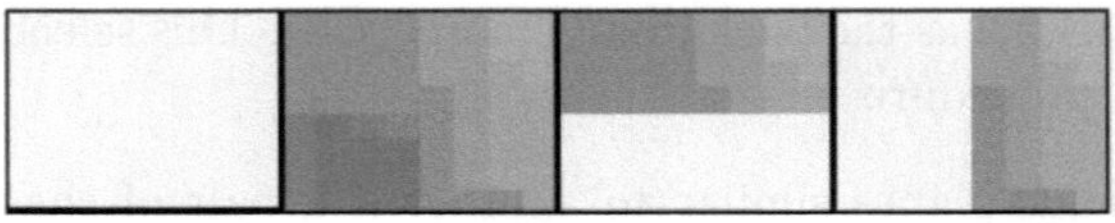

Fig. 16. Experimental results for the sample image#1

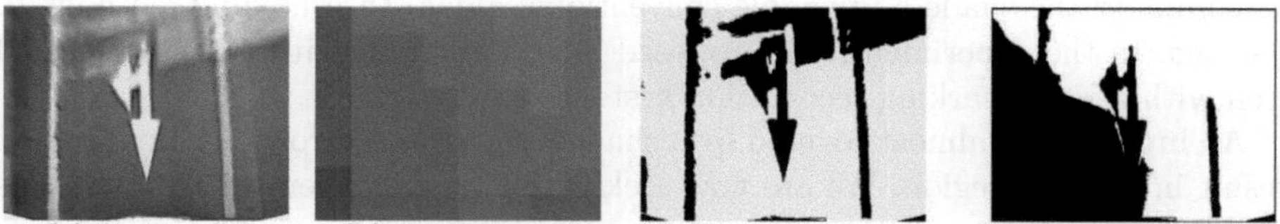

(a) Image #2 (b) Background (c) Binarized image (d) Niblack's method

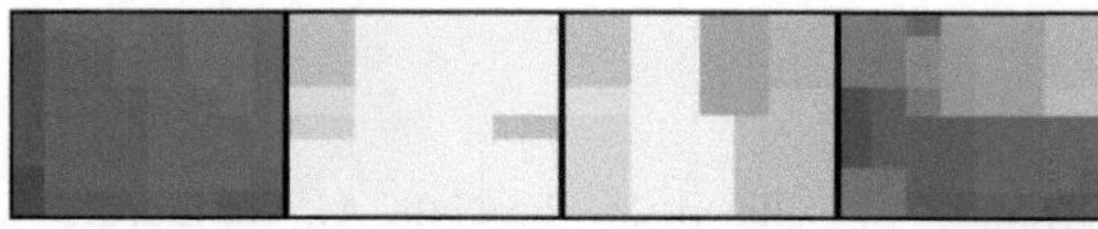

(e) The shade-planes produced by this algorithm

Fig. 17. Experimental results for the sample image#2

Table 1. Recognition performance

Movie No.	Frames	Markings	Detected markings	Errors	Precision	Recall rate
1	27032	64	53	0	100%	83%
2	29898	131	110	0	100%	84%
3	63941	84	65	0	100%	77%
total	120871	279	228	0	100%	82%

The binarization error observed in the upper part in Fig.17(c) is caused by selecting "Plane#1", which corresponds to the shadow region that covers the most area in the image. This led to the binarization error in the sunlit region, for which "Plane#4" would be better.

We implemented the road marking recognition system with the proposed binarization algorithm on a PC with 800MHz P3 processor as an experimental system. The recognition system described above was tested with the QVGA movies taken on the street. The processing time per frame was 20msec on average, and was fast enough to process movie sequences by 30fps. Table 1 shows the recognition performance of these movies in this experiment. The average recall rate of marking-recognition was 82% and no false positives were observed throughout 120,871 frames.

6 Conclusion

A binarization algorithm that tolerates both shade and shadows without color information is described in this paper. In this algorithm, shade-planes associated to gradual changes of intensity are introduced. The shade-planes are produced by a quasi-optimization algorithm based on the divide and conquer approach. Consequently, one of the shade-planes is selected as an estimated background to eliminate the shade and enable conventional global thresholding methods to be used. In the experiment, the proposed binarization algorithm has performed well with a road marking recognition system.

An input image almost covered by a shadow showed an erroneous binarization result in a sunlit region. We are now seeking for an enhancement to mend this problem.

References

1. Bertozzi, M., Broggi, A., Cellario, M., Fascioli, A., Lombardi, P., Porta, M.: Artificial Vision in Road Vehicles. Proc. IEEE 90(7), 1258–1271 (2002)
2. Charbonnier, P., Diebolt, F., Guillard, Y., Peyret, F.: Road markings recognition using image processing. In: IEEE Conference on Intelligent Transportation System (ITSC 1997), November 9-12, 1997, pp. 912–917 (1997)
3. Ohta, H., Shiono, M.: An Experiment on Extraction and Recognition of Road Markings from a Road Scene Image, Technical Report of IEICE, PRU95-188, 1995-12, pp. 79–86 (in Japanese)
4. Niblack: An Introduction to Digital Image Processing, pp. 115–116. Prentice-Hall, Englewood Cliffs (1986)
5. Yanowitz, S.D., Bruckstein, A.M.: A new method for image segmentation. Comput.Vision Graphics Image Process. 46, 82–95 (1989)
6. Blayvas, I., Bruckstein, A., Kimmel, R.: Efficient computation of adaptive threshold surfaces for image binarization. In: Proceedings of the 2001 IEEE Computer Society Conference on Computer Vision and Pattern Recognition, December 2001, vol. 1, pp. 737–742 (2001)
7. Finlayson, G.D., Hordley, S.D., Cheng Lu Drew, M.S.: On the removal of shadows from images. IEEE Transactions on Pattern Analysis and Machine Intelligence 28, 59–68 (2006)
8. Nielsen, M., Madsen, C.B.: Graph Cut Based Segmentation of Soft Shadows for Seamless Removal and Augmentation. In: Ersbøll, B.K., Pedersen, K.S. (eds.) SCIA 2007. LNCS, vol. 4522, pp. 918–927. Springer, Heidelberg (2007)
9. Forsyth, D.A., Ponce, J.: Computer Vision A Modern Approach, pp. 20–37. Prentice Hall, Englewood Cliffs (2003)
10. Nakayama, H., et al.: White line detection by tracking candidates on a reverse projection image, Technical report of IEICE, PRMU 2001-87, pp. 15–22 (2001) (in Japanese)
11. Oja, E.: Subspace Methods of Pattern Recognition. Research Studies Press Ltd. (1983)
12. Otsu, N.: A threshold selection method from gray-level histograms. IEEE Trans. Sys. Man Cyber. 9(1), 62–66 (1979)

Rotation Invariant Image Description with Local Binary Pattern Histogram Fourier Features

Timo Ahonen[1], Jiří Matas[2], Chu He[3,1], and Matti Pietikäinen[1]

[1] Machine Vision Group, University of Oulu, Finland
{tahonen,mkp}@ee.oulu.fi
[2] Center for Machine Percpetion, Dept. of Cybernetics,
Faculty of Elec. Eng., Czech Technical University in Prague
matas@cmp.felk.cvut.cz
[3] School of Electronic Information, Wuhan University, P.R. China
chuhe.whu@gmail.com

Abstract. In this paper, we propose Local Binary Pattern Histogram Fourier features (LBP-HF), a novel rotation invariant image descriptor computed from discrete Fourier transforms of local binary pattern (LBP) histograms. Unlike most other histogram based invariant texture descriptors which normalize rotation locally, the proposed invariants are constructed globally for the whole region to be described. In addition to being rotation invariant, the LBP-HF features retain the highly discriminative nature of LBP histograms. In the experiments, it is shown that these features outperform non-invariant and earlier version of rotation invariant LBP and the MR8 descriptor in texture classification, material categorization and face recognition tests.

1 Introduction

Rotation invariant texture analysis is a widely studied problem [1], [2], [3]. It aims at providing with texture features that are invariant to rotation angle of the input texture image. Moreover, these features should typically be robust also to image formation conditions such as illumination changes.

Describing the appearance locally, e.g., using co-occurrences of gray values or with filter bank responses and then forming a global description by computing statistics over the image region is a well established technique in texture analysis [4]. This approach has been extended by several authors to produce rotation invariant features by transforming each local descriptor to a canonical representation invariant to rotations of the input image [2], [3], [5]. The statistics describing the whole region are then computed from these transformed local descriptors.

Even though such approaches have produced good results in rotation invariant texture classification, they have some weaknesses. Most importantly, as each local descriptor (e.g., filter bank response) is transformed to canonical representation independently, the relative distribution of different orientations is lost. Furthermore, as the transformation needs to be performed for each texton, it must be computationally simple if the overall computational cost needs to be low.

A.-B. Salberg, J.Y. Hardeberg, and R. Jenssen (Eds.): SCIA 2009, LNCS 5575, pp. 61–70, 2009.

In this paper, we propose novel Local Binary Pattern Histogram Fourier features (LBP-HF), a rotation invariant image descriptor based on uniform Local Binary Patterns (LBP) [2]. LBP is an operator for image description that is based on the signs of differences of neighboring pixels. It is fast to compute and invariant to monotonic gray-scale changes of the image. Despite being simple, it is very descriptive, which is attested by the wide variety of different tasks it has been successfully applied to. The LBP histogram has proven to be a widely applicable image feature for, e.g., texture classification, face analysis, video background subtraction, interest region description, etc[1].

Unlike the earlier local rotation invariant features, the LBP-HF descriptor is formed by first computing a non-invariant LBP histogram over the whole region and then constructing rotationally invariant features from the histogram. This means that rotation invariance is attained globally, and the features are thus invariant to rotations of the whole input signal but they still retain information about relative distribution of different orientations of uniform local binary patterns.

2 Rotation Invariant Local Binary Pattern Descriptors

The proposed rotation invariant local binary pattern histogram Fourier features are based on uniform local binary pattern histograms. First, the LBP methodology is briefly reviewed and the LBP-HF features are then introduced.

2.1 The Local Binary Pattern Operator

The local binary pattern operator [2] is a powerful means of texture description. The original version of the operator labels the image pixels by thresholding the 3x3-neighborhood of each pixel with the center value and summing the thresholded values weighted by powers of two.

The operator can also be extended to use neighborhoods of different sizes [2] (See Fig.1). To do this, a circular neighborhood denoted by (P, R) is defined. Here P represents the number of sampling points and R is the radius of the neighborhood. These sampling points around pixel (x, y) lie at coordinates $(x_p, y_p) = (x + R\cos(2\pi p/P), y - R\sin(2\pi p/P))$. When a sampling point does not fall at integer coordinates, the pixel value is bilinearly interpolated. Now the LBP label for the center pixel (x, y) of image $f(x, y)$ is obtained through

$$LBP_{P,R}(x,y) = \sum_{p=0}^{P-1} s(f(x,y) - f(x_p, y_p))2^p, \tag{1}$$

where $s(z)$ is the thresholding function

$$s(z) = \begin{cases} 1, & z \geq 0 \\ 0, & z < 0 \end{cases} \tag{2}$$

[1] See LBP bibliography at http://www.ee.oulu.fi/mvg/page/lbp_bibliography

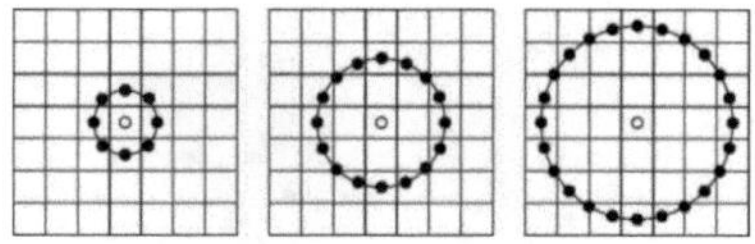

Fig. 1. Three circular neighborhoods: (8,1), (16,2), (24,3). The pixel values are bilinearly interpolated whenever the sampling point is not in the center of a pixel.

Further extensions to the original operator are so called *uniform* patterns [2]. A local binary pattern is called uniform if the binary pattern contains at most two bitwise transitions from 0 to 1 or vice versa when the bit pattern is considered circular. In the computation of the LBP histogram, uniform patterns are used so that the histogram has a separate bin for every uniform pattern and all non-uniform patterns are assigned to a single bin. The 58 possible uniform patterns in neighborhood of 8 sampling points are shown in Fig. 2.

The original rotation invariant LBP operator, denoted here as LBP^{riu2}, is achieved by circularly rotating each bit pattern to the minimum value. For instance, the bit sequences 1000011, 1110000 and 0011100 arise from different rotations of the same local pattern and they all correspond to the normalized sequence 0000111. In Fig. 2 this means that all the patterns from one row are replaced with a single label.

2.2 Invariant Descriptors from LBP Histograms

Let us denote a specific uniform LBP pattern by $U_P(n,r)$. The pair (n,r) specifies an uniform pattern so that n is the number of 1-bits in the pattern (corresponds to row number in Fig. 2) and r is the rotation of the pattern (column number in Fig. 2).

Now if the neighborhood has P sampling points, n gets values from 0 to $P+1$, where $n = P+1$ is the special label marking all the non-uniform patterns. Furthermore, when $1 \leq n \leq P-1$, the rotation of the pattern is in the range $0 \leq r \leq P-1$.

Let $I^{\alpha^\circ}(x,y)$ denote the rotation of image $I(x,y)$ by α degrees. Under this rotation, point (x,y) is rotated to location (x',y'). If we place a circular sampling neighborhood on points $I(x,y)$ and $I^{\alpha^\circ}(x',y')$, we observe that it also rotates by α°. See Fig. 3.

If the rotations are limited to integer multiples of the angle between two sampling points, *i.e.* $\alpha = a\frac{360^\circ}{P}, a = 0,1,\ldots,P-1$, this rotates the sampling neighborhood exactly by a discrete steps. Therefore the uniform pattern $U_P(n,r)$ at point (x,y) is replaced by uniform pattern $U_P(n, r+a \mod P)$ at point (x',y') of the rotated image.

Now consider the uniform LBP histograms $h_I(U_P(n,r))$. The histogram value h_I at bin $U_P(n,r)$ is the number of occurrences of uniform pattern $U_P(n,r)$ in image I.

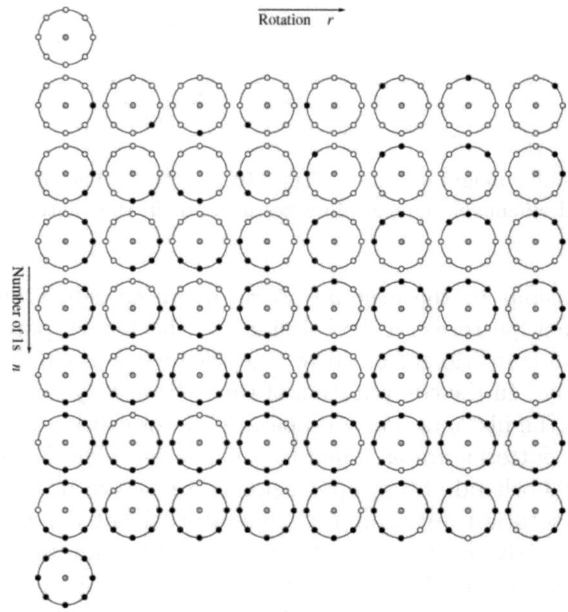

Fig. 2. The 58 different uniform patterns in (8,R) neighborhood

If the image I is rotated by $\alpha = a\frac{360^\circ}{P}$, based on the reasoning above, this rotation of the input image causes a cyclic shift in the histogram along each of the rows,

$$h_{I^{\alpha^\circ}}(U_P(n, r + a)) = h_I(U_P(n, r)) \tag{3}$$

For example, in the case of 8 neighbor LBP, when the input image is rotated by 45°, the value from histogram bin $U_8(1, 0) = 000000001b$ moves to bin $U_8(1, 1) = 00000010b$, value from bin $U_8(1, 1)$ to bin $U_8(1, 2)$, etc.

Based on the property, which states that rotations induce shift in the polar representation (P, R) of the neighborhood, we propose a class of features that are invariant to rotation of the input image, namely such features, computed along the input histogram rows, that are invariant to cyclic shifts.

We use the Discrete Fourier Transform to construct these features. Let $H(n, \cdot)$ be the DFT of nth row of the histogram $h_I(U_P(n, r))$, *i.e.*

$$H(n, u) = \sum_{r=0}^{P-1} h_I(U_P(n, r))e^{-i2\pi ur/P}. \tag{4}$$

Now for DFT it holds that a cyclic shift of the input vector causes a phase shift in the DFT coefficients. If $h'(U_P(n, r)) = h(U_P(n, r - a))$, then

$$H'(n, u) = H(n, u)e^{-i2\pi ua/P}, \tag{5}$$

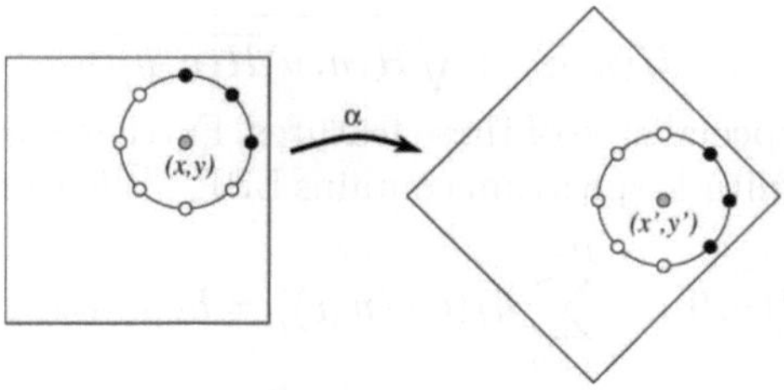

Fig. 3. Effect of image rotation on points in circular neighborhoods

and therefore, with any $1 \leq n_1, n_2 \leq P-1$,

$$H'(n_1,u)\overline{H'(n_2,u)} = H(n_1,u)e^{-i2\pi ua/P}\overline{H(n_2,u)e^{i2\pi ua/P}} = H(n_1,u)\overline{H(n_2,u)}, \quad (6)$$

where $\overline{H(n_2,u)}$ denotes the complex conjugate of $H(n_2,u)$.

This shows that with any $1 \leq n_1, n_2 \leq P-1$ and $0 \leq u \leq P-1$, the features

$$\text{LBP}^{u2}\text{-HF}(n_1,n_2,u) = H(n_1,u)\overline{H(n_2,u)}, \quad (7)$$

are invariant to cyclic shifts of the rows of $h_I(U_P(n,r))$ and consequently, they are invariant also to rotations of the input image $I(x,y)$. The Fourier magnitude spectrum

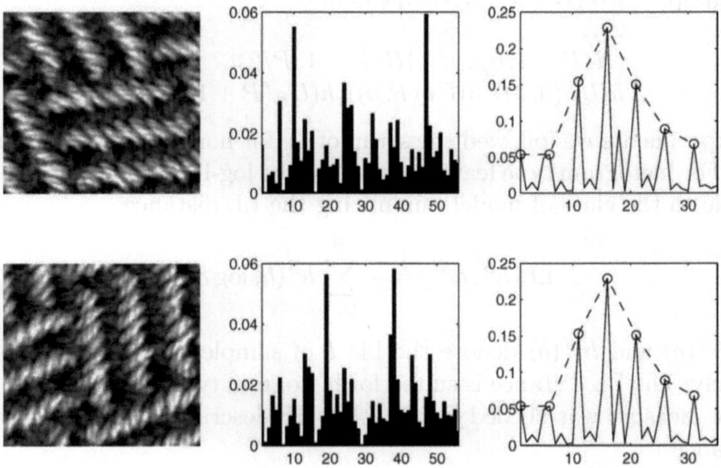

Fig. 4. 1st column: Texture image at orientations 0° and 90°. 2nd column: bins 1–56 of the corresponding LBP^{u2} histograms. 3rd column: Rotation invariant features $|H(n,u)|, 1 \leq n \leq 7, 0 \leq u \leq 5$, (solid line) and LBP^{riu2} (circles, dashed line). Note that the LBP^{u2} histograms for the two images are markedly different, but the $|H(n,u)|$ features are nearly equal.

$$|H(n,u)| = \sqrt{H(n,u)\overline{H(n,u)}} \tag{8}$$

can be considered a special case of these features. Furthermore it should be noted that the Fourier magnitude spectrum contains LBP^{riu2} features as a subset, since

$$|H(n,0)| = \sum_{r=0}^{P-1} h_I(U_P(n,r)) = h_{\mathrm{LBP}^{riu2}}(n). \tag{9}$$

An illustration of these features is in Fig. 4

3 Experiments

We tested the performance of the proposed descriptor in three different scenarios: texture classification, material categorization and face description. The proposed rotation invariant LBP-HF features were compared against non-invariant LBP^{u2} and the older rotation invariant version LBP^{riu2}. In the texture classification and material categorization experiments, the MR8 descriptor [3] was used as an additional control method. The results for the MR8 descriptor were computed using the setup from [6].

In preliminary tests, the Fourier magnitude spectrum was found to give most consistent performance over the family of different possible features (Eq. (7)). Therefore, in the following we use feature vectors consisting of three LBP histogram values (all zeros, all ones, non-uniform) and Fourier magnitude spectrum values. The feature vectors are of the following form:

$$\begin{aligned} fv_{\text{LBP-HF}} = [&|H(1,0)|, \ldots, |H(1,P/2)|, \\ &\ldots, \\ &|H(P-1,0)|, \ldots, |H(P-1,P/2)|, \\ &h(U_P(0,0)), h(U_P(P,0)), h(U_P(P+1,0))]_{1\times((P-1)(P/2+1)+3)}. \end{aligned}$$

In experiments we followed the setup of [2] for nonparametric texture classification. For histogram type features, we used the log-likelihood statistic, assigning a sample to the class of model minimizing the LL distance

$$\mathrm{LL}(h^S, h^M) = -\sum_{b=1}^{B} h^S(b) \log h^M(b), \tag{10}$$

where $h^S(b)$ and $h^M(b)$ denote the bin b of sample and model histograms, respectively. The LL distance is suited for histogram type features, thus a different distance measure was needed for the LBP-HF descriptor. For these features, the L_1 distance

$$\mathrm{L}_1(fv^S_{\text{LBP-HF}}, fv^M_{\text{LBP-HF}}) = \sum_{k=1}^{K} |fv^S_{\text{LBP-HF}}(k) - fv^M_{\text{LBP-HF}}(k)| \tag{11}$$

was selected. We derived from the setup of [2] by using nearest neighbor (NN) classifier instead of 3NN because no significant performance difference between the two was observed and in the setup for the last experiment we had only 1 training sample per class.

Table 1. Texture recognition rates on Outex_TC_0012 dataset

	LBP^{u2}	LBP^{riu2}	LBP-HF
$(8, 1)$	0.566	0.646	0.773
$(16, 2)$	0.578	0.791	0.873
$(24, 3)$	0.45	0.833	0.896
$(8, 1) + (16, 2)$	0.595	0.821	0.894
$(8, 1) + (24, 3)$	0.512	0.883	0.917
$(16, 2) + (24, 3)$	0.513	0.857	0.915
$(8, 1) + (16, 2) + (24, 3)$	0.539	0.87	**0.925**
MR8		0.761	

3.1 Experiment 1: Rotation Invariant Texture Classification

In the first experiment, we used the Outex_TC_0012 [7] test set intended for testing rotation invariant texture classification methods. This test set consists of 9120 images representing 24 different textures imaged under different rotations and lightings. The test set contains 20 training images for each texture class. The training images are under single orientation whereas different orientations are present in the total of 8640 testing images. We report here the total classification rates over all test images.

The results of the first experiment are in Table 1. As it can be observed, the both rotation invariant features provide better classification rates than non-invariant features. The performance of LBP-HF features is clearly higher than that of MR8 and LBP^{riu2}. This can be observed at all tested scales, but the difference between LBP-HF and LBP^{riu2} is particularly large at the smallest scale $(8, 1)$.

3.2 Experiment 2: Material Categorization

In next experiments, we aimed to test how well the novel rotation invariant features retain the discriminativeness of the original LBP features. This was tested using two challenging problems, namely material categorization and illumination invariant face recognition

In Experiment 2, we tested the performance of the proposed features in material categorization using the KTH-TIPS2 database[2]. For this experiment, we used the same setup as in Experiment 1. This test setup resembles the most difficult setup used in [8].

The KTH-TIPS2 database contains 4 samples of 11 different materials, each sample imaged at 9 different scales and 12 lighting and pose setups, totaling 4572 images. Using each of the descriptors to be tested, a nearest neighbor classifier was trained with one sample (i.e. 9*12 images) per material category. The remaining 3*9*12 images were used for testing. This was repeated with 10000 random combinations as training and testing data and the mean categorization rate over the permutations is used to assess the performance.

[2] http://www.nada.kth.se/cvap/databases/kth-tips/

Table 2. Material categorization rates on KTH TIPS2 dataset

	LBP^{u2}	LBP^{riu2}	LBP-HF
$(8, 1)$	0.528	0.482	0.525
$(16, 2)$	0.511	0.494	0.533
$(24, 3)$	0.502	0.481	0.513
$(8, 1) + (16, 2)$	0.536	0.502	0.542
$(8, 1) + (24, 3)$	0.542	0.507	0.542
$(16, 2) + (24, 3)$	0.514	0.508	0.539
$(8, 1) + (16, 2) + (24, 3)$	0.536	0.514	**0.546**
MR8		0.455	

Results of material categorization experiments are in Table 2. LBP-HF reaches, or with most scales even exceeds the performance of LBP^{u2}. The performance of LBP^{riu2} is consistently lower than that of the other two, and the MR8 descriptor gives the lowest recognition rate. The reason for LBP-HF not performing significantly better then non-invariant LBP is most likely that different orientations are present in the training data so rotational invariance does not benefit much here. Unlike with LBP^{riu2}, no information is lost either, but a slight improvement over the non-invariant descriptor is achieved instead.

3.3 Experiment 3: Face Recognition

The third experiment was aimed to further assess whether useful information is lost due to the transformation making the features rotation invariant. For this test, we chose the face recognition problem where the input images have been manually registered, so rotation invariance is not actually needed.

The CMU PIE (Pose, Illumination, and Expression) database [9] was used for this experiment. Totally, the database contains 41368 images of 68 subjects taken at different angles, lighting conditions and with varying expression. For our experiments, we selected a set of 23 images of each of the 68 subjects. 2 of these are taken with the room lights on and the remaining 21 each with a flash at varying positions.

In obtaining a descriptor for the facial image, the procedure of [10] was followed. The faces were first normalized so that the eyes are at fixed positions. The uniform LBP operator at chosen scale was then applied and the resulting label image was cropped to size 128×128 pixels. The cropped image was further divided into blocks of size of 16×16 pixels and histograms were computed in each block individually. In case of LBP-HF descriptor, the rotation invariant transform was applied to the histogram, and finally the features obtained within each block were concatenated to form the spatially enhanced histogram describing the face.

Due to the sparseness of the resulting histograms, Chi square distance was used with histogram type features in this experiments. With LBP-HF descriptor, L1 distance was used as in the previous experiment.

Table 3. Face recognition rates on CMU PIE dataset

	LBP^{u2}	LBP^{riu2}	LBP-HF
(8, 1)	0.726	0.649	0.716
(8, 2)	0.744	0.699	**0.748**
(8, 3)	0.727	0.680	0.726

On each test round, one image per person was used for training and the remaining 22 images for testing. Again, 10000 random selections into training and testing data were used.

Results of the face recognition experiment are in Table 3. Surprisingly, the performance of rotation invariant LBP-HF is almost equal to non-invariant LBP^{u2} even though there are no global rotations present in the images.

4 Discussion and Conclusion

In this paper, we proposed rotation invariant LBP-HF features based on local binary pattern histograms. It was shown that rotations of the input image cause cyclic shifts of the values in the uniform LBP histogram. Relying on this observation we proposed discrete Fourier transform based features that are invariant to cyclic shifts of input vector and, when computed from uniform LBP histograms, hence invariant to rotations of input image.

Several other histogram based rotation invariant texture features have been discussed in the literature, e.g., [2], [3], [5]. The method proposed here differs from those since LBP-HF features are computed from the histogram representing the whole region, i.e. the invariants are constructed globally instead of computing invariant independently at each pixel location. The major advantage of this approach is that the relative distribution of local orientations is not lost.

Another benefit of constructing invariant features globally is that invariant computation needs not to be performed at every pixel location. This allows for using computationally more complex invariant functions still keeping the total computational cost reasonable. In case of LBP-HF descriptor, the computational overhead is negligible. After computing the non-invariant LBP histogram, only $P - 1$ Fast Fourier Transforms of P points need to be computed to construct the rotation invariant LBP-HF descriptor.

In the experiments, it was shown that in addition to being rotation invariant, the proposed features retain the highly discriminative nature of LBP histograms. The LBP-HF descriptor was shown to outperform the MR8 descriptor and the non-invariant and earlier version of rotation invariant LBP in texture classification, material categorization and face recognition tests.

Acknowledgements. This work was supported by the Academy of Finland and the EC project IST-214324 MOBIO. JM was supported by EC project ICT-215078 DIPLECS.

References

1. Zhang, J., Tan, T.: Brief review of invariant texture analysis methods. Pattern Recognition 35(3), 735–747 (2002)
2. Ojala, T., Pietikäinen, M., Mäenpää, T.: Multiresolution gray-scale and rotation invariant texture classification with local binary patterns. IEEE Transactions on Pattern Analysis and Machine Intelligence 24(7), 971–987 (2002)
3. Varma, M., Zisserman, A.: A statistical approach to texture classification from single images. International Journal of Computer Vision 62(1–2), 61–81 (2005)
4. Tuceryan, M., Jain, A.K.: Texture analysis. In: Chen, C.H., Pau, L.F., Wang, P.S.P. (eds.) The Handbook of Pattern Recognition and Computer Vision, 2nd edn., pp. 207–248. World Scientific Publishing Co., Singapore (1998)
5. Arof, H., Deravi, F.: Circular neighbourhood and 1-d dft features for texture classification and segmentation. IEE Proceedings - Vision, Image and Signal Processing 145(3), 167–172 (1998)
6. Ahonen, T., Pietikäinen, M.: Image description using joint distribution of filter bank responses. Pattern Recognition Letters 30(4), 368–376 (2009)
7. Ojala, T., Mäenpää, T., Pietikäinen, M., Viertola, J., Kyllönen, J., Huovinen, S.: Outex - new framework for empirical evaluation of texture analysis algorithms. In: Proc. 16th International Conference on Pattern Recognition (ICPR 2002), vol. 1, pp. 701–706 (2002)
8. Caputo, B., Hayman, E., Mallikarjuna, P.: Class-specific material categorisation. In: 10th IEEE International Conference on Computer Vision (ICCV 2005), pp. 1597–1604 (2005)
9. Sim, T., Baker, S., Bsat, M.: The cmu pose, illumination, and expression database. IEEE Transactions on Pattern Analysis and Machine Intelligence 25(12), 1615–1618 (2003)
10. Ahonen, T., Hadid, A., Pietikäinen, M.: Face description with local binary patterns: Application to face recognition. IEEE Transactions on Pattern Analysis and Machine Intelligence 28(12), 2037–2041 (2006)

Weighted DFT Based Blur Invariants for Pattern Recognition

Ville Ojansivu and Janne Heikkilä

Machine Vision Group, Department of Electrical and Information Engineering,
University of Oulu, PO Box 4500, 90014, Finland
{vpo,jth}@ee.oulu.fi

Abstract. Recognition of patterns in blurred images can be achieved without deblurring of the images by using image features that are invariant to blur. All known blur invariants are based either on image moments or Fourier phase. In this paper, we introduce a method that improves the results obtained by existing state of the art blur invariant Fourier domain features. In this method, the invariants are weighted according to their reliability, which is proportional to their estimated signal-to-noise ratio. Because the invariants are non-linear functions of the image data, we apply a linearization scheme to estimate their noise covariance matrix, which is used for computation of the weighted distance between the images in classification. We applied similar weighting scheme to blur and blur-translation invariant features in the Fourier domain. For illustration we did experiments also with other Fourier and spatial domain features with and without weighting. In the experiments, the classification accuracy of the Fourier domain invariants was increased by up to 20 % through the use of weighting.

1 Introduction

Recognition of objects and patterns in images is a fundamental part of computer vision with numerous applications. The task is difficult as the objects rarely look similar in different conditions. Images may contain various artefacts such as geometrical and convolutional degradations. In an ideal situation, an image analysis system should be invariant to the degradations.

We are specifically interested in invariance to image blurring, which is one type of image degradation. Typically, blur is caused by motion between the camera and the scene, an out of focus of the lens, or atmospheric turbulence. Although most of the research on invariants has been devoted to geometrical invariance [1], there are also papers considering blur invariance [2,3,4,5,6]. An alternative approach to blur insensitive recognition would be deblurring of the images, followed by recognition of the sharp pattern. However, deblurring is an ill-posed problem which often results in new artefacts in images [7].

All of the blur invariant features introduced thus far are invariant to uniform centrally symmetric blur. In an ideal case, the point spread functions (PSF) of linear motion, out of focus, and atmospheric turbulence blur for a long exposure

A.-B. Salberg, J.Y. Hardeberg, and R. Jenssen (Eds.): SCIA 2009, LNCS 5575, pp. 71–80, 2009.

are centrally symmetric [7]. The invariants are computed either in the spatial domain [2,3,4] or in the Fourier domain [5,6], and have also geometrical invariance properties.

For blur and blur-translation invariants, the best classification results are obtained using the invariants proposed in [5], which are computed from the phase spectrum or bispectrum phase of the images. The former are called *phase blur invariants* (PBI) and the latter, which are also translation invariant, are referred to as *phase blur-translation invariants* (PBTI). These methods are less sensitive to noise compared to image moment based blur-translation invariants [2] and are also faster to compute using FFT. Also other Fourier domain blur invariants have been proposed, which are based on a tangent of the Fourier phase [2] and are referred as the phase-tangent invariants in this paper. However, these invariants tend to be very unstable due to the properties of the tangent-function. PBTIs are also the only combined blur-translation invariants in the Fourier domain. Because all the Fourier domain invariants utilize only the phase, they are additionally invariant to uniform illumination changes.

The stability of the phase-tangent invariants was greatly improved in [8] by using a statistical weighting of the invariants based on the estimated effect of image noise. Weighting improved also the results of moment invariants slightly. In this paper, we utilize a similar weighting scheme for the PBI and PBTI features. We also present comparative experiments between all the blur and blur-translation invariants, with and without weighting.

2 Blur Invariant Features Based on DFT Phase

The blur invariant features introduced in [5] assume that the blurred images $g(\mathbf{n})$ are generated by a linear shift invariant (LSI) process which is given by the convolution of the ideal image $f(\mathbf{n})$ with a point spread function (PSF) of the blur $h(\mathbf{n})$, namely

$$g(\mathbf{n}) = (f * h)(\mathbf{n}) \ , \tag{1}$$

where $\mathbf{n} = [n_1, n_2]^T$ denotes discrete spatial coordinates. It is further assumed that $h(\mathbf{n})$ is centrally symmetric, that is $h(\mathbf{n}) = h(-\mathbf{n})$. In practice, images contain also noise, whereupon the observed image becomes

$$\hat{g}(\mathbf{n}) = g(\mathbf{n}) + w(\mathbf{n}) \ , \tag{2}$$

where $w(\mathbf{n})$ denotes additive noise.

In the Fourier domain, the same blurring process is given by a multiplication. By neglecting the noise term, this is expressed by

$$G(\mathbf{u}) = F(\mathbf{u}) \cdot H(\mathbf{u}) \ , \tag{3}$$

where $G(\mathbf{u})$, $F(\mathbf{u})$, and $H(\mathbf{u})$ are the 2-D discrete Fourier transforms (DFT) of the observed image, the ideal image, and the PSF of the blur, and where

$\mathbf{u} = [u_1, u_2]^T$ is a vector of frequencies. The DFT phase $\phi_g(\mathbf{u})$ of the observed image is given by the sum of the phases of the ideal image and the PSF, namely

$$\phi_g(\mathbf{u}) = \phi_f(\mathbf{u}) + \phi_h(\mathbf{u}) . \tag{4}$$

Because $h(\mathbf{n}) = h(-\mathbf{n})$, the $H(\mathbf{u})$ is real valued and $\phi_h(\mathbf{u}) \in \{0, \pi\}$. Thus, $\phi_g(\mathbf{u})$ may deviate from $\phi_f(\mathbf{u})$ by angle π. This effect of $\phi_h(\mathbf{u})$ can be cancelled by doubling the phase modulo 2π, resulting to the *phase blur invariants* (PBI)

$$\begin{aligned} \mathcal{B}(\mathbf{u}_i) \equiv \mathcal{B}(\mathbf{u}_i, G) &= 2\phi_g(\mathbf{u}_i) \bmod 2\pi \\ &= 2\arctan(\frac{p_i^0}{p_i^1}) \bmod 2\pi , \end{aligned} \tag{5}$$

where $\mathbf{p}_i = [p_i^0, p_i^1] = [\mathrm{Im}\{G(\mathbf{u}_i)\}, \mathrm{Re}\{G(\mathbf{u}_i)\}]$, and where $\mathrm{Im}\{\cdot\}$ and $\mathrm{Re}\{\cdot\}$ denote the real and imaginary parts of a complex number.

In [5], a shift invariant bispectrum slice of the observed image, defined by

$$\Psi(\mathbf{u}) = G(\mathbf{u})^2 G^*(2\mathbf{u}) , \tag{6}$$

was used to obtain blur and translation invariants. The phase of the bispectrum slice is expressed by

$$\phi_\Psi(\mathbf{u}) = 2\phi_g(\mathbf{u}) - \phi_g(2\mathbf{u}) . \tag{7}$$

Also the phase of the bispectrum slice is made invariant to blur by doubling it modulo 2π. This results in combined *phase blur-translation invariants* (PBTI), given by

$$\begin{aligned} \mathcal{T}(\mathbf{u}_i) \equiv \mathcal{T}(\mathbf{u}_i, G) &= 2[2\phi_g(\mathbf{u}_i) - \phi_g(2\mathbf{u}_i)] \bmod 2\pi \\ &= 2\Big[2\arctan(\frac{p_i^0}{p_i^1}) - \arctan(\frac{q_i^0}{q_i^1})\Big] \bmod 2\pi , \end{aligned} \tag{8}$$

where $\mathbf{p}_i$ is as above while $\mathbf{q}_i = [q_i^0, q_i^1] = [\mathrm{Im}\{G(2\mathbf{u}_i)\}, \mathrm{Re}\{G(2\mathbf{u}_i)\}]$.

3 Weighting of the Blur Invariant Features

For image recognition purposes, the similarity between two blurred and noisy images $\hat{g}_1(\mathbf{n})$ and $\hat{g}_2(\mathbf{n})$ can be deduced based on some distance measure between the vectors of PBI or PBTI features computed for the images. Because the values of the invariants are affected by the image noise, the image classification result can be improved if the contribution of the individual invariants to the distance measure is weighted according to their noisiness. In this section, we introduce a method for computation of a weighted distance between the PBI or PBTI feature vectors based on the estimated signal-to-noise ratio of the features. The method is similar to the one given in paper [8] for the moment invariants and phase-tangent invariants. The weighting is done by computing a Mahalanobis distance between the feature vectors of distorted images $\hat{g}_1(\mathbf{n})$ and $\hat{g}_2(\mathbf{n})$ as shown in Sect. 3.1. For the computation of the Mahalanobis distance, we need the covariance matrices of the PBI and PBTI features, which are derived in Sects. 3.2 and 3.3, respectively.

It is assumed that invariants (5) and (8) are computed for noisy N-by-N image $\hat{g}(\mathbf{n})$ of which DFT is given by

$$\begin{aligned}\hat{G}(\mathbf{u}) &= \sum_{\mathbf{n}} \{g(\mathbf{n}) + w(\mathbf{n})\} e^{-2\pi j(\mathbf{u}^T\mathbf{n})/N} \\ &= G(\mathbf{u}) + \sum_{\mathbf{n}} w(\mathbf{n}) e^{-2\pi j(\mathbf{u}^T\mathbf{n})/N} \ ,\end{aligned} \tag{9}$$

where noise $w(\mathbf{n})$ is assumed to be zero-mean independent and identically distributed with variance σ^2. These noisy invariants are denoted by $\hat{\mathcal{B}}(\mathbf{u}_i) \equiv \mathcal{B}(\mathbf{u}_i, \hat{G})$ and $\hat{\mathcal{T}}(\mathbf{u}_i) \equiv \mathcal{T}(\mathbf{u}_i, \hat{G})$. We use also the following notation: $\hat{\mathbf{p}}_i = [\hat{p}_i^0, \hat{p}_i^1] = [\text{Im}\{\hat{G}(\mathbf{u}_i)\}, \text{Re}\{\hat{G}(\mathbf{u}_i)\}]$ and $\hat{\mathbf{q}}_i = [\hat{q}_i^0, \hat{q}_i^1] = [\text{Im}\{\hat{G}(2\mathbf{u}_i)\}, \text{Re}\{\hat{G}(2\mathbf{u}_i)\}]$. As only the relative effect of noise is considered, σ^2 does not have to be known.

3.1 Weighted Distance between the Feature Vectors

Weighting of the invariant features is done by computing a Mahalanobis distance between the feature vectors. Mahalanobis distance is then used as a similarity measure in classification of the images. Mahalanobis distance is computed by using the sum $\mathbf{C}_S = \mathbf{C}_T^{(\hat{g}_1)} + \mathbf{C}_T^{(\hat{g}_2)}$ of the covariance matrices of the PBI or PBTI features of images $\hat{g}_1(\mathbf{n})$ and $\hat{g}_2(\mathbf{n})$, and is given by

$$\text{distance} = \mathbf{d}^T \mathbf{C}_S^{-1} \mathbf{d} \ , \tag{10}$$

where $\mathbf{d} = [d_0, d_1, \ldots, d_{N_T-1}]^T$, contains the unweighted differences of the invariants for images $\hat{g}_1(\mathbf{n})$ and $\hat{g}_2(\mathbf{n})$ in the range $[-\pi, \pi]$, which are expressed by

$$d_i = \begin{cases} \alpha_i - 2\pi & \text{if } \alpha_i > \pi \\ \alpha_i & \text{otherwise, and} \end{cases} \tag{11}$$

where $\alpha_i = [\hat{\mathcal{B}}(\mathbf{u}_i)^{(\hat{g}_1)} - \hat{\mathcal{B}}(\mathbf{u}_i)^{(\hat{g}_2)} \bmod 2\pi]$ for PBIs and $\alpha_i = [\hat{\mathcal{T}}(\mathbf{u}_i)^{(\hat{g}_1)} - \hat{\mathcal{T}}(\mathbf{u}_i)^{(\hat{g}_2)} \bmod 2\pi]$ for PBTIs. $\hat{\mathcal{B}}(\mathbf{u})^{(\hat{g}_k)}$ and $\hat{\mathcal{T}}(\mathbf{u})^{(\hat{g}_k)}$ denote invariants (5) and (8), respectively, for image $\hat{g}_k(\mathbf{n})$.

Basically the modulo operator in (5) and (8) can be omitted due to the use of the same operator in computation of α_i. The modulo operator of (5) and (8) can be neglected also in the computation of the covariance matrices in Sects. 3.2 and 3.3.

3.2 The Covariances of the PBI Features

The covariance matrix of the PBIs (5) can not be computed directly as they are a non-linear function of the image data. Instead, we approximate the N_T-by-N_T covariance matrix $\mathbf{C}_T$ of N_T invariants $\hat{\mathcal{B}}(\mathbf{u}_i)$, $i = 0, 1, \ldots, N_T - 1$, using linearization

$$\mathbf{C}_T \approx \mathbf{J} \cdot \mathbf{C} \cdot \mathbf{J}^T \ , \tag{12}$$

where $\mathbf{C}$ is $2N_T$-by-$2N_T$ covariance matrix of the elements of vector $\mathbf{P} = [\hat{p}_0^0, \hat{p}_0^1, \hat{p}_1^0, \hat{p}_1^1, \cdots, \hat{p}_{N_T-1}^0, \hat{p}_{N_T-1}^1]$, and $\mathbf{J}$ is a Jacobian matrix. It can be shown, that due to the orthogonality of the Fourier transform, the covariance terms of $\mathbf{C}$ are zero and the $2N_T$-by-$2N_T$ covariance matrix is diagonal resulting in

$$\mathbf{C}_T \approx \frac{N^2}{2}\sigma^2 \mathbf{J} \cdot \mathbf{J}^T \ . \tag{13}$$

The Jacobian matrix is block diagonal and given by

$$\mathbf{J} = \begin{bmatrix} \mathbf{J}_0 & 0 & \cdots & 0 \\ 0 & \mathbf{J}_1 & \cdots & 0 \\ \vdots & \vdots & \ddots & \vdots \\ 0 & 0 & \cdots & \mathbf{J}_{N_T-1} \end{bmatrix} , \tag{14}$$

where $\mathbf{J}_i$, $i = 0, \ldots, N_T - 1$ contains the partial derivatives of the invariants $\mathcal{B}(\mathbf{u}_i)$ with respect to $\hat{p}_i^0$ and $\hat{p}_i^1$, namely

$$\begin{aligned} \mathbf{J}_i &= \left[\frac{\partial \hat{\mathcal{B}}(\mathbf{u}_i)}{\partial \hat{p}_i^0}, \frac{\partial \hat{\mathcal{B}}(\mathbf{u}_i)}{\partial \hat{p}_i^1} \right] \\ &= \left[\frac{2\hat{p}_i^1}{c_i}, \frac{-2\hat{p}_i^0}{c_i} \right] , \end{aligned} \tag{15}$$

where $c_i = [\hat{p}_i^0]^2 + [\hat{p}_i^1]^2$. Notice that the modulo operator in (5) does not have any effect on the derivatives of $\mathcal{B}(\mathbf{u})$, and it can be omitted.

3.3 The Covariances of the PBTI Features

For PBTIs (8) the covariance matrix $\mathbf{C}_T$ is computed also using linearization (12). $\mathbf{C}$ is now a $4N_T$-by-$4N_T$ covariance matrix of the elements of vector $\mathbf{R} = [\mathbf{P}, \mathbf{Q}]$, where $\mathbf{Q} = [\hat{q}_0^0, \hat{q}_0^1, \hat{q}_1^0, \hat{q}_1^1, \cdots, \hat{q}_{N_T-1}^0, \hat{q}_{N_T-1}^1]$. The Jacobian matrix can be expressed by

$$\mathbf{J} = [\mathbf{K}, \mathbf{L}] = \begin{bmatrix} \mathbf{K}_0 & 0 & \cdots & 0 & \mathbf{L}_0 & 0 & \cdots & 0 \\ 0 & \mathbf{K}_1 & \cdots & 0 & 0 & \mathbf{L}_1 & \cdots & 0 \\ \vdots & \vdots & \ddots & \vdots & \vdots & \vdots & \ddots & \vdots \\ 0 & 0 & \cdots & \mathbf{K}_{N_T-1} & 0 & 0 & \cdots & \mathbf{L}_{N_T-1} \end{bmatrix} . \tag{16}$$

$\mathbf{K}_i$ contains partial derivatives of the invariants $\hat{\mathcal{T}}(\mathbf{u}_i)$ with respect to $\hat{p}_i^0$ and $\hat{p}_i^1$ and is given by

$$\begin{aligned} \mathbf{K}_i \equiv \mathbf{K}_{i,i} &= \left[\frac{\partial \hat{\mathcal{T}}(\mathbf{u}_i)}{\partial \hat{p}_i^0}, \frac{\partial \hat{\mathcal{T}}(\mathbf{u}_i)}{\partial \hat{p}_i^1} \right] \\ &= \left[\frac{4\hat{p}_i^1}{c_i}, \frac{-4\hat{p}_i^0}{c_i} \right] , \end{aligned} \tag{17}$$

while $\mathbf{L}_i$ contains partial derivatives with respect to $\hat{q}_i^0$ and $\hat{q}_i^1$, namely

$$\mathbf{L}_i \equiv \mathbf{L}_{i,i} = \left[\frac{\partial \hat{T}(\mathbf{u}_i)}{\partial \hat{q}_i^0}, \frac{\partial \hat{T}(\mathbf{u}_i)}{\partial \hat{q}_i^1} \right] = \left[\frac{-2\hat{q}_i^1}{e_i}, \frac{2\hat{q}_i^0}{e_i} \right], \tag{18}$$

where $e_i = [\hat{q}_i^0]^2 + [\hat{q}_i^1]^2$.

Equation (12) simplifies to (13) also for PBTIs when discarding redundant coefficients $\hat{\mathbf{q}}_i$ from $\mathbf{R}$ that correspond to frequencies $\hat{\mathbf{q}}_i = \hat{\mathbf{p}}_j$ for some $i, j \in \{0, 1, \ldots, N_T - 1\}$. The Jacobian matrix (16) has to be organized accordingly: $\mathbf{L}_i$ corresponding to redundant coefficients are replaced by $\mathbf{K}_{i,j}$ given by

$$\mathbf{K}_{i,j} = \left[\frac{\partial \hat{T}(\mathbf{u}_i)}{\partial \hat{p}_j^0}, \frac{\partial \hat{T}(\mathbf{u}_i)}{\partial \hat{p}_j^1} \right] = \left[\frac{-2\hat{p}_j^1}{c_j}, \frac{2\hat{p}_j^0}{c_j} \right]. \tag{19}$$

4 Experiments

In the experiments, we compared the performance of the weighted and unweighted PBI and PBTI features in classification of blurred and noisy images using nearest neighbour classification. For comparison, we present similar results with and without weighting for the central moment invariants and the phase-tangent invariants [2]. As the phase-tangent invariants are not shift invariant, they are used only in the first experiment. For the moment invariants, we used invariants up to the order 7, as proposed in [2], which results in 18 invariants. For all the frequency domain invariants, we used the invariants for which $\sqrt{u_1^2 + u_2^2} \leq \sqrt{10}$, but without using the conjugate symmetric or zero frequency invariants. This results also in $N_T = 18$ invariants.

In the first experiment, the invariants only for blur were considered, namely the PBIs, the phase-tangent invariants, and the central moment invariants (invariant also to shift, but give better results than regular moment invariants [5]).

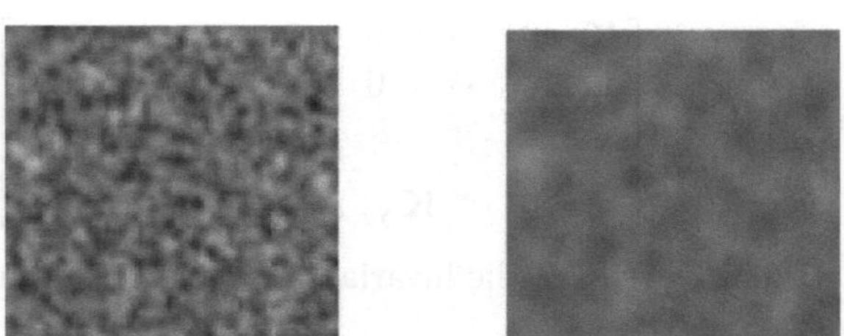

(a) (b)

Fig. 1. (a) An example of the 40 filtered noise images used in the first experiment, and (b) a degraded version of it with blur radius 5 and PSNR 30 dB

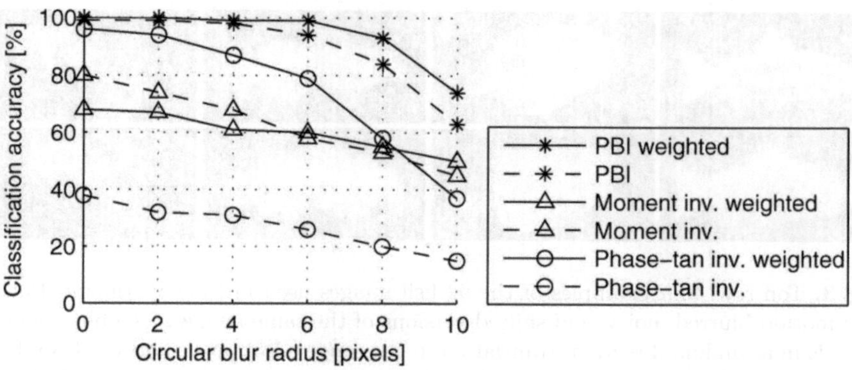

Fig. 2. The classification accuracy of the nearest neighbour classification of the out of focus blurred and noisy (PSNR 20 dB) images using various blur invariant features

As test images, we had 40 computer generated images of uniformly distributed noise, which were filtered using a Gaussian low pass filter of size 10-by-10 with the standard deviation σ=1 to acquire an image, as in Fig. 3.3, that resembles some natural texture. One image at a time was degraded by blur and noise, and was classified as one of the 40 original images using the invariants. The blur was generated by convolving the images with a circular PSF with a radius varying from 0 to 10 pixels with steps of 2 pixels, which models out of focus blur. The PSNR was 20 dB. The image size was cropped finally to 80-by-80 containing only the valid part of the convolution. The experiment was repeated 20 times for each blur size and for each of the 40 images.

All the tested methods are invariant to circular blur, but there are differences in robustness to noise and boundary error caused by convolution that extends beyond the borders of the observed image. The percentage of correct classification for the three methods, the PBIs, the moment invariants, and the phase-tangent invariants, is shown in Fig. 2 with and without weighting. Clearly, the non-weighted phase-tangent invariants are the most sensitive to disturbances. Their classification result is also improved most by the weighting. The non-weighted moment invariants are known to be more robust to distortions than the corresponding phase-tangent invariants, and this is confirmed by the results. However, the weighting improves the result for moment invariants much less, and only for a blur radii up to 5 pixels making the phase-tangent invariants preferable. Clearly, the best classification results are obtained with the PBIs. Although the PBIs result in the best classification accuracy without weighting, the result is still improved up to 10 % if the weighting is used.

In the second experiment, we tested the blur-translation invariant methods, the PBTIs and the central moment invariants. The test material consisted of 94 100 × 100 fish images. These original images formed the target classes into which the distorted versions of the images were to be classified. Some

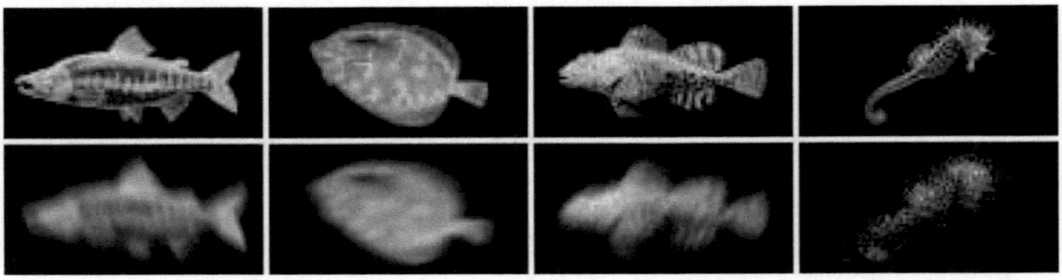

Fig. 3. Top row: four examples of the 94 fish images used in the experiment. Bottom row: motion blurred, noisy, and shifted versions of the same images. The blur length is 6 pixels in a random direction, translation in the range [-5,5] pixels and the PSNRs are from left to right 50, 40, 30, and 20 dB. (45 × 90 images are cropped from 100 × 100 images.)

original and distorted fish images are shown in Fig. 3. The distortion included linear motion blur of six pixels in a random direction, noise with PSNR from 50 to 10 dB, and random displacement in the horizontal and vertical direction in the range [-5,5] pixels. The objects were segmented from the noisy background before classification using a threshold and connectivity analysis. At the same time, this results in realistic distortion at the boundaries of the objects as some information is lost. The distance between the images of the fish image database was computed using $\mathbf{C}_T^{(\hat{g}_1)}$ or $\mathbf{C}_T^{(\hat{g}_2)}$ separately instead of their sum $\mathbf{C}_S = \mathbf{C}_T^{(\hat{g}_1)} + \mathbf{C}_T^{(\hat{g}_2)}$, and selecting the larger of the resulting distances, namely distance $= \max\{\mathbf{d}^T[\mathbf{C}_T^{(\hat{g}_1)}]^{-1}\mathbf{d}, \mathbf{d}^T[\mathbf{C}_T^{(\hat{g}_2)}]^{-1}\mathbf{d}\}$. This resulted in significantly better classification accuracy for PBTI features (and also for PBI features without displacement of the images), and the result was slightly better also for moment invariants.

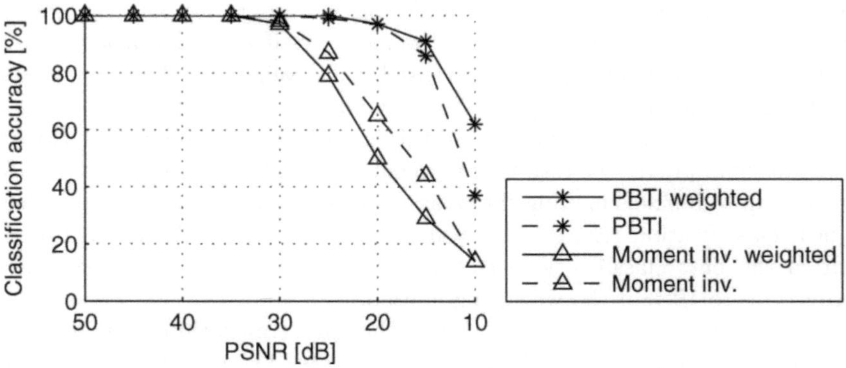

Fig. 4. The classification accuracy of nearest neighbour classification of motion blurred and noisy images using the PBTIs and the moment invariants

The classification results are shown in the diagram of Fig. 4. Both methods classify images correctly when the noise level is low. When the noise level increases, after 35 dB the PBTIs perform clearly better than the moment invariants. It can be observed that the weighting does not improve the result of the moment invariants, which is probably due to strong nonlinearity of the moment invariants that cannot be well linearized by (12). However, for the PBTIs the result is improved by up to 20 % through the use of weighting.

5 Conclusions

Only few blur invariants have been introduced in the previous literature, and they are based either on image moments or Fourier transform phase. We have shown that the Fourier phase based blur invariants and blur-translation invariants, namely the PBIs and PBTIs, are more robust to noise compared to the moment invariants. In this paper, we introduced a weighting scheme that still improves the results of the Fourier domain blur invariants in classification of blurred images and objects. For the PBIs, the improvement in classification accuracy was up to 10 % and for the PBTIs, the improvement was up to 20 %. For comparison, we also showed the results for a similar weighting scheme applied to the moment invariants and the phase-tangent based invariants. The experiments clearly indicated that the weighted PBIs and PBTIs are superior in terms of classification accuracy to other existing methods.

Acknowledgments

The authors would like to thank the Academy of Finland (project no. 127702), and Prof. Petrou and Dr. Kadyrov for providing us with the fish image database.

References

1. Wood, J.: Invariant pattern recognition: A review. Pattern Recognition 29(1), 1–17 (1996)
2. Flusser, J., Suk, T.: Degraded image analysis: An invariant approach. IEEE Trans. Pattern Anal. Machine Intell. 20(6), 590–603 (1998)
3. Flusser, J., Zitová, B.: Combined invariants to linear filtering and rotation. Int. J. Pattern Recognition and Artificial Intelligence 13(8), 1123–1136 (1999)
4. Suk, T., Flusser, J.: Combined blur and affine moment invariants and their use in pattern recognition. Pattern Recognition 36(12), 2895–2907 (2003)
5. Ojansivu, V., Heikkilä, J.: Object recognition using frequency domain blur invariant features. In: Ersbøll, B.K., Pedersen, K.S. (eds.) SCIA 2007. LNCS, vol. 4522, pp. 243–252. Springer, Heidelberg (2007)
6. Ojansivu, V., Heikkilä, J.: A method for blur and similarity transform invariant object recognition. In: Proc. International Conference on Image Analysis and Processing (ICIAP 2007), Modena, Italy, September 2007, pp. 583–588 (2007)

7. Lagendijk, R.L., Biemond, J.: Basic methods for image restoration and identification. In: Bovik, A. (ed.) Handbook of Image and Video Processing, pp. 167–182. Academic Press, London (2005)
8. Ojansivu, V., Heikkilä, J.: Motion blur concealment of digital video using invariant features. In: Blanc-Talon, J., Philips, W., Popescu, D., Scheunders, P. (eds.) ACIVS 2006. LNCS, vol. 4179, pp. 35–45. Springer, Heidelberg (2006)

The Effect of Motion Blur and Signal Noise on Image Quality in Low Light Imaging

Eero Kurimo[1], Leena Lepistö[2], Jarno Nikkanen[2], Juuso Grén[2], Iivari Kunttu[2], and Jorma Laaksonen[1]

[1] Helsinki University of Technology
Department of Information and Computer Science
P.O. Box 5400, FI-02015 TKK, Finland
jorma.laaksonen@tkk.fi
http://www.tkk.fi
[2] Nokia Corporation
Visiokatu 3, FI-33720 Tampere, Finland
{leena.i.lepisto,jarno.nikkanen,juuso.gren,
iivari.kunttu}@nokia.com
http://www.nokia.com

Abstract. Motion blur and signal noise are probably the two most dominant sources of image quality degradation in digital imaging. In low light conditions, the image quality is always a tradeoff between motion blur and noise. Long exposure time is required in low illumination level in order to obtain adequate signal to noise ratio. On the other hand, risk of motion blur due to tremble of hands or subject motion increases as exposure time becomes longer. Loss of image brightness caused by shorter exposure time and consequent underexposure can be compensated with analogue or digital gains. However, at the same time also noise will be amplified. In relation to digital photography the interesting question is: What is the tradeoff between motion blur and noise that is preferred by human observers? In this paper we explore this problem. A motion blur metric is created and analyzed. Similarly, necessary measurement methods for image noise are presented. Based on a relatively large testing material, we show experimental results on the motion blur and noise behavior in different illumination conditions and their effect on the perceived image quality.

1 Introduction

The development in the area of digital imaging has been rapid during recent years. The camera sensors have become smaller whereas the number of pixels has increased. Consequently the pixel sizes are nowadays much smaller than before. This is particularly the case in the digital pocket cameras and mobile phone cameras. Due to the smaller size, one pixel is able to receive smaller number of photons within the same exposure time. On the other hand, the random noise caused by various sources is present in the obtained signal. The most effective way to reduce the relative amount of noise in the image (i.e. signal to noise ratio, SNR) is to use longer exposure times, which allows more photons to be observed by the sensor. However, in the case of long exposure times, the risk of motion blur increases.

A.-B. Salberg, J.Y. Hardeberg, and R. Jenssen (Eds.): SCIA 2009, LNCS 5575, pp. 81–90, 2009.

Motion blur occurs when the camera or the subject moves during the exposure period. When this happens, the image of the subject moves to different area of the camera sensor photosensitive surface during the exposure time. Small camera movements soften the image and diminish the details whereas larger movements can make the whole image incomprehensible [8]. This way, either the camera movement or the movement of the object in the scene are likely to become visible in the image, when the exposure time is long. This obviously is dependent on the manner how the images are taken, but usually this problem is recognized in low light conditions in which long exposure times are required to collect enough photons to the sensor pixels. The decision on the exposure time is typically made by using an automatic exposure algorithm. An example of this kind algorithm can be found in e.g. [11]. A more sophisticated exposure control algorithm presented in [12] tries to optimize the ratio between signal noise and motion blur.

The perceived image quality is always subjective. Some people prefer somewhat noisy but detailed images over smooth but blurry images, and some tolerate more blur than noise. The image subject and the purpose of the image also affect on the perceived image quality. For example, images containing text may be a bit noisy but still readable, similarly e.g. images of landscapes can sometimes be a bit blurry. In this paper, we analyze the effect of motion blur and noise on the perceived image quality and try to find the relationship of these two with respect to the camera parameters such as exposure time. The analysis is based on the measured motion blur and noise and the image quality perceived by human observers.

Although both image noise and motion blur have been intensively investigated in the past, their relationship and their relative effect on the image quality has not been studied in the same extent. Especially the effect of the motion blur on the image quality has not received much attention. In [16], a model to estimate the tremble of hands was presented and it was measured, but it was not compared to noise levels in the image. Also the subjective image quality was not studied. In this paper, we analyze the effects of the motion blur and noise to the perceived image quality in order to optimize the exposure time in different levels of image quality, motion blur, noise and illumination. For this purpose, a motion blur metric is created and analyzed. Similarly, necessary measurement methods for image noise are presented. In a quite comprehensive testing part, we created a set of test images captured by several test persons. The relationship between the motion blur and noise is measured by means of these test images. The subjective image quality of the test set images is evaluated and the results are compared to the measured motion blur and noise in different imaging circumstances.

The organization of this paper is the following: Sections 2 and 3 present the framework for the motion blur and noise measurements, respectively. In section 4, we present the experiments made to validate the framework presented in this paper. The results are discussed and conclusions drawn in section 5.

2 Motion Blur Measurements

Motion blur is one of the most significant reasons for image quality decrease. Noise is also influential, but it increases gradually and can be accurately estimated from the signal values. Motion blur, on the other hand, has no such benefits. It is very difficult to estimate the amount of motion blur either a priori or a posteriori. It is even more difficult to estimate the motion blur a priori from the exposure time because motion blur only follows a random distribution based on the exposure time and the characteristics of the camera and the photographer. The expected amount of motion blur can be estimated a priori if the knowledge on the photographer behavior is available, but because of the high variance of the motion blur distribution of the exposure time, the estimation is very imprecise at best.

The framework for motion blur inspection has been presented in [8], in which types of motion blur are presented. In [8], a three-dimensional model, in which the camera may move along or spin around three different axes, was presented. Motion blur is typically modeled as angular blur, which is not necessarily always the case. It has been shown that camera motion should be considered as straight linear motion when the exposure time is less than 0.125 seconds [16]. If the point spread function (PSF) is known, or it is possible to estimate, then it is possible to correct the blur by using Wiener filtration [15]. The amount of blur can be estimated in many manners. A basic approach is to detect the blur in the image by using an edge detector, such as Canny method, or the local scale control method proposed by James and Steven [6], and measure the edge width at each edge point [10]. Another more practical method was proposed in [14], which uses the characteristics of sharp and dull edges after Haar wavelet transform. It is clear that the motion blur analysis is more reliable in the cases where two or more consequent frames are available [13]. In [9], the strength and direction of the motion was estimated this way, and this information was used to reduce the motion blur. Also in [2], a method for estimating and removing blur from two blurry images was presented. A two camera approach was presented also in [1]. The methods based on several frames, however, are not always practical in all mobile devices due to their memory requirements.

2.1 Blur Metric

An efficient and simple way of measuring the blur from the image is to use laser spots projected to the image subject. The motion blur can be estimated from the size of the laser spot area [8]. To get a more reliable motion blur measurement result and also include the camera rotation around the optical axes (roll) into measurement, the use of multiple laser spots is preferable. In the experiments related to this paper, we have used three laser spots located in center, and two corners of the scene. To make the identification process faster and easier, a smaller image is cropped from the image, and the blur pattern is extracted by means of adaptive thresholding, in which the laser spot threshold could be determined by keeping the ratio between the threshold and the exposure time at a constant level. This method produced roughly the same size laser spot regions of no motion blur with varying exposure times.

Once the laser spot regions in each image are located, the amount of motion blur in the images can be estimated. First, a skeleton is created by thinning the thresholded binary laser spot region image. The thinning algorithm, proposed as Algorithm A1 in [4] and implemented in the Image processing toolbox of the Matlab software, is iterated until the final homotopic skeleton is reached. After the skeletonization, the centroid, orientation and major and minor axis lengths of the best-fit ellipse fit to the skeleton pixels can be calculated. The major axis length is then used as a scalar measure for the blur of the laser spot.

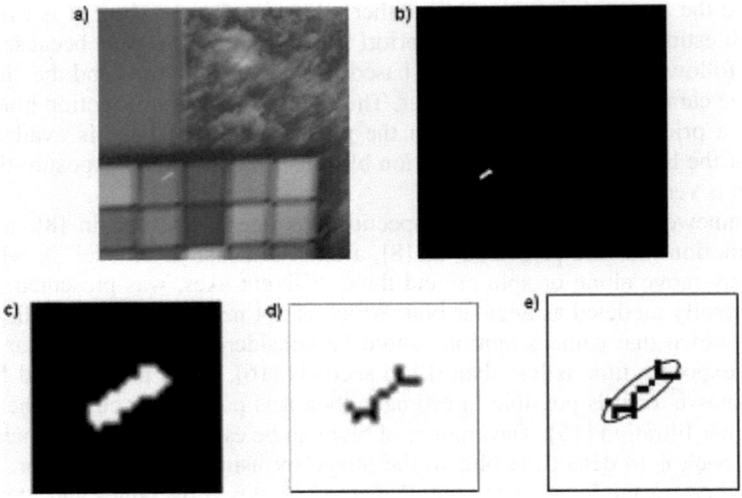

Fig. 1. a) Blur measurement process: a) piece extracted from the original image, b) the thresholded binary image c) enlarged laser spot, d) its extracted homotropic skeleton and e) the ellipse fitted around the skeleton

Figure 1 illustrates the blur measurement process. First, subfigures 1a and 1b show a piece extracted from the original image and the corresponding thresholded binary image of the laser spot. Then, subfigures 1c, 1d and 1e display the enlarged laser spot, its extracted homotopic skeleton and finally the best-fit ellipse, respectively. In the case of this illustration, the blur was measured to be 15.7 pixels in length.

3 Noise Measurement

During the decades, digital camera noise research has identified many additive and multiplicative noise sources, especially inside the image sensor transistors. Some noise sources have even been completely eliminated. Dark current is the noise generated by the photosensor voltage leaks independent of the received photons. The amount of dark current noise depends on the temperature of the sensors, the exposure time and the physical properties of the sensors. Shot noise comes from the random arrival of photons to a sensor pixel. It is the dominant noise source at the lower signal values just above the dark current noise. The arrivals of photons to the sensor pixel are uncorrelated events. This means that the number of captured photons by a sensor pixel during a time interval can be described as a Poisson process. It follows that the SNR of a signal that follows the Poisson distribution has the SNR that is proportional to the number of photons captured by the sensor. Consequently, the effects of shot noise can be reduced only by increasing the number of captured photons. Fixed pattern noise (FPN) comes from the nonuniformity of the image sensor pixels. It is caused by imperfections and other variations between the pixels, which result in slightly different pixel sensitivities. The FPN is the dominant noise source with high signal values. It is to be noticed that the SNR of fixed pattern noise is independent of signal level and remains at a constant level. This means that the SNR cannot be affected by increasing the light or exposure time, but only by using a more uniform pixel sensor array.

The total noise of the camera system is a quadrature sum of its dark current, shot and fixed pattern noise components. These can be studied by using the photon transfer curve (PTC) method [7]. Signal and noise levels are measured from sample images of a uniformly illuminated uniform white subject in different exposure times. The measured noise is plotted against the measured signal on a log-log scale. The plotted curve will have three distinguishable sections as illustrated in figure 2a.

With the lowest signals the signal noise is constant, which indicates the read noise consisting of the noise sources independent of the signal level, such as the dark current and on-chip noise. As the signal value increases, the shot noise becomes the dominant noise source. Finally the fixed pattern noise becomes the dominant noise source, and indicating the full well of the image sensor.

3.1 Noise Metric

For a human observer, it is possible to intuitively approximate how much visual noise there is present in the image. However, measuring this algorithmically has proven to be a difficult task. Measuring noise directly from the image without any a priori knowledge on the camera noise behavior is a challenging task but has not received much attention. Foi et al [3] have proposed an approach, in which the image is segmented into regions of different signal values $y \pm \delta$ where y is the signal value of the segment and δ is a small variability allowed inside the segment.

Signal noise is in practice generally considered as the standard deviation of subsequent measurements of some constant signal. An accurate image noise measurement method would be to measure the standard deviation of a group of pixels inside an area of uniform luminosity. An old and widely used camera performance analysis method is based on the photon transfer curve (PTC) [7]. Methods similar to the one used in this study have been applied in [5]. The PTC method generates a curve showing the standard deviation of an image sensor pixel value in different signal levels. The noise σ should grow monotonically with the signal S according to:

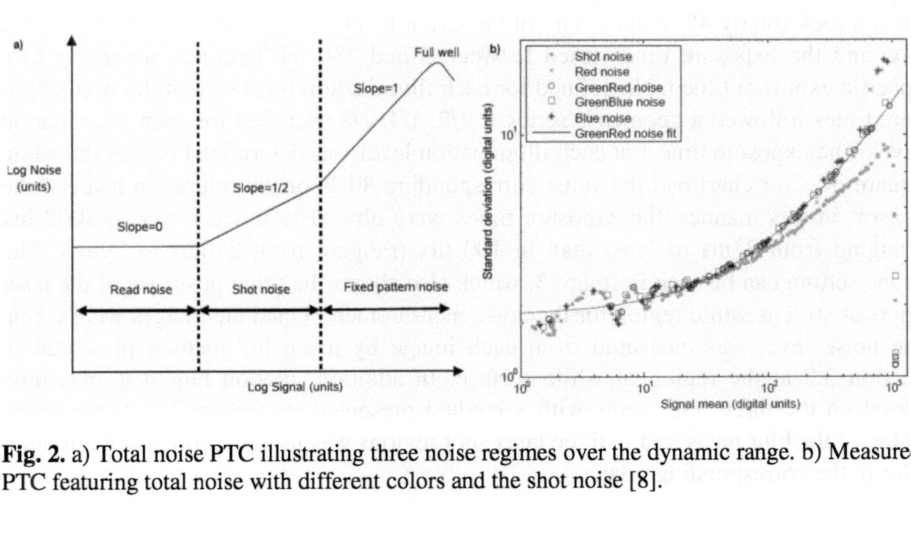

Fig. 2. a) Total noise PTC illustrating three noise regimes over the dynamic range. b) Measured PTC featuring total noise with different colors and the shot noise [8].

$$\sigma = aS^b + c \tag{1}$$

before reaching the full well. If the noise monotonicity hypothesis holds for the camera, the noisiness of each image pixel could be directly estimated from the curve when knowing the signal value.

In our calibration procedure, the read noise floor was first determined using dark frames by capturing images without any exposure to light. Dark frames were taken with varying exposure times to determine also the effect of longer exposure times. Figure 2b shows noise measurements made for experimental image data. The noise measurement was carried out in three color channels and shot noise from images when fixed pattern noise is removed. The noise model was created by fitting the equation (1) to the green pixel values using values a = 0.04799, b = 0.798 and c = 1.819.

For the signal noise measurement, a uniform white surface was located into the scene, and the noise level of the test images was estimated as a local standard deviation on this surface. Similarly, the signal value estimate was the local average of the signal on this region. The signal to noise ratio (SNR) can be calculated as a ratio between these two.

4 Experiments

The goal of the experiments was to obtain sample images with good spectrum of different motion blurs and noise levels. The noise, motion blur and the image had to be able to be measured from the sample images. All the experiments were carried out in an imaging studio in which the illumination levels can be accurately controlled.

All the experiments were made by using a standard mobile camera device containing a CMOS sensor with 1151x864 pixel resolution. There were totally four test persons with varying amount of experience on photography. Each person captured hand held camera photographs in four different illumination levels and with four different exposure times. At each setting, three images were taken, which means that each test person took totally 48 images. The illumination levels were 1000, 500, 250, and 100 lux, and the exposure time varied between 3 and 230 milliseconds according to a specific exposure time table defined for each illumination level so that the used exposure times followed a geometric series 1, 1/2, 1/4, 1/8 specified for each illumination level. The exposure time 1 at each illumination level was determined so that the white square in color chart had the value corresponding 80 % of the saturation level of the sensor. in this manner, the exposure times were obviously much lower in 1000 lux (ranging from 22ms to 3ms) than in 100 lux (ranging from 230ms to 29ms). The scene setting can be seen in figure 3, which also shows the three positions of the laser spots as well as white region for the noise measurement. Once the images were taken, the noise level was measured from each image by using the method presented in section 3.2 at the region of white surface. In addition, motion blur was measured based on the three laser spots with a method presented in section 2.1. The average value of the blur measured in three laser spot regions was used to represent the motion blur in the corresponding image.

Fig. 3. Two example images from the testing in 100 lux illumination. The exposure times in left and right are 230 and 29 ms, respectively. This causes motion blur in left and noise in right side image. The subjective brightness of the images is adjusted to the same level by using appropriate gain factors. The three laser spots are clearly visible in both images.

After that, the subjective visual image quality evaluation was carried out. For the evaluation, the images were processed by using adjusted gain factors so that the brightness of all the images was at the same level. There were totally 5 persons who independently evaluated the image quality. This was made in terms of overall quality, blur and noise. The evaluating persons gave a grade in scale between zero and five for each image, zero meaning poor and five meaning excellent image quality with no apparent quality degradations. The image quality was evaluated in three manners, in terms of overall quality, motion blur as well as noise. The evaluating persons gave the grades for each image in these three manners.

4.1 Noise and Motion Blur Analysis

To evaluate the perceived image quality against the noise and motion blur metrics presented in this paper, we compared them to the subjective evaluation results. This was made by taking the average subjective image quality evaluation results for each sample image, and plotting them against the measurements calculated to these images. The result of this comparison is shown in figure 4. As presented in this figure, both noise and motion blur metrics follow well the subjective interpretation of these two image characteristics. In the case of SNR, the perceived image quality smoothly rises with increasing SNR in the cases where there is no motion blur. On the other hand, it is essential to note that if there is significant motion in the image, the image quality grade is poor even if the noise level is relatively low. When considering the motion blur, however, an image is considered a relatively good quality even though there was some noise in it. This supports a conclusion that human observers find motion blur more disturbing than noise.

4.2 Exposure Time and Illumination Analysis

The second part of the analysis considered the relationship of exposure time and motion blur versus the perceived image quality. This analysis is essential in terms of the scope of this paper, since the risk of tremble of hands increases with increasing

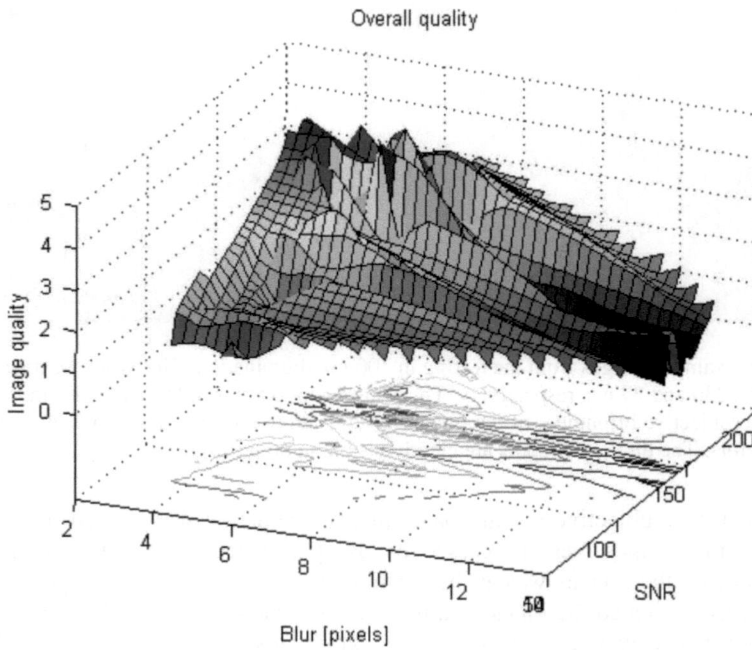

Fig. 4. Average overall evaluation results for the image set plotted versus measured blur and SNR

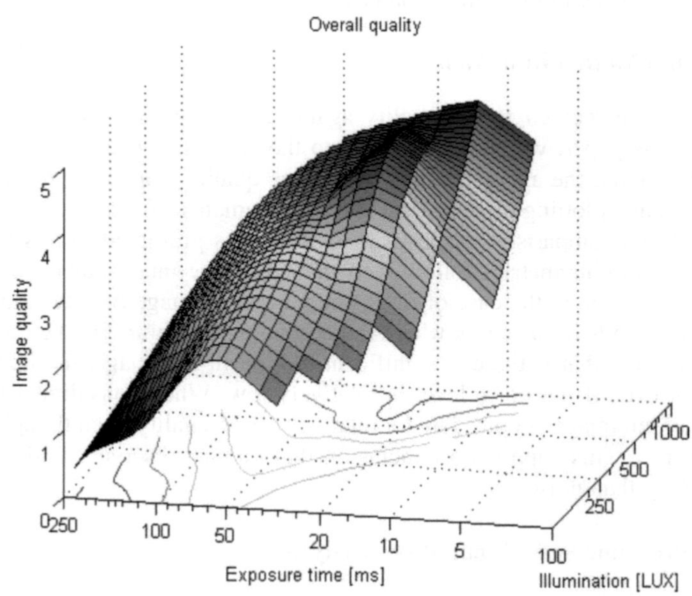

Fig. 5. Average overall evaluation results for the image set plotted versus illumination and exposure time

exposure time. Therefore, the analysis of optimal exposure times is a key factor in this study. Figure 5 shows the average grades given by the evaluating persons as a function of exposure time and illumination. The plot presented in figure 5 shows that image quality is clearly the best with high illumination levels, but it slowly decreases when illumination or exposure time decreases. This is an obvious result in general. However, the value of this kind of analysis is the fact that it can be used to optimize the exposure time at different illumination levels.

5 Discussion and Conclusions

Automatically determining the optimal exposure time using a priori knowledge is an important step in many digital imaging applications, but has not much been publicly studied. Because signal noise and motion blur are the most severe reasons for digital image quality degradations, and both are heavily affected by the exposure time, their effects on the image quality were the focus of this paper. Motion blur distribution and camera noise in different exposure times should be automatically estimated from the sample images taken just before the actual shot using recent advances in image processing. Using these estimates, the expected image quality for different exposure times can be determined using the methods of the framework presented in this paper.

In this paper, we have presented a framework for the analysis of the relationship between noise and motion blur. In addition, the information given by the tools provided in this paper is able to steer the optimization of the exposure time in different lighting conditions. It is obvious that a proper method for the estimation of the camera motion is needed to make this kind of optimization more accurate, but even a rough understanding of the risk of the motion blur on each lighting level greatly helps e.g. the development of more accurate exposure algorithms.

To make the model of the motion blur and noise relationship more accurate, an extensive testing with a covering test person group of different types of people is needed. However, the contribution of this paper is clear: a simple and robust method for the motion blur measurement and related metrics are developed, and the ratio between measured motion blur and measured noise could be determined in different lighting conditions. The effect of this on the perceived image quality was evaluated. Hence the work presented in this paper is a framework that can be used in the development of methods for the optimization of the ratio between noise and motion blur.

One aspect that is not considered in this paper is the impact of noise reduction algorithms. It is obvious that by utilizing a very effective noise reduction algorithm it is possible to use shorter exposure times and higher digital or analogue gains. This is because the resulting amplified noise can be reduced in the final image, hence improving the perceived image quality. An interesting topic for further study would be to quantify the difference between simple and more advanced noise reduction methods in this respect.

References

1. Ben-Ezra, M., Nayat, S.K.: Motion based motion deblurring. IEEE Transactions on Pattern Analysis and Machine Intelligence 26(6), 689–698 (2004)
2. Cho, S., Matsushita, Y., Lee, S.: Removing non-uniform motion blur from images (2007)

3. Foi, A., Alenius, S., Katkovnik, V., Egiazatrian, K.: Noise measurement for raw-data of digital imaging sensors by automatic segmentation of non-uniform targets. IEEE Sensors Journal 7(10), 1456–1461 (2007)
4. Guo, Z., Hall, R.W.: Parallel Thinning with Two-Subiteration Algorithms. Communications of the ACM 32(3), 359–373 (1989)
5. Hytti, H.T.: Characterization of digital image noise properties based on RAW data. In: Proceedings of SPIE, vol. 6059, pp. 86–97 (2006)
6. James, H., Steven, W.: Local scale control for edge detection and blur estimation. IEEE Transactions on Pattern Analysis and Machine Intelligence, 699–716 (1996)
7. Janesick, J.: Scientific Charge Coupled Devices, vol. PM83 (2001)
8. Kurimo, E.: Motion blur and signal noise in low light imaging, Master Thesis, Helsinki University of Technology, Faculty of Electronics, Communications and Automation, Department of Information and Computer Science (2008)
9. Liu, X., Gamal, A.E.: Simultaneous image formation and motion blur restoration via multiple capture,....
10. Marziliano, P., Dufaux, F., Winkler, S., Ebrahimi, T., Genimedia, S.A., Lausanne, S.: A no-reference perceptual blur metric. In: Proceedings of International Conference on Image Processing, vol. 3 (2002)
11. Nikkanen, J., Kalevo, O.: Menetelmä ja järjestelmä digitaalisessa kuvannuksessa valotuksen säätämiseksi ja vastaava laite. Patent FI 116246 B (2003)
12. Nikkanen, J., Kalevo, O.: Exposure of digital imaging. Patent application PCT/FI2004/050198 (2004)
13. Rav-Acha, A., Peleg, S.: Two motion blurred images are better than one. Pattern Recognition letters 26, 311–317 (2005)
14. Tong, H., Li, M., Zhang, H., Zhang, C.: Blur detection for digital images using wavelet transform. In: Proceedings of IEEE International Conference on Multimedia and Expo., vol. 1 (2004)
15. Wiener, N.: Extrapolation, interpolation, and smoothing of stationary time series (1992)
16. Xiao, F., Silverstein, A., Farrell, J.: Camera-motion and effective spatial resolution. In: International Congress of Imaging Science, Rochester, NY (2006)

A Hybrid Image Quality Measure for Automatic Image Quality Assessment

Atif Bin Mansoor[1], Maaz Haider[1], Ajmal S. Mian[2], and Shoab A. Khan[1]

[1] National University of Sciences and Technology, Pakistan
[2] Computer Science and Software Engineering,
The University of Western Australia, Australia
atif-cae@nust.edu.pk, smaazhaider@yahoo.com, ajmal@csse.uwa.edu.au,
kshoab@yahoo.com

Abstract. Automatic image quality assessment has many diverse applications. Existing quality measures are not accurate representatives of the human perception. We present a hybrid image quality (HIQ) measure, which is a combination of four existing measures using an 'n' degree polynomial to accurately model the human image perception. First we undertook time consuming human experiments to subjectively evaluate a given set of training images, and resultantly formed a Human Perception Curve (HPC). Next we define a HIQ measure that closely follows the HPC using curve fitting techniques. The HIQ measure is then validated on a separate set of images by similar human subjective experiments and is compared to the HPC.The coefficients and degree of the polynomial are estimated using regression on training data obtained from human subjects. Validation of the resultant HIQ was performed on a separate validation data. Our results show that HIQ gives an RMS error of 5.1 compared to the best RMS error of 5.8 by a second degree polynomial of an individual measure HVS (Human Visual System) absolute norm (H_1) amongst the four considered metrics. Our data contains subjective quality assessment (by 100 individuals) of 174 images with various degrees of fast fading distortion. Each image was evaluated by 50 different human subjects using double stimulus quality scale, resulting in an overall 8,700 judgements.

1 Introduction

The aim of image quality assessment is to provide a quantitative metric that can automatically and reliably predict how an image will be perceived by humans. However, human visual system is a complex entity, and despite all advancements in the opthalmology, the phenomenon of image perception by humans is not clearly understood. Understanding the human visual perception is a challenging task, encompassing the complex areas of biology, psychology, vision etc. Likewise, developing an automatic quantitative measure that accurately correlates with the human perception of images is a challenging assignment [1]. An effective quantitative image quality measure finds its use in different image processing applications including image quality control systems, benchmarking and optimizing of image processing systems and algorithms [1]. Moreover, it

A.-B. Salberg, J.Y. Hardeberg, and R. Jenssen (Eds.): SCIA 2009, LNCS 5575, pp. 91–98, 2009.

can facilitate in evaluating the performance of imaging sensors, compression algorithms, image restoration and denoising algorithms etc. In the absence of a well defined mathematical model, researchers have attempted to find a quantitative metric based upon various heuristics to model the human image perception [2], [3]. These heuristics are based upon frequency contents, statistics, structure and Human Visual System. Miyahara et al [4] proposed a Picture Quality Scale (PQS), as a combination of three essential distortion factors; namely the amount, location and structure of error. Mean squared error (MSE) or its identical measure, peak signal to noise ratio (PSNR) has often been used as a quality metric. In [5], Guo and Meng have tried to evaluate the effectiveness of MSE as a quality measure. As per their findings, MSE alone cannot be a reliable quality index. Wang and Bovik [6] proposed a new universal image quality index Q, by modeling any image distortion as the combination of loss of correlation, luminance distortion and contrast distortion. The experimental results have been compared with MSE, demonstrating superiority of Q index over MSE. Wang et al [7] proposed a quality assessment named Structural Similarity Index based upon degradation of structural information. The approach was further improved by them to incorporate the multi scale structural information [8]. Shnayderman et al [9] explored the feasibility of Singular Value Decomposition (SVD) for quality measurement. They compared their results with PSNR, Universal Quality Index [6] and Structural Similarity Index [7] to demonstrate the effectiveness of the proposed measure. Sheikh et al. [10] gave a survey and statistical evaluation of full reference image quality measures. They included PSNR (Peak Signal to Noise Ratio), JND Metrix [11], DCTune [12], PQS [4], NQM [13], fuzzy S7 [14], BSDM (Block Spectral Distance Meausurement) [15], MSSIM (Multiscale Structural Similarity Index Measure) [8], IFC (Information Fidelity Criteria) [16], VIF (Visual Information Fidelity) [17] in the study and concluded that VIF performs the best among these parameters. Chandler and Hemami proposed a two staged wavelet based visual signal to noise ratio based on near-threshold and supra-threshold properties of human vision [18].

2 Hybrid Image Quality Measure

2.1 Choice of Individual Quality Measures

Researchers have devised various image quality measures following different approaches, and showed their effectiveness in respective domains. These measures prove effective in certain conditions and show restricted performance otherwise. In our approach, instead of proposing a new quality metric, we suggest an apt combinational metric benefiting from the strength of individual measures. Therefore, the choice of constituent measures has a direct bearing on the performance of the proposed hybrid metric. Avcibas et al. [15] performed a statistical evaluation of 26 image quality measures. They categorized these quality measures into six distinct groups based on the used type of information. More importantly, they clustered these 26 measures using a Self-Organizing Map (SOM) of distortion measures. Based on the clustering results, Analysis of variance (ANOVA) and subjective mean opinion score they concluded that five of the quality measures are most discriminating. These measures are edge stability measure (E_2), spectral phase magnitude error (S_2), block spectral phase magnitude error (S_5), HVS (Human Visual System) absolute norm (H_1) and HVS L2 norm (H_2). We chose four (H_1, H_2, S_2, S_5) of these five prominent quality measures due to their mutual non redundancy. E_2 was dropped due to its close proximity to H_2 in the SOM.

2.2 Experiment Setup

A total of 174 color images, obtained from LIVE image quality assessment database [19] representing diverse contents, were used in our experiments. These images have been degraded by using varying levels of fast fading distortion by inducing bit errors during transmission of compressed JPEG 2000 bitstream over a simulated wireless channel. The different levels of distortion resulted in a wide variation in the quality of these images. We carried out our own perceptual tests on these images. The tests were administered as per the guidelines specified in the ITU-Recommendations for subjective assessment of quality for television pictures [20]. We used three identical workstations with 17-inch CRT displays of approximately the same age. The resolution of displays were identical, 1024 x 768. External light effects were minimized, and all tests were carried out under the same indoor illumination. All subjects viewed the display from a distance of 2 to 2.5 screen heights. We employed Double stimulus quality scale method, keeping in view its more precise image quality assessments. A matlab based graphical user interface was designed to show the assessors a pair of pictures i.e. original and degraded. The images were rated using a five point quality scale; excellent, good, fair, poor and bad. The corresponding rating was scaled on a 1-100 score.

2.3 Human Subjects

The human subjects were screened and then trained according to the ITU-Recommendations [20]. The subjects of the experiment were male and female undergraduate students with no experience in image quality assessment. All participants were tested for vision impairments e.g., colour blindness. The aim of the test was communicated to each assessor. Before each session, a demonstration was given using the developed GUI with images different from the actual test images.

2.4 Training and Validation Data

Each of the 174 test images was evaluated by 50 different human subjects, resulting in 8,700 judgements. This data was divided into training and validation sets. The training set comprised 60 images, whereas the remaining 114 images were used for validation of the proposed HIQ.

A mean opinion score was formulated from the Human Perception Values (HPVs) adjudged by the human subjects for various distortion levels. As expected, it was observed that different humans subjectively evaluated the same image differently. To cater this effect, we further normalized the distortion levels and plotted the average MOS against these levels. It means that average mean opinion score of different human subjects against all the images with a certain level of degradation was plotted. As the images of a wide variety with different levels of degradation are used, therefore in this manner, we achieved an image independent Human Perception Curve (HPC).

Similarly, average values were calculated for H_1, H_2, S_2 and S_5 for the normalized distortion levels using code from [19]. All these quality measures were regressed upon HPC by using a polynomial of 'n' degree. The general form of the HIQ is given by Eqn. 1.

$$\mathrm{HIQ} = a_0 + \sum_{i=1}^{n}(a_i H_1^i) + \sum_{j=1}^{n}(b_j H_2^j) + \sum_{k=1}^{n}(c_k S_2^k) + \sum_{l=1}^{n}(d_l S_5^l) \tag{1}$$

We tested different combinations of these measures taking one, two, three and four measures at a time. All these combinations were tested up to fourth degree polynomial.

Table 1. RMS errors for various combination of Quality Measures. First block gives RMS error for individual measures, second, third and fourth blocks for combination of two, three and four measures respectively.

	Polynomial of degree 1		Polynomial of degree 2		Polynomial of degree 3		Polynomial of degree 4	
Comb. of Measures	Training RMS error	Validation RMS error	Training RMS error	Validation RMS error	Training RMS error	Validation RMS error	Training RMS error	Validation RMS error
S2	12.9	9.2	9.2	6.6	9.7	6.2	10.5	6.1
S5	13.2	10.2	6.9	7.3	7.2	6.9	7.7	7.1
H1	10.1	6.8	8.4	5.8	8.8	6.0	9.5	6.2
H2	14.8	10.8	15.4	10.0	14.4	20.4	10.5	75.7
S2–S5	11.7	9.0	5.6	8.1	4.9	8.5	4.8	8.8
S2–H1	7.2	5.8	4.2	6.3	4.0	6.2	3.9	6.6
S2–H2	9.4	7.5	6.6	7.2	6.5	7.5	6.8	6.4
S5–H1	7.2	6.2	2.9	6.4	2.9	6.4	2.4	6.3
S5–H2	9.4	8.3	4.2	8.0	4.1	8.9	4.0	9.1
H1–H2	4.4	5.4	3.1	6.5	2.8	9.9	2.2	23.1
S2–S5–H1	7.2	5.8	2.2	6.7	0.2	12	0.3	16.9
S2–S5–H2	9.4	8.0	2.9	9.3	1.0	15.8	0.4	21.5
S2–H1–H2	4.0	5.1	1.5	5.6	1.3	7.6	1.9	5.5
S5–H1–H2	4.2	5.1	1.9	5.4	1.1	6.0	0.0	22.9
S2–S5–H1–H2	3.7	5.5	1.3	7.2	0.0	14.1	0.3	16.9

3 Results

We performed a comparison of the mean square error for individual and various combinations of the quality measures for fast fading degradation. Table 1 shows the RMS errors obtained after regression on the training data and then verified on the validation data. The minimum RMS errors (approx equal to zero) on the training data were achieved using a third degree polynomial combination of all the four measures and a fourth degree polynomial combination of S_5, H_1, H_2. However, using the same combinations resulted in unexpected RMS errors of 14.1 and 22.9 respectively during validation indicating cases of overfitting on the training data. The most optimal results are given by a linear combination of H_1, H_2, S_2 which provide RMS errors of 4.0 and 5.1 on the training and validation data respectively. Therefore, we concluded that a linear combination of these measures gives the best estimate of human perception. Resultantly, regressing the values of these quality measures against HPC of the training data, the coefficients a_0, a_1, b_1, c_1 as given in Eqn. 1 were found. Thus, the HIQ measure achieved is given by:

$$\text{HIQ} = 85.33 - 529.51H_1 - 2164.50H_2 - 0.0137S_2 \tag{2}$$

Fig. 1 shows the HPV curve and the regressed HIQ measure plot for the training data. The HPV curve was calculated by averaging the HPVs of all images

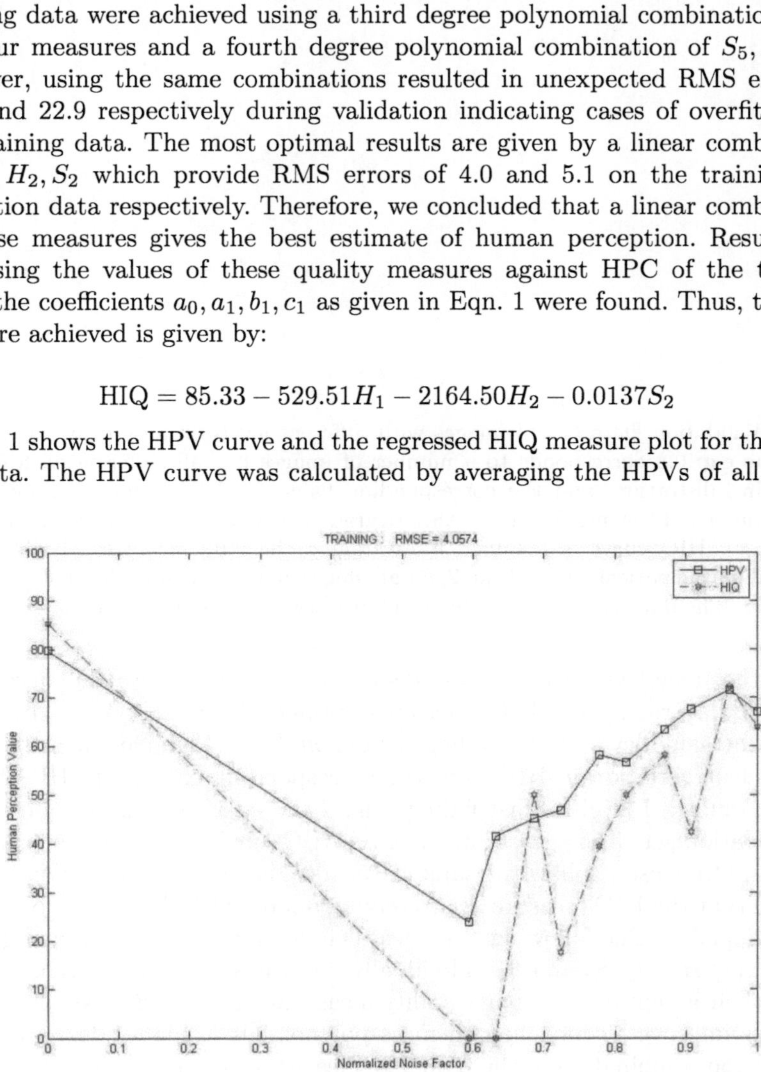

Fig. 1. Training Data of 60 images with different levels of noise degradation. Any one value e.g. 0.2 corresponds to a number of images but all suffering with 0.2% of fast fading distortion, and the corresponding value of HPV is mean opinion score of all human judgements for these 0.2% degraded images (50 human judgements for one image). HIQ curve is obtained by averaging the HIQ measures obtained from proposed mathematical model, Eqn. 2, for all images having the same level of fast fading distortion. The data is made available at http://www.csse.uwa.edu.au/ ~ ajmal/.

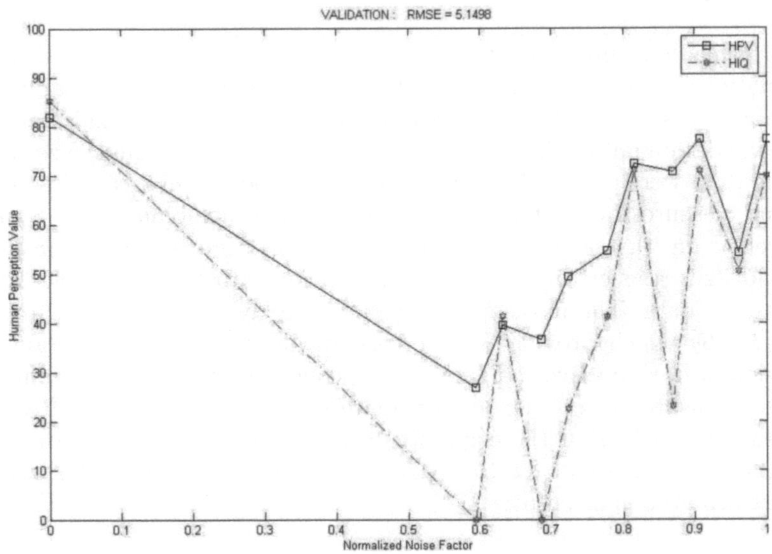

Fig. 2. Validation Data of 114 images with different levels of noise degradation. Any one value e.g. 0.8 corresponds to a number of images but all suffering with 0.8% of fast fading distortion, and the corresponding value of HPV is mean opinion score of all human judgements for these 0.8% degraded images (50 human judgements for one image). HIQ curve is obtained by averaging the HIQ measures obtained from proposed mathematical model, Eqn. 2, for all images having the same level of fast fading distortion. The data is made available at http://www.csse.uwa.edu.au/ ~ ajmal/.

having the same level of fast fading distortion. Similarly, the HIQ curve is calculated by averaging the HIQ measures obtained from Eqn. 2 for all images having the same level of fast fading distortion. Thus Fig. 1 depicts the image independent variation in HPV and the corresponding changes in HIQ for different normalized levels of fast fading. Fig. 2 shows similar curves obtained on the validation set of images. Note that the HIQ curves, in both the cases (i.e. Fig. 1 and 2), closely follow the same pattern of the HPV curves which is an indication that the HIQ measure accurately correlates with the human perception of image quality. The following inferences can be made from our results given in Table 1. (1) H_1, H_2, S_2 and S_5 individually perform satisfactorily which demonstrates their acceptance as image quality measures. (2) The effectiveness of these measures improve by modeling them as polynomials of higher degrees. (3) Increasing the combination of these quality measures e.g., using all four measures does not necessarily increase their effectiveness, as this may suffer from overfitting on training data. (4) An important finding is validation of the fact that HIQ measure closely follows the human perception curve, as evident from Fig. 2 where HIQ curve has similar trend as of HPV, though both are calculated independently. (5) Finally, a linear combination of H_1, H_2, S_2 gives the best estimate of the human perception of an image quality.

4 Conclusion

We presented a hybrid image quality measure, HIQ, consisting of a first order polynomial combination of three different quality metrics. We demonstrated its effectiveness by evaluating it over a separate validation data consisting of a separate set of 114 different images. HIQ proved to closely follow the human perception curve and gave an error improvement over the individual measures. In the future, we plan to investigate the HIQ for other degradation models like white noise, JPEG compression, gaussian blur etc.

References

1. Wang, Z., Bovik, A.C., Lu, L.: Why is Image Quality Assessment so difficult. In: IEEE International Conference on Acoustics, Speech and Signal Processing, vol. 4, pp. 3313–3316 (2002)
2. Eskicioglu, A.M.: Quality measurement for monochrome compressed images in the past 25 years. In: IEEE International Conference on Acoustics, Speech and Signal Processing, vol. 4, pp. 1907–1910 (2000)
3. Eskicioglu, A.M., Fisher, P.S.: Image Quality Measures and their Performance. IEEE Transaction on Communications 43, 2959–2965 (1995)
4. Miyahara, M., Kotani, K., Algazi, V.R.: Objective Picture Quality Scale (PQS) for image coding. IEEE Transaction on Communications 9, 1215–1225 (1998)
5. Guo, L., Meng, Y.: What is Wrong and Right with MSE. In: Eighth IASTED International Conference on Signal and Image Processing, pp. 212–215 (2006)
6. Wang, Z., Bovik, A.C.: A universal image quality index. IEEE Signal Processing Letters 9, 81–84 (2002)
7. Wang, Z., Bovik, A.C., Sheikh, H.R., Simoncelli, E.P.: Image quality assessment: From error measurement to structural similarity. IEEE Transaction on Image Processing 13 (January 2004)
8. Wang, Z., Simoncelli, E.P., Bovik, A.C.: Multi-scale structural similarity for image quality assessment. In: 37th IEEE Asilomar Conference on Signals, Systems, and Computers (2003)
9. Shnayderman, A., Gusev, A., Eskicioglu, A.M.: An SVD-Based Gray-Scale Image Quality Measure for Local and Global Assessment. IEEE Transaction on Image Processing 15 (February 2006)
10. Sheikh, H.R., Sabir, M.F., Bovik, A.C.: A statistical evaluation of recent full reference image quality assessment algorithms. IEEE Transaction on Image Processing 15, 3440–3451 (2006)
11. Sarnoff Corporation, JNDmetrix Technology, `http://www.sarnoff.com`
12. Watson, A.B.: DC Tune: A technique for visual optimization of DCT quantization matrices for individual images, Society for Information Display Digest of Technical Papers, vol. XXIV, pp. 946–949 (1993)
13. Damera-Venkata, N., Kite, T.D., Geisler, W.S., Evans, B.L., Bovik, A.C.: Image Quality Assessment based on a Degradation Model. IEEE Transaction on Image Processing 9, 636–650 (2000)
14. Weken, D.V., Nachtegael, M., Kerre, E.E.: Using similarity measures and homogeneity for the comparison of images. Image and Vision Computing 22, 695–702 (2004)

15. Avcibas, I., Sankur, B., Sayood, K.: Statistical Evaluation of Image Quality Measures. Journal of Electronic Imaging 11, 206–223 (2002)
16. Sheikh, H.R., Bovik, A.C., de Veciana, G.: An information fidelity criterion for image quality assessment using natural scene statistics. IEEE Transaction on Image Processing 14, 2117–2128 (2005)
17. Sheikh, H.R., Bovik, A.C.: Image information and Visual Quality. IEEE Transaction on Image Processing 15, 430–444 (2006)
18. Chandler, D.M., Hemami, S.S.: VSNR: A Wavelet base Visual Signla-to-Noise Ratio for Natural Images. IEEE Transaction on Image Processing 16, 2284–2298 (2007)
19. Sheikh, H.R., Wang, Z., Cormack, L., Bovik, A.C.: LIVE image quality assessment database, `http://live.ece.utexas.edu/research/quality`
20. ITU-R Rec. BT. 500-11, Methodology for the Subjective Assessment of the Quality for Television Pictures

Framework for Applying Full Reference Digital Image Quality Measures to Printed Images

Tuomas Eerola, Joni-Kristian Kämäräinen*, Lasse Lensu, and Heikki Kälviäinen

Machine Vision and Pattern Recognition Research Group (MVPR)
*MVPR/Computational Vision Group, Kouvola
Department of Information Technology
Lappeenranta University of Technology (LUT), Finland
firstname.lastname@lut.fi

Abstract. Measuring visual quality of printed media is important as printed products play an essential role in every day life, and for many "vision applications", printed products still dominate the market (e.g., newspapers). Measuring visual quality, especially the quality of images when the original is known (full-reference), has been an active research topic in image processing. During the course of work, several good measures have been proposed and shown to correspond with human (subjective) evaluations. Adapting these approaches to measuring visual quality of printed media has been considered only rarely and is not straightforward. In this work, the aim is to reduce the gap by presenting a complete framework starting from the original digital image and its hard-copy reproduction to a scanned digital sample which is compared to the original reference image by using existing quality measures. The proposed framework is justified by experiments where the measures are compared to a subjective evaluation performed using the printed hard copies.

1 Introduction

The importance of measuring visual quality is obvious from the viewpoint of limited data communications bandwidth or feasible storage size: an image or video compression algorithm is chosen based on which approach provides the best (average) visual quality. The problem should be well-posed since it is possible to compare the compressed data to the original (full-reference measure). This appears straightforward, but it is not because the underlying process how humans perceive quality or its deviation is unknown. Some physiological facts are know, e.g., the modulation transfer function of the human eye, but the accompanying cognitive process is still unclear. For digital media (images), it has been possible to devise heuristic full-reference measures, which have been shown to correspond with the average human evaluation at least for a limited number of samples, e.g., the visible difference predictor [1], structural similarity metric [2], and visual information fidelity [3]. Despite the fact that "analog" media (printed images) have been used for a much longer time, they cannot overcome certain limitations, which on the other hand, can be considered as the strengths of

A.-B. Salberg, J.Y. Hardeberg, and R. Jenssen (Eds.): SCIA 2009, LNCS 5575, pp. 99–108, 2009.

digital reproduction. For printed images, it has been considered to be impossible to utilise a similar full-reference strategy since the information undergoes various non-linear transformations (printing, scanning) before its return to the digital form. Therefore, the visual quality of printed images has been measured with various low-level measures which represent some visually relevant characteristic of the reproduced image, e.g., mottling [4] and the number of missing print dots [5]. However, since the printed media still dominate in many reproduction forms of visual information (journals, newspapers, etc.), it is intriguing to enable the use of well-studied full-reference digital visual quality measures in the context of printed media.

For digital images, the relevant literature consists of full-reference (FR) and no-reference (NR) quality measures according to whether a reproduced image is compared to a known reference image (FR), or a reference does not exist (NR). Where the NR measures stand out as a very challenging research problem [6], the FR measures are based on a more stronger rationale. The current FR measures make use of various heuristics and their correlation to the human quality experience is tested usually with a limited set for pre-defined types of distortions. The FR measures, however, posses an almost unexplored topic for printed images where the subjective human evaluation trials are often much more general. By closing the gap, completely novel research results can be achieved. An especially intriguing study where a very comprehensive comparison between the state-of-the-art FR measures was performed for digital images was published by Sheikh et al. [7]. How could this experiment be replicated for the printed media?

The main challenges in enabling the use of the FR measures with the printed media are actually those completely missing from the digital reproduction: image correspondence by accurate registration and removal of reproduction distortions (e.g., halftone patterns). In this study, we address these problems with known computer vision techniques. Finally, we present a complete framework for applying the FR digital image quality measures to printed images. The framework contains the full flow from a digital original and printed hard-copy sample to a single scalar representing the overall quality computed by comparing the corresponding re-digitised and aligned image to the original digital reference. The stages of the framework, the registration stage in particular, are studied in detail to solve the problems and provide as accurate results as possible. Finally, we justify our approach by comparing the computed quality measure values to an extensive set of subjective human evaluations.

The article is organised as follows. In Sec. 2, the whole framework is presented. In Sec. 3, the framework is tested and improved, as well as, some full reference measures are tested. Future work is discussed in Sec. 4, and finally, conclusions are devised in Sec. 5.

2 The Framework

When the quality of a compressed image is analysed by comparing it to an original (reference) image, the FR measures can be straightforwardly computed, cf., computing "distance measures". This is possible as digital representations are in correspondence, i.e., there exists no rigid, partly rigid or non-rigid (elastic) spatial shifts between the images and compression should retain photometric equivalence. This is not the case with printed media. In modern digital printing, a digital reference exists, but it will undergo various irreversible transforms, especially in printing and scanning, until another digital image for the comparison is established. The first important consideration is the scanning process. Since we are not interested in the scanning but printing quality, a scanner must be an order of magnitude better than a printing system. Fortunately, this is not difficult to achieve with the available top-quality scanners in which sub-pixel accuracy of the original can be used. It is important to use sub-pixel accuracy because this prevents the scanning distortions to affect the registration. Furthermore, to prevent photometric errors from occurring, the scanner colour mapping should be adjusted to correspond to the original colour map.This can be achieved by using a scanner profiling software that comes along with the high-quality scanners. Secondly, a printed image contains halftone patterns, and therefore, descreening is needed to remove high halftone frequencies and form a continuous tone image comparable to the reference image. Thirdly, the scanned image needs to be very accurately registered with the original image before the FR image quality measures or dissimilarity between the images can be computed. The registration can be assumed to be rigid since non-rigidity is a reproduction error and partly-rigid correspondence should be avoided by using the high scanning resolution.

Based on the above general discussion, it is possible to sketch the main structure for our framework of computing FR image quality measures from printed images. The framework structure and data flow are illustrated in Fig. 1. First, the printed halftone image is scanned using a colour-profiled scanner. Second, the descreening is performed using a Gaussian low-pass filter (GLPF) which produces a continuous tone image. To perform the descreening in a more psychophysically plausible way, the image is converted to the CIE L*a*b* colour space where all the channels are filtered separately. The purpose of CIE L*a*b* is to span a perceptually uniform colour space and not suffer from the problems related to, e.g., RGB where the colour differences do not correspond to the human visual system [8]. Moreover, the filter cut-off frequency is limited by the printing resolution (frequency of the halftone pattern) and should not be higher than 0.5 mm which is the smallest detail visible to human eyes when unevenness of a print is evaluated from the viewing distance of 30 cm [4]. To make the input and reference images comparable, the reference image needs to be filtered with the identical cut-off frequency.

2.1 Rigid Image Registration

Rigid image registration was considered as a difficult problem until the invention of general interest point detectors and their rotation and scale invariant descriptors. These methods provide unparametrised methods which yield accurate and robust correspondence essential for the registration. The most popular method which combines both the interest point detection and description is David Lowe's SIFT [9]. Registration based on the SIFT features has been utilised, for example,

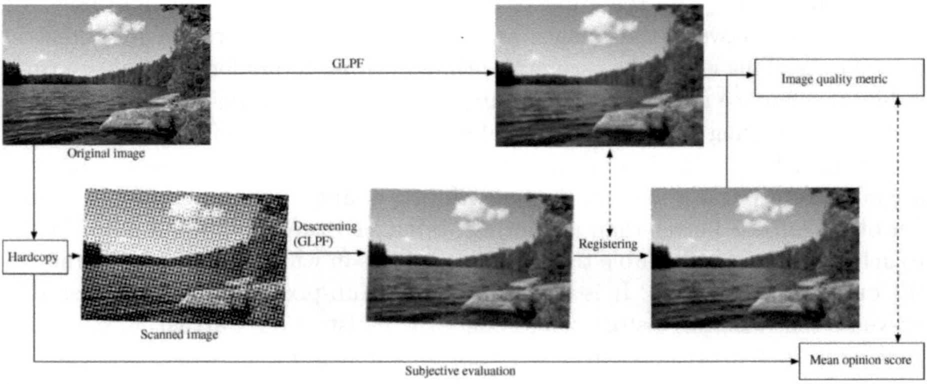

Fig. 1. The structure of the framework and data flow for computing full-reference image quality measures for printed images

in mosaicing panoramic views [10]. The registration consists of 4 stages: extract local features from both images, match the features (correspondence), find a 2D homography between correspondence and finally transform one image to another for comparison.

Our method performs a scale and rotation invariant extraction of local features using the scale-invariant feature transform (SIFT) by Lowe [9]. The SIFT method includes also the descriptor part which can be used for matching, i.e., the correspondence search. As a standard procedure, the random sample consensus (RANSAC) principle presented in [11] is used to find the best homography using exact homography estimation for the minimum number of points and linear estimation methods for all "inliers". The linear methods are robust and accurate also for the final estimation since the number of correspondences is typically quite large (several hundreds of points). The implemented linear homography estimation methods are Umeyama for isometry and similarity [12], a restricted direct linear transform (DLT) for affinity and the standard normalised DLT for projectivity [13]. The only adjustable parameters in our method are the number of random iterations and the inlier distance threshold for the RANSAC which can be safely set to 2000 and 0.7 mm, respectively. This makes the whole registration algorithm parameter free. In image transformation, we utilise standard remapping using bicubic interpolation.

2.2 Full Reference Quality Measures

Simplest FR quality measures are mathematical formulae for computing element-wise similarity or dissimilarity between two matrices (images), such as, the mean squared error (MSE) or peak signal-to-noise ratio (PSNR). These methods are widely used in signal processing since they are computationally efficient and have a clear physical meaning. These measures should, however, be restricted by the known physiological facts to bring them in correspondence with the human visual system. For example, the MSE can be generalised to colour images by computing Euclidean distances in the perceptually uniform CIE L*a*b* colour space as

$$LabMSE = \frac{1}{MN} \sum_{i=0}^{M-1} \sum_{j=0}^{N-1} [\Delta L^*(i,j)^2 + \Delta a^*(i,j)^2 + \Delta b^*(i,j)^2] \quad (1)$$

where $\Delta L^*(i,j)$, $\Delta a^*(i,j)$ and $\Delta b^*(i,j)$ are differences of the colour components at point (i,j) and M and N are the width and height of the image. This measure is known as the L*a*b* perceptual error [14]. There are several more exotic and more plausible methods surveyed, e.g., in [7], but since our intention here is only to introduce and study our framework, we utilise the standard MSE and PSNR measures in the experimental part of this study. Using any other FR quality measure in our framework is straightforward.

3 Experiments

Our "ground truth", i.e., the dedicatedly selected test targets (prepared independently by a media technology research group) and their extensive subjective evaluations (performed independently by a vision psychophysics research group) were recently introduced in detail in [15,16,17]. The test set consisted of natural images printed with a high quality inkjet printer on 16 different paper grades. The printed samples were scanned using a high quality scanner with 1250 dpi resolution and 48-bit RGB colours. A colour management profile was derived for the scanner before scanning, scanner colour correction, descreening and other automatic settings were disabled, and the digitised images were saved using lossless

Fig. 2. The reference image

compression. Descreening was performed using the cut-off frequency of 0.1 mm which was selected based on the resolution of the printer (360 dpi). The following experiments were conducted using the reference image in Fig. 2, which contains different objects generally considered as most important for quality inspection: natural solid regions, high texture frequencies and a human face. The size of the original (reference) image was 2126 × 1417 pixels.

3.1 Registration Error

The success of the registration was studied by examining error magnitudes and orientations in different parts of the image. For a good registration result in general, the magnitudes should be small (sub-pixel) and random, and similarly their orientations should be randomly distributed. The registration error was estimated by setting the inlier threshold, used by the RANSAC, to relatively loose and by studying the relative locations of accepted local features (matches) between the reference and input images after registration. This should be a good estimate of the geometrical error of the registration. Despite the fact that the loose inlier threshold causes a lot of false matches, the most of the matches are still correct, and the trend of distances between the correspondence in different parts of the image describes the real geometrical registration error.

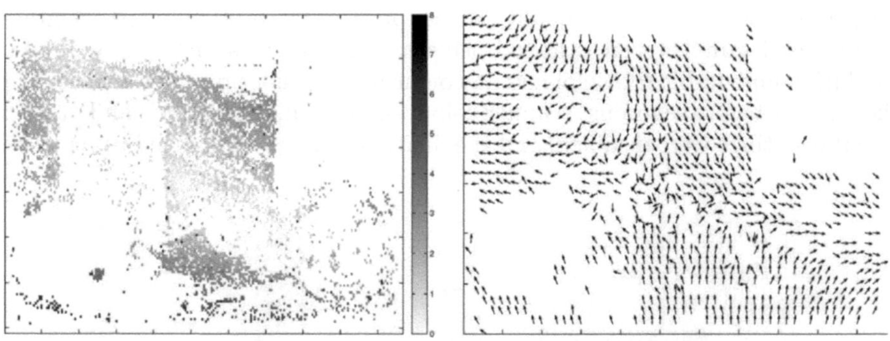

(a) (b)

Fig. 3. Registration error of similarity transformation: (a) error magnitudes; (b) error orientations

In Fig. 3, the registration errors are visualised for similarity as the selected homography. Similarity should be the correct homography as in the ideal case, the homography between the original image and its printed reproduction should be similarity (translation, rotation and scaling). However, as it can be seen in Fig. 3(a), the registration is accurate to sub-pixel accuracy only in the centre of the image where the number of local features is high. However, the error magnitudes increase to over 10 pixels near the image borders which is far from sufficient for the FR measures. The reason for the spatially varying inaccuracy

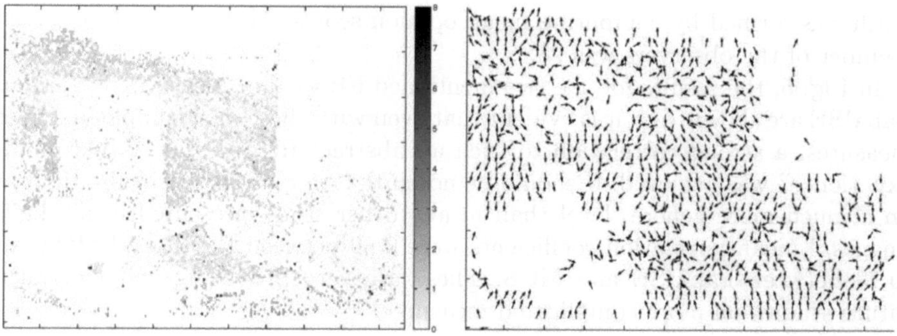

(a) (b)

Fig. 4. Registration error of affine transformation: (a) error magnitudes; (b) error orientations

can be seen from Fig. 3(b), where the error orientations are away from the centre on the left- and right side of the image, and towards the centre on the top and at the bottom. The correct interpretation is that there exists small stretching in the printing direction. This stretching is not fatal for the human eye, but it causes a transformation which does not follow similarity. Similarity must be replaced with another more general transformation, affinity being the most intuitive. In Fig. 4, the registration errors for affine transformation are visualised. Now, the registration errors are very small over the whole image (Fig. 4(a)) and the error orientations correspond to a uniform random distribution (Fig. 4(b)).

In some cases, e.g., if the paper in the printer or imaging head of the scanner do not move at constant speed, registration may need to be performed in a piecewise manner to get accurate registration results. One noteworthy benefit of the piecewise registration is that after joining the registered image parts, the falsely registered images are clearly visible and can be either re-registered or eliminated from biasing further studies. In the following experiments, the images are registered in two parts.

3.2 Full Reference Quality Measures

The above presented experiment was already a proof-of-concept for our framework, but we wanted to briefly apply some simple FR quality measures to test the framework in practise.

The performance of the FR quality measures was studied against the subjective evaluation results (ground truth) introduced in [15]. In brief, all samples (same image content) were placed on a table in random order. Also the numbers from 1 to 5 were presented on the table. An observer was asked to select the sample representing the worst quality of the sample set and place it on the number 1. Then, the observer was asked to select the best sample and place it on the number 5. After that, the observer was asked to place the remaining samples on numbers 1 to 5 so that the quality grows regularly from 1 to 5. The final ground truth was formed by computing mean opinion scores (MOS) over all observers. Number of the observers was 28.

In Fig. 5, the results for the two mentioned FR quality measures, PSNR and LabMSE are shown, and it is evident that even with these most simple pixel-wise measures, a strong correlation to such an abstract task as the "visual quality experience" was achieved. It should be noted that our subjective evaluations are on a much more general level than in any other study presented using digital images. Linear correlation coefficients were 0.69 between PSNR and MOS, and -0.79 between LabMSE and MOS. These are very promising and motivating future studies on more complicated measures.

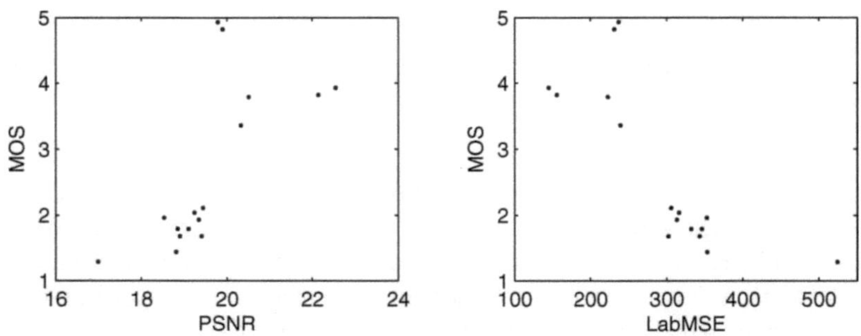

(a) (b)

Fig. 5. Scatter plots between simple FR measures computed in our framework and subjective MOS: (a) PSNR; (b) LabMSE

4 Discussion and Future Work

The most important consideration in the future work is to find FR measures which are more appropriate for printed media. Although our registration method works very well, sub-pixel errors still appear and they always affect simple pixel-wise distance formula, such as the MSE. In other words, we need FR measures which are less sensitive to small registration errors. Another notable problem arises from the nature of subjective tests with printed media: The experiments are carried out using printed (hard-copy) samples and the actual digital reference (original) is not available for the observers and not even interesting; the visual quality experience is not a task of finding differences between the reproduction and original, but a more complex process of what is seen as excellent, good, moderate or poor quality. This point has been wrongly omitted in many digital image quality studies, but it must be embedded in FR measures.

In the literature, several approaches have been proposed to enhance the FR algorithms to be more consistent with the human perception: mathematical distance formulations (e.g., fuzzy similarity measures [18]), human visual system (HVS) model based (e.g., Sarnoff JNDmetrix [19]), HVS models combined application specific modelling (DCTune [20]), structural (structural similarity metric [2]), and information theoretic (visual information fidelity [3]). It will be interesting to evaluate these more advanced methods in our framework. Proper statistical evaluation, however, requires a larger amount of samples and several different image contents. Another important aspect is the effect of the cut-off frequency in the descreening stage. What is the suitable cut-off frequency and does it depend on the used FR measure?

5 Conclusions

In this work, we presented a framework to compute full reference (FR) image quality measures, common in digital image quality research field, for printed natural images. The work was first of its kind in this extent and generality, and it will provide a new basis for future studies on evaluating visual quality of printed products using methods common in the field of computer vision and digital image processing.

Acknowledgement

The authors would like to thank Raisa Halonen from the Department of Media Technology in Helsinki University of Technology for providing the test material and Tuomas Leisti from the Department of Psychology in University of Helsinki for providing the subjective evaluation data. The authors would like to thank also the Finnish Funding Agency for Technology and Innovation (TEKES) and partners of the DigiQ project (No. 40176/06) for support.

References

1. Daly, S.: Visible differences predictor: an algorithm for the assessment of image fidelity. In: Proc. SPIE, San Jose, USA. Human Vision, Visual Processing, and Digital Display III, vol. 1666, pp. 2–15 (1992)
2. Wang, Z., Bovik, A.C., Sheikh, H.R., Simoncelli, E.P.: Image quality assessment: From error visibility to structural similarity. IEEE Transactions on Image Processing 13(4), 600–612 (2004)
3. Sheikh, H.R., Bovik, A.C.: Image information and visual quality. IEEE Transactions On Image Processing 15(2), 430–444 (2006)
4. Sadovnikov, A., Salmela, P., Lensu, L., Kamarainen, J., Kalviainen, H.: Mottling assessment of solid printed areas and its correlation to perceived uniformity. In: 14th Scandinavian Conference of Image Processing, Joensuu, Finland, pp. 411–418 (2005)
5. Vartiainen, J., Sadovnikov, A., Kamarainen, J.K., Lensu, L., Kalviainen, H.: Detection of irregularities in regular patterns. Machine Vision and Applications 19(4), 249–259 (2008)
6. Sheikh, H.R., Bovik, A.C., Cormack, L.: No-reference quality assessment using natural scene statistics: JPEG 2000. IEEE Transactions on Image Processing 14(11), 1918–1927 (2005)
7. Sheikh, H.R., Sabir, M.F., Bovik, A.C.: A statistical evaluation of recent full reference image quality assessment algorithms. IEEE Transactions On Image Processing 15(11), 3440–3451 (2006)

8. Wyszecki, G., Stiles, W.S.: Color science: concepts and methods, quantitative data and formulae, 2nd edn. Wiley, Chichester (2000)
9. Lowe, D.: Distinctive image features from scale-invariant keypoints. International Journal of Computer Vision 60(2), 91–110 (2004)
10. Brown, M., Lowe, D.G.: Automatic panoramic image stitching using invariant features. International Journal of Computer Vision 74(1), 59–73 (2007)
11. Fischler, M., Bolles, R.: Random sample consensus: A paradigm for model fitting with applications to image analysis and automated cartography. Graphics and Image Processing 24(6) (1981)
12. Umeyama, S.: Least-squares estimation of transformation parameters between two point patterns. IEEE-TPAMI 13(4), 376–380 (1991)
13. Hartley, R., Zisserman, A.: Multiple View Geometry in Computer Vision, 2nd edn. Cambridge University Press, Cambridge (2003)
14. Avcibaş, I., Sankur, B., Sayood, K.: Statistical evaluation of image quality measures. Journal of Electronic Imaging 11(2), 206–223 (2002)
15. Oittinen, P., Halonen, R., Kokkonen, A., Leisti, T., Nyman, G., Eerola, T., Lensu, L., Kälviäinen, H., Ritala, R., Pulla, J., Mettänen, M.: Framework for modelling visual printed image quality from paper perspective. In: SPIE/IS&T Electronic Imaging 2008, Image Quality and System Performance V, San Jose, USA (2008)
16. Eerola, T., Kamarainen, J.K., Leisti, T., Halonen, R., Lensu, L., Kälviäinen, H., Nyman, G., Oittinen, P.: Is there hope for predicting human visual quality experience? In: Proc. of the IEEE International Conference on Systems, Man, and Cybernetics, Singapore (2008)
17. Eerola, T., Kamarainen, J.K., Leisti, T., Halonen, R., Lensu, L., Kälviäinen, H., Oittinen, P., Nyman, G.: Finding best measurable quantities for predicting human visual quality experience. In: Proc. of the IEEE International Conference on Systems, Man, and Cybernetics, Singapore (2008)
18. van der Weken, D., Nachtegael, M., Kerre, E.E.: Using similarity measures and homogeneity for the comparison of images. Image and Vision Computing 22(9), 695–702 (2004)
19. Lubin, J., Fibush, D.: Contribution to the IEEE standards subcommittee: Sarnoff JND vision model (August 1997)
20. Watson, A.B.: DCTune: A technique for visual optimization of DCT quantization matrices for individual images. Society for Information Display Digest of Technical Papers XXIV, 946–949 (1993)

Colour Gamut Mapping as a Constrained Variational Problem

Ali Alsam[1] and Ivar Farup[2]

[1] Sør-Trøndelag University College, Trondheim, Norway
[2] Gjøvik University College, Gjøvik, Norway

Abstract. We present a novel, computationally efficient, iterative, spatial gamut mapping algorithm. The proposed algorithm offers a compromise between the colorimetrically optimal gamut clipping and the most successful spatial methods. This is achieved by the iterative nature of the method. At iteration level zero, the result is identical to gamut clipping. The more we iterate the more we approach an optimal, spatial, gamut mapping result. Optimal is defined as a gamut mapping algorithm that preserves the hue of the image colours as well as the spatial ratios at all scales. Our results show that as few as five iterations are sufficient to produce an output that is as good or better than that achieved in previous, computationally more expensive, methods. Being able to improve upon previous results using such low number of iterations allows us to state that the proposed algorithm is $O(N)$, N being the number of pixels. Results based on a challenging small destination gamut supports our claims that it is indeed efficient.

1 Introduction

To accurately define a colour three independent variables need to be fixed. In a given three dimensional colour-space the colour gamut is the volume which encloses all the colour values that can be reproduced by the reproduction device or present in the image. Colour gamut mapping is the problem of representing the colour values of an image in the space of a reproduction device: Typically, a printer or a monitor. Furthermore, in the general case, when an image gamut is larger than the destination gamut some image-information will be lost. We therefore redefine gamut mapping as: The problem of representing the colour values of an image in the space of a reproduction device with minimum information loss.

Unlike single colours, images are represented in a higher dimensional space than three, i.e. knowledge of the exact colour values is not, on its own, sufficient to reproduce an unknown image. In order to fully define an image, the spatial location of each colour pixel needs to be fixed. Based on this, we define two categories of gamut mapping algorithms: In the first, colours are mapped independent of their spatial location [1]. In the second, the mapping is influenced by

A.-B. Salberg, J.Y. Hardeberg, and R. Jenssen (Eds.): SCIA 2009, LNCS 5575, pp. 109–118, 2009.

the location of each colour value [2,3,4,5]. The latter category is referred to as spatial gamut mapping.

Eschbach [6] stated that: Although the accuracy of mapping a single colour is well defined, the reproduction accuracy of images isn't. To elucidate this claim, with which we agree, we consider a single colour that is defined by its hue, saturation and lightness. Assuming that such a colour is outside the target gamut, we can modify its components independently. That is to say, if the colour is lighter or more saturated than what can be achieved inside the reproduction gamut, we shift its lightness and saturation to the nearest feasible values. Further, in most cases it is possible to reproduce colours without shifting their hue.

Taking the spatial location of colours into account presents us with the challenge of defining the spatial components of a colour pixel and incorporating this information into the gamut mapping algorithm. Generally speaking, we need to define rules that would result in mapping two colours with identical hue, saturation and lightness to two different locations depending on their location in the image plane. The main challenge is, thus, defining the spatial location of an image pixel in a manner that results in an improved gamut mapping. By improved we mean that the appearance of the resultant, in gamut, image is visually preferred by a human observer. Further, from a practical point of view, the new definition needs to result in an algorithm that is fast and does not result in image artifacts.

It is well understood that the human visual system is more sensitive to spatial ratios than absolute values [7]. This knowledge is at the heart of all spatial gamut mapping algorithms. A definition of spatial gamut mapping is then: The problem of representing the colour values of an image in the space of a reproduction device while preserving the spatial ratios between different colour pixels. In an image spatial ratios are the difference, given some difference metric, between a pixel and its surround. This can be the difference between one pixel and its adjacent neighbors or pixels far away from it. Thus, we face the problem that: Spatial ratios are defined in different scales and dependent on the chosen difference metric.

McCann suggested to preserve the spatial gradients at all scales while applying gamut mapping [8]. Meyer and Barth [9] suggested to compress the lightness of the image using a low-pass filter in the Fourier domain. As a second step the high-pass image information is added back to the gamut compressed image. Many spatial gamut mapping algorithms have been based upon this basic idea [2,10,11,12,4].

A completely different approach was taken by Nakauchi et al. [13]. They defined gamut mapping as an optimization problem of finding the image that is perceptually closest to the original and has all pixels inside the gamut. The perceptual difference was calculated by applying band-pass filters to Fourier-transformed CIELab images and then weighing them according to the human contrast sensitivity function. Thus, the best gamut mapped image is the image having contrast (according to their definition) as close as possible to the original.

Kimmel et al. [3] presented a variational approach to spatial gamut mapping where it was shown that the gamut mapping problem leads to a quadratic programming formulation, which is guaranteed to have a unique solution if the gamut of the target device is convex.

The algorithm presented in this paper adheres to our, previously, stated definition of spatial gamut mapping in that we aim to preserve the spatial ratios between pixels in the image. We start by calculating the gradients of the original image in CIELab colour space. The image is then gamut mapped by projecting the colour values to the nearest, in gamut, point along hue-constant lines. The difference between the gradient of the gamut mapped image and that of the original is then iteratively minimized with the constraint that the resultant colour is a convex combination of its gamut mapped representation and the center of the destination gamut. Imposing the convexity constraint ensures that the resultant colour is inside the reproduction gamut and has the same hue as the original. Further, if the convexity constraint is removed then the result of the gradient minimization is the original image. The scale at which the gradient is preserved is related to the number of iterations and the extent to which we can fit the original gradients into the destination gamut.

The main contributions of this work are as follows: We first present a mathematically elegant formulation of the gamut mapping problem in colour space. Our formulation can be extended to a higher dimensional space than three. Secondly, our algorithm offers a compromise between the colorimetrically optimal gamut clipping and the most successful spatial methods. This latter aspect is achieved by the iterative nature of the methods. At zero iteration level, the result is identical to gamut clipping. The more we iterate the more we approach McCann's definition of an optimal gamut mapping result. The calculations are performed in the three-dimensional colour space, thus, the goodness of the hue preservation is dependent not upon our formulation but the extent to which the hue lines in the colour space are linear. Finally, our results show that as few as five iterations are sufficient to produce an output that is similar or better than previous methods. Being able to improve upon previous results using such low number of iterations allows us to state that the proposed algorithm is: Fast.

2 Spatial Gamut Mapping: A Mathematical Definition

Let's say we have an original image with pixel values $\mathbf{p}(x, y)$ (bold face to indicate vector) in CIELab or any similarly structured colour space. A gamut clipped image can be obtained by leaving in-gamut colours untouched, and moving out-of-gamut colours along staight lines towards $\mathbf{g}$, the center of the gamut on the L axis until they hit the gamut surface. Let's denote the gamut clipped image $\mathbf{p_c}(x, y)$. From the original image and the gamut clipped one, we can define

$$\alpha_c(x,y) = \frac{||\mathbf{p}_c(x,y) - \mathbf{g}||}{||\mathbf{p}(x,y) - \mathbf{g}||}, \tag{1}$$

where $||\cdot||$ denotes the L_2 norm of the colour space. Since $\mathbf{p}_c(x,y) - \mathbf{g}$ is parallel to $\mathbf{p}(x,y) - \mathbf{g}$, this means that the gamut clipped image can be obtained as a linear convex combination of the original image and the gamut clipped one,

$$\mathbf{p}_c(x,y) = \alpha_c(x,y)\mathbf{p}(x,y) + (1 - \alpha_c(x,y))\mathbf{g}. \tag{2}$$

Given that we want to perform the gamut mapping in this direction: This is the least amount of gamut mapping we can do. If we want to impose some more gamut mapping in addition to the clipping, e.g., in order to preserve details, this can be obtained by multiplying $\alpha_c(x,y)$ with some number $\alpha_s(x,y) \in [0,1]$ (s for spatial). With this introduced, the final spatial gamut mapped image can be written as the linear convex combination

$$\mathbf{p}_s(x,y) = \alpha_s(x,y)\alpha_c(x,y)\mathbf{p}(x,y) + (1 - \alpha_s(x,y)\alpha_c(x,y))\mathbf{g}. \tag{3}$$

Now, we assume that the best spatially gamut mapped image is the one having gradients as close as possible to the original image. This means that we want to find

$$\min \int ||\nabla \mathbf{p}_s(x,y) - \nabla \mathbf{p}(x,y)||_F^2 \, dA \quad \text{subject to} \quad \alpha_s(x,y) \in [0,1]. \tag{4}$$

where $||\cdot||_F$ denotes the Frobenius norm on $\mathbb{R}^{3\times 2}$.

In Equation (3), everything exept $\alpha_s(x,y)$ can be determined in advance. Let's therefore rewrite $\mathbf{p}_s(x,y)$ as

$$\mathbf{p}_s(x,y) = \alpha_s(x,y)\alpha_c(x,y)(\mathbf{p}(x,y) - \mathbf{g}) + \mathbf{g} \equiv \alpha_s(x,y)\mathbf{d}(x,y) + \mathbf{g}, \tag{5}$$

where $\mathbf{d}(x,y) = \alpha_c(\mathbf{p}(x,y) - \mathbf{g})$ has been introduced. Then, since $\mathbf{g}$ is constant,

$$\nabla \mathbf{p}_s(x,y) = \nabla(\alpha_s(x,y)\mathbf{d}(x,y)), \tag{6}$$

and the optimisition problem at hand reduces to finding

$$\min \int ||\nabla(\alpha_s(x,y)\mathbf{d}(x,y)) - \nabla \mathbf{p}(x,y)||_F^2 \, dA \quad \text{subject to} \quad \alpha_s(x,y) \in [0,1]. \tag{7}$$

This corresponds to solving the Euler–Lagrange equation:

$$\nabla^2(\alpha_s(x,y)\mathbf{d}(x,y) - \mathbf{p}(x,y)) = 0. \tag{8}$$

Finally, in Figure (1) we present a graphical representation of the spatial gamut problem. $p(x,y)$ is the original colour at image pixel (x,y), this value is clipped to the gamut boundary resulting in a new colour $p_c(x,y)$ which is compressed based on the gradient information to a new value $p_s(x,y)$.

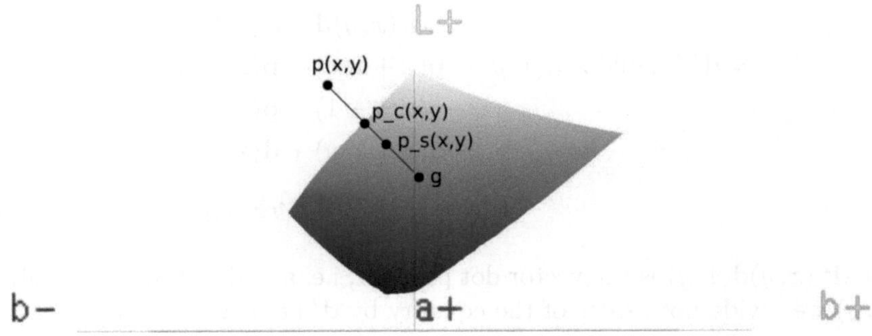

Fig. 1. A representation of the spatial gamut mapping problem. $p(x, y)$ is the original colour at image pixel (x, y), this value is clipped to the gamut boundary resulting in a new colour $p_c(x, y)$ which is compressed based on the gradient information to a new value $p_s(x, y)$.

3 Numerical Implementation

In this section, we present a numerical implementation to solve the minimization problem described in Equation (8) using finite difference. For each image pixel $\mathbf{p}(x, y)$, we calculate forward-facing and backward-facing derivatives. That is: $[\mathbf{p}(x,y)-\mathbf{p}(x+1,y)]$, $[\mathbf{p}(x,y)-\mathbf{p}(x-1,y)]$, $[\mathbf{p}(x,y)-\mathbf{p}(x,y+1)]$, $[\mathbf{p}(x,y)-\mathbf{p}(x,y-1)]$. Based on that, the discrete version of Equation (8) can be expressed as:

$$
\begin{aligned}
\alpha_s(x,y)\mathbf{d}(x,y) - \mathbf{d}(x+1,y) + \alpha_s(x,y)\mathbf{d}(x,y) - \mathbf{d}(x-1,y) \\
+\alpha_s(x,y)\mathbf{d}(x,y) - \mathbf{d}(x,y+1) + \alpha_s(x,y)\mathbf{d}(x,y) - \mathbf{d}(x,y-1) \\
= \mathbf{p}(x,y) - \mathbf{p}(x+1,y) + \mathbf{p}(x,y) - \mathbf{p}(x-1,y) \\
+\mathbf{p}(x,y) - \mathbf{p}(x,y+1) + \mathbf{p}(x,y) - \mathbf{p}(x,y-1)
\end{aligned}
\qquad (9)
$$

where $\alpha_s(x, y)$ is a scalar. Note that in Equation (9) we assume that $\alpha_s(x+1, y)$, $\alpha_s(x-1, y)$, $\alpha_s(x, y+1)$, $\alpha_s(x, y-1)$ are equal to one. This simplifies the calculation, but makes the convergence of the numerical scheme slightly slower.

We rearrange Equation (9) to get:

$$
\begin{aligned}
&\alpha_s(x,y)\mathbf{d}(x,y) \\
&= [4 \times \mathbf{p}(x,y) - \mathbf{p}(x+1,y) - \mathbf{p}(x-1,y) \\
&\quad -\mathbf{p}(x,y+1) - \mathbf{p}(x,y-1) \\
&\quad +\mathbf{d}(x+1,y) + \mathbf{d}(x-1,y) \\
&\quad +\mathbf{d}(x,y+1) + \mathbf{d}(x,y-1)] \times \frac{1}{4}
\end{aligned}
\qquad (10)
$$

To solve for $\alpha_s(x, y)$, we use least squares. To do that we multiply both sides of the equality by $\mathbf{d}^T(x, y)$ where T denotes vector transpose operator.

$$\begin{aligned}
&\alpha_s(x,y)\mathbf{d}^T(x,y)\mathbf{d}(x,y) \\
&= \mathbf{d}^T(x,y)[4\times\mathbf{p}(x,y)-\mathbf{p}(x+1,y)-\mathbf{p}(x-1,y) \\
&\quad -\mathbf{p}(x,y+1)-\mathbf{p}(x,y-1) \\
&\quad +\mathbf{d}(x+1,y)+\mathbf{d}(x-1,y) \\
&\quad +\mathbf{d}(x,y+1)+\mathbf{d}(x,y-1)]\times\frac{1}{4}
\end{aligned} \tag{11}$$

where $\mathbf{d}^T(x,y)\mathbf{d}(x,y)$ is the vector dot product, i.e. a scalar. Finally, to solve for $\alpha_s(x,y)$ we divide both sides of the equality by $\mathbf{d}^T(x,y)\mathbf{d}(x,y)$, i.e.:

$$\begin{aligned}
&\alpha_s(x,y) \\
&= \mathbf{d}^T(x,y)[4\times\mathbf{p}(x,y)-\mathbf{p}(x+1,y)-\mathbf{p}(x-1,y) \\
&\quad -\mathbf{p}(x,y+1)-\mathbf{p}(x,y-1) \\
&\quad +\mathbf{d}(x+1,y)+\mathbf{d}(x-1,y) \\
&\quad +\mathbf{d}(x,y+1)+\mathbf{d}(x,y-1)]\times\frac{1}{4}\times\frac{1}{\mathbf{d}^T(x,y)\mathbf{d}(x,y)}
\end{aligned} \tag{12}$$

To insure that $\alpha_s(x,y)$ has values in the range [0 1], we clip values greater than one or less than zero to one, i.e. $if\ \alpha_s(x,y) > 1 \quad \alpha_s(x,y) = 1$ and $if\ \alpha_s(x,y) < 0 \quad \alpha_s(x,y) = 1$, the last one to reset the calculation if the iterative scheme overshoots the gamut compensation.

At each iteration level we update $\mathbf{d}(x,y)$, i.e.:

$$\mathbf{d}(x,y)^{i+1} = \alpha_s(x,y)^i \times \mathbf{d}(x,y)^i \tag{13}$$

The result of the optimization is a map, $\alpha_s(x,y)$, that has values in the range [0 1], where zero takes the value of the clipped pixel $\mathbf{d}(x,y)$ to the average of the gamut and one results in no change.

Clearly, the description given in Equation (12) is an extension of the spatial domain solution of a Poisson equation. It is an extension because we introduce the weights $\alpha_s(x,y)$ with the [0 1] constraint. We solve the optimization problem using Jacobi iteration, with homogenous Neumann boundary conditions to ensure zero derivative at the image boundary.

4 Results

Figures 2 and 3 shows the result when gamut mapping two images. From the α_s maps shown on the right hand side of the figures, the inner workings of the algorithm can be seen. At the first stages, only small details and edges are corrected. Iterating further, the local changes are propagated to larger regions in order to maintain the spatial ratios. Already at two iterations, the result resembles closely those presented in [4], which is, according to Dugay et al. [14] a state-of-the-art algorithm. For many of the images tried, an optimum seems to be found around five iterations. Thus, the algorithm is very fast, the complexity of each iteration being $O(N)$ for an image with N pixels.

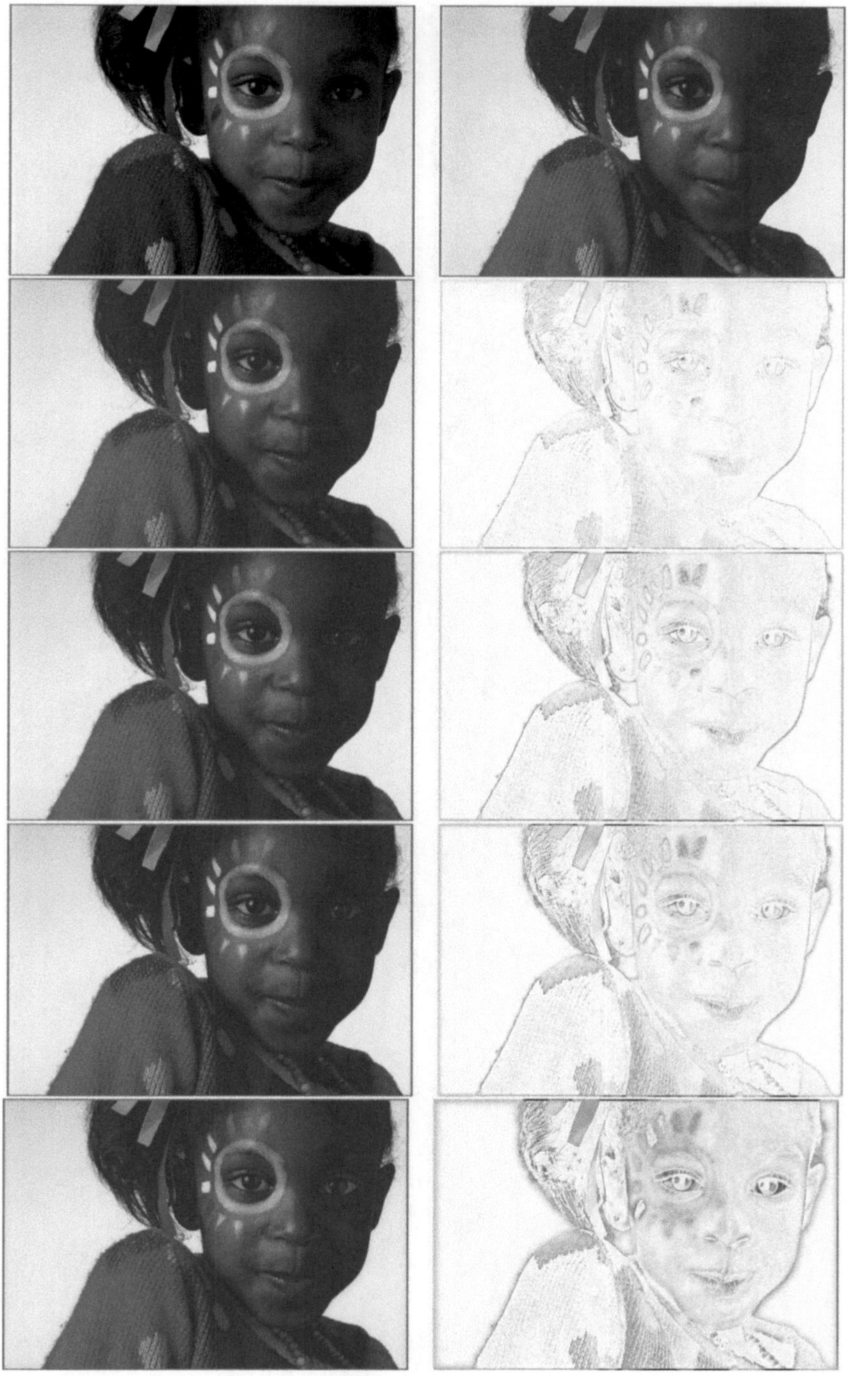

Fig. 2. Original (top left) and gamut clipped (top right) image, resulting image (left column) and α_s (right column) for running the proposed algorithm with 2, 5, 10, and 50 iterations of the algorithm (top to bottom)

Fig. 3. Original (top left) and gamut clipped (top right) image, resulting image (left column) and α_s (right column) for running the proposed algorithm with 2, 5, 10, and 50 iterations of the algorithm (top to bottom)

As part of this work, we have experimented with 20 images which we mapped to a small destination gamut. Our results shows that keeping the iteration level below twenty results in improved gamut mapping with no visible artifacts. Using a higher number of iterations results in the creation of halos at strong edges and the desaturation of flat regions. A trade-off between these tendencies can be made by keeping the number of iterations below twenty. Further, a larger destination gamut would allow us to recover more lost information without artifacts. We thus recommend that the number of iterations is calculated as a function of the size of the destination gamut.

5 Conclusion

Using a variational approach, we have developed a spatial colour gamut mapping algorithm that performs, at least, as well as state-of-the-art algorithms. The algorithm presented is, however, computationally very efficient and lends itself to implementation as part of an imaging pipeline for commercial applications. Unfortunately, it also shares some of the minor disadvantages of other spatial gamut mapping algorithms: halos and desaturation of flat regions for particularly difficult images. Currently, we working on a modification of the algorithm that incorporates knowledge of the strength of the edge. We believe that this modification will solve or at least reduce strongly these minor problems. This is, however, left as future work.

References

1. Morovič, J., Ronnier Luo, M.: The fundamentals of gamut mapping: A survey. Journal of Imaging Science and Technology 45(3), 283–290 (2001)
2. Bala, R., de Queiroz, R., Eschbach, R., Wu, W.: Gamut mapping to preserve spatial luminance variations. Journal of Imaging Science and Technology 45(5), 436–443 (2001)
3. Kimmel, R., Shaked, D., Elad, M., Sobel, I.: Space-dependent color gamut mapping: A variational approach. IEEE Trans. Image Proc. 14(6), 796–803 (2005)
4. Farup, I., Gatta, C., Rizzi, A.: A multiscale framework for spatial gamut mapping. IEEE Trans. Image Proc. 16(10) (2007), doi:10.1109/TIP.2007.904946
5. Giesen, J., Schubert, E., Simon, K., Zolliker, P.: Image-dependent gamut mapping as optimization problem. IEEE Trans. Image Proc. 6(10), 2401–2410 (2007)
6. Eschbach, R.: Image reproduction: An oxymoron? Colour: Design & Creativity 3(3), 1–6 (2008)
7. Land, E.H., McCann, J.J.: Lightness and retinex theory. Journal of the Optical Society of America 61(1), 1–11 (1971)
8. McCann, J.J.: A spatial colour gamut calculation to optimise colour appearance. In: MacDonald, L.W., Luo, M.R. (eds.) Colour Image Science, pp. 213–233. John Wiley & Sons Ltd., Chichester (2002)
9. Meyer, J., Barth, B.: Color gamut matching for hard copy. SID Digest, 86–89 (1989)
10. Morovič, J., Wang, Y.: A multi-resolution, full-colour spatial gamut mapping algorithm. In: Proceedings of IS&T and SID's 11th Color Imaging Conference: Color Science and Engineering: Systems, Technologies, Applications, Scottsdale, Arizona, pp. 282–287 (2003)

11. Eschbach, R., Bala, R., de Queiroz, R.: Simple spatial processing for color mappings. Journal of Electronic Imaging 13(1), 120–125 (2004)
12. Zolliker, P., Simon, K.: Retaining local image information in gamut mapping algorithms. IEEE Trans. Image Proc. 16(3), 664–672 (2007)
13. Nakauchi, S., Hatanaka, S., Usui, S.: Color gamut mapping based on a perceptual image difference measure. Color Research and Application 24(4), 280–291 (1999)
14. Dugay, F., Farup, I., Hardeberg, J.Y.: Perceptual evaluation of color gamut mapping algorithms. Color Research and Application 33(6), 470–476 (2008)

Geometric Multispectral Camera Calibration

Johannes Brauers and Til Aach

Institute of Imaging & Computer Vision, RWTH Aachen University,
Templergraben 55, D-52056 Aachen, Germany
Johannes.Brauers@lfb.rwth-aachen.de
http://www.lfb.rwth-aachen.de

Abstract. A large number of multispectral cameras uses optical bandpass filters to divide the electromagnetic spectrum into passbands. If the filters are placed between the sensor and the lens, the different thicknesses, refraction indices and tilt angles of the filters cause image distortions, which are different for each spectral passband. On the other hand, the lens also causes distortions which are critical in machine vision tasks. In this paper, we propose a method to calibrate the multispectral camera geometrically to remove all kinds of geometric distortions. To this end, the combination of the camera with each of the bandpass filters is considered as single camera system. The systems are then calibrated by estimation of the intrinsic and extrinsic camera parameters and geometrically merged via a homography. The experimental results show that our algorithm can be used to compensate for the geometric distortions of the lens and the optical bandpass filters simultaneously.

1 Introduction

Multispectral imaging considerably improves the color accuracy in contrast to conventional three-channel RGB imaging [1]: This is because RGB color filters exhibit a systematic color error due to production conditions and thus violate the Luther rule [2]. The latter states that, for a human-like color acquisition, the color filters have to be a linear combination of the human observer's ones. Additionally, multispectral cameras are able to differentiate metameric colors, i.e., colors with different spectra but whose color impressions are the same for a human viewer or an RGB camera. Furthermore, different illuminations can be simulated with the acquired spectral data after acquisition. A well-established multispectral camera type, viz., the one with a filter wheel, has been patented by Hill and Vorhagen [3] and is used by several research groups [4,5,6,7].

One disadvantage of the multispectral filter wheel camera are the different optical properties of the bandpass filters. Since the filters are positioned in the optical path, their different thicknesses, refraction indices and tilt angles cause a different path of rays for each passband when the filter wheel index position is changed. This causes both longitudinal and transversal aberrations in the acquired images: Longitudinal aberrations produce a blurring or defocusing effect

A.-B. Salberg, J.Y. Hardeberg, and R. Jenssen (Eds.): SCIA 2009, LNCS 5575, pp. 119–127, 2009.

in the image as shown in our paper in [8]. In the present paper, we consider the transversal aberrations, causing a geometric distortion. A combination of the uncorrected passband images leads to color fringes (see Fig. 3a). We presented a detailed physical model and compensation algorithm in [9]. Other researchers reported heuristic algorithms to correct the distortions [10,11,12] caused by the bandpass filters. A common method is the geometric warping of all passband images to a selected reference passband, which eliminates the color fringes in the final reconstructed image.

However, the reference passband image also exhibits distortions caused by the lens. To overcome this limitation, we have developed an algorithm to compensate both types of aberrations, namely the ones caused by the different optical properties of the bandpass filters *and* the aberrations caused by the lens. Our basic idea is shown in Fig. 1: We interpret the combination of the camera with each optical bandpass filter as a separate camera system. We then use camera calibration techniques [13] in combination with a checkerboard test chart to estimate calibration parameters for the different optical systems. Afterwards, we warp the images geometrically according to a homography.

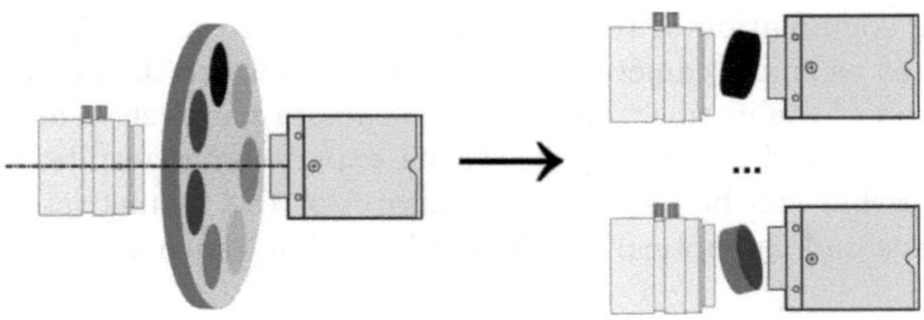

Fig. 1. With respect to camera calibration, our multispectral camera system can be interpreted as multiple camera systems with different optical bandpass filters

We have been inspired by two publications from Gao et. al [14,15], who used a plane-parallel plate in front of a camera to acquire stereo images. To a certain degree, our bandpass filters are optically equivalent to a plane-parallel plate. In our case, we are not able to estimate depth information because the base width of our system is close to zero. Additionally, our system exhibits seven different optical filters, whereas Gao uses only one plate. Furthermore, our optical filters are placed between optics and sensor, whereas Gao used the plate in front of the camera.

In the following section we describe our algorithm, which is subdivided into three parts: First, we compute the intrinsic and extrinsic camera parameters for all multispectral passbands. Next, we compute a homography between points in the image to be corrected and a reference image. In the last step, we finally compensate the image distortions. In the third section we present detailed practical results and finish with the conclusions in the fourth section.

2 Algorithm

2.1 Camera Calibration

A pinhole geometry camera model [13] serves as the basis for our computations. We use

$$\mathbf{x}_n = \frac{1}{Z}\begin{pmatrix} X \\ Y \end{pmatrix} \tag{1}$$

to transform the world coordinates $\mathbf{X} = (X, Y, Z)^T$ to normalized image coordinates $\mathbf{x}_n = (x_n, y_n)^T$. Together with the radius

$$r_n^2 = x_n^2 + y_n^2 \tag{2}$$

we derive the distorted image coordinates $\mathbf{x}_d = (x_d, y_d)^T$ with

$$\begin{aligned} \mathbf{x}_d &= \left(1 + k_1 r_n^2 + k_2 r_n^4\right) \mathbf{x}_n + \begin{pmatrix} 2k_3 x_n y_n + k_4 \left(r_n^2 + 2x_n^2\right) \\ k_3 \left(r_n^2 + 2y_n^2\right) + 2k_4 x_n y_n \end{pmatrix} \\ &= f\left(\mathbf{x}_n, \mathbf{k}\right) . \end{aligned} \tag{3}$$

The coefficients k_1, k_2 account for radial distortions and the coefficients k_3, k_4 for tangential ones. The function $f()$ describes the distortions and takes a normalized, undistorted point $\mathbf{x}_n$ and a coefficient vector $\mathbf{k} = (k_1, k_2, k_3, k_4)^T$ as parameters.

The mapping of the distorted, normalized image coordinates $\mathbf{x}_d$ to the pixel coordinates $\mathbf{x}$ is computed by

$$\mathbf{x}' = \begin{pmatrix} x' \\ y' \\ z' \end{pmatrix} = \mathbf{K}\begin{pmatrix} \mathbf{x}_d \\ 1 \end{pmatrix} \qquad \mathbf{K} = \begin{pmatrix} f/s_x & 0 & c_x \\ 0 & f/s_y & c_y \\ 0 & 0 & 1 \end{pmatrix} \tag{4}$$

and

$$\mathbf{x} = \begin{pmatrix} x \\ y \end{pmatrix} = \frac{1}{z'}\begin{pmatrix} x' \\ y' \end{pmatrix}, \tag{5}$$

where f denotes the focal length of the lens and s_x, s_y the size of the sensor pixels. The parameters c_x and c_y specify the image center, i.e., the point where the optical axis hits the sensor layer. In brief, the intrinsic parameters of the camera are given by the camera matrix $\mathbf{K}$ and the distortion parameters $\mathbf{k} = (k_1, k_2, k_3, k_4)^T$.

As mentioned in the introduction, each filter wheel position of the multispectral camera is modeled as a single camera system with specific intrinsic parameters. For instance, the parameters for the filter wheel position using an optical bandpass filter with the selected wavelength $\lambda_{\text{sel}} = 400\,\text{nm}$ is described by the intrinsic parameters $\mathbf{K}_{\lambda_{\text{sel}}}$ and $\mathbf{k}_{\lambda_{\text{sel}}}$.

2.2 Computing the Homography

In addition to lens distortions, which are mainly characterized by the intrinsic parameters $\mathbf{k}_{\lambda_{\text{sel}}}$, the perspective geometry for each passband is slightly different because of the different optical properties of the bandpass filters: As shown in more detail in [9], a variation of the tilt angle causes an image shift, whereas changes in the thickness or refraction index causes the image to be enlarged or shrunk. Therefore, we have to compute a relation between the image pixel coordinates of the selected passband and the reference passband. The normalized and homogeneous coordinates are derived by

$$\mathbf{x}_{n,\lambda_{\text{sel}}} = \frac{\mathbf{X}_{\lambda_{\text{sel}}}}{Z_{\lambda_{\text{sel}}}} = \frac{\mathbf{X}_{\lambda_{\text{sel}}}}{\mathbf{e}_z^T \mathbf{X}_{\lambda_{\text{sel}}}} \quad \text{and} \quad \mathbf{x}_{n,\lambda_{\text{ref}}} = \frac{\mathbf{X}_{\lambda_{\text{ref}}}}{Z_{\lambda_{\text{ref}}}} = \frac{\mathbf{X}_{\lambda_{\text{ref}}}}{\mathbf{e}_z^T \mathbf{X}_{\lambda_{\text{ref}}}}, \tag{6}$$

respectively, where $\mathbf{X}_{\lambda_{\text{sel}}}$ and $\mathbf{X}_{\lambda_{\text{ref}}}$ are coordinates for the selected and the reference passband. The normalization transforms $\mathbf{X}_{\lambda_{\text{sel}}}$ and $\mathbf{X}_{\lambda_{\text{ref}}}$ to a plane in the position $z_{n,\lambda_{\text{sel}}} = 1$ and $z_{n,\lambda_{\text{ref}}} = 1$, respectively. In the following, we treat them as homogeneous coordinates, i.e., $\mathbf{x}_{n,\lambda_{\text{sel}}} = (x_{n,\lambda_{\text{sel}}}, y_{n,\lambda_{\text{sel}}}, 1)^T$.

According to our results in [9], where we proved that an affine transformation matrix is well suited to characterize the distortions caused by the bandpass filters solely, we estimate a matrix

$$\mathbf{H}\mathbf{x}_{n,\lambda_{\text{ref}}} = \mathbf{x}_{n,\lambda_{\text{sel}}} . \tag{7}$$

The matrix $\mathbf{H}$ transforms coordinates $\mathbf{x}_{n,\lambda_{\text{ref}}}$ from the reference passband to coordinates $\mathbf{x}_{n,\lambda_{\text{sel}}}$ of the selected passband. In practice, we use a set of coordinates from the checkerboard crossing detection during the calibration for reliable estimation of $\mathbf{H}$ and apply a least squares algorithm to solve the overdetermined problem.

2.3 Performing Rectification

Finally, the distortions of all passband images have to be compensated and the images have to be adapted geometrically to the reference passband as described in the previous section. Doing this straightforwardly, we would transform the coordinates of a selected passband to the ones of the reference passband. To keep an equidistant sampling in the resulting image this is in practice done the other way round: We start out from the destination coordinates of the final image and compute the coordinates in the selected passband, where the pixel values have to be taken from.

The undistorted, homogeneous pixel coordinates in the target passband are here denoted by $(x_{\lambda_{\text{ref}}}, y_{\lambda_{\text{ref}}}, 1)^T$, the ones of the selected passband are computed by

$$\begin{pmatrix} u' \\ v' \\ w' \end{pmatrix} = \mathbf{H}\mathbf{K}_{\lambda_{\text{ref}}}^{-1} \begin{pmatrix} x_{\lambda_{\text{ref}}} \\ y_{\lambda_{\text{ref}}} \\ 1 \end{pmatrix}, \tag{8}$$

where $\mathbf{K}_{\lambda_{\text{ref}}}^{-1}$ transforms from pixel coordinates to normalized camera coordinates and $\mathbf{H}$ performs the affine transformation introduced in section 2.2. The normalized coordinates $(u, v)^T$ in the selected passband are then computed by

$$u = \frac{u'}{w'} \qquad v = \frac{v'}{w'} . \tag{9}$$

Furthermore, the distorted coordinates are determined using

$$\begin{pmatrix} \tilde{u} \\ \tilde{v} \end{pmatrix} = f\left(\begin{pmatrix} u \\ v \end{pmatrix}, \mathbf{k}_{\lambda_{\text{sel}}} \right) , \tag{10}$$

where $f()$ is the distortion function introduced above and $\mathbf{k}_{\lambda_{\text{sel}}}$ are the distortion coefficients for the selected spectral passband. The camera coordinates in the selected passband are then derived by

$$\mathbf{x}_{\lambda_{\text{sel}}} = \mathbf{K}_{\lambda_{\text{sel}}} \begin{pmatrix} \tilde{u} \\ \tilde{v} \\ 1 \end{pmatrix} , \tag{11}$$

where $\mathbf{K}_{\lambda_{\text{sel}}}$ is the camera matrix for the selected passband.

The final warping for a passband image with the wavelength λ_{sel} is done by taking a pixel at the position $\mathbf{x}_{\lambda_{\text{sel}}}$ from the image using bilinear interpolation and storing it at position $\mathbf{x}_{\lambda_{\text{ref}}}$ in the corrected image. This procedure is repeated for all image pixels and passbands.

3 Results

A sketch of our multispectral camera is shown in Fig. 1. The camera features a filter wheel with seven optical filters in the range from 400 nm to 700 nm in steps of 50 nm and a bandwidth of 40 nm. The internal grayscale camera is a Sony XCD-SX900 with a resolution of 1280 × 960 pixel and a cell size of 4.65 μm × 4.65 μm. While the internal camera features a C-mount, we use F-mount lenses to be able to place the filter-wheel between sensor and lens. In our experiments, we use a Sigma 10-20mm F4-5.6 lens. Since the sensor is much smaller than a full frame sensor (36 mm × 24 mm), the focal lengths of the lens has to be multiplied with the crop factor of 5.82 to compute the apparent focal length. This also means that only the center part of the lens is really used for imaging and therefore the distortions are reduced compared to a full frame camera.

For our experiments, we used the calibration chart shown in Fig. 2, which comprises a checkerboard pattern with 9 × 7 squares and a unit length of 30 mm. We acquired multispectral images for 20 different poses of the chart. Since each multispectral image consists of seven grayscale images representing the passbands, we acquired a total of 140 images. We performed the estimation of intrinsic and extrinsic parameters with the well-known Bouguet toolbox [16] for each passband separately, i.e., we obtain seven parameter datasets. The calibration is then done using the equations in section 2. In this paper, the multispectral images, which

Fig. 2. Exemplary calibration image; distortions have been compensated with the proposed algorithm. The detected checkerboard pattern is marked with a grid. The small rectangle marks the crop area shown enlarged in Fig. 3.

(a) Without geometric calibration color fringes are not compensated.

(b) Calibration shown in [9]: color fringes are removed but lens distortions remain.

(c) Proposed calibration scheme: both color fringes and lens distortions are removed.

Fig. 3. Crops of the area shown in Fig. 2 for different calibration algorithms

consist of multiple grayscale images, are transformed to the sRGB color space for visualization. Details of this procedure are, e.g., given in [17].

When the geometric calibration is omitted, the final RGB image shows large color fringes as shown in Fig. 3a. Using our previous calibration algorithm in [9], the color fringes vanish (see Fig. 3b), but lens distortions still remain: The undistorted checkerboard squares are indicated by thin lines in the magnified image; the corner of the lines is not aligned with the underlying image, and thus shows the distortion of the image. Small distortions might be acceptable for several imaging tasks, where geometric accuracy is rather unimportant. However, e.g., industrial machine vision tasks often require a distortion-free image, which can be computed by our algorithm. The results are shown in Fig. 3c, where the edge of the overlayed lines is perfectly aligned with the checkerboard crossing of the underlying image.

Table 1. Reprojection errors in pixels for all spectral passbands. Each entry shows the mean of Euclidean length and maximum pixel error, separated with a slash. For a detailed explanation see text.

	400 nm	450 nm	500 nm	550 nm	600 nm	650 nm	700 nm	all
no calib.	2.0 / 4.9	1.2 / 2.6	0.6 / 2.2	0.0 / 0.0	5.0 / 5.4	2.2 / 3.3	3.8 / 7.0	2.11 / 6.97
intra-band	0.1 / 0.6	0.1 / 0.6	0.1 / 0.6	0.1 / 0.6	0.1 / 0.6	0.1 / 0.5	0.1 / 0.6	0.10 / 0.61
inter-band	0.1 / 0.7	0.1 / 0.6	0.2 / 0.9	0.1 / 0.6	0.2 / 0.8	0.1 / 0.7	0.2 / 0.7	0.14 / 0.91

Fig. 4. Distortions caused by the bandpass filters; calibration pattern pose 11 for passband 550 nm (reference passband); *scaled* arrows indicate distortions between this passband and the 500 nm passband

Table 1 shows reprojection errors for all spectral passbands from 400 nm to 700 nm and a summary in the last column "all". The second row lists the deviations when no calibration is performed at all. For instance, the fourth column denotes the mean and maximum distances (separated with a slash) of checkerboard crossings between the 500 nm and the 550 nm passband: This means, in the worst case, the checkerboard crossing in the 500 nm passband is located 2.2 pixel away from the corresponding crossing in the 550 nm passband. In other words, the color fringe in the combined image has a width of 2.2 pixel at this location, which is not acceptable. The distortions are also shown in Fig. 4.

The third row "intra-band" indicates the reprojection errors between the projection of 3D points to pixel coordinates via Eqs. (1)-(5) and their corresponding measured coordinates. We call these errors "intra-band" because only differences in the same passband are taken into account; the differences show how well the passband images can be calibrated themselves, without considering the geometrical connection between them. Since the further transformation via a homography introduces additional errors, the errors given in the third row mark a theoretical limit for the complete calibration (fourth row).

In contrast to the "intra-band" errors, the "inter-band" errors denoted in the fourth row include errors caused by the homography between different spectral passbands. More precisely, we computed the difference between a projection of 3D points in the *reference* passband to pixel coordinates in the selected passband and compared them to measured coordinates in the selected passband. These numbers show how well the overall model is suited to model the multispectral camera, i.e., the deviation which remains after calibration. The mean overall error of 0.14 pixels for all passbands lies in the subpixel range. Therefore, our algorithm is well suited to model the distortions of the multispectral camera. The intra and inter band errors (third and fourth row) for the 550 nm reference passband are identical because no homography is required here and thus no additional errors are introduced.

Compared to our registration algorithm presented in [9], the algorithm shown in this paper is able to compensate for lens distortions as well. As a side-effect, we also gain information about the focal length and the image center, since both properties are computed implicitly by the camera calibration. However, the advantage of [9] is that almost every image can be used for calibration – there is no need to perform an explicit calibration with a dedicated test chart, which might be time consuming and not possible in all situations. Also, the algorithms for camera calibration mentioned in this paper are more complex, although most of them are provided in toolboxes. Finally, for our specific configuration, the lens distortions are very small. This is due to a high-quality lens and because we use a smaller sensor (C-mount size) than the lens is designed for (F-mount size); therefore, only the center part of the lens is used.

4 Conclusions

We have shown that both color fringes caused by the different optical properties of the color filters in our multispectral camera as well as geometric distortions caused by the lens can be corrected with our algorithm. The mean absolute calibration error for our multispectral camera is 0.14 pixel, and the maximum error is 0.91 pixel for all passbands. Without calibration, mean and maximum errors are 6.97 and 2.11, respectively. Our framework is based on standard tools for camera calibration; with these tools, our algorithm can be implemented easily.

Acknowledgments

The authors are grateful to Professor Bernhard Hill and Dr. Stephan Helling, RWTH Aachen University, for making the wide angle lens available.

References

1. Yamaguchi, M., Haneishi, H., Ohyama, N.: Beyond Red-Green-Blue (RGB): Spectrum-based color imaging technology. Journal of Imaging Science and Technology 52(1), 010201–1–010201–15 (2008)

2. Luther, R.: Aus dem Gebiet der Farbreizmetrik. Zeitschrift für technische Physik 8, 540–558 (1927)
3. Hill, B., Vorhagen, F.W.: Multispectral image pick-up system, U.S.Pat. 5,319,472, German Patent P 41 19 489.6 (1991)
4. Tominaga, S.: Spectral imaging by a multi-channel camera. Journal of Electronic Imaging 8(4), 332–341 (1999)
5. Burns, P.D., Berns, R.S.: Analysis multispectral image capture. In: IS&T Color Imaging Conference, Springfield, VA, USA, vol. 4, pp. 19–22 (1996)
6. Mansouri, A., Marzani, F.S., Hardeberg, J.Y., Gouton, P.: Optical calibration of a multispectral imaging system based on interference filters. SPIE Optical Engineering 44(2), 027004.1–027004.12 (2005)
7. Haneishi, H., Iwanami, T., Honma, T., Tsumura, N., Miyake, Y.: Goniospectral imaging of three-dimensional objects. Journal of Imaging Science and Technology 45(5), 451–456 (2001)
8. Brauers, J., Aach, T.: Longitudinal aberrations caused by optical filters and their compensation in multispectral imaging. In: IEEE International Conference on Image Processing (ICIP 2008), San Diego, CA, USA, pp. 525–528. IEEE, Los Alamitos (2008)
9. Brauers, J., Schulte, N., Aach, T.: Multispectral filter-wheel cameras: Geometric distortion model and compensation algorithms. IEEE Transactions on Image Processing 17(12), 2368–2380 (2008)
10. Cappellini, V., Del Mastio, A., De Rosa, A., Piva, A., Pelagotti, A., El Yamani, H.: An automatic registration algorithm for cultural heritage images. In: IEEE International Conference on Image Processing, Genova, Italy, September 2005, vol. 2, pp. II-566–9 (2005)
11. Kern, J.: Reliable band-to-band registration of multispectral thermal imager data using multivariate mutual information and cyclic consistency. In: Proceedings of SPIE, November 2004, vol. 5558, pp. 57–68 (2004)
12. Helling, S., Seidel, E., Biehlig, W.: Algorithms for spectral color stimulus reconstruction with a seven-channel multispectral camera. In: IS&Ts Proc. 2nd European Conference on Color in Graphics, Imaging and Vision CGIV 2004, Aachen, Germany, April 2004, vol. 2, pp. 254–258 (2004)
13. Hartley, R.I., Zisserman, A.: Multiple View Geometry in Computer Vision, 2nd edn. Cambridge University Press, Cambridge (2004)
14. Gao, C., Ahuja, N.: Single camera stereo using planar parallel plate. In: Ahuja, N. (ed.) Proceedings of the 17th International Conference on Pattern Recognition, vol. 4, pp. 108–111 (2004)
15. Gao, C., Ahuja, N.: A refractive camera for acquiring stereo and super-resolution images. In: Ahuja, N. (ed.) IEEE Computer Society Conference on Computer Vision and Pattern Recognition, New York, USA, vol. 2, pp. 2316–2323 (2006)
16. Bouguet, J.Y.: Camera Calibration Toolbox for Matlab
17. Brauers, J., Schulte, N., Bell, A.A., Aach, T.: Multispectral high dynamic range imaging. In: IS&T/SPIE Electronic Imaging, San Jose, California, USA, January 2008, vol. 6807 (2008)

A Color Management Process for Real Time Color Reconstruction of Multispectral Images

Philippe Colantoni[1,2] and Jean-Baptiste Thomas[3,4]

[1] Université Jean Monnet, Saint-Étienne, France
[2] Centre de recherche et de restauration des musées de France, Paris, France
[3] Université de Bourgogne, LE2I, Dijon, France
[4] Gjøvik university College, The Norwegian color research laboratory, Gjøvik, Norway

Abstract. We introduce a new accurate and technology independent display color characterization model for color rendering of multispectral images. The establishment of this model is automatic, and does not exceed the time of a coffee break to be efficient in a practical situation. This model is a part of the color management workflow of the new tools designed at the C2RMF for multispectral image analysis of paintings acquired with the material developed during the CRISATEL European project. The analysis is based on color reconstruction with virtual illuminants and use a GPU (Graphics processor unit) based processing model in order to interact in real time with a virtual lighting.

1 Introduction

The CRISATEL European Project [4] opened the possibility to the C2RMF of acquiring multispectral images through a convenient framework. We are now able to scan in one shot a much larger surface than before (resolution of 12000×20000) in 13 different bands of wavelengths from ultraviolet to near infrared, covering all the visible spectrum.

The multispectral analysis of paintings via a very complex image processing pipeline, allows us to investigate a painting in ways that were totally unknown until now [6].

Manipulating these images is not easy considering the amount of data (about 4GB by image). We can either use a pre-computation process, which will produce even bigger files, or compute everything on the fly.

The second method is complex to implement because it requires an optimized (cache friendly) representation of data and a large amount of computations. This second point is not anymore a problem if we use parallel processors like graphic processor units (GPU) for the computation. For the data we use a traditional multi-resolution tiled representation of an uncorrelated version of the original multispectral image.

The computational capabilities of GPU have been used for other applications such as numerical computations and simulations [7]. The work of Colantoni and

A.-B. Salberg, J.Y. Hardeberg, and R. Jenssen (Eds.): SCIA 2009, LNCS 5575, pp. 128–137, 2009.

al. [2] demonstrated that a graphic card can be suitable for color image processing and multispectral image processing.

In this article, we present a part of the color flow used in our new software (PCASpectralViewer): the color management process. As constraints, we want the display color characterization model to be as accurate as possible on any type of display and we want the color correction to be in real time (no pre-processing). Moreover, we want the model establishment not to exceed the time of a coffee break.

We first introduce a new accurate display color characterization method. We evaluate this method and then describe its GPU implementation for real time rendering.

2 Color Management Process

The CRISATEL project produces 13 planes multispectral images which correspond to the following wavelengths: 400, 440, 480, 520, 560, 600, 640, 680, 720, 760, 800, 900 and 1000nm. Only the 10 first planes interact with the visible part of the light. Considering this, we can estimate the corresponding $XYZ()$ tri-stimulus values for each pixel of the source image using Equation 1:

$$\begin{cases} X = \sum_{\lambda=400}^{\lambda=760} x(\lambda)R(\lambda)L(\lambda) \\ Y = \sum_{\lambda=400}^{\lambda=760} y(\lambda)R(\lambda)L(\lambda) \\ Z = \sum_{\lambda=400}^{\lambda=760} z(\lambda)R(\lambda)L(\lambda) \end{cases} \tag{1}$$

where $R(\lambda)$ is the reflectance spectrum and $L(\lambda)$ is the light spectrum (the illuminant).

Using a GPU implementation of this formula we can compute in real-time the XYZ and the corresponding $L^*a^*b^*$ values for each pixel of the original multispectral image with a virtual illuminant provided by the user (standard or custom illuminants).

If we want to provide a correct color representation of these computed XYZ values, we must apply a color management process, based on the color characterization of the display device used, in our color flow. We then have to find which *RGB* values to input to the display in order to produce the same color stimuli than the retrieved XYZ values represents, or at least the closest color stimuli (according to the display limits).

In the following, we introduce a color characterization method which gives accurate color rendering on all available display technologies.

2.1 Display Characterization

A display color characterization model aims to provide a function which estimates the displayed color stimuli for a given 3-tuple *RGB* input to the display. Different approaches can be used for this purpose [5], based on measurements of input values (i.e. *RGB* input values to a display device) and output values (i.e.

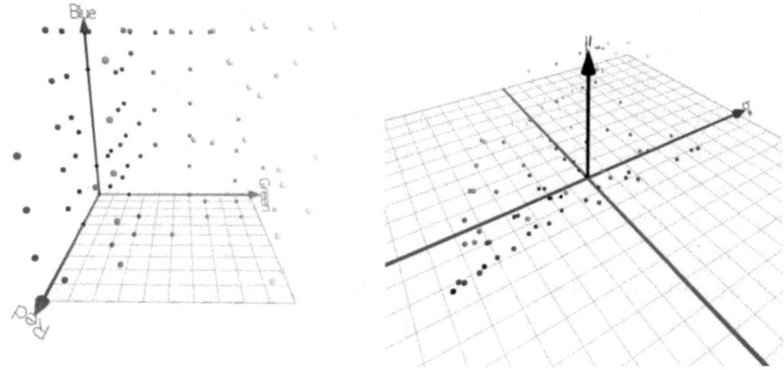

Fig. 1. Characterization process from *RGB* to $L^*a^*b^*$

XYZ or $L^*a^*b^*$ values measured on the screen by a colorimeter or spectrometer) (see figure.1).

The method we present here is based on the generalization of measurements at some position in the color space. It is an empirical method which does not consider any assumptions based on display technology. The forward direction (*RGB* to $L^*a^*b^*$), is based on RBF interpolation on an optimal set of measured patches. The backward model ($L^*a^*b^*$ to *RGB*) is based on tetrahedral interpolation. An overview of this model is shown in Figure 2.

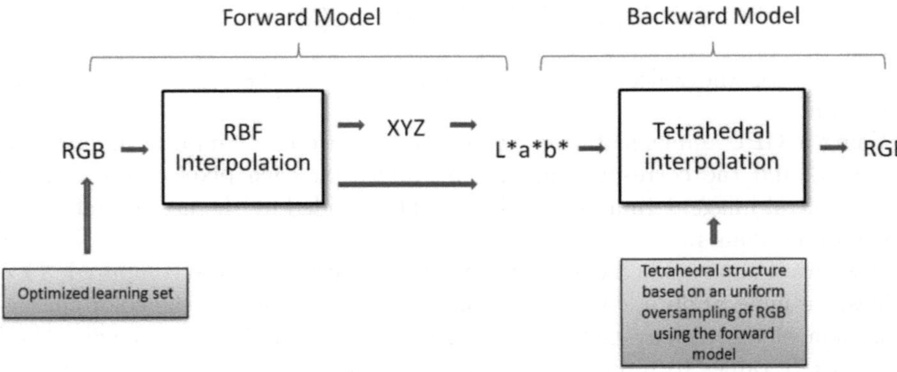

Fig. 2. Overview of the display color characterization model

2.2 Forward Model

Traditionally a characterization model (or forward model) is based on an interpolation or an approximation method. We found that radial basis function interpolation (RBFI) was the best model for our purpose.

RBF Interpolation. is an interpolation/approximation [1] scheme for arbitrarily distributed data. The idea is to build a function f whose graph passes through the data and minimizes a bending energy function. For a general M-dimensional case, we want to interpolate a valued function $f(X) = Y$ given by the set of values $f = (f_1, ..., f_N)$ at the distinct points $X = x_1, ..., x_N \subset \Re^M$. We choose $f(X)$ to be a Radial Basis Function of the shape:

$$f(x) = p(x) + \sum_{i=1}^{N} \lambda_i \phi(||x - x_i||) \quad x \in \Re^M$$

where p is a polynomial, λ_i is a real-valued weight, ϕ is a basis function, $\phi : \Re^M \to \Re$, and $||x - x_i||$ is the euclidean norm between x and x_i. Therefore, a RBF is a weighted sum of translations of a radially symetric basis function augmented by a polynomial term. Different basis functions (kernel) $\phi(x)$ can by used.

Considering the color problem, we want to establish three three-dimensional functions $f_i(x, y, z)$. The idea is to build a function $f(x, y, z)$ whose graph passes through the tabulated data and minimizes the following bending energy function:

$$\iiint_{\Re^3} (f_{xxx}^3 + f_{yyy}^3 + f_{zzz}^3 + 3f_{xxy}^3 + 3f_{xxz}^3 + 3f_{xyy}^3 + 3f_{xzz}^3 + 3f_{yyz}^3 + 3f_{yzz}^3 + 6f_{xyz}^3)dxdydz \quad (2)$$

For a set of data $\{(x_i, y_i, z_i, w_i)\}_{i=1}^{n}$ (where $w_i = f(x_i, y_i, z_i)$) the minimizing function is such as:

$$f(x, y, z) = b_0 + b_1 x + b_2 y + b_3 z + \sum_{j=1}^{n} a_j \phi(||(x - x_j, y - y_j, z - z_j)||) \quad (3)$$

where the coefficients a_j and $b_{0,1,2,3}$ are determined by requiring exact interpolation using the following equation

$$w_i = \sum_{j=1}^{n} \phi_{ij} a_j + b_0 + b_1 x_i + b_2 y_i + b_3 z_i \quad (4)$$

for $1 \leq n$ where $\phi_{ij} = \phi(||(x_i - x_j, y_i - y_j, z_i - z_j)||)$. In matrix form this is

$$h = Aa + Bb \quad (5)$$

where $A = [\phi_{ij}]$ is an $n \times n$ matrix and where B is an $n \times 4$ matrix whose rows $[1 \quad x_i \quad y_i \quad z_i]$. An additional implication is that

$$B^T a = 0 \quad (6)$$

These two vector equations can be solved to obtain

$$a = A^{-1}(h - Bb) \text{ and } b = (B^T A^{-1} B)^{-1} B^T A^{-1} h.$$

It is possible to provide a smoothing term. In this case the interpolation is not exact and becomes an approximation. The modification is to use the equation

$$h = (A + \lambda I)a + Bb \quad (7)$$

$$a = (A + \lambda I)^{-1}(h - Bb) \text{ and } b = (B^T(A + \lambda I)^{-1}B)^{-1}B^T(A + \lambda I)^{-1}h.$$

where $\lambda > 0$ is a smoothing parameter and I is the $n \times n$ identity matrix.

In our context we used a set of 4 real functions as kernel, the biharmonic ($\phi(x) = x$), triharmonic ($\phi(x) = x^3$), thin-plate spline 1 ($\phi(x) = x^2 log(x)$) and thin-plate spline 2 ($\phi(x) = x^2 log(x^2)$), with x the distance from the origin. The use of a given basis function depends on the display device which is characterized, and gives some freedom to the model.

Color Space Target. Our forward model uses $L^*a^*b^*$ as target ($L^*a^*b^*$ is a target well adapted for the gamut clipping that we use). This does not imply that we have to use $L^*a^*b^*$ as target for the RBF interpolation. In fact we have two choices. We can use either $L^*a^*b^*$ which seems to be the most logical target or XYZ associated with a XYZ to $L^*a^*b^*$ color transformation.

The use of different color spaces as target gives us another degree of freedom.

Smooth Factor Choice. Once the kernel and the color space target are fixed, the smooth factor, includes in the RBFI model used here, is the only parameter which can be used to change the properties of the transformation. With a zero value the model is a pure interpolation. With a different smooth factor, the model becomes an approximation. This is an important feature because it helps us to deal with the measurement problems due to the display stability (a color rendering for a given *RGB* value can change with the time) and to the measure repeatability of the measurement device.

2.3 Backward Model Using Tetrahedral Interpolation

While the forward model defines the relationship between the device "color space" and the CIE system of color measurement, we present in this section the inversion of this transform. Our problem is to find, for a $L^*a^*b^*$ values computed by the GPU from the multispectral image and the chosen illuminant, the corresponding *RGB* values (for a display device previously characterized).

This backward model could use the same interpolation methods previously presented but we used a new and more accurate method [3]. This new method uses the fact that if our forward model is very good then it is associated with an optimal patch database (see 2.4). Basically, we use a hybrid method; a tetrahedral interpolation associated with an over-sampling of the *RGB* cube (see Figure 3). We have chosen the tetrahedral interpolation method because of its geometrical aspect (this method is associated with our gamut clipping algorithm).

We build the initial tetrahedral structure using an uniform over sampling of the *RGB* cube ($n \times n \times n$ samples). This over sampling process uses the forward model to compute the corresponding structure in the $L^*a^*b^*$ color space. Once this structure is built, we can compute, for an unknown C_{Lab} color, the associated C_{RGB} color in two steps: First, the tetrahedron which encloses the point C_{Lab} to be interpolated should be found (the scattered point set is tetrahedrized); and then, an interpolation scheme is used within each tetrahedron.

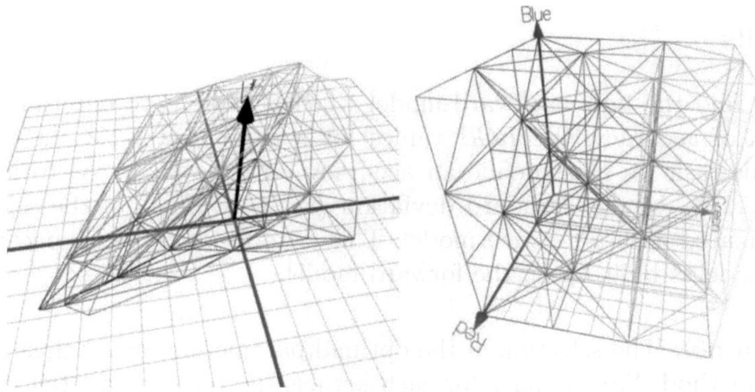

Fig. 3. Tetrahedral structure in $L^*a^*b^*$ and the correponding structure in *RGB*

More precisely, the color value C of the point is interpolated from the color values C_i of the tetrahedron vertices. A tri-linear interpolation within a tetrahedron can be performed as follows:

$$C = \sum_{i=0}^{3} w_i C_i$$

The weights can be calculated by $w_i = \frac{V_i}{V}$ with V the volume of the tetrahedron and V_i the volume of the sub-tetrahedron according to:

$$V_i = \frac{1}{6}(P_i - P)[(P_{i+1} - P)(P_{i+2} - P)]; i = 0, ..., 3$$

where P_i are the vertices of the tetrahedron and the indices are taken modulo 4.

The over-sampling used is not the same for each axis of *RGB*. It is computed according to the shape of the display device gamut in the $L^*a^*b^*$ color space. We found that than an equivalent to $36 \times 36 \times 36$ samples was a good choice. Using such a tight structure linearizes locally our model which becomes perfectly compatible with the used of a tetrahedral interpolation.

2.4 Optimized Learning Data Set

In order to increase the reliability of the model, we introduce a new way to determine the learning data set for the RBF based interpolation (e.g. the set of color patches measured on the screen). We found that our interpolation model was most efficient when the learning data set used to initialize the interpolation was regularly distributed in our destination color space ($L^*a^*b^*$). This new method is based on a regular 3D sampling of $L^*a^*b^*$ color space combined with a forward - backward refinement process after the selection of each patch. This algorithm allows us to find the optimal set of RGB colors to measure.

This technique needs to select incrementally the RGB color patches that will be integrated into the learning database. For this reason it has been integrated into a custom software tool which is able to drive a colorimeter. This software also measures a set of 100 random test patches equiprobably distributed in *RGB* used in order to determine the accuracy of the model.

2.5 Results

We want to find the best backward model which allows us to determine, with a maximum of accuracy, the *RGB* values for a computed *XYZ*. In order to complete this task we must define an accuracy criteria. We chose to multiply the average ΔE_{76} by the standard deviation (STD) of ΔE_{76} of the set of 100 patches evaluated with a forward model. This criteria makes sense because the backward model is built up on the forward model.

Optimal Model. The selection of the optimal parameters can be done using a brute force method. We compute for each kernels (ie. biharmonic, triharmonic, thin-plate spline 1, thin-plate spline 2), each color space target ($L^*a^*b^*$, XYZ and several smooth factors (0, 1e-005, 0.0001, 0.0005, 0.001, 0.005, 0.01, 0.05, 0.1) the values of this criteria and we select the minimum.

For example the following table shows the report obtains for a SB2070 Mitsubishi DiamondPro with a triharmonic kernel for $L^*a^*b^*$ (Table 1) and XYZ (Table 2) as color space target (using a learning data set of 216 patches):

According to our criteria the best kernel is the triharmonic with a smooth factor of 0.01 and XYZ as target.

Table 1. Part of the report obtained in order to evaluate the best model parameters. The presented results are considering $L^*a^*b^*$ as target color space, and a triharmonic kernel for a CRT monitor SB2070 Mitsubishi DiamondPro.

smooth factor	0	0.0001	0.001	0.01	0.1
ΔE Mean	0.379	0.393	0.376	0.386	0.739
ΔE STD	0.226	0.218	0.201	0.224	0.502
ΔE Max	1.374	1.327	1.132	1.363	2.671
ΔE 95%	0.882	0.848	0.856	0.828	1.769
ΔRGB Mean	0.00396	0.00459	0.00438	0.00421	0.00826
ΔRGB STD	0.00252	0.00323	0.00316	0.00296	0.00728
ΔRGB Max	0.01567	0.02071	0.01768	0.01554	0.05859
ΔRGB 95%	0.00886	0.01167	0.01162	0.01051	0.01975

Table 2. Part of the report obtained in order to evaluate the best model parameters. The presented results are considering XYZ as target color space, and a triharmonic kernel for a CRT monitor SB2070 Mitsubishi DiamondPro.

smooth factor	0	0.0001	0.001	0.01	0.1
ΔE Mean	0.495	0.639	0.539	0.332	0.616
ΔE STD	0.293	0.424	0.360	0.179	0.691
ΔE Max	1.991	2.931	2.548	1.075	4.537
ΔE 95%	1.000	1.427	1.383	0.7021	1.751
ΔRGB Mean	0.00674	0.00905	0.00720	0.00332	0.00552
ΔRGB STD	0.00542	0.00740	0.00553	0.00220	0.00610
ΔRGB Max	0.02984	0.03954	0.03141	0.01438	0.04036
ΔRGB 95%	0.01545	0.02081	0.01642	0.00597	0.01907

The measurement process took about 5 minutes and the optimization process took 2 minutes (with a 4 cores processor). We reached our goal which was to provide an optimal model during a coffee break of the user.

Our different experimentation showed that a 216 patches learning set was a good compromise (equivalent to a $6 \times 6 \times 6$ sampling of the RGB cube). A smaller data set gives us a degraded accuracy, a bigger gives us similar results because we are facing the measurement problems introduced previously.

Optimized Learning Data Set. Table 3 and Table 4 show the results obtained with our model for two displays of different technologies. These tables show clearly how the optimized learning data set can produce better results with the same number of patches.

Table 3. Accuracy of the model established with 216 patches in forward and backward direction for a LCD Wide Gamut display (HP2408w). The distribution of the patches plays a major role for the model accuracy.

	Forward model		Backward model	
	ΔE Mean	ΔE Max	ΔRGB Mean	ΔRGB Max
Optimized	1.057	4.985	0.01504	0.1257
Uniform	1.313	9.017	0.01730	0.1168

Table 4. Accuracy of the model established with 216 patches in forward and backward direction for a CRT display (Mitsubishi SB2070). The distribution of the patches plays a major role for the model accuracy.

	Forward model		Backward model	
	ΔE Mean	ΔE Max	ΔRGB Mean	ΔRGB Max
Optimized	0.332	1.075	0.00311	0.01267
Uniform	0.435	1.613	0.00446	0.01332

Table 5. Accuracy of the model established with 216 patches in forward and backward direction for three other displays. The model performs well on all monitors.

	Forward model		Backward model	
	ΔE Mean	ΔE Max	ΔRGB Mean	ΔRGB Max
EIZO CG301W (LCD)	0.783	1.906	0.00573	0.01385
Sensy 24KAL (LCD)	0.956	2.734	0.01308	0.06051
DiamondPlus 230 (CRT)	0.458	2.151	0.00909	0.06380

Results for Different Displays. Table 5 presents different results obtained for 3 others displays (2 LCD and 1 CRT).

Considering that non trained humans can not discriminate ΔE less than 2, we can see here that our model gives very good results on a wide range of display.

2.6 Gamut Mapping

The aim of gamut mapping is to ensure a good correspondence of overall color appearance between the original and the reproduction by compensating for the mismatch in the size, shape and location between the original and reproduction gamuts.

The $L^*a^*b^*$ computed color can be out of gamut (i.e. the destination display cannot generate the corresponding color). To ensure an accurate colorimetric rendering, considering $L^*a^*b^*$ color space, and low computational requirements, we used a geometrical gamut clipping method based on the pre-computed tetrahedral structure (generated in our backward model) and more especially on the surface of this geometrical structure (see figure.3).

The clipped color is defined by the intersection of the gamut boundaries and the segment between a target point and the input color. The target point used here is an achromatic $L^*a^*b^*$ color with a luminance of 50.

3 GPU-Based Implementation

Our color management method is based on a conversion process which will compute for a *XYZ* values the corresponding *RGB*.

It is possible to implement the presented algorithm with a specific GPU language, like CUDA, but our application will only works with CUDA compatible GPU (*n*VIDIA^{TM} G80, G90 and GT200). Our goal was to have a working application on a large number of GPU (*AMD* and *n*VIDIA^{TM} GPUs), for this reason we choose to implement a classical method using a 3D lookup table.

During an initialization process we build a three dimensional *RGBA* floating point texture which cover the $L^*a^*b^*$ color space. The alpha channel of the *RGBA* values saves the distance between the initial $L^*a^*b^*$ value and $L^*a^*b^*$ value obtained after the gamut mapping process. If this value is 0 the $L^*a^*b^*$ color which will have to be converted is in the gamut of the display otherwise this color is out gamut and we are displaying the closest color (according to our gamut mapping process). This allows us to display in real time the color errors due to the screen inability to display every visible colors.

Finaly our complete color pipeline includes: a reflectance to *XYZ* conversion then a *XYZ* to $L^*a^*b^*$ conversion (using the white of the screen as reference) and our color management process based on the 3D lookup table associated with a tri-linear interpolation process.

4 Conclusion

We presented a part of a large multispectral application used at the C2RMF. It has been shown that it is possible to implement an accurate color management process even for a real time color reconstruction. We showed a color management process based only on colorimetric consideration. The next step is to introduce a color appearance model in our color flow. The use of such color appearance model, built up on our accurate color management process, will allows us to do virtual exhibition of painting.

References

[1] Carr, J.C., Beatson, R.K., Cherrie, J.B., Mitchell, T.J., Fright, W.R., McCallum, B.C., Evans, T.R.: Reconstruction and Representation of 3D Objects with Radial Basis Functions. In: SIGGRAPH, pp. 12–17 (2001)

[2] Colantoni, P., Boukala, N., Da Rugna, J.: Fast and Accurate Color Image Processing Using 3D Graphics Cards. In: Vision Modeling and Visualization, VMV 2003, pp. 383–390 (2003)

[3] Colantoni, P., Stauder, J., Blond, L.: Device and method for characterizing a colour device Thomson Corporate Research, European Patent, EP 05300165.7 (2005)

[4] Ribés, A., Schmitt, F., Pillay, R., Lahanier, C.: Calibration and Spectral Reconstruction for CRISATEL: an Art Painting Multispectral Acquisition System. Journal of Imaging Science and Technology 49, 563–573 (2005)

[5] Bastani, B., Cressman, B., Funt, B.: An evaluation of methods for producing desired colors on CRT monitors. Color Research & Application 30, 438–447 (2005)

[6] Colantoni, P., Pitzalis, D., Pillay, R., Aitken, G.: GPU Spectral Viewer: analysing paintings from a colorimetric perspective. In: The 8th International Symposium on Virtual Reality, Archaeology and Cultural Heritage, Brighton, United Kingdom (2007)

[7] `http://www.gpgpu.org`

Precise Analysis of Spectral Reflectance Properties of Cosmetic Foundation

Yusuke Moriuchi, Shoji Tominaga, and Takahiko Horiuchi

Graduate School of Advanced Integration Science, Chiba University,
1-33, Yayoi-cho, Inage-ku, Chiba 263-8522, Japan

Abstract. The present paper describes the detailed analysis of the spectral reflection properties of skin surface with make-up foundation, based on two approaches of a physical model using the Cook-Torrance model and a statistical approach using the PCA. First, we show how the surface-spectral reflectances changed with the observation conditions of light incidence and viewing, and also the material compositions. Second, the Cook-Torrance model is used for describing the complicated reflectance curves by a small number of parameters, and rendering images of 3D object surfaces. Third, the PCA method is presented the observed spectral reflectances analysis. The PCA shows that all skin surfaces have the property of the standard dichromatic reflection, so that the observed reflectances are represented by two components of the diffuse reflectance and a constant reflectance. The spectral estimation is then reduced to a simple computation using the diffuse reflectance, some principal components, and the weighting coefficients. Finally, the feasibility of the two methods is examined in experiments. The PCA method performs reliable spectral reflectance estimation for the skin surface from a global point of view, compared with the model-based method.

Keywords: Spectral reflectance analysis, cosmetic foundation, color reproduction, image rendering.

1 Introduction

Foundation has various purposes. Basically, foundation makes skin color and skin texture appears more even. Moreover, it can be used to cover up blemishes and other imperfections, and reduce wrinkles. The essential role is to improve the appearance of skin surfaces. Therefore it is important to evaluate the change of skin color by foundation. However, there was not enough scientific discussion on the spectral analysis of foundation material and skin with make-up foundations [1]. In a previous report [2], we discussed the problem of analyzing the reflectance properties of skin surface with make-up foundation. We presented a new approach based on the principal-component analysis (PCA), useful for describing the measured spectral reflectances, and showed the possibility of estimating the reflectance under any lighting and viewing conditions.

The present paper describes the detailed analysis of the spectral reflection properties of skin surface with make-up foundation by using two approaches based on a

A.-B. Salberg, J.Y. Hardeberg, and R. Jenssen (Eds.): SCIA 2009, LNCS 5575, pp. 138–148, 2009.

physical model approach and a statistical approach. Foundations with different material compositions are painted on a bio-skin. Light reflected from the skin surface is measured using a gonio-spectrophotometer.

First, we show how appearances of the surface, including specularity, gloss, and matte appearance, change with the observation conditions of light incidence and viewing, and also the material compositions. Second, we use the Cook-Torrance model as a physical reflection model for describing the three-dimensional (3D) reflection properties of the skin surface with foundation. This model is effective for image rendering of 3D object surfaces. Third, we use the PCA as a statistical approach for analyzing the reflection properties. The PCA is effective for statistical analysis of the complicated spectral curves of the skin surface reflectance. We present an improved algorithm for synthesizing the spectral reflectance. Finally, the feasibility of both approaches is examined in experiments from the point of view of spectral reflectance analysis and color image rendering.

2 Foundation Samples and Reflectance Measurements

Although the make-up foundation is composed of different materials such as mica, talc, nylon, titanium, and oil, the two materials of mica and talc are the important components which affect the appearance of skin surface painted with the foundation. So many foundations were made by changing the quantity and the ratio of two materials. For instance, the combination ratio of mica (M) and talc (T) was changed as (M=0, T=60), (M=10, T=50), …, (M=60, T=0), the ratio of mica was changed with a constant T as (M=0, T=40), (M=10, T=40), …, (M=40, T=40), and the size of mica was also changed in the present study. Table 1 shows typical foundation samples used for spectral reflectance analysis. Powder foundations with the above compositions were painted on a flat bio-skin surface with the fingers. The bio-skin is made of urethane which looks like human skin. Figure 1 shows a board sample of bio-skin with foundation. The foundation layer is very thin as 5-10 microns in thickness on the skin.

Table 1. Foundation samples with different composition of mica and talc

Samples	IKD-0	IKD-10	IKD-20	IKD-40	IKD-54	IKD-59
Mica	0	10	20	40	54	59
Talc	59	49	39	19	5	0

A gonio-spectrophotometer is used for observing surface-spectral reflections of the skin surface with foundations under different lighting and viewing conditions. This instrument has two degrees of freedom on the light source position and the sensor position as shown in Fig. 2, although in the real system, the sensor position is fixed, and both light source and sample object can rotate. The ratio of the spectral radiance from the sample to the one from the reference white diffuser, called the spectral radiance factor, is output as spectral reflectance. The spectral reflectances of all samples were measured at 13 incidence angles of 0, 5, 10, …, 60 degrees and 81 viewing angles of -80, -78, …, -2, 0, 2, …, 78, 80 degrees.

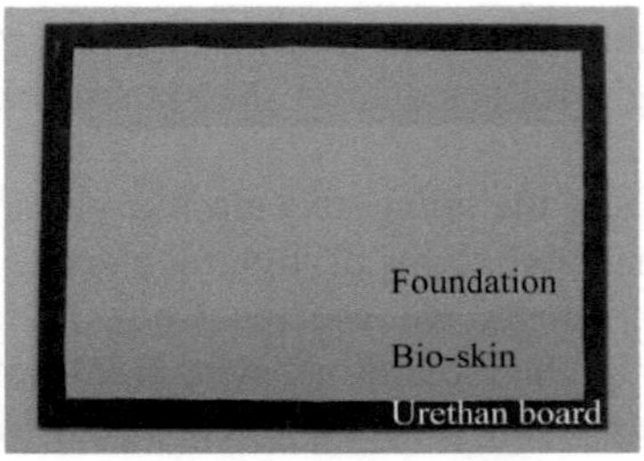

Fig. 1. Sample of bio-skin with foundation

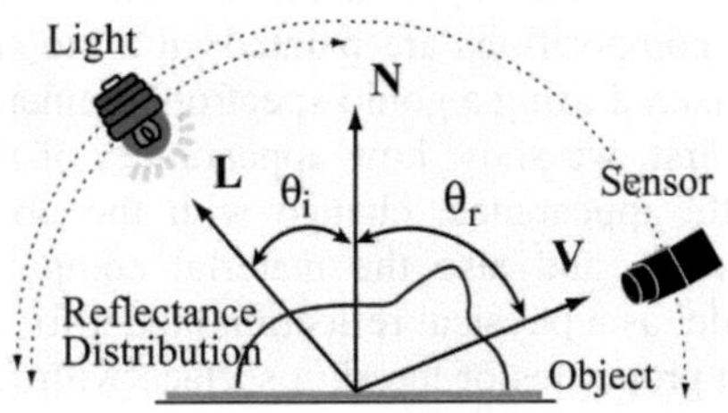

Fig. 2. Measuring system of surface reflectance

Figure 3(a) shows a 3D perspective view of spectral radiance factors measured from the bio-skin itself and the skin with a foundation sample IKD-54 at the incidence angle of 20 degrees. This figure suggests how the foundation changes effectively the spectral reflectance of the skin surface. In Fig. 3(a), solid mesh and broken mesh indicate the spectral radiance factors from bio-skin and IKD-54 itself, respectively, where the spectral curves are depicted as a function of viewing angle.

The spectral reflectance depends not only on the viewing angle, but also on the incidence angle. In order to make this point clear, we average the radiance factors on wavelength in the visible range. Figure 3(b) depicts a set of the average curves at different incidence angles as a function of viewing angle for both bio-skin and IKD-54.

A comparison between solid curves and broken curves in Fig. 3 suggests several typical features of skin surface reflectance with foundation as follows:

(1) Reflectance hump at around the vertical viewing angle,
(2) Back-scattering at around -70 degrees, and
(3) Specular reflectance with increasing viewing angle.

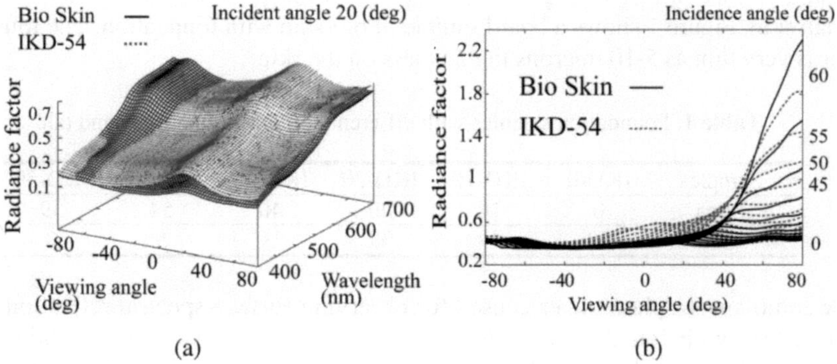

Fig. 3. Reflectance measurements from a sample IKD-54 and bio-skin. (a) 3D view of spectral reflectances at θ_i =20, (b) Average reflectances as a function of viewing angle.

Moreover we have investigated how the surface reflectance depends on the material composition of foundation. Figure 4 shows the average reflectances for three cases among difference material compositions. As a result, we find the following two basic properties:

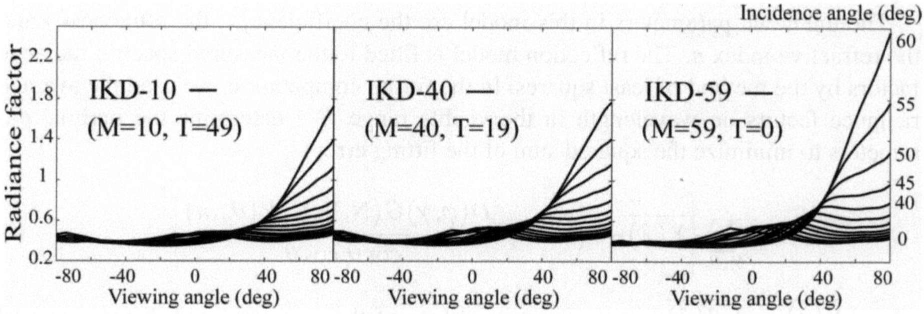

Fig. 4. Reflectance measurements from different make-up foundations

(1) When the quantity of mica increases, the whole reflectance of skin surface increases at all angles of incidence and viewing.
(2) When the quantity of talc increases, the surface reflectance decreases at large viewing angles, but increases at matte regions.

3 Model-Based Analysis of Spectral Reflectance

In the field of computer graphics and vision, the Phong model [3] and the Cook-Torrance model [4] are known as a 3D reflection model used for describing light reflection of an object surface. The former model is convenient for inhomogeneous dielectric object like plastics, although the mathematical expression is simple, and the number of model parameters is small. The latter model is a physically precise model which is available for both dielectrics and metals. In this paper, we analyze the spectral reflectances of the skin surface based on the Cook-Torrance model.

The Cook-Torrance model can be written in terms of the spectral radiance factor as

$$Y(\lambda)=S(\lambda)+\ \beta\frac{D(\varphi,\gamma)G(\mathbf{N},\mathbf{V},\mathbf{L})F(\theta_Q,n)}{\cos\theta_i\cos\theta_r}, \qquad (1)$$

where the first and second terms represent, respectively, the diffuse and specular reflection components. β is the specular reflection coefficient. A specular surface is assumed to be an isotropic collection of planar microscopic facets by Torrance and Sparrow [5]. The area of each microfacet is much smaller than the pixel size of an image. Note that the surface normal vector **N** represents the normal vector of a macroscopic surface. Let **Q** be the vector bisector of an **L** and **V** vector pair, that is, the normal vector of a microfacet. The symbol θ_i is the incidence angle, θ_r is the viewing angle, φ is the angle between **N** and **Q**, and θ_Q is the angle between **L** and **Q**.

The specular reflection component consists of several terms: *D* is the distribution function of the microfacet orientation, and *F* represents the Fresnel spectral reflectance [6] of the microfacets. *G* is the geometrical attenuation factor. *D* is assumed as a Gaussian distribution function with rotational symmetry about the surface normal **N** as $D(\varphi,\gamma)=\exp\{-\log(2)\varphi^2/\gamma^2\}$, where the parameter γ is a constant that represents surface roughness. The Fresnel reflectance *F* is described as a nonlinear function with the parameter of the refractive index *n*.

The unknown parameters in this model are the coefficient β, the roughness γ and the refractive index n. The reflection model is fitted to the measured spectral radiance factors by the method of least squares. In the fitting computation, we used the average radiance factors on wavelength in the visible range. We determine the optimal parameters to minimize the squared sum of the fitting error

$$e = \min \sum_{\theta_i,\theta_r} \left\{ \overline{Y(\lambda)} - \overline{S(\lambda)} - \beta \frac{D(\varphi,\gamma) G(\mathbf{N},\mathbf{V},\mathbf{L}) F(\theta_Q,n)}{\cos\theta_i \cos\theta_r} \right\}^2, \tag{2}$$

where $\overline{Y(\lambda)}$ and $\overline{S(\lambda)}$ are the average values of the measured and diffuse spectral reference factors, respectively. The diffuse reflectance $S(\lambda)$ is chosen as a minimum of the measured spectral reflectance factors. The above error minimization is done over all angles of θ_i and θ_r. For simplicity of the fitting computation, we determine the refractive index n to 1.90 because the skin surface with foundation is considered as inhomogeneous dielectric.

Figure 5(b) shows the results of model fitting to the sample IKD-54 shown in Fig. 3, where solid curves indicate the fitted reflectances, and a broken curve indicates the original measurements. Figure 5(a) shows the fitting results for spectral reflectances at the incidence angle of 20 degrees. The model parameters were estimated as β =0.74 and γ=0.20. The squared error was e=4.97. These figures suggest that the model describes the surface-spectral reflectances at the low range of viewing angle with relatively good accuracy. However the fitting error tends to increase with the viewing angle.

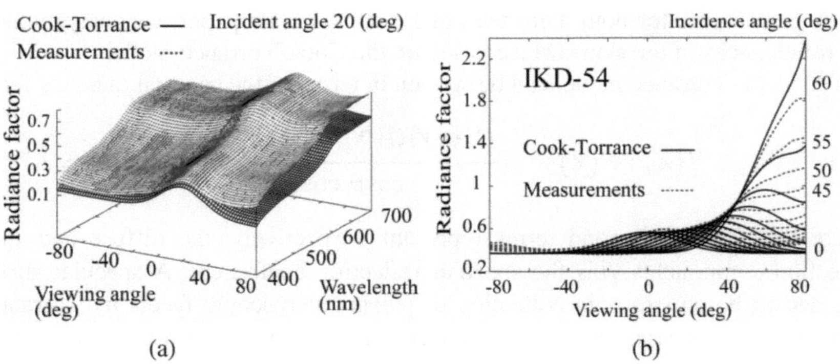

Fig. 5. Fitting results of the Cook-Torrance model to IKD-54. (a) 3D view of spectral reflectances at θ_i =20, (b) Average reflectances as a function of viewing angle.

We have repeated the same fitting experiment of the model to many skin samples with different material compositions for foundation. Then a relationship between the material compositions and the model parameters was found as follows:

(1) As the quantity of mica increases, both parameters β and γ increase.

(2) As the size of mica increases, β decreases and γ increases.

(3) As the quantity of talc increases, β decreases abruptly and γ increases gradually.

Table 2 shows a list of the estimated model parameters for the foundation IKD-0 - IKD-59 with different material compositions. Thus, a variety of skin surface with different make-up foundations is described by the Cook-Torrance model with a small number of parameters.

Table 2. Composition and model parameters of a human hand with different foundations

Samples	Composition (M, T)	Parameters β	γ	n
IKD-0	(0, 59)	0.431	0.249	1.90
IKD-10	(10, 49)	0.426	0.249	1.90
IKD-20	(20, 39)	0.485	0.220	1.90
IKD-40	(40, 19)	0.570	0.191	1.90
IKD-54	(54, 5)	0.744	0.170	1.90
IKD-59	(59, 0)	0.736	0.180	1.90

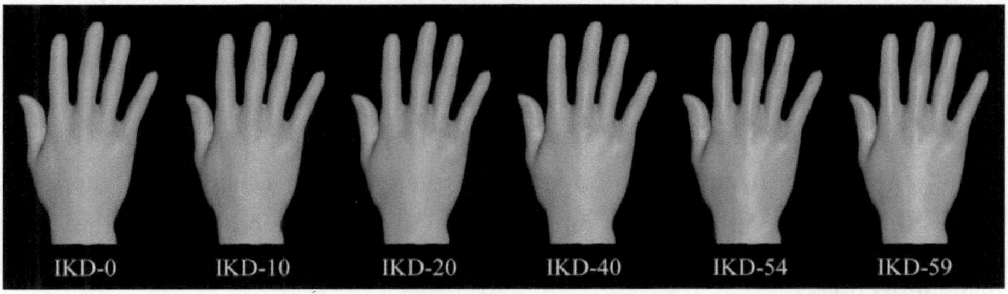

Fig. 6. Image rendering results for a human hand with different make-up foundations

For application to image rendering, we render color images of the skin surface of a human hand by using the present model fitting results. The 3D shape of the human hand was acquired separately by using a laser range finder system. Figure 6 demonstrates the image rendering results of the 3D skin surface with different make-up foundations. A ray-tracing algorithm was used for rendering realistic images, which performed wavelength-based color calculation precisely. Only the Cook-Torrance model was used for spectral reflectance computation of IKD-0 - IKD-59. We assume that the light source is D65 and the illumination direction is the normal direction to the hand.

In the rendered images, the appearance changes such that the gloss of skin surface increases with the quantity of mica. These rendered images show the feasibility of the model-based approach. A detailed comparison between spectral reflectance curves such as Fig. 5, however, suggests that there is a certain discrepancy between the measured reflectances and the estimated ones by the model. The similar discrepancy occurs for all the other samples.

4 PCA-Based Analysis of Spectral Reflectance

Let us consider another approach to describing spectral reflectance of the skin surface with make-up foundation. The PCA is effective for statistical analysis of the complicated spectral curves of the skin surface reflectance.

First, we have to know the basic reflection property of the skin surface. In the previous report [2], we showed that the skin surface could be described by the standard dichromatic reflection model [6]. The standard model assumes that the surface reflection consists of two additive components, the body (diffuse) reflection and the interface (specular) reflection, which is independent of wavelength. The spectral reflectance (radiance factor) $Y(\theta_i,\theta_r,\lambda)$ of the skin surface is a function of the wavelength and the geometric parameters of incidence angle θ_i and viewing angle θ_r. Therefore the reflectance is expressed in a linear combination of the diffuse reflectance $S(\lambda)$ and the constant reflectance as

$$Y(\theta_i,\theta_r,\lambda)=C_1(\theta_i,\theta_r)S(\lambda)+C_2(\theta_i,\theta_r), \tag{3}$$

where the weights $C_1(\theta_i,\theta_r)$ and $C_2(\theta_i,\theta_r)$ are the geometric scale factors.

To confirm the adequacy of this model, the PCA was applied to the whole set of spectral reflectance curves observed under different geometries of θ_i and θ_r with an equal 5nm interval in the range 400-700nm. A singular value decomposition (SVD) is used for the practical PCA computation of spectral reflectances. The SVD shows two-dimensionality of the set of spectral reflectance curves. Therefore, all spectral reflectances of skin surface can be represented by only two principal-component vectors $\mathbf{u}_1$ and $\mathbf{u}_2$. Moreover, $\mathbf{u}_1$ and $\mathbf{u}_2$ can be fitted to a unit vector $\mathbf{i}$ using linear regression, that is, the constant reflectance is represented by the two components. By the above reason, we can conclude that the skin surface has the property of the standard dichromatic reflection.

Next, let us consider the estimation of spectral reflectances for various angles of incidence and viewing without observation. Note that the observed spectral reflectances from the skin surface are described using the two components of the diffuse reflectance $S(\lambda)$ and the constant specular reflectance. Hence we expect that any unknown spectral reflectances are described in terms of the same components. Then the reflectances can be estimated by the following function with two parameters,

$$Y(\theta_i,\theta_r,\lambda)=\hat{C}_1(\theta_i,\theta_r)S(\lambda)+\hat{C}_2(\theta_i,\theta_r), \tag{4}$$

where $\hat{C}_1(\theta_i,\theta_r)$ and $\hat{C}_2(\theta_i,\theta_r)$ denote the estimates of the weighting coefficients on a pair of angles (θ_i,θ_r).

In order to develop the estimation procedure, we analyze the weighting coefficients $C_1(\theta_i,\theta_r)$ and $C_2(\theta_i,\theta_r)$ based on the observed data. Again the SVD is applied to the data set of those weighting coefficients. When we consider an approximate representation of the weighting coefficients in terms of several principal components, the performance index of the chosen principle components is given by the percent variance $P(K)=\sum_{i=1}^{K}\mu_i^2 / \sum_{i=1}^{n}\mu_i^2$. The performance indices are $P(2)$=0.994 for the first two components and $P(3)$=0.996 for the first three components in both coefficient data $C_1(\theta_i,\theta_r)$ and $C_2(\theta_i,\theta_r)$ from IKD-59. Then, the weighting coefficients can be decomposed into two basis functions with a single parameter as

$$C_1(\theta_i,\theta_r)=\sum_{j=1}^{K} w_{1j}(\theta_i)v_{1j}(\theta_r),\quad C_2(\theta_i,\theta_r)=\sum_{j=1}^{K} w_{2j}(\theta_i)v_{2j}(\theta_r)\,,\ (K\text{=2 or 3}) \tag{5}$$

where (v_{1j})and (v_{2j}) are two sets of principal components as a function of viewing angle θ_r, and (w_{1j}) and (w_{2j}) are two sets of the corresponding weights to those principal components, which are a function of incidence angle θ_i. $\hat{w}$ is the principal components and v is the weights determined by interpolating the coefficients at observation points.

The performance values $P(2)$ and $P(3)$ are close each other. We examine the accuracy in the two cases for describing the surface-spectral reflectances under all observation conditions. Figure 7 depicts the root-mean squared errors (RMSE) of the reflectance approximation for K=2, 3. In the case of K=2, although the absolute error of overall fitting is relatively small, noticeable errors occur at the incident angles of around 0, 40, and 60 degrees. In particular, it should be emphasized that the errors at the incident and viewing angles of around 0 degree deteriorate seriously the image rendering results of 3 D objects. We find that K=3 improves much to express the surface-spectral reflectances by only one additional component.

Therefore the estimation of $\hat{C}_1(\theta_i,\theta_r)$ and $\hat{C}_2(\theta_i,\theta_r)$ for any unknown reflectance can be reduced into a simple form

$$\begin{aligned}\hat{C}_1(\theta_i,\theta_r)&=\hat{w}_{11}(\theta_i)v_{11}(\theta_r)+\hat{w}_{12}(\theta_i)v_{12}(\theta_r)+\hat{w}_{13}(\theta_i)v_{13}(\theta_r)\\ \hat{C}_2(\theta_i,\theta_r)&=\hat{w}_{21}(\theta_i)v_{21}(\theta_r)+\hat{w}_{22}(\theta_i)v_{22}(\theta_r)+\hat{w}_{23}(\theta_i)v_{23}(\theta_r)\end{aligned}, \tag{6}$$

where $\hat{w}_{ij}(\theta_i)(i=1,2; j=1,2,3)$ are determined by interpolating the coefficients at observation points such as $w_{ij}(0)$, $w_{ij}(5)$, …, $w_{ij}(60)$. Thus, the spectral reflectance of the skin surface at arbitrary angular conditions is generated using the diffuse spectral reflectance $S(\lambda)$, the principal component $v_{ij}(\theta_r)(i=1,2; j=1,2,3)$, and three pairs of weights $\hat{w}_{ij}(\theta_i)(i=1,2; j=1,2,3)$. Note that these basis data are all one-dimensional.

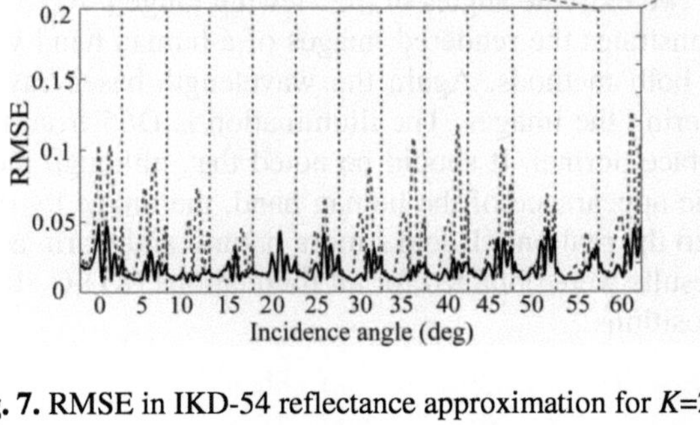

Fig. 7. RMSE in IKD-54 reflectance approximation for K=2, 3

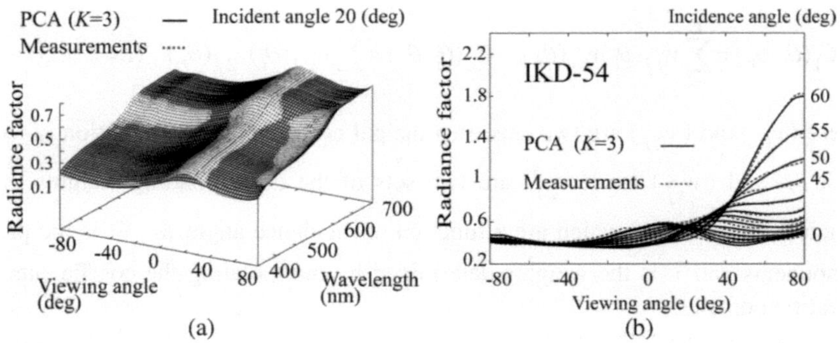

Fig. 8. Estimation results of surface-spectral reflectances for IKD-54. (a) 3D view of spectral reflectances at θ_i =20, (b) Average reflectances as a function of viewing angle.

Figure 8 shows the estimation results to the sample IKD-54, where solid curves indicates the reflectances by the proposed method, and broken curves indicate the original measurements. We should note that the surface spectral reflectances of the skin with make-up foundation are recovered with sufficient accuracy.

5 Performance Comparisons and Applications

A comparison between Fig. 8 by the PCA method and Fig. 5 by the Cook-Torrance model suggests clearly that the estimated surface-spectral reflectances with K=3 are almost coincident with the measurements at all angles. The estimated spectral curves represent accurately the whole features of skin reflectance, including not only reflectance hump at around the vertical viewing angle, but also back-scattering at around -70 degrees, and increasing reflectance at around 70 degrees.

Figure 9 shows the typical estimation results of surface-spectral reflectance of IKD-54 at the incidence of 20 degrees. The estimated reflectance by the PCA method is more closely coincident with the measurements at all angles, while clear discrepancy occurs for the Cook-Torrance model at large viewing angles. Figure 10 summarizes the RMSE of both methods for IKD-54. The solid mesh indicates the estimation results by the Cook-Torrance method and the broken mesh indicates the estimates by the PCA method. The PCA method with K=3 provides much better performance than the Cook-Torrance model method. Note that the Cook-Torrance method has large discrepancy at the two extreme angles of the viewing range [-70, 70].

Figure 11 demonstrates the rendered images of a human hand with the foundation IKD-54 by using both methods. Again the wavelength-based ray-tracing algorithm was used for rendering the images. The illumination is D65 from the direction of 45 degrees to the surface normal. It should be noted that, although the rendered images represent a realistic appearance of the human hand, the image by the PCA method is sufficiently close to the real one. It looks more natural and warm atmosphere like our skins. The same results were obtained for all foundations IKD-0 - IKD-50 with different material compositions.

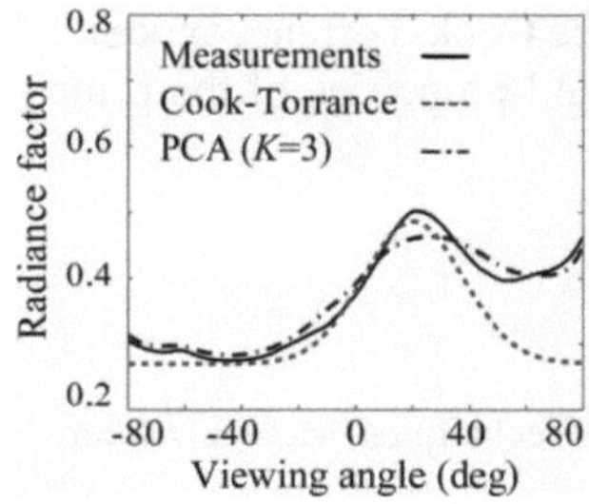

Fig. 9. Reflectance estimates for IKD-54 as a function of viewing angle

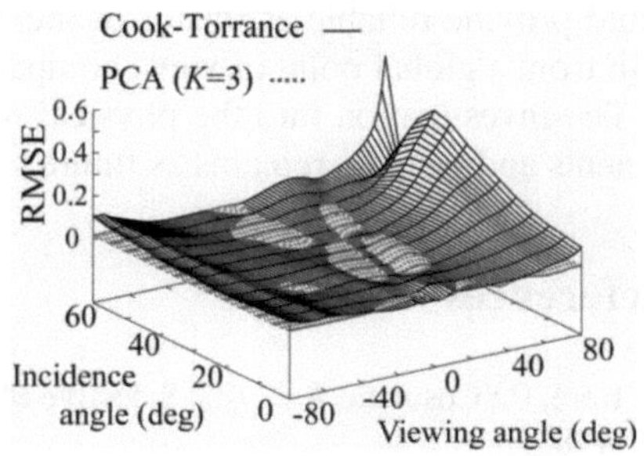

Fig. 10. RMSE in IKD-54 reflectance estimates

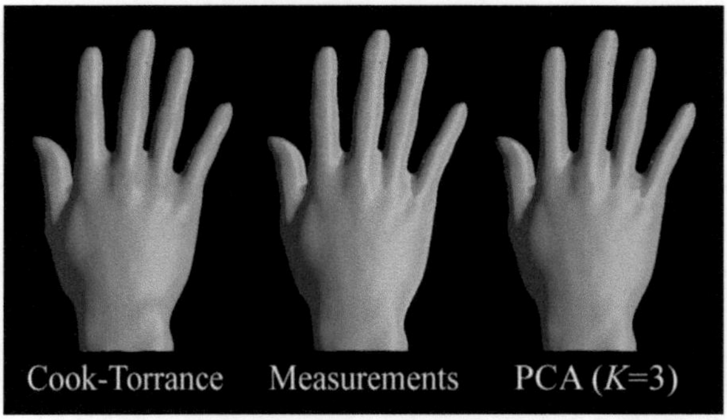

Fig. 11. Image rendering results of a human hand with make-up foundation IKD-54

6 Conclusions

This paper has described the detailed analysis of the spectral reflection properties of skin surface with make-up foundation, based on two approaches of a physical model using the Cook-Torrance model and a statistical approach using the PCA.

First, we showed how the surface-spectral reflectances changed with the observation conditions of light incidence and viewing, and also the material compositions. Second, the Cook-Torrance model was useful for describing the complicated reflectance curves by a small number of parameters, and rendering images of 3D object surfaces. We showed that parameter β increased as the quantity of mica increased. However, the model did not have sufficient accuracy for describing the surface reflection under some geometry conditions. Third, the PCA of the observed spectral reflectances suggested that all skin surfaces satisfied the property of the standard dichromatic reflection. Then the observed reflectances were represented by two spectral components of a diffuse reflectance and constant reflectance. The spectral estimation was reduced to a simple computation using the diffuse reflectance, some principal components, and the weighting coefficients. The PCA method could describe the surface reflection properties with foundation with sufficient accuracy. Finally, the feasibility was examined in experiments. It was shown that the PCA method could provide reliable estimates of the surface-spectral reflectance for the foundation skin from a global point of view, compared with the Cook-Torrance model.

The investigation into the physical meanings and properties of the principal components and weights remains as future works.

References

1. Boré, P.: Cosmetic Analysis: Selective Methods and Techniques. Marcel Dekker, New York (1985)
2. Tominaga, S., Moriuchi, Y.: PCA-based reflectance analysis/synthesis of cosmetic foundation. In: CIC 16, pp. 195–200 (2008)
3. Phong, B.T.: Illumination for computer-generated pictures. Comm. ACM 18(6), 311–317 (1975)
4. Cook, R., Torrance, K.: A reflection model for computer graphics. In: Proc. SIGGRAPH 1981, vol. 15(3), pp. 307–316 (1981)
5. Torrance, K.E., Sparrow, E.M.: Theory for off-specular reflection from roughened surfaces. J. of Optical Society of America 57, 1105–1114 (1967)
6. Born, M., Wolf, E.: Principles of Optics, pp. 36–51. Pergamon Press, Oxford (1987)

Extending Diabetic Retinopathy Imaging from Color to Spectra

Pauli Fält[1], Jouni Hiltunen[1], Markku Hauta-Kasari[1], Iiris Sorri[2], Valentina Kalesnykiene[2], and Hannu Uusitalo[2,3]

[1] InFotonics Center Joensuu, Department of Computer Science and Statistics, University of Joensuu, P.O. Box 111, FI-80101 Joensuu, Finland
{pauli.falt,jouni.hiltunen,markku.hauta-kasari}@ifc.joensuu.fi
http://spectral.joensuu.fi

[2] Department of Ophthalmology, Kuopio University Hospital and University of Kuopio, P.O. Box 1777, FI-70211 Kuopio, Finland
iiris.sorri@kuh.fi, valentina.kalesnykiene@uku.fi

[3] Department of Ophthalmology, Tampere University Hospital, Tampere, Finland
hannu.uusitalo@uta.fi

Abstract. In this study, spectral images of 66 human retinas were collected. These spectral images were measured in vivo from 54 voluntary diabetic patients and 12 control subjects using a modified ophthalmic fundus camera system. This system incorporates the optics of a standard fundus microscope, 30 narrow bandpass interference filters ranging from 400 to 700 nanometers at 10 nm intervals, a steady-state broadband lightsource and a monochrome digital charge-coupled device camera. The introduced spectral fundus image database will be expanded in the future with professional annotations and will be made public.

Keywords: Spectral image, human retina, ocular fundus camera, interference filter, retinopathy, diabetes mellitus.

1 Introduction

Retinal image databases have been important for scientists developing improved pattern recognition methods and algorithms for the detection of retinal structures – such as vascular tree and optic disk – and retinal abnormalities (e.g. microaneurysms, exudates, drusens, etc.). Examples of such publicly available databases are DRIVE [1,2] and STARE [3]. Also, retinal image databases including markings made by eye care professionals exist: e.g. DiaRetDB1 [4].

Traditionally, these databases contain only three-channel RGB-images. Unfortunately, the amount of information in images with only three channels is very limited (red, green and blue channel). In an RGB-image, each channel is an integrated sum over a broad spectral band. Thus, depending on application, an RGB-image can contain useless information that obscures the actual desired data. Better alternative is to take multi-channel spectral images of the retina, because with different wavelengths, different objects of the retina can be emphasized and researchers

A.-B. Salberg, J.Y. Hardeberg, and R. Jenssen (Eds.): SCIA 2009, LNCS 5575, pp. 149–158, 2009.

have indeed started to show growing interest in applications based on spectral color information. Fundus reflectance information can be used in various applications: e.g. in non-invasive study of the ocular media and retina [5,6,7], retinal pigments [8,9,10], oxygen saturation in the retina [11,12,13,14,15], etc.

For example, Styles et al. measured multi-spectral images of the human ocular fundus using an ophthalmic fundus camera equipped with a liquid crystal tunable filter (LCTF) [16]. In their approach, the LCTF-based spectral camera measured spectral color channels from 400 to 700 nm at 10 nm intervals. The constant unvoluntary eye movement is problematic, since the LCTF requires separate lengthy non-stop procedures to acquire exposure times for the color channels and to perform the actual measurement. In general, human ocular fundus is a difficult target to measure in vivo due to the constant eye movements, optical aberrations and reflections from the cornea and optical media (aqueous humor, crystalline lens, and vitreous body), possible medical conditions (e.g. cataract), and the fact that the fundus must be illuminated and measured through a dilated pupil.

To overcome the problems of non-stop measurements, Johnson et al. introduced a snapshot spectral imaging apparatus which used a diffractive optical element to separate a white light image into several spectral channel images [17]. However, this method required complicated calibration and data post-processing to produce the actual spectral image.

In this study, an ophthalmic fundus camera system was modified to use 30 narrow bandpass interference filters, an external steady-state broadband lightsource and a monochrome digital charge-coupled device (CCD) camera. Using this system, spectral images of 66 human ocular fundi were recorded. The voluntary human subjects included 54 persons with abnormal retinal changes caused by diabetes mellitus (diabetic retinopathy) and 12 non-diabetic control subjects. Subject's fundus was illuminated with light filtered through an interference filter and an 8-bit digital image was captured from the light reflected from the retina. This procedure was repeated using each of the 30 filters one by one. Resulting images were normalized to a unit exposure time and registered using an automatic GDB-ICP algorithm by Stewart et al. [18,19]. The registered spectral channel images were then "stacked" into a spectral image. The final 66 spectral retinal images were gathered in a database which will be further expanded in the future. In the database, the 12 control spectral images are necessary for identifying normal and abnormal retinal features. Spectra from these images could be used, for example, as a part of a test set for an automatic detection algorithm.

The ultimate goal of the study was to create a spectral image database of diabetic ocular fundi with additional annotations made by eye care professionals. The database will be made public for all researchers, and it can be used e.g. for teaching, or for creating and testing new and improved methods for manual and automatic detection of diabetic retinopathy. To authors' knowledge, similar public spectral image database with professional annotations does not yet exist.

2 Equipment and Methods

2.1 Spectral Fundus Camera

An ophthalmic fundus camera system is a standard tool in health care systems for the inspection and documentation of the ocular fundus. Normally, such system consists of xenon flash light source, microscope optics for guiding the light into the eye, and optics for guiding the reflected light to a standard RGB-camera. For focusing, there usually exists a separate aiming-light and a video camera.

In this study, a Canon CR5-45NM fundus camera system (Canon, Inc.) was modified for spectral imaging (see Figs. 1 and 2). All unneeded components of the system (including the internal light source) were removed – only the basic fundus microscope optics were left inside the device body – and appropriate openings were cut for the filter holders and the fiber optic cable. Four filter holders and a rail for them were fabricated from acrylic glass, and the rail was installed inside the fundus camera body. Each of the four filter holders could hold up to eight filters and the 30 narrow bandpass interference filters (Edmund Optics, Inc.) were attached to them in a sequence from 400 to 700 nm leaving the two last of the 32 positions empty. The transmittances of the filters are shown in Fig. 3.

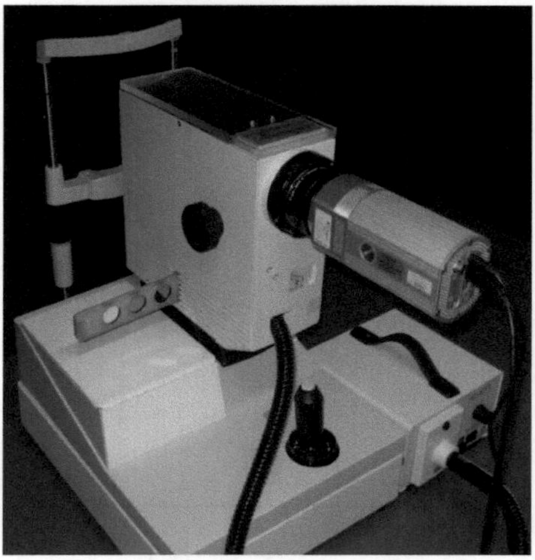

Fig. 1. The modified fundus camera system used in this study

The rail and the identical openings on both sides of the fundus camera allowed the filter holders to be slided through the device manually. A spring-based mechanical stopper locked the holder (and a filter) always in the correct place on the optical path of the system. As a broadband light source, an external Schott Fostec DCR III lightbox (SCHOTT North America, Inc.) with a 150 W OSRAM halogen lamp (OSRAM Corp.) and a daylight-simulating filter was used. Light

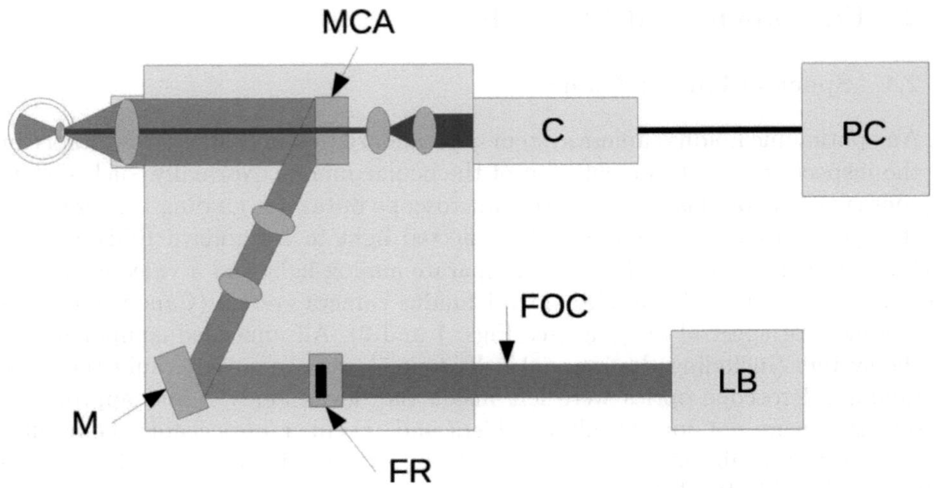

Fig. 2. Simplified structure and operation of the modified ophthalmic fundus camera in Fig. 1: a light box (*LB*), a fiber optic cable (*FOC*), a filter rail (*FR*), a mirror (*M*), a mirror with a central aperture (*MCA*), a CCD camera (*C*), a personal computer (*PC*), and lenses (*ellipses*)

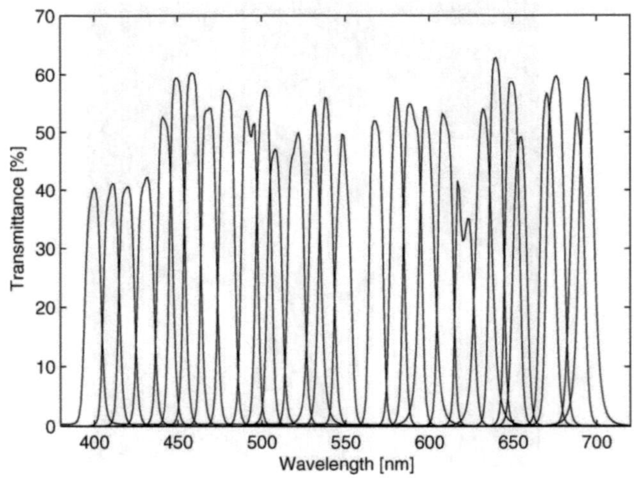

Fig. 3. The spectral transmittances of the 30 narrow bandpass interference filters

was guided into the fundus camera system via a fiber optic cable of the Schott lightbox. In the same piece as the rail was also a mount for the optical cable, which held the end of the cable tightly in place. The light source was allowed to warm up and stabilize for 30 minutes before the beginning of the measurements.

The light exiting the cable was immediately filtered by narrow bandpass filter and the filtered light was guided inside the subject's eye through a dilated pupil. Light reflecting back from the retina was captured with a QImaging Retiga-4000RV digital monochrome CCD camera (QImaging Corp.), which had a 2048 × 2048 pixel detector array and was attached to the fundus camera with a C-mount adapter. The camera was controlled via a Firewire port with a standard desktop PC running QImaging's QCapture Pro 6.0 software. The live preview function of the software allowed the camera-operator to monitor the subject's ocular fundus in real time, which was important for positioning and focusing of the fundus camera, and also for determining the exposure time. Exposure times were calculated from a small area in the retina with the highest reflectivity (typically the optic disk). The typical camera parameters – gain, offset and gamma – were set to 6, 0 and 1, respectively. Gain-value was increased to shorten the exposure time.

The camera was programmed to capture five images as fast as possible and to save the resulting images to the PC's harddrive automatically. Five images per filter were needed because of the constant involuntary movements of the eye: usually at least one of the images was acceptable; if not, a new new set of five images was taken. Image acquisition produced 8-bit grayscale TIFF-images sized 1024×1024 pixels (using 2×2 binning). For each of the 30 filters, a set of five images were captured, and from each set only one image was selected for spectral image formation.

The selected images were co-aligned using the efficient automatic image registration algorithm by Stewart et al. called the generalized dual-bootstrap iterative closest point (GDB-ICP) algorithm [18,19]. Some difficult image pairs had to be registered manually with MATLAB's Control Point Selection Tool [20]. The registered spectral channel images were then normalized to unit exposure time, i.e. 1 second, and stacked in wavelength-order into a 1024×1024×30 spectral image.

2.2 Spectral Image Corrections

Let us derive a formula for the reflectance spectrum $\boldsymbol{r}_{\text{final}}$ at point (x, y) in the final registered and white-corrected reflectance spectral image: The digital signal output v_i for the interference filter i, $i = 1, \ldots, 30$, from one pixel (x, y) of the one-sensor CCD detector array is of the form

$$v_i = \int_{\lambda} s(\lambda) t_i(\lambda) t_{\text{FC}}(\lambda) t_{\text{OM}}^2(\lambda) r_{\text{retina}}(\lambda) h_{\text{CCD}}(\lambda) d\lambda + n_i \ , \qquad (1)$$

where $s(\lambda)$ is the spectral power distribution of the light coming out of the fiber optic cable, λ is the wavelength of the electromagnetic radiation, $t_i(\lambda)$ is the spectral transmittance of the ith interference filter, $t_{\text{FC}}(\lambda)$ is the spectral transmittance of the fundus camera optics, $t_{\text{OM}}(\lambda)$ is the spectral transmittance of the ocular media of the eye, $r_{\text{retina}}(\lambda)$ is the spectral reflectance of the retina, $h_{\text{CCD}}(\lambda)$ is the spectral sensitivity of the detector, and n_i is noise. In Eq. (1), the second power of $t_{\text{OM}}(\lambda)$ is used, because reflected light goes through these media twice.

Let us write the above spectra for pixel (x, y) as discrete m-dimensional vectors (in this application $m = 30$) $\boldsymbol{s}$, $\boldsymbol{t}_i$, $\boldsymbol{t}_{\text{FC}}$, $\boldsymbol{t}_{\text{OM}}$, $\boldsymbol{r}_{\text{retina}}$, $\boldsymbol{h}_{\text{CCD}}$ and $\boldsymbol{n}$. Now, from (1) one gets the spectrum $\boldsymbol{v}$ for each pixel (x, y) in the non-white-corrected spectral image as a matrix-equation

$$\boldsymbol{v} = \boldsymbol{W}\boldsymbol{T}_{\mathrm{OM}}^2\boldsymbol{r}_{\mathrm{retina}} + \boldsymbol{n} \ , \tag{2}$$

where $\boldsymbol{W} = \mathrm{diag}(\boldsymbol{w})$,

$$\boldsymbol{w} = \boldsymbol{S}\boldsymbol{T}_{\mathrm{FC}}\boldsymbol{H}_{\mathrm{CCD}}\boldsymbol{T}_{\mathrm{filters}}\boldsymbol{1}_{30} \tag{3}$$

and $\boldsymbol{T}_{\mathrm{OM}} = \mathrm{diag}\,(\boldsymbol{t}_{\mathrm{OM}})$, $\boldsymbol{S} = \mathrm{diag}\,(\boldsymbol{s})$, $\boldsymbol{T}_{\mathrm{FC}} = \mathrm{diag}\,(\boldsymbol{t}_{\mathrm{FC}})$, $\boldsymbol{H}_{\mathrm{CCD}} = \mathrm{diag}\,(\boldsymbol{h}_{\mathrm{CCD}})$, and $\boldsymbol{T}_{\mathrm{filters}}$ is a matrix that has the spectra $\boldsymbol{t}_i$ on its columns. Finally, $\boldsymbol{1}_{30}$ denotes a 30-vector of ones.

Here $\boldsymbol{w}$ is a 30-vector that describes the effect of the entire fundus imaging system, and it was measured by using a diffuse non-fluorescent Spectralon white reflectance standard (Labsphere, Inc.) as a imaging target instead of an eye. In this case

$$\boldsymbol{v}_{\mathrm{white}} = \boldsymbol{W}\boldsymbol{r}_{\mathrm{white}} + \boldsymbol{n}_{\mathrm{white}} \ . \tag{4}$$

Spectralon-coating reflects $> 99\%$ of all the wavelengths in the visual range (380–780 nm). Hence, by assuming the reflectance $\boldsymbol{r}_{\mathrm{white}}(\lambda) \approx 1$, $\forall \lambda \in [380, 780]$ nm in (4), and that the backround noise is minimal, i.e. $\boldsymbol{n} \approx \boldsymbol{n}_{\mathrm{white}} \approx \boldsymbol{0}_{30}$, one gets (3). Now, (2) and (3) yield

$$\boldsymbol{r}_{\mathrm{final}} = \boldsymbol{T}_{\mathrm{OM}}^2\boldsymbol{r}_{\mathrm{retina}} = \boldsymbol{W}^{-1}\boldsymbol{v} \ . \tag{5}$$

As usual, the superscript -1 denotes matrix (pseudo)inverse. In Eq. (5), $\boldsymbol{r}_{\mathrm{final}}$ describes the "pseudo-reflectance" of the retina at point (x, y) of the spectral image, because, in practice, it is not possible to measure the transmittance of the ocular media $t_{\mathrm{OM}}(\lambda)$ in vivo. One gets $\boldsymbol{W}$ and $\boldsymbol{v}$ by measuring the white reflectance sample and the actual retina with the spectral fundus camera, respectively. Another thing to consider is that a fundus camera is designed to take images of a curved surface, but no appropriate curved white reflectance standards exist. The Labsphere standard used in this study was flat, so the light was unevenly distributed on its surface. Because of this, using the 30 spectral channel images taken from the standard to make the corrections directly would have resulted in unrealistic results. Instead, a mean-spectrum from a 100×100 pixel spatial area in the middle of the white standard's spectral image was used as $\boldsymbol{w}$.

3 Voluntary Human Subjects

Using the spectral fundus camera system described above, spectral images of 66 human ocular fundi were recorded in vivo from 54 diabetic patients and 12 healthy volunteers. This study was approved by the local ethical committee of the University of Kuopio and was designed and performed in accordance with the ethical standards of the Declaration of Helsinki. Fully informed consent was obtained from each participant prior to his or her inclusion into the study.

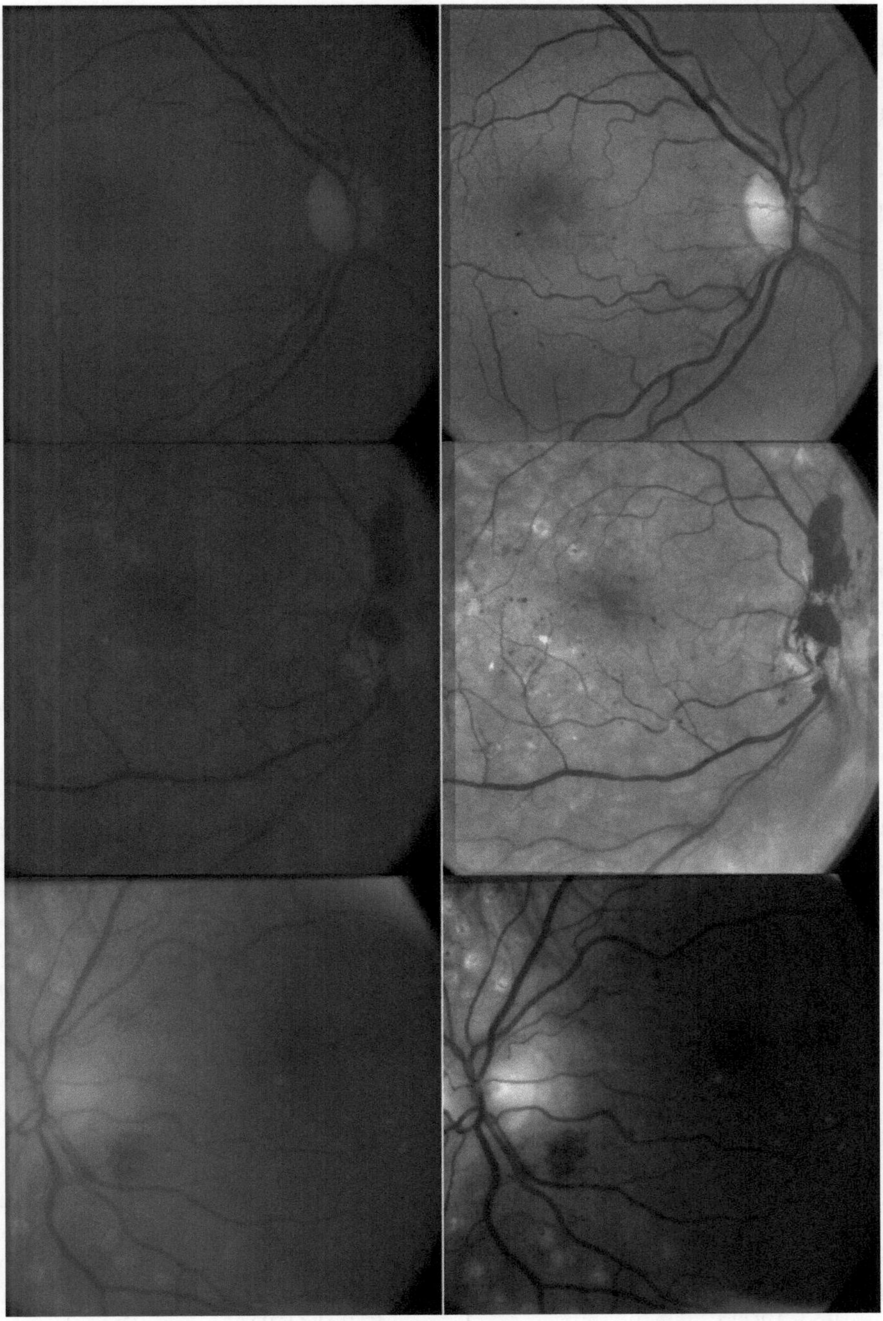

Fig. 4. RGB-images calculated from three of the 66 spectral fundus images for the CIE 1931 standard observer and D65 illumination (*left column*), and three-channel images the same fundi using specified registered spectral color channels (*right column*). No image processing (e.g. contrast enhancement) was applied to any of the images.

Imaging of the diabetic subjects was conducted in the Department of Ophthalmology in the Kuopio University Hospital (Kuopio, Finland). The control subjects were imaged in the color research laboratory of the University of Joensuu (Joensuu, Finland). The subjects' pupils were dilated using tropicamide eye drops (Oftan Tropicamid, Santen Oy, Finland), and only one eye was imaged from each subject. The database doesn't yet contain any follow-up spectral images of individual patients.

Subject's fundus was illuminated with 30 different filtered lights and images were captured in each case. Usually, due to the light source's poor emission of violet light, the very first spectral channels contained no useful information and were thus omitted from the spectral images. Also, the age-related yellowing of the crystalline lens of the eye [21] and other obstructions (mostly cataract) played a significant role in this.

4 Results and Discussion

Total of 66 spectral fundus images were collected using equipment and methods descriped above. These spectral images were then saved with MATLAB to a custom file format called "spectral binary", which stores the spectral data and their wavelength range is a lossless, uncompressed form. In this study, a typical size for one spectral binary file with 27 spectral channels (the first three channels contained no information) was approx. 108 MB, and the total size of the database was approx. 7 GB.

From the spectral images, normal RGB-images were calculated for visualization (see three example images in Fig. 4, left column). Spectral-to-RGB calculations were performed for the CIE 1931 standard colorimetric observer and illuminant D65 [22]. The 54 diabetes images showed typical findings for background and proliferative diabetic retinopathy, such as microaneurysms, small hemorrhages, hard lipid exudates, soft exudates (microinfarcts), intra-retinal microvascular abnormalities (IRMA), preretinal bleeding, neovascularization, and fibrosis. Due to the spectral channel image registration process, the colors on the outer edges of the images were distorted. On the right column of Fig. 4, some preliminary results of using specified spectral color channels are shown.

5 Conclusions

A database of spectral images of 66 human ocular fundi were presented. Also the methods of image acquisition and post-processing were described. A modified version of a standard ophthalmic fundus camera system was used with 30 narrow bandpass interference filters (400–700 nm at 10 nm intervals), a steady-state broadband light source and a monochrome digital CCD camera. Final spectral images had a 1024×1024 pixel spatial resolution and a varying number of spectral color channels (usually 27, since the first three channels beginning from 400 nm contained practically no information). Spectral images were saved in an uncompressed "spectral binary" format.

The database consists of fundus spectral images taken from 54 diabetic patients demonstrating different signs and severities of diabetic retinopathy and from 12 healthy volunteers. In the future we aim to establish a full spectral benchmarking database including both spectral images and manually annotated ground truth similarly to DiaRetDB1 [4]. Due to the special attention and solutions needed in capturing and processing the spectral data, the image acquisition and data post-processing were described in detail in this study. The augmentation of the database with annotations and additional data will be future work. The database will be made public for all researchers.

Acknowledgments. The authors would like to thank Tekes – the Finnish Funding Agency for Technology and Innovation – for funding (FinnWell program, funding decision 40039/07, filing number 2773/31/06).

References

1. DRIVE: Digital Retinal Images for Vessel Extraction, `http://www.isi.uu.nl/Research/Databases/DRIVE/`
2. Staal, J.J., Abramoff, M.D., Niemeijer, M., Viergever, M.A., van Ginneken, B.: Ridge based vessel segmentation in color images of the retina. IEEE Trans. Med. Imag. 23, 501–509 (2004)
3. STARE: STructured Analysis of the Retina, `http://www.parl.clemson.edu/stare/`
4. Kauppi, T., Kalesnykiene, V., Kämäräinen, J.-K., Lensu, L., Sorri, I., Raninen, A., Voutilainen, R., Uusitalo, H., Kälviäinen, H., Pietilä, J.: DIARETDB1 diabetic retinopathy database and evaluation protocol. In: Proceedings of the 11th Conference on Medical Image Understanding and Analysis (MIUA 2007), pp. 61–65 (2007)
5. Delori, F.C., Burns, S.A.: Fundus reflectance and the measurement of crystalline lens density. J. Opt. Soc. Am. A 13, 215–226 (1996)
6. Savage, G.L., Johnson, C.A., Howard, D.L.: A comparison of noninvasive objective and subjective measurements of the optical density of human ocular media. Optom. Vis. Sci. 78, 386–395 (2001)
7. Delori, F.C.: Spectrophotometer for noninvasive measurement of intrinsic fluorescence and reflectance of the ocular fundus. Appl. Opt. 33, 7439–7452 (1994)
8. Van Norren, D., Tiemeijer, L.F.: Spectral reflectance of the human eye. Vision Res. 26, 313–320 (1986)
9. Delori, F.C., Pflibsen, K.P.: Spectral reflectance of the human ocular fundus. Appl. Opt. 28, 1061–1077 (1989)
10. Bone, R.A., Brener, B., Gibert, J.C.: Macular pigment, photopigments, and melanin: Distributions in young subjects determined by four-wavelength reflectometry. Vision Res. 47, 3259–3268 (2007)
11. Beach, J.M., Schwenzer, K.J., Srinivas, S., Kim, D., Tiedeman, J.S.: Oximetry of retinal vessels by dual-wavelength imaging: calibration and influence of pigmentation. J. Appl. Physiol. 86, 748–758 (1999)
12. Ramella-Roman, J.C., Mathews, S.A., Kandimalla, H., Nabili, A., Duncan, D.D., D'Anna, S.A., Shah, S.M., Nguyen, Q.D.: Measurement of oxygen saturation in the retina with a spectroscopic sensitive multi aperture camera. Opt. Express 16, 6170–6182 (2008)

13. Khoobehi, B., Beach, J.M., Kawano, H.: Hyperspectral Imaging for Measurement of Oxygen Saturation in the Optic Nerve Head. Invest. Ophthalmol. Vis. Sci. 45, 1464–1472 (2004)
14. Hirohara, Y., Okawa, Y., Mihashi, T., Amaguchi, T., Nakazawa, N., Tsuruga, Y., Aoki, H., Maeda, N., Uchida, I., Fujikado, T.: Validity of Retinal Oxygen Saturation Analysis: Hyperspectral Imaging in Visible Wavelength with Fundus Camera and Liquid Crystal Wavelength Tunable Filter. Opt. Rev. 14, 151–158 (2007)
15. Hammer, M., Thamm, E., Schweitzer, D.: A simple algorithm for in vivo ocular fundus oximetry compensating for non-haemoglobin absorption and scattering. Phys. Med. Biol. 47, N233–N238 (2002)
16. Styles, I.B., Calcagni, A., Claridge, E., Orihuela-Espina, F., Gibson, J.M.: Quantitative analysis of multi-spectral fundus images. Med. Image Anal. 10, 578–597 (2006)
17. Johnson, W.R., Wilson, D.W., Fink, W., Humayun, M., Bearman, G.: Snapshot hyperspectral imaging in ophthalmology. J. Biomed. Opt. 12, 014036 (2007)
18. Stewart, C.V., Tsai, C.-L., Roysam, B.: The dual-bootstrap iterative closest point algorithm with application to retinal image registration. IEEE Trans. Med. Imag. 22, 1379–1394 (2003)
19. Yang, G., Stewart, C.V., Sofka, M., Tsai, C.-L.: Registration of challenging image pairs: initialization, estimation, and decision. IEEE Trans. Pattern Anal. Mach. Intell. 29, 1973–1989 (2007)
20. MATLAB: MATrix LABoratory, The MathWorks, Inc., `http://www.mathworks.com/matlab`
21. Gaillard, E.R., Zheng, L., Merriam, J.C., Dillon, J.: Age-related changes in the absorption characteristics of the primate lens. Invest. Ophthalmol. Vis. Sci. 41, 1454–1459 (2000)
22. Wyszecki, G., Stiles, W.S.: Color Science: Concepts and Methods, Quantitative Data and Formulae, 2nd edn. John Wiley & Sons, Inc., New York (1982)

Fast Prototype Based Noise Reduction

Kajsa Tibell[1], Hagen Spies[1], and Magnus Borga[2]

[1] Sapheneia Commercial Products AB,
Teknikringen 8, 583 30 Linkoping, Sweden
[2] Department of Biomedical Engineering,
Linkoping University, Linkoping, Sweden
{kajsa.tibell,hagen.spies}@scpab.eu,
magnus@imt.liu.se

Abstract. This paper introduces a novel method for noise reduction in medical images based on concepts of the Non-Local Means algorithm. The main objective has been to develop a method that optimizes the processing speed to achieve practical applicability without compromising the quality of the resulting images. A database consisting of prototypes, composed of pixel neighborhoods originating from several images of similar motif, has been created. By using a dedicated data structure, here Locality Sensitive Hashing (LSH), fast access to appropriate prototypes is granted. Experimental results show that the proposed method can be used to provide noise reduction with high quality results in a fraction of the time required by the Non-local Means algorithm.

Keywords: Image Noise Reduction, Prototype, Non-Local.

1 Introduction

Noise reduction without removing fine structures is an important and challenging issue within medical imaging. The ability to distinguish certain details is crucial for confident diagnosis and noise can obscure these details. To dissolve this problem some noise reduction method is usually applied. However, many of the existing algorithms assume that noise is dominant for high frequencies and that the image is smooth or piecewise smooth when, unfortunately, many fine structures in images correspond to high frequencies and regular white noise has smooth components. This can cause unwanted loss of detail in the image.

The Non-Local Means algorithm, first proposed in 2005, addresses this problem and has been proven to produce state-of-the-art results compared to other common techniques. It has been applied to medical images (MRI, 3D-MRI images) [12] [1] with excellent results. Unlike existing techniques, which rely on local statistics to suppress noise, the Non-Local Means algorithm processes the image by replacing every pixel by the weighted average of all pixels in that image having similar neighborhoods. However, its complexity implies a huge computational burden which makes the processing take unreasonably long time. Several improvements have been proposed (see for example [1] [3] [13]) to increase the speed, but they are still too slow for practical applications. Other related methods include Discrete Universal Denoising (DUDE) proposed by *Weissman et al*

A.-B. Salberg, J.Y. Hardeberg, and R. Jenssen (Eds.): SCIA 2009, LNCS 5575, pp. 159–168, 2009.

[11] and Unsupervised Information-Theoretic, Adaptive filtering (UINTA) by *Awate and Whitaker* [10].

This work presents a method for reducing noise based on concepts of the Non-Local Means algorithm with dramatically reduced processing times. The central idea is to take advantage of the fact that medical images are limited in the matter of motif and that there already exists a huge amount of images for different kinds of examinations, and perform as much of the computations as possible prior to the actual processing.

These ideas are implemented by creating a database of pixel neighborhood averages, called prototypes, originating from several images of a certain type of examination. This database is then used to process any new image of that type of examination. Different databases can be created to provide the possibility to process different images. During processing, the prototypes of interest can be rapidly accessed, in the appropriate database, using a fast nearest neighbor search algorithm, here the Locality Sensitive Hashing (LSH) is used. Thus, the time spent on processing an image is dramatically reduced. Other benefits of this approach are that a lot more neighborhoods can contribute to the estimation of a pixel and the algorithm is more likely to find at least one neighborhood in the more unusual cases.

The outline of this paper is as follows. The theory of the Non-Local Means algorithm is described in Section 2 and the proposed method is described in Section 3. The experimental results are presented and discussed in Section 4 and finally conclusions are drawn in Section 5.

2 Summary of the Non-local Means Algorithm

This chapter recalls the basic concept upon which the proposed method is based.

The Non-Local means algorithm was first proposed by *Buades et al.* [2] in 2005 and is based on the idea that the redundancy of information in the image under study can be used to remove noise. For each pixel in the image the algorithm selects a square window of surrounding pixels with size $(2d+1)^2$ where d is the radius. This window is called the neighborhood of that pixel. The restored value of a pixel, i, is then estimated by taking the average of all pixels in the image weighted depending on the similarity between their neighborhood and the neighborhood of i.

Each neighborhood is described by a vector $v(N_i)$ containing the gray level values of the pixels of which it consists. The similarity between two pixels i and j will then depend on the similarity of the intensity gray level vectors $v(N_i)$ and $v(N_j)$. This similarity is computed as a Gaussian weighted Euclidean distance $\|v(N_i) - v(N_j)\|_{2,a}^2$ which is a standard L_2-norm convolved with a Gaussian kernel of standard deviation a.

As described earlier the pixels need to be weighted so that pixels with a similar neighborhood to $v(N_i)$ are assigned larger weights on the average. Given the distance between the neighborhood vectors $v(N_i)$ and $v(N_j)$, the weight, $w(i, j)$ is computed as follows:

$$w(i,j) = \frac{1}{Z(i)} e^{-\frac{\|v(N_i) - v(N_j)\|_{2,a}^2}{h^2}} \quad (1)$$

where $Z(i)$ is the normalizing factor $Z(i) = \sum_j e^{-\frac{\|v(N_i)-v(N_j)\|_{2,a}^2}{h^2}}$. The decay of the weights, is controlled by the parameter h.

Given a noisy image $v = v(i)$ defined on the discrete grid I, where $i \in I$, the Non-Local Means filtered image is given by:

$$NL(v)(i) = \sum_{j \in I} w(i,j)v(j), \quad (2)$$

where $v(j)$ is the intensity of the pixel j and $w(i,j)$ is the weights assigned to $v(j)$ in the restoration of the pixel i.

Several attempts have been made to reduce the computational burden related to the Non-Local Means. Already when introducing the algorithm in the original paper [2], the authors emphasized the problem and proposed some improvements. For example, they suggested to limit the comparison of neighborhoods to a so called "search window" centered at the pixel under study. Another suggestion they had was "Blockwise implementation" where the image is divided into overlapping blocks. A Non-Local Means-like restoration of these blocks is then performed and finally the pixel values are restored based on the restored values of the blocks that they belong to. Examples of other improvements are "Pixel selection" proposed by *Mahmoudi and Sapiro* in [3] and "Parallel computation" and a combination of several optimizations proposed by *Coup et al* in [1].

3 Noise Reduction Using Non-local Means Based Prototype Databases

Inspired by the previously described Non-Local Means algorithm and using some favorable properties of medical images a method for fast noise reduction of CT images has been developed. The following key aspects were used:

1. Create a database of pixel neighborhoods originating from several similar images.
2. Perform as much of the computations as possible during preprocessing, i.e. during the creation of the database.
3. Create a data structure that provides fast access to prototypes in the database.

3.1 Neighborhood Database

As described earlier, CT images are limited in terms of motif due to the technique of the acquisition and the restricted number of examination types. Furthermore, several images of similar motif already exist in medical archiving systems. This implies that it is possible to create a system that uses neighborhoods of pixels from several images.

A database of neighborhoods that can be searched when processing an image is constructed as follows.

As in the Non-Local Means algorithm, the neighborhood $n(i)$ of a pixel i is defined as a window of arbitrary radius surrounding the pixel i.

Let N_I be a number of images of similar motif with size I^2. For every image $I_{1...N_I}$ extract the neighborhoods $n(i)_{1,...,I^2}$ of all pixels $i_{1,...,I^2}$ in the image. Store each extracted neighborhood as a vector $v(n)$ in a database. The database $D(v)$ will then consist of $S_D = N_I * I^2$ neighborhood vectors $v(n)_{1,...,S_D}$:

$$D(v) = v(n)_{1,...,S_D} \tag{3}$$

3.2 Prototypes

Similar to the blockwise implementation suggested in [2] the idea is to reduce the number of distance and average computations performed during processing by combining neighborhoods. The combined neighborhoods are called prototypes. Then the pixel values can be restored based on the values of these prototypes.

If $q(n)$ is a random neighborhood vector stored in the database $D(v)$ a prototype is created by computing the average of the neighborhood vectors $v(n)$ at distance at most w from $q(n)$.

By randomly selecting N_p number of neighborhood vectors from the database and compute the weighted average for each of them the entire database can be altered so that all neighborhood vectors are replaced by prototypes. The prototypes are given by:

$$P(v)_{1,...,N_p} = \frac{1}{C_i} \sum_{i \in D} v(n)_i \; if \; \|q(n) - v(n)_i\|_2^2 < w \tag{4}$$

where $C_i = \sum_{i \in D} v(n)_i$. Clearly, the number of prototypes in the database will be much smaller than the number of neighborhood vectors. Thus, the number of similarity comparisons during processing is decreased. However, for fast processing the relevant prototypes need to be accessed without having to search through the whole database.

3.3 Similarity

The neighborhood vectors can be considered to be feature vectors of each pixel of an image. Thus, they can be represented as points in a feature space with the same dimensionality as the size of the neighborhood. The points that are closest to each other in that feature space are also the most similar neighborhoods. Finding a neighborhood similar to a query neighborhood then becomes a Near Neighbor problem (see [9] [5] for definition).

The prototypes are, as described earlier, restored neighborhoods and thereby also points living in the same feature space as the neighborhood vectors. They are simply points representing a collection of the neighborhood vector points that lie closest to each other in the feature space.

As mentioned before, the Near Neighbor problem can be solved by using a dedicated data structure. In that way linear search can be avoided and replaced by fast access to the prototypes of interest.

3.4 Data Structure

The data structure chosen is the Locality Sensitive Hashing (LSH) scheme proposed by *Datar et al* [6] in 2003 which uses p-stable distributions [8] [7] and works directly on points in Euclidean space. Their version is a further development of the original scheme introduced by *P. Indyk and R. Motwani* [5] in 1998 whose key idea was to hash the points in a data set using hash functions such that the probability of collision is much higher for points which are close to each other than for points that are far apart. Points that collide are collected in "buckets" and stored in hash tables. The type of functions used to hash the points belong to what is called a locality-sensitive hash (LSH) family.

For a domain S of the point set with distance D, an locality-sensitive hash (LSH) family is defined as:

Definition 1. *A family* $H = h : S \rightarrow U$ *is called locality-sensitive (or* (r_1, r_2, p_1, p_2)*-sensitive) for* D *if for any* $v, q \in S$

- *if* $v \in B(q, r_1)$ *then* $Pr_H[h(q) = h(v)] \geq p_1$
- *if* $v \notin B(q, r_2)$ *then* $Pr_H[h(q) = h(v)] \leq p_2$

where $r_1 = R$ and $r_2 = c * R$, $B(q, r)$ is a ball of radius r centered in q and $Pr_H[h(q) = h(v)]$ is the probability that a point q and a point v will collide if using a hash function $h \in H$. The LSH family has to satisfy the inequalities $p_1 > p_2$ and $r_1 < r_2$ in order to be useful. By using functions from the LSH family the set of points can be preprocessed so that adjacent points are stored in the same bucket. When searching for the neighbors of a query point q the same functions are used to compute which "bucket" shall be considered. Instead of the whole set of points, only the points inside that "bucket" need to be searched.

The LSH algorithm was chosen since it has proven to have better query time than spatial data structures, the dependency on dimension and data size is sublinear and it is somewhat easy to implement.

3.5 Fast Creation of the Prototypes

As described in 3.2 a prototype is created by finding all neighborhood vectors similar to a randomly chosen neighborhood in the database and computing their average. To achieve fast creation of the prototypes the LSH data structure is applied. Given a number N_I of similar images the procedure is as follows: First all neighborhoods $n(i)_{1,\ldots,I^2}$ of the first image are stored using the LSH data structure described above. Next, a random vector is chosen and used as a query q to find all similar neighborhood vectors. The average of all neighborhood vectors at distance at most w from the query is computed producing the prototype $P(v)_i$. The procedure is repeated until an chosen number N_p of prototypes is created. Finally all neighborhood vectors are deleted from the hash tables and the prototypes $P(v)_{1,\ldots,N_p}$ are inserted instead. For all subsequent images every neighborhood vector is used as a query searching for similar prototypes. If a prototype is found the neighborhood vector is added to that by computing the average of the prototype and the vector itself. Since a prototype $P(v)_i$ most often is created of several neighborhood vectors and the query vector q is single, the query vector should not have equal impact on the average. Thus, the average has to be weighted by the number of neighborhood vectors included.

$$P(v)_{iNew} = \frac{P(v)_i * N_v + q}{N_v + 1} \tag{5}$$

where N_v is the number of neighborhood vectors that the prototype $P(v)_i$ is composed of. If for some query vector no prototype is found that query vector will constitute a new prototype itself. Thereby, unusual neighborhoods will still be represented.

3.6 The Resulting Pipeline

The resulting pipeline of the proposed method consist of two phases. The pre-processing phase where a database is created and stored using the LSH scheme and the processing phase where the algorithm reduces the noise in an image using the information stored in the database.

Creating the Database. First the framework of the data structure is constructed. Using this framework the neighborhood vectors $v(n)_i$ of N_I similar images are transformed into prototypes. The prototypes $P(v)_{iNew}$, which constitutes the database, are stored in "buckets" depending on their location in the high dimensional space in which they live. The "buckets" are then stored in hash tables $T_1, ..., T_L$ using a universal hash function, see fig 1.

Processing an Image. For every pixel in the image to be processed a new value is estimated using the prototypes stored in the database. By utilizing the data structure the prototypes to be considered can be found simply by calculating the "buckets" $g_1, ..., g_L$ corresponding to the neighborhood vector of the pixel under process and the indexes of those "buckets" in the hash tables $T_1, ..., T_L$. If more than one prototype is found the distance to each prototype is computed. The intensity value $p(i)$ of the pixel i is then estimated by interpolating the prototypes $P(v)_k$ that lies within radius s from the neighborhood $v(n)_i$ of i using inverse distance weighting (IDW).

Applying the general form of the IDW using a weight function defined by *Shepard* in [4] gives the expression for the interpolated value $p(i)$ of the point i:

$$p(i) = \frac{\sum_{k \in N_p} w(i)_k P(v)_k}{\sum_{k \in N_p} w(i)_k} \tag{6}$$

where $w(i)_k = \frac{1}{(\|v(n)_i - P(v)_k\|_2^2)^t}$, N_p is the number of prototypes in the database and t is a positive real number, called the power parameter. Greater values of t emphasizes the influence of the values closest to the interpolated point and the most common value of t is 2. If no prototype is found the original value of the pixel will remain unmodified.

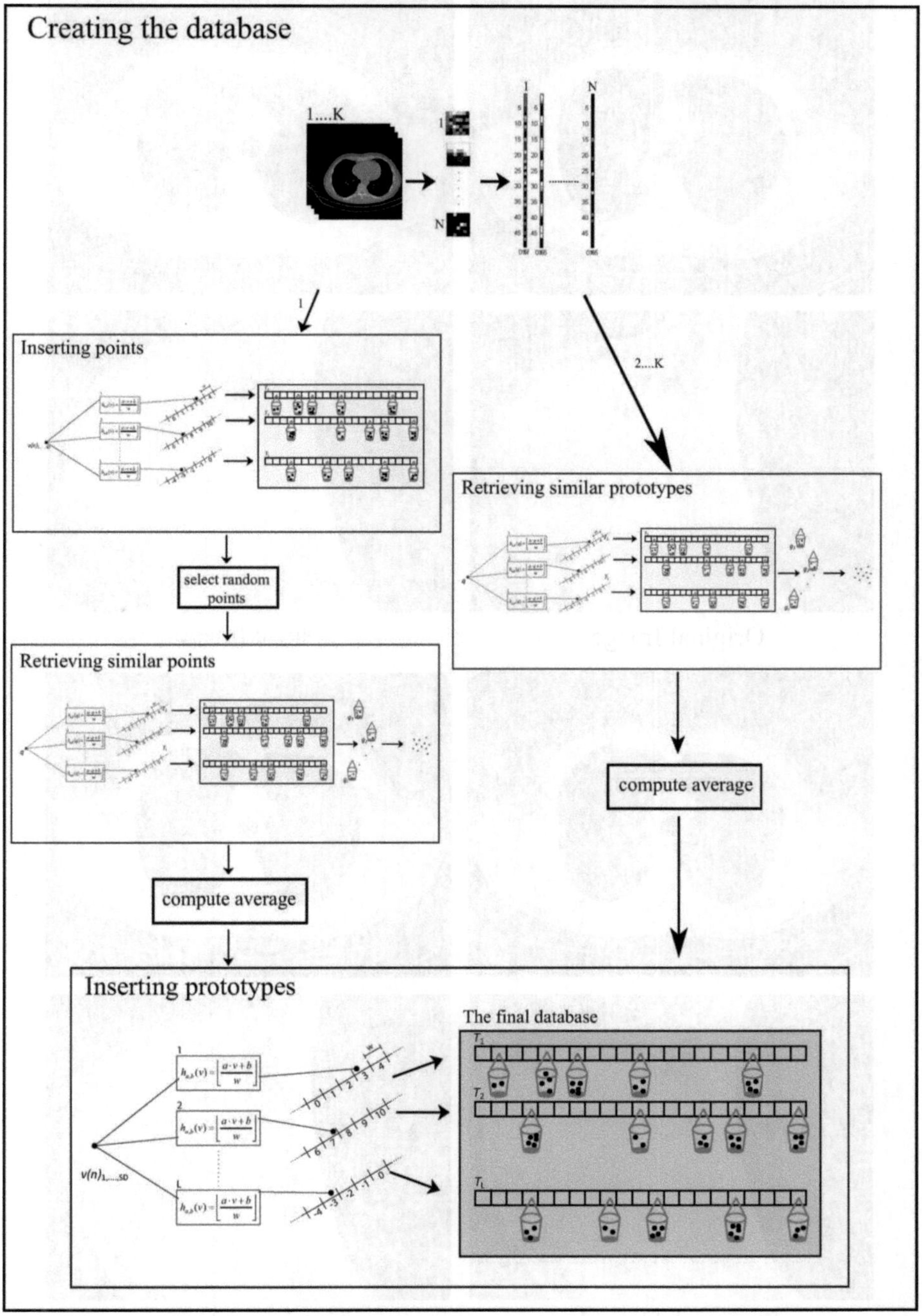

Fig. 1. A schematic overview of the creation of a database

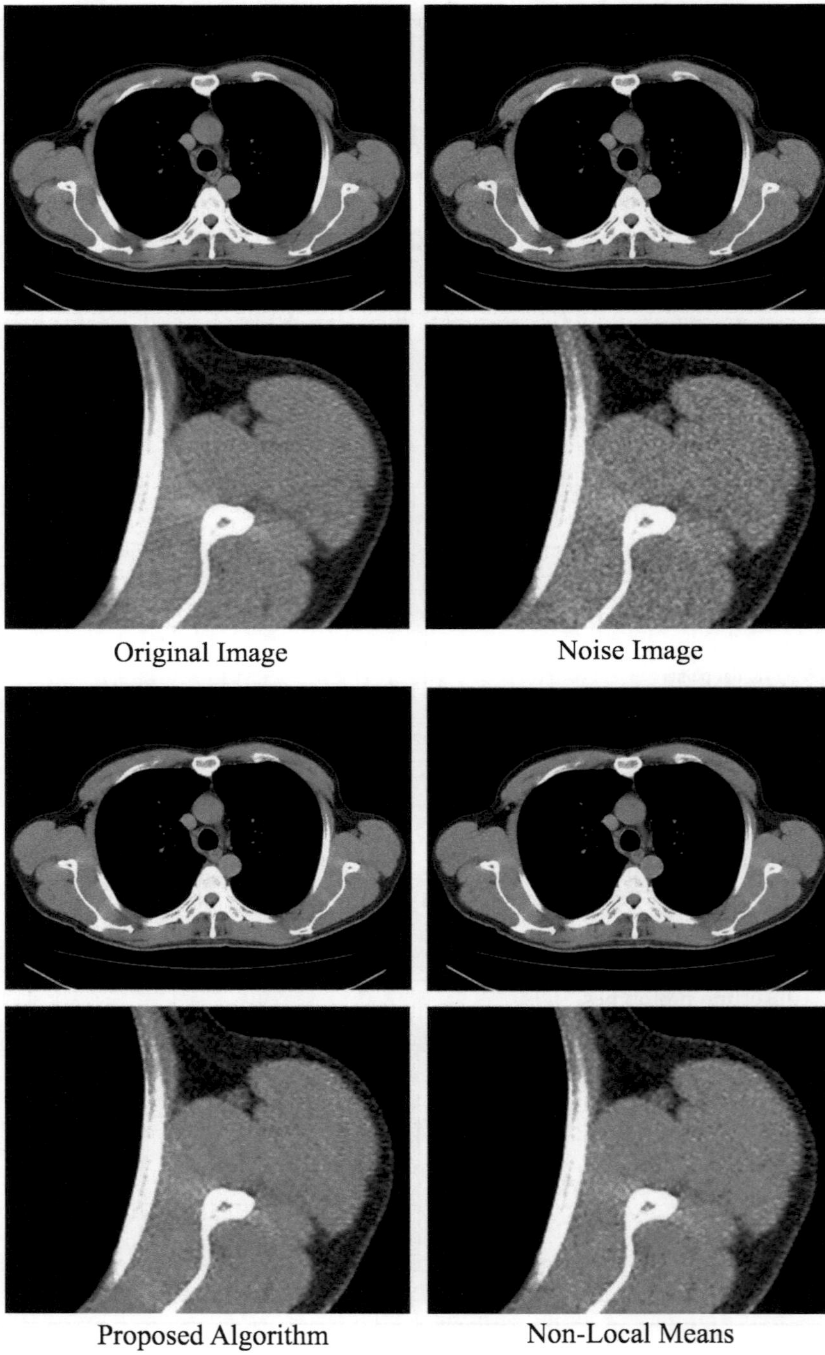

Fig. 2. CT image from lung with enlarged section below

4 Experimental Results

To test the performance of the proposed algorithm several databases have been created using different numbers of images. As expected, increasing the number of images used also increased the quality of the resulting images. The database used for processing the images in fig 2 consisted of 48 772 prototypes obtained from the neighborhoods of 17 similar images. Two sets of images were tested one of which is presented here. White Gaussian noise was applied to all images in one of the test sets (presented here) and the size of the neighborhoods was set to $7 * 7$ pixels.

The results was compared to The Non-Local Means algorithm and to evaluate the performance of the algorithms, quantitatively, the peak-to-peak signal to noise ratio (PSNR) was computed.

Table 1. PSNR and processing times for the test images

Method	*PSNR*	*Time(s)*
Non-Local Means	126.9640	34576
Proposed method	129.9270	72

The results in fig 2 shows that the proposed method produces an improved visual result compared to the Non-Local Means. The details in the resulting image are better preserved while a high level of noise reduction is still maintained. Table 1 shows the PSNR and processing times obtained.

5 Conclusions and Future Work

This paper introduced a noise reduction approach based on concepts of the Non-Local Means algorithm. By creating a well-adjusted database of prototypes that can be rapidly accessed using a dedicated data structure it was shown that a noticeably improved result can be achieved in a small fraction of the time required by the existing Non-Local Means algorithm. Some further improvement in the implementation will enable using the method for practical purposes and the presented method is currently being integrated in the Sapheneia Clarity product line for low dose CT applications.

Future work will include investigation of alternative features, of the neighborhoods, replacing the currently used intensity values. Furthermore, the dynamic capacity of the chosen data structure will be utilized for examining the possibility to continuously integrate the neighborhoods, of the images being processed, into the database for making it adaptive.

References

1. Coupe, P., Yger, P., Prima, S., Hellier, P., Kervrann, C., Barillot, C.: An Optimized Blockwise Nonlocal Means Denoising Filter for 3-D Magnetic Resonance Images. IEEE Transactions on Medical Imaging 27(4), 425–441 (2008)

2. Buades, A., Coll, B., Morel, J.M.: A review of image denoising algorithms, with a new one. Multiscale Modeling & Simulation 4(2), 490–530 (2005)
3. Mahmoudi, M., Sapiro, G.: Fast image and video denoising via nonlocal means of similar neighborhoods. IEEE Signal Processing Letters 12(12), 839–842 (2005)
4. Shepard, D.: A two-dimensional interpolation function for irregularly-spaced data. In: Proceedings of the 1968 ACM National Conference, pp. 517–524 (1968)
5. Indyk, P., Motwani, R.: Approximate nearest neighbor: towards removing the curse of dimensionality. In: Proceedings of the 30th Symposium on Theory of Computing, pp. 604–613 (1998)
6. Datar, M., Immorlica, N., Indyk, P., Mirrokni, V.: Locality-sensitive hashing scheme based on p-stable distributions. In: DIMACS Workshop on Streaming Data Analysis and Mining (2003)
7. Nolan, J.P.: Stable Distributions - Models for Heavy Tailed Data. Birkhäuser, Boston (2007)
8. Zolotarev, V.M.: One-Dimensional Stable Distributions. Translations of Mathematical Monographs 65 (1986)
9. Andoni, A., Indyk, P.: Near-Optimal hashing algorithm for approximate nearest neighbor in high dimensions. Communications of the ACM 51(1) (2008)
10. Awate, S.A., Whitaker, R.T.: Image denoising with unsupervised, information-theoretic, adaptive filtering. In: Proceedings of the IEEE International Conference on Computer Vision and Pattern Recognition (2005)
11. Weissman, T., Ordentlich, E., Seroussi, G., Verdu, S., Weinberger, M.: Universal discrete denoising: Known channel. IEEE Transactions on Information Theory 51, 5–28 (2005)
12. Manjón, J.V., Carbonell-Caballero, J., Lull, J.J., García-Martía, G., Martí-Bonmatíb, L., Robles, M.: MRI denoising using Non-Local Means. Medical Image Analysis 12, 514–523 (2008)
13. Wong, A., Fieguth, P., Clausi, D.: A Perceptually-adaptive Approach to Image Denoising using Anisotropic Non-Local Means. In: The Proceedings of IEEE International Conference on Image Processing (ICIP) (2008)

Towards Automated TEM for Virus Diagnostics: Segmentation of Grid Squares and Detection of Regions of Interest

Gustaf Kylberg[1], Ida-Maria Sintorn[1,2], and Gunilla Borgefors[1]

[1] Centre for Image Analysis, Uppsala University,
Lägerhyddsvägen 2, SE-751 05 Uppsala, Sweden
[2] Vironova AB, Smedjegatan 6, SE-131 34 Nacka, Sweden
{gustaf,ida.sintorn,gunilla}@cb.uu.se

Abstract. When searching for viruses in an electron microscope the sample grid constitutes an enormous search area. Here, we present methods for automating the image acquisition process for an automatic virus diagnostic application. The methods constitute a multi resolution approach where we first identify the grid squares and rate individual grid squares based on content in a grid overview image and then detect regions of interest in higher resolution images of good grid squares. Our methods are designed to mimic the actions of a virus TEM expert manually navigating the microscope and they are also compared to the expert's performance. Integrating the proposed methods with the microscope would reduce the search area by more than 99.99 % and it would also remove the need for an expert to perform the virus search by the microscope.

Keywords: TEM, virus diagnostics, automatic image acquisition.

1 Introduction

Ocular analysis of transmission electron microscopy (TEM) images is an essential virus diagnostic tool in infectious disease outbreaks as well as a means of detecting and identifying new or mutated viruses [1,2]. In fact, virus taxonomy, to a large extent, still uses TEM to classify viruses based on their morphological appearance, as it has since it was first proposed in 1943 [3]. The use of TEM as a virus diagnostic tool in an infectious emergency situation was, for example, shown in both the SARS pandemic and the human monkey pox outbreak in the US 2003 [4,5]. The viral pathogens were identified using TEM before any other method provided any results or information. It can provide an initial identification of the viral pathogen faster than the molecular diagnostic methods more commonly used today.

The main problems with ocular TEM analysis are the need of an expert to perform the analysis by the microscope and that the result is highly dependent on the expert's skill and experience. To make virus diagnostic using TEM more useful, automated image acquisition combined with automatic analysis would hence be desirable. The method presented in this paper focuses on the first part,

A.-B. Salberg, J.Y. Hardeberg, and R. Jenssen (Eds.): SCIA 2009, LNCS 5575, pp. 169–178, 2009.

i.e., enabling automation of the image acquisition process. It is part of a project with the aim to develop a fully automatic system for virus diagnostics based on TEM in combination with automatic image analysis.

Modern transmission electron microscopes are, to a large extent, controlled via a computer interface. This opens up the possibility to add on software to automate the image acquisition procedure. For other biological sample types and applications (mainly 3D reconstructions of proteins and protein complexes), procedures for fully automated or semi automated image acquisition already exist as commercially available software or as in house systems in specific labs, i.e., [6,7,8,9,10].

For the application of automatically diagnosing viral pathogens, a pixel size of about 0.5 nm is necessary to capture the texture on the viral surfaces. If images with such high spatial resolution would be acquired over the grid squares of a TEM grid with a diameter of 3 mm, one would end up with about 28.3 terapixels of image data, where only a small fraction might actually contain viruses. Consequently, to be able to create a rapid and automatic detection system for viruses on TEM grids the search area has to be narrowed down to areas where the probability of finding viruses is high.

In this paper we present methods for a multi resolution approach, using low resolution images to guide the acquisition of high resolution images, mimicking the actions of an expert in virus diagnosis using TEM. This allows for efficient acquisition of high resolution images of regions of an TEM grid likely to contain viruses.

2 Methods

The main concept in the method is to:

1. segment grid squares in overview images of an TEM grid,
2. rate the segmented grid squares in the overview images,
3. identify regions of interest in images with higher spatial resolution of single good squares.

2.1 Segmenting Grid Squares

An EM grid is a thin-foil mesh of usually 3.05 mm in diameter. They can be made from a number of different metals such as copper, gold or nickel. The mesh is covered with a thin film or membrane of carbon and on top of this sits the biological material. Overview images of 400-Mesh EM grids at magnifications between 190× and 380× show a number of bright squares which are the carbon membrane in the holes of the metal grid, see Fig. 1(a). One assumption is made about the EM grid in this paper; the shape of the grid squares is quadratic or rectangular with parallel edges. Consequently there should exist two main directions of the grid square edges.

Detecting Main Directions. The main directions in these overview images are detected in images that are downsampeled to half the original size, simply to save

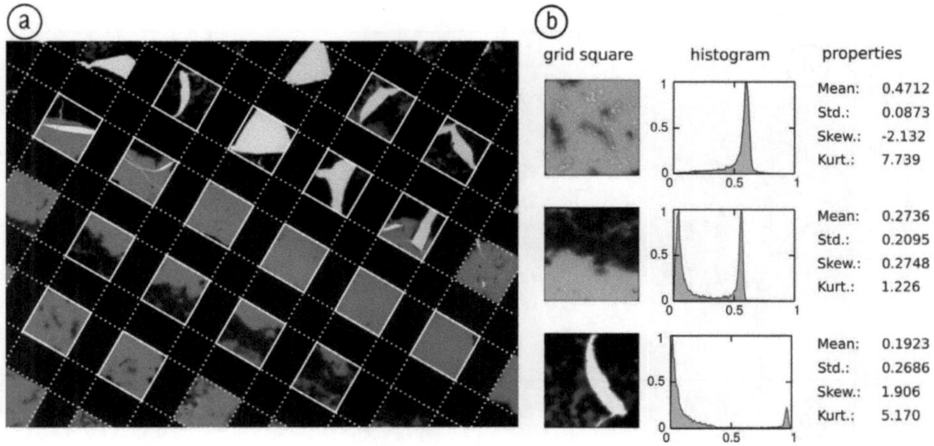

Fig. 1. a) One example overview image of an TEM grid with a sample containing rotavirus. The detected lines and grid square edges are marked with overlaid white dashed and continuous lines respectively. **b)** Three grid squares with corresponding gray level histograms and some properties.

computational time. The gradient magnitude of the image is calculated using the first order derivative of a Gaussian kernel. This is equivalent to computing the derivative in a pixel-wise fashion of an image smoothed with a Gaussian. This can be expressed in one dimension as:

$$\frac{\partial}{\partial x}\{f(x) \otimes G(x)\} = f(x) \otimes \frac{\partial}{\partial x}G(x), \qquad (1)$$

where $f(x)$ is the image function and $G(x)$ is a Gaussian kernel. The smoothing properties makes this method less noise sensitive compared to calculating derivatives with Prewitt or Sobel operators [11].

The Radon transform [12], with parallel beams, is applied on the gradient magnitude image to create projections in angles from 0 to 180 degrees. In 2D the Radon transform integrates the gray-values along straight lines in the desired directions. The Radon space is hence a parameter space of the radial distance from the image center and angle between the image x-axis and the normal of the projection direction. To avoid the image proportions to bias the Radon transform only a circular disc in the center of the gradient magnitude image is used.

Figure 2(a) shows the Radon transform for the example overview image in Fig. 1(a). A distinct pattern of local maxima can be seen at two different angles. These two angles correspond to the two main directions of the grid square edges. These two main directions can be separated from other angles by analyzing the variance of the integrated gray-values for the angles. Figure 2(b) shows the variance in the Radon image for each angle. The two local maxima correspond to the angles of the main directions of the grid square borders. These angles can be even better identified by finding the two lowest minima in the second derivative, also shown in Fig. 2(b). If there are several broken grid squares with edges in the same direction analyzing the second derivative of the variance is necessary.

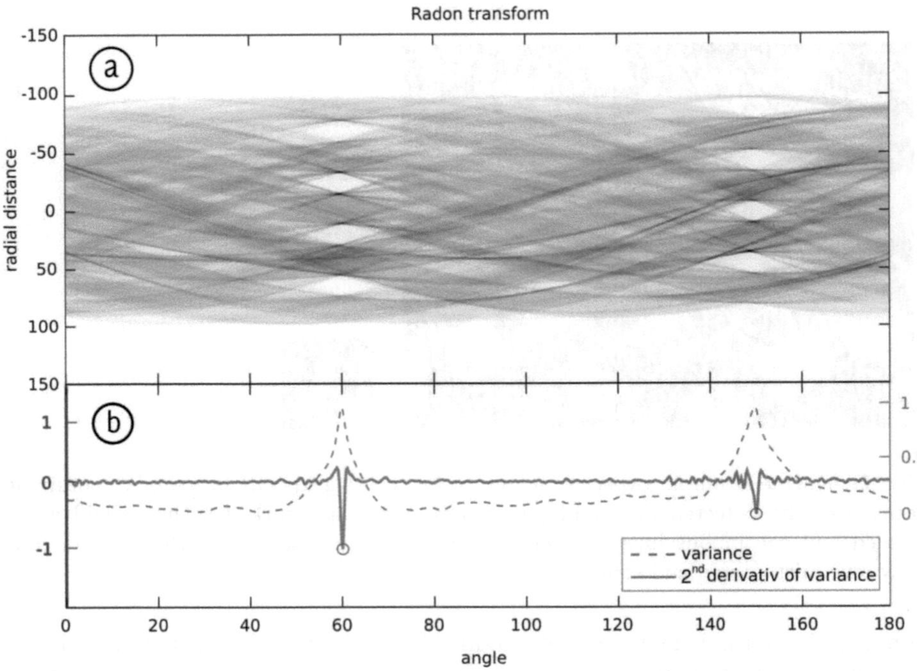

Fig. 2. a) The Radon transform of the central disk of the gradient magnitude image of the downsampled overview image. **b)** The variance, normalized to [0,1], of the angular values of the Radon transform in a) and its second derivative. The detected local minima are marked with red circles.

Detecting Edges in Main Directions. To find the straight lines connecting the edges in the gradient magnitude image the Radon transform is applied once more, but now only in the two main directions. Figure 3(a) shows the Radon transform for one of the main directions. These functions are fairly periodic, corresponding to the repetitive pattern of grid square edges. The periodicity can be calculated using autocorrelation. The highest correlation occurs when the function is aligned with itself, the second highest peak in the correlation occurs when the function is shifted one period etc., see Fig. 3(b). In Fig. 3(c) the function is split into its periods and stacked (cumulatively summed). These summed periods have one high and one low plateau separated by two local maxima which we want to detect. By using Otsu's method for binary thresholding [13] these plateaux are detected. Thereafter, the two local maxima surrounding the low plateau are found. The high and low plateaux correspond to the inside and outside of the squares, respectively. Knowing the distance between the peaks (the length of the high plateau) and the period length the peak positions can be propagated in the Radon transform. This enables filling in missing lines, due to damaged grid square edges. The distance between the lines, representing the square edges, may vary a few units throughout the function, therefore, the peak positions are fine tuned by finding the local maxima in a small region around the

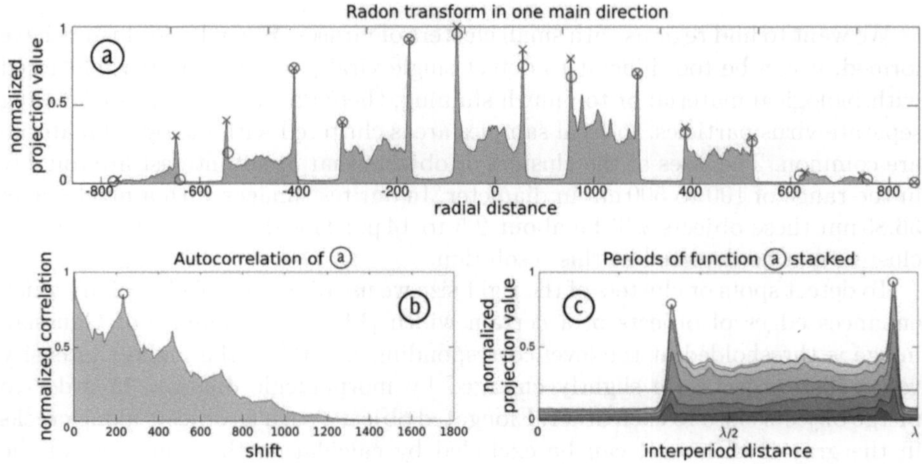

Fig. 3. a) The Radon transform in one of the main directions of the gradient magnitude image of the grid overview image. The red circles are the peaks detected in b) and c). Red crosses are the peak positions after fine tuning. **b)** The autocorrelation of the function in a). The peak used to calculate the period length is marked with a red circle. The horizontal axis is the shift starting with full overlap. **c)** The periods of the function in a) stacked. The red horizontal line is the threshold used to separate the high and the low plateaux and the peaks detected are marked with red circles.

peak position, shown as red circles and crosses in Fig. 3(a). This step completes the grid square segmentation.

2.2 Rating Grid Squares

The segmented grid squares are rated on a five level scale from 'good' to 'bad'. The rating system mimics the performance of an expert operator. The rating is based on whether a square is broken, empty or too cluttered with biological material. Statistical properties of the gray level histogram such as mean and the central moments variance, skewness and kurtosis are used to differentiate between squares with broken membranes, cluttered squares and squares suitable for further analysis. To get comparable mean gray values of the overview images their intensities are normalized to $[0, 1]$.

A randomly selected set of 53 grid squares rated by a virologist was used to train a naive Bayes classifier with a quadratic discriminant function. The rest of the segmented grid squares was rated with this classifier and compared with the rating done by the virologist, see Sec. 4.

2.3 Detecting Regions of Interest

In order to narrow down the search area further, only the top rated grid squares should be imaged at higher resolution at an approximate magnification of 2000× to allow detection of areas more likely to contain viruses.

We want to find regions with small clusters of viruses. When large clusters have formed, it can be too difficult to detect single viral particles. In areas cluttered with biological material or too much staining, there are small chances of finding separate virus particles. In fecal samples areas cluttered with biological material are common. The sizes of the clusters or objects that are of interest are roughly in the range of 100 to 500 nm in diameter. In our test images with a pixel size of 36.85 nm these objects will be about 2.5 to 14 pixels wide. This means that the clusters can be detected at this resolution.

To detect spots or clusters of the right size we use difference of Gaussians which enhances edges of objects of a certain width [14]. The difference of Gaussian image is thresholded at the level corresponding to 50 % of the highest intensity value. The objects are slightly enlarged by morphologic dilation, in order to merge objects close to each other. Elongated objects, such as objects along cracks in the gray level image, can be excluded by calculating the roundness of the objects. The roundness measure used is defined as follows:

$$roundness = \frac{4\pi \times area}{perimeter^2}, \tag{2}$$

where the area is the number of pixels in the object and the perimeter is the sum of the local distances of neighbouring pixels on the eight connected border of the object. The remaining objects correspond to regions with a higher probability to contain small clusters of viruses.

3 Material and Implementation

Human fecal samples and domestic dog oral samples were used, as well as cell-cultured viruses. A standard sample preparation protocol for biological material with negative staining was used. The samples were diluted in 10% phosphate buffered saline (PBS) before being applied to carbon coated 400-Mesh TEM grids and let to adhere for 60 seconds before excess sample were blotted of with filter paper. Next, the samples were stained with the negative staining phosphotungstic acid (PTA). To avoid PTA crystallization the grids were tilted 45 °. Excess of PTA was blotted off with filter paper, and left to air dry.

The different samples contained adenovirus, rotavirus, papillomavirus and semliki forest virus. These are all viruses with icosahedral capsids.

A Tecnai 10 electron microscope was used and it was controlled via Olympus AnalySIS software. The TEM camera used was a CCD based side-mounted Olympus MegaView III camera. The images were acquired in 16 bit gray scale resolution TIFF format with a size of 1376×1032 pixels. For grid square segmentation overview images in magnifications between 190× and 380× were acquired. To decide the size of the sigmas used for the Gaussian kernels in the difference of Gaussian in Sec. 2.3 image series with decreasing magnification of manually detected regions with virus were acquired. To verify the method image series with increasing magnification of manually picked regions were taken. Magnification steps in the image series used were between 650× and 73000×.

The methods described in Sec. 2 were implemented in MATLAB[15]. The computer used was a HP xw6600 Workstation running the Red Hat Linux distribution with the GNOME desktop environment.

4 Results

Segmenting and Rating Grid Squares. The method described in Sec. 2.1 was applied on 24 overview images. One example is shown in Fig. 1. The sigma for the Gaussian used in the calculation of the gradient magnitude was set to 1 and the filter size was 9×9. The Radon transform was used with an angular resolution of 0.25 degrees. The fine tuning of peaks was done within ten units of the radial distance. All the 159 grid squares completely within the borders of the 24 overview images were correctly segmented. The segmentation of the example overview image is shown in Fig. 1(a).

The segmented grid squares were classified according to the method in Sec. 2.2. One third, 53 squares, of the manually classified squares were randomly picked as training data and the other two thirds, 106 squares, were automatically classified. This procedure was repeated twenty times. The resulting average confusion matrix is shown in Table 1. When rating the grid squares they were on the average, 73.1 % correctly classified according to the rating done by the virologist. Allowing the classification to deviate ± 1 from the true rating 97.2 % of the grid squares were correctly classified. The best preforming classifier in these twenty training runs was selected as the classifier of choice.

Table 1. Confusion matrix comparing the automatic classification result and the classification done by the expert virologist. The numbers are the rounded mean values from 20 training and classification runs. The scale goes from bad (1) to good (5). The tridiagonal and diagonal are marked in the matrix.

automatic classification \ classification by expert	1	2	3	4	5
1	33	7	0	0	0
2	3	22	4	0	0
3	0	2	21	2	1
4	0	0	5	2	2
5	0	0	1	1	1

Detecting Regions of Interest. Eight resolution series of images with decreasing resolutions on regions with manually detected virus clusters were used to choose suitable sigmas for the Gaussian kernels in the method in Sec. 2.3. The sigmas were set to 2 and 3.2 for images with a pixel size of 36.85 nm and scaled accordingly for images with other pixel sizes. The method was tested on the eight resolution series with increasing magnification available. The limit for roundness

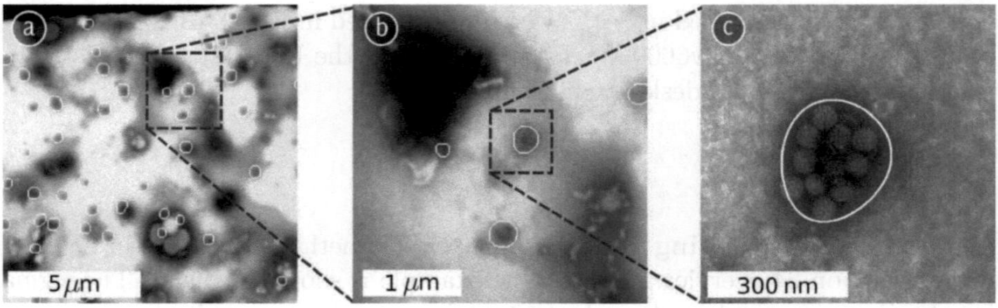

Fig. 4. Section of a resolution series with increasing resolution. The borders of the detected regions are shown in white. **a)** image with a pixel size of 36.85 nm. **b)** Image with a pixel size of 2.86 nm of the virus cluster in a). **c)** Image with a pixel size of 1.05 nm of the same virus cluster as in a) and b). The round shapes are individual viruses.

of objects was set to 0.8. Figure 4 shows a section of one of the resolution series for one detected virus cluster at three different resolutions.

5 Discussion and Conclusions

In this paper we have presented a method that enables reducing the search area considerably when looking for viruses in TEM grids. The segmentation of grid squares, followed by rating of individual squares, resembles how a virologist operates the microscope to find regions with high probability to have virus content.

The segmentation method utilizes information from several squares and their regular patterns to be able detect damaged squares. If overview images are acquired with a very low contrast between the grid and the membrane or if all squares in the image are lacking the same edges, the segmentation method might be less successfull. This is, however, an unlikely event. By decreasing the magnification, more squares can be fit in a single image and the probability that all squares have the same defects will decrease. Another solution is to use information from adjacent images from the same grid. This grid-square segmentation method can be used in in other TEM applications using the same kind of grids.

The classification result when rating grid squares shows that the size of the training data is adequate. Resuts when using different sets of 53 manually rated grid squares to train the naive Bayes classifier indicates that the choise of training set is sufficient as long as each class is represented in the training set.

The detection of regions of interest narrows down the search area within good grid squares. For the images at a magnification of 1850×, showing a large part of one grid square, the decrease in search area was calculated to be on average a factor 137. In other terms on average 99.3 % of the area of each analyzed grid square was discarded. The remaining regions have higher probability of containing small clusters of viruses.

By combining the segmentiation and rating of grid squares with the detection of regions of interest in the ten highest rated grid squares (usually more than ten good grid squares are never visually analyzed by an expert) the search area can be decreased with a factor of about 4000, assuming a standard 400 mesh TEM grid is used. This means that about 99.99975 % of the original search area can be descarded, assuming a standard 400 mesh TEM grid is used.

Parallel to this work we are developing automatic segmentation and classification methods for viruses in TEM images. Future work includes integration of these methods and those presented in this paper with softwares for controlling electron microscopes.

Acknowledgement. We would like to thank Dr. Kjell-Olof Hedlund at the Swedish Institute for Infectious Disease Control for providing the samples and being our model expert, and Dr. Tobias Bergroth and Dr. Lars Haag at Vironova AB for acquiring the image. The work presented in this paper is part of a project funded by the Swedish Agency for Innovative systems (VINNOVA), Swedish Defence Materiel Administration (FMV), and the Swedish Civil Contingencies Agency (MSB). The project aims to combine TEM and automated image analysis to develop a rapid diagnostic system for screening and identification of viral pathogens in humans and animals.

References

1. Hazelton, P.R., Gelderblom, H.R.: Electron microscopy for rapid diagnosis of infectious agents in emergent situations. Emerg. Infect. Dis. 9(3), 294–303 (2003)
2. Gentile, M., Gelderblom, H.R.: Rapid viral diagnosis: role of electron microscopy. New Microbiol. 28(1), 1–12 (2005)
3. Kruger, D.H., Schneck, P., Gelderblom, H.R.: Helmut ruska and the visualisation of viruses. Lancet 355, 1713–1717 (2000)
4. Reed, K.D., Melski, J.W., Graham, M.B., Regnery, R.L., Sotir, M.J., Wegner, M.V., Kazmierczak, J.J., Stratman, E.J., Li, Y., Fairley, J.A., Swain, G.R., Olson, V.A., Sargent, E.K., Kehl, S.C., Frace, M.A., Kline, R., Foldy, S.L., Davis, J.P., Damon, I.K.: The detection of monkeypox in humans in the western hemispher. N. Engl. J. Med. 350(4), 342–350 (2004)
5. Ksiazek, T.G., Erdman, D., Goldsmith, C.S., Zaki, S.R., Peret, T., Emery, S., Tong, S., Urbani, C., Comer, J.A., Lim, W., Rollin, P.E., Ngheim, K.H., Dowell, S., Ling, A.E., Humphrey, C., Shieh, W.J., Guarner, J., Paddock, C.D., Rota, P., Fields, B., DeRisi, J., Yang, J.Y., Cox, N., Hughes, J., LeDuc, J.W., Bellini, W.J., Anderson, L.J.: A novel coronavirus associated with severe acute respiratory syndrome. N. Engl. J. Med. 348, 1953–1966 (2003)
6. Suloway, C., Pulokas, J., Fellmann, D., Cheng, A., Guerra, F., Quispe, J., Stagg, S., Potter, C.S., Carragher, B.: Automated molecular microscopy: The new Leginon system. J. Struct. Biol. 151, 41–60 (2005)
7. Lei, J., Frank, J.: Automated acquisition of cryo-electron micrographs for single particle reconstruction on an fei Tecnai electron microscope. J. Struct. Biol. 150(1), 69–80 (2005)
8. Lefman, J., Morrison, R., Subramaniam, S.: Automated 100-position specimen loader and image acquisition system for transmission electron microscopy. J. Struct. Biol. 158(3), 318–326 (2007)

9. Zhang, P., Beatty, A., Milne, J.L.S., Subramaniam, S.: Automated data collection with a tecnai 12 electron microscope: Applications for molecular imaging by cryomicroscopy. J. Struct. Biol. 135, 251–261 (2001)
10. Zhu, Y., Carragher, B., Glaeser, R.M., Fellmann, D., Bajaj, C., Bern, M., Mouche, F., de Haas, F., Hall, R.J., Kriegman, D.J., Ludtke, S.J., Mallick, S.P., Penczek, P.A., Roseman, A.M., Sigworth, F.J., Volkmann, N., Potter, C.S.: Automatic particle selection: results of a comparative study. J. Struct. Biol. 145, 3–14 (2004)
11. Gonzalez, R.C., Woods, R.E.: Ch. 10.2.6. In: Digital Image Processing, 3rd edn. Pearson Education Inc., London (2006)
12. Gonzalez, R.C., Woods, R.E.: Ch. 5.11.3. In: Digital Image Processing, 3rd edn. Pearson Education Inc., London (2006)
13. Otsu, N.: A threshold selection method from gray-level histograms. IEEE Trans. Syst. Man Cybern. 9(1), 62–66 (1979)
14. Sonka, M., Hlavac, V., Boyle, R.: Ch. 5.3.3. In: Image Processing, Analysis, and Machine Vision, 3rd edn. Thomson Learning (2008)
15. The MathWorks Inc., MATLAB: system for numerical computation and visualization. R2008b edn. (2008-12-05), `http://www.mathworks.com`

Unsupervised Assessment of Subcutaneous and Visceral Fat by MRI

Peter S. Jørgensen[1,2], Rasmus Larsen[1], and Kristian Wraae[3]

[1] Department of Informatics and Mathematical Modelling, Technical University of Denmark, Denmark
[2] Fuel Cells and Solid State Chemistry Division, National Laboratory for Sustainable Energy, Technical University of Denmark, Denmark
[3] Odense University Hospital, Denmark

Abstract. This paper presents a method for unsupervised assessment of visceral and subcutaneous adipose tissue in the abdominal region by MRI. The identification of the subcutaneous and the visceral regions were achieved by dynamic programming constrained by points acquired from an active shape model. The combination of active shape models and dynamic programming provides for a both robust and accurate segmentation. The method features a low number of parameters that give good results over a wide range of values.The unsupervised segmentation was compared with a manual procedure and the correlation between the manual segmentation and unsupervised segmentation was considered high.

Keywords: Image processing, Abdomen, Visceral fat, Dynamic programming, Active shape model.

1 Introduction

There is growing evidence that obesity is related to a number of metabolic disturbances such as diabetes and cardiovascular disease [1]. It is of scientific importance to be able to accurately measure both visceral adipose tissue (VAT) and subcutaneous adipose tissue (SAT) distributions in the abdomen. This is due to the metabolic disturbances being closely correlated with particularly visceral fat [2].

Different techniques for fat assessment is currently available including anthropometry (waist-hip ratio, Body Mass Index), computed tomography (CT) and magnetic resonance imaging (MRI) [3].

These methods differ in terms of cost, reproducibility, safety and accuracy. The anthropometric measures are easy and inexpensive to obtain but do not allow quantification of visceral fat. Other techniques like CT will allow for this distinction in an accurate and reproducible way but are not safe to use due to the ionizing radiation [4]. MRI on the other hand does not have this problem and will also allow a visualization of the adipose tissue.

The potential problems with MRI measures are linked to the technique by which images are obtained. MRI does not have the advantage of CT in terms of

A.-B. Salberg, J.Y. Hardeberg, and R. Jenssen (Eds.): SCIA 2009, LNCS 5575, pp. 179–188, 2009.

direct classification of tissues based on Hounsfield units and will therefore usually require an experienced professional to visually mark and measure the different tissues on each image making it a time consuming and expensive technique.

The development of a robust and accurate method for unsupervised segmentation of visceral and subcutaneous adipose tissue would be a both inexpensive and fast way of assessing abdominal fat.

The validation of MRI to assess adipose tissue has been done by [5]. A high correlation was found between adipose tissue assessed by segmentation of MR images and dissection in human cadavers. A number of approaches have been developed for abdominal assessment of fat by MRI. A semi automatic method that fits Gaussian curves to the histogram of intensity levels and uses manual delineation of the visceral area has been developed by [6]. [7] uses fuzzy connectedness and Voronoi diagrams in a semi automatic method to segment adipose tissue in the abdomen. An Unsupervised method has been developed by [8] using active contour models to delimit the subcutaneous and visceral areas and fuzzy c-mean clustering to perform the clustering. [9] has developed an unsupervised method for assessment of abdominal fat in minipigs. The method performs a bias correction on the MR data and uses active contour models and dynamic programming to delimit the subcutaneous and visceral regions.

In this paper we present an unsupervised method that is robust to the poor image quality and large bias field that is present on older low field scanners. The method features a low number of parameters that are all non critical and give good results over a wide range of values. This is opposed to active contour models where accurate parameter tuning is required to yield good results. Furthermore, active contour models are not robust to large variations in intensity levels.

2 Data

The test data consisted of MR images from 300 subjects. The subjects were all human males with highly varying levels of obesity. Thus both very obese and very slim subjects were included in the data. Volume data was recorded for each subject in an anatomically bounded unit ranging from the bottom of the second lumbar vertebra to the bottom of the 5th lumbar vertebra. In this unit slices were acquired with a spacing of 10 mm. Only the T1 modality of the MRI data was used for further processing. A low field scanner was used for the image acquisition and images were scanned at a resolution of 256×256. The low field scanners generally have poor image quality compared to high field scanners. This is due to the presence of a stronger bias field and the extended amount of time needed for the image acquisition process thus not allowing breath-hold techniques to be used.

3 Method

3.1 Bias Field Correction

The slowly varying bias field present on all the MR images was corrected using a new way of sampling same tissue voxels evenly distributed over the subjects anatomy. The method works by first computing all local intensity maxima inside the subjects anatomy (the Region Of Interest - ROI) on a given slice. The ROI is then subdivided into a number of overlapping rectangular regions and the voxel with the highest intensity is stored for each region. We assume that this local maximum intensity voxel is a fat voxel. A threshold percentage is defined and all voxels with intensities below this percentage of the highest intensity voxel in each region is removed. We use a 85 % threshold for all images. However, this parameter is not critical and equally good results are obtained over a range of values (80-90 %).

The dimensions of the regions are determined so that it is impossible to place such a rectangle within the ROI without it overlapping at least one high intensity fat voxel. We subdivide the ROI into 8 rectangles vertically and 12 rectangles horizontally for all images. Again these parameters are not critical and equally good results are obtained for subdivisions 6−10 vertically and 6−12 horizontally. The acquired sampling locations are spatially trimmed to get evenly distributed samples across the subjects anatomy.

We assume an image model where the observed original biased image is the product of the unbiased image and the bias field

$$I_{biased} = I_{unbiased} \cdot bias \ . \tag{1}$$

The estimation of the bias field was done by fitting a 3 dimensional Thin Plate Spline to the sampled points in each subject volume. We apply a smoothing spline penalizing bending energy.

Assume N observations in $\mathbb{R}^3$, with each observation $\mathbf{s}$ having coordinates $[s_1 \quad s_2 \quad s_3]^T$ and values y. Instead of using the sampling points as knots a regular grid of n knots $\mathbf{t}$ is defined with coordinates $[t_1 \quad t_2 \quad t_3]^T$. We seek to find a function f, that describes a 3-dimensional hypersurface that provides an optimal fit to the observation points with minimal bending energy. The problem is formulated as minimizing the function S subject to f.

$$S(f) = \sum_{i=1}^{N} \{y_i - f(\mathbf{s}_i)\}^2 + \alpha J(f) \tag{2}$$

where $J(f)$ is a function for the curvature of f:

$$J(f) = \int\int\int_{\mathbb{R}^3} \sum_{i=1}^{3}\sum_{j=1}^{3} \left(\frac{\partial^2 f}{\partial x_i x_j}\right)^2 dx_1 dx_2 dx_3 \tag{3}$$

and f is of the form [10]:

$$f(\mathbf{t}) = \beta_0 + \beta_1^T \mathbf{t} + \sum_{j}^{n} \delta_j ||\mathbf{t} - \mathbf{t_j}||^3 \ . \tag{4}$$

α is a parameter that penalizes for curvature. With $\alpha = 0$ there is no penalty for curvature, this corresponds to an interpolating surface function where the function passes through each observation point. At higher α values the surface becomes more and more smooth since curvature is penalized. For α going towards infinity the surface will go towards the plane with the least squares fit, since no curvature is allowed.

To solve the system of equations we write the system on matrix form. First a coordinate matrix for the knots and the data points are defined.

$$\mathbf{T_k} = \begin{bmatrix} 1 & \cdots & 1 \\ \mathbf{t}_1 & \cdots & \mathbf{t}_n \end{bmatrix}_{[4\times n]} \tag{5}$$

$$\mathbf{T_d} = \begin{bmatrix} 1 & \cdots & 1 \\ \mathbf{s}_1 & \cdots & \mathbf{s}_N \end{bmatrix}_{[4\times N]} \quad . \tag{6}$$

Matrices containing all pairwise evaluations of the cubed distance measure from Equation 4 are defined as

$$\{\mathbf{E_k}\}_{ij} = ||\mathbf{t}_i - \mathbf{t}_j||^3 \quad i,j = 1,\cdots,n \tag{7}$$

$$\{\mathbf{E_d}\}_{ij} = ||\mathbf{s}_i - \mathbf{t}_j||^3 \quad i = 1,\cdots,N \quad j = 1,\cdots,n \tag{8}$$

$J(f)$ can be written as

$$J(f) = \delta^T \mathbf{E_k} \delta \ . \tag{9}$$

We can now write equation 2 on matrix form, incorporating the constraints $\mathbf{T}_k\delta = \mathbf{0}$ by the method of Lagrange multipliers.

$$S(f) = \left[\mathbf{Y} - \mathbf{E_d}\delta - \mathbf{T_d}^T\beta\right]^T \left[\mathbf{Y} - \mathbf{E_d}\delta - \mathbf{T_d}^T\beta\right] + \alpha\delta\mathbf{E_k}\delta + \lambda^T\mathbf{T_k}\delta \tag{10}$$

where λ is the Lagrange multiplier vector and $\beta = [\beta_0; \beta_1]_{[4\times 1]}$. By setting the 3 partial derivatives $\frac{\partial S}{\partial \delta} = \frac{\partial S}{\partial \beta} = \frac{\partial S}{\partial \lambda} = 0$ we get the following linear system

$$\begin{bmatrix} \mathbf{E_d}^T\mathbf{E_d} + \alpha\mathbf{E_k} & \mathbf{E_d}^T\mathbf{T_d}^T & \mathbf{T_k}^T \\ \mathbf{T_d}\mathbf{E_d} & \mathbf{T_d}\mathbf{T_d}^T & \mathbf{0} \\ \mathbf{T_k} & \mathbf{0} & \mathbf{0} \end{bmatrix} \begin{bmatrix} \delta \\ \beta \\ \lambda \end{bmatrix} = \begin{bmatrix} \mathbf{E_d}^T\mathbf{Y} \\ \mathbf{T_d}\mathbf{Y} \\ \mathbf{0} \end{bmatrix} \ . \tag{11}$$

An example result of the bias correction can be seen on Fig. 1.

Fig. 1. (right) The MR image before the bias correction. (center) The sample points from which the bias field is estimated. (left) The MR image after the bias correction.

3.2 Identifying Image Structures

Automatic outlining of 3 image structures was necessary in order to determine the regions for subcutaneous adipose tissue (SAT) and visceral adipose tissue (VAT): The external SAT outline, the internal SAT outline and the VAT area outline. First, a rough identification of the location of each outline was found using an active shape model trained on a small sample. Outlines found using this rough model were then used as constraints to drive a simple dynamic programming through polar transformed images.

3.3 Active Shape Models

The Active Shape Models approach developed by [12] is able to fit a point model of an image structure to image structures in an unknown image. The model is constructed from a set of 11 2D slices from different individuals at different vertical positions. This training set consists of images selected to represent the variation of the image structures of interest across all data. We have annotated the outer and inner subcutaneous outlines as well as the posterior part of the inner abdominal outline with a total of 99 landmarks. Fig. 2 shows an example of annotated images in the training set.

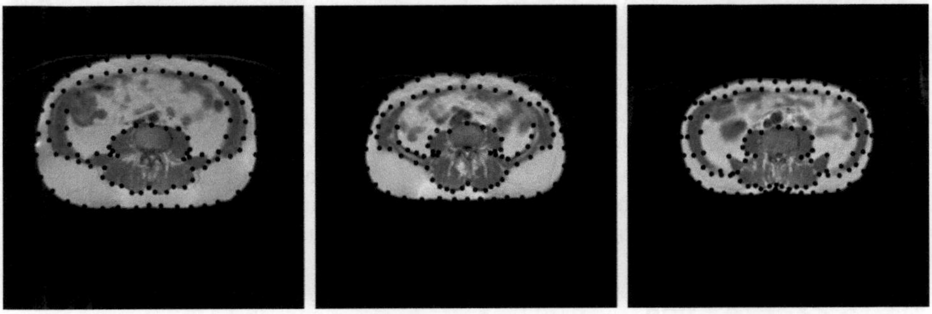

Fig. 2. 3 examples of annotated images from the training set

The 3 outlines are jointly aligned using a generalized Procrustes analysis [13,14], and principal components accounting for 95% of the variation are retained.

The search for new points in the unknown image is done by searching along a profile normal to the shape boundary through each shape point. Samples are taken in a window along the sampled profile. A statistical model of the grey-level structure near the landmark points in the training examples is constructed. To find the best match along the profile the Mahalanobis distance between the sampled window and the model mean is calculated. The Mahalanobis distance is linearly related to the log of the probability that the sample is drawn from a Gaussian-model. The best fit is found where the Mahalanobis distance is lowest and thus the probability that the sample comes from the model distribution is highest.

3.4 Dynamic Programming

The shape points acquired from the active shape models were used as constraints for dynamic programming. First a polar transformation was applied to the images to give the images a form suitable for dynamic programming [15]. A difference filter was applied radially to give edges from the original image a ridge representation in the transformed image. The same transformation was applied to the shape points of the ASM. The shape points were then used as constraints for the optimal path of the dynamic programming, only allowing the path to pass within a band of width 7 pixels centered on the ASM outline.

The optimal paths were then transformed back into the original image format to yield the outline of the external SAT border, the internal SAT border and the VAT area border. The method is illustrated on Fig. 3.

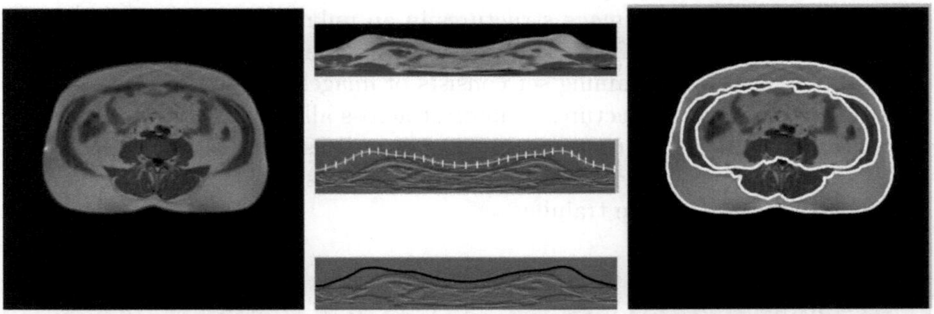

Fig. 3. Dynamic programming with ASM acquired constraints. (left) The bias corrected MR image. (center top) The polar transformed image. (center middle) The vertical difference filter applied on the transformed image with the constraint ranges superimposed (in white). (center bottom) The optimal path (in black) found through the transformed image for the external SAT border. (right) The 3 optimal paths from the constrained dynamic programming superimposed on the bias corrected image.

3.5 Post Processing

A set of voxels were defined for each of the 3 image structure outlines and set operations were applied to form the regions for SAT and VAT. Fuzzy c-mean clustering was used inside the VAT area to segment adipose tissue from other tissue. 3 classes were used: one for adipose tissue, one for other tissue and one for void. The class with the highest intensity voxels was assumed to be adipose tissue. Finally the connectivity of adipose tissue from the fuzzy c-mean clustering was used to correct a number of minor errors in regions where no clear border between SAT and VAT was available. A few examples of the final segmentation can be seen on Fig. 4.

4 Results

The amount of voxels in each class for each slice in the subjects were counted and measures for the total volume of the anatomically bounded unit were calculated.

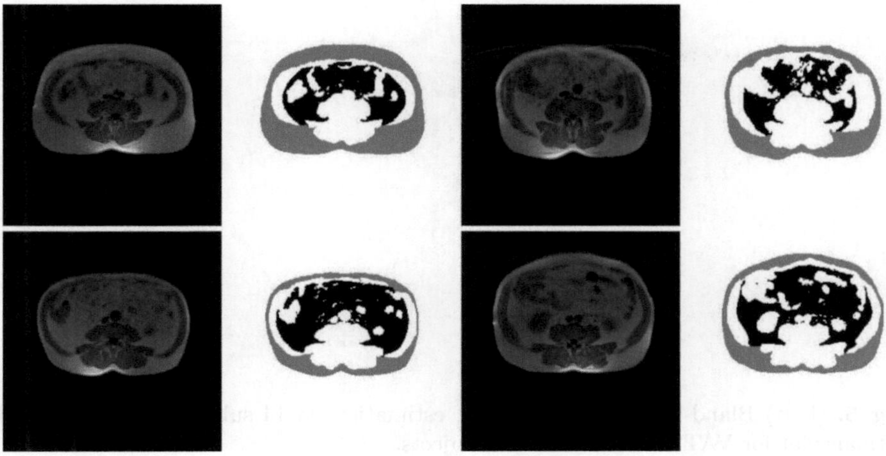

Fig. 4. 4 examples of the final segmentation. The segmented image is shown to the right of the original biased image. Grey: SAT; black:VAT; White:Other.

For each subject the distribution of tissue on the 3 classes: SAT, VAT and other tissue was computed. The results of the segmentation have been assessed by medical experts on a smaller subset of data and no significant aberrations between manual and unsupervised segmentation were found.

The unsupervised method was compared with manual segmentation. The manual method consist of manually segmenting the SAT by drawing the outlines of the internal and external SAT outlines. The VAT is estimated by drawing an outline around the visceral area and setting an intensity threshold that separates adipose tissue from muscle tissue.

A total of 14 subject volumes were randomly selected and segmented both automatic and manually. The correlation between the unsupervised and manual segmentation is high for both VAT ($r = 0.9599$, $P < 0.0001$) and SAT ($r = 0.9917$, $P < 0.0001$).

Figure 5(a) shows the Bland-Altman plot for SAT. The automatic method generally slightly overestimates compared to the manual method. The very blurry area near the umbilicus caused by the infeasibility of the breath-hold technique will have intensities that are very close to the threshold intensity between muscle and fat. This makes very slight differences between the automatic and manual threshold have large effects on the result.

The automatic estimates of the VAT also suffers from overestimation compared to the manual estimates, as seen on Figure 5(b). The partial volume effect is particularly significant in the visceral area and the adipose tissue estimate is thus very sensitive to small variations of the voxel intensity classification threshold.

Generally, the main source of disparity between the automatic and manual methods is the difference in the voxel intensity classification threshold. The manual method generally sets the threshold higher than the automatic method, which causes the automatic method to systematically overestimate compared to the manual method.

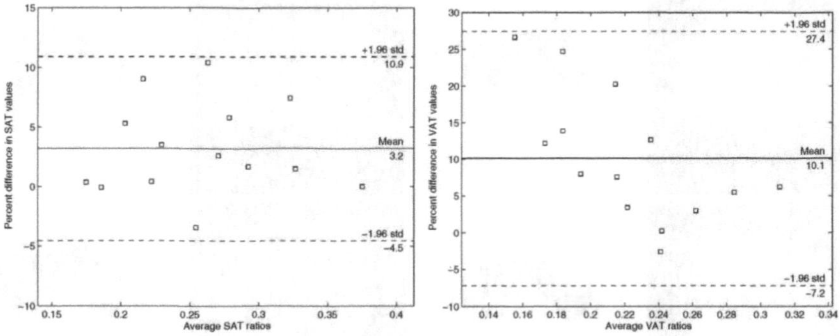

Fig. 5. (Left) Bland-Altman plot for SAT estimation on 14 subjects. (Right) Bland-Altman plot for VAT estimation on 14 subjects.

Fat in the visceral area is hard to estimate due to the partial volume effect. The manual estimate might thus not be more correlated with the true amount of fat in the region than the automatic estimate. The total truncus fat on the 14 subjects was estimated using DEXA and the correlation was found to be higher between the estimated total fat of automatic segmentation ($r = 0.8455$) than the manual segmentation ($r = 0.7913$).

5 Discussion

The described bias correction procedure allows the segmentation method to be used on low field scanners. The method will improve in accuracy on images scanned by newer high field scanners with better image quality using the breath-hold technique.

The use of ASM to find the general location of image structures makes the method robust to blurry areas (especially near the umbilicus) where a snake implementation is prone to failure [9]. Our method yields good results even on images acquired over an extended time period where the breath-hold technique is not applied.

The combination of ASM with DP makes the method both robust and accurate by combining the robust but inaccurate high level ASM method with the more fragile but accurate low level DP method.

The method proposed here is fully automated and has a very low amount of adjustable parameters. The low amount of parameters makes the method easily adaptable to new data, such as images acquired from other scanners. Furthermore, all parameters yield good results over a wide range of values.

The use of an automated unsupervised method has the potential to be much more precise than manual segmentation. A large amount of slices can be analyzed at a low cost thus minimizing the effect of errors on individual slices. The increased feasible amount of slices to segment with an unsupervised method allows for anatomically bounded units to be segmented with full volume information.

Using manual segmentation it is only feasible to segment a low number of slices in the subjects anatomy. The automatic volume segmentation will be less vulnerable to varying placement of organs on specific slices that could greatly bias single slice adipose tissue assessments. Furthermore the unsupervised segmentation method is not affected by intra- or inter-observer variability.

In conclusion, the presented methodology provides a both robust and accurate segmentation with only a small number of easily adjustable parameters.

Acknowledgements. We would like to thank Torben Leo Nielsen, MD Odense University Hospital, Denmark for allowing us access to the image data from the Odense Androgen Study and for valuable inputs during the course of this work.

References

1. Vague, J.: The degree of masculine differentiation of obesity: a factor determining predisposition to diabetes, atherosclerosis, gout, and uric calculous disease. Obes. Res. 4 (1996)
2. Bjorntorp, P.P.: Adipose tissue as a generator of risk factors for cardiovascular diseases and diabetes. Arteriosclerosis 10 (1990)
3. McNeill, G., Fowler, P.A., Maughan, R.J., McGaw, B.A., Gvozdanovic, D., Fuller, M.F.: Body fat in lean and obese women measured by six methods. Prof. Nutr. Soc. 48 (1989)
4. Van der Kooy, K., Seidell, J.C.: Techniques for the measurement of visceral fat: a practical guide. Int. J. Obes. 17 (1993)
5. Abate, N., Burns, D., Peshock, R.M., Garg, A., Grundy, S.M.: Estimation of adipose tissue by magnetic resonance imaging: validation against dissection in human cadavers. Journal of Lipid Research 35 (1994)
6. Poll, L.W., Wittsack, H.J., Koch, J.A., Willers, R., Cohnen, M., Kapitza, C., Heinemann, L., Mödder, U.: A rapid and reliable semiautomated method for measurement of total abdominal fat volumes using magnetic resonance imaging. Magnetic Resonance Imaging 21 (2003)
7. Jin, Y., Imielinska, C.Z., Laine, A.F., Udupa, J., Shen, W., Heymsfield, S.B.: Segmentation and evaluation of adipose tissue from whole body MRI scans. In: Ellis, R.E., Peters, T.M. (eds.) MICCAI 2003. LNCS, vol. 2878, pp. 635–642. Springer, Heidelberg (2003)
8. Positano, V., Gastaldelli, A., Sironi, A.M., Santarelli, M.F., Lmobardi, M., Landini, L.: An accurate and robust method for unsupervised assessment of abdominal fat by MRI. Journal of Magnetic Resonance Imaging 20 (2004)
9. Engholm, R., Dubinskiy, A., Larsen, R., Hanson, L.G., Christoffersen, B.Ø.: An adipose segmentation and quantification scheme for the abdominal region in minipigs. In: International Symposium on Medical Imaging 2006, San Diego, CA, USA. The International Society for Optical Engineering, SPIE (February 2006)
10. Green, P.J., Silverman, B.W.: Nonparametric regression and generalized linear models, a roughness penalty approach. Chapman & Hall, Boca Raton (1994)
11. Hastie, T., Tibshirani, R., Friedman, J.: The elements of statistical learning. Springer, Heidelberg (2001)

12. Cootes, T.F., Taylor, C.J.: Statistical models of appearence for medical image analysis and computer vision. In: Proc. SPIE Medical Imaging (2001) (to appear)
13. Gower, J.C.: Generalized procrustes analysis. Psychometrika 40 (1975)
14. Ten Berge, J.M.F.: Orthogonal procrustes rotation for two or more matrices. Psychometrika 42 (1977)
15. Glasbey, C.A., Young, M.J.: Maximum a posteriori estimation of image boundaries by dynamic programming. Journal of the Royal Statistical Society - Series C Applied Statistics 51(2), 209–222 (2002)

Decomposition and Classification of Spectral Lines in Astronomical Radio Data Cubes

Vincent Mazet[1], Christophe Collet[1], and Bernd Vollmer[2]

[1] LSIIT (UMR 7005 University of Strasbourg–CNRS), Bd Sébastien Brant, BP 10413, 67412 Illkirch Cedex, France
[2] Observatoire Astronomique de Strasbourg (UMR 7550 University of Strasbourg–CNRS), 11 rue de l'Université, 67000 Strasbourg, France
{vincent.mazet,c.collet,bernd.vollmer}@unistra.fr

Abstract. The natural output of imaging spectroscopy in astronomy is a 3D data cube with two spatial and one frequency axis. The spectrum of each image pixel consists of an emission line which is Doppler-shifted by gas motions along the line of sight. These data are essential to understand the gas distribution and kinematics of the astronomical object. We propose a two-step method to extract coherent kinematic structures from the data cube. First, the spectra are decomposed into a sum of Gaussians using a Bayesian method to obtain an estimation of spectral lines. Second, we aim at tracking the estimated lines to get an estimation of the structures in the cube. The performance of the approach is evaluated on a real radio-astronomical observation.

Keywords: Bayesian inference, MCMC, spectrum decomposition, multicomponent image, spiral galaxy NGC 4254.

1 Introduction

Astronomical data cubes are 3D images with spatial coordinates as the two first axis and the frequency (velocity channels) as third axis. We consider in this paper 3D observations of galaxies made at different wavelengths, typically in the radio (> 1 cm) or near-infrared bands ($\approx 10\,\mu$m). Each pixel of these images contains an atomic or molecular line spectrum which we called in the sequel *spexel*. The spectral lines contain information about the gas distribution and kinematics of the astronomical object. Indeed, due to the Doppler effect, the lines are shifted according to the radial velocity of the observed gas. A coherent physical gas structure gives rise to a coherent structure in the cube.

The standard method for studying cubes is the visual inspection of the channel maps and the creation of moment maps (see figure 1 a and b): moment 0 is the integrated intensity or the emission distribution and moment 1 is the velocity field. As long as the intensity distribution is not too complex, these maps give a fair impression of the 3D information contained in the cube. However, when the 3D structure becomes complex, the inspection by eye becomes difficult and important information is lost in the moment maps because they are produced

A.-B. Salberg, J.Y. Hardeberg, and R. Jenssen (Eds.): SCIA 2009, LNCS 5575, pp. 189–198, 2009.

by integrating the spectra, and thus do not reflect the individual line profiles. Especially, the analysis becomes extremely difficult when the spexels contain two or more components. Anyway, the need of an automatic method for the analysis of data cube is justified by the fact that eye inspection is subjective and time-consuming.

If the line components were static in position and widths, the problem would come down to be a source separation problem from which a number of works have been proposed in the context of astrophysical source maps from 3D cubes in the last years [2]. However, theses techniques cannot be used in our application where the line components (*i.e.* the sources) may vary between two spatial locations. Therefore, Flitti *et al.* [5] have proposed a Bayesian segmentation carried out on reduced data. In this method, the spexels are decomposed into Gaussian functions yielding reduced data feeding a Markovian segmentation algorithm to cluster the pixels according to similar behaviors (figure 1 c).

We propose in this paper a two-step method to isolate coherent kinematic structures in the cube by first decomposing the spexels to extract the different line profiles and then to classify the estimated lines. The first step (section 2) decomposes each spexel in a sum of Gaussian components whose number, positions, amplitudes and widths are estimated. A Bayesian model is presented: it aims at using all the available information since pertinent data are too few. The major difference with Flitti's approach is that the decomposition is not set on a unique basis: line positions and widths may differ between spexels. The second step (section 3) classifies each estimated component line assuming that two components in two neighbouring spexels are considered in the same class if their parameters are close. This is a new supervised method allowing the astronomer to set a threshold on the amplitudes. The information about the spatial dependence between spexels is introduced in this step. Performing the decomposition and classification steps separately is simpler that performing them together. It also allows the astronomer to modify the classification without doing again the decomposition step which is time consuming. The method proposed in this paper is intended to help astronomers to handle complex data cubes and to be complementary to the standard method of analysis. It provides a set of spatial zones corresponding to the presence of a coherent kinematic structure in the cube, as well as spectral characteristics (section 4).

2 Spexel Decomposition

2.1 Spexel Model

Spexel decomposition is typically an object extraction problem consisting here in decomposing each spexel as a sum of spectral component lines. A spexel is a sum of spectral lines which are different in wavelength and intensity, but also in width. Besides, the usual model in radioastronomy assumes that the lines are Gaussian. Therefore, the lines are modeled by a parametric function $\boldsymbol{f}$ with unknown parameters (position $\boldsymbol{c}$, intensity $\boldsymbol{a}$ and width $\boldsymbol{w}$) which are estimated as well as the component number. We consider in the sequel that the cube contains S spexels. Each spexel $s \in \{1, \ldots, S\}$ is a signal $\boldsymbol{y}_s$ modeled as a noisy sum of K_s components:

$$\boldsymbol{y}_s = \sum_{k=1}^{K_s} \boldsymbol{a}_{sk} \boldsymbol{f}(\boldsymbol{c}_{sk}, \boldsymbol{w}_{sk}) + \boldsymbol{e}_s = \boldsymbol{F}_s \boldsymbol{a}_s + \boldsymbol{e}_s, \tag{1}$$

where $\boldsymbol{f}$ is a vector function of length N, $\boldsymbol{e}_s$ is a $N \times 1$ vector modeling the noise, $\boldsymbol{F}_s$ is a $N \times K_s$ matrix and $\boldsymbol{a}_s$ is a $K_s \times 1$ vector:

$$\boldsymbol{F}_s = \begin{pmatrix} \boldsymbol{f}_1(\boldsymbol{c}_{s1}, \boldsymbol{w}_{s1}) & \cdots & \boldsymbol{f}_1(\boldsymbol{c}_{sK_s}, \boldsymbol{w}_{sK_s}) \\ \vdots & & \vdots \\ \boldsymbol{f}_N(\boldsymbol{c}_{s1}, \boldsymbol{w}_{s1}) & \cdots & \boldsymbol{f}_N(\boldsymbol{c}_{sK_s}, \boldsymbol{w}_{sK_s}) \end{pmatrix} \qquad \boldsymbol{a}_s = \begin{pmatrix} \boldsymbol{a}_{s1} \\ \vdots \\ \boldsymbol{a}_{sK_s} \end{pmatrix}.$$

The vector function $\boldsymbol{f}$ for component $k \in \{1, \ldots, K_s\}$ in pixel $s \in \{1, \ldots, S\}$ at frequency channel $n \in \{1, \ldots, N\}$ equals:

$$\boldsymbol{f}_n(\boldsymbol{c}_{sk}, \boldsymbol{w}_{sk}) = \exp\left(-\frac{(n - \boldsymbol{c}_{sk})^2}{2\boldsymbol{w}_{sk}^2}\right).$$

For simplicity, the expression of a Gaussian function was multiplied by $\sqrt{2\pi \boldsymbol{w}_{sk}^2}$ so that $\boldsymbol{a}_{sk}$ corresponds to the maximum of the line. In addition, we have $\forall s, k,\ \boldsymbol{a}_{sk} \geq 0$ because the lines are supposed to be non-negative. A perfect Gaussian shape is open to criticism because in reality the lines may be asymmetric, but modelling the asymmetry needs to consider one (or more) unknown and appears to be unnecessary complex.

Spexel decomposition is set in a Bayesian framework because it is clearly an ill-posed inverse problem [8]. Moreover, the posterior being a high dimensional complex density, usual optimisation techniques fail to provide a satisfactory solution. So, we propose to use Monte Carlo Markov chain (MCMC) methods [12] which are efficient techniques for drawing samples X from the posterior distribution π by generating a sequence of realizations $\{X^{(i)}\}$ through a Markov chain having π as its stationary distribution.

Besides, we are interesting in this step to decompose the whole cube, so the spexels are not decomposed independently each other. This allows to consider some global hyperparameters (such as a single noise variance allover the spexels).

2.2 Bayesian Model

The chosen priors are described hereafter for all s and k. A hierarchical model is used since it allows to set priors rather than a constant on hyperparameters. Some priors are conjugate so as to get usual conditional posteriors. We also try to work with usual priors for which simulation algorithms are available [12].

- the prior model is specified by supposing that K_s is drawn from a Poisson distribution with expected component number λ [7];
- the noise $\boldsymbol{e}_s$ is supposed to be white, zero-mean Gaussian, independent and identically distributed with variance $r_{\boldsymbol{e}}$;

- because we do not have any information about the component locations $\boldsymbol{c}_{sk}$, they are supposed uniformly distributed on $[1; N]$;
- component amplitudes $\boldsymbol{a}_{sk}$ are positive, so we consider that they are distributed according to a (conjugate) Gaussian distribution with variance $r_{\boldsymbol{a}}$ and truncated in zero to get positive amplitudes. We note: $\boldsymbol{a}_{sk} \sim \mathcal{N}^+(0, r_{\boldsymbol{a}})$ where $\mathcal{N}^+(\mu, \sigma^2)$ stands for a Gaussian distribution with positive support defined as (erf is the error function):

$$p(x|\mu,\sigma) = \sqrt{\frac{2}{\pi\sigma^2}} \left[1 + \operatorname{erf}\left(\frac{\mu}{\sqrt{2\sigma^2}}\right)\right]^{-1} \exp\left(-\frac{(x-\mu)^2}{2\sigma^2}\right) \mathbb{1}_{[0;+\infty[}(x);$$

- we choose an inverse gamma prior $\mathcal{IG}(\alpha_{\boldsymbol{w}}, \beta_{\boldsymbol{w}})$ for component width $\boldsymbol{w}_{sk}$ because this is a positive-support distribution whose parameters can be set according to the approximate component width known a priori. This is supposed to equal 5 for the considered data but, because this value is very approximative, we also suppose a large variance (equals to 100), yielding $\alpha_{\boldsymbol{w}} \approx 2$ and $\beta_{\boldsymbol{w}} \approx 5$;
- the hyperparameter $r_{\boldsymbol{a}}$ is distributed according to an (conjugate) inverse gamma prior $\mathcal{IG}(\alpha_{\boldsymbol{a}}, \beta_{\boldsymbol{a}})$. We propose to set the mean to the approximate real line amplitude (say μ) which can be roughly estimated, and to assign a large value to the variance. This yields: $\alpha_{\boldsymbol{a}} = 2 + \varepsilon$ and $\beta_{\boldsymbol{a}} = \mu + \varepsilon$ with $\varepsilon \ll 1$;
- Again, we adopt an inverse gamma prior $\mathcal{IG}(\alpha_{\boldsymbol{e}}, \beta_{\boldsymbol{e}})$ for $r_{\boldsymbol{e}}$, whose parameters are both set close to zero ($\alpha_{\boldsymbol{e}} = \beta_{\boldsymbol{e}} = \zeta$, with $\zeta \ll 1$). The limit case corresponds to the common Jeffreys prior which is unfortunately improper.

The posterior has to be integrable to ensure that the MCMC algorithm is valid. This cannot been checked mathematically because of the posterior complexity but, since the priors are integrable, a sufficient condition is fulfilled. The conditional posterior distributions of each unknown is obtained thanks to the prior defined above:

$$\begin{aligned}
&\boldsymbol{c}_{sk}|\cdots \propto \exp\Big(-\|\boldsymbol{y}_s - \boldsymbol{F}_s\boldsymbol{a}_s\|^2/2r_{\boldsymbol{e}}\Big)\mathbb{1}_{[1,N]}(\boldsymbol{c}_{sk}),\\
&\boldsymbol{a}_{sk}|\cdots \sim \mathcal{N}^+(\mu_{sk}, \rho_{sk}),\\
&\boldsymbol{w}_{sk}|\cdots \propto \exp\left(-\frac{1}{2r_{\boldsymbol{e}}}\|\boldsymbol{y}_s - \boldsymbol{F}_s\boldsymbol{a}_s\|^2 - \frac{\beta_{\boldsymbol{w}}}{\boldsymbol{w}_{sk}}\right)\frac{1}{\boldsymbol{w}_{sk}^{\alpha_w+1}}\mathbb{1}_{[0;+\infty[}(\boldsymbol{w}_{sk}),\\
&r_{\boldsymbol{a}}|\cdots \sim \mathcal{IG}\left(\frac{L}{2} + \alpha_{\boldsymbol{a}}, \frac{1}{2}\sum_s\sum_{k=1}^{K_s}\boldsymbol{a}_{sk}^2 + \beta_{\boldsymbol{w}}\right),\\
&r_{\boldsymbol{e}}|\cdots \sim \mathcal{IG}\left(\frac{NS}{2} + \alpha_{\boldsymbol{e}}, \frac{1}{2}\sum_s\|\boldsymbol{y}_s - \boldsymbol{F}_s\boldsymbol{a}_s\|^2 + \beta_{\boldsymbol{e}}\right)
\end{aligned}$$

where $x|\cdots$ means x conditionally to $\boldsymbol{y}$ and the other variables, N is the spectum length, S is the spexel number, $L = \sum_s K_s$ denotes the component number and

$$\mu_{sk} = \frac{\rho_{sk}}{r_e} \boldsymbol{z}_{sk}^T \boldsymbol{F}_{sk}, \qquad \rho_{sk} = \frac{r_a r_e}{r_e + r_a \boldsymbol{F}_{sk}^T \boldsymbol{F}_{sk}}, \qquad \boldsymbol{z}_{sk} = \boldsymbol{y}_s - \boldsymbol{F}_s \boldsymbol{a}_s + \boldsymbol{F}_{sk} \boldsymbol{a}_{sk}$$

where $\boldsymbol{F}_{sk}$ corresponds to the kth column of matrix $\boldsymbol{F}_s$.

The conditional posterior expressions for $\boldsymbol{c}_{sk}$, $\boldsymbol{w}_{sk}$ and the hyperparameters are straightforward, contrary to the conditional posterior for $\boldsymbol{a}_{sk}$ whose detail of computation can be found in [10, Appendix B].

2.3 MCMC Algorithm and Estimation

MCMC methods dealing with variable dimension models are known as trans-dimensional MCMC methods. Among them, the reversible jump MCMC algorithm [7] appears to be popular, fast and flexible [1]. At each iteration of this algorithm, a move which can either change the model dimension or generate a random variable is randomly performed. We propose these moves:

- B_s "birth in s": a component is created in spexel s;
- D_s "death in s": a component is deleted in spexel s;
- U_s "update in s": variables $\boldsymbol{c}_s$, $\boldsymbol{a}_s$ and $\boldsymbol{w}_s$ are updated;
- H "hyperparameter update": hyperparameters r_a and r_e are updated.

The probabilities b_s, d_s, u_s and h of moves B_s, D_s, U_s and H are chosen so that:

$$b_s = \frac{\gamma}{S+1} \min\left\{1, \frac{p(K_s+1)}{p(K_s)}\right\} \qquad d_s = \frac{\gamma}{S+1} \min\left\{1, \frac{p(K_s-1)}{p(K_s)}\right\}$$

$$u_s = \frac{1}{S+1} - b_s - d_s \qquad h = \frac{1}{S+1}$$

with γ such that $b_s + d_s \leq 0.9/(S+1)$ (we choose $\gamma = 0.45$) and $d_s = 0$ if $K_s = 0$.

We now discuss the simulation of the posteriors. Many methods available in literature are used for sampling positive normal [9] and inverse gamma distributions [4,12]. Besides, $\boldsymbol{c}_{sk}$ and $\boldsymbol{w}_{sk}$ are sampled using a random-walk Metropolis-Hastings algorithm [12]. To improve the speed of the algorithm, they are sampled jointly avoiding to compute the likelihood twice. The proposal distribution is a (separable) truncated Gaussian centered on the current values:

$$\tilde{\boldsymbol{c}}_{sk} \sim \mathcal{N}\left(\boldsymbol{c}_{sk}^*, r_c\right) \mathbb{1}_{[1,N]}(\tilde{\boldsymbol{c}}_{sk}), \qquad \tilde{\boldsymbol{w}}_{sk} \sim \mathcal{N}^+\left(\boldsymbol{w}_{sk}^*, r_w\right)$$

where $\tilde{\cdot}$ stands for the proposal and $\cdot^*$ denotes the current value. The algorithm efficiency depends on the scaling parameters r_c and r_w chosen by the user (generally with heuristics methods, see for example [6]).

The estimation is computed by picking in each Markov chain the sample which minimises the mean square error: it is a very simple estimation of the maximum a posteriori which does not need to save the chains. Indeed, the number of unknowns, and as a result, the number of Markov chains to save, is prohibitive.

3 Component Classification

3.1 New Proposed Approach

The decomposition method presented in section 2 provides for each spexel K_s components with parameter $\boldsymbol{x}_{sk} = \{\boldsymbol{c}_{sk}, \boldsymbol{a}_{sk}, \boldsymbol{w}_{sk}\}$. The goal of component classification is to assign to each component (s,k) a class $\boldsymbol{q}_{sk} \in \mathbb{N}^*$. One class corresponds to only one structure, so that components with the same class belong to the same structure. We also impose that, in each pixel, a class is present once at the most.

First of all, the components whose amplitude is lower than a predefined threshold τ are vanished in the following procedure (this condition helps the astronomer to analyse the gas location with respect to the intensity). To perform the classification, we assume that the component parameters exhibit weak variation between two neighbouring spexels, *i.e.* two components in two neighbouring spexels are considered in the same class if their parameters are close. The spatial dependency is introduced by defining a Gibbs field over the decomposed image [3]:

$$p(\boldsymbol{q}|\boldsymbol{x}) = \frac{1}{Z}\exp\left(-U(\boldsymbol{q}|\boldsymbol{x})\right) = \frac{1}{Z}\exp\left(-\sum_{c\in\mathcal{C}} U_c(\boldsymbol{q}|\boldsymbol{x})\right) \tag{2}$$

where Z is the partition function, $\mathcal{C}$ gathers the cliques of order 2 in a 4-connexity system and the potential function is defined as the total cost of the classification.

Let consider one component (s,k) located in spexel $s \in \{1,\ldots,S\}$ ($k \in \{1,\ldots,K_s\}$), and a neighbouring pixel $t \in \{1,\ldots,S\}$. Then, the component (s,k) may be classified with a component (t,l) ($l \in \{1,\ldots,K_t\}$) if their parameters are similar. In this case, we define the cost of component (s,k) equals to a distance $D(\boldsymbol{x}_{sk}, \boldsymbol{x}_{tl})^2$ computed with the component parameters (we see further why we choose the square of the distance). On the contrary, if no component in spexel t is close enough to component (s,k), we choose to set the cost of the component to a threshold σ^2 which codes the weaker similarity allowed. Indeed, if the two components (s,k) and (t,l) are too different (that is $D(\boldsymbol{x}_{sk}, \boldsymbol{x}_{tl})^2 > \sigma^2$), it would be less costly to let them in different classes. Finally, the total cost of the classification (*i.e.* the potential function) corresponds to the sum of the component costs.

Formally, these considerations read in the following manner. The potential function is defined as:

$$U_c(\boldsymbol{q}|\boldsymbol{x}) = \sum_{k=1}^{K_s} \varphi(\boldsymbol{x}_{sk}, \boldsymbol{q}_{sk}, \boldsymbol{x}_t, \boldsymbol{q}_t) \tag{3}$$

where s and t are the two spexels involved in the clique c, and $\varphi(\boldsymbol{x}_{sk}, \boldsymbol{q}_{sk}, \boldsymbol{x}_t, \boldsymbol{q}_t)$ represents the cost associated for the component (s,k) and defined as:

$$\varphi(\boldsymbol{x}_{sk}, \boldsymbol{q}_{sk}, \boldsymbol{x}_t, \boldsymbol{q}_t) = \begin{cases} D(\boldsymbol{x}_{sk}, \boldsymbol{x}_{tl})^2 & \text{if } \exists\, l \text{ such that } \boldsymbol{q}_{sk} = \boldsymbol{q}_{tl}, \\ \sigma^2 & \text{otherwise.} \end{cases} \tag{4}$$

In some ways, $\varphi(\boldsymbol{x}_{sk}, \boldsymbol{q}_{sk}, \boldsymbol{x}_t, \boldsymbol{q}_t)$ can be seen as a truncated quadratic function which is known to be very appealing in the context of outliers detection [13].

We choose for the distance $D(\boldsymbol{x}_{sk}, \boldsymbol{x}_{tl})$ a normalized Euclidean distance:

$$D(\boldsymbol{x}_{sk}, \boldsymbol{x}_{tl}) = \sqrt{\left(\frac{\boldsymbol{c}_{sk} - \boldsymbol{c}_{tl}}{\delta_c}\right)^2 + \left(\frac{\boldsymbol{a}_{sk} - \boldsymbol{a}_{tl}}{\delta_a}\right)^2 + \left(\frac{\boldsymbol{w}_{sk} - \boldsymbol{w}_{tl}}{\delta_w}\right)^2}. \quad (5)$$

The distance is normalized because the three parameters have not the same unity. δ_c and δ_w are the normalizing factors in the frequency domain whereas δ_a is the one in the intensity domain. We consider that two components are similar if their positions or widths do not differ for more than 1.2 wavelength channel, or if the difference between the amplitudes do not exceed 40% of the maximal amplitude. So, we set $\delta_c = \delta_w = 1.2$, $\delta_a = \max(\boldsymbol{a}_{sk}, \boldsymbol{a}_{s'k'}) \times 40\%$ and $\sigma = 1$.

To resume, we look for:

$$\hat{\boldsymbol{q}} = \arg\max_{\boldsymbol{q}} p(\boldsymbol{q}|\boldsymbol{x}) \quad \Leftrightarrow \quad \hat{\boldsymbol{q}} = \arg\min_{\boldsymbol{q}} \sum_{c \in C} \sum_{k=1}^{K_s} \varphi(\boldsymbol{x}_{sk}, \boldsymbol{q}_{sk}, \boldsymbol{x}_t, \boldsymbol{q}_t) \quad (6)$$

subject to the uniqueness of each class in each pixel.

3.2 Algorithm

We propose a greedy algorithm to perform the classification because it yields good results in an acceptable computation time (≈ 36 s on the cube considered in section 4 containing 9463 processed spexels). The algorithm is presented below. The main idea consists in tracking the components through the image by starting from an initial component and looking for the components with similar parameters spexel by spexel. These components are then classified in the same class, and the algorithm starts again until every estimated component is classified. We note z^* the increasing index coding the class, and the set $\mathcal{L}$ gathers the estimated components to classify.

1. set $z^* = 0$
2. while it exists some components that are not yet classified:
   3. $z^* = z^* + 1$
   4. choose randomly a component (s, k)
   5. set $\mathcal{L} = \{(s, k)\}$
   6. while $\mathcal{L}$ is not empty:
      7. set (s, k) as the first element of $\mathcal{L}$
      8. set $\boldsymbol{q}_{sk} = z^*$
      9. delete component (s, k) from $\mathcal{L}$
      10. among the 4 neighbouring pixels t of s, choose the components l that satisfy the following conditions:
          (C1) they are not yet classified
          (C2) they are similar to component (s, k) that is $D(\boldsymbol{x}_{sk}, \boldsymbol{x}_{tl})^2 < \sigma^2$
          (C3) $D(\boldsymbol{x}_{sk}, \boldsymbol{x}_{tl}) = \arg\min_{m \in \{1,\ldots,K_t\}} D(\boldsymbol{x}_{sk}, \boldsymbol{x}_{tm})$
          (C4) their amplitude is greater than τ
      11. Add (t, l) to $\mathcal{L}$

4 Application to a Modified Data Cube of NGC 4254

The data cube is a modified radio line observations made with the VLA of NGC 4254, a spiral galaxy located in the Virgo cluster [11]. It is a well-suited test case because it contains mainly only one single line (the HI 21 cm line). For simplicity, we keep in this paper pixel numbers for the spatial coordinates axis and channel numbers for the frequency axis (the data cube is a $512 \times 512 \times 42$ image, figures show only the relevant region). In order to investigate the ability of the proposed method to detect regions of double line profiles, we added an artificial line in a circular region north of the galaxy center. The intensity of the artificial line follows a Gaussian profile. Figure 1 (a and b) shows the maps of the first two moments integrated over the whole data cube and figure 1 c shows the estimation obtained with Flitti's method [5]. The map of the HI emission distribution (figure 1 a) shows an inclined gas disk with a prominent one-armed spiral to the west, and the additional line produces a local maximum. Moreover, the velocity field (figure 1 b) is that of a rotating disk with perturbations to the north-east and to the north. In addition, the artifical line produces a pronounced asymmetry. The double-line nature of this region cannot be recognized in the moment maps.

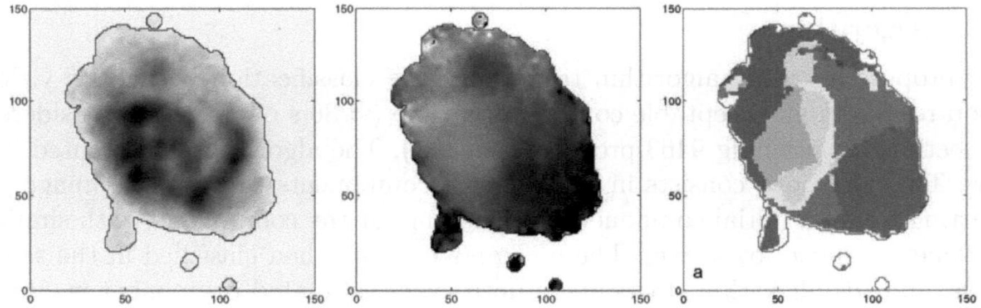

a b c

Fig. 1. Spiral galaxy NGC 4254 with a double line profile added: emission distribution (left) and velocity field (center); the figures are shown in inverse video (black corresponds to high values). Right: Flitti's estimation [5] (gray levels denote the different classes). The mask is displayed as a thin black line. The x-axis corresponds to right ascension, the y-axis to declination, the celestial north is at the top of the images and the celestial east at the left.

To reduce the computation time, a mask is determined to process only the spexels whose maximum intensity is greater than three times the standard deviation of the channel maps. A morphological dilation is then applied to connect close regions in the mask (a disk of diameter 7 pixels is chosen for structuring element).

The algorithm ran for 5000 iterations with an expected component number $\lambda = 1$ and a threshold $\tau = 0$. The variables are initialized by simulating them from the priors. The processing was carried out using Matlab on a double core (each 3.8 GHz) PC and takes 5h43. The estimation is very satisfactory because the difference between the original and the estimated cubes is very small; this is confirmed by inspecting by eye some spexel decomposition. The estimated components are then classified into 9056 classes, but the majority are very small and, consequently, not significant. In fact, only three classes, gathering more than 650 components each, are relevant (see figure 2): the large central structure (a & d), the "comma" shape in the south-east (b & e) and the artificially added component (c & f) which appears clearly as a third relevant class. Thus, our approach operates successfully since it is able to distinguish clearly the three main structures in the galaxy.

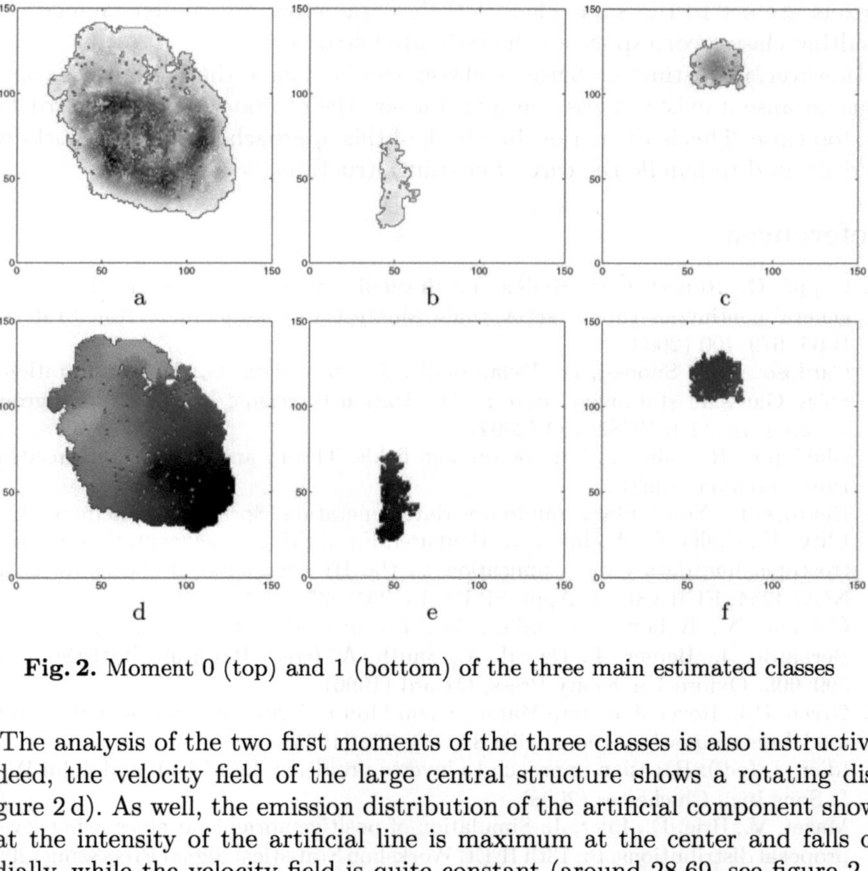

Fig. 2. Moment 0 (top) and 1 (bottom) of the three main estimated classes

The analysis of the two first moments of the three classes is also instructive. Indeed, the velocity field of the large central structure shows a rotating disk (figure 2 d). As well, the emission distribution of the artificial component shows that the intensity of the artificial line is maximum at the center and falls off radially, while the velocity field is quite constant (around 28.69, see figure 2, c and f). This is in agreement with the data since the artificial component is a Gaussian profile in intensity and has a center velocity at channel number 28.

Flitti *et al.* propose a method that clusters the pixels according to the six most representative components. Then, it is able to distinguish two structures that crosses while our method cannot because it exists at least one spexel where the components of each structure are too close. However, Flitti's method is unable to distinguish superimposed structures (since each pixel belongs to a single class) and a structure may be split into different kinematic zones if the spexels inside are evoluting too much: these drawbacks are clearly shown in figure 1 c. Finally, our method is more flexible and can better fit complex line profiles.

5 Conclusion and Perspectives

We proposed in this paper a new method for the analysis of astronomical data cubes and their decomposition into structures. In a first step, each spexel is decomposed into a sum of Gaussians whose number and parameters are estimated via a Bayesian framework. Then, the estimated components are classified with respect to their shape similarity: two components located in two neighbouring spexels are set in the same class if their parameters are similar enough. The resulting classes correspond to the estimated structures.

However, no distinction between classes can be done if the structure is continuous because it exists at less one spexel where the components of each structure are too close. This is the major drawback of this approach, and future works will be dedicated to handle the case of crossing structures.

References

1. Cappé, O., Robert, C.P., Rydèn, T.: Reversible jump, birth-and-death and more general continuous time Markov chain Monte Carlo samplers. J. Roy. Stat. Soc. B 65, 679–700 (2003)
2. Cardoso, J.-F., Snoussi, H., Delabrouille, J., Patanchon, G.: Blind separation of noisy Gaussian stationary sources. Application to cosmic microwave background imaging. In: 11th EUSIPCO (2002)
3. Chellappa, R., Jain, A.: Markov random fields. Theory and application. Academic Press, London (1993)
4. Devroye, L.: Non-uniform random variate generation. Springer, Heidelberg (1986)
5. Flitti, F., Collet, C., Vollmer, B., Bonnarel, F.: Multiband segmentation of a spectroscopic line data cube: application to the HI data cube of the spiral galaxy NGC 4254. EURASIP J. Appl. Si. Pr. 15, 2546–2558 (2005)
6. Gelman, A., Roberts, G., Gilks, W.: Efficient Metropolis jumping rules. In: Bernardo, J., Berger, J., Dawid, A., Smith, A. (eds.) Bayesian Statistics 5, pp. 599–608. Oxford University Press, Oxford (1996)
7. Green, P.J.: Reversible jump Markov chain Monte Carlo computation and Bayesian model determination. Biometrika 82, 711–732 (1995)
8. Idier, J. (ed.): Bayesian approach to inverse problems. ISTE Ltd. and John Wiley & Sons Inc., Chichester (2008)
9. Mazet, V., Brie, D., Idier, J.: Simulation of positive normal variables using several proposal distributions. In: 13th IEEE Workshop Statistical Signal Processing (2005)
10. Mazet, V.: Développement de méthodes de traitement de signaux spectroscopiques : estimation de la ligne de base et du spectre de raies. PhD. thesis, Nancy University, France (2005)
11. Phookun, B., Vogel, S.N., Mundy, L.G.: NGC 4254: a spiral galaxy with an $m = 1$ mode and infalling gas. Astrophys. J. 418, 113–122 (1993)
12. Robert, C., Casella, G.: Monte Carlo statistical methods. Springer, Heidelberg (2002)
13. Rousseeuw, P., Leroy, A.: Robust Regression and Outlier Detection. Series in Applied Probability and Statistics. Wiley-Interscience, Hoboken (1987)

Segmentation, Tracking and Characterization of Solar Features from EIT Solar Corona Images

Vincent Barra[1], Véronique Delouille[2], and Jean-Francois Hochedez[2]

[1] LIMOS, UMR 6158, Campus des Cézeaux, 63173 Aubière, France
vincent.barra@isima.fr

[2] Royal Observatory of Belgium, Circular Avenue 3, B-1180 Brussels, Belgium
{verodelo,hochedez}@sidc.com

Abstract. With the multiplication of sensors and instruments, size, amount and quality of solar image data are constantly increasing, and analyzing this data requires defining and implementing accurate and reliable algorithms. In the context of solar features analysis, it is particularly important to accurately delineate their edges and track their motion, to estimate quantitative indices and analyse their evolution through time. Herein, we introduce an image processing pipeline that segment, track and quantify solar features from a set of multispectral solar corona images, taken with eit EIT instrument. We demonstrate the method on the automatic tracking of Active Regions from EIT images, and on the analysis of the spatial distribution of coronal bright points. The method is generic enough to allow the study of any solar feature, provided it can be segmented from EIT images or other sources.

Keywords: Segmentation, tracking, EIT Images.

1 Introduction

With the multiplication of both ground-based and onboard satellites sensors and instruments, size, amount and quality of solar image data are constantly increasing, and analyzing this data requires the mandatory definition and implementation of accurate and reliable algorithms. Several applications can benefit from such an analysis, from data mining to the forecast of solar activity or space weather. More particularly, solar features, such as sunspots, filaments or solar flares partially express energy transfer processes in the Sun, and detecting, tracking and quantifying their characteristics can provide information about how these processes occur, evolve and affect total and spectral solar irradiance or photochemical processes in the terrestrial atmosphere.

The problem of solar image segmentation in general and the detection and tracking of these solar features in particular has thus been addressed in many ways in the last decade. The detection of sunspots [18,22,27], umbral dots [21] active regions [4,13,23], filaments [1,7,12,19,25], photospheric [5,17] or chromospheric structures [26], solar flares [24], bright points [8,9] or coronal holes [16] mainly use classical image processing techniques, from region-based to edge-based segmentation methods.

A.-B. Salberg, J.Y. Hardeberg, and R. Jenssen (Eds.): SCIA 2009, LNCS 5575, pp. 199–208, 2009.

In this article we propose an image processing pipeline that segment, track and quantify solar features from a set of multispectral solar corona images, taken with eit EIT instrument. The EIT telescope [10] onboard the SoHO ESA-NASA solar mission takes daily several data sets composed of four images (17.1 nm, 19.5 nm, 28.4 nm and 30.4 nm), all acquired within 30 minutes. They are thus well spatially registered and provide for each pixel a collection of 4 intensities that potentially permit to recognize the standard solar atmosphere region, or more generally solar features, to which it belongs..

This paper is organized as follows : section 2 introduces the general segmentation method. It basically recalls the original *SPoCA* algorithm, then specializes it to the automatic segmentation and tracking of solar features, and finally introduces some solar features properties suitable for the characterization of such objects. Section 3 demonstrate some results on some EIT images of a 9-year images dataset spanning solar cycle 23, and section 4 sheds lights on perspectives and conclusion.

2 Method

2.1 Segmentation

We introduced in [2] and refined in [3] *SPoCA*, an unsupervised fuzzy clustering algorithm allowing the fast and automatic segmentation of coronal holes, active regions and quiet sun from multispectral EIT images. In the following, we only recall the basic principle of this algorithm, and we more particularly focus on its application for the segmentation of solar features.

SPoCA. Let $I = (I^i)_{\{1\leq i\leq p\}}$, $I^i = (I^i_j)_{\{1\leq j\leq N\}}$, be the set of p images to be processed. Pixel j, $1 \leq j \leq N$ is described by a feature vector x_j. x_j can be the p-dimensional vector $(I^1_j \cdots I^p_j)^T$ or any r-dimensional vector describing local properties (textures, egdes,...) of j. In the following, the size of x_j will be denoted as r. Let $\mathcal{N}_j$ denote the neighborhood of pixel j, containing j, and Card($\mathcal{N}_j$) be the number of elements in $\mathcal{N}_j$. In the following, we note $X = \{\boldsymbol{x_j}, 1 \leq j \leq N, \boldsymbol{x_j} \in \mathbb{R}^r\}$ the set of feature vectors describing pixels j of I.

SPoCA is an iterative algorithm that searches for C compact clusters in X by computing both a fuzzy partition matrix $U = (u_{ij}), 1 \leq i \leq C, 1 \leq j \leq N$, $u_{i,j} = u_i(\boldsymbol{x_j}) \in [0,1]$ being the membership degree of $\boldsymbol{x_j}$ to class i, and unknown cluster centers $B = (b_i \in \mathbb{R}^r, 1 \leq i \leq C)$. It uses iterative optimizations to find the minimum of a constrained objective function:

$$J_{SPoCA}(B,U,X) = \sum_{i=1}^{C}\left(\sum_{j=1}^{N} u_{ij}^m \sum_{k\in\mathcal{N}_j} \beta_k d(\boldsymbol{x_k},\boldsymbol{b_i}) + \eta_i \sum_{j=1}^{N}(1-u_{ij})^m\right) \quad (1)$$

subject for all $i \in \{1 \cdots C\}$ to $\sum_{j=1}^{N} u_{ij} < N$, for all $j \in \{1 \cdots N\}$ to $\max_i u_{ij} > 0$, where $m > 1$ is a fuzzification parameter [6], and

$$\beta_k = \begin{cases} 1 & \text{if} \quad k = j \\ \frac{1}{\text{Card}(\mathcal{N}_j)-1} & \text{otherwise} \end{cases} \tag{2}$$

Parameter η_i can be interpreted as the mean distance of all feature vectors $\boldsymbol{x_j}$ to $\boldsymbol{b_i}$ such that $u_{ij} = 0.5$. η_i can be computed as the intra-class mean fuzzy distance [14]:

$$\eta_i = \frac{\sum_{j=1}^{N} u_{ij}^m d(\boldsymbol{x_j}, \boldsymbol{b_i})}{\sum_{j=1}^{N} u_{ij}^m}$$

The first term in (1) is the total fuzzy intra-cluster variance, while the second term prevents the trivial solution $U = 0$ and relaxes the probabilistic constraint $\sum_{j=1}^{N} u_{ij} = 1, 1 \leq i \leq C$, stemming from the classical Fuzzy-C-means (FCM) algorithm [6]. SPoCa is a spatially-constrained version of the possibilistic clustering algorithm proposed by Krishnapuram and Keller [14], which allows memerships to be interpreted as true degrees of belonging, and not as degrees of sharing pixels amongst all classes, which is the case in the FCM method.
We showed in [2] that U and B could be computed as

$$u_{ij} = \left[1 + \left(\frac{\sum_{k \in \mathcal{N}_j} \beta_k d(\boldsymbol{x_k}, \boldsymbol{b_i})}{\eta_i} \right)^{\frac{1}{m-1}} \right]^{-1} \quad \text{and} \quad \boldsymbol{b_i} = \frac{\sum_{j=1}^{N} u_{ij}^m \sum_{k \in \mathcal{N}_j} \beta_k \boldsymbol{x_k}}{2 \sum_{j=1}^{N} u_{ij}^m}$$

SPoCA provides thus coronal holes (CH), Active regions (AR) and Quiet Sun (QS) fuzzy maps $U_i = (u_{ij})$ for $i \in \{CH, QS, AR\}$, modeled as distributions of possibility π_i [11] and represented by fuzzy images. Figure 1 presents an example of such fuzzy maps, processed on a 19.5 nm EIT image taken on August 3, 2000.

To this original algorithm, we added [3] some pre and post processings (temporal stability, limb correction, edge smoothing, optimal clustering based on a sursegmentation), which dramatically improved the results.

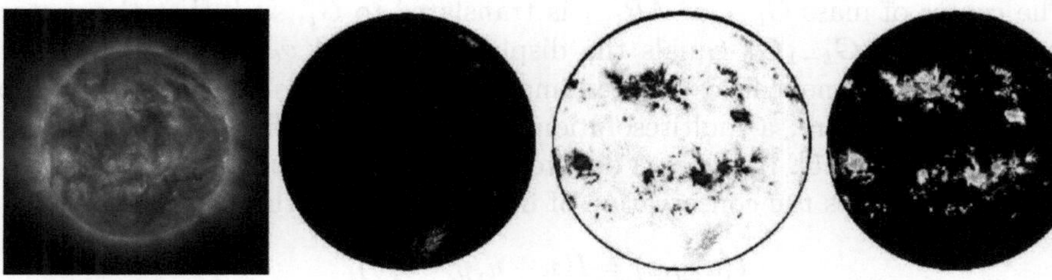

Original Image CH map π_{CH} QS map π_{QS} AR map π_{AR}

Fig. 1. Fuzzy segmentation of a 19.5 nm EIT image taken on August 3, 2000

Segmentation of Solar Features. From coronal holes (CH), Active regions (AR) and Quiet Sun (QS) fuzzy maps, solar features can then be segmented using both memberships and expert knowledge provided by solar physicists. The basic principle is to find connected components in a fuzzy map being homogeneous with respect to some statistical criteria, related to the physical properties of the features, and/or having some predefined geometrical properties. Some region growing techniques and mathematical morphology are thus used here to achieve this segmentation process. Typical solar features that can directly be extracted from EIT images only include coronal bright points (figure 2(a)) or active regions (figure 2(b)).

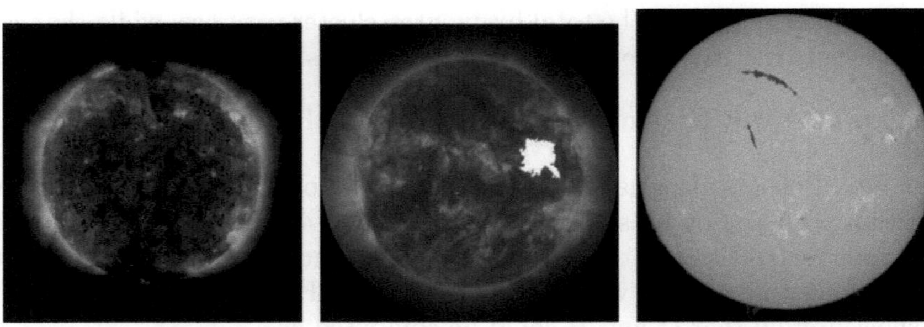

(a) Bright points from EIT image (1998-02-03) (b) Active regions from EIT image (2000-08-04) (c) Filaments from H-α image

Fig. 2. Several solar features

Additional information can also be added to these maps to allow the segmentation of other solar features. We for example processed in [3] the segmentation of filaments from the fusion of EIT and H-α images, from Kanzelhoehe observatory (figure 2(c)).

2.2 Tracking

In this article, we propose to illustrate the method on the automatic tracking of Active Regions. We more particularly focus on the largest active region, and algorithm 3 gives an overview of the method.

The center of mass G_{t-1} of AR_{t-1} is translated to G_t, such that the vector with start point $\boldsymbol{G_{t-1}G_t}$ equals the displacement field ν_G observed at pixel G_{t-1}. The displacement field between images I_{t-1} and I_t is estimated with the *opticalFlow* procedure, a multiresolution version of the differential Lucas and Kanade algorithm [15]. If $I(x, y, t)$ denote the gray-level of pixel (x, y) at date t, the method assumes the conservation of image intensities through time:

$$I(x, y, t) = I(x - u, y - v, 0)$$

where $\nu = (u, v)$ is the velocity vector. Under the hypothesis of small displacements, a Taylor expansion of this expression gives the gradient constraint equation:

```
Data: (I_1 ··· I_N) N EIT images
Result: Timeseries of parameters of the tracked AR
// Find the Largest connected component on the AR fuzzy map of I_1
AR_1=FindLargestCC(I_1^{AR})
// Compute the Center of mass of AR_1
G_1=ComputeCenterMass(AR_1)
for t=2 to N do
    // Compute the Optical flow between I_{t-1} and I_t
    F_{t-1}=opticalFlow(I_{t-1}, I_t)
    // Compute the New center of mass, given the velocity field
    G_t = Forecast(G_{t-1}, F_{t-1})
    // Find the Connected component in AR fuzzy map of I_t, centered
       on G_t
    AR_t = FindCC(G_t)
// Timeseries analysis of regions AR_1 ··· AR_t
return Timeseries(AR_1 ··· AR_N)
```

Fig. 3. Active region tracking

$$\nabla I(x,y,t)^T \nu + \frac{\partial I}{\partial t}(x,y,t) = 0 \tag{3}$$

Equation (3) allows to compute the projection of ν in the direction of ∇I, and the other component of ν is found by regularizing the estimation of the vector field, through a weighted least squares fit of (3) to a constant model for ν in each of small spatial neighborhood Ω:

$$Min \sum_{(x,y)\in\Omega} W^2(x,y)\left[\nabla I(x,y,t)^T \nu + \frac{\partial I}{\partial t}(x,y,t)\right]^2 \tag{4}$$

where $W(x,y)$ denotes a window function that gives more influence to constraints at the center of the neighborhood than those at the surroundings. The solution of (4) is given by solving

$$A^T W^2 A \nu = A^T W^2 b$$

where for n points $(x_i, y_i) \in \Omega$ at time t

$$\begin{aligned} A &= (\nabla I(x_1,y_1,t) \cdots \nabla I(x_n,y_n,t))^T \\ W &= diag(W(x_1,y_1) \cdots W(x_n,y_n)) \\ b &= \left(-\frac{\partial I}{\partial t}(x_1,y_1,t) \cdots - \frac{\partial I}{\partial t}(x_n,y_n,t)\right)^T \end{aligned}$$

A classical calculus of linear algebra directly gives $\nu = (A^T W^2 A)^{-1} A^T W^2 b$.

In this work, we applied a multiresolution version of this algorithm : the images were downsampled to a given lowest resolution, then the optical flow algorithm was computed for this resolution, and serves as an initialization for the computation of optical flow at the next resolution. This process was iteratively applied until the initial resolution was reached. This allows a coarse-to-fine estimation of velocities. This procedure is simple and fast, and hence allows for a real-time tracking of AR.

Although we can suppose here that because of the slow motion between I_{t-1} and I_t, G_t will lie in the trace of AR_{t-1} in I_t (and thus a region growing technique may be sufficient, directly starting from G_t in I_t), we use the optical flow for handling non successive images I_t and $I_{t+j}, j >> 1$, but also for computing some velocity parameters of the active regions such as the magnitude, the phase, etc, and to allow the tracking of any solar feature, whatever its size (*cf.* section 3.3).

2.3 Quantifying Solar Features

Several quantitative indices can finally be computed on a given solar feature, given the previous segmentation. We investigate here both geometric and photometric (irradiance) indices for a solar feature S_t segmented from image I_t at time t:

- location L_t, given as as function of the latitude on the solar disc
- area $a_t = \int\int_{S_t} dxdy$,
- Integrated and mean intentities: $i_t = \int\int_{S_t} I(x,y,t)dxdy$ and $m(t) = i_t/a_t$
- fractal dimension, estimated using a box counting method

All of these numerical indices give relevant information on S_t, and more important, the analysis of the timeseries of these indices can reveal important facts on the birth, the evolution and the dead of solar features.

3 Results

3.1 Data

We apply our segmentation procedure on subsets of 1024×1024 EIT images taken from 14 February 1997 up till 30 April 2005, thus spanning more than 8 years of the 11-year solar cycle. During the 8 years period, there were two extended periods without data: from 25 June up to 12 October 1998, and during the whole month of January 1999. Almost each day during this period, EIT images taken with less than 30 min apart were considered. These images did not contain telemetry missing blocks, and were preprocessed using the standard *eit_prep* procedure of the *solar software (ssw)* library. Image intensities were moreover normalized by their median value.

3.2 First Example: Automatic Tracking of the Biggest Active Region

Active regions (AR) are areas on the Sun where magnetic fields emerge through the photosphere into the chromosphere and corona. Active regions are the source of intense solar flares and coronal mass ejections. Studying their birth, their evolution and their impact on total solar irradiance is of great importance for several applications, such as space weather.

We illustrate our method with the tracking and the quantification of the largest AR of the solar disc, during the first 15 days of August, 2000. Figure 4 presents an example on a sequence of images, taken from 2000-08-01 to 2000-08-10. Active Regions segmented from SPoCA are highlighted with red edges, the biggest one being labeled in white. From this segmentation, we computed and plotted several quantitative indices, and we illustrate the timeseries of area, maximum intensity and fractal dimension over the period showed in figure 4.

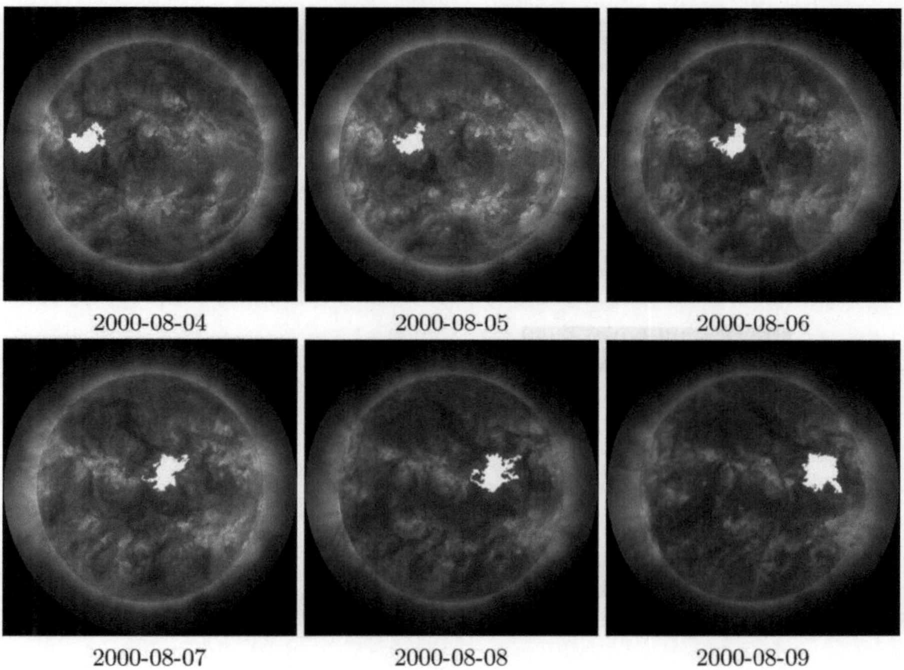

Fig. 4. Example of an AR tracking process. The tracking was performed on an active region detected on 2000-08-04, up to 2000-08-09.

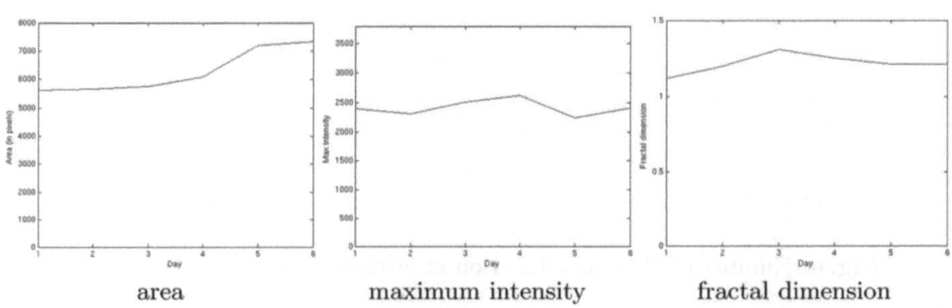

Fig. 5. Example of AR quantification indices for the period 2000-08-04 - 2000-08-09

Such results demonstrate the ability of the method to track and quantify active regions. It is now important not only to track such a solar feature over a solar rotation period, but also to record its birth and capture its evolution through several solar rotations. For this, we now plan to characterized solar features with their vector of quantification indices, and to recognize new features appearing on the limb, among the set of solar feature already been registered, using an unsupervised pattern recognition algorithm.

3.3 Second Example: Distribution of Coronal Bright Points

Coronal Bright Points (CBP) are of great importance for the analysis of the structure and dynamics of solar corona. They are identified as small and short-lived (< 2 days) coronal features with enhanced emission, mostly located in quiet-Sun regions and coronal holes. Figure 6 presents a segmentation of CBP of an image taken on February, 2nd, 1998. This image was chosen so as to compare the results with the one provided by [20] Several other indices can be computed from this analysis, such as N/S assymetry, timeseries of the number of CBP, intensity analysis of CBP...

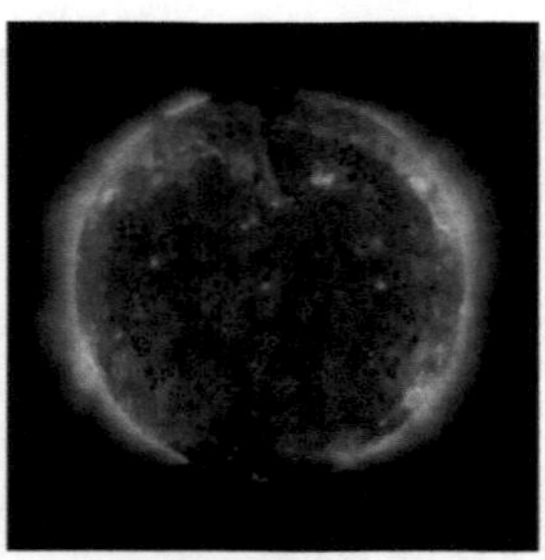

Sgmentation of CBP using 19.5 nm EIT image

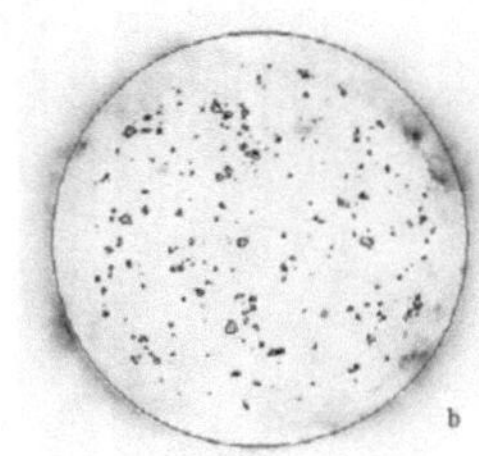

CBP [20]

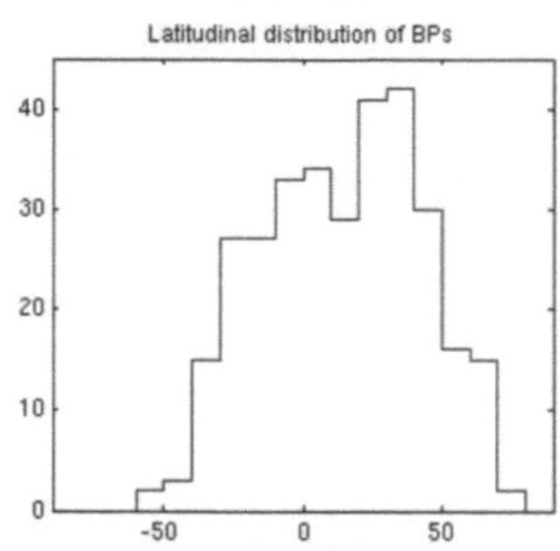

Number of CBP as a function of latitude

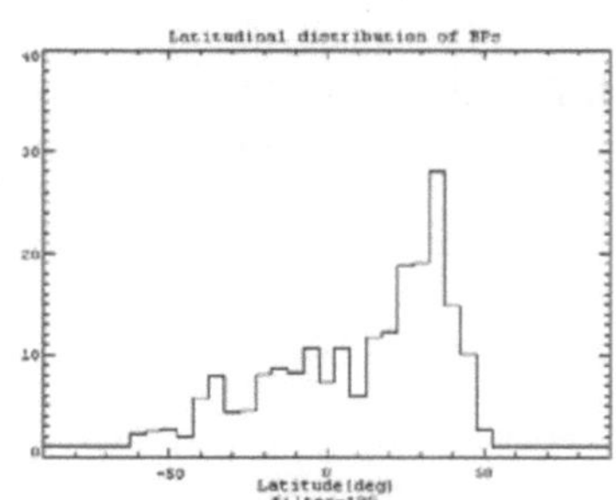

same from [20]

Fig. 6. Number of CBP as a function of latitude: comparison with [20]

4 Conclusion

We proposed in this article an image processing pipeline that segment, track and quantify solar features from a set of multispectral solar corona images, taken with eit EIT instrument. Based on a validated segmentation scheme, the method is fully described and illustrated on two preliminary studies: the automatic tracking of Active Regions from EIT images taken during solar cycle 23, and the analysis of spatial distribution of coronal bright points on the sular surface. The method is generic enough to allow the study of any solar feature, provided it can be segmented from EIT images or other sources. As stated above, our main perspective is to follow solar feature and to track their reappearance after a solar rotation S. We plan to use the quantification indices computed on a given solar feature F to characterize it and to find, over new solar features appearing on the solar limb at time $t + S/2$, the one closest to F. We also intend to implement a multiple activity region tracking, using a natural extension of our method.

References

1. Aboudarham, J., Scholl, I., Fuller, N.: Automatic detection and tracking of filaments for a solar feature database. Annales Geophysicae 26, 243–248 (2008)
2. Barra, V., Delouille, V., Hochedez, J.F.: Segmentation of extreme ultraviolet solar images via multichannel Fuzzy Clustering Algorithm. Adv. Space Res. 42, 917–925 (2008)
3. Barra, V., Delouille, V., Hochedez, J.F.: Fast and robust segmentation of solar EUV images: algorithm and results for solar cycle 23. A&A (submitted)
4. Benkhalil, A., Zharkova, V., Zharkov, S., Ipson, S.: Proceedings of the AISB 2003 Symposium on Biologically-inspired Machine Vision, Theory and Application, ed. S. L. N. in Computer Science, pp. 66–73 (2003)
5. Berrili, F., Moro, D.D., Russo, S.: Spatial clustering of photospheric structures. The Astrophysical Journal 632, 677–683 (2005)
6. Bezdek, J.C., Hall, L.O., Clark, M., Goldof, D., Clarke, L.P.: Medical image analysis with fuzzy models. Stat. Methods Med. Res. 6, 191–214 (1997)
7. Bornmann, P., Winkelman, D., Kohl, T.: Automated solar image processing for flare forecasting. In: Proc. of the solar terrestrial predictions workshop, Hitachi, Japan, pp. 23–27 (1996)
8. Brajsa, R., Whöl, H., Vrsnak, B., Ruzdjak, V., Clette, F., Hochedez, J.F.: Solar differential rotation determined by tracing coronal bright points in SOHO-EIT images. Astronomy and Astrophysics 374, 309–315 (2001)
9. Brajsa, R., Wöhl, H., Vrsnak, B., Ruzdjak, V., Clette, F., Hochedez, J.F., Verbanac, G., Temmer, M.: Spatial Distribution and North South Asymmetry of Coronal Bright Points from Mid-1998 to Mid-1999. Solar Physics 231, 29–44 (2005)
10. Delaboudinière, J.P., Artzner, G.E., Brunaud, J., et al.: EIT: Extreme-Ultraviolet Imaging Telescope for the SOHO Mission. Solar Physics 162, 291–312 (1995)
11. Dubois, D., Prade, H.: Possibility theory, an approach to the computerized processing of uncertainty. Plenum Press (1985)
12. Fuller, N., Aboudarham, J., Bentley, R.D.: Filament Recognition and Image Cleaning on Meudon Hα Spectroheliograms. Solar Physics 227, 61–75 (2005)

13. Hill, M., Castelli, V., Chu-Sheng, L.: Solarspire: querying temporal solar imagery by content. In: Proc. of the IEEE International Conference on Image Processing, pp. 834–837 (2001)
14. Krishnapuram, R., Keller, J.M.: A possibilistic approach to clustering. IEEE Trans. Fuzzy Sys. 1, 98–110 (1993)
15. Lucas, B.D., Kanade, T.: An iterative image registration technique with an application to stereovision. In: Proc. Imaging Understanding Workshop, pp. 121–130 (1981)
16. Nieniewski, M.: Segmentation of extreme ultraviolet (SOHO) sun images by means of watershed and region growing. In: Wilson, A. (ed.) Proc. of the SOHO 11 Symposium on From Solar Min to Max: Half a Solar Cycle with SOHO, Noordwijk, pp. 323–326 (2002)
17. Ortiz, A.: Solar cycle evolution of the contrast of small photospheric magnetic elements. Advances in Space Research 35, 350–360 (2005)
18. Pettauer, T., Brandt, P.: On novel methods to determine areas of sunspots from photoheliograms. Solar Physics 175, 197–203 (1997)
19. Qahwaji, R.: The Detection of Filaments in Solar Images. In: Proc. of the Solar Image Recognition Workshop, ed. Brussels, Belgium (2003)
20. Sattarov, I., Pevtsov, A., Karachek, N.: Proc. of the International Astronomical Union, pp. 665–666. Cambridge University Press, Cambridge (2004)
21. Sobotka, M., Brandt, P.N., Simon, G.W.: Fine structures in sunspots. I. Sizes and lifetimes of umbral dots. Astronomy and astrophysics 2, 682–688 (1997)
22. Steinegger, M., Bonet, J., Vazquez, M.: Simulation of seeing influences on the photometric determination of sunspot areas. Solar Physics 171, 303–330 (1997)
23. Steinegger, M., Bonet, J., Vazquez, M., Jimenez, A.: On the intensity thresholds of the network and plage regions. Solar Physics 177, 279–286 (1998)
24. Veronig, A., Steinegger, M., Otruba, W.: Automatic Image Segmentation and Feature Detection in solar Full-Disk Images. In: Wilson, N.E.P.D.A. (ed.) Proc. of the 1st Solar and Space Weather Euroconference, Noordwijk, p. 455 (2000)
25. Wagstaff, K., Rust, D.M., LaBonte, B.J., Bernasconi, P.N.: Automated Detection and Characterization of Solar Filaments and Sigmoids. In: Proc. of the Solar image recognition workshop, ed. Brussels, Belgium (2003)
26. Worden, J., Woods, T., Neupert, W., Delaboudiniere, J.: Evolution of Chromospheric Structures: How Chromospheric Structures Contribute to the Solar He ii 30.4 Nanometer Irradiance and Variability. The Astrophysical Journal, 965–975 (1999)
27. Zharkov, S., Zharkova, V., Ipson, S., Benkhalil, A.: Automated Recognition of Sunspots on the SOHO/MDI White Light Solar Images. In: Negoita, M.G., Howlett, R.J., Jain, L.C. (eds.) KES 2004. LNCS, vol. 3215, pp. 446–452. Springer, Heidelberg (2004)

Galaxy Decomposition in Multispectral Images Using Markov Chain Monte Carlo Algorithms

Benjamin Perret[1], Vincent Mazet[1], Christophe Collet[1], and Éric Slezak[2]

[1] LSIIT (UMR CNRS-Université de Strasbourg 7005), France
{perret,mazet,collet}@lsiit.u-strasbg.fr

[2] Laboratoire Cassiopée (UMR CNRS-Observatoire de la Côte d'Azur 6202), France
eric.slezak@oca.eu

Abstract. Astronomers still lack a multiwavelength analysis scheme for galaxy classification. In this paper we propose a way of analysing multispectral observations aiming at refining existing classifications with spectral information. We propose a global approach which consists of decomposing the galaxy into a parametric model using physically meaningful structures. Physical interpretation of the results will be straightforward even if the method is limited to regular galaxies. The proposed approach is fully automatic and performed using Markov Chain Monte Carlo (MCMC) algorithms. Evaluation on simulated and real 5-band images shows that this new method is robust and accurate.

Keywords: Bayesian inference, MCMC, multispectral image processing, galaxy classification.

1 Introduction

Galaxy classification is a necessary step in analysing and then understanding the evolution of these objects in relation to their environment at different spatial scales. Current classifications rely mostly on the De Vaucouleurs scheme [1] which is an evolution of the original idea by Hubble. These classifications are based only on the visible aspect of galaxies and identifies five major classes: ellipticals, lenticulars, spirals with or without bar, and irregulars. Each class is characterized by the presence, with different strengths, of physical structures such as a central bright bulge, an extended fainter disc, spiral arms,... and each class and the intermediate cases are themselves divided into finer stages.

Nowadays wide astronomical image surveys provide huge amount of multiwavelength data. For example, the Sloan Digital Sky Survey (SDSS[1]) has already produced more than 15 Tb of 5-band images. Nevertheless, most classifications still do not take advantage of colour information, although this information gives important clues on galaxy evolution allowing astronomers to estimate the star formation history, the current amount of dust, etc. This observation motivates the research of a more efficient classification including spectral information over all available bands. Moreover due to the quantity of available data (more than

[1] http://www.sdss.org/

A.-B. Salberg, J.Y. Hardeberg, and R. Jenssen (Eds.): SCIA 2009, LNCS 5575, pp. 209–218, 2009.

930,000 galaxies for the SDSS), it appears relevant to use an automatic and unsupervised method.

Two kinds of methods have been proposed to automatically classify galaxies following the Hubble scheme. The first one measures galaxy features directly on the image (*e.g.* symmetry index [2], Pétrosian radius [3], concentration index [4], *clumpiness* [5], ...). The second one is based on decomposition techniques (shapelets [6], the basis extracted with principal component analysis [7], and the pseudo basis modelling of the physical structures: bulge and disc [8]). Parameters extracted from these methods are then used as the input to a traditional classifier such as a support vector machine [9], a multi layer perceptron [10] or a Gaussian mixture model [6].

These methods are now able to reach a good classification efficiency (equal to the experts' agreement rate) for major classes [7]. Some attempts have been made to use decomposition into shapelets [11] or feature measurement methods [12] on multispectral data by processing images band by band. Fusion of spectral information is then performed by the classifier. But the lack of physical meaning of data used as inputs for the classifiers makes results hard to interpret. To avoid this problem we propose to extend the decomposition method using physical structures to multiwavelength data. This way we expect that the interpretation of new classes will be straightforward.

In this context, three 2D galaxy decomposition methods are publicly available. Gim2D [13] performs bulge and disc decomposition of distant galaxies using MCMC methods, making it robust but slow. Budda [14] handles bulge, disc, and stellar bar, while Galfit [15] handles any composition of structures using various brightness profiles. Both of them are based on deterministic algorithms which are fast but sensitive to local minima. Because these methods cannot handle multispectral data, we propose a new decomposition algorithm. This works with multispectral data and any parametric structures. Moreover, the use of MCMC methods makes it robust and allows it to work in a fully automated way.

The paper is organized as follows. In Sec. 2, we extend current models to multispectral images. Then, we present in Sec. 3 the Bayesian approach and a suitable MCMC algorithm to estimate model parameters from observations. The first results on simulated and raw images are discussed in Sec. 4. Finally some conclusions and perspectives are drawn in Sec. 5.

2 Galaxy Model

2.1 Decomposition into Structures

It is widely accepted by astronomers that spiral galaxies for instance can be decomposed into physically significant structures such as bulge, disc, stellar bar and spiral arms (Fig. 4, first column). Each structure has its own particular shape, populations of stars and dynamic. The bulge is a spheroidal population of mostly old red stars located in the centre of the galaxy. The disc is a planar structure with different scale heights which includes most of the gas and dust if any and populations of stars of various ages and colour from old red to younger and bluer ones. The stellar bar is an elongated structure composed of old red stars across the galaxy centre. Finally, spiral arms are over-bright regions in the disc that are the principal regions of star formation. The visible aspect of these structures are the fundamental criterion in the Hubble classification. It is noteworthy that this model only concerns regular galaxies and that no model for irregular or peculiar galaxies is available.

We only consider in this paper bulge, disc, and stellar bar. Spiral arms are not included because no mathematical model including both shape and brightness informations is available; we are working at finding such a suitable model.

2.2 Structure Model

We propose in this section a multispectral model for bulge, disc, and stellar bar. These structures rely on the following components: a generalized ellipse (also known as super ellipse) is used as a shape descriptor and a Sérsic law is used for the brightness profile [16]. These two descriptors are flexible enough to describe the three structures.

The major axis r of a generalized ellipse centred at the origin with axis parallel to coordinate axis and passing trough point $(x, y) \in \mathbb{R}^2$ is given by:

$$r(x, y) = \left(|x|^{c+2} + \left|\frac{y}{e}\right|^{c+2}\right)^{\frac{1}{c+2}} \tag{1}$$

where e is the ratio of the minor to the major axis and c controls the misshapenness: if $c = 0$ the generalized ellipse reduces to a simple ellipse, if $c < 0$ the ellipse is said to be *disky* and if $c > 0$ the ellipse is said to be *boxy* (Fig. 1). Three more parameters are needed to complete shape information: the centre (c_x, c_y) and the position angle α between abscissa axis and major axis.

The Sérsic law [16] is generally used to model the brightness profile. It is a generalization of the traditional exponential and De Vaucouleurs laws usually used to model disc and bulge brightness profiles. Its high flexibility allows it to vary continuously from a nearly flat curve to a very piked one (Fig. 2). The brightness at major axis r is given by:

$$I(r) = I\,\mathrm{e}^{-k_n\left(\left(\frac{r}{R}\right)^{\frac{1}{n}} - 1\right)} \tag{2}$$

where R is the effective radius, n is the Sérsic index, and I the brightness at the effective radius. k_n is an auxiliary function such that $\Gamma(2n) = 2\gamma(2n, k_n)$ to ensure that half of the total flux is contained in the effective radius (Γ and γ are respectively the complete and incomplete gamma function).

Then, the brightness at pixel (x, y) is given by:

$$F(x, y) = (F_1(x, y), \ldots, F_B(x, y)) \tag{3}$$

with B the number of bands and the brightness in band b is defined as:

$$F_b(x, y) = I_b\,\mathrm{e}^{-k_{n_b}\left(\left(\frac{r(x,y)}{R_b}\right)^{\frac{1}{n_b}} - 1\right)} \tag{4}$$

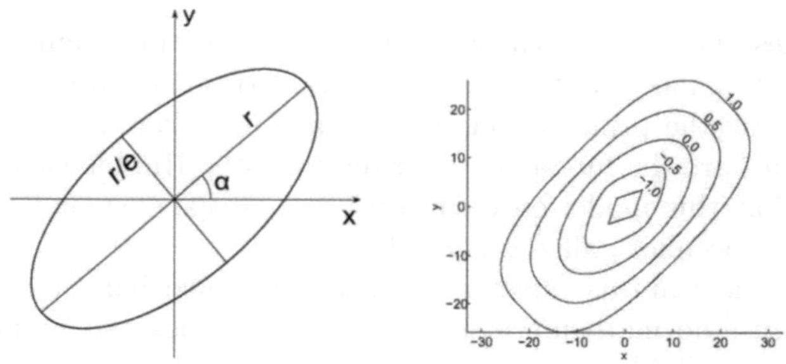

Fig. 1. Left: a simple ellipse with position angle α, major axis r and minor axis r/e. Right: generalized ellipse with variations of parameter c (displayed near each ellipse).

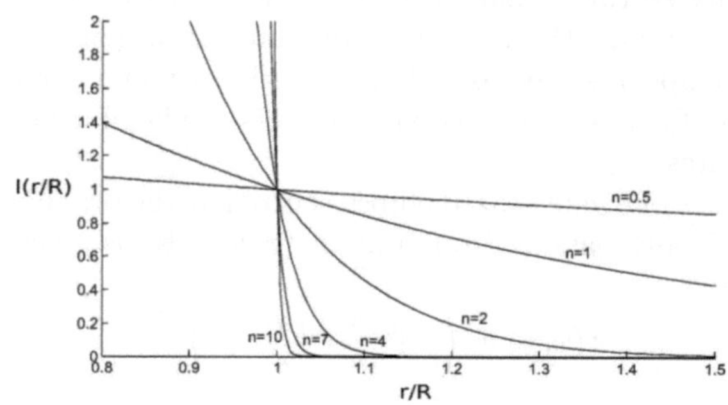

Fig. 2. The Sérsic law for different Sérsic index n. $n = 0.5$ yields a Gaussian, $n = 1$ yields an exponential profile and for $n = 4$ we obtain the De Vaucouleurs profile.

As each structure is supposed to represent a particular population of stars and galactic environment, we also assume that shape parameters do not vary between bands. This strong assumption seems to be verified in observations suggesting that shape variations between bands is negligible compared with deviation induced by noise. Moreover, this assumption reduces significantly the number of unknowns. The stellar bar has one more parameter which is the cut-off radius R_{max}; its brightness is zero beyond this radius. For the bulge (respectively the stellar bar), all Sérsic parameters are free which leads to a total number of $5+3B$ (respectively $6+3B$) unknowns. For the disc, parameter c is set to zero and Sérsic index is set to one leading to $4+2B$ free parameters. Finally, we assume that the centre is identical for all structures yielding a total of $11+8B$ unknowns.

2.3 Observation Model

Atmospheric distortions can be approximated by a spatial convolution with a Point Spread Function (PSF) H given as a parametric function or an image. Other noises are a composition of several sources and will be approximated by a Gaussian noise $\mathcal{N}(0, \Sigma)$. Matrix Σ and PSF H are not estimated as they can be measured using a deterministic procedure. Let Y be the observations and e the noise, we then have:

$$Y = Hm + e \quad \text{with} \quad m = F_{\mathfrak{B}} + F_{\mathfrak{D}} + F_{\mathfrak{Ba}} \tag{5}$$

with $\mathfrak{B}$, $\mathfrak{D}$, and $\mathfrak{B}a$ denoting respectively the bulge, the disc, and the stellar bar.

3 Bayesian Model and Monte Carlo Sampling

The problem being clearly ill-posed, we adopt a Bayesian approach. Priors assigned to each parameter are summarized in Table 1; they were determined from literature when possible and empirically otherwise. Indeed experts are able to determine limits for parameters but no further information is available: that is why Probability Density Functions (pdf) of chosen priors are uniformly distributed. However we expect to be able to determine more informative priors in future work. The posterior reads then:

$$P(\phi|Y) = \frac{1}{(2\pi)^{\frac{N}{2}} \det(\Sigma)^{\frac{1}{2}}} e^{-\frac{1}{2}(Y-Hm)^T \Sigma^{-1} (Y-Hm)} P(\phi) \tag{6}$$

where $P(\phi)$ denotes the priors and ϕ the unknowns. Due to its high dimensionality it is intractable to characterize the posterior pdf with sufficient accuracy. Instead, we aim at finding the Maximum A Posteriori (MAP).

Table 1. Parameters and their priors. All proposal distributions are Gaussians whose covariance matrix (or deviation for scalars) are given in the last column.

Structure	Parameter	Prior Support	Algorithm
$\mathfrak{B}$, $\mathfrak{B}a$, $\mathfrak{D}$	centre (c_x, c_y)	Image domain	RWHM with $\begin{pmatrix} 1 & 0 \\ 0 & 1 \end{pmatrix}$
$\mathfrak{B}$	major to minor axis (e)	$[1; 10]$	RWHM with 1
	position angle (α)	$[0; 2\pi]$	RWHM with 0.5
	ellipse misshapenness (c)	$[-0.5; 1]$	RWHM with 0.1
	brightness factor (I)	$\mathbb{R}^+$	direct with $\mathcal{N}^+(\mu, \sigma^2)$
	radius (R)	$[0; 200]$	ADHM with $\begin{pmatrix} 0.16 & -0.02 \\ -0.02 & 0.01 \end{pmatrix}$
	Sérsic index (n)	$[1; 10]$	
$\mathfrak{D}$	major to minor axis (e)	$[1; 10]$	RWHM with 0.2
	position angle (α)	$[0; 2\pi]$	RWHM with 0.5
	brightness factor (I)	$\mathbb{R}^+$	direct with $\mathcal{N}^+(\mu, \sigma^2)$
	radius (R)	$[0; 200]$	RWHM with 1
$\mathfrak{B}a$	major to minor axis (e)	$[4; 10]$	RWHM with 1
	position angle (α)	$[0; 2\pi]$	RWHM with 0.5
	ellipse misshapenness (c)	$[0.6; 2]$	RWHM with 0.1
	brightness factor (I)	$\mathbb{R}^+$	direct with $\mathcal{N}^+(\mu, \sigma^2)$
	radius (R)	$[0; 200]$	ADHM with $\begin{pmatrix} 0.16 & -0.02 \\ -0.02 & 0.01 \end{pmatrix}$
	Sérsic index (n)	$[0.5; 10]$	
	cut-off radius (R_{max})	$[10; 100]$	RWHM with 1

Because of the posterior complexity, the need for a robust algorithm leads us to choose MCMC methods [17]. MCMC algorithms are proven to converge in infinite time, and in practice the time needed to obtain a good estimation may be quite long. Thus several methods are used to improve convergence speed: simulated annealing, adaptive scale [18] and direction [19] Hastings Metropolis (HM) algorithm. As well, highly correlated parameters like Sérsic index and radius are sampled jointly to improve performance.

The main algorithm is a Gibbs sampler consisting in simulating variables separately according to their respective conditional posterior. One can note that the brightness factors posterior reduces to a truncated positive Gaussian $\mathcal{N}^+\left(\mu,\sigma^2\right)$ which can be efficiently sampled using an accept-reject algorithm [20]. Other variables are generated using the HM algorithm.

Some are generated with a Random Walk HM (RWHM) algorithm whose proposal is a Gaussian. At each iteration a random move from the current value is proposed. The proposed value is accepted or rejected with respect to the posterior ratio with the current value. The parameters of the proposal have been chosen by examining several empirical posterior distributions to find preferred directions and optimal scale. Sometimes the posterior is very sensitive to input data and no preferred directions can be found. In this case we decided to use the Adaptive Direction HM (ADHM). ADHM algorithm uses a sample of already simulated points to find preferred directions. As it needs a group of points to start with we choose to initialize the algorithm using simple RWHM. When enough points have been simulated by RWHM, the ADHM algorithm takes over. Algorithm and parameters of proposal distributions are summarized in Table 1.

Also, parameters I_b, R_b, and n_b are jointly simulated. R_b, n_b are first sampled according to $P\left(R_b, n_b \mid \phi_{\setminus\{R_b,n_b,I_b\}}\right)$ where I_b has been integrated and then I_b is sampled [21]. Indeed, the posterior can be decomposed in:

$$P\left(R_b, n_b, I_b \mid \phi_{\setminus\{R_b,n_b,I_b\}}, Y\right) = P\left(R_b, n_b \mid \phi_{\setminus\{R_b,n_b,I_b\}}, Y\right) P\left(I_b \mid \phi_{\setminus\{I_b\}}, Y\right) \tag{7}$$

4 Validation and Results

We measured two values for each parameter: the MAP and the variance of the chain in the last iterations. The latter gives an estimation of the uncertainty on the estimated value. A high variance can have different interpretations. In case of an observation with a low SNR, the variance naturally increases. But the variance can also be high when a parameter is not relevant. For example, the position angle is significant if the structure is not circular, the radius is also significant if the brightness is strong enough. We have also checked visually the residual image (the difference between the observation and the simulated image) which should contain only noise and non modelled structures.

Parameters are initialized by generating random variables according to their priors. This procedure ensures that the algorithm is robust so that it will not be fooled by a bad initialisation, even if the burn-in period of the Gibbs sampler is quite long (about 1,500 iterations corresponding to 1.5 hours).

4.1 Test on Simulated Images

We have validated the procedure on simulated images to test the ability of the algorithm to recover input parameters. The results showed that the algorithm is able to provide a solution leading to a residual image containing only noise (Fig. 3). Some parameters like elongation, position angle, or centre are retrieved with a very good precision (relative error less than 0.1%). On the other hand, Sérsic parameters are harder to estimate. Thanks to the extension of the disc, its radius and its brightness are estimated with a relative error of less than 5%. For the bulge and the stellar bar, the situation is complex because information is held by only a few pixels and an error in the estimation of Sérsic parametres does not lead to a high variation in the likelihood. Although the relative error increases to 20%, the errors seem to compensate each other.

Another problem is the evaluation of the presence of a given structure. Because the algorithm seeks at minimizing the residual, all the structures are always used. This can lead to solutions where structures have no physical significance. Therefore, we tried to introduce a Bernoulli variable coding the structure occurrence. Unfortunately, we were not able to determine a physically significant Bernoulli parameter. Instead we could use a pre- or post-processing method to determine the presence of each structure. These questions are highly linked to the astrophysical meaning of the structures we are modelling and we have to ask ourselves why some structures detected by the algorithm should in fact not be used. As claimed before, we need to define more informative joint priors.

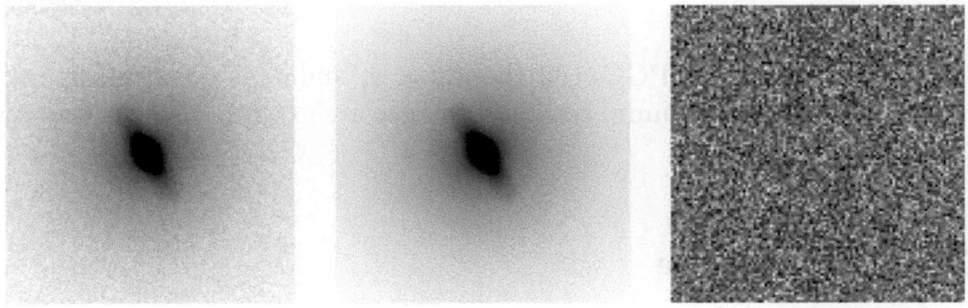

Fig. 3. Example of estimation on a simulated image (only one band on five is shown). Left: simulated galaxy with a bulge, a disc and a stellar bar. Centre: estimation. Right: residual. Images are given in inverse gray scale with enhanced contrast.

4.2 Test on Real Images

We have performed tests on about 30 images extracted from the EFIGI database [7] which is composed of thousands of galaxy images extracted from the SDSS. Images are centred on the galaxy but may contain other objects (stars, galaxies, artefacts, ...). Experiments showed that the algorithm performs well as long as no other bright object is present in the image (see Fig. 4 for example). As there is no ground truth available on real data we compared the results of our algorithm on monospectral images with those provided by Galfit. This shows a very good agreement since Galfit estimations are within the confidence interval proposed by our method.

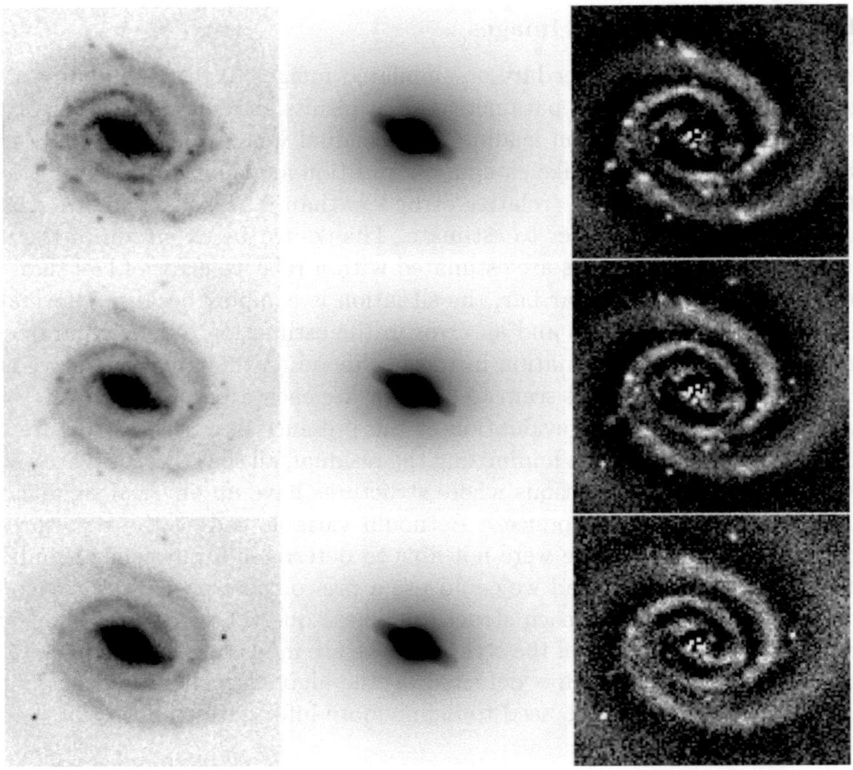

Fig. 4. Left column: galaxy PGC2182 (bands g, r, and i) is a barred spiral. Centre column: estimation. Right column: residual. Images are given in inverse gray scale with enhanced contrast.

4.3 Computation Time

Most of the computation time is used to evaluate the likelihood. Each time a parameter is modified, this implies the recomputation of the brightness of each affected structure for all pixels. Processing 1,000 iterations on a 5-band image of 250×250 pixels takes about 1 hour with a Java code running on an Intel Core 2 processor (2,66 GHz). We are exploring several ways to improve performance such as providing a good initialisation using fast algorithms or finely tuning the algorithm to simplify exploration of the posterior pdf.

5 Conclusion

We have proposed an extension of the traditional bulge, disc, stellar bar decomposition of galaxies to multiwavelength images and an automatic estimation process based on Bayesian inference and MCMC methods. We aim at using the decomposition results to provide an extension of the Hubble's classification to multispectral data. The proposed approach decomposes multiwavelength observations in a global way. The chosen model relies on some physically significant structures and can be extended with other structures such as spiral arms. In agreement with the experts, some parameters are identical in every band while others are specific to each band. The algorithm is non-supervised in order to obtain a fully automatic method. The model and estimation process have been validated on simulated and real images.

We are currently enriching the model with a parametric multispectral description of spiral arms. Other important work being carried out with experts is to determine joint priors that would ensure the significance of all parameters. Finally we are looking for an efficient initialisation procedure that would greatly increase convergence speed and open the way to a fast and fully unsupervised algorithm for multiband galaxy classification.

Acknowledgements

We would like to thank É. Bertin from the Institut d'Astrophysique de Paris for giving us a full access to the EFIGI image database.

References

1. De Vaucouleurs, G.: Classification and Morphology of External Galaxies. Handbuch der Physik 53, 275 (1959)
2. Yagi, M., Nakamura, Y., Doi, M., Shimasaku, K., Okamura, S.: Morphological classification of nearby galaxies based on asymmetry and luminosity concentration. Monthly Notices of Roy. Astr. Soc. 368, 211–220 (2006)
3. Petrosian, V.: Surface brightness and evolution of galaxies. Astrophys. J. Letters 209, L1–L5 (1976)
4. Abraham, R.G., Valdes, F., Yee, H.K.C., van den Bergh, S.: The morphologies of distant galaxies. 1: an automated classification system. Astrophys. J. 432, 75–90 (1994)
5. Conselice, C.J.: The Relationship between Stellar Light Distributions of Galaxies and Their Formation Histories. Astrophys. J. Suppl. S. 147, 1–28 (2003)
6. Kelly, B.C., McKa, T.A.: Morphological Classification of Galaxies by Shapelet Decomposition in the Sloan Digital Sky Survey. Astron. J. 127, 625–645 (2004)
7. Baillard, A., Bertin, E., Mellier, Y., McCracken, H.J., Géraud, T., Pelló, R., Leborgne, F., Fouqué, P.: Project EFIGI: Automatic Classification of Galaxies. In: Astron. Soc. Pac. Conf. ADASS XV, vol. 351, p. 236 (2006)
8. Allen, P.D., Driver, S.P., Graham, A.W., Cameron, E., Liske, J., de Propris, R.: The Millennium Galaxy Catalogue: bulge-disc decomposition of 10095 nearby galaxies. Monthly Notices of Roy. Astr. Soc. 371, 2–18 (2006)
9. Tsalmantza, P., Kontizas, M., Bailer-Jones, C.A.L., Rocca-Volmerange, B., Korakitis, R., Kontizas, E., Livanou, E., Dapergolas, A., Bellas-Velidis, I., Vallenari, A., Fioc, M.: Towards a library of synthetic galaxy spectra and preliminary results of classification and parametrization of unresolved galaxies for Gaia: Astron. Astrophys. 470, 761–770 (2007)

10. Bazell, D.: Feature relevance in morphological galaxy classification. Monthly Notices of Roy. Astr. Soc. 316, 519–528 (2000)
11. Kelly, B.C., McKay, T.A.: Morphological Classification of Galaxies by Shapelet Decomposition in the Sloan Digital Sky Survey. II. Multiwavelength Classification. Astron. J. 129, 1287–1310 (2005)
12. Lauger, S., Burgarella, D., Buat, V.: Spectro-morphology of galaxies: A multi-wavelength (UV-R) classification method. Astron. Astrophys. 434, 77–87 (2005)
13. Simard, L., Willmer, C.N.A., Vogt, N.P., Sarajedini, V.L., Phillips, A.C., Weiner, B.J., Koo, D.C., Im, M., Illingworth, G.D., Faber, S.M.: The DEEP Groth Strip Survey. II. Hubble Space Telescope Structural Parameters of Galaxies in the Groth Strip. Astrophys. J. Suppl. S. 142, 1–33 (2002)
14. de Souza, R.E., Gadotti, D.A., dos Anjos, S.: BUDDA: A New Two-dimensional Bulge/Disk Decomposition Code for Detailed Structural Analysis of Galaxies. Astrophys. J. Suppl. S. 153, 411–427 (2004)
15. Peng, C.Y., Ho, L.C., Impey, C.D., Rix, H.-W.: Detailed Structural Decomposition of Galaxy Images. Astron. J. 124, 266–293 (2002)
16. Sérsic, J.L.: Atlas de galaxias australes. Cordoba, Argentina: Observatorio Astronomico (1968)
17. Gilks, W.R., Richardson, S., Spiegelhalter, D.J.: Markov Chain Monte Carlo In Practice. Chapman & Hall/CRC, Washington (1996)
18. Gilks, W.R., Roberts, G.O., Sahu, S.K.: Adaptive Markov chain Monte Carlo through regeneration. J. Amer. Statistical Assoc. 93, 1045–1054 (1998)
19. Roberts, G.O., Gilks, W.R.: Convergence of adaptive direction sampling. J. of Multivariate Ana. 49, 287–298 (1994)
20. Mazet, V., Brie, D., Idier, J.: Simulation of positive normal variables using several proposal distributions. In: IEEE Workshop on Statistical Sig. Proc., pp. 37–42 (2005)
21. Devroye, L.: Non-Uniforme Random Variate Generation. Springer, New York (1986)

Head Pose Estimation from Passive Stereo Images

M.D. Breitenstein[1], J. Jensen[2], C. Høilund[2], T.B. Moeslund[2], and L. Van Gool[1]

[1] ETH Zurich, Switzerland
[2] Aalborg University, Denmark

Abstract. We present an algorithm to estimate the 3D pose (location and orientation) of a previously unseen face from low-quality range images. The algorithm generates many pose candidates from a signature to find the nose tip based on local shape, and then evaluates each candidate by computing an error function. Our algorithm incorporates 2D and 3D cues to make the system robust to low-quality range images acquired by passive stereo systems. It handles large pose variations (of $\pm 90°$ yaw and $\pm 45°$ pitch rotation) and facial variations due to expressions or accessories. For a maximally allowed error of 30°, the system achieves an accuracy of 83.6%.

1 Introduction

Head pose estimation is the problem of finding a human head in digital imagery and estimating its orientation. It can be required explicitly (*e.g.*, for gaze estimation in driver-attentiveness monitoring [11] or human-computer interaction [9]) as well as during a preprocessing step (*e.g.*, for face recognition or facial expression analysis).

A recent survey [12] identifies the assumptions of many state-of-the-art methods to simplify the pose estimation problem: small pose changes between frames (*i.e.*, continuous video input), manual initialization, no drift (*i.e.*, short duration of the input), 3D data, limited pose range, rotation around one single axis, permanent existence of facial features (*i.e.*, no partial occlusions and limited pose variation), previously seen persons, and synthetic data. The vast majority of previous approaches are based on 2D data and suffer from several of those limitations [12]. In general, purely image-based approaches are sensitive to illumination, shadows, lack of features (due to self-occlusion), and facial variations due to expressions or accessories like glasses and hats (*e.g.*, [14,6]). However, recent work indicates that some of these problems could be avoided by using *depth information* [2,15].

In this paper, we present a method for robust and automatic head pose estimation from *low-quality range images*. The algorithm relies only on 2.5D range images and the assumption that the nose of a head is visible in the image. Both assumptions are weak. Two color images (instead of one) are sufficient to compute depth information in a passive stereo system, thus, passive stereo imagery is

A.-B. Salberg, J.Y. Hardeberg, and R. Jenssen (Eds.): SCIA 2009, LNCS 5575, pp. 219–228, 2009.

cheap and relatively easy to obtain. Secondly, the nose is normally visible whenever the face is (in contrast to the corners of both eyes, as required by other methods, *e.g.*, [17]). Furthermore, our method particularly does not require any manual initialization, is robust to very large pose variations (of $\pm 90^\circ$ yaw and $\pm 45^\circ$ pitch rotation), and is identity-invariant.

Our algorithm is an extension of earlier work [1] that relies on high-quality range data (from an active stereo system) and does not work for low-quality passive stereo input. Unfortunately, the need for high-quality data is a strong limitation for real-world applications. With active stereo systems, users are often blinded by the bright light from a projector or suffer from unhealthy laser light. In this work, we generalize the original method and extend it for the use of low-quality range image data (captured, *e.g.*, by an off-the-shelf passive stereo system).

Our algorithm works as follows: First, a region of interest (ROI) is found in the color image to limit the area for depth reconstruction. Second, the resulting range image is interpolated and smoothed to close holes and remove noise. Then, the following steps are performed for each input range image. A pixel-based signature is computed to identify regions with high curvature, yielding a set of *candidates for the nose position*. From this set, we generate *head pose candidates*. To evaluate each candidate, we compute an error function that uses pre-computed *reference pose range images*, the ROI detector, motion direction estimation, and favors temporal consistency. Finally, the candidate with the lowest error yields the *final pose estimation* and a confidence value.

In comparison to our earlier work [1], we substantially changed the error function and added preprocessing steps. The presented algorithm works on single range images, making it possible to overcome drift and complete frame drop-outs in case of occlusions. The result is a system that can directly be used together with a low-cost stereo acquisition system (*e.g.*, passive stereo).

Although a few other face pose estimation algorithms use stereo input or multi-view images [8,17,21,10], most do not explicitly exploit depth information. Often, they need manual initialization, have limited pose range, or do not generalize to arbitrary faces. Instead of 2.5D range images, most systems using depth information are based on complete 3D information [7,4,3,20], the acquisition of which is complex and thus of limited use for most real-world applications. Most similar to our algorithm is the work of Seemann *et al.* [18], where the disparity and grey values are directly used in Neural Networks.

2 Range Image Acquisition and Preprocessing

Our head pose estimation algorithm is based on depth, color and intensity information. The data is extracted using an off-the-shelf stereo system (the Point Grey Bumblebee XB3 stereo system [16]), which provides color images with a resolution of 640×480 pixels. The applied stereo matching algorithm is a sum-of-absolute-differences correlation method that is relatively fast but produces mediocre range images. We speed it up further by limiting the allowed disparity range (*i.e.*, reducing the search region for the correlation).

(a) Input. (b) ROI only. (c) Interpolated.

Fig. 1. a) The range image, b) after background noise removal, c) after interpolation

The data is acquired in a common office setup. Two standard desk lamps are placed near the camera to ensure sufficient lighting. However, shadows and specularities on the face cause a considerable amount of noise and holes in the resulting depth images.

To enhance the quality of the range images, we remove background and foreground noise. The former can be seen in Fig. 1(a) in form of the large, isolated objects around the head. These objects originate from physical objects behind the user's head or due to erroneous 3D estimation. We handle such background noise by computing a region of interest (ROI) and ignoring all computed 3D points outside (see result in Fig. 1(b)). For this purpose, we apply a frontal 2D face detector [6]. As long as both eyes are visible, it detects the face reliably. When no face is detected we keep the ROI from the previous frame. In Fig. 1(b), foreground noise is visible, caused by the stereo matching algorithm. If the stereo algorithm fails to compute depth values, *e.g.*, in regions that are visible for one camera only, or due to specularities, holes appear in the resulting range image. We fill such holes by linear interpolation to remove large discontinuities on the surface (see Fig. 1(c)).

3 Finding Pose Candidates

The overall strategy of our algorithm is to find good candidates for the face pose (location and orientation) and then to evaluate them (see Sec 4). To find pose candidates, we try to locate the nose tip and estimate its orientation around object-centered rotation axes as local positional extremities. This step needs only local computations and thus can be parallelized for implementation on the GPU.

3.1 Finding Nose Tip Candidates

One strategy to find the nose tip is to compute the curvature of the surface, and then to search for local maxima (like previous methods, *e.g.*, [3]). However, curvature computation is very sensitive to noise, which is prominent especially in passively acquired range data. Additionally, nose detection in profile views based on curvature is not reliable because the curvature of the visible part of the nose significantly changes for different poses. Instead, our algorithm is based on a signature to approximate the local shape of the surface.

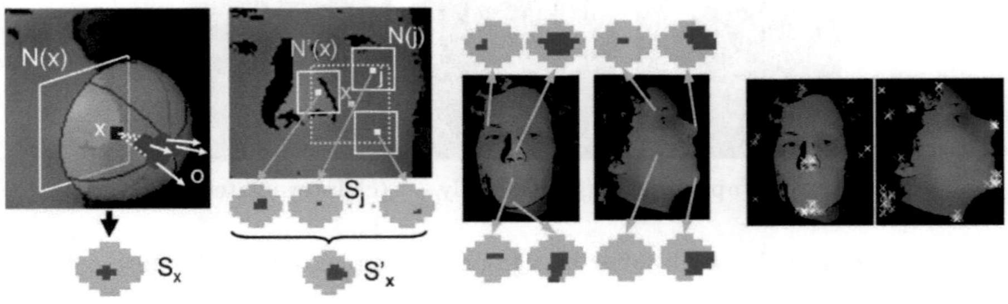

(a) (b) (c) (d)

Fig. 2. a) The *single signature* $S_{\boldsymbol{x}}$ is the set of orientations $\boldsymbol{o}$ for which the pixel's position $\boldsymbol{x}$ is a maximum along $\boldsymbol{o}$ compared to pixels in the neighborhood $N(\boldsymbol{x})$. b) Single signatures $S_{\boldsymbol{j}}$ of points $\boldsymbol{j}$ in $N'(\boldsymbol{x})$ are merged into the final signature $S'_{\boldsymbol{x}}$. c) The resulting signatures for different facial regions are similar across different poses. The signatures at nose and chin indicate high curvature areas compared to those at cheek and forehead. d) Nose candidates (white), generated based on selected signatures.

To locate the nose, we compute a 3D shape signature that is distinct for regions with high curvature. In a first step, we search for pixels $\boldsymbol{x}$ whose 3D position is a maximum along an orientation $\boldsymbol{o}$ compared to pixels in a local neighborhood $N(\boldsymbol{x})$ (see Fig. 2(a)). If such a pixel (called a *local directional maximum*) is found, a *single* signature $S_{\boldsymbol{x}}$ is stored (as a boolean matrix). In $S_{\boldsymbol{x}}$, one cell corresponds to one orientation $\boldsymbol{o}$, which is marked (red in Fig. 2(a)) if the pixel is a local directional maximum along this orientation. We only compute $S_{\boldsymbol{x}}$ for the orientations on the half sphere towards the camera, because we operate on range data (2.5D).

The resulting *single* signatures typically contain only a few marked orientations. Hence, they are not distinctive enough yet to reliably distinguish between different facial regions. Therefore, we merge single signatures $S_{\boldsymbol{j}}$ in a neighborhood $N'(\boldsymbol{x})$ to get signatures that are characteristic for the local shape of a whole region (see Fig. 2(b)).

Some resulting signatures for different facial areas are illustrated in Fig. 2(c). As can be seen, the resulting signatures reflect the characteristic local curvature of facial areas. The signatures are distinct for large, convex extremities, such as the nose tip and the chin. Their marked cells typically have a compact shape and cover many adjacent cells compared to those of facial regions that are flat, such as the cheek or forehead. Furthermore, the signature for a certain facial region looks similar if the head is rotated.

3.2 Generating Pose Candidates

Each *pose candidate* consists of the location of a *nose tip candidate* and its respective *orientation*. We select points as nose candidates based on the signatures using two criteria: first, the whole area around the point has a convex shape, *i.e.*, a large amount of the cells in the signature has to be marked. Secondly, the

(a) (b)

Fig. 3. The final output of the system: a) the range image with the estimated face pose and the signature of the best nose candidate, b) the color image with the output of the face ROI (red box), the nose ROI (green box), the KLT feature points (green), and the final estimation (white box). *(Best viewed in color)*

point is a "typical" point for the area represented by the signature (*i.e.*, it is in the center of the convex area). This is guaranteed if the cell in the center of all marked cells (*i.e.*, the *mean orientation*) is part of the pixel's *single* signature. Fig. 2(d) shows the resulting nose candidates based on the signatures of Fig. 2(c). Finally, the 3D positions and mean orientations of selected nose tip candidates form the set of final *head pose candidates* $\{P\}$.

4 Evaluating Pose Candidates

To evaluate each pose candidate P_{cur} corresponding to the nose candidate N_{cur}, we compute an error function. Finally, the candidate with the lowest error yields the final pose estimation:

$$P_{final} = \arg\min_{P_{cur}}(\alpha e_{nroi} + \beta e_{feature} + \gamma e_{temp} + \delta e_{align} + \theta e_{com}) \quad (1)$$

The error function consists of several error terms e (and their respective weights), which are described in the following subsections. The final error value can also be used as a (inverse) confidence value.

4.1 Error Term Based on Nose ROI

The face detector used in the preprocessing step (Sec. 2) yields a ROI containing the face. Our experiments have shown that the ROI is always centered close to the position of the nose in the image, independent of the head pose. Thus, we compute ROI_{nose}, a region of interest around the nose, using 50% of the size of the original ROI (see Fig. 3(b)). Since we are interested in pose candidates corresponding to nose candidates inside ROI_{nose}, we ignore all the other candidates.

In practice, instead of a hard pruning, we introduce a penalty value χ for candidates outside and no penalty value for candidates inside the nose ROI:

$$e_{nroi} = \begin{cases} \chi & \text{if } N_{cur} \notin ROI_{nose} \\ 0 & \text{otherwise} \end{cases} \quad (2)$$

This effectively prevents candidates outside of the nose ROI from being selected as long as there is one other candidate within the nose ROI.

4.2 Error Term Based on Average Feature Point Tracking

Usually, the poses in consecutive frames don't change dramatically. Therefore, we further evaluate pose candidates by checking the temporal correlation between two frames. The change of the nose position between the position in the last frame and the current candidate is defined as a motion vector V_{nose} and should be similar to the overall head movement in the current frame, denoted as V_{head}. However, this depends on the accuracy of the pose estimation in the previous frame. Therefore, we apply this check only if the confidence value of the last estimation is high (*i.e.*, if the respective final error value is below a threshold).

To implement this error term, we introduce the penalty function

$$e_{feature} = \begin{cases} |V_{head} - V_{nose}| & \text{if } |V_{head} - V_{nose}| > T_{feature} \\ 0 & \text{otherwise.} \end{cases} \quad (3)$$

We estimate V_{head} as the average displacement of a number of feature points from the previous to the current frame. Therefore, we use the Kanade-Lucas-Tomasi (KLT) tracker [19] on color images to find feature points and to track them (see Fig. 3(b)). The tracker is configured to select around 50 feature points. In case of an uncertain tracking result, the KLT tracker is reinitialized (*i.e.*, new feature points are identified). This is done if the number of feature points is too low (in our experiments, 15 was a good threshold).

4.3 Error Term Based on Temporal Pose Consistency

We introduce another error term e_{temp}, which punishes large differences between the estimated head pose P_{prev} from the last time step and the current pose candidate P_{cur}. Therefore, the term enforces temporal consistency. Again, this term is only introduced if the confidence value of the estimation in the last frame was high.

$$e_{temp} = \begin{cases} |P_{prev} - P_{cur}| & \text{if } |P_{prev} - P_{cur}| > T_{temp} \\ 0 & \text{otherwise.} \end{cases} \quad (4)$$

4.4 Error Term Based on Alignment Evaluation

The current pose candidate is further assessed by evaluating the alignment of the corresponding *reference pose range image*. Therefore, an average 3D face model was generated from the mean of an eigenvalue decomposition of laser scans from 97 male and 41 female adults (the subjects are not contained in our test dataset for the pose estimation). In an offline step, this average model (see Fig. 4(a)) is then rendered for all possible poses, and the resulting *reference pose range images* are directly stored on the graphics card. The possible number of poses depends on the memory size of the graphics card; in our case, we can

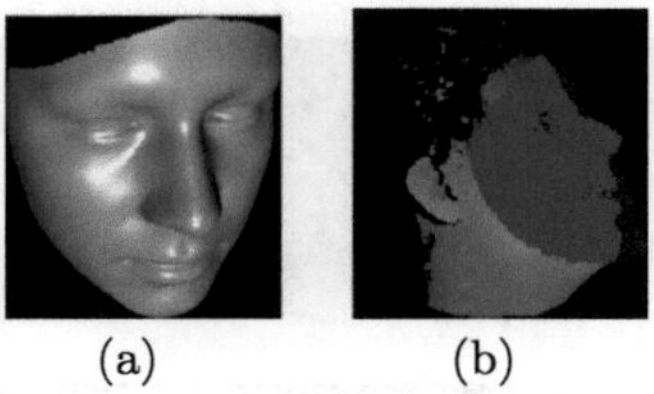

Fig. 4. a) The 3D model. b) An alignment of one reference image and the input.

store reference pose range images with a step size of 6 ° steps within ±90 ° yaw and ±45 ° pitch rotation. The error e_{align} consists of two error terms, the depth difference error e_d and the coverage error e_c

$$e_{align} = e_d(M_{\boldsymbol{O}}, I_{\boldsymbol{x}}) + \lambda \cdot e_c(M_{\boldsymbol{O}}, I_{\boldsymbol{x}}), \tag{5}$$

where e_{align} is identical with [1]; we refer to this paper for details. Because e_{align} only consists of pixel-wise operations, the alignment of all pose hypotheses is evaluated in parallel on the GPU.

The term e_d is the normalized sum of squared depth differences between reference range image $M_{\boldsymbol{O}}$ and input range image $I_{\boldsymbol{x}}$ for all foreground pixels (*i.e.*, pixels where a depth was captured), without taking into account the actual number of pixels. Hence, it does not penalize small overlaps between input and model (*e.g.*, the model could be perfectly aligned to the input but the overlap consists only of one pixel). Therefore, the second error term e_c favors those alignments where all pixels of the reference model fit to foreground pixels of the input image.

4.5 Error Term Based on Rough Head Pose Estimate

The KLT feature point tracker used for the error term $e_{feature}$ relies on motion, but does not help in static situations. Therefore, we introduce a penalty function that compares the current pose candidate P_{cur} with the result P_{com} from a simple head pose estimator.

We apply the idea of [13], where the center of the bounding box around the head (we use the ROI from preprocessing) is compared with the center of mass *com* of the face region. Therefore, the face pixels S are found using an ad-hoc skin color segmentation algorithm ($x_{r,g,b}$ are the values in the color channels)

$$S = \{\mathbf{x} | x_r > x_g \wedge x_r > x_b \wedge x_g > x_b \wedge x_r > 150 \wedge x_g > 100\}. \tag{6}$$

The error term e_{com} is then computed as follows:

$$e_{com} = \begin{cases} |P_{com} - P_{cur}| & \text{if } |P_{com} - P_{cur}| > T_{com} \\ 0 & \text{otherwise} \end{cases} \tag{7}$$

The pose estimation P_{com} is only valid for the horizontal direction and not very precise. However, it provides a rough estimate of the overall viewing direction that can be used to make the algorithm more robust.

Fig. 5. Pose estimation results: good (top), acceptable (middle), bad (bottom)

5 Experiments and Results

The different parameters for the algorithm are determined experimentally and set to $[T_{feature}, T_{temp}, T_{com}, \chi, \lambda] = [40, 25, 30, 10000, 10000]$. The weights of the error terms are chosen as $[\alpha, \beta, \gamma, \delta, \theta] = [1, 10, 50, 1, 20]$. None of them is particularly critical. To obtain test data with ground truth, a magnetic tracking system [5] is applied with a receiver mounted on a headband each test person wears. Each test person used to evaluate the system is first asked to look straight ahead to calibrate the magnetic tracking system for the ground truth. However, this initialization phase is not necessary for our algorithm. Then, each person is asked to freely move the head from frontal up to profile poses, while recording 200 frames. We use 15 test persons yielding 3000 frames in total[1].

We first evaluate the system qualitatively by inspecting each frame and judging whether the estimated pose (superimposed as illustrated in Fig. 5) is acceptable. We define acceptable as whether the estimated pose has correctly captured the general direction of the head. In Fig. 5 the first two rows are examples of acceptable poses in contrast to the last row. This test results in around 80% correctly estimated poses. In a second run, we looked at the ground truth for the acceptable frames and found that our instinctive notion of acceptable corresponds to a maximum pose error of about $\pm 30°$. We used this error condition in a quantitative test, where we compared the pose estimation in each frame with the ground truth. This results in a recognition rate of 83.6%.

We assess the isolated effects of the different error terms (Sec. 4) in Table 1, which shows the recognition rates when only the alignment term and one other

[1] Note that outliers (*e.g.*, a person looks backwards *w.r.t.*the calibration direction) are removed before testing. Therefore, the effect of some of the error terms is reduced due to missing frames, hence the recognition rate is lowered – but more realistic.

Table 1. The result of using different combinations of error terms

Error term	Error ≤ 15°	Error ≤ 30°
Alignment	29.0%	61.4%
Nose ROI	36.7%	75.7%
Feature	36.4%	68.7%
Temporal	37.7%	73.4%
Center of Mass	34.0%	66.4%
All	47.3%	**83.6%**

term is used. In [1], a success rate of 97.8% is reported, while this algorithm achieves only 29.0% in our setup. The main reason is the very bad quality of the passively acquired range images. In most error cases, a large part of the face is not reconstructed at all. Hence, special methods are required to account for the quality difference, as done in this work by using complementary error terms.

There are mainly two reasons for the algorithm to fail. First, when the nose ROI is incorrect, nose tip candidates far from the nose could be selected (especially those at the boundary, since such points are local directional maxima for many directions); see middle image of last row in Fig. 5. The nose ROI is incorrect when the face detector breaks for a longer time period (and the last accepted ROI is used). Secondly, if the depth reconstruction of the face surface is too flawed, the alignment evaluation will not be able to distinguish the different pose candidates correctly (see right and left image of the last row in Fig. 5). This is mostly the case if there are very large holes in the surface, which is mainly due to specularities or uniformly textured and colored regions.

The whole system runs with a frame-rate of several fps. However, it could be optimized for real-time performance, *e.g.*, by consistently using the GPU.

6 Conclusion

We presented an algorithm for estimating the pose of unseen faces from low-quality range images acquired by a passive stereo system. It is robust to very large pose variations and for facial variations. For a maximally allowed error of 30°, the system achieves an accuracy of 83.6%. For most applications from surveillance or human-computer interaction, such a coarse head orientation estimation system can be used directly for further processing.

The estimation errors are mostly caused by a bad depth reconstruction. Therefore, the simplest way to improve the accuracy would be to improve the quality of the range images. Although better reconstruction methods exist, there is a tradeoff between accuracy and speed. Further work will include experiments with different stereo reconstruction algorithms.

Acknowledgments. Supported by the EU project HERMES (IST-027110).

References

1. Breitenstein, M.D., Kuettel, D., Weise, T., Van Gool, L., Pfister, H.: Real-time face pose estimation from single range images. In: CVPR (2008)
2. Chang, K.I., Bowyer, K.W., Flynn, P.J.: An evaluation of multimodal 2D+3D face biometrics. PAMI 27(4), 619–624 (2005)
3. Chang, K.I., Bowyer, K.W., Flynn, P.J.: Multiple nose region matching for 3d face recognition under varying facial expression. PAMI 28(10), 1695–1700 (2006)
4. Colbry, D., Stockman, G., Jain, A.: Detection of anchor points for 3d face verification. In: A3DISS, CVPR Workshop (2005)
5. Fastrak, `http://www.polhemus.com`
6. Jones, M., Viola, P.: Fast multi-view face detection. Technical Report TR2003-096, Mitsubishi Electric Research Laboratories (2003)
7. Lu, X., Jain, A.K.: Automatic feature extraction for multiview 3D face recognition. In: FG (2006)
8. Matsumoto, Y., Zelinsky, A.: An algorithm for real-time stereo vision implementation of head pose and gaze direction measurement. In: FG (2000)
9. Morency, L.-P., Sidner, C., Lee, C., Darrell, T.: Head gestures for perceptual interfaces: The role of context in improving recognition. Artificial Intelligence 171(8-9) (2007)
10. Morency, L.-P., Sundberg, P., Darrell, T.: Pose estimation using 3D view-based eigenspaces. In: FG (2003)
11. Murphy-Chutorian, E., Doshi, A., Trivedi, M.M.: Head pose estimation for driver assistance systems: A robust algorithm and experimental evaluation. In: Intelligent Transportation Systems Conference (2007)
12. Murphy-Chutorian, E., Trivedi, M.M.: Head pose estimation in computer vision: A survey. PAMI (2008) (to appear)
13. Nasrollahi, K., Moeslund, T.: Face quality assessment system in video sequences. In: Workshop on Biometrics and Identity Management (2008)
14. Osadchy, M., Miller, M.L., LeCun, Y.: Synergistic face detection and pose estimation with energy-based models. In: NIPS (2005)
15. Phillips, P.J., Flynn, P.J., Scruggs, T., Bowyer, K.W., Chang, J., Hoffman, K., Marques, J., Min, J., Worek, W.: Overview of the face recognition grand challenge. In: CVPR (2005)
16. Point Grey Research, `http://www.ptgrey.com/products/bumblebee/index.html`
17. Sankaran, P., Gundimada, S., Tompkins, R.C., Asari, V.K.: Pose angle determination by face, eyes and nose localization. In: FRGC, CVPR Workshop (2005)
18. Seemann, E., Nickel, K., Stiefelhagen, R.: Head pose estimation using stereo vision for human-robot interaction. In: FG (2004)
19. Tomasi, C., Kanade, T.: Detection and tracking of point features. Technical report, Carnegie Mellon University (April 1991)
20. Xu, C., Tan, T., Wang, Y., Quan, L.: Combining local features for robust nose location in 3D facial data. Pattern Recognition Letters 27(13), 1487–1494 (2006)
21. Yao, J., Cham, W.K.: Efficient model-based linear head motion recovery from movies. In: CVPR (2004)

Multi-band Gradient Component Pattern (MGCP): A New Statistical Feature for Face Recognition

Yimo Guo[1,2], Jie Chen[1], Guoying Zhao[1], Matti Pietikäinen[1], and Zhengguang Xu[2]

[1] Machine Vision Group, Department of Electrical and Information Engineering, University of Oulu, P.O. Box 4500, FIN-90014, Finland
[2] School of Information Engineering, University of Science and Technology Beijing, Beijing, 100083, China

Abstract. A feature extraction method using multi-frequency bands is proposed for face recognition, named as the Multi-band Gradient Component Pattern (MGCP). The MGCP captures discriminative information from Gabor filter responses in virtue of an orthogonal gradient component analysis method, which is especially designed to encode energy variations of Gabor magnitude. Different from some well-known Gabor-based feature extraction methods, MGCP extracts geometry features from Gabor magnitudes in the orthogonal gradient space in a novel way. It is shown that such features encapsulate more discriminative information. The proposed method is evaluated by performing face recognition experiments on the FERET and FRGC ver 2.0 databases and compared with several state-of-the-art approaches. Experimental results demonstrate that MGCP achieves the highest recognition rate among all the compared methods, including some well-known Gabor-based methods.

1 Introduction

Face recognition receives much attention from both research and commercial communities, but it remains challenging in real applications. The main task of face recognition is to represent object appropriately for identification. A well designed representation method should extract discriminative information effectively and improve recognition performance. This depends on deep understanding of the object and recognition task itself. Especially, there are two problems involved: (i) what representation is desirable for pattern recognition; (ii) how to represent the information contained in both neighborhood and global structure. In the last decades, numerous face recognition methods and their improvements have been proposed. These methods can be generally divided into two categories: holistic matching methods and local matching methods. Some representative methods are Eigenfaces [1], Fisherfaces [2], Independent Component Analysis [3], Bayesian [4], Local Binary Pattern (LBP) [5,6], Gabor features [7,12,13], gradient magnitude and orientation maps [8], Elastic Bunch Graph Matching [9] and so on. All these methods exploit the idea to obtain features using an operator and build up a global representation or local neighborhood representation.

Recently, some Gabor-based methods that belong to local matching methods have been proposed, such as the local Gabor binary pattern (LGBPHS) [10], enhanced local

A.-B. Salberg, J.Y. Hardeberg, and R. Jenssen (Eds.): SCIA 2009, LNCS 5575, pp. 229–238, 2009.

Gabor binary pattern (ELGBP) [11] and the histogram of Gabor phase patterns (HGPP) [12]. LGBPHS and ELGBP explore information from Gabor magnitude, which is a commonly used part of the Gabor filter response, by applying local binary pattern to Gabor filter responses. Similarly, HGPP introduced LBP for further feature extraction from Gabor phase that was demonstrated to provide useful information. Although LBP is an efficient descriptor for image representation, it is good at capturing neighborhood relationships from original images in the spatial domain. To process multi-frequency bands responses using LBP would increase complexity and lose information.

Therefore, to improve the recognition performance and efficiency, we propose a new method to extract discriminative information especially from Gabor magnitude. Useful information would be extracted from Gabor filter responses in an elaborate way by making use of the characteristics of Gabor magnitude. In detail, based on Gabor function and gradient theory, we design a Gabor energy variation analysis method to extract discriminative information. This method encodes Gabor energy variations to represent images for face recognition. The gradient orientations are selected in a hierarchical fashion, which aims to improve the capability of capturing discriminative information from Gabor filter responses. The spatially enhanced representation is finally described as the combination of these histogram sequences at different scales and orientations. From experiments conducted on the FERET database and FRGC ver 2.0 database, our method is shown to be more powerful than many other methods, including some well-known Gabor-based methods.

The rest of this paper is organized as follows. In Section 2, the image representation method for face recognition is presented. Experiments and result analysis are reported in Section 3. Conclusions are drawn in Section 4.

2 Multi-band Gradient Component Pattern (MGCP)

Gabor filters have been widely used in pattern recognition because of their multi-scale, multi-orientation, multi-frequency and processing capability. Most of the proposed Gabor-based methods take advantage of Gabor magnitude to represent face images. Although Gabor phase was demonstrated to be a good compensation to the magnitude, information should be exploited elaborately from the phase in order to avoid the sensitivity to local variations [11]. Considering that the Gabor magnitude part varies slowly with spatial position and contains enough discriminative information for classification, we extract features from this part of Gabor filter responses. In detail, features are obtained from Gabor responses using an energy variation analysis method. The gradient component is adopted here because: (i) gradient magnitudes contain intensity variation information; (ii) gradient orientations of neighborhood pixels contain rich directional information and are insensitive to illumination and pose variations [15]. In this way, features are described as histogram sequences explored from Gabor filter responses at each scale and orientation.

2.1 Multi-frequency Bands Feature Extraction Method Using Gabor Filters

Gabor function is biologically inspired, since Gabor like receptive fields have been found in the visual cortex of primates [16]. It acts as low-level oriented edge and texture discriminator and is sensitive to different frequencies and scale information.

These characteristics raise considerable interests for researchers to extensively exploit its properties. Gabor wavelets are biologically motivated convolution kernels in the shape of plane waves restricted by a Gaussian envelope function [17]. The general form of a 2D Gabor wavelet is defined as:

$$\Psi_{u,v}(z) = \left(\|k_{u,v}\|^2/\sigma^2\right)\exp\left(-\|k_{u,v}\|^2\|z\|^2/2\sigma^2\right)\left[\exp(ik_{u,v}z) - \exp(-\sigma^2/2)\right], \quad (1)$$

where u and v define the orientation and scale of Gabor kernels. σ is a parameter to control the scale of Gaussian. $\vec{k}_{u,v}$ is a 2D wave vector whose magnitude and angle determine the scale and orientation of Gabor kernel respectively. In most cases, Gabor wavelets at five different scales $v:\{0,...4\}$ and eight orientations $u:\{0,...7\}$ are used [18,19,20]. The Gabor wavelet transformation of an image is the convolution of the image with a family of Gabor kernels, as defined by:

$$G_{u,v}(z) = I(z) * \Psi(z), \quad (2)$$

where $z = (x, y)$. The operator $*$ is the convolution operator. $G_{u,v}(z)$ is the convolution corresponding to Gabor kernels at different scales and orientations. The Gabor magnitude is defined as:

$$M_{u,v}(z) = \sqrt{\mathrm{Re}(G_{u,v}(z))^2 + \mathrm{Im}(G_{u,v}(z))^2}, \quad (3)$$

where $\mathrm{Re}(\cdot)$ and $\mathrm{Im}(\cdot)$ denote the real and imaginary part of Gabor transformed image respectively, as shown in Fig. 1. In this way, 40 Gabor magnitudes are calculated to form the representation. The visualization of Gabor magnitudes are shown in Fig. 2.

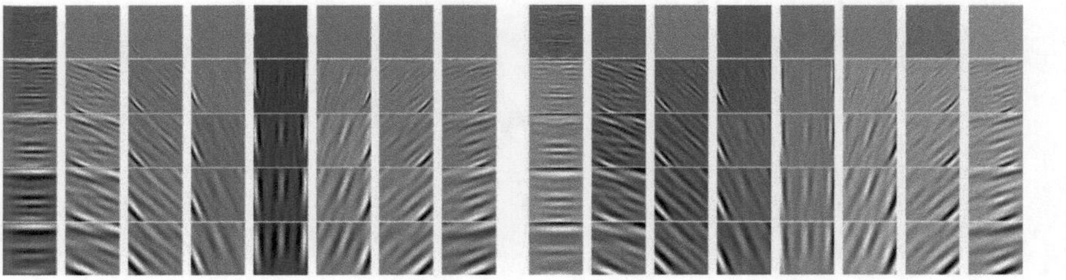

(a) (b)

Fig. 1. The visualization of a) the real part and b) imaginary part of a Gabor transformed image

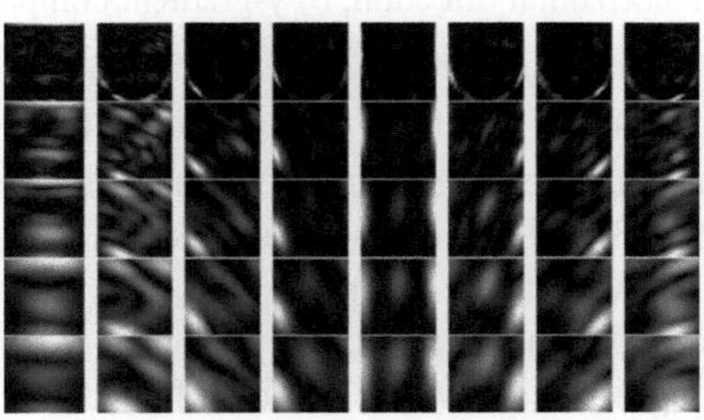

Fig. 2. The visualization of Gabor magnitudes

2.2 Orthogonal Gradient Component Analysis

There has been some recent work makes use of gradient information in object representation [21,22]. As Gabor magnitude part varies slowly with spatial position and embodies energy information, we explore Gabor gradient components for representation. Motivated by using the Three Orthogonal Planes to encode texture information [23], we select orthogonal orientations (horizontal and vertical) here. This is mainly because Gabor gradient is defined based on Gaussian function, which is not declining at exponential speed as in Gabor wavelets. These two orientations are selected as: (i) the gradient of orthogonal orientations could encode more variations with less correlation; (ii) less time is needed to calculate two orientations than in some other Gabor-based methods, such as LGBPHS and ELGBP, which calculate eight neighbors to capture discriminative information from Gabor magnitude.

Given an image $I(z)$, where $z=(x,y)$ indicates the pixel location. $G_{u,v}(z)$ is the convolution corresponding to the Gabor kernel at scale v and orientation u. The gradient of $G_{u,v}(z)$ is defined as:

$$\left|\nabla_d G_{u,v}(z)\right| = \left(\partial G_{u,v}/\partial x\right)\hat{i} + \left(\partial G_{u,v}/\partial y\right)\hat{j} \ . \qquad (4)$$

Equation 4 is the set of vectors pointing at appointed directions of increasing values of $G_{u,v}(z)$. The $\partial G_{u,v}/\partial x$ corresponds to differences in the horizontal (row) direction, while the $\partial G_{u,v}/\partial y$ corresponds to differences in the vertical (column) direction. The $x-$and $y-$ gradient components of Gabor filter responses are calculated at each scale and orientation. The gradient components are shown in Fig. 3.

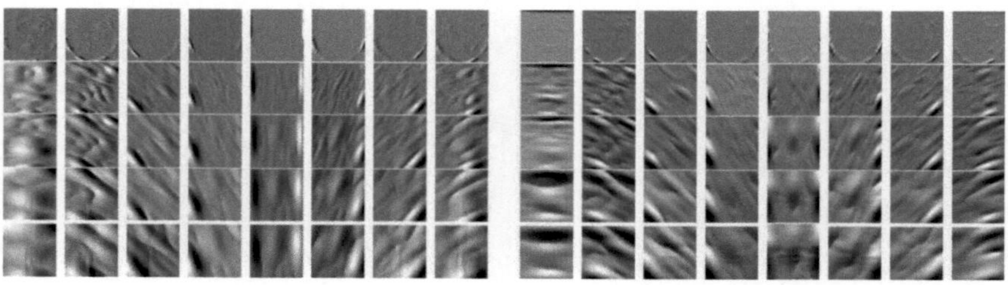

(a) (b)

Fig. 3. The gradient components of Gabor filter responses at different scales and orientations. a) x-gradient components in horizontal direction; b) y-gradient components in vertical direction.

The histograms (256 bins) of $x-$and $y-$gradient components of Gabor responses at different scales and orientations are calculated and concatenated to form the representation. From Equations 3 and 4, we can see that MGCP actually encodes the information of Gabor energy variations in orthogonal orientations, which contains very discriminative information as shown in Section 4.

Considering Gabor magnitude provides useful information for face recognition, we propose MGCP to encode Gabor energy variations for face representation. However, a single histogram suffers from losing spatial structure information. Therefore, images are decomposed into non-overlapping sub-regions, from which local features are extracted. To capture both the global and local information, all these histograms are concatenated to an extended histogram for each scale and orientation. Examples of concatenated histograms are illustrated in Fig. 4 (c) when images are divided into non-overlapping 4×4 sub-regions. The 4×4 decomposition will result in a little weak feature but can further demonstrate the performance of our method. Fig. 4 (b) illustrates the MGCP ($u=90$, $v=5.47$) of four face images for two subjects. The u and v are selected randomly. The capability of these discriminative patterns could be observed from histogram distances, listed in Table 1.

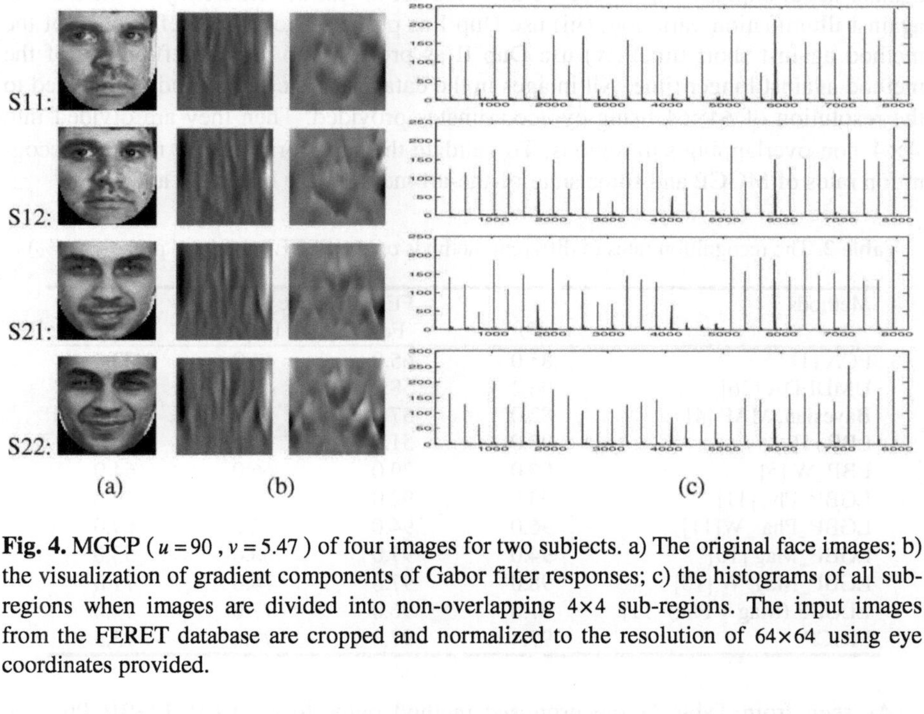

Fig. 4. MGCP ($u=90$, $v=5.47$) of four images for two subjects. a) The original face images; b) the visualization of gradient components of Gabor filter responses; c) the histograms of all sub-regions when images are divided into non-overlapping 4×4 sub-regions. The input images from the FERET database are cropped and normalized to the resolution of 64×64 using eye coordinates provided.

Table 1. The histogram distances of four images for two subjects using MGCP

Subjects	S11	S12	S21	S22
S11	0	4640	5226	5536
S12	--	0	4970	5266
S21	--	--	0	4708
S22	--	--	--	0

3 Experiments

The proposed method is tested on the FERET database and FRGC ver 2.0 database [24,25]. The classifier is the simplest classification scheme: nearest neighbour classifier in image space with Chi square statistics as the similarity measure.

3.1 Experiments on the FERET Database

To conduct experiments on the FERET database, we use the same Gallery and Probe sets as the standard FERET evaluation protocol. For the FERET database, we use Fa as gallery, which contains 1196 frontal images of 1196 subjects. The probe sets consist of Fb, Fc, Dup I and Dup II. Fb contains 1195 images of expression variations, Fc contains 194 images taken under different illumination conditions, Dup I has 722 images taken later in time and Dup II (a subset of Dup I) has 234 images taken at least one year after the corresponding Gallery images. Using Fa as the gallery, we design the following experiments: (i) use Fb as probe set to test the efficiency of the method against facial expression; (ii) use Fc as probe set to test the efficiency of the method against illumination variation; (iii) use Dup I as probe set to test the efficiency of the method against short time; (iv) use Dup II as probe set to test the efficiency of the method against longer time. All images in the database are cropped and normalized to the resolution of 64×64 using eye coordinates provided. Then they are divided into 4×4 non-overlapping sub-regions. To validate the superiority of our method, recognition rates of MGCP and some state-of-the-art methods are listed in Table 2.

Table 2. The recognition rates of different methods on the FERET database probe sets (%)

Methods	FERET Probe Sets			
	Fb	Fc	Dup I	Dup II
PCA [1]	85.0	65.0	44.0	22.0
UMDLDA [26]	96.2	58.8	47.2	20.9
Bayesian, MAP [4]	82.0	37.0	52.0	32.0
LBP [5]	93.0	51.0	61.0	50.0
LBP_W [5]	97.0	79.0	66.0	64.0
LGBP_Pha [11]	93.0	92.0	65.0	59.0
LGBP_Pha _W[11]	96.0	94.0	72.0	69.0
LGBP_Mag [10]	94.0	97.0	68.0	53.0
LGBP_Mag_W [10]	98.0	97.0	74.0	71.0
ELGBP (Mag + Pha) [11]	97.0	96.0	77.0	74.0
MGCP	97.4	97.3	77.8	73.5

As seen from Table 2, the proposed method outperforms LBP, LGBP_Pha and their corresponding methods with weights. The MGCP also outperforms LGBP_Mag that represents images using Gabor magnitude information. Moreover, from experimental results of Fa-X (X: Fc, Dup I and Dup II), MGCP without weights performs better than LGBP_Mag with weights. From experimental results of Fa-Y (Y: Fb, Fc and Dup I), MGCP performs even better than ELGBP that combines both the magnitude and phase patterns of Gabor filter responses.

3.2 Experiments on the FRGC Ver 2.0 Database

To further evaluate the performance of the proposed method, we conduct experiments on the FRGC version 2.0 database which is one of the most challenging databases [25]. The face images are normalized and cropped to the size of 120×120 using eye coordinates provided. Some samples are shown in Fig. 5.

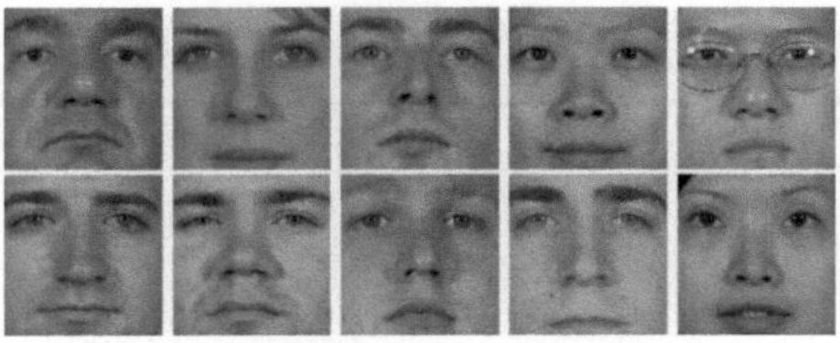

Fig. 5. Face images from FRGC 2.0 database

In FRGC 2.0 database, there are 12776 images taken from 222 subjects in the training set and 16028 images in the target set. We conduct Experiment 1 and Experiment 4 protocols to evaluate the performance of different approaches. In Experiment 1, there are 16028 query images taken under the controlled illumination condition. The goal of Experiment 1 is to test the basic recognition ability of approaches. In Experiment 4, there are 8014 query images taken under the uncontrolled illumination condition. Experiment 4 is the most challenging protocol in FRGC because the uncontrolled large illumination variations bring significant difficulties to achieve high recognition rate. The experimental results on the FRGC 2.0 database in Experiment 1 and 4 are evaluated by Receiving Operator Characteristics (ROC), which is face verification rate (FVR) versus false accept rate (FAR). Tables 3 and 4 list the performance of different approaches on face verification rate (FVR) at false accept rate (FAR) of 0.1% in Experiment 1 and 4.

From experimental results listed in Table 3, MGCP achieves the best performance, which demonstrates its basic abilities in face recognition. Table 4 exhibits results of MGCP and two well-known approaches: BEE Baseline and LBP. MGCP is also compared with some recently proposed methods and the results are listed in Table 5. The database used in experiments for Gabor + FLDA, LGBP, E-GV-LBP, GV-LBP-TOP are reported to be a subset of FRGC 2.0, while the whole database is used in experiments for UCS and MGCP. It is observed from Table 4 and 5 that MGCP could overcome uncontrolled condition variations effectively and improve face recognition performance.

Table 3. The FVR value of different approaches at FAR = 0.1% in Experiment 1 of the FRGC 2.0 database

Methods	FVR at FAR = 0.1% (in %)		
	ROC 1	ROC 2	ROC 3
BEE Baseline [25]	77.63	75.13	70.88
LBP [5]	86.24	83.84	79.72
MGCP	97.52	94.08	92.57

Table 4. The FVR value of different approaches at FAR = 0.1% in Experiment 4 of the FRGC 2.0 database

Methods	FVR at FAR = 0.1% (in %)		
	ROC 1	ROC 2	ROC 3
BEE Baseline [25]	17.13	15.22	13.98
LBP [5]	58.49	54.18	52.17
MGCP	76.08	75.79	74.41

Table 5. ROC 3 on the FRGC 2.0 in Experiment 4

Methods	ROC 3, FVR at FAR = 0.1% (in %)
BEE Baseline [25]	13.98
Gabor + FLDA [27]	48.84
LBP [27]	52.17
LGBP [27]	52.88
E-GV-LBP [27]	53.66
GV-LBP-TOP [27]	54.53
UCS [28]	69.92
MGCP	74.41

4 Conclusions

To extend traditional use of multi-band responses, the proposed feature extraction method encodes Gabor magnitude gradient component in an elaborate way, which is different from some previous Gabor-based methods that directly apply some proposed feature extraction methods on Gabor filter responses. Especially, the gradient orientations are organized in a hierarchical fashion. Experimental results show that orthogonal orientations could improve the capability to capture energy variations of Gabor responses. The spatial histograms of multi-frequency bands gradient component pattern at each scale and orientation are finally concatenated to represent face images, which could encode both the structure and local information. From experimental results conducted on the FERET and FRGC 2.0, it is observed that the proposed method is insensitive to many variations, such as illumination and pose. The experimental results also demonstrate its efficiency and validity in face recognition.

Acknowledgments. The authors would like to thank the Academy of Finland for their support to this work.

References

1. Turk, M., Pentland, A.: Eigenfaces for recognition. Journal of Cognitive Neuroscience 3(1), 71–86 (1991)
2. Belhumeur, P.N., Hespanha, J.P., Kriegman, D.J.: Eigenfaces vs. Fisherfaces: Recognition using class specific linear projection. IEEE Transactions on Pattern Analysis and Machine Intelligence 19(7), 711–720 (1997)
3. Bartlett, M.S., Movellan, J.R., Sejnowski, T.J.: Face recognition by independent component analysis. IEEE Transactions on Neural Networks 13(6), 1450–1464 (2002)
4. Phillips, P., Syed, H., Rizvi, A., Rauss, P.: The FERET evaluation methodology for face-recognition algorithms. IEEE Transactions on Pattern Analysis and Machine Intelligence 22(10), 1090–1104 (2000)
5. Ahonen, T., Hadid, A., Pietikäinen, M.: Face recognition with local binary patterns. In: Pajdla, T., Matas, J(G.) (eds.) ECCV 2004. LNCS, vol. 3021, pp. 469–481. Springer, Heidelberg (2004)
6. Ahonen, T., Hadid, A., Pietikäinen, M.: Face description with local binary pattern. IEEE Transactions on Pattern Analysis and Machine Intelligence 28, 2037–2041 (2006)

7. Daugman, J.G.: Two-dimensional spectral analysis of cortical receptive field problems. Vision Research (20), 847–856 (1980)
8. Lowe, D.: Object recognition from local scale-invariant features. In: Conference on Computer Vision and Pattern Recognition, pp. 1150–1157 (1999)
9. Wiskott, L., Fellous, J.-M., Kruger, N., Malsburg, C.v.d.: Face recognition by Elastic Bunch Graph Matching. IEEE Transactions on Pattern Analysis and Machine Intelligence 19(7), 775–779 (1997)
10. Zhang, W., Shan, S., Gao, W., Chen, X., Zhang, H.: Local Gabor Binary Pattern Histogram Sequence (LGBPHS): a novel non-Statistical model for face representation and recognition. In: International Conference on Computer Vision, pp. 786–791 (2005)
11. Zhang, W., Shan, S., Chen, X., Gao, W.: Are Gabor phases really useless for face recognition? In: International Conference on Pattern Recognition, vol. 4, pp. 606–609 (2006)
12. Zhang, B., Shan, S., Chen, X., Gao, W.: Histogram of Gabor Phase Pattern (HGPP): A novel object representation approach for face recognition. IEEE Transactions on Image Processing 16(1), 57–68 (2007)
13. Lyons, M.J., Budynek, J., Plante, A., Akamatsu, S.: Classifying facial attributes using a 2-D Gabor wavelet representation and discriminant analysis. In: Conference on Automatic Face and Gesture Recognition, pp. 1357–1362 (2000)
14. Liu, C., Wechsler, H.: Gabor feature based classification using the enhanced fisher linear discriminant model for face recognition. IEEE Transactions on Image Processing 11, 467–476 (1997)
15. Chen, H., Belhumeur, P., Jacobs, D.W.: In search of illumination invariants. In: Conference on Computer Vision and Pattern Recognition, pp. 254–261 (2000)
16. Daniel, P., Whitterridge, D.: The representation of the visual field on the cerebral cortex in monkeys. Journal of Physiology 159, 203–221 (1961)
17. Wiskott, L., Fellous, J.-M., Kruger, N., Malsburg, C.v.d.: Face recognition by Elastic Bunch Graph Matching. In: Intelligent Biometric Techniques in Fingerprint and Face Recognition, ch. 11, pp. 355–396 (1999)
18. Field, D.: Relations between the statistics of natural images and the response properties of cortical cells. Journal of the Optical Society of America A: Optics Image Science and Vision 4(12), 2379–2394 (1987)
19. Jones, J., Palmer, L.: An evaluation of the two-dimensional Gabor filter model of simple receptive fields in cat striate cortex. Journal of Neurophysiology 58(6), 1233–1258 (1987)
20. Burr, D., Morrone, M., Spinelli, D.: Evidence for edge and bar detectors in human vision. Vision Research 29(4), 419–431 (1989)
21. Lowe, D.: Distinctive Image Features from Scale-Invariant Keypoints. International Journal of Computer Vision 60(2), 91–110 (2004)
22. Dalal, N., Triggs, B.: Histograms of oriented gradients for human detection. In: Conference on Computer Vision and Pattern Recognition, vol. 1, pp. 886–893 (2005)
23. Zhao, G., Pietikäinen, M.: Dynamic texture recognition using local binary patterns with an application to facial expressions. IEEE Transactions on Pattern Analysis and Machine Intelligence 29(6), 915–928 (2007)
24. Phillips, P., Wechsler, H., Huang, J., Rauss, P.: The FERET database and evaluation procedure for face recognition algorithms. Image and Vision Computing 16(5), 295–306 (1998)
25. Phillips, P.J., Flynn, P.J., Scruggs, T., Bowyer, K.W., Chang, J., Hoffman, K., Marques, J., Min, J., Worek, W.: Overview of the face recognition grand challenge. In: IEEE Conference on Computer Vision and Pattern Recognition, pp. 947–954 (2005)

26. Ravela, S., Manmatha, R.: Retrieving images by appearance. In: International Conference on Computer Vision, pp. 608–613 (1998)
27. Lei, Z., Liao, S., He, R., Pietikäinen, M., Li, S.: Gabor volume based local binary pattern for face representation and recognition. In: IEEE conference on Automatic Face and Gesture Recognition (2008)
28. Liu, C.: Learning the uncorrelated, independent, and discriminating color spaces for face recognition. IEEE Transactions on Information Forensics and Security 3(2), 213–222 (2008)

Weight-Based Facial Expression Recognition from Near-Infrared Video Sequences

Matti Taini, Guoying Zhao, and Matti Pietikäinen

Machine Vision Group, Infotech Oulu and Department of Electrical and Information Engineering,
P.O. Box 4500 FI-90014 University of Oulu, Finland
{mtaini,gyzhao,mkp}@ee.oulu.fi

Abstract. This paper presents a novel weight-based approach to recognize facial expressions from the near-infrared (NIR) video sequences. Facial expressions can be thought of as specific dynamic textures where local appearance and motion information need to be considered. The face image is divided into several regions from which local binary patterns from three orthogonal planes (LBP-TOP) features are extracted to be used as a facial feature descriptor. The use of LBP-TOP features enables us to set different weights for each of the three planes (appearance, horizontal motion and vertical motion) inside the block volume. The performance of the proposed method is tested in the novel NIR facial expression database. Assigning different weights to the planes according to their contribution improves the performance. NIR images are shown to deal with illumination variations comparing with visible light images.

Keywords: Local binary pattern, region based weights, illumination invariance, support vector machine.

1 Introduction

Facial expression is natural, immediate and one of the most powerful means for human beings to communicate their emotions and intentions, and to interact socially. The face can express emotion sooner than people verbalize or even realize their feelings. To really achieve effective human-computer interaction, the computer must be able to interact naturally with the user, in the same way as human-human interaction takes place. Therefore, there is a growing need to understand the emotions of the user. The most informative way for computers to perceive emotions is through facial expressions in video.

A novel facial representation for face recognition from static images based on local binary pattern (LBP) features divides the face image into several regions (blocks) from which the LBP features are extracted and concatenated into an enhanced feature vector [1]. This approach has been used successfully also for facial expression recognition [2], [3], [4]. LBP features from each block are extracted only from static images, meaning that temporal information is not taken into consideration. However, according to psychologists, analyzing a sequence of images leads to more accurate and robust recognition of facial expressions [5].

A.-B. Salberg, J.Y. Hardeberg, and R. Jenssen (Eds.): SCIA 2009, LNCS 5575, pp. 239–248, 2009.

Psycho-physical findings indicate that some facial features play more important roles in human face recognition than other features [6]. It is also observed that some local facial regions contain more discriminative information for facial expression classification than others [2], [3], [4]. These studies show that it is reasonable to assign higher weights for the most important facial regions to improve facial expression recognition performance. However, weights are set only based on the location information. Moreover, similar weights are used for all expressions, so there is no specificity for discriminating two different expressions.

In this paper, we use local binary pattern features extracted from three orthogonal planes (LBP-TOP), which can describe appearance and motion of a video sequence effectively. Face image is divided into overlapping blocks. Due to the LBP-TOP operator it is furthermore possible to divide each block into three planes, and set individual weights for each plane inside the block volume. To the best of our knowledge, this constitutes novel research on setting weights for the planes. In addition to the location information, the plane-based approach obtains also the feature type: appearance, horizontal motion or vertical motion, which makes the features more adaptive for dynamic facial expression recognition.

We learn weights separately for every expression pair. This means that the weighted features are more related to intra- and extra-class variations of two specific expressions. A support vector machine (SVM) classifier, which is exploited in this paper, separates two expressions at a time. The use of individual weights for each expression pair makes the SVM more effective for classification.

Visible light (VL) (380-750 nm) usually changes with locations, and can also vary with time, which can cause significant variations in image appearance and texture. Those facial expression recognition methods that have been developed so far perform well under controlled circumstances, but changes in illumination or light angle cause problems for the recognition systems [7]. To meet the requirements of real-world applications, facial expression recognition should be possible in varying illumination conditions and even in near darkness. Near-infrared (NIR) imaging (780-1100 nm) is robust to illumination variations, and it has been used successfully for illumination invariant face recognition [8]. Our earlier work shows that facial expression recognition accuracies in different illuminations are quite consistent in the NIR images, while results decrease much in the VL images [9]. Especially for illumination cross-validation, facial expression recognition from the NIR video sequences outperforms VL videos, which provides promising performance for real applications.

2 Illumination Invariant Facial Expression Descriptors

LBP-TOP features, which are appropriate for describing and recognizing dynamic textures, have been used successfully for facial expression recognition [10]. LBP-TOP features describe effectively appearance (XY plane), horizontal motion (XT plane) and vertical motion (YT plane) from the video sequence. For each pixel a binary code is formed by thresholding its neighborhood in a circle to the center pixel value. LBP code is computed for all pixels in XY, XT and YT planes or slices separately. LBP histograms are computed to all three planes or slices in order to collect up the occurrences of different binary patterns. Finally those histograms are concatenated into one feature histogram [10].

For facial expressions, an LBP-TOP description computed over the whole video sequence encodes only the occurrences of the micro-patterns without any indication about their locations. To overcome this effect, a face image is divided into overlapping blocks. A block-based approach combines pixel-, region- and volume-level features in order to handle non-traditional dynamic textures in which image is not homogeneous and local information and its spatial locations need to be considered. LBP histograms for each block volume in three orthogonal planes are formed and concatenated into one feature histogram. This operation is demonstrated in Fig. 1. Finally all features extracted from each block volume are connected to represent the appearance and motion of the video sequence.

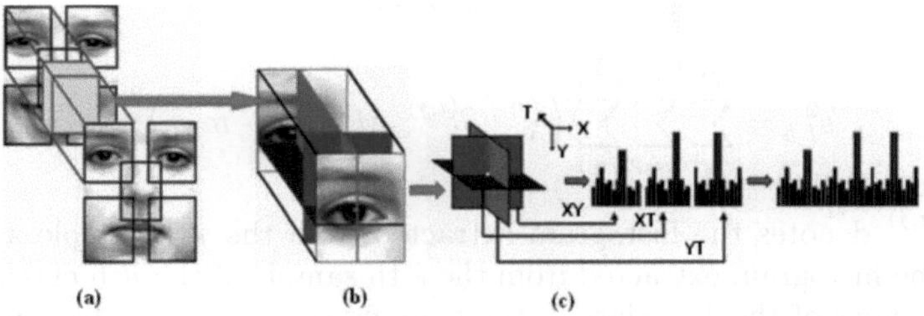

Fig. 1. Features in each block volume. (a) block volumes, (b) LBP features from three orthogonal planes, (c) concatenated features for one block volume.

For LBP-TOP, it is possible to change the radii in axes X, Y and T, which can be marked as R_X, R_Y and R_T. Also a different number of neighboring points can be used in the XY, XT and YT planes or slices, which can be marked as P_{XY}, P_{XT} and P_{YT}. Using these notations, LBP-TOP features can be denoted as $\text{LBP-TOP}_{P_{XY},P_{XT},P_{YT},R_X,R_Y,R_T}$.

Uncontrolled environmental lighting is an important issue to be solved for reliable facial expression recognition. An NIR imaging is robust to illumination changes. Because of the changes in the lighting intensity, NIR images are subject to a monotonic transform. LBP-like operators are robust to monotonic gray-scale changes [10]. In this paper, the monotonic transform in the NIR images is compensated for by applying the LBP-TOP operator to the NIR images. This means that illumination invariant representation of facial expressions can be obtained by extracting LBP-TOP features from the NIR images.

3 Weight Assignment

Different regions of the face have different contribution for the facial expression recognition performance. Therefore it makes sense to assign different weights to different face regions when measuring the dissimilarity between expressions. In this section, methods for weight assignment are examined in order to improve facial expression recognition performance.

3.1 Block Weights

In this paper, a face image is divided into overlapping blocks and different weights are set for each block, based on its importance. In many cases, weights are designed empirically, based on the observation [2], [3], [4]. Here, the Fisher separation criterion is used to learn suitable weights from the training data [11].

For a C class problem, let the similarities of different samples of the same expression compose the intra-class similarity, and those of samples from different expressions compose the extra-class similarity. The mean ($m_{I,b}$) and the variance ($s^2_{I,b}$) of intra-class similarities for each block can be computed by as follows:

$$m_{I,b} = \frac{1}{C}\sum_{i=1}^{C}\frac{2}{N_i(N_i-1)}\sum_{k=2}^{N_i}\sum_{j=1}^{k-1}\chi^2\left(S_b^{(i,j)}, M_b^{(i,k)}\right) \; , \tag{1}$$

$$s^2_{I,b} = \sum_{i=1}^{C}\sum_{k=2}^{N_i}\sum_{j=1}^{k-1}\left(\chi^2\left(S_b^{(i,j)}, M_b^{(i,k)}\right) - m_{I,b}\right)^2 \; , \tag{2}$$

where $S_b^{(i,j)}$ denotes the histogram extracted from the *j*-th sample and $M_b^{(i,k)}$ denotes the histogram extracted from the *k*-th sample of the *i*-th class, *Ni* is the sample number of the *i*-th class in the training set, and the subsidiary index *b* means the *b*-th block. In the same way, the mean ($m_{E,b}$) and the variance ($s^2_{E,b}$) of the extra-class similarities for each block can be computed by as follows:

$$m_{E,b} = \frac{2}{C(C-1)}\sum_{i=1}^{C-1}\sum_{j=i+1}^{C}\frac{1}{N_iN_j}\sum_{k=1}^{N_i}\sum_{l=1}^{N_j}\chi^2\left(S_b^{(i,k)}, M_b^{(j,l)}\right) \; , \tag{3}$$

$$s^2_{E,b} = \sum_{i=1}^{C-1}\sum_{j=i+1}^{C}\sum_{k=1}^{N_i}\sum_{l=1}^{N_j}\left(\chi^2\left(S_b^{(i,k)}, M_b^{(j,l)}\right) - m_{E,b}\right)^2 \; . \tag{4}$$

The Chi square statistic is used as dissimilarity measurement of two histograms

$$\chi^2(S,M) = \sum_{i}^{L}\frac{(S_i - M_i)^2}{S_i + M_i} \; , \tag{5}$$

where S and M are two LBP-TOP histograms, and L is the number of bins in the histogram.

Finally, the weight for each block can be computed by

$$w_b = \frac{(m_{I,b} - m_{E,b})^2}{s^2_{I,b} + s^2_{E,b}} \; . \tag{6}$$

The local histogram features are discriminative, if the means of intra and extra classes are far apart and the variances are small. In that case, a large weight will be assigned to the corresponding block. Otherwise the weight will be small.

3.2 Slice Weights

In the block-based approach, weights are set only to the location of the block. However, different kinds of features do not contribute equally in the same location. In LBP-TOP representation, the LBP code is extracted from three orthogonal planes, describing appearance in the XY plane and temporal motion in the XT and YT planes. The use of LBP-TOP features enables us to set different weights for each plane or slice inside the block volume. In addition to the location information, the slice-based approach obtains also the feature type: appearance, horizontal motion or vertical motion, which makes the features more suitable and adaptive for classification.

In the slice-based approach, the similarity within class and diversity between classes can be formed when every slice histogram from different samples is compared separately. $\chi^2_{i,j}(XY)$, $\chi^2_{i,j}(XT)$ and $\chi^2_{i,j}(YT)$ are the similarity of the LBP-TOP features in three slices from samples i and j. With this kind of approach, the dissimilarity for three kinds of slices can be obtained. In the slice-based approach, different weights can be set based on the importance of the appearance, horizontal motion and vertical motion features. Equation (5) can be used to compute weights also for each slice, when S and M are considered as two slice histograms.

3.3 Weights for Expression Pairs

In the weight computation above, the similarities of different samples of the same expression composed the intra-class similarity, and those of samples from different expressions composed the extra-class similarity. In that kind of approach, similar weights are used for all expressions and there is no specificity for discriminating two different expressions. To deal with this problem, expression pair learning is utilized. This means that the weights are learned separately for every expression pair, so extra-class similarity can be considered as a similarity between two different expressions.

Every expression pair has different and specific features which are of great importance when expression classification is performed on expression pairs [12]. Fig. 2 demonstrates that for different expression pairs, {E(I), E(J)} and {E(I), E(K)}, different appearance and temporal motion features are the most discriminative ones. The symbol "/" inside each block expresses the appearance, symbol "-" indicates horizontal motion and symbol "|" indicates vertical motion. As we can see from Fig. 2, for class pair {E(I), E(J)}, the appearance feature in block (1,3), the horizontal motion feature in block (3,1) and the appearance feature in block (4,4) are more discriminative and be assigned bigger weights, while for pair {E(I), E(K)}, the horizontal motion feature in block (1,3) and block (2,4), and the vertical motion feature in block (4,2) are more discriminative.

The aim in expression pair learning is to learn the most specific and discriminative features separately for each expression pair, and to set bigger weights for those features. Learned features are different depending on expression pairs, and they are in that way more related to intra- and extra-class variations of two specific expressions. The SVM classifier, which is exploited in this paper, separates

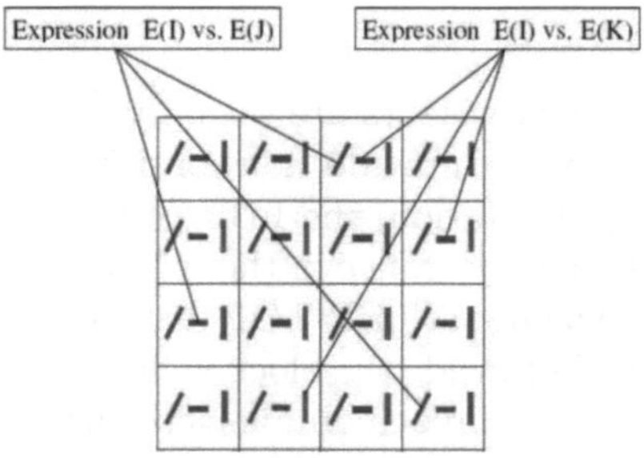

Fig. 2. Different features are selected for different class pairs

two expressions at a time. The use of individual weights for each expression pair can make the SVM more effective and adaptive for classification.

4 Weight Assignment Experiments

1602 video sequences from the novel NIR facial expression database [9] were used to recognize six typical expressions: anger, disgust, fear, happiness, sadness and surprise. Video sequences came from 50 subjects, with two to six expressions per subject. All of the expressions in the database were captured with both NIR camera and VL camera in three different illumination conditions: Strong, weak and dark. Strong illumination means that good normal lighting is used. Weak illumination means that only computer display is on and subject sits on the chair in front of the computer. Dark illumination means near darkness.

The positions of the eyes in the first frame were detected manually and these positions were used to determine the facial area for the whole sequence. 9×8 blocks, eight neighbouring points and radius three are used as the LBP-TOP parameters. SVM classifier separates two classes, so our six-expression classification problem is divided into 15 two-class problems, then a voting scheme is used to perform the recognition. If more than one class gets the highest number of votes, 1-NN template matching is applied to find out the best class [10].

In the experiments, the subjects are separated into ten groups of roughly equal size. After that a "leave one group out" cross-validation, which can also be called a "ten-fold cross-validation" test scheme, is used for evaluation. Testing is therefore performed with novel faces and it is subject-independent.

4.1 Learning Weights

Fig. 3 demonstrates the learning process of the weights for every expression pair. Fisher criterion is adopted to compute the weights from the training samples for each expression pair according to (6). This means that testing is subject-independent also when weights are used. Obtained weights were so small that they needed to be scaled from one to six. Otherwise the weights would have been meaningless.

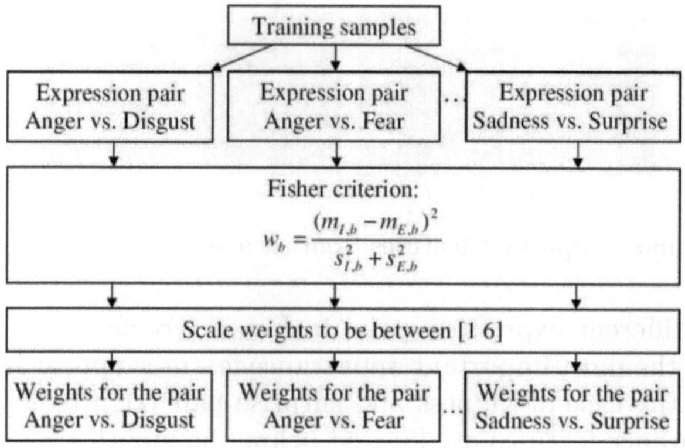

Fig. 3. Learning process of the weights

In Fig. 4, images are divided into 9 × 8 blocks, and expression pair specific block and slice weights are visualized for the pair fear and happiness. Weights are learned from the NIR images in strong illumination. Darker intensity means smaller weight and brighter intensity means larger weight. It can be seen from Fig. 4 (middle image in top row) that the highest block-weights for the pair fear and happiness are in the eyes and in the eyebrows. However, the most important appearance features (leftmost image in bottom row) are in the mouth region. This means that when block-weights are used, the appearance features are not weighted correctly. This emphasizes the importance of the slice-based approach, in which separate weights can be set for each slice based on its importance.

The ten most important features from each of the three slices for the expression pairs fear-happiness and sadness-surprise are illustrated in Fig. 5. The symbol "/" expresses appearance, symbol "-" indicates horizontal motion and symbol "|" indicates vertical motion features. The effectiveness of expression pair learning can be seen by comparing the locations of appearance features (symbol

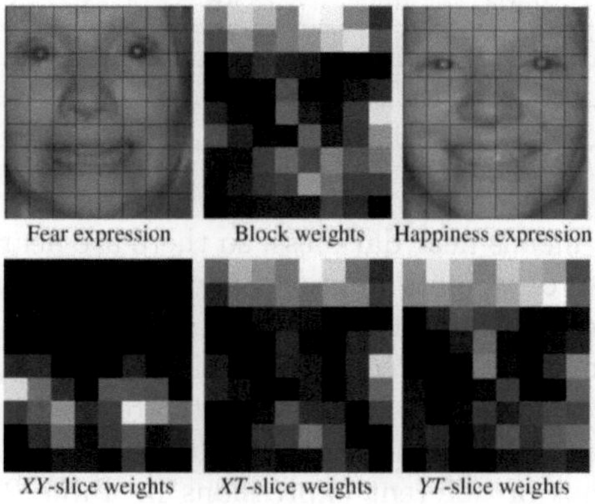

Fig. 4. Expression pair specific block and slice weights for the pair fear and happiness

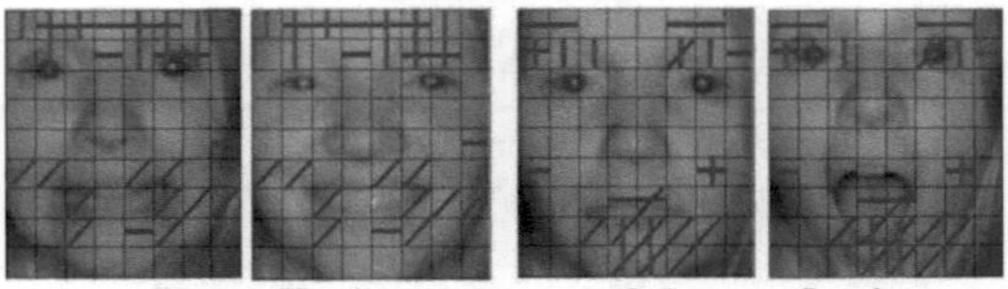

Fig. 5. The ten most important features from each slice for different expression pairs

"/") between different expression pairs in Fig. 5. For fear and happiness pair (leftmost pair) the most important appearance features appear in the corners of the mouth. In the case of sadness and surprise pair (rightmost pair) the most essential appearance features are located below the mouth.

4.2 Using Weights

Table 1 shows the recognition accuracies when different weights are assigned for each expression pair. The use of weighted blocks decreases the accuracy because weights are based only on the location information. However, different feature types are not equally important. When weighted slices are assigned to expression pairs, accuracies in the NIR images in all illumination conditions are improved, and the increase is over three percent in strong illumination. In the VL images, the recognition accuracies are decreased in strong and weak illuminations because illumination is not always consistent in those illuminations. In addition to facial features, there is also illumination information in the face area, and this makes the training of the strong and weak illumination weights harder.

Table 1. Results (%) when different weights are set for each expression pair

	Without weights	With weighted blocks	With weighted slices
NIR_Strong	79.40	77.15	82.77
NIR_Weak	73.03	76.03	75.28
NIR_Dark	76.03	74.16	76.40
VL_Strong	79.40	77.53	76.40
VL_Weak	74.53	69.66	71.16
VL_Dark	58.80	61.80	62.55

Dark illumination means near darkness, so there are nearly no changes in the illumination. The use of weights improves the results in dark illumination, so it was decided to use dark illumination weights also in strong and weak illuminations in the VL images. The recognition accuracy is improved from 71.16% to 74.16% when dark illumination slice-weights are used in weak illumination, and from 76.40% to 76.78% when those weights are used in strong illumination. Recognition accuracies of different expressions in Table 2 are obtained using weighted slices. In the VL images, dark illumination slice-weights are used also in the strong and weak illuminations.

Table 2. Recognition accuracies (%) of different expressions

	Anger	Disgust	Fear	Happiness	Sadness	Surprise	Total
NIR_Strong	84.78	90.00	73.17	84.00	72.50	90.00	82.77
NIR_Weak	73.91	70.00	68.29	84.00	55.00	94.00	75.28
NIR_Dark	76.09	80.00	68.29	82.00	55.00	92.00	76.40
VL_Strong	76.09	80.00	68.29	84.00	67.50	82.00	76.78
VL_Weak	76.09	67.50	60.98	88.00	57.50	88.00	74.16
VL_Dark	67.39	55.00	43.90	72.00	47.50	82.00	62.55

Table 3 illustrates subject-independent illumination cross-validation results. Strong illumination images are used in training, and strong, weak or dark illumination images are used in testing. The results in Table 3 show that the use of weighted slices is beneficial in the NIR images, and that different illumination between training and testing videos does not affect much on overall recognition accuracies in the NIR images. Illumination cross-validation results in the VL images are poor because of significant illumination variations.

Table 3. Illumination cross-validation results (%)

Training	NIR_Strong	NIR_Strong	NIR_Strong	VL_Strong	VL_Strong	VL_Strong
Testing	NIR_Strong	NIR_Weak	NIR_Dark	VL_Strong	VL_Weak	VL_Dark
No weights	79.40	72.28	74.16	79.40	41.20	35.96
Slice weights	82.77	71.54	75.66	76.40	39.70	29.59

5 Conclusion

We have presented a novel weight-based method to recognize facial expressions from the NIR video sequences. Some local facial regions were known to contain more discriminative information for facial expression classification than others, so higher weights were assigned for the most important facial regions. The face image was divided into overlapping blocks. Due to the LBP-TOP operator, it was furthermore possible to divide each block into three slices, and set individual weights for each of the three slices inside the block volume. In the slice-based approach, different weights can be set not only for the location, as in the block-based approach, but also for the appearance, horizontal motion and vertical motion. To the best of our knowledge, this constitutes novel research on setting weights for the slices. Every expression pair has different and specific features which are of great importance when expression classification is performed on expression pairs, so we learned weights separately for every expression pair.

The performance of the proposed method was tested in the novel NIR facial expression database. Experiments show that slice-based approach performs better than the block-based approach, and that expression pair learning provides more specific information between two expressions. It was also shown that NIR imaging can handle illumination changes. In the future, the database will be extended with 30 people using more different lighting directions in video capture. The advantages of NIR are likely to be even more obvious for videos taken under different lighting directions. Cross-imaging system recognition will be studied.

Acknowledgments. The financial support provided by the European Regional Development Fund, the Finnish Funding Agency for Technology and Innovation and the Academy of Finland is gratefully acknowledged.

References

1. Ahonen, T., Hadid, A., Pietikäinen, M.: Face Description with Local Binary Patterns: Application to Face Recognition. IEEE PAMI 28(12), 2037–2041 (2006)
2. Feng, X., Hadid, A., Pietikäinen, M.: A Coarse-to-Fine Classification Scheme for Facial Expression Recognition. In: Campilho, A.C., Kamel, M.S. (eds.) ICIAR 2004. LNCS, vol. 3212, pp. 668–675. Springer, Heidelberg (2004)
3. Shan, C., Gong, S., McOwan, P.W.: Robust Facial Expression Recognition Using Local Binary Patterns. In: 12th IEEE ICIP, pp. 370–373 (2005)
4. Liao, S., Fan, W., Chung, A.C.S., Yeung, D.-Y.: Facial Expression Recognition Using Advanced Local Binary Patterns, Tsallis Entropies and Global Appearance Features. In: 13rd IEEE ICIP, pp. 665–668 (2006)
5. Bassili, J.: Emotion Recognition: The Role of Facial Movement and the Relative Importance of Upper and Lower Areas of the Face. Journal of Personality and Social Psychology 37, 2049–2059 (1979)
6. Zhao, W., Chellappa, R., Phillips, P.J., Rosenfeld, A.: Face Recognition: A Literature Survey. ACM Computing Surveys 35(4), 399–458 (2003)
7. Adini, Y., Moses, Y., Ullman, S.: Face Recognition: The Problem of Compensating for Changes in Illumination Direction. IEEE PAMI 19(7), 721–732 (1997)
8. Li, S.Z., Chu, R., Liao, S., Zhang, L.: Illumination Invariant Face Recognition Using Near-Infrared Images. IEEE PAMI 29(4), 627–639 (2007)
9. Taini, M., Zhao, G., Li, S.Z., Pietikäinen, M.: Facial Expression Recognition from Near-Infrared Video Sequences. In: 19th ICPR (2008)
10. Zhao, G., Pietikäinen, M.: Dynamic Texture Recognition Using Local Binary Patterns with an Application to Facial Expressions. IEEE PAMI 29(6), 915–928 (2007)
11. Duda, R., Hart, P., Stork, D.: Pattern Classification. Wiley & Sons, New York (2001)
12. Zhao, G., Pietikäinen, M.: Principal Appearance and Motion from Boosted Spatiotemporal Descriptors. In: 1st IEEE Workshop on CVPR4HB, pp. 1–8 (2008)

Stereo Tracking of Faces for Driver Observation

Markus Steffens[1,2], Stephan Kieneke[1,2], Dominik Aufderheide[1,2], Werner Krybus[1], Christine Kohring[1], and Danny Morton[2]

[1] South Westphalia University of Applied Sciences, Luebecker Ring 2, 59494 Soest, Germany
{steffens,krybus,kohring}@fh-swf.de
[2] University of Bolton, Deane Road, Bolton BL3 5AB UK
d.morton@bolton.ac.uk

Abstract. This report contributes a coherent framework for the robust tracking of facial structures. The framework comprises aspects of structure and motion problems, as there are feature extraction, spatial and temporal matching, re-calibration, tracking, and reconstruction. The scene is acquired through a calibrated stereo sensor. A cue processor extracts invariant features in both views, which are spatially matched by geometric relations. The temporal matching takes place via prediction from the tracking module and a similarity transformation of the features' 2D locations between both views. The head is reconstructed and tracked in 3D. The re-projection of the predicted structure limits the search space of both the cue processor as well as the re-construction procedure. Due to the focused application, the instability of calibration of the stereo sensor is limited to the relative extrinsic parameters that are re-calibrated during the re-construction process. The framework is practically applied and proven. First experimental results will be discussed and further steps of development within the project are presented.

1 Introduction and Motivation

Advanced Driver Assistance Systems (ADAS) are investigated today. The European Commission states their capabilities to weakening and avoiding heavy accidents to approx. 70% [1]. According to an investigation of German insurance companies, a quarter of all deadly car accidents are caused by tiredness [2]. The aim of all systems is to deduce characteristic states like the spatial position and orientation of head or face and the eyeballs as well as the clamping times of the eyelids. The environmental conditions and the variability of person-specific appearances put high demands on the methods and systems. Past developments were unable to achieve the necessary robustness and usability needed to gain acceptance by the automotive industry and consumers. Current prognoses, as in [2] and [3], expect rudimental but reliable approaches after 2011. It is expected, that those products will be able to reliably detect certain lines of sight, e.g. into the mirrors or instrument panel. A broad analysis on this topic can be found in a former paper [4].

A.-B. Salberg, J.Y. Hardeberg, and R. Jenssen (Eds.): SCIA 2009, LNCS 5575, pp. 249–258, 2009.

In this report a new concept for spatio-temporal modeling and tracking of partially rigid objects (Figures 1) is presented as was generally proposed in [4]. It is based on methods for spatio-temporal scene acquisition, graph theory, adaptive information fusion and multi-hypotheses-tracking (section 3). In this paper parts of this concept will be designed into a complete system (section 4) and examined (section 5). Future work and further systems will be discussed (section 6).

2 Previous Work

Methodically, the presented contributions are originated in former works about structure and stereo motion like [11, 12, 13], about spatio-temporal tracking of faces such as [14, 15], evolution of cues [16], cue fusion and tracking like in [17, 18], and graph-based modeling of partly-rigid objects such as [19, 20, 21, 22]. The underlying scheme of all concepts is summarized in Figure 1.

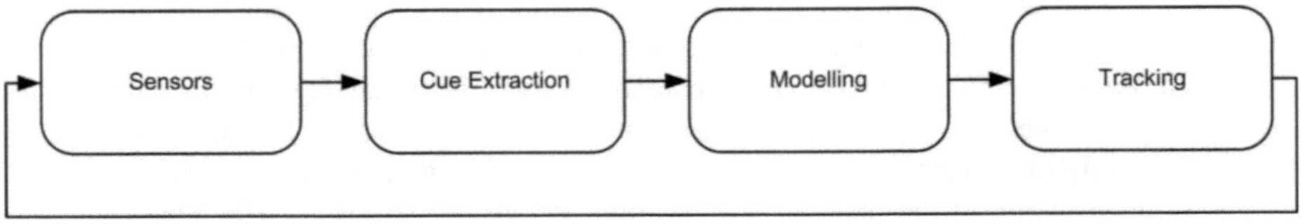

Fig. 1. General concept of spatio-temporal scene analysis for stereo tracking of faces

However, in all previously and further studied publications no coherent framework was developed like the one originally proposed here. The scheme was firstly discussed in [4]. This report originally contributes a more detailed and exact structure of the approach (section 3), a complete design of a real-world system (section 4), and first experimental results (section 5).

3 Spatio-temporal Scene Analysis for Tracking

The overall framework (Figure 1) utilizes information from a stereo sensor. In both views cues are to be detected and extracted by a cue processor. All cues are modeled in a scene graph, where the spatial (e.g. position and distance) and temporal relations (e.g. appearance and spatial dynamics) are organized. All cues are tracked over time. Information from the graph, the cue processor, and the tracker are utilized to evolve a robust model of the scene in terms of features' positions, dynamics, and cliques of features which are rigidly connected. Since all these modules are generally independent of a concrete object, a semantic model links information from the above modules into a certain context such as the T-shape of the facial features from eyes and nose. The re-calibration or auto-calibration, being a rudimental part of all systems in this field, performs a calibration of the sensors, either partly or in complete. The underlying idea is that besides utilizing an object model, facial cues are observed without a-priori semantic relations.

4 System Design and Outline

4.1 Preliminaries

The system will incorporate a stereo head with verged cameras which are strongly calibrated as described in [23]. The imagers can be full-spectrum or infrared sensors. During operation, it is expected that only the relative camera motion becomes un-calibrated, that is, it is assumed that the sensors reside intrinsically calibrated.

The general framework as presented in Figure 1 will be implemented with one cue type, a simple graph covering the spatial positions and dynamics (i.e. velocities), tracking will be performed with a Kalman filter and a linear motion model, re-calibration is performed via an overall skew measure of the corresponding rays. The overall process chain is covered in Figure 2. Currently, the rigidity constraint is implicitly met by the feature detector and no partitioning of the scene graph takes place. Consequently, the applicability of the framework is demonstrated while the overall potentials are part of further publications.

4.2 Feature Detection and Extraction

Detecting cues of interest is one significant task in the framework. Of special interest in this context is the observation of human faces. Invariant characteristics of human

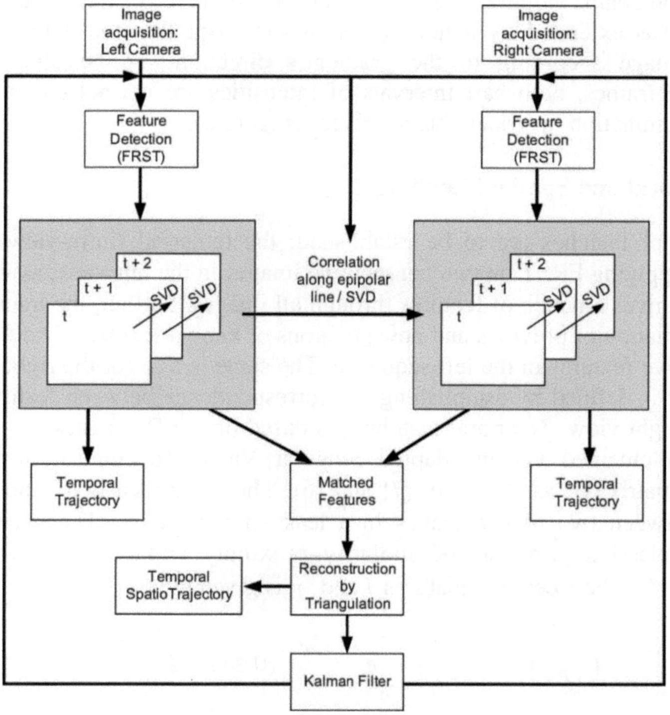

Fig. 2. Applied concept for tracking of faces

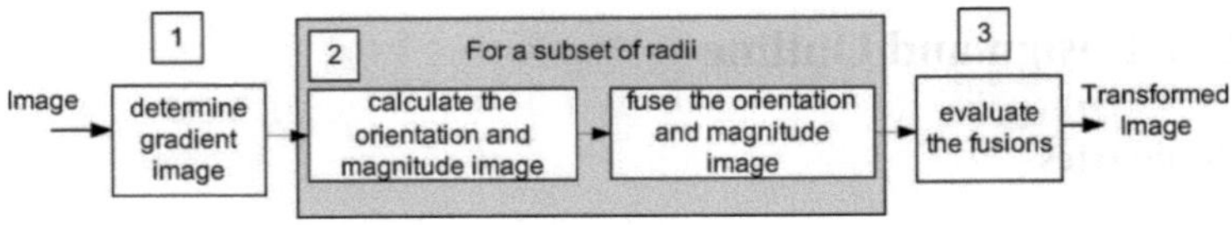

Fig. 3. Data flow of the Fast Radial Symmetry Transform (FRST)

faces are the pupils, eye corners, nostrils, top of the nose, or mouth corners. All offer an inherent characteristic, namely the presence of radial symmetric properties. For example a pupil has a shape as a circle and also nostrils have a circle-like shape. The Fast Radial Symmetry Transform (FRST) [5] is well suited for detecting such cues.

To reduce the search space in the images, an elliptic mask indicating the area of interest is evolved over the time [24]. Consequently, all subsequent steps are limited to this area and no further background model is needed.

The FRST further developed in [5] determines radial symmetric elements in an image. This algorithm is based on evaluating the gradient image to infer the contribution of each pixel to a certain centre of symmetry. The transform can be split into three parts (Figure 3). From a given image the gradient image is produced (1). Based on this gradient image, a magnitude and orientation image is built for a defined radii subset (2). Based on the resultant orientation and magnitude image, a resultant image is assembled, which encodes the radial symmetric components (3). The mathematical details would exceed the current scope; therefore have a look at [5]. The transform was extended by a normalization step such that the output is a signed intensity image according to the gradient's direction. To be able to compare consecutive frames, both half intervals of intensities are normalized independently yielding illumination invariant characteristics (Figure 6).

4.3 Temporal and Spatial Matching

Two cases of matches are to be established: the temporal (intra-view) and stereo matches. Applying FRST on two consecutive images in the left view, as well as in the right view, gives a bunch of features through all images. Further, the tracking module gives information of previous and new positions of known features. The first task is to find repetitive features in the left sequence. The same is true for the right stream. The second task is defined by establishing the correspondence between features from the left in the right view. Temporal matching is based on the Procrustes Analysis, which can be implemented via an adapted Singular Value Decomposition (SVD) of a proximity matrix $\mathbf{G}$ as shown in [7] and [6]. The basic idea is to find a rotational relation between two planar shapes in a least-squares sense. The pairing problem fulfills the classical principles of similarity, proximity, and exclusion. The similarity (proximity) $G_{i,j}$ between two features i and j is given by:

$$G_{i,j} = \left[e^{(-C_{i,j}-1)^2/2\gamma^2} \right] e^{-r_{i,j}^2/2\sigma^2} \quad (0 \le G_{i,j} \le 1) \tag{1}$$

where r is the distance between any two features in 2D and σ is a free parameter to be adapted. To account for the appearance, in [6] the normalized areal correlation index $C_{i,j}$ was introduced. The output of the algorithm is a feature pairing according to their locations in 2D between two consecutive frames in time from one view. The similarity factor indicates the quality of fit between two features.

Spatial matching takes place via a correlation method combined with epipolar properties to accelerate the entire search process by shrinking the search space to epipolar lines. Some authors like in [6] also apply SVD-based matching for the stereo correspondence, but this method only works well under strict setups, that are fronto-parallel retinas, so that both views show similar perspectives. Therefore, a rectification into the fronto-parallel setup is needed. But since no dense matching is needed [23], the correspondence search along epipolar lines is suitable. The process of finding a corresponding feature in the other view is carried out in three steps: First a window around the feature is extracted giving a template. Usually, the template shape is chosen as a square. Good results for matching are gained here for edge length between 8 and 11 pixel. Seconldy, the template is searched for along the corresponding epipolar line (Figure 5). According to the cost function (correlation score) the matched feature is found, otherwise none is found, e.g. due to occlusions. Taking only features from one view into account lead to less matches since each view may cover features which are not detected in the other view. Therefore, the previous process is also performed from the right to the left view.

4.4 Reconstruction

The spatial reconstruction takes place via triangulation with the found consistent correspondences in both views. In a fully calibrated system, the solution of finding the world coordinates of a point can be formulated as a least-square problem which can be solved via singular value decomposition (SVD). In Figure 9, the graph of a reconstructed pair of views is shown.

4.5 Tracking

This approach is characterized by feature position estimation in 3D, which is carried out by a Kalman filter currently [8] as shown in Figure 4. A window around the estimated feature, back-projected into 2D, reduces the search space for the temporal as well as the spatial search in the successive images (Figure 5). Consequently, computational costs for detecting the corresponding features are limited. Furthermore, features which are temporarily occluded can be tracked over time in case they can be classified as belonging to a group of rigidly connected features. The graph and the cue processor estimate their states from the state of the clique to which the occluded feature belongs.

The linear Kalman filter comprises a simple process model. The features move in 3D, so the state vector contains the current X-, Y- and Z-position as well as the feature's velocity. Thus, the state is the 6-vector $\mathbf{x} = [X, Y, Z, V_X, V_Y, V_Z]$. The process matrix $\mathbf{A}$ maps the previous position with the velocity multiplied by the time step to the new position $\mathbf{P}_{t+1} = \mathbf{P}_t + \mathbf{V}_t \Delta t$. The velocities are mapped identically. The measurement matrix $\mathbf{H}$ maps the positions from $\mathbf{x}$ identically to the world coordinates in $\mathbf{z}$.

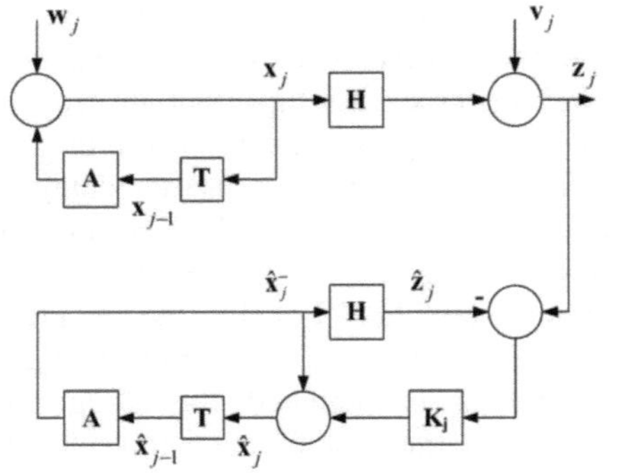

Fig. 4. Kalman Filter as block diagram [10]

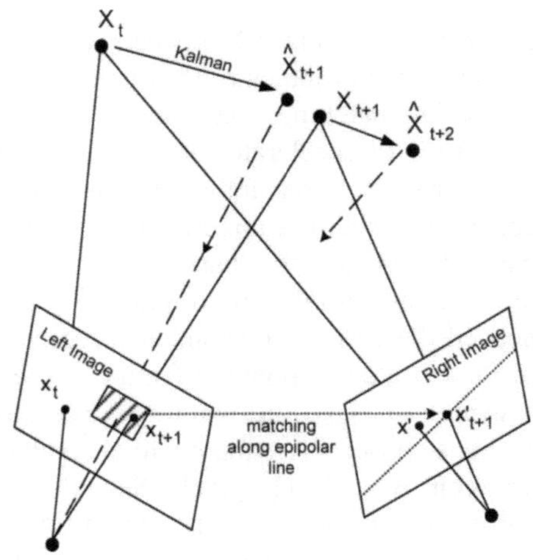

Fig. 5. Spatio-Temporal Tracking using Kalman-Filter

5 Experimental Results

An image sequence of 40 frames is taken exemplarily here. The face moves from the left to the right and back. The eyes are directed into the cameras, while in some frames the gaze is shifting away.

5.1 Feature Detection

The first part of the evaluation proves the announced property and verifies the robust ability of locating radial symmetric elements. The radius is varied by a fixed radial strictness parameter α . The algorithm yields the transformed images in Figure 6. The parameter for the FRST is a radii subset of one up to 15 pixels. The radial strictness parameter is 2.4. With exceeding a radius of 15 pixels, the positions of the pupils are highlighted uniquely. The same is true for the nostrils. By exceeding the radius of 6, the nostrils are extracted accurately. The influence of the strictness parameter α yields comparably significant results. The higher the strictness parameter, the more contour fading can be noticed. The transform was further examined under varying illumination and line-of-sights. The internal parameters were optimized accordingly with different sets of face images. The results obtained are conforming to those in [5].

Fig. 6. Performing FRST by varying the subset of radii and fixed strictness parameter (radius increases). Dark and bright pixels are features with a high radial symmetric property.

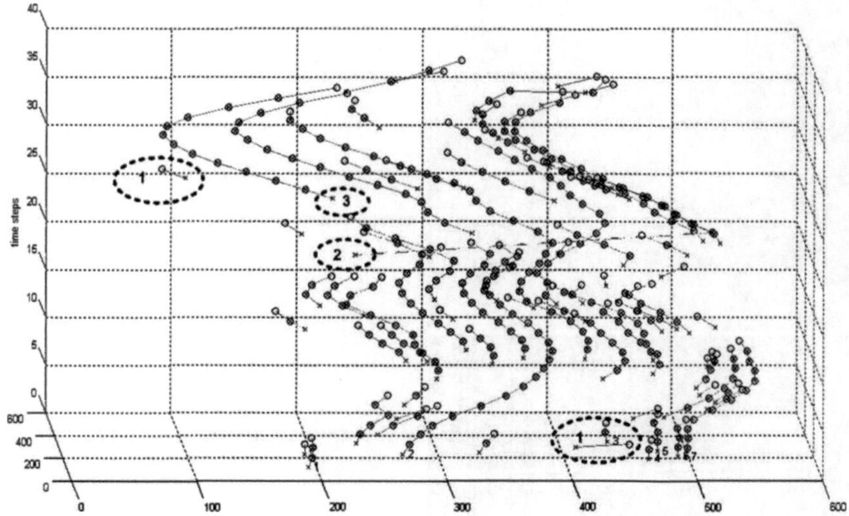

Fig. 7. Trajectory of the temporal tracking of the 40-frame sequence in one view. A single cross indicates the first occurrence of a feature, while a single circle indicates the last occurrence.

5.2 Matching

The temporal matching is performed as described. Figure 7 presents the trajectory of the sequence with the mentioned FRST parameters. A trajectory is represented by a line. Time is passing along the third axis from the bottom up. A cross without a circle indicates a feature appearing the first time in this view. A circle without cross encodes the last frame in which a certain feature appeared. A cross combined with a circle declares a successful matching of a feature in the current frame with the previous and following frame. Temporarily matched correspondences are connected by a line.

At first one is able to recognize an upstanding similar movement of most of the features. This movement has a shape similar to a wave. This correlates exactly to the real movement of the face in the observed image sequence. In Figure 10, there are four positions marked, which highlight some characteristics of the temporal matching. The first mark is a feature which was not traceable for more than one frame. The third mark is the starting point of a feature which is track-able for a longer time. In particular, this feature was observed in 14 frames. Noteworthy is the fact, that in this sequence no feature is tracked over the full sequence. It is not unusual due to the matter of the radial symmetric feature characteristic in faces. For example a recorded eye blink leads to a feature loss. Also, due to head rotations, certain features are rotated out of the image plane. The second mark shows a bad matching. Due to the rigid object and coherent movement, such a feature displacement is not realistic. The correlation threshold was chosen relatively low to 0.6, while it is working fine for this image sequence. For demonstrating the spatial matching, 21 characteristic features are selected. Figure 8 represents the results for an exemplary image pair.

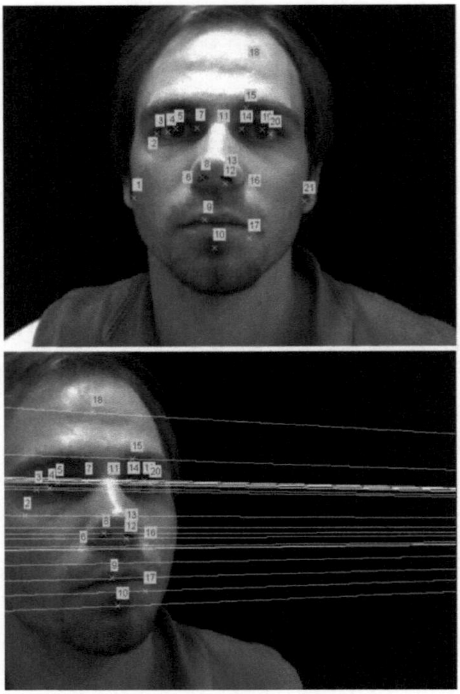

Fig. 8. Left Image with applied FRST, serves as basis for reconstruction (top); the corresponding right image (bottom)

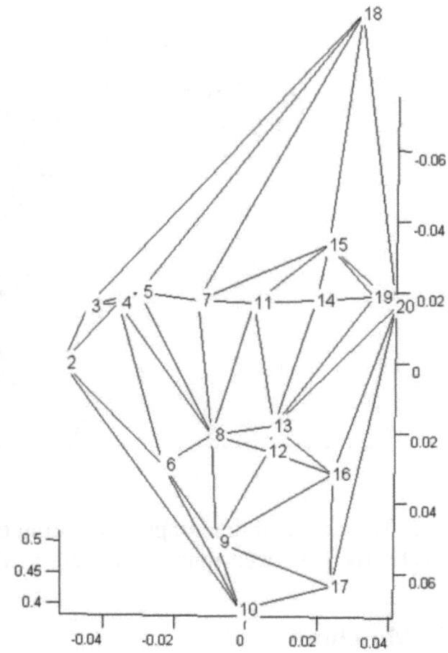

Fig. 9. Reconstructed scene graph of world points from a pair of views selected for reconstruction (scene dynamics excluded for brevity). Best viewed in color.

5.3 Reconstruction

The matching process on the corresponding right image is performed by applying areal correlation along epipolar lines [9]. The reconstruction is based on least-squares triangulation, instead of taking the mean of the closest distance between two skew rays.

Figure 8 shows the left and right view, which is the basis for reconstruction. Applying the FRST algorithm, 21 features are detected in the left view. The reconstruction based on the corresponding right view is shown in Figure 9. As one can see, almost the entire bunch of features from the left view (Figure 8, top) is detected in the right view. Due to the different camera positions, features 1 and 21 are not covered in the right image and consequently not matched. Although the correlation assignment criteria is quite simple, namely the maximum correlation along an epipolar line, this method yields a robust matching as shown in Figures 8 and 9. All features, except feature 18, are assigned correctly. Due to the wrong correspondence, a wrong triangulation and consequently a wrong reconstruction of feature 18 is the outcome as can be inspected in Figure 9.

5.4 Tracking

In this subsection the tracking approach will be evaluated. The previous sequence of 40 frames was used for tracking. The covariance matrices are currently deduced experimentally. This way the filter works stable over all frames. The predictions by the filter and the measurements lie on common trajectories. However, the chosen motion model is only suitable for relatively smooth motions. The estimates of the filter were further used during fitting of the facial regions in the images. The centroid of all features in 2D was used as an estimate of the center of the ellipse.

6 Future Work

At the moment there are different areas under research. Here, only some important should be named: robust dense stereo matching, cue processor incorporating fusion, graphical models, model fusion of semantic and structure models, auto- and re-calibration, and particle filters in Bayesian networks.

7 Summary and Discussion

This report introduces current issues on driver assistance systems and presents a novel framework designed for this kind of application. Different aspects of a system for spatio-temporal tracking of faces are demonstrated. Methods for feature detection, for tracking in the 3D world, and reconstruction utilizing a structure graph were presented. While all methods are at a simple level, the overall potentials of the approach could be demonstrated. All modules are incorporated into a working system and future work is indicated.

References

[1] European Commission, Directorate General Information Society and Media: Use of Intelligent Systems in Vehicles. Special Eurobarometer 267 / Wave 65.4. 2006

[2] Büker, U.: Innere Sicherheit in allen Fahrsituationen. Hella KGaA Hueck & Co., Lippstadt (2007)

[3] Mak, K.: Analyzes Advanced Driver Assistance Systems (ADAS) and Forecasts 63M Systems For 2013, UK (2007)

[4] Steffens, M., Krybus, W., Kohring, C.: Ein Ansatz zur visuellen Fahrerbeobachtung, Sensorik und Algorithmik zur Beobachtung von Autofahrern unter realen Bedingungen. In: VDI-Konferenz BV 2007, Regensburg, Deutschland (2007)

[5] Lay, G., Zelinsky, A.: A fast radial symmetry transform for detecting points of interest. Technical report, Australien National University, Canberra (2003)

[6] Pilu, M.: Uncalibrated stereo correspondence by singular valued decomposition. Technical report, HP Laboratories Bristol (1997)

[7] Scott, G., Longuet-Higgins, H.: An algorithm for associating the features of two patterns. In: Proceedings of the Royal Statistical Society of London, vol. B244, pp. 21–26 (1991)

[8] Welch, G., Bishop, G.: An introduction to the kalman filter (July 2006)

[9] Steffens, M.: Polar Rectification and Correspondence Analysis. Technical Report Laboratory for Image Processing Soest, South Westphalia University of Applied Sciences, Germany (2008)
[10] Cheever, E.: Kalman filter (2008)
[11] Torr, P.H.S.: A structure and motion toolkit in matlab. Technical report, Microsoft Research (2002)
[12] Oberle, W.F.: Stereo camera re-calibration and the impact of pixel location uncertainty. Technical Report ARL-TR-2979, U.S. Army Research Laboratory (2003)
[13] Pollefeys, M.: Visual 3Dmodeling from images. Technical report, University of North Carolina - Chapel Hill, USA (2002)
[14] Newman, R., Matsumoto, Y., Rougeaux, S., Zelinsky, A.: Real-Time Stereo Tracking for Head Pose and Gaze Estimation. In: FG 2000, pp. 122–128 (2000)
[15] Heinzmann, J., Zelinsky, A.: 3-D Facial Pose and Gaze Point Estimation using a Robust Real-Time Tracking Paradigm, Canberra, Australia (1997)
[16] Seeing Machines: WIPO Patent WO/2004/003849
[17] Loy, G., Fletcher, L., Apostoloff, N., Zelinsky, A.: An Adaptive Fusion Architecture for Target Tracking, Canberra, Australia (2002)
[18] Kähler, O., Denzler, J., Triesch, J.: Hierarchical Sensor Data Fusion by Probabilistic Cue Integration for Robust 3-D Object Tracking, Passau, Deutschland (2004)
[19] Mills, S., Novins, K.: Motion Segmentation in Long Image Sequences, Dunedin, New Zealand (2000)
[20] Mills, S., Novins, K.: Graph-Based Object Hypothesis. Dunedin, New Zealand (1998)
[21] Mills, S.: Stereo-Motion Analysis of Image Sequences. Dunedin, New Zealand (1997)
[22] Kropatsch, W.: Tracking with Structure in Computer Vision TWIST-CV. Project Proposal, Pattern Recognition and Image Processing Group, TU Vienna (2005)
[23] Steffens, M.: Close-Range Photogrammetry. Technical Report Laboratory for Image Processing Soest, South Westphalia University of Applied Sciences, Germany (2008)
[24] Steffens, M., Krybus, W.: Analysis and Implementation of Methods for Face Tracking. Technical Report Laboratory for Image Processing Soest, South Westphalia University of Applied Sciences, Germany (2007)

Camera Resectioning from a Box

Henrik Aanæs[1], Klas Josephson[2], François Anton[1],
Jakob Andreas Bærentzen[1], and Fredrik Kahl[2]

[1] DTU Informatics, Technical University of Denmark, 2800 Kgs. Lyngby, Denmark
[2] Centre For Mathematical Sciences, Lund University, Lund, Sweden

Abstract. In this paper we describe how we can do camera resectioning from a box with unknown dimensions, i.e. determine the camera model, assuming that image pixels are square. This assumption is equivalent to assuming that the camera has an aspect ratio of one and zero skew, and this holds for most — if not all — digital cameras. Our proposed method works by first deriving 9 linear constraints on the projective camera matrix from the box, leaving a 3-dimensional subspace in which the projective camera matrix can lie. A single solution in this 3D subspace is then found via a method by Triggs in 1999, which uses the square pixel assumption to set up a 4th degree polynomial to which the solution is the desired model. This approach is, however, numerically challenging, and we use several means to tackle this issue. Lastly the solution is refined in an iterative manner, i.e. using bundle adjustment.

1 Introduction

With the ever increasing use of interactive 3D environments for online social interaction, computer gaming and online shopping, there is also an ever increasing need for 3D modelling. And even though there has been a tremendous increase in our ability to process and display such 3D environments, the creation of such 3D content is still mainly a manual — and thus expensive — task. A natural way of automating 3D content creation is via image based methods, where several images are taken of a real world object upon which a 3D model is generated, c.f. e.g. [9,12]. However, such fully automated image based methods do not yet exist for general scenes. Hence, we are contemplating doing such modelling in a semi-automatic fashion, where 3D models are generated from images with a minimum of user input, inspired e.g. by Hengel et al. [18].

For many objects, especially man made, boxes are a natural building blocks. Hence, we are contemplating a system where a user can annotate the bounding box of an object in several images, and from this get a rough estimate of the geometry, see Figure 1. However, we do not envision that the user will supply the dimensions (even relatively) of that box. Hence, in order to get a correspondence between the images, and thereby refine the geometry, we need to be able to do camera resectioning from a box. That is, given an annotation of a box, as seen in Figure 1, we should be able to determine the camera geometry. At present, to the best of our knowledge, no solution is available for this particular resectioning

A.-B. Salberg, J.Y. Hardeberg, and R. Jenssen (Eds.): SCIA 2009, LNCS 5575, pp. 259–268, 2009.

Fig. 1. A typical man made object, which at a coarse level is approximated well by a box. It is the annotation of such a box, that we assume the user is going to do in a sequence of images.

problem, and such a solution is what we present here. Thus, taking the first step towards building a semi-automatic image based 3D modelling system.

Our proposed method works by first extracting 9 linear constraints from the geometry of the box, as explained in Section 2, and thereupon resolving the ambiguity by enforcing the constraint that the pixels should be square. Our method extends the method of Triggs [16] from points to boxes, does not require elimination of variables, and is numerically more stable. Moreover, the complexity of our method is polynomial by opposition to the complexity of the method of Triggs, which is doubly exponential. It results in solving a 4th degree polynomial system in 2 variables. This is covered in Section 3. There are however some numerical issues which need attention as described in Section 4. Lastly our solution is refined via Bundle adjustment c.f. e.g. [17].

1.1 Relation to Other Work

Solutions to the camera resectioning problem are by no means novel. For the uncalibrated pinhole camera model the resectioning problem can be solved from 6 or more points via a direct linear transform from 6 or more points c.f. e.g. [9], using so called algebraic methods. If the camera is calibrated, in the sense that the internal parameters are known, solutions exist for 3 or more known 3D points c.f. e.g. [8], given that the camera is a pinhole camera. In the general case – the camera is not assumed to obey the pinhole camera model – of a calibrated camera and 3 or more points, Nister et al. [11] have provided a solution. In the rest of this paper, a pinhole camera model is assumed. A linear algorithm for resectioning of a calibrated camera from 4 or more points or lines [1] exists.

If parts of the intrinsic camera parameters are known, e.g. that the pixels are square, solutions also exist c.f. e.g. [16]. Lastly, we would like to mention that from a decent initial estimate we can solve any – well posed – resection problem via bundle adjustment c.f. e.g. [17].

Most of the methods above require the solution to a system of multivariate polynomials, c.f. [5,6]. And also many of these problems end up being numerically challenging as addressed within a computer vision context in [3].

2 Basic Equations

Basically, we want to do camera resectioning from the geometry illustrated in Figure 2, where a and b are unknown. The two remaining corners are fixed to $(0,0,0)$ and $(1,0,0)$ in order to fix a frame of reference, and thereby remove the ambiguity over all scale rotations and translations. Assuming a projective or pinhole camera model, $\mathbf{P}$, the relationship between a 3D point Q_i and it's corresponding 2D point q_i is given by

$$q_i = \mathbf{P}Q_i \ , \tag{1}$$

where Q_i and q_i are in homogeneous coordinates, and $\mathbf{P}$ is a 3 by 4 matrix. It is known that Q_i and q_i induces the following linear constraint on $\mathbf{P}$, c.f. [9]

$$0 = [q_i]_x \mathbf{P} Q_i = Q_i^T \otimes [q_i]_x \bar{P} \ , \tag{2}$$

where $[q_i]_x$ is the 3 by 3 matrix corresponding to taking the cross product with q_i, $\otimes$ is the Kronecker product and $\bar{P}$ is the elements of $\mathbf{P}$ arranged as a vector. Setting $c_i = Q_i^T \otimes [q_i]_x$, and arranging the c_i in a matrix $\mathbf{C} = [c_1^T, \ldots, c_n^T]^T$, we have a linear system of equations

$$\mathbf{C}\bar{P} = 0 \tag{3}$$

constraining $\mathbf{P}$. This is the method used here.

To address the issue that we do not know a and b, we assume that the box has right angles, in which case the box defines points at infinity. These points at infinity are, as illustrated in Figure 2, independent of the size of a and b, and can be derived by calculating the intersections of the lines composing the edges of the box.[1] We thus calculate linear constraints, c_i, based on $[0,0,0,1]^T$ and $[1,0,0,1]^T$ and the three points at infinity $[1,0,0,0]^T$, $[0,1,0,0]^T$, $[0,0,1,0]^T$. This, however, only yields 9 constraints on $\mathbf{P}$, i.e. the rank of $\mathbf{C}$ is 9. Usually a 3D to 2D point correspondence gives 2 constraints, and we should have 10 constraints. The points $[0,0,0,1]^T$, $[1,0,0,1]^T$ and $[1,0,0,0]^T$ are, however, located on a line making them partly linearly dependant, and thus giving an extra degree of freedom, leaving us with our 9 constraints.

To define $\mathbf{P}$ completely we need 11 constraints, in that it has 12 parameters and is independent of scale. The null space of $\mathbf{C}$ is thus (by the dimension

[1] Note that in projective space infinity is a point like any other.

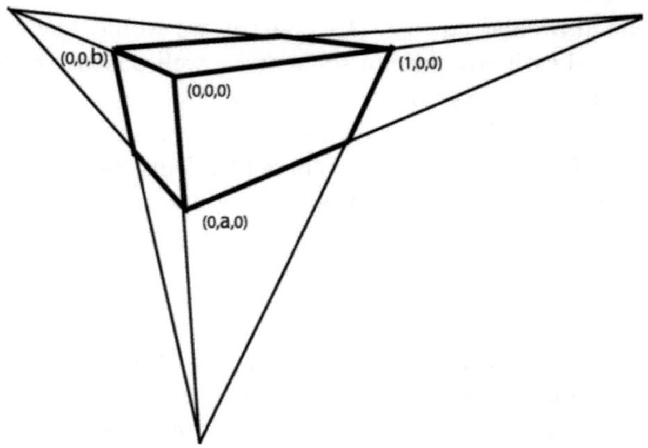

Fig. 2. The geometric outline of the box, from which we want to do the resectioning, along with the associated points at infinity denoted. Here a and b are the unknowns.

theorem for subspaces) 3-dimensional instead of 1-dimensional. We are thus 2 degrees short. By requiring that the images are taken by a digital camera the pixels should be perfectly square. This assumption gives us the remaining two degrees of freedom, in that a pinhole camera model has a parameters for skewness of the pixels as well as one for their aspect ratio. The issue is, however, how to incorporate these two constraints in a computationally feasible way. In order to do this, we will let the 3D right-null space of $\mathbf{C}$ be spanned by v_1, v_2, v_3. The usual way to find v_1, v_2, v_3 is via singular value decomposition (SVD) of $\mathbf{C}$. But during our experiments we found that it does not yield the desired result. Instead, one of the equations in $\mathbf{C}$ corresponding to the point $[0, 0, 0, 1]^T$ was removed, and by that, we can calculate the null space of the remaining nine equations. This turned out to be a crucial step to get the proposed method to work. We have also tried to remove any of the theoretically linearly dependent equations, and the result proved not to be dependent on the equations that were removed. Then, $\mathbf{P}$ is seen to be a linear combination of v_1, v_2, v_3, i.e.

$$\bar{P} = \mu_1 v_1 + \mu_2 v_2 + \mu_3 v_3 \ . \tag{4}$$

For computational reasons, we will set $\mu_3 = 1$, and if this turns out to be numerically unstable, we will set one of the other coefficients to one.

3 Polynomial Equation

Here we are going to find the solution to (4), by using the method proposed by Triggs in [16]. To do this, we decompose the pinhole camera into intrinsic parameters $\mathbf{K}$, rotation $\mathbf{R}$ and translation t, such that

$$\mathbf{P} = \mathbf{K}[\mathbf{R}|t] \ . \tag{5}$$

The dual image of the absolute quadric, ω is given by [9,16]

$$\omega = \mathbf{P}\Omega\mathbf{P}^T = \mathbf{K}\mathbf{K}^T \ , \tag{6}$$

where Ω is the absolute dual quadric,

$$\Omega = \begin{bmatrix} \mathbf{I} & 0 \\ 0 & 0 \end{bmatrix} \ .$$

Here $\mathbf{P}$ and thus $\mathbf{K}$ and ω are functions of $\mu = [\mu_1, \mu_2]^T$. Assuming that the pixels are square is equivalent to $\mathbf{K}$ having the form

$$\mathbf{K} = \begin{bmatrix} f & 0 & \Delta x \\ 0 & f & \Delta y \\ 0 & 0 & 1 \end{bmatrix} \ , \tag{7}$$

where f is the focal length and $(\Delta x, \Delta y)$ is the optical center of the camera. In this case the the upper 2 by 2 part of ω^{-1} is proportional to an identity matrix. Using the matrix of cofactors, it is seen that this coresponds to the minor of ω_{11} equals the minor of ω_{22} and that the minor of ω_{12} equals 0, i.e.

$$\omega_{22}\omega_{33} - \omega_{23}^2 = \omega_{11}\omega_{33} - \omega_{13}^2 \tag{8}$$

$$\omega_{21}\omega_{33} - \omega_{23}\omega_{31} = 0 \tag{9}$$

This corresponds to a fourth degree polynomial in the elements of $\mu = [\mu_1, \mu_2]^T$. Solving this polynomial equation will give us the linear combination in (4), corresponding to a camera with square pixels, and thus the solution to our resectioning problem.

3.1 Polynomial Equation Solver

To solve the system of polynomial equations Gröbner basis methods are used. These methods compute the basis of the vector space (called the quotient algebra) of all the unique representatives of the residuals of the (Euclidean) multivariate division of all polynomials by the polynomials of the system to be solved, without relying on elimination of variables, nor performing the doubly exponential time computation of the Gröbner basis. Moreover, this computation of the Gröbner basis, which requires the successive computation of remainders in floating point arithmetic, would induce an explosion of the errors. This approach has been a successful method used to solve several systems of polynomial equations in computer vision in recent years e.g. [4,13,14]. The pros of Gröbner basis methods is that they give a fast way to solve systems of polynomial equations, and that they reduce the problem of the computation of these solutions to a linear algebra (eigenvalue) problem, which is solvable by radicals if the size of the matrix does not exceed 4, yielding a closed form in such cases. On the other hand the numerical accuracy can be a problem [15]. A simple introduction to Gröbner bases and the field of algebraic geometry (which is the theoretical basis of the Gröbner basis) can be found in the two books by Cox et al. [5,6].

The numerical Gröbner basis methods we are using here require that the number of solutions to the problem needs to be known beforehand, because we do not actually compute the Gröbner basis. An upper bound to a system is given by Bézout's theorem [6]. It states that the number of solutions of a system of polynomial equations is generically the product of the degrees of the polynomials. The upper bound is reached only if the decompositions of the polynomials into irreducible factors do not have any (irreducible) factor in common. In this case, since there are two polynomials of degree four in the system to be solved, the maximal number of solutions is 16. This is also the true number of complex solutions of the problem. The number of solutions is later used when the action (also called the multiplication map in algebraic geometry) matrix is constructed, it is also the size of the minimal eigenvalue problem necessary to solve. We are using a threshold to determine whether monomials are certainly standard monomials (which are the elements of the basis of the quotient algebra) or not. The monomials for which we are not sure whether they are standard are added to the basis, yielding a higher dimensional representation of the quotient algebra.

The first step when a system of polynomial equations is solved with such a numerical Gröbner basis based quotient algebra representation is to put the system in matrix form. A homogenous system can be written,

$$\mathbf{C}X = 0. \tag{10}$$

In this equation $\mathbf{C}$ holds the coefficients in the equations and X the monomials. The next step is to expand the number of equation. This is done by multiplying the original equations by a handcrafted set of monomials in the unknown variables. This is done to get more linearly independent equations with the same set of solutions. For the problem in this paper we multiply with all monomials up to degree 3 in the two unknown variables μ_1 and μ_2. The result of this is twenty equations with the same solution set as the original two equations. Once again we put this on matrix form,

$$\mathbf{C}_{exp} X_{exp} = 0, \tag{11}$$

in this case $\mathbf{C}_{exp}$ is a 20×36 matrix. From this step the method of [3] is used. By using those methods with truncation and automatic choice of the basis monomials the numeric stability is considerably improved. The only parameters that are left to choose is the variable used to construct the action matrix and the truncation threshold. We choose μ_1 as action variable and the truncation threshold is fixed to 10^{-8}.

An alternative way to solve the polynomial equation is to use the automatic generator for minimal problems presented by Kukelova et al. [10]. A solver generated this way doesn't use the methods of basis selection, which will reduce the numerical stability. We could also use exact arithmetic for computing the Gröbner basis exactly, but this would yield in the tractable cases a much longer computation time, and in the other cases an aborted computation due to a memory shortage.

3.2 Resolving Ambiguity

It should be expected that there are more than one real valued solution to the polynomial equations. To determine which of those solutions are correct, an alternative method to calculate the calibration matrix, **K**, is used. After that, the solution from the polynomial equations with a calibration matrix closest to the alternatively calculated calibration matrix is used. The method used is described in [9]. It uses that in the case of square pixels and zero skew the image of the absolute conic has the form

$$\omega^{-1} = \begin{bmatrix} \omega_1 & 0 & \omega_2 \\ 0 & \omega_1 & \omega_3 \\ \omega_2 & \omega_3 & \omega_4 \end{bmatrix} \tag{12}$$

and that for each pair of orthogonal vanishing points v_i, v_j the relation $v_i^T \omega^{-1} v_j = 0$ holds. The three orthogonal vanishing points known from the drawn box in the image thus gives three constraints on ω^{-1} that can be expressed on matrix form according to $\mathbf{A}\bar{\omega}^{-1} = 0$ where **A** is a 3×4 matrix. The vector $\bar{\omega}^{-1}$ can then be found as the null space of **A**. The calibration matrix is then obtained by calculating the Cholesky factorization of ω as described in equation 6.

The use of the above method also has an extra advantage. Since it doesn't enforce ω to be positive definite it can be used as a method to detect uncertainty in the data. If ω isn't positive definite, the Cholesky factorization can't be performed and, hence, the result will not be good in the solution of the polynomial equations. To nevertheless have something to compare with, we substitute ω with $\omega - \delta\mathbf{I}$, where δ equals the smallest eigenvalue of ω times 1.1.

To decide which solution from the polynomial equations to use the extra constraints that the two points $[0, 0, 0]$ and $[1, 0, 0]$ are in front of the camera is enforced. Among those solutions fulfilling this constraint the solution with smallest difference in matrix norm between the calibration matrix from the method described above and those from the solutions of the polynomial equations is used.

4 Numerical Considerations

The most common use of Gröbner basis solvers is in the core of a RANSAC engine[7]. In those cases there is no problem if the numerical errors gets large in a few setups since the problem is calculated for many instances and only the best is used. In the problem of this paper this is not the case instead we need a good solution for every null space used in the polynomial equation solver. To find the best possible solution the accuracy of the solution is measured by the condition number of the matrix that is inverted when the Gröbner basis is calculated. This has been shown to be a good marker of the quality of the solution [2]. Since the order of the vectors in the null space is independent we choose to try a new ordering if this condition number is larger than 10^5. If all orderings gives a condition number larger than 10^5 we choose the solution with the smallest condition number. By this we can eliminate the majority of the large errors.

To even further improve the numerical precision the first step in the calculation is to change the scale of the images. The scale is chosen so that the largest absolute value of any image coordinate of the drawn box equals one. By doing this the condition number of ω decreases from approximately 10^6 to one for an image of size 1000 by 1000.

5 Experimental Results

To evaluate the proposed method we went to the local furniture store and took several images of their furniture, e.g. Figure 1. On this data set we manually annotated 30 boxes, outlining furniture, see e.g. Figure 3, and ran our proposed method on the annotated data to get an initial result, and refined the solution with a bundle adjuster. In all but one of these we got acceptable results, in the

Fig. 3. Estimated boxes. The annotated boxes from furniture images denoted blue lines. The initial estimate is denoted by green lines, and the final result is denoted by a dashed magenta line.

last example, there were no real solutions to the polynomial equations. As seen from Figure 3, the results are fully satisfactory, and we are now working on using the proposed method in a semi-automatic modelling system. As far as we can see, the reason that we can refine the initial results is that there are numerical inaccuracies in our estimation. To push the point, that fact that we can find a good fit of a box, implies that we have been able to find a model, consisting of camera position and internal parameters as well as values for the unknown box sides a and b, that explains the data well. Thus, from the given data, we have a good solution to the camera resectioning problem.

6 Conclusion

We have proposed a method for solving the camera resectioning problem from an annotated box, assuming only that the box has right angles, and that the camera's pixels are square. Once several numerical issues have been addressed, the method produces good results.

Acknowledgements

We wish to thank ILVA A/S in Kgs. Lyngby for helping us gather the furniture images used in this work. This work has been partly funded by the European Research Council (GlobalVision grant no. 209480), the Swedish Research Council (grant no. 2007-6476) and the Swedish Foundation for Strategic Research (SSF) through the programme Future Research Leaders.

References

1. Ansar, A., Daniilidis, K.: Linear pose estimation from points or lines. IEEE Transactions on Pattern Analysis and Machine Intelligence 25(5), 578–589 (2003)
2. Byröd, M., Josephson, K., Åström, K.: Improving numerical accuracy of gröbner basis polynomial equation solvers. In: International Conference on Computer Vision (2007)
3. Byröd, M., Josephson, K., Åström, K.: A column-pivoting based strategy for monomial ordering in numerical gröbner basis calculations. In: The 10th European Conference on Computer Vision (2008)
4. Byröd, M., Kukelova, Z., Josephson, K., Pajdla, T., Åström, K.: Fast and robust numerical solutions to minimal problems for cameras with radial distortion. In: Conference on Computer Vision and Pattern Recognition (2008)
5. Cox, D., Little, J., O'Shea, D.: Using Algebraic Geometry, 2nd edn. Springer, Heidelberg (2005)
6. Cox, D., Little, J., O'Shea, D.: Ideals, Varieties, and Algorithms. Springer, Heidelberg (2007)
7. Fischler, M.A., Bolles, R.C.: Random sample consensus: a paradigm for model fitting with applications to image analysis and automated cartography. Communications of the ACM 24(6), 381–395 (1981)

8. Haralick, R.M., Lee, C.-N., Ottenberg, K., Nolle, M.: Review and analysis of solutions of the three point perspective pose estimation problem. International Journal of Computer Vision 13(3), 331–356 (1994)
9. Hartley, R.I., Zisserman, A.: Multiple View Geometry, 2nd edn. Cambridge University Press, Cambridge (2003)
10. Kukelova, M., Bujnak, Z., Pajdla, T.: Automatic generator of minimal problem solvers. In: The 10th European Conference on Computer Vision, pp. 302–315 (2008)
11. Nister, D., Stewenius, H.: A minimal solution to the generalised 3-point pose problem. Journal of Mathematical Imaging and Vision 27(1), 67–79 (2007)
12. Seitz, S.M., Curless, B., Diebel, J., Scharstein, D., Szeliski, R.: A comparison and evaluation of multi-view stereo reconstruction algorithms. In: 2006 IEEE Computer Society Conference on Computer Vision and Pattern Recognition, vol. 1, pp. 519–528 (2006)
13. Stewénius, H., Engels, C., Nistér, D.: Recent developments on direct relative orientation. ISPRS Journal of Photogrammetry and Remote Sensing 60(4), 284–294 (2006)
14. Stewenius, H., Nister, D., Kahl, F., Schaffilitzky, F.: A minimal solution for relative pose with unknown focal length. Image and Vision Computing 26(7), 871–877 (2008)
15. Stewénius, H., Schaffalitzky, F., Nistér, D.: How hard is three-view triangulation really? In: Proc. Int. Conf. on Computer Vision, Beijing, China, pp. 686–693 (2005)
16. Triggs, B.: Camera pose and calibration from 4 or 5 known 3D points. In: Proc. 7th Int. Conf. on Computer Vision, pp. 278–284. IEEE Computer Society Press, Los Alamitos (1999)
17. Triggs, B., McLauchlan, P.F., Hartley, R.I., Fitzgibbon, A.W.: Special sessions - bundle adjustment - a modern synthesis. In: Triggs, B., Zisserman, A., Szeliski, R. (eds.) ICCV-WS 1999. LNCS, vol. 1883, pp. 298–372. Springer, Heidelberg (2000)
18. van den Hengel, A., Dick, A., Thormahlen, T., Ward, B., Torr, P.H.S.: Videotrace: rapid interactive scene modelling from video. ACM Transactions on Graphics 26(3), 86–1–5 (2007)

Appearance Based Extraction of Planar Structure in Monocular SLAM

José Martínez-Carranza and Andrew Calway

Department of Computer Science
University of Bristol, UK
{csjmc,csadc}@bristol.ac.uk

Abstract. This paper concerns the building of enhanced scene maps during real-time monocular SLAM. Specifically, we present a novel algorithm for detecting and estimating planar structure in a scene based on both geometric and appearance and information. We adopt a hypothesis testing framework, in which the validity of planar patches within a triangulation of the point based scene map are assessed against an appearance metric. A key contribution is that the metric incorporates the uncertainties available within the SLAM filter through the use of a test statistic assessing error distribution against predicted covariances, hence maintaining a coherent probabilistic formulation. Experimental results indicate that the approach is effective, having good detection and discrimination properties, and leading to convincing planar feature representations[1].

1 Introduction

Several systems now exist which are capable of tracking the 3-D pose of a moving camera in real-time using feature point depth estimation within previously unseen environments. Advances in both structure from motion (SFM) and simultaneous localisation and mapping (SLAM) have enabled both robust and stable tracking over large areas, even with highly agile motion, see e.g. [1,2,3,4,5]. Moreover, effective relocalisation strategies also enable rapid recovery in the event of tracking failure [6,7]. This has opened up the possibility of highly portable and low cost real-time positioning devices for use in a wide range of applications, from robotics to wearable computing and augmented reality.

A key challenge now is to take these systems and extend them to allow real-time extraction of more complex scene structure, beyond the sparse point maps upon which they are currently based. As well as providing enhanced stability and reducing redundancy in representation, deriving richer descriptions of the surrounding environment will significantly expand the potential applications, notably in areas such as augmented reality in which knowledge of scene structure is an important element. However, the computational challenges of inferring both geometric and topological structure in real-time from a single camera are highly

[1] Example videos can be found at http://www.cs.bris.ac.uk/home/carranza/scia09/

A.-B. Salberg, J.Y. Hardeberg, and R. Jenssen (Eds.): SCIA 2009, LNCS 5575, pp. 269–278, 2009.

demanding and will require the development of alternative strategies to those that have formed the basis of current off-line approaches, which in the main are based on optimization over very large numbers of frames.

Most previous work on extending scene descriptions in real-time systems has been done in the context of SLAM. This includes several approaches in which 3-D edge and planar patch features are used for mapping [8,9,10,11]. However, the motivation in these cases was more to do with gaining greater robustness in localisation, rather than extending the utility of the resulting scene maps. More recently, Gee *et al* [12] have demonstrated real-time plane extraction in which planar structure is inferred from the geometry of subsets of mapped point features and then parameterised within the state, allowing simultaneous update alongside existing features. However, the method relies solely on geometric information and thus planes may not correspond to physical scene structure. In [13], Castle *et al* detect the presence of planar objects for which appearance knowledge has been learned *a priori* and then use the known geometric structure to allow insertion of the objects into the map. This gives direct relationship to physical structure but at the expense of prior user interaction.

The work reported in this paper aims to extend these methods. Specifically, we describe a novel approach to detecting and extracting planar structure in previously unseen environments using both geometric and appearance information. The latter provides direct correspondence to physical structure. We adopt a hypothesis testing strategy, in which the validity of planar patch structures derived from triangulation of mapped point features is tested against appearance information within selected frames. Importantly, this is based on a test statistic which compares matching errors against the predicted covariance derived from the SLAM filter, giving a probabilistic formulation which automatically takes account of the inherent uncertainty within the system. Results of experiments indicate that this gives both robust and consistent detection and extraction of planar structure.

2 Monocular SLAM

For completeness we start with an overview of the underlying monocular SLAM system. Such systems are now well documented, see e.g. [14], and thus we present only brief details. They provide estimates of the 3-D pose of a moving camera whilst simultaneously estimating the depth of feature points in the scene. This is based on measurements taken from the video stream captured by the camera and is done in real-time, processing the measurements sequentially as each video frame is captured. Stochastic filtering provides an ideal framework for this and we use the version based on the Kalman filter (KF) [15].

The system state contains the current camera pose $\mathbf{v} = (\mathbf{q}, \mathbf{t})$, defined by position $\mathbf{t}$ and orientation quarternion $\mathbf{q}$, and the positions of M scene points, $\mathbf{m} = (\mathbf{m}_1, \mathbf{m}_2, \ldots, \mathbf{m}_M)$. The system is defined by a process and an observation model. The former defines the assumed evolution of the camera pose (we use a constant velocity model), whilst the latter defines the relationship between the state and the measurements. These are 2-D points $(\mathbf{z}_1, \mathbf{z}_2, \ldots, \mathbf{z}_M)$, assumed to be noisy versions of the projections of a subset of 3-D map points. Both of these models are non-linear and hence the extended KF (EKF) is used to obtain sub-optimal estimates of the state mean and covariance at each time step.

This probabilistic formulation provides a coherent framework for modeling the uncertainties in the system, ensuring the proper maintenance of correlations amongst the estimated parameters. Moreover, the estimated covariances, when projected through the observation model, provide search regions for the locations of the 2-D measurements, aiding the data association task and hence minimising image processing operations. As described below, they also play a key role in the work presented in this paper.

For data association, we use the multi-scale descriptor developed by Chekhlov *et al* [4], combined with a hybrid implementation of FAST and Shi and Tomasi feature detection integrated with non-maximal suppression [5]. The system operates with a calibrated camera and feature points are initialised using the inverse depth formulation [16].

3 Detecting Planar Structure

The central theme of our work is the robust detection and extraction of planar structure in a scene as SLAM progresses. We aim to do so with minimal caching of frames, sequentially processing measurements, and taking into account the uncertainties in the system.

We adopt a hypothesis testing strategy in which we take triplets of mapped points and test the validity of the assertion that the planar patch defined by the points corresponds to a physical plane in the scene. For this we use a metric based on appearance information within the projections of the patches in the camera frames. Note that unlike the problem of detecting planar homographies in uncalibrated images [17], in a SLAM system we have access to estimates of the camera pose and hence can utilise these when testing planar hypotheses.

Consider the case illustrated in Fig. 1, in which the triangular patch defined by the mapped points $\{\mathbf{m}_1, \mathbf{m}_2, \mathbf{m}_3\}$ - we refer to these as 'control points' - is projected into two frames. If the patch corresponds to a true plane, then we could test validity simply by comparing pixel values in the two frames after transforming to take account of the relative camera positions and the plane normal. Of course, such an approach is fraught with difficulty: it ignores the uncertainty about our knowledge of the camera motion and the position of the control points, as well as the inherent ambiguity in comparing pixel values caused by lighting effects, lack of texture, etc.

Instead, we base our method on matching salient points within the projected patches and then analysing the deviation of the matches from that predicted by the filter state, taking into account the uncertainty in the estimates. We refer to these as 'test points'. The use of salient points is important since it helps to minimise ambiguity as well as reducing computational load. The algorithm can be summarised as follows:

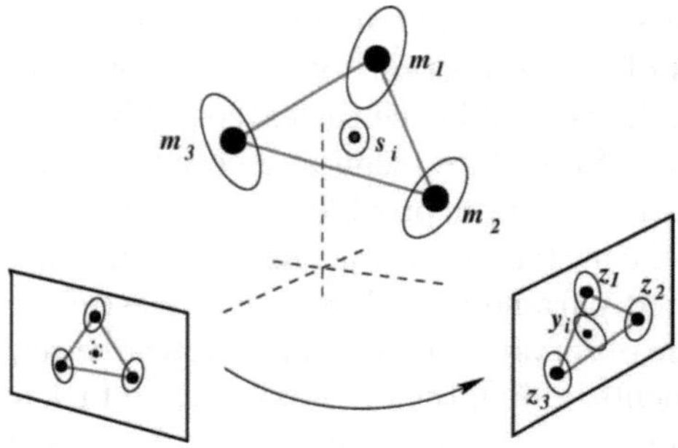

Fig. 1. Detecting planar structure: errors in matching test points y_i are compared with the predicted covariance obtained from those predicted for the control points z_i, hence taking account of estimation uncertainty within the SLAM filter

1. Select a subset of test points within the triangular patch within the reference view;
2. Find matching points within the triangular patches projected into subsequent views;
3. Check that the set of corresponding points are consistent with the planar hypothesis and the estimated uncertainty in camera positions and control points.

For (1), we use the same feature detection as that used for mapping points, whilst for (2) we use warped normalised cross correlation between patches about the test points, where the warp is defined by the mean camera positions and plane orientation. The method for checking correspondence consistency is based on a comparison of matching errors with the predicted covariances using a χ^2 test statistic as described below.

3.1 Consistent Planar Correspondence

Our central idea for detecting planar structure is that if a set of test points do indeed lie on a planar patch in 3-D, then the matching errors we observe in subsequent frames should agree with our uncertainty about the orientation of the patch. We can obtain an approximation for the latter from the uncertainty about the position of the control points derived from covariance estimates within the EKF.

Let $\mathbf{s} = (\mathbf{s}_1, \mathbf{s}_2, \ldots, \mathbf{s}_K)$ be a set of K test points within the triangular planar patch defined by control points $\mathbf{m} = (\mathbf{m}_1, \mathbf{m}_2, \mathbf{m}_3)$ (see Fig. 1). From the planarity assumption we have

$$\mathbf{s}_k = \sum_{i=1}^{3} a_{ki} \mathbf{m}_i \tag{1}$$

where the weights a_{ki} define the positions of the points within the patch and $\sum_i a_{ki} = 1$. In the image plane, let $\mathbf{y} = (\mathbf{y}_1, \ldots, \mathbf{y}_K)$ denote the perspective projections of the $\mathbf{s}_k$ and then define the following measurement model for the kth test point using linearisation about the mean projection

$$\mathbf{y}_k \approx P(\mathbf{v})\mathbf{s}_k + \mathbf{e}_k \approx \sum_{i=1}^{3} a_{ki}\mathbf{z}_i + \mathbf{e}_k \tag{2}$$

where $P(\mathbf{v})$ is a matrix representing the linearised projection operator defined by the current estimate of the camera pose, $\mathbf{v}$, and $\mathbf{z}_i$ is the projection of the control point $\mathbf{m}_i$. The vectors $\mathbf{e}_k$ represent the expected noise in the matching process and we assume these to be independent with zero mean and covariance R.

Thus we have an expression for the projected test points in terms of the projected control points, and we can obtain a prediction for the covariance of the former in terms of those for the latter, i.e. from (2)

$$C_y = \begin{bmatrix} C_y(1,1) \cdots & C_y(1,K) \\ \cdots \quad \cdots & \cdots \\ C_y(K,1) \cdots & C_y(K,K) \end{bmatrix} \tag{3}$$

in which the block terms $C_y(k,l)$ are 2×2 matrices given by

$$C_y(k,l) = \sum_{i=1}^{3}\sum_{j=1}^{3} a_{ki}a_{lj}C_z(i,j) + \delta_{kl}R \tag{4}$$

where $\delta_{kl} = 1$ for $k = l$ and 0 otherwise, and $C_z(i,j)$ is the 2×2 cross covariance of $\mathbf{z}_i$ and $\mathbf{z}_j$. Note that we can obtain estimates for the latter from the predicted innovation covariance within the EKF [15].

The above covariance indicates how we should expect the matching errors for test points to be distributed under the hypothesis that they lie on the planar patch[2]. We can therefore assess the validity of the hypothesis using the χ^2 test [15]. In a given frame, let $\mathbf{u}$ denote the vector containing the positions of the matches obtained for the set of test points $\mathbf{s}$. Assuming Gaussian statistics, the Mahalanobis distance given by

$$\epsilon = (\mathbf{u} - \mathbf{y})' C_y^{-1} (\mathbf{u} - \mathbf{y}) \tag{5}$$

then has a χ^2 distribution with $2K$ degrees of freedom. Hence, ϵ can be used as a test statistic, and comparing it with an appropriate upper bound allows assessment of the planar hypothesis. In other words, if the distribution of the errors exceeds that of the predicted covariance, then we have grounds based on appearance for concluding that the planar patch does not correspond to a physical plane in the scene. The key contribution here is that the test explicitly and rigorously takes account of the uncertainty within the filter, both in terms of the mapped points and the current estimate of the camera pose. As we show in the experiments, this yields an *adaptive test*, allowing greater variation in matching error of the test points during uncertain operation and tightening up the test when state estimates improve.

[2] Note that by 'matching errors' we refer to the difference in position of the detected matches and those predicted by the hypothesised positions on the planar patch.

We can extend the above to allow assessment of the planar hypothesis over multiple frames by considering the following time-averaged statistic over N frames

$$\bar{\epsilon}_N = \frac{1}{N} \sum_{n=1}^{N} \upsilon(n)' C_y^{-1}(n) \upsilon(n) \tag{6}$$

where $\upsilon(n) = \mathbf{u}(n) - \mathbf{y}(n)$ is the set of matching errors in frame n and $C_y^{-1}(n)$ is the prediction for its covariance derived from the current innovation covariance in the EKF. In this case, the statistic $N\bar{\epsilon}_N$ is χ^2 distributed with $2KN$ degrees of freedom [15]. Note again that this formulation is adaptive, with the predicted covariance, and hence the test statistic, adapting from frame to frame according to the current level of uncertainty. In practice, sufficient parallax between frames is required to gain meaningful measurements, and thus in the experiments we computed the above time averaged statistic at intervals corresponding to approximately 2° degrees of change in camera orientation (the 'parallax interval').

4 Experiments

We evaluated the performance of the method during real-time monocular SLAM in an office environment. A calibrated hand-held web-cam was used with a resolution of 320 × 240 pixels and a wide-angled lens with 81° FOV. Maps of around 30-40 features were built prior to turning on planar structure detection.

We adopted a simple approach for defining planar patches by computing a Delaunay triangulation [18] over the set of visible mapped features in a given reference frame. The latter was selected by the user at a suitable point. For each patch, we detected salient points within its triangular projection and patches were considered for testing if a sufficient number of points were detected and that they were sufficiently distributed. The back projections of these points onto the 3-D patch were then taken as the test points $\mathbf{s}_k$ and these were then used to compute the weights a_{ki} in (1).

The validity of the planar hypothesis for each patch was then assessed over subsequent frames at parallax intervals using the time averaged test statistic in (6). We set the measurement error covariance R to the same value as that used in the SLAM filter, i.e. isotropic with a variance of 2 pixels. A patch remaining in the 95% upper bound for the test over 15 intervals (corresponding to 30° of parallax) was then accepted as a valid plane, with others being rejected when the statistic exceeded the upper bound. The analysis was then repeated, building up a representation of planar structure in the scene. Note that our emphasis in these experiments was to assess the effectiveness of the planarity test statistic, rather than building complete representations of the scene. Future work will look at more sophisticated ways of both selecting and linking planar patches.

Figure 2 shows examples of detected and rejected patches during a typical run. In this example we used 10 test points for each patch. The first column shows the view through the camera, whilst the other two columns show two different views of the 3-D representation within the system, showing the estimates of camera pose and mapped point features, and the Delaunay triangulations. Covariances

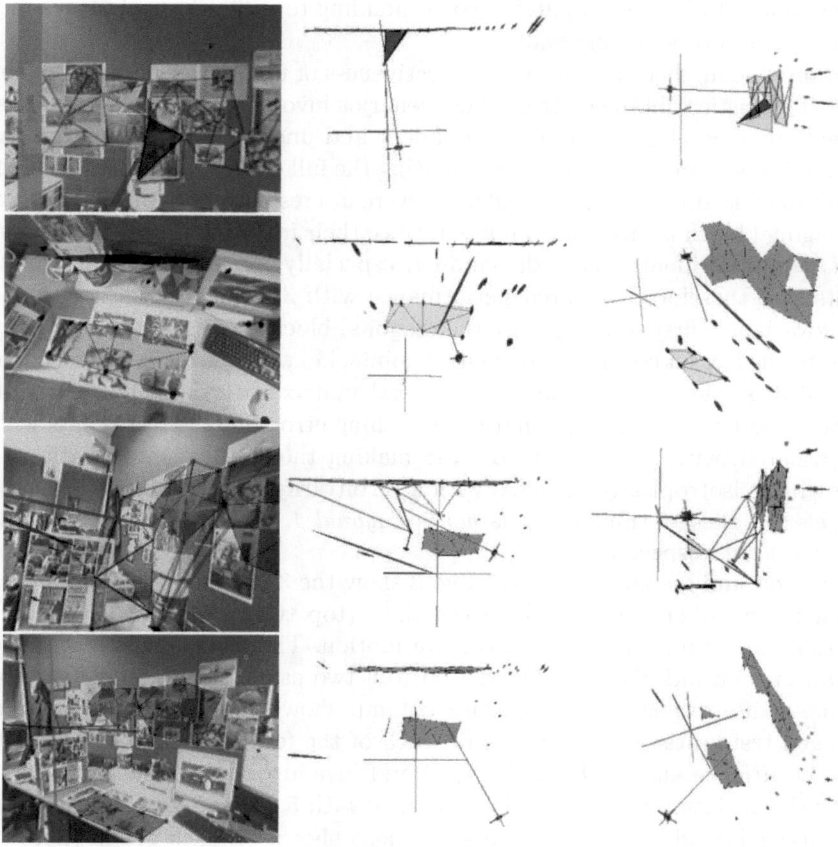

Fig. 2. Examples from a typical run of real time planar structure detection in an office environment: yellow/green patches indicate detected planes; red patches indicate rejected planes; pink patches indicate near rejection. Note that the full video for this example is available via the web link given in the abstract.

for the pose and mapped points are also shown as red ellipsoids. The first row shows the results of testing the statistic after the first parallax interval. Note that only a subset of patches are being tested within the triangulation; those not tested were rejected due to a lack of salient points. The patches in yellow indicate that the test statistic was well below the 95% upper bound, whilst those in red or pink were over or near the upper bound.

As can be seen from the 3-D representations and the image in the second row, the two red patches and the lower pink patch correspond to invalid planes, with vertices on both the background wall and the box on the desk. All three of these are subsequently rejected. The upper pink patch corresponds to a valid plane and this is subsequently accepted. The vast majority of yellow patches correspond to valid planes, the one exception being that below the left-hand red patch, but this is subsequently rejected at later parallax intervals. The other yellow patches are all accepted. Similar comments apply to the remainder of the sequence, with all the final set of detected patches corresponding to valid physical planes in the scene on the box, desk and wall.

To provide further analysis of the effectiveness of the approach, we considered the test statistics obtained for various scenarios involving both valid and invalid single planar patches during both confident and uncertainty periods of SLAM. We also investigated the significance of using the full covariance formulation in (4 within the test statistic. In particular, we were interested in the role played by the off diagonal block terms, $C_y(k,l)$, $k \neq l$, since their inclusion makes the inversion of C_y computationally more demanding, especially for larger numbers of test points. We therefore compared performance with 3 other formulations for the test covariance: first, keeping only the diagonal block terms; second, setting the latter to the largest covariance of control points, i.e. with the largest determinant; and third, setting it to a constant diagonal matrix with diagonal values of 4. These formulation all assume that the matching errors for the test points will be uncorrelated, with the last version also making the further simplification that they will be isotropically bounded with a (arbitrarily fixed) variance of 4 pixels. We refer to these formulations as *block diagonal 1*, *block diagonal 2* and *block diagonal fixed*, respectively.

The first and second columns of Fig. 3 show the 3-D representation and view through the camera for both high certainty (top two rows) and low certainty (bottom two rows) estimation of camera motion. The top two cases show both a valid and invalid plane, whilst the bottom two cases show a single valid and invalid plane, respectively. The third column shows the variation of the time averaged test statistic over frames for each of the four formulations of the test point covariance and for both the valid and invalid patches. The forth column shows the variation using the full covariance with 5, 10 and 20 test points. The 95% upper bound on the test statistic is also shown on each graph (note that this varies with frame as we are using the time averaged statistic).

The key point to note from these results is that the full covariance method performs as expected for all cases. It remains approximately constant and well below the upper bound for valid planes and rises quickly above the bound for invalid planes. Note in particular that its performance is not adversely affected by uncertainty in the filter estimates. This is in contrast to the other formulations, which, for example, rise quickly with increasing parallax in the case of the valid plane being viewed with low certainty (3rd row). Thus, with these formulations, the valid plane would eventually be rejected. Note also that the full covariance method has higher sensitivity to invalid planes, correctly rejecting them at lower parallax than all the other formulations. This confirms the important role played by the cross terms, which encode the correlations amongst the test points. Note also that the full covariance method performs well even for smaller numbers of test points. The notable difference is a slight reduction in sensitivity to invalid planes when using fewer points (3rd row, right). This indicates a trade off between sensitivity and the computational cost involved in computing the inverse covariance. In practice, we found that the use of 10 points was a good compromise.

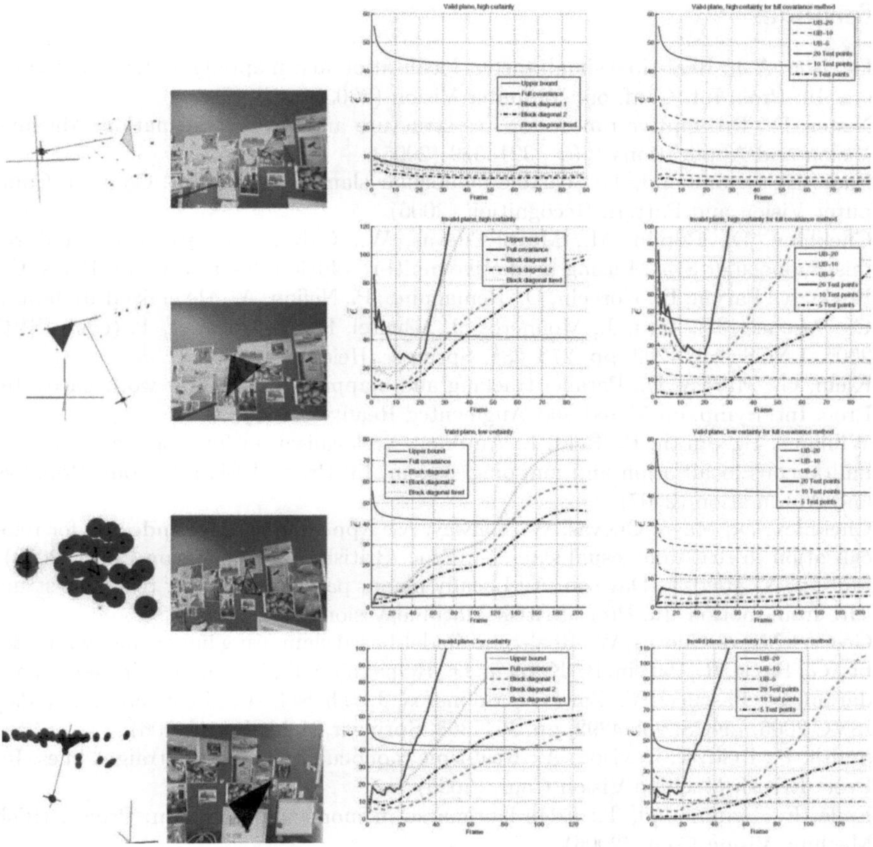

Fig. 3. Variation of the time averaged test statistic over frames for cases of valid and invalid planes during high and low certainty operation of the SLAM filter

5 Conclusions

We have presented a novel method that uses appearance information to validate planar structure hypotheses in a monocular SLAM system using a full probabilistic approach. The key contribution is that the statistic underlying the hypothesis test adapts to the uncertainty in camera pose and depth estimation within the system, giving reliable assessment of valid and invalid planar structure even in conditions of high uncertainty. Our future work will look at more sophisticated methods of selecting and combining planar patches, with a view to building more complete scene representations. We also intend to investigate the use of the resulting planar patches to gain greater stability in SLAM, as advocated in [12] and [19].

Acknowledgements. This work was funded by CONACYT Mexico under the grant 189903.

References

1. Davison, A.J.: Real-time simultaneous localisation and mapping with a single camera. In: Proc. Int. Conf. on Computer Vision (2003)
2. Nister, D.: Preemptive ransac for live structure and motion estimation. Machine Vision and Applications 16(5), 321–329 (2005)
3. Eade, E., Drummond, T.: Scalable monocular slam. In: Proc. Int. Conf. on Computer Vision and Pattern Recognition (2006)
4. Chekhlov, D., Pupilli, M., Mayol-Cuevas, W., Calway, A.: Real-time and robust monocular SLAM using predictive multi-resolution descriptors. In: Bebis, G., Boyle, R., Parvin, B., Koracin, D., Remagnino, P., Nefian, A., Meenakshisundaram, G., Pascucci, V., Zara, J., Molineros, J., Theisel, H., Malzbender, T. (eds.) ISVC 2006. LNCS, vol. 4292, pp. 276–285. Springer, Heidelberg (2006)
5. Klein, G., Murray, D.: Parallel tracking and mapping for small ar workspaces. In: Proc. Int. Symp. on Mixed and Augmented Reality (2007)
6. Williams, B., Smith, P., Reid, I.: Automatic relocalisation for a single-camera simultaneous localisation and mapping system. In: Proc. IEEE Int. Conf. Robotics and Automation (2007)
7. Chekhlov, D., Mayol-Cuevas, W., Calway, A.: Appearance based indexing for relocalisation in real-time visual slam. In: Proc. British Machine Vision Conf. (2008)
8. Molton, N., Ried, I., Davison, A.: Locally planar patch features for real-time structure from motion. In: Proc. British Machine Vision Conf. (2004)
9. Gee, A., Mayol-Cuevas, W.: Real-time model-based slam using line segments. In: Bebis, G., Boyle, R., Parvin, B., Koracin, D., Remagnino, P., Nefian, A., Meenakshisundaram, G., Pascucci, V., Zara, J., Molineros, J., Theisel, H., Malzbender, T. (eds.) ISVC 2006. LNCS, vol. 4292, pp. 354–363. Springer, Heidelberg (2006)
10. Smith, P., Reid, I., Davison, A.: Real-time monocular slam with straight lines. In: Proc. British Machine Vision Conf. (2006)
11. Eade, E., Drummond, T.: Edge landmarks in monocular slam. In: Proc. British Machine Vision Conf. (2006)
12. Gee, A., Chekhlov, D., Calway, A., Mayol-Cuevas, W.: Discovering higher level structure in visual slam. IEEE Trans. on Robotics 24(5), 980–990 (2008)
13. Castle, R.O., Gawley, D.J., Klein, G., Murray, D.W.: Towards simultaneous recognition, localization and mapping for hand-held and wearable cameras. In: Proc. Int. Conf. Robotics and Automation (2007)
14. Davison, A., Reid, I., Molton, N., Stasse, O.: Monoslam: Real-time single camera slam. IEEE Trans. on Pattern Analysis and Machine Intelligence 29(6), 1052–1067 (2007)
15. Bar-Shalom, Y., Kirubarajan, T., Li, X.: Estimation with Applications to Tracking and Navigation (2002)
16. Civera, J., Davison, A., Montiel, J.: Inverse depth to depth conversion for monocular slam. In: Proc. Int. Conf. Robotics and Automation (2007)
17. Hartley, R.I., Zisserman, A.: Multiple View Geometry in Computer Vision. Cambridge University Press, Cambridge (2000)
18. Renka, R.J.: Algorithm 772: Stripack: Delaunay triangulation and voronoi diagram on the surface of a sphere. In: ACM Trans. Math. Softw., vol. 23, pp. 416–434. ACM, New York (1997)
19. Pietzsch, T.: Planar features for visual slam. In: Dengel, A.R., Berns, K., Breuel, T.M., Bomarius, F., Roth-Berghofer, T.R. (eds.) KI 2008. LNCS, vol. 5243. Springer, Heidelberg (2008)

A New Triangulation-Based Method for Disparity Estimation in Image Sequences

Dimitri Bulatov, Peter Wernerus, and Stefan Lang

Research Institute for Optronics and Pattern Recognition,
Gutleuthausstr. 1, 76275 Ettlingen, Germany
{bulatov,wernerus,lang}@fom.fgan.de

Abstract. We give a simple and efficient algorithm for approximating computation of disparities in a pair of rectified frames of an image sequence. The algorithm consists of rendering a sparse set of correspondences, which are triangulated, expanded and corrected in the areas of occlusions and homogeneous texture by a color distribution algorithm. The obtained approximations of the disparity maps are refined by a semi-global algorithm. The algorithm was tested for three data sets with rather different data quality. The results of the performance of our method are presented and areas of applications and future research are outlined.

Keywords: Color, dense, depth map, disparity map, histogram, matching, reconstruction, semi-global, surface, triangulation.

1 Introduction

Retrieving dense three-dimensional point clouds from monocular images is the key-issue in a large number of computer vision applications. In the areas of navigation, civilian emergency and military missions, the need for fast, accurate and robust retrieving of disparity maps from small and inexpensive cameras is rapidly growing. However, the matching process is usually complicated by low resolution, occlusion, weakly textured regions and image noise. In order to compensate these negative effects, robust state-of-the-art methods such as [2], [10], [13], [20], are usually global or semi-global, i.e. the computation of matches is transformed into a global optimization problem. Therefore all these methods require high computational costs. On the other hand, the local methods, such as [3], [12], are able to obtain dense sets of correspondences, but the quality of the disparity maps obtained by these methods is usually below the quality achieved by global methods.

In our applications, image sequences are recorded with handheld or airborne cameras. Characteristic points are found by means of [8] or [15] and the fundamental matrices are computed from the point correspondences by robust algorithms (such as a modification of RANSAC [16]). As a further step, the structure and motion can be reconstructed using tools described in [9]. If the cameras are not calibrated, the reconstruction can be carried out in a projective coordinate system and afterwards upgraded to a metric reconstruction using methods

A.-B. Salberg, J.Y. Hardeberg, and R. Jenssen (Eds.): SCIA 2009, LNCS 5575, pp. 279–290, 2009.

of auto-calibration ([9], Chapter 19). The point clouds thus obtained have extremely irregular density: Areas with a sparse density of points arising from homogeneous regions in the images are usually quite close to areas with high density resulting from highly textured areas. In order to reconstruct the surface of the unknown terrain, it is extremely important to obtain a homogeneous density of points. In this paper, we want to enrich the sparse set of points by a dense set, i.e. to predict the position in space of (almost) every pixel in every image. It is always useful to consider *all* available information in order to facilitate the computation of such dense sets. Beside methods cited above and those which were tested in the survey due to Scharstein and Szeliski [21], there are several methods which combine the approaches of disparity estimation and surface reconstruction. In [1], for example, the authors propose to initialize layers in the images which correspond to (almost) planar surfaces in space. The correspondences of layers in different images are thus given by homographies induced by these surfaces. Since the surface is not really piecewise planar, the authors introduce the distances between the point on the surface and its planar approximation at each pixel as additional parameters. However, it is difficult to initialize the layers without prior knowledge. In addition, the algorithm could have problems in the regions which belong to the same segment but have depth discontinuities. In [19], the Delaunay triangulation of points already determined is obtained; [18] proposes using edge-flip algorithms in order to obtain a better triangulation since the edges of Delaunay-triangles in the images are not likely to correspond to the object edges. Unfortunately, the sparse set of points usually produces a rather coarse estimation of disparity maps; also, this method can not detect occlusions. In this paper, we will investigate to what extent disparity maps can be initialized by triangular meshes in the images.

In the method proposed here, we will use the set of sparse point correspondences $\mathbf{x} = \mathbf{x}_1 \leftrightarrow \mathbf{x}_2$ to create initial disparity maps from the support planes for the triangles with vertices in $\mathbf{x}$. The set $\mathbf{x}$ will then be iteratively enriched. Furthermore, in the areas of weak texture and gradient discontinuities, we will investigate to what extent the color distribution algorithms can detect the outliers and occlusions among the triangle vertices and edges. Finally, we will use the result of the previous steps as an initial value for the global method [10], which uses a random disparity map as input. The necessary theoretical background will be described in Sec. 2.1 and the three steps mentioned above in Sec. 2.2, 2.3, and 2.4. The performance of our method is compared with semi-global algorithms without initial estimation of disparities in Sec. 3. Finally, Sec. 4 provides the conclusions and the research fields of the future work.

2 Our Method

2.1 Preliminaries

Suppose that we have obtained the set of sparse point correspondences and the set of camera matrices in a projective coordinate system, for several images of an airborne or handheld image sequence. The fundamental matrix can be extracted from any pair of cameras according to the formula (9.1) of [9]. In order to facilitate the search for correspondences in a pair of images, we perform *image rectification*, i.e. we transform the images and points by two homographies to make the corresponding points (denoted by $\mathbf{x}_1, \mathbf{x}_2$) have the same y-coordinates. In the rectification method we chose, [14], the epipoles $\mathbf{e}_1, \mathbf{e}_2$ must be transformed to the point at infinity $(1,0,0)^T$, therefore $\mathbf{e}_1, \mathbf{e}_2$ must be bounded away from the image domain in order to avoid significant distortion of the images. We can assume that such a pair of images with enough overlap can be chosen from the entire sequence. We also assume that the percentage of outliers among the points in $\mathbf{x} = \mathbf{x}_1$ is low because most of the outliers are supposed to be eliminated by robust methods. Finally, we remark that we are not interested to compute correspondences of all points inside of the overlap of both rectified images (which will be denoted by I_1 respectively I_2) but restrict ourselves to the convex hull of the points in $\mathbf{x}$. Computing point correspondences of pixels outside of the convex hulls does not make much sense since they often do not lie in the overlap area and, especially in the case of uncalibrated cameras, suffer more from the lens distortion effects. One should better use another pair of images to compute disparities for these points.

Now suppose we have a partition of $\mathbf{x}$ into triangles. Hereafter, $\breve{\mathbf{p}}$ denotes the homogeneous representation of a point $\mathbf{p}$; T represents a triple of integer numbers; thus, $\mathbf{x}_{1,T}$ are the columns of $\mathbf{x}_1$ specified by T. By $\mathbf{p}_1 \in T$, we will denote that the pixel $\mathbf{p}_1$ in the first rectified image lies in triangle $\mathbf{x}_{1,T}$. Given such a partition, every triangle can be associated with its support plane which induces a triangle-to-triangle homography. This homography only possesses three degrees of freedom which are stored in its first row since the displacement of a point in a rectified image only concerns its x-coordinate.

Result 1: Let $\mathbf{p}_1 \in T$ and let $\mathbf{x}_{1,T}, \mathbf{x}_{2,T}$ be the coordinates of the triangle vertices in the rectified images. The homography induced by T maps $\mathbf{x}_1$ onto the point $\mathbf{p}_2 = (X_2, Y)$, where $X_2 = \mathbf{v}\breve{\mathbf{p}}_1$, $\mathbf{v} = x_{2,T}\left(\breve{\mathbf{x}}_{1,T}\right)^{-1}$, and $x_{2,T}$ is the row vector consisting of x-coordinates of $\mathbf{x}_{2,T}$.

Proof: Since triangle vertices $\mathbf{x}_{1,T}, \mathbf{x}_{2,T}$ are corresponding points, their correct locations are on the corresponding epipolar lines. Therefore they have pairwise the same y-coordinates. Moreover, the epipole is given by $\mathbf{e}_2 = (1,0,0)^T$ and the fundamental matrix is $F = [\mathbf{e}_2]_\times$. Inserting this information into Result 13.6 of [9], p. 331 proves, after some simplifications, the statement of **Result 1**.

Determining and storing the entries of $\mathbf{v} = \mathbf{v}_T$ for each triangle, optionally refining $\mathbf{v}$ for the triangles in the big planar regions by error minimization and calculating disparities according to **Result 1** provide, in many cases, a coarse approximation for the disparity map in the areas where the surface is approximately piecewise planar and does not have many self-occlusions.

2.2 Initialization of Disparity Maps Given from Triangulations

Starting from the Delaunay-Triangulation obtained from several points in the image, we want to expand, because the first approximation is too coarse, the quantity of points. Since the fundamental matrix obtained from structure-from-motion algorithms is noisy, it is necessary to search for correspondences not only in the direction along the epipolar lines but also in the vertical direction. We suppose that the distance of a pair of corresponding points to the corresponding epipolar lines to be bounded by 1 pel. Therefore, given a point $\mathbf{p}_1 = (X_1, Y_1) \in T$, we consider the search window in the second image given by:

$$\begin{aligned} W_s &= [X_1 + X_{\min}; X_1 + X_{\max}] \times [Y-1; Y+1], \\ X_{\min} &= \max(d_{\min} - \varepsilon, \min(s_T)), X_{\max} = \min(d_{\max} + \varepsilon, \max(s_T)) \end{aligned} \tag{1}$$

where $\varepsilon = 3$ is a fixed scalar, s_T are the x-coordinates of at most six intersection points between the epipolar lines at $Y, Y-1, Y+1$ and the edges of $\mathbf{x}_{1,T}$ and $d_{\min}, d_{\max}$ are the estimates of smallest and biggest possible disparities which can be obtained from the point coordinates.

The search for correspondent points succeeds by means of the *normalized cross correlation* (NCC) algorithm between the quadratic window $I_1(W(\mathbf{p}_1))$ of size between 5 and 10 pixels and $I_2(W_s)$. However, in order to avoid including mismatches into the set of correspondences, we impose three filters on the result of the correlation. A pair of points $\mathbf{p}_1 = (X_1, Y)$ and $\mathbf{p}_2 = (X_2, Y)$ is added to the set of correspondences if and only if: 1. the correlation coefficient c_0 of the winner exceeds a user-specified value $c_{\min}$ (= 0.7-0.9 in our experiments), 2. the windows have approximately the same luminance, i. e. $\|I_1(W(\mathbf{p}_1)) - I_2(W(\mathbf{p}_2))\|_1 < |W|u_{\max}$ where $|W|$ is the number of pixels in the window and $u_{\max} = 15$ in our experiments, and, 3. in order to avoid erroneous correspondences along epipolar lines which coincide with edges in the images, we eliminate the matches where the ratio of the maximal correlation coefficient in the sub-windows

$$([X_{\min}; X_2 - 1] \cup [X_2 + 1; X_{\max}]) \times [Y-1; Y+1], \tag{2}$$

and c_0 (second-best to best) exceeds a threshold γ, which is usually 0.9. Here $X_{\min}, X_{\max}$ in (2), are specified according to (1). An alternative way to handle the mismatches is using more cameras, as described, for example, in [7]. Further research on this topic will be part of our future work.

Three concluding remarks will be given at the end of present subsection:

1. It is not necessary to use *every* point in *every* triangle for determining corresponding points. It is recommendable not to search corresponding points in lowly textured areas but to take the points with a maximal (within a small window) response of a suitable point-detector. In our implementation, it is the *Harris-operator*, see [8], so the *structural tensor* A for a given image as well as the "cornerness" term $\text{trace}(A) - 0.04\det(A)$ can be precomputed and stored once for all.

2. It also turned out to be helpful to subdivide only triangles with area exceeding a reasonable threshold (100-500 pel^2 in our experiments) and *non-compatible* with the surface, which means that the highest correlation coefficient for the barycenter $\mathbf{p}_1$ of the triangle T was obtained at X_2 and for $\mathbf{v} = \mathbf{v}_T$ computed according to **Result 1**, we have $|\mathbf{v}\check{\mathbf{p}} - X_2| > 1$. After obtaining correspondences, the triangulation could be refined by using edge-flipping algorithms, but in the current implementation, we do not follow this approach.
3. The coordinates of corresponding points can be refined to subpixel values, according to one of four methods discussed in [23]. For the sake of computation time, subpixel coordinates for correspondences are computed according to correlation parabolas. We denote by c_- and c_+ the correlation values in the pixels left and right from X_2. The correction term $\hat{X}_2$ in x-direction is then given by:
$$\hat{X}_2 = X_2 - \frac{c_+ - c_-}{2(c_- + c_+ - 2c_0)}.$$
Also the value of X_2 is corrected for triangles compatible with the surface according to **Result 1**.

2.3 Color-Distribution Algorithms for Occlusion Detection

The main drawback of the initialization with an (expanded) set of disparities are the outliers in the data as well as the occlusions since the sharp edge of depth in the triangle on the left and on the right of edge with disparity discontinuities will be blurred. While the outliers can be efficiently eliminated by means of disparities of their neighbors (a procedure which we apply once before and once after the expansion), in the case of occlusions, we shall investigate how the color-distribution algorithms can restore the disparities at the edges of discontinuities.

At present, we mark all triangles for which the standard deviation of disparities at the vertexes exceeds a user-specified threshold ($\sigma_0 = 2$ in our experiments) as unfeasible. Given a list of unfeasible triangles, we want to find similar triangles in the neighborhood. In our approach this similarity is based on color distribution represented by three histograms, each for a different color in the color space RGB (red, green and blue).

A histogram is defined over the occurrence of different color values of the pixels inside the considered triangle T. Each color contains values from 0 to 255, thus each color histogram has b bins with a bin size of $256/b$. Let the number of pixels in a triangle be n. In order to obtain the probability of this distribution and to make it independent of the size of the triangle, we obtain for the i-th bin of the *normalized histogram*

$$H_T(i) = \frac{1}{n} \cdot \# \left\{ \mathbf{p} \,\middle|\, \mathbf{p} \in T \text{ and } \frac{256 \cdot i}{b} \leq I_1(\mathbf{p}) < \frac{256 \cdot (i+1)}{b} \right\}.$$

The three histograms H_T^R, H_T^G, H_T^B represent the color distribution of the considered triangle. It is also useful to split big, inhomogeneous, unfeasible triangles into smaller ones. To perform splitting, characteristic edges ([4]) are found in every candidate triangle and saved in form of a binary image $G(\mathbf{p})$.

To find the line with maximum support, we apply the radon transformation ([6]) to $G(\mathbf{p})$:

$$\breve{G}(u,\varphi) = R\{G(\mathbf{p})\} = \int_{-\infty}^{\infty}\int_{-\infty}^{\infty} G(\mathbf{p})\delta(\mathbf{p}^T\mathbf{e}_\varphi - u)d\mathbf{p}$$

with the Dirac delta function $\delta(x) = \infty$ if $x = 0$ and 0 otherwise and line parameters $\mathbf{p}^T\mathbf{e}_\varphi - u$, where $\mathbf{e}_\varphi = (cos\varphi, sin\varphi)^T$ is the normal vector and u the distance to origin. The strongest edge in the triangle is found if the maximum of $\breve{G}(u,\varphi)$ is over a certain threshold for the minimum line support. This line intersects the edges of the considered triangle T in two intersection points. We disregard intersection points too close to a vertex of T. If new points were found, the original triangle is split in two or three smaller triangles. These new smaller triangles consider the edges in the image.

Next the similarity of two neighboring triangles has to be calculated by means of the color distribution. Two triangles are called neighbors if they share at least one vertex. There are a lot of different approaches measuring the distance between histograms [5]. In our case we define the distance of two neighboring triangles T_1 and T_2 as follows:

$$d(T_1,T_2) = w_R \cdot d\left(H^R_{T_1}, H^R_{T_2}\right) + w_G \cdot d\left(H^G_{T_1}, H^G_{T_2}\right) + w_B \cdot d\left(H^B_{T_1}, H^B_{T_2}\right) \quad (3)$$

where w_R, w_G, w_B are different weights for the colors. The distance between two histograms in (3) is the sum of absolute differences of their bins.

In the next step, the disparity in the vertices of unfeasible triangles will be corrected. Given an unfeasible triangle T_1, we define

$$T_2 = \text{argmin}_T \{d(T_1,T)|\text{area } (T) > A_0, d(T_1,T) < c_0 \text{ and } T \text{ is not unfeasible}\},$$

where $c_0 = 2$, $A_0 = 30$ and $d(T_1,T)$ is computed according to (3). If such T_2 does exist, we recompute the disparities of pixels in T_1 with $\mathbf{v}_{T_2}$ according to **Result 1**. Usually this method performs rather well as long as the assumption holds that neighboring triangles with similar color information lie indeed in the same planar region of the surface.

2.4 Refining of the Results with a Global Algorithm

Many dense stereo correspondence algorithms improve their disparity map estimation by minimizing disparity discontinuities. The reason is that neighboring pixels probably map to the same surface in the scene, and thus their disparity should not differ much. This could be achieved by minimizing the energy

$$E(D) = \sum_{\mathbf{p}} \left\{ C(\mathbf{p}, d_{\mathbf{p}}) + P_1 \cdot N_{\mathbf{p}}(1) + P_2 \cdot \sum_{i=2}^{\infty} N_{\mathbf{p}}(i) \right\}, \quad (4)$$

where $C(\mathbf{p}, d)$ is the cost function for disparity $d_{\mathbf{p}}$ at pixel $\mathbf{p}$; P_1, P_2, with $P_1 < P_2$ are penalties for disparity discontinuities and $N_{\mathbf{p}}(i)$ is the number of pixels $\mathbf{q}$ in the neighborhood of $\mathbf{p}$ for which $|d_{\mathbf{p}} - d_{\mathbf{q}}| = i$. Unfortunately, the minimization of (4) is NP-hard. Therefore an approximation is needed. One approximation method yielding good results, while simultaneously being computational fast compared to many other approaches, was developed by Hirschmüller [10].

This algorithm, called Semi-Global Matching (SGM), uses mutual information for matching cost estimation and a path approach for energy minimization. The matching cost method is an extension of the one suggested in [11]. The accumulation of corresponding intensities to a probability distribution from an initial disparity map is the input for the cost function to be minimized. The original approach is to start using a random map and iteratively calculate improved maps, which are used for a new cost calculation. To speed up this process, Hirschmüller first iteratively halves the original image by downsampling it, thus creating image pyramids. The random initialization and first disparity approximation take place at lowest scale and are iteratively upscaled until the original scale is achieved.

To approximate the energy functional $E(D)$, paths from 16 different directions leading into one pixel are accumulated. The cost for one path in direction $\mathbf{r}$ ending in pixel $\mathbf{p}$ is recursively defined as: $L_{\mathbf{r}}(\mathbf{p}, d) = C(\mathbf{p}, d)$ for $\mathbf{p}$ near image border and

$$L_{\mathbf{r}}(\mathbf{p}, d) = C(\mathbf{p}, d) + \min[L_{\mathbf{r}}(\mathbf{p}-\mathbf{r}, d), L_{\mathbf{r}}(\mathbf{p}-\mathbf{r}, d\pm 1) + P_1, \min_i (L_{\mathbf{r}}(\mathbf{p}-\mathbf{r}, i)) + P_2]$$

otherwise. The optimal disparity for pixel $\mathbf{p}$ is then determined by summing up costs of all paths of the same disparity and choosing the disparity with the lowest result. Our method comes in as a substitution for the random initialization and iterative improvement of the matching cost. The disparity map achieved by our algorithm is simply used to compute the cost function once without iterations.

In the last step, the disparity map in the opposite direction is calculated. Pixels with corresponding disparities are considered correctly estimated, the remaining pixels occluded.

3 Results

In this section, results from three data sets will be presented. The first data set is taken from the well known *Tsukuba* benchmark-sequence. No camera rectification was needed since the images are already aligned. Although we do not consider this image sequence as characteristic for our applications, we decided to demonstrate the performance of our algorithm for a data set with available ground truth. In the upper row of Fig. 1, we present the ground truth, the result of our implementation of [10] and the result of depth maps estimation initialized with ground truth. In the bottom row, one sees from left to right, the result of Step 1 of our algorithm described in Sec. 2.2, the correction of the result as described in Step 2 (Sec. 2.3) and the result obtained by Hirschmüller algorithm as described in Sec. 2.4 with initialization. The disparities are drawn in pseudo-colors and with occlusions marked in black.

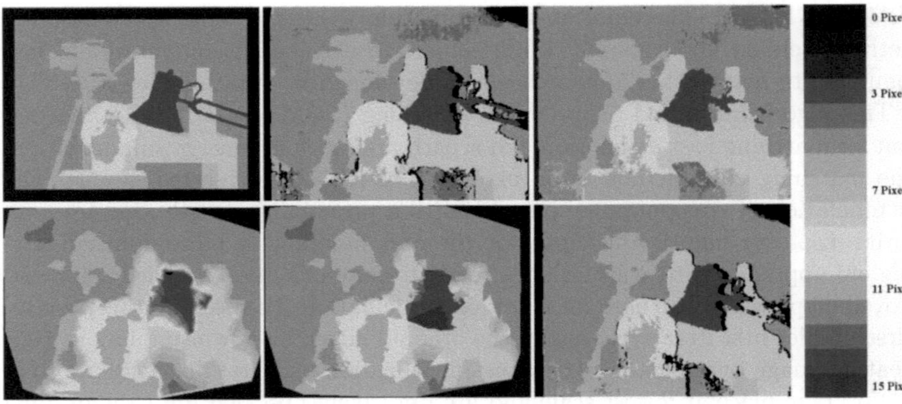

Fig. 1. Top row, left to right: the ground truth from the sequence *Tsukuba*, the result of disparity map rendered by [10], the result of disparity map rendered by [10] initialized with ground truth. Bottom row, left to right: initialization of the disparity map created Step 1 by our algorithm, initialization of the disparity map created Step 2 by our algorithm and the result of [10] with initialization. Right: color scale representing different disparity values.

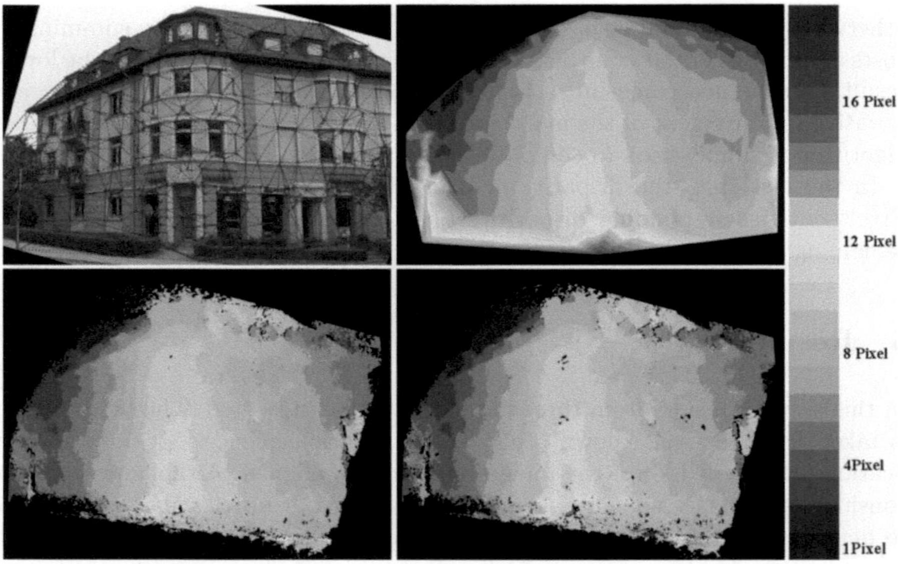

Fig. 2. Top row: left: a rectfied image from the sequence *Old House* with the mesh from the point set in the rectified image; right: initialization of the disparity map created by our algorithm. Bottom row: results of [10] with and without initialization. Right: color scale representing disparity values.

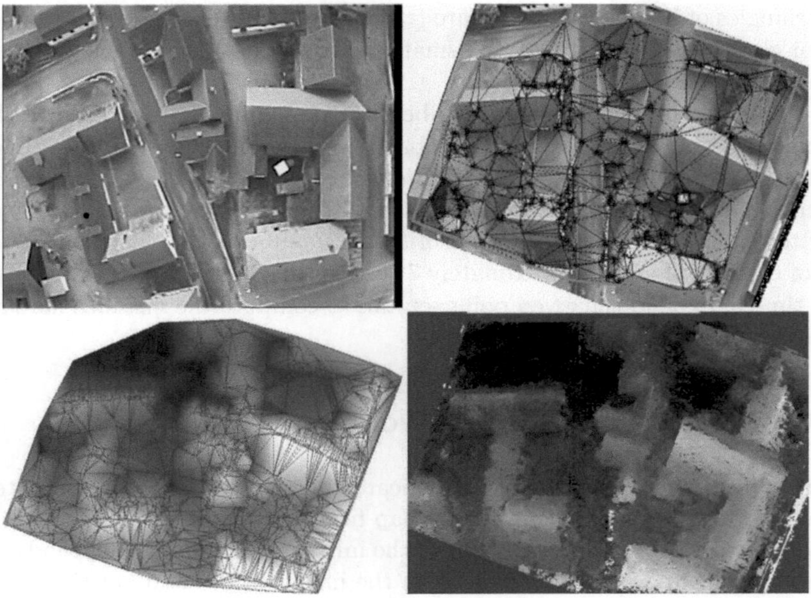

Fig. 3. Top row: left: a frame from the sequence *Bonnland*; right: the rectified image and mesh from the point set. Bottom row: initialization of the disparity map created by our algorithm with the expanded point set and the result of [10] with initialization.

The data set *Old House* shows a view of a building in Ettlingen, Germany, recorded by a handheld camera. In the top row of Fig. 2, the rectified image with the triangulated mesh of points detected with [8] as well as the disparity estimation by our method is shown. The bottom row shows the results of the disparity estimation with (left) and without (right) initialization drawn with pseudo-colors and with occlusions marked in black.

The data set *Bonnland* was taken from a small unmanned aerial vehicle which carries a small inexpensive camera on board. The video therefore suffers from reception disturbances, lens distortion effects and motion blur. However, obtaining fast and feasible depth information from these kinds of sequences is very important for practical applications. In the top row of Fig. 3, we present a frame of the sequence and the rectified image with triangulated mesh of points. The convex hull of the points is indicated by a green line. In the bottom row, we present the initialization obtained from the expanded point set as well as the disparity map computed by [10] with initialization and occlusions marked in red.

The demonstrated results show that in many practical applications, the initialization of disparity maps from already available point correspondences is a feasible tool for disparity estimation. The results are the more feasible, the more the surface is piecewise planar and the less occlusions as well as segments of the same color lying in different support planes there are. The algorithm maps well triangles of homogeneous texture (compatible with the surface), while even a semi-global method produces mismatches in these areas, as one can see in the areas in front of the house in Fig. 2 and in some areas of Fig. 3. The results obtained with the method described in Sec. 2.2 and 2.3 usually provide an acceptable initialization for a semi-global algorithm. The computation time for our implementation of [10] *without initialization* was around 80 seconds for the sequence *Bonnland* (two frames of size 823 × 577 pel, the algorithm run twice in order to detect occlusions) and *with initialization* about 10% faster. The difference of elapsed times is approximately 7 seconds and it takes approximately the same time to expand the given point set and to compute the distance matrix for correcting unfeasible triangles.

4 Conclusions and Future Work

The results presented in this paper indicate that it is possible to compute acceptable initialization of the disparity map from a pair of images by means of a sparse point set. The computing time of the initialization does not depend on the disparity range and is less dependent on the image size as state-of-the-art local and global algorithms since a lower point density not necessarily means worse results. Given an appropriate point detector, our method is able to consider pairs of images with different radiometric information. In this contribution, for instance, we extract depths maps from different frames of the same video sequence, so the correspondences of points are likely to be established from intensity differences; but in the case of pictures with significantly different radiometry, one can take the SIFT-operator ([15]) as a robust point detector and the cost function will be given by the scalar product of the descriptors.

The enriched point clouds may be used as input for scene and surface reconstruction algorithms. These algorithms benefit from a regular density of points, which makes the task of fast and accurate retrieving additional 3D-points (especially) in the areas of low texture extremely important. It is therefore necessary to develop robust color distribution algorithms to perform texture analysis and to correct unfeasible triangles, as we have indicated in Sec. 2.3.

The main drawback of Sec. 2.2 are outliers among the new correspondences as well as occlusions which are not always corrected at later stages. Since the initialization of disparities is spanned from triangles, the complete regions around these points will be given wrong disparities. It has been shown that using redundant information given from more than two images ([22], [7]) can significantly improve the performance; therefore we will concentrate our future efforts on integration of multi-view-systems into our triangulation networks. Another interesting aspect will be the integration of 3D-information given from calibrated cameras into the process of robust determination of point correspondences, as described, for example, in [17], [7]. Moreover, we want to investigate how the expanded point clouds can improve the performance of the state-of-the-art surface reconstruction algorithms.

References

1. Baker, S., Szeliski, R., Anandan, P.: A layered approach to stereo reconstruction. In: Computer Vision and Pattern Recognition (CVPR), pp. 434–441 (1998)
2. Bleyer, M., Gelautz, M.: Simple but Effective Tree Structures for Dynamic Programming-based Stereo Matching. In: International Conference on Computer Vision Theory and Applications (VISAPP), (2), pp. 415–422 (2008)
3. Boykov, Y., Veksler, O., Zabih, R.: A variable window approach to early vision. IEEE Transactions on Pattern Analysis and Machine Intelligence (TPAMI) 20(12), 1283–1294 (1998)
4. Canny, J.A.: Computational approach to edge detection. IEEE Transactions on Pattern Analysis and Machine Intelligence (TPAMI) 8(6), 679–698 (1986)
5. Cha, S.-H., Srihari, S.N.: On measuring the distance between histograms. Pattern Recognition 35(6), 1355–1370 (2002)
6. Deans, S.: The Radon Transform and Some of Its Applications. Wiley, New York (1983)
7. Furukawa, Y., Ponce, J.: Accurate, Dense, and Robust Multi-View Stereopsis. In: Proc. of the IEEE Conference on Computer Vision and Pattern Recognition (CVPR), Anchorage, USA, pp. 1–8 (2008)
8. Harris, C.G., Stevens, M.J.: A Combined Corner and Edge Detector. In: Proc. of 4th Alvey Vision Conference, pp. 147–151 (1998)
9. Hartley, R., Zisserman, A.: Multiple View Geometry in Computer Vision. Cambridge University Press, Cambridge (2000)
10. Hirschmüller, H.: Accurate and Efficient Stereo Processing by Semi-Global Matching and Mutual Information. In: Proc. of the IEEE Conference on Computer Vision and Pattern Recognition (CVPR), (2), San Diego, USA, pp. 807–814 (2005)
11. Kim, J., Kolmogorov, V., Zabih, R.: Visual correspondence using energy minimization and mutual information. In: Proc. of International Conference on Computer Vision (ICCV), (2), pp. 1033–1040 (2003)
12. Klaus, A., Sormann, M., Karner, K.: Segment-Based Stereo Matching Using Belief Propagation and a Self-Adapting Dissimilarity Measure. In: Proc. of International Conference on Pattern Recognition, (3), pp. 15–18 (2006)
13. Kolmogorov, V., Zabih, R.: Computing visual correspondence with occlusions using graph cuts. In: Proc. of International Conference on Computer Vision (ICCV), (2), pp. 508–515 (2001)
14. Loop, C., Zhang, Z.: Computing rectifying homographies for stereo vision. Technical Report MSR-TR-99-21, Microsoft Research (1999)
15. Lowe, D.G.: Distinctive Image Features from Scale-Invariant Keypoints. International Journal of Computer Vision (IJCV) 60(2), 91–110 (2004)
16. Matas, J., Chum, O.: Randomized Ransac with $T_{d,d}$-test. Image and Vision Computing 22(10), 837–842 (2004)
17. Mayer, H., Ton, D.: 3D Least-Squares-Based Surface Reconstruction. In: Photogrammetric Image Analysis (PIA 2007), (3), Munich, Germany, pp. 69–74 (2007)
18. Morris, D., Kanade, T.: Image-Consistent Surface Triangulation. In: Proc. of the IEEE Conference on Computer Vision and Pattern Recognition (CVPR), (1), Los Alamitos, pp. 332–338 (2000)
19. Nistér, D.: Automatic dense reconstruction from uncalibrated video sequences. PhD Thesis, Royal Institute of Technology KTH, Stockholm, Sweden (2001)
20. Scharstein, D., Szeliski, R.: Stereo matching with nonlinear diffusion. International Journal of Computer Vision (IJCV) 28(2), 155–174 (1998)

21. Scharstein, D., Szeliski, R.: A taxonomy and evaluation of dense two-frame stereo correspondence algorithms. International Journal of Computer Vision (IJCV) 47(1), 7–42 (2002)
22. Stewart, C.V., Dyer, C.R.: The Trinocular General Support Algorithm: A Three-camera Stereo Algorithm For Overcoming Binocular Matching Errors. In: Second International Conference on Computer Vision (ICCV), pp. 134–138 (1988)
23. Tian, Q., Huhns, M.N.: Algorithms for subpixel registration. In: Graphical Models and Image Processing (CVGIP), vol. 35, pp. 220–233 (1986)

Sputnik Tracker: Having a Companion Improves Robustness of the Tracker

Lukáš Cerman, Jiří Matas, and Václav Hlaváč

Czech Technical University
Faculty of Electrical Engineering, Center for Machine Perception
121 35 Prague 2, Karlovo naměstí 13, Czech Republic
{cermal1,hlavac}@fel.cvut.cz, matas@cmp.felk.cvut.cz

Abstract. Tracked objects rarely move alone. They are often temporarily accompanied by other objects undergoing similar motion. We propose a novel tracking algorithm called Sputnik[1] Tracker. It is capable of identifying which image regions move coherently with the tracked object. This information is used to stabilize tracking in the presence of occlusions or fluctuations in the appearance of the tracked object, without the need to model its dynamics. In addition, Sputnik Tracker is based on a novel template tracker integrating foreground and background appearance cues. The time varying shape of the target is also estimated in each video frame, together with the target position. The time varying shape is used as another cue when estimating the target position in the next frame.

1 Introduction

One way to approach the tracking and scene analysis is to represent an image as a collection of independently moving planes [1,2,3,4]. One plane (layer) is assigned to the background, the remaining layers are assigned to the individual objects. Each layer is represented by its appearance and support (segmentation mask). After initialization, the motion of every layer is estimated in each step of the video sequence together with the changes of its appearance and support.

The layer-based approach has found its applications in video insertion, sprite-based video compression, and video summarization [2]. For the purpose of a single object tracking, we propose a similar method using only one foreground layer attached to the object and one background layer. Other objects, if present, are not modelled explicitly. They become parts of the background outlier process. Such approach can be also viewed as a generalized background subtraction combined with an appearance template tracker.

Unlike background subtraction based techniques [5,6,7,8], which model only background appearance, or appearance template trackers, which usually model only the foreground appearance [9,10,11,12], the proposed tracker uses the complete observation model which makes it more robust to appearance changes in both foreground and background.

The image-based representation of both foreground and background, inherited from the layer-based approaches, contrasts with statistical representations used by classifiers [13] or discriminative template trackers [14,15], which do not model the spatial structure of the layers. The inner structure of each layer can be useful source of information for localizing the layer.

[1] Sputnik, pronounced \'sput-nik in Russian, was the first Earth-orbiting satellite launched in 1957. According to Merriam-Webster dictionary, the English translation of the Russian word *sputnik* is a travelling companion.

A.-B. Salberg, J.Y. Hardeberg, and R. Jenssen (Eds.): SCIA 2009, LNCS 5575, pp. 291–300, 2009.

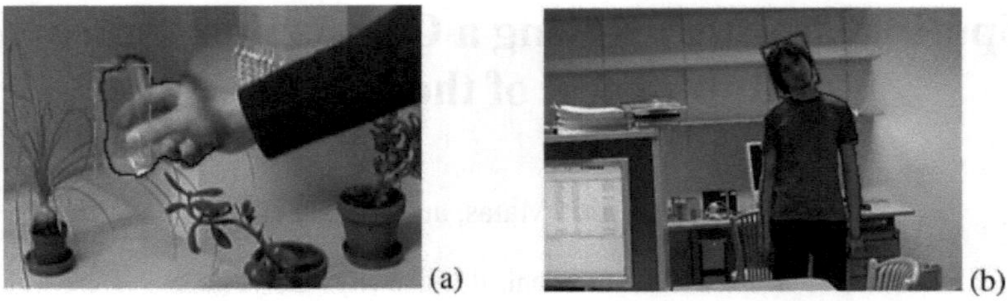

Fig. 1. Objects with a companion. Foreground includes not just the main object, e.g., (a) a glass or (b) a head, but also other image regions, such as (a) hand or (b) body.

The foreground layer often includes not just the object of interest but also other image regions which move coherently with the object. The connection of the object to the companion may be temporary, e.g., a glass can be picked up by hand and dragged from the table, or it may be permanent, e.g., a head of a man always moves together with his torso, see Figure 1 for examples. As the core contribution of this paper, we show how the companion, i.e., the non-object part of the foreground motion layer, contributes to robust tracking and expands situations in which successful tracking is possible, e.g, when the object of interest is not visible or abruptly changes its appearance. Such situations would distract the trackers that look only for the object itself.

The task of tracking a single object can be then decomposed to several sub-problems: (1) On-line learning of the foreground layer appearance, support and motion, i.e., "What is the foreground layer?". (2) Learning of the background layer appearance, support and motion. In our current implementation, the camera is fixed and the background appearance is learned off-line from the training sequence. However, the principle of the proposed tracker allows us to estimate the background motion and its appearance changes on-line in the future versions. (3) Separating the object from its companion, i.e., "Where is the object?". (4) Modelling appearance of the object.

The proposed Sputnik Tracker is based on this reasoning. It learns and is able to estimate which parts of the image area accompany the object, be it temporarily or permanently, and which parts together with the object form the foreground layer. In this paper we do not deal with tracker initialization and re-initialization after failure.

The Sputnik Tracker requires the foreground to be modelled as a structure of connected, independently moving parts, unlike approaches based on the pictorial structures [7,16,17]. The foreground layer is represented by a plane containing only image regions which perform similar movement. To track a part of an object, the Sputnik Tracker does not need to have a prior knowledge of the object structure, i.e., the number of parts and their connections.

The rest of the paper is structured as follows: In *Section 2*, the probabilistic model implemented in Sputnik Tracker will be explained together with the on-line learning of the model parameters. The tracking algorithm will be described. In *Section 3*, it will be demonstrated on several challenging sequences how the estimated companion contributes to robust tracking. The contributions will be concluded in *Section 4*.

2 The Sputnik Tracker

2.1 Integrating Foreground and Background Cues

We pose the object tracking probabilistically as finding the foreground position $l^{\star}$, in which the likelihood of the observed image I is maximized over all possible locations l given the foreground model ϕ_F and the background model ϕ_B

$$l^{\star} = \operatorname*{argmax}_{l} P(I|\phi_F, \phi_B, l) \,. \tag{1}$$

When the foreground layer has the position l then the observed image can be divided in two disjoint areas – $I_{F(l)}$ containing pixels associated with foreground layer and $I_{B(l)}$ containing pixels belonging to the background layer. Assuming that pixel intensities observed on the foreground are independent of those observed on the background, the likelihood of observing the image I can be rewritten as

$$P(I|\phi_F, \phi_B, l) = P(I_{F(l)}, I_{B(l)}|\phi_F, \phi_B) = P(I_{F(l)}|\phi_F)P(I_{B(l)}|\phi_B) \,. \tag{2}$$

Ignoring dependencies on the foreground-background boundary:

$$P(I|\phi_B) = P(I_{F(l)}|\phi_B)P(I_{B(l)}|\phi_B) \,, \tag{3}$$

Equation (2) can be rewritten as

$$P(I|\phi_F, \phi_B, l) = \frac{P(I_{F(l)}|\phi_F)}{P(I_{F(l)}|\phi_B)} P(I|\phi_B) \,. \tag{4}$$

The last term in Equation (4) does not depend on l. It follows that likelihood of the whole image (with respect to l) is maximized by maximizing the likelihood ratio of the image region $I_{F(l)}$ with the respect to the foreground ϕ_F and background model ϕ_B. The optimal position $l^{\star}$ is then

$$l^{\star} = \operatorname*{argmax}_{l} \frac{P(I_{F(l)}|\phi_F)}{P(I_{F(l)}|\phi_B)} \,. \tag{5}$$

Note that by modelling $P(I_{F(l)}|\phi_B)$ as the uniform distribution with respect to $I_{F(l)}$, one gets, as a special case, a standard template tracker which maximizes likelihood of $I_{F(l)}$ with respect to the foreground model only.

2.2 Object and Companion Models

Very often some parts of the visible scene undergo the same motion as the object of interest. The foreground layer, the union of such parts, is modelled by the *companion model* ϕ_C. The companion model is adapted on-line in each step of tracking. It is gradually extended by the neighboring image areas which exhibit the same movement as the tracked object. The involved areas are not necessarily connected.

Should such a group of objects split later, it must be decided which image area contains the object of interest. Sputnik Tracker maintains another model for this reason, the *object model* ϕ_O, which describes the appearance of the main object only. Unlike the companion model ϕ_C, which adapts on-line very quickly, the object model ϕ_O adapts slowly, with lower risk of drift.

In the current implementation, both models are based on the same pixel-wise representation:

$$\phi_C = \{(\mu_j^C, s_j^C, m_j^C);\ j \in \{1 \dots N\}\} \,, \tag{6}$$

$$\phi_O = \{(\mu_j^O, s_j^O, m_j^O);\ j \in \{1 \dots N_O\}\} \,, \tag{7}$$

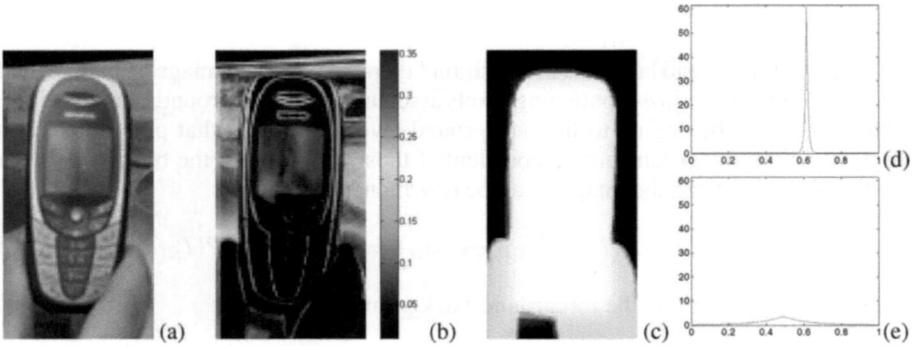

Fig. 2. Illustration of the model parameters: (a) median, (b) scale and (c) mask. Right side displays the pixel intensity PDF which is parametrized by its median and scale, see Equation (8) and (9). There are two examples, one of pixel with (d) low variance and other with (e) high variance.

where N and N_O denote the number of pixels in the template, which is illustrated in Figure 2. In the probabilistic model, each individual pixel is represented by the probability density function (PDF) based on the mixture of Laplace distribution

$$f(x|\mu, s) = \frac{1}{2s} \exp\left(-\frac{|x-\mu|}{s}\right) \tag{8}$$

restricted to the interval $\langle 0, 1 \rangle$, and uniform distribution over the interval $\langle 0, 1 \rangle$:

$$p(x|\mu, s) = \omega U_{\langle 0,1 \rangle}(x) + (1-\omega) f_{\langle 0,1 \rangle}(x|\mu, s) \,, \tag{9}$$

where $U_{\langle 0,1 \rangle}(x) = 1$ represents the uniform distribution and

$$f_{\langle 0,1 \rangle}(x|\mu, s) = \left[f(x|\mu, s) + \frac{\int_{R-\langle 0,1 \rangle} f(x'|\mu, s)\, \mathrm{d}x'}{\int_{\langle 0,1 \rangle} 1 \, \mathrm{d}x'} \right] \tag{10}$$

represents the restricted Laplace distribution. The parameter $\omega \in (0, 1)$ weighs the mixture. It has the same value for all pixels and represents the probability of an unexpected measurement. The individual pixel PDFs are parametrized by their median μ and scale s.

The mixture of the Laplace distribution with the uniform distribution provides distribution with heavier tails which is more robust to unpredicted disturbances. Examples of PDF in the form of Equation (9) are shown in Figure 2d,e. The distribution in the form of Equation (10) has the desirable property that it approaches uniform distribution by increasing the uncertainty in the model. This is likely to happen in fast and unpredictably changing object areas that would otherwise disturb the tracking.

The models ϕ_C and ϕ_O also include segmentation mask (support) which assigns each pixel j in the model the value m_j representing a probability that the pixel belongs to the object.

2.3 Evolution of the Models

At the end of each tracking step at time t, after the new position of the object has been estimated, the model parameters μ, s and the segmentation mask are updated. For each pixel in the model, its median is updated using the exponential forgetting principle,

$$\mu^{(t)} = \alpha\,\mu^{(t-1)} + (1-\alpha)\,x\,, \tag{11}$$

where x is the observed intensity of the corresponding image pixel in the current frame and α is the parameter controlling the speed of exponential forgetting. Similarly, the scale is updated as

$$s^{(t)} = \max\{\alpha\,s^{(t-1)} + (1-\alpha)|x^{(t)} - \mu^{(t)}|, s_{\min}\}\,. \tag{12}$$

The scale values are limited by the manually chosen lower bound $s_{\min}$ to prevent overfitting and to enforce robustness to a sudden change of the previously stable object area.

The segmentation mask of the companion model ϕ_C is updated at each step of the tracking following updates of μ and s. First, a binary segmentation $A = \{a_j; a_j \in \{0,1\}, j \in 1\ldots N\}$ is calculated using Graph Cuts algorithm [18]. An update to the object segmentation mask is then obtained as

$$m_j^{C,(t)} = \alpha\,m_j^{C,(t-1)} + (1-\alpha)\,a_j\,. \tag{13}$$

2.4 Background Model

The background is modelled using the same distribution as the foreground. Background pixels are considered independent and are represented by PDF expressed by formula (9). Each pixel of the background is then characterized by its median μ and scale s:

$$\phi_B = \{(\mu_i^B, s_i^B); i \in \{1\ldots\|I\|\}\}\,, \tag{14}$$

This model is suitable for a fixed camera. However, by geometrically registering consecutive frames in the video sequence, it might be used with pan-tilt-zoom (PTZ) cameras, which have a lot of applications in surveillance, or even with freely moving camera, provided that the movement is not large so that the robust tracker will overcome the model error caused by the change of the parallax. Cases with almost planar background, like aerial images of the Earth surface, can be also handled by the rigid geometrical image registration.

In the current implementation, the background parameters μ and scale s are learned off-line from a training sequence using EM algorithm. The sequence does not necessarily need to exhibit empty scene. It might also contain objects moving in the foreground. The foreground objects are detected as outliers and are robustly filtered out by the learning algorithm. Description of the learning algorithm is out of the scope of this paper.

2.5 The Tracking Algorithm

The state of the tracker is characterized by object appearance model ϕ_O, companion model ϕ_C and object location l. In the current implementation, we model the affine rigid motion of the object. This does not restrict us to track rigid objects only, it only limits the space of possible locations l such that the coordinate transform $j = \psi(i|l)$ is affine. The transform maps indices i in the image pixel to indices j in the model, see Figure 3. Appearance changes due to a non-rigid object or its non-affine motion are handled by adapting on-line the companion appearance model ϕ_C.

The tracker is initialized by marking the area covered by the object to be tracked in the first image of the sequence. The size of the companion model ϕ_C is set to cover a

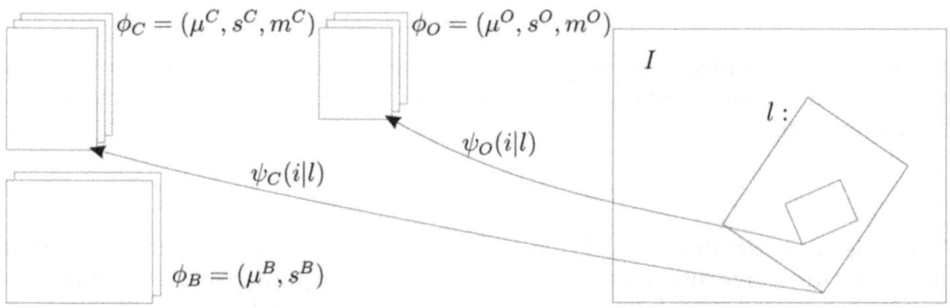

Fig. 3. Transforms between image and model coordinates

rectangular area larger than the object. That area has potential to become a companion of the object. Initial values of μ_j^C are set to image intensities observed in the corresponding image pixels, s_j^C are set to $s_{\min}$. Mask values m_j^C are set to 1 in areas corresponding to the object and to 0 elsewhere.

Object model ϕ_O is initialized in the similar way, but it covers only the object area. Only the scale of the object model, s_j^O, is updated during tracking.

Tracking is approached as minimization of the cost based on the negative logarithm of the likelihood ratio, Equation (5),

$$\mathcal{C}(l, M) = -\sum_{i \in F(l)} p(I(i)|\mu^M_{\psi_M(i|l)}, s^M_{\psi_M(i|l)}) + \sum_{i \in F(l)} p(I(i)|\mu_i^B, s_i^B)], \tag{15}$$

where $F(l)$ are indices of image pixels covered by the object/companion if it were at the location l, the assignment is determined by the model segmentation mask and $\psi_M(i|l)$. The model selector (companion or object) is denoted $M \in \{O, C\}$. The following steps are executed for each image in the sequence.

1. Find the optimal object position induced by the companion model by minimizing the cost $l_C^\star = \operatorname{argmin} \mathcal{C}(l, C)$. The minimization is performed using the gradient descent method starting at the previous location.
2. Find the optimal object position induced by the object model $l_O^\star = \operatorname{argmin} \mathcal{C}(l, O)$ using the same approach.
3. If $\mathcal{C}(l_O^\star, O)$ is high then continue from step 5.
4. If the location $l_O^\star$ gives better fit to the object model, $\mathcal{C}(l_O^\star, O) < \mathcal{C}(l_C^\star, O)$, then set the new object location to $l = l_O^\star$ and continue from step 6.
5. The object may be occluded or its appearance may be changed. Set the new object location to $l = l_C^\star$.
6. Update model parameters μ_j^C, s_j^C, m_j^C and s_j^O using method described in *Section 2.3*.

The above described algorithm is controlled by several manually chosen parameters which were described in the previous sections. To recapitulate, those are: ω – the probability of unexpected pixel intensity, it controls the amount of uniform distribution in the mixture PDF; α – the speed of the exponential forgetting; $s_{\min}$ the lover bound on the scale s. The unoptimized MATLAB implementation of the process takes 1 to 10 seconds per image on a standard PC.

3 Results

To show the strengths of the Sputnik Tracker, a successful tracking on some challenging sequences will be demonstrated. In all following illustrations, the red rectangle is used

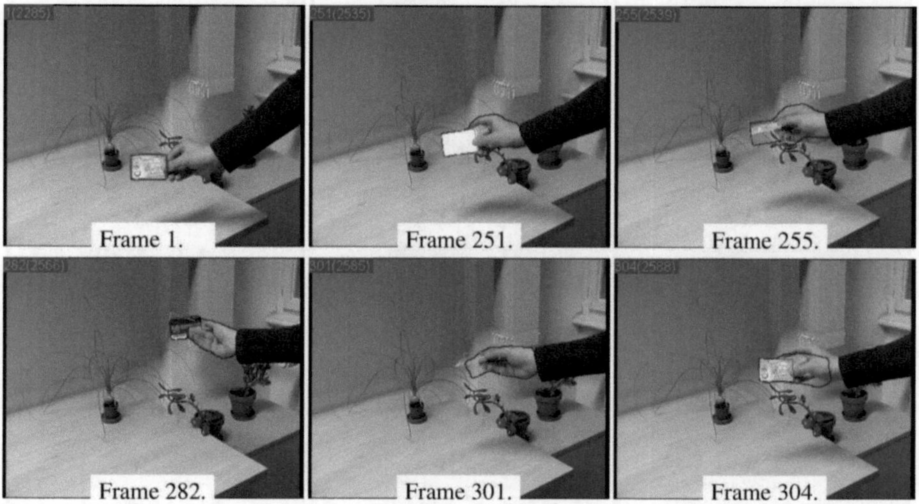

Fig. 4. Tracking a card carried by the hand. The strong reflection in frame 251 or flipping the card later does not cause the Sputnik Tracker to fail.

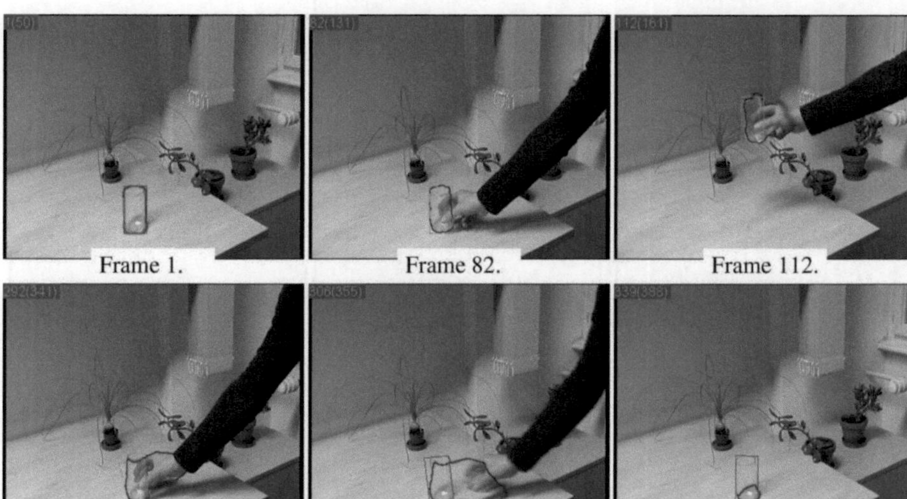

Fig. 5. Tracking a glass after being picked by a hand and put back later. The glass moves with the hand which is recognized as companion and stabilizes the tracking.

Fig. 6. Tracking the head of a man. The body is correctly recognized as a companion (the blue line). This helped to keep tracking the head while the man turns around between frames 202 and 285 and after the head gets covered with a picture in the frame 495 and the man hides behind the sideboard. In those moments, an occlusion was detected, see the green rectangle in place of the red one, but the head position was still tracked, given the companion.

to illustrate a successful object detection, a green rectangle corresponds to the recognized occlusion or the change of object appearance. The blue line shows the contour of the foreground layer including the estimated companion. The thickness of the line is proportional to the uncertainty in the layer segmentation. The complete sequences can be downloaded from http://cmp.felk.cvut.cz/~cermal1/supplements-scia/ as video files.

The first sequence shows the tracking of an ID card, see Figure 4 for several frames selected from the sequence. After initialization with the region belonging to the card, the Sputnik Tracker learns that the card is accompanied by the hand. This prevents it from failing in the frame 251 where the card reflects strong light source and its image is oversaturated. Any tracker that looks only for the object itself would have a very hard time at this moment. Similarly, the knowledge of the companion helps to keep a successful tracking even when the card is flipped in the frame 255. The appearance on the backside differs from the frontside. The tracker recognizes this change and reports an occlusion. However, the rough position of the card is still maintained with respect to the companion. When the card is flipped back it is redetected in the frame 304.

Figure 5 shows tracking of a glass being picked by a hand in the frame 82. At this point, the tracker reports an occlusion that is caused by the fingers and the hand is becoming a companion. This allows the tracking of the glass while it is being carried around the view. The glass is dropped back to the table in the frame 292 and when the hand moves away it is recognized back in the frame 306.

Figure 6 shows head tracking through occlusion. After initialization to the head area in the first image, the Sputnik Tracker estimates the body as a companion, see frame 118. While the man turns around between frames 202 and 285 the tracker reports occlusion of the tracked object (head) and maintains its position relative to the companion. The tracking is not lost even when the head gets covered with a picture and the man moves behind a sideboard and only the picture covering the head remains visible. This would be very difficult to achieve without learning the companion. After the picture is removed in the frame 635, the head is recognized again in the frame 735. The man then leaves the view while his head is still being successfully tracked.

4 Conclusion

We have proposed a novel approach to tracking based on the observation that objects rarely move alone and their movement can be coherent with other image regions. Learning which image regions move together with the object can help to overcome occlusions or unpredictable changes in the object appearance.

To demonstrate this we have implemented a Sputnik Tracker and presented a successful tracking in several challenging sequences. The tracker learns on-line which image regions accompany the object and maintain an adaptive model of the companion appearance and shape. This makes it robust to situations that would be distractive to trackers focusing only on the object alone.

Acknowledgments

The authors wish to thank Libor Špaček for careful proofreading. The authors were supported by Czech Ministry of Education project 1M0567 and by EC project ICT-215078 DIPLECS.

References

1. Tao, H., Sawhney, H.S., Kumar, R.: Dynamic layer representation with applications to tracking. In: Proceedings of the International Conference on Computer Vision and Pattern Recognition, vol. 2, pp. 134–141. IEEE Computer Society, Los Alamitos (2000)
2. Tao, H., Sawhney, H.S., Kumar, R.: Object tracking with Bayesian estimation of dynamic layer representations. IEEE Transactions on Pattern Analysis and Machine Intelligence 24(1), 75–89 (2002)
3. Weiss, Y., Adelson, E.H.: A unified mixture framework for motion segmentation: Incorporating spatial coherence and estimating the number of models. In: Proceedings of the International Conference on Computer Vision and Pattern Recognition, pp. 321–326. IEEE Computer Society, Los Alamitos (1996)
4. Wang, J.Y.A., Adelson, E.H.: Layered representation for motion analysis. In: Proceedings of the International Conference on Computer Vision and Pattern Recognition, pp. 361–366. IEEE Computer Society, Los Alamitos (1993)
5. Stauffer, C., Grimson, W.E.L.: Adaptive background mixture models for real-time tracking. In: Proceedings of the International Conference on Computer Vision and Pattern Recognition, vol. 2, p. 252 (1999)
6. Stauffer, C., Grimson, W.E.L.: Learning patterns of activity using real-time tracking. IEEE Transactions on Pattern Analysis and Machine Intelligence 22(8), 747–757 (2000)
7. Felzenschwalb, P.F., Huttenlocher, D.P.: Pictorial structures for object recognition. International Journal of Computer Vision 61(1), 55–79 (2005)
8. Korč, F., Hlaváč, V.: Detection and tracking of humans in single view sequences using 2D articulated model. In: Human Motion, Understanding, Modelling, Capture and Animation, vol. 36, pp. 105–130. Springer, Heidelberg (2007)
9. Comaniciu, D., Ramesh, V., Meer, P.: Kernel-based object tracking. IEEE Transactions on Pattern Analysis and Machine Intelligence 25(5), 564–575 (2003)
10. Babu, R.V., Pérez, P., Bouthemy, P.: Robust tracking with motion estimation and local kernel-based color modeling. Image and Vision Computing 25(8), 1205–1216 (2007)
11. Georgescu, B., Comaniciu, D., Han, T.X., Zhou, X.S.: Multi-model component-based tracking using robust information fusion. In: Comaniciu, D., Mester, R., Kanatani, K., Suter, D. (eds.) SMVP 2004. LNCS, vol. 3247, pp. 61–70. Springer, Heidelberg (2004)
12. Jepson, A.D., Fleet, D.J., El-Maraghi, T.F.: Robust online appearance models for visual tracking. IEEE Transactions on Pattern Analysis and Machine Intelligence 25(10), 1296–1311 (2003)
13. Grabner, H., Grabner, M., Bischof, H.: Real-time tracking via on-line boosting. In: Proceedings of the British Machine Vision Conference, vol. 1, pp. 47–56 (2006)
14. Collins, R., Liu, Y., Leordeanu, M.: Online selection of discriminative tracking features. IEEE Transactions on Pattern Analysis and Machine Intelligence 27(10), 1631–1643 (2005)
15. Kristan, M., Pers, J., Perse, M., Kovacic, S.: Closed-world tracking of multiple interacting targets for indoor-sports applications. Computer Vision and Image Understanding (in press, 2008)
16. Ramanan, D.: Learning to parse images of articulated bodies. In: Schölkopf, B., Platt, J., Hoffman, T. (eds.) Advances in Neural Information Processing Systems, pp. 1129–1136. MIT Press, Cambridge (2006)
17. Ramanan, D., Forsyth, D.A., Zisserman, A.: Tracking people by learning their appearance. IEEE Transactions on Pattern Analysis and Machine Intelligence 29(1), 65–81 (2007)
18. Boykov, Y., Funka-Lea, G.: Graph cuts and efficient n-d image segmentation. Int. J. Comput. Vision 70(2), 109–131 (2006)

A Convex Approach to Low Rank Matrix Approximation with Missing Data

Carl Olsson and Magnus Oskarsson

Centre for Mathematical Sciences
Lund University, Lund, Sweden
{calle,magnuso}@maths.lth.se

Abstract. Many computer vision problems can be formulated as low rank bilinear minimization problems. One reason for the success of these problems is that they can be efficiently solved using singular value decomposition. However this approach fails if the measurement matrix contains missing data.

In this paper we propose a new method for estimating missing data. Our approach is similar to that of L_1 approximation schemes that have been successfully used for recovering sparse solutions of under-determined linear systems. We use the nuclear norm to formulate a convex approximation of the missing data problem. The method has been tested on real and synthetic images with promising results.

1 Bilinear Models and Factorization

Bilinear models have been applied successfully to several computer vision problems such as structure from motion [1,2,3], nonrigid 3D reconstruction [4,5], articulated motion [6], photometric stereo [7] and many other. In the typical application, the observations of the system are collected in a measurement matrix which (ideally) is known to be of low rank due to the bilinearity of the model. The successful application of these models is mostly due to the fact that if the entire measurement matrix is known, singular value decomposition (SVD) can be used to find a low rank factorization of the matrix.

In practice, it is rarely the case that all the measurements are known. Problems with occlusion and tracking failure lead to missing data. In this case SVD can not be employed, which motivates the search for methods that can handle incomplete data.

To our knowledge there is, as of yet, no method that can solve this problem optimally. One approach is to use iterative local methods. A typical example is to use a two step procedure. Here the parameters of the model are divided into two groups where each one is chosen such that the model is linear when the other group is fixed. The optimization can then be performed by alternating the optimization over the two groups [8]. Other local approaches such as non-linear Newton methods have also been applied [9]. There are however no guarantee of convergence and therefore these methods are in need of good initialization. This

A.-B. Salberg, J.Y. Hardeberg, and R. Jenssen (Eds.): SCIA 2009, LNCS 5575, pp. 301–309, 2009.

is typically done with a batch algorithm (e.g. [1]) which usually optimizes some algebraic criterion.

In this paper we propose a different approach. Since the original problem is difficult to solve due to its non convexity we derive a simple convex approximation. Our solution is independent of initialization, however batch algorithms can still be used to strengthen the approximation. Further more, since our program is convex it is easy to extend it to other error measures or to include prior information.

2 Low Rank Approximations and the Nuclear Norm

In this section we will present the nuclear norm. It has previously been used in applications such as image compression, system identification and similar problems that can be stated as low rank approximation problems (see [10,11,12]). The theory largely parallels that of L_1 approximation (see [13,14,15]) which has been used successfully in various applications.

Let M the matrix with entries m_{ij} containing the measurements. The typical problem of finding a low rank matrix X that describes the data well can be posed as

$$\min_{X} \; ||X - M||_F^2 \tag{1}$$
$$\text{s.t rank}(X) \leq r, \tag{2}$$

where $||\cdot||_F$ denotes the Frobenius norm, and r is the given rank. This problem can be solved optimally with SVD even though the rank constraint is highly non-convex (see [16]). The SVD-approach does however not extend to the case when the measurement matrix is incomplete. Let W be a matrix with entries $w_{ij} = 1$ if the value of m_{ij} has been observed and zeros otherwise. Note that the values of W can also be chosen to represent weights modeling the confidence of the measurements. The new problem can be formulated as

$$\min_{X} ||W \odot (X - M)||_F^2 \tag{3}$$
$$\text{s.t} \quad \text{rank}(X) \leq r \tag{4}$$

where $\odot$ denotes element-wise multiplication. In this case SVD can not be directly applied since the whole matrix M is not known. Various approaches for estimating the missing data exist and the most simple one (which is commonly used for initializing different iterative methods) is simply to let the missing entries be zeros. In terms of optimization this corresponds to finding the minimum Frobenius norm solution X such that $W \odot (X - M) = 0$. In effect what we are minimizing is

$$||X||_F^2 = \sum_{i=1}^{m} \sigma_i(X)^2, \tag{5}$$

where $\sigma_i(X)$ is the i'th largest singular value of the $m \times n$ matrix X. It is easy to see that this function penalizes larger values proportionally more than small values (see figure 1). Hence, this function favors solutions with many small singular values as opposed to a small number of large singular values, which is exactly the opposite of what we want.

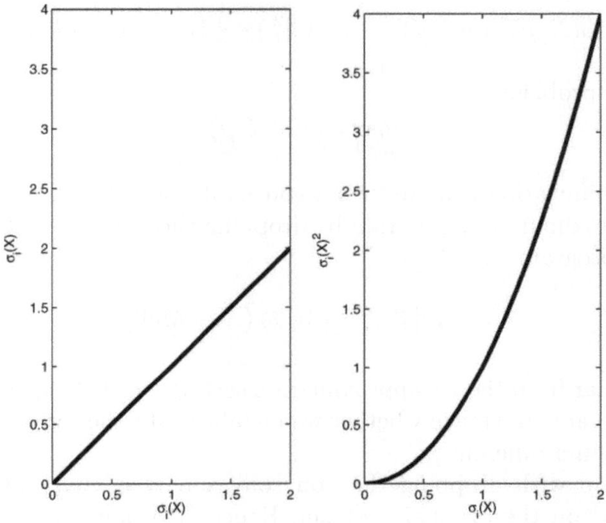

Fig. 1. Comparison between the Frobenius norm and the nuclear norm, showing on the left: $\sigma_i(X)$ and on the right: $\sigma_i(X)^2$

Since we cannot minimize the rank function directly, because of its non-convexity, we will use the so called nuclear norm which is given by

$$||X||_* = \sum_{i=1}^{m} \sigma_i(X). \tag{6}$$

The nuclear norm can also be seen as the dual norm of the operator norm $||\cdot||_2$, that is

$$||X||_* = \max_{||Y||_2 \leq 1} \langle X, Y \rangle \tag{7}$$

where the inner product is defined by $\langle X, Y \rangle = \text{tr}(X^T Y)$, see [10]. By the above characterization it is easy to see that $||X||_*$ is convex, since a maximum of functions linear in X is always convex (see [17]).

The connection between the rank function and the nuclear norm can be seen via the following inequality (see [16]), which holds for any matrix of at most rank r

$$||X||_* \leq \sqrt{r}||X||_F. \tag{8}$$

In fact it turns out that the nuclear norm is the convex envelope of the rank function on the set $\{X; ||X||_F \leq 1\}$ (see [17]). In view of (8) we can try to solve the following program

$$\min_{X} \quad ||W \odot (X - M)||_F^2 \tag{9}$$

$$\text{s.t } ||X||_*^2 - r||X||_F^2 \leq 0. \tag{10}$$

The Lagrangian of this problem is

$$L(X, \mu) = \mu(||X||_*^2 - r||X||_F^2) + ||W \odot (X - M)||_F^2, \tag{11}$$

with the dual problem

$$\max_{\mu > 0} \min_{X} L(X, \mu). \tag{12}$$

The inner minimization is however not convex if μ is not zero. Therefore we are forced to approximate this program by dropping the non convex term $-r||X||_F^2$, yielding the program

$$\min_{X} \mu||X||_*^2 + ||W \odot (X - M)||_F^2. \tag{13}$$

which is familiar from the L_1-approximation setting (see [13,14,15]). Note that it does not make any difference whether we penalize with the term $||X||_*$ or $||X||_*^2$, it just results in a different μ.

The problem with dropping the non convex part is that (13) is no longer a lower bound on the original problem. Hence (13) does not tell us anything about the global optimum, it can only be used as a heuristic for generating good solutions. An interesting exception is when the entire measurement matrix is known. In this case we can write the Lagrangian as

$$L(x, \mu) = \mu||X||_*^2 + (1 - \mu r)||X||_F^2 + 2\langle X, M\rangle + ||M||_F^2. \tag{14}$$

Thus, here L will be convex if $0 \leq \mu \leq 1/r$. Note that if $\mu = 1/r$ then the term $||X||_F^2$ is completely removed. In fact this offers some insight as to why the problem can be solved exactly when M is completely known, but we will not pursue this further.

2.1 Implementation

In our experiments we use (13) to fill in the missing data of the measurement matrix. If the resulting matrix is not of sufficiently low rank then we use SVD to approximate it. In this way it is possible to use methods such as [5] that work when the entire measurement matrix is known. The program (13) can be implemented in various ways (see [10]). The easiest way (which we use) is to reformulate it as a semidefinite program, and use any standard optimization software to solve it. The semidefinite formulation can be obtained from the dual norm (see equation (7)). Suppose the matrix X (and Y) has size $m \times n$, and let I_m, I_n denote the identity matrices of size $m \times m$ and $n \times n$ respectively. That the matrix Y has operator-norm $||Y||_2 \leq 1$ means that all the eigenvalues of Y^TY are smaller than 1, or equivalently that $I_m - Y^TY \succeq 0$. Using the Schur complement [17] and (7) it is now easy to see that minimizing the nuclear norm can be formulated as

$$\min_{X} \max_{Y} \quad \mathrm{tr}(Y^T X) \tag{15}$$

$$\begin{bmatrix} I_m & Y \\ Y^T & I_n \end{bmatrix} \succeq 0 \tag{16}$$

Taking the dual of this program, we arrive at the linear semidefinite program

$$\min_{X, Z_{11}, Z_{22}} \quad \mathrm{tr}(Z_{11} + Z_{22}), \tag{17}$$

$$\begin{bmatrix} Z_{11} & \frac{X}{2} \\ \frac{X^T}{2} & Z_{22} \end{bmatrix} \succeq 0. \tag{18}$$

Linear semidefinite programs have been extensively studied in the optimization literature and there are various softwares for solving them. In our experiments we use SeDuMi [18] (which is freely available) but any solver that can handle the semidefinite program and the Frobenius-norm term in (13) will work.

3 Experiments

Next we present two simple experiments for evaluating the performance of the approximation. In both experiments we select the observation matrix W randomly. Not a realistic scenario for most real applications, however we do this since we want to evaluate the performance for different levels of missing data with respect to ground truth. It is possible to strengthen the relaxation by using batch algorithms. However, since we are only interested in the performance of (13) itself we do not do this.

In the first experiment points on a shark are tracked in a sequence of images. The same sequence has been used before, see e.g. [19]. The shark undergoes a deformation as it moves. In this case the deformation can be described by two shape modes S_0 and S_1. Figure 2 shows three images from the sequence (with no missing data). To generate the measurement matrix we added noise and randomly selected W for different levels of missing data. Figure 3 shows the

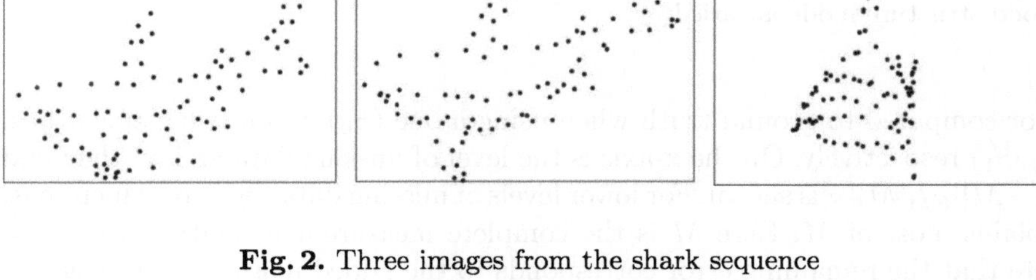

Fig. 2. Three images from the shark sequence

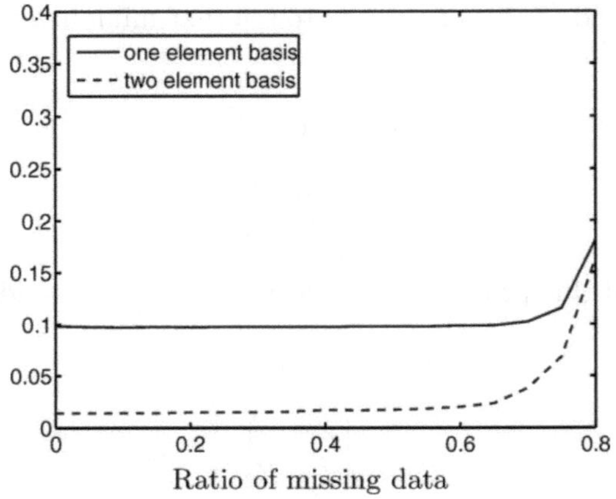

Fig. 3. Reconstruction error for the Shark experiment, for a one and two element basis, as a function of the level of missing data. On the x-axis is the level of missing data and on the y-axis is $||X - M||_F/||M||_F$.

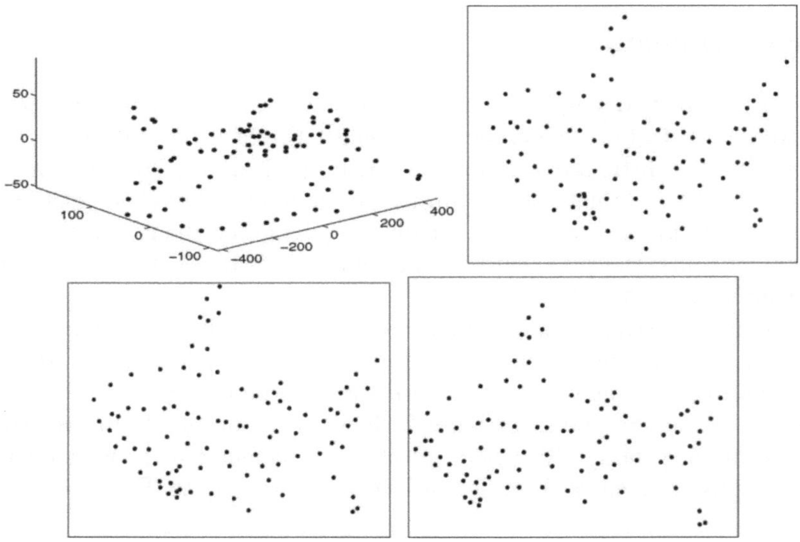

Fig. 4. A 3D-reconstruction of the shark. The first shape mode in 3D and three generated images. The camera is the same for the three images but the coefficient of the second structure mode is varied.

error compared to ground truth when using a one (S_0) and a two element basis (S_0, S_1) respectively. On the x-axis is the level of missing data and on the y-axis $||X-M||_F/||M||_F$ is shown. For lower levels of missing data the two element basis explains most of M. Here M is the complete measurement matrix with noise. Note that the remaining error corresponds to the added noise. For missing data

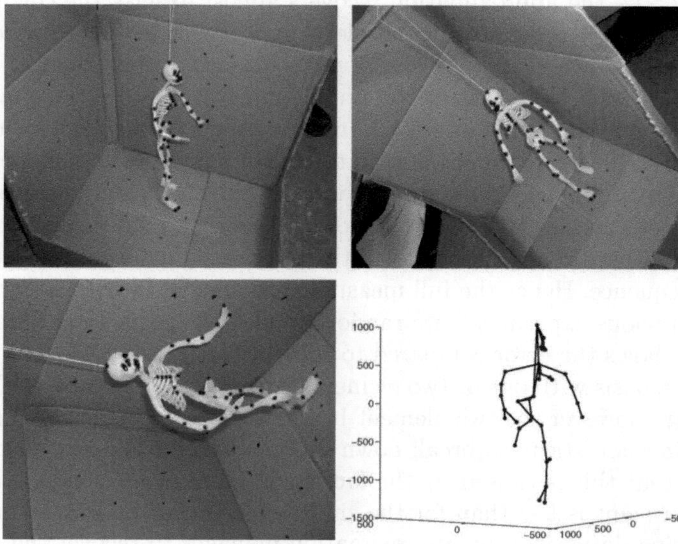

Fig. 5. Three images from the skeleton sequence, with tracked image points, and the 1st mode of reconstructed nonrigid-structure

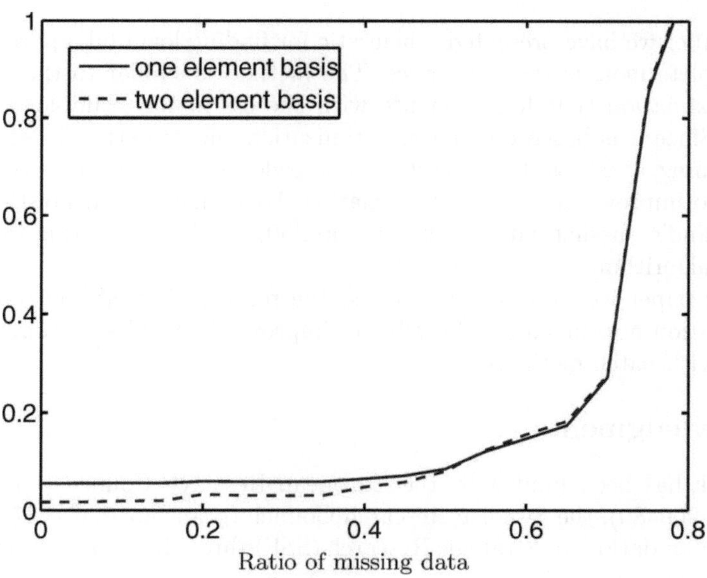

Fig. 6. Reconstruction error for the Skeleton experiment, for a one and two element basis, as a function of the level of missing data. On the y-axis $||X - M||_F/||M||_F$ is shown.

levels below 50% the approximation recovers almost exactly the correct matrix (without noise). When the missing data level approaches 70%, the approximation starts to break down. Figure 4 shows the obtained reconstruction when the missing data is 40%. Note that we are not claiming to improve the quality of the reconstructions; We are only interested in recovering M. The reconstructions are just included to illustrate the results. To the upper left is the first shape mode S_0, and the others are images generated by varying the coefficient corresponding to the second mode S_1 (see [4]). Figure 5 shows the setup for the second experiment. In this case we used real data where all the interest points were tracked through the entire sequence. Hence the full measurement matrix M with noise is known. As in the previous experiment, we randomly selected the missing data.

Figure 6 shows the error compared to ground truth (i.e. $||X - M||_F/||M||_F$) when using a basis with one or two elements. In this case the rank of the motion is not known, however the two element basis seems to be sufficient. In this case the approximation starts to break down sooner than for the shark experiment. We believe that this is caused by the fact that the number of points and views in this experiment is less than for the shark experiment, making it more sensitive to missing data. Still the approximation manages to recover the matrix M well, for noise levels up to 50% without any knowledge other than the low rank assumption.

4 Conclusions

In this paper we have presented a heuristic for finding low rank approximations of incomplete measurement matrices. The method is similar to the concept of L_1-approximation that has been use with success in for example compressed sensing. Since it is based on convex optimization and in particular semidefinite programming, it is possible to add more knowledge in the form of convex constraints to improve the resulting estimation. Experiments indicate that we are able to handle missing data levels of around 50% without resorting to any type of batch algorithm.

In this paper we have merely studied the relaxation itself and it is still an open question how much it is possible to improve the results by combining our method with batch methods.

Acknowledgments

This work has been funded by the European Research Council (GlobalVision grant no. 209480), the Swedish Research Council (grant no. 2007-6476) and the Swedish Foundation for Strategic Research (SSF) through the programme Future Research Leaders.

References

1. Tardif, J., Bartoli, A., Trudeau, M., Guilbert, N., Roy, S.: Algorithms for batch matrix factorization with application to structure-from-motion. In: Int. Conf. on Computer Vision and Pattern Recognition, Minneapolis, USA (2007)

2. Sturm, P., Triggs, B.: A factorization bases algorithm for multi-image projective structure and motion. In: European Conference on Computer Vision, Cambridge, UK (1996)
3. Tomasi, C., Kanade, T.: Shape and motion from image sttreams under orthography: a factorization method. Int. Journal of Computer Vision 9 (1992)
4. Bregler, C., Hertzmann, A., Biermann, H.: Recovering non-rigid 3D shape from image steams. In: Int. Conf. on Computer Vision and Pattern Recognition, Hilton Head, SC, USA (2000)
5. Xiao, J., Kanade, T.: A closed form solution to non-rigid shape and motion recovery. International Journal of Computer Vision 67, 233–246 (2006)
6. Yan, J., Pollefeys, M.: A factorization approach to articulated motion recovery. In: IEEE Conf. on Computer Vision and Pattern Recognition, San Diego, USA (2005)
7. Basri, R., Jacobs, D., Kemelmacher, I.: Photometric stereo with general, unknown lighting. Int. Journal of Computer Vision 72, 239–257 (2007)
8. Hartley, R., Schaffalitzky, F.: Powerfactoriztion: An approach to affine reconstruction with missing and uncertain data. In: Australia-Japan Advanced Workshop on Computer Vision, Adelaide, Australia (2003)
9. Buchanan, A., Fitzgibbon, A.: Damped newton algorithms for matrix factorization with missing data. In: IEEE Computer Society Conference on Computer Vision and Pattern Recognition, CVPR 2005, June 20-25, 2005, vol. 2, pp. 316–322 (20)
10. Recht, B., Fazel, M., Parrilo, P.: Guaranteed minimum-rank solutions of linear matrix equations via nuclear norm minimization (2007), `http://arxiv.org/abs/0706.4138v1`
11. Fazel, M., Hindi, H., Boyd, S.: A rank minimization heuristic with application to minimum order system identification. In: Proceedings of the American Control Conference (2003)
12. El Ghaoui, L., Gahinet, P.: Rank minimization under lmi constraints: A framework for output feedback problems. In: Proceedings of the European Control Conference (1993)
13. Tropp, J.: Just relax: convex programming methods for identifying sparse signals in noise. IEEE Transactions on Information Theory 52, 1030–1051 (2006)
14. Donoho, D., Elad, M., Temlyakov, V.: Stable recovery of sparse overcomplete representations in the presence of noise. IEEE Transactions on Information Theory 52, 6–18 (2006)
15. Candes, E., Romberg, J., Tao, T.: Stable signal recovery from incomplete and inaccurate measurments. Communications of Pure and Applied Mathematics 59, 1207–1223 (2005)
16. Golub, G., van Loan, C.: Matrix Computations. The Johns Hopkins University Press (1996)
17. Boyd, S., Vandenberghe, L.: Convex Optimization. Cambridge University Press, Cambridge (2004)
18. Sturm, J.F.: Using sedumi 1.02, a matlab toolbox for optimization over symmetric cones (1998)
19. Torresani, L., Hertzmann, A., Bregler, C.: Non-rigid structure-from-motion: Estimating shape and motion with hierarchical priors. IEEE Transactions on Pattern Analysis and Machine Intelligence 30 (2008)
20. Raiko, T., Ilin, A., Karhunen, J.: Principal component analysis for sparse high-dimensional data. In: 14th International Conference on Neural Information Processing, Kitakyushu, Japan, pp. 566–575 (2007)

Multi-frequency Phase Unwrapping from Noisy Data: Adaptive Local Maximum Likelihood Approach

José Bioucas-Dias[1], Vladimir Katkovnik[2], Jaakko Astola[2], and Karen Egiazarian[2]

[1] Instituto de Telecomunicações, Instituto Superior Técnico, TULisbon, 1049-001 Lisboa, Portugal
bioucas@lx.it.pt
[2] Signal Processing Institute, University of Technology of Tampere, P.O. Box 553, Tampere, Finland
{katkov,jta,karen}@cs.tut.fi

Abstract. The paper introduces a new approach to absolute phase estimation from frequency diverse wrapped observations. We adopt a discontinuity preserving nonparametric regression technique, where the phase is reconstructed based on a local maximum likelihood criterion. It is shown that this criterion, applied to the multifrequency data, besides filtering the noise, yields a $2\pi Q$-periodic solution, where $Q > 1$ is an integer. The filtering algorithm is based on local polynomial (LPA) approximation for the design of nonlinear filters (estimators) and the adaptation of these filters to the unknown spatially smoothness of the absolute phase. Depending on the value of Q and of the original phase range, we may obtain complete or partial phase unwrapping. In the latter case, we apply the recently introduced robust (in the sense of discontinuity preserving) PUMA unwrapping algorithm [1]. Simulations give evidence that the proposed method yields state-of-the-art performance, enabling phase unwrapping in extraordinary difficult situations when all other algorithms fail.

Keywords: Interferometric imaging, phase unwrapping, diversity, local maximum-likelihood, adaptive filtering.

1 Introduction

Many remote sensing systems exploit the phase coherence between the transmitted and the scattered waves to infer information about physical and geometrical properties of the illuminated objects such as shape, deformation, movement, and structure of the object's surface. Phase estimation plays, therefore, a central role in these coherent imaging systems. For instance, in synthetic aperture radar interferometry (InSAR), the phase is proportional to the terrain elevation height; in magnetic resonance imaging, the phase is used to measure temperature, to map the main magnetic field inhomogeneity, to identify veins in the tissues, and to segment water from fat. Other examples can be found in adaptive optics,

A.-B. Salberg, J.Y. Hardeberg, and R. Jenssen (Eds.): SCIA 2009, LNCS 5575, pp. 310–320, 2009.

diffraction tomography, nondestructive testing of components, and deformation and vibration measurements (see, *e.g.*, [2], [4], [3], [5]). In all these applications, the observation mechanism is a 2π-periodic function of the true phase, hereafter termed *absolute phase.* The inversion of this function in the interval $[-\pi, \pi)$ yields the so-called *principal* phase values, or *wrapped* phases, or *interferogram*; if the true phase is outside the interval $[-\pi, \pi)$, the associated observed value is wrapped into it, corresponding to the addition/subtraction of an integer number of 2π. It is thus impossible to unambiguously reconstruct the absolute phase, unless additional assumptions are introduced into this inference problem.

Data acquisition with diversity has been exploited to eliminate or reduce the ambiguity of absolute phase reconstruction problem. In this paper, we consider multichannel sensors, each one operating at a different frequency (or wavelengths).

Let ψ_s, for $s = 1, \ldots, L$, stand for the wrapped phase acquired by a L-channel sensor. In the absence of noise, the wrapped phase is related with the true absolute phase, φ, as $\mu_s\varphi = \psi_s + 2\pi k_s$, where k_s is an integer, $\psi_s \in [-\pi, \pi)$, and μ_s is a channel depending scale parameter, to which we attach the meaning of relative frequency. This parameter establishes a link between the absolute phase φ and the wrapped phase ψ_s measured at the s-channel:

$$\psi_s = W(\mu_s\varphi_s) \equiv \mathrm{mod}\{\mu_s\varphi + \pi, 2\pi\} - \pi, \quad s = 1, \ldots, L, \tag{1}$$

where $W(\cdot)$ is the so-called wrapping operator, which decomposes the absolute phase φ into two parts: the fractional part ψ_s and the integer part defined as $2\pi k_s$. The integers k_s are known in interferometry as fringe orders. We assume that the frequencies for the different channels are strictly decreasing, *i.e.*, $\mu_1 > \mu_2 > ... > \mu_L$, or, equivalently, the corresponding wavelengths $\lambda_s = 1/\mu_s$ are strictly increasing, $\lambda_1 < \lambda_2, \ldots \lambda_L$.

Let us mention some of the techniques used for the multifrequency phase unwrap. Multi-frequency interferometry (see, *e.g.*, [16]) provides a solution for fringe order identification using the method of *excess fractions.* This technique computes a set of integers k_s compatible with the simultaneous set of equations $\mu_s\varphi = \psi_s + 2\pi k_s$, for $s = 1, \ldots, L$. It is assumed that the frequencies μ_s do not share common factors, *i.e.*, they are pair-wise relatively prime. The solution is obtained by maximizing the interval of possible absolute phase values.

A different approach formulates the phase unwrapping problem in terms of the Chinese remainder theorem, where the absolute phase φ is reconstructed from the remainders ψ_s, given the frequencies μ_s. This formulation assumes that all variables known and unknown are scaled to be integral. An accurate theory and results, in particular concerning the existence of a unique solution, is a strong side of this approach [18].

The initial versions of the excess fraction and Chinese remainder theorem based methods are highly sensitive to random errors. Efforts have been made to make these methods resistant to noise. The works [19] and [17], based on he Chinese remainder approach, are results of these efforts.

Statistical modeling for multi-frequency phase unwrapping based on the maximum likelihood approach is proposed in [13]. This work addresses the surface reconstruction from the multifrequency InSAR data. The unknown surface is approximated by local planes. The optimization problem therein formulated is tackled with simulated annealing.

An obvious idea that comes to mind to attenuate the damaging effect of the noise is prefiltering the wrapped observations. We would like, however, to emphasize that prefiltering, although desirable, is a rather delicate task. In fact, if prefiltering is too strong, the essential pattern of the absolute phase coded in the wrapped phase is damaged, and the reconstruction of absolute phase is compromised. On the other hand, if we do not filter, the unwrapping may be impossible because of the noise. A conclusion is, therefore, that filtering is crucial but should be designed very carefully. One of the ways to ensure efficiency is to adapt the strength of the prefiltering according to the phase surface smoothness and the noise level. In this paper, we use the wrapped phase prefiltering technique developed in [20] for a single frequency phase unwrapping.

2 Proposed Approach

We introduce a novel phase unwrapping technique based on local polynomial approximation (LPA) with varying adaptive neighborhood used in reconstruction. We assume that the absolute phase is a piecewise smooth function, which is well approximated by a polynomial in a neighborhood of the estimation point. Besides the wrapped phase, also the size and possibly the shape of this neighborhood are estimated. The adaptive window selection is based on two independent ideas: local approximation for design of nonlinear filters (estimators) and adaptation of these filters to the unknown spatially varying smoothness of the absolute phase. We use LPA for approximation in a sliding varying size window and intersection of confidence intervals (ICI) for window size adaptation. The proposed technique is a development of the PEARLS algorithm proposed for the single wavelength phase reconstruction from noisy data [20].

We assume that the frequencies μ_s can be represented as ratios

$$\mu_s = p_s/q_s, \tag{2}$$

where p_s, q_s are positive integers and the pairs (p_s, q_t), for $s, t \in \{1, \ldots, L\}$ do not have common multipliers, *i.e.*, p_s and q_t are pair-wise relatively prime.

Let

$$Q = \prod_{s=1}^{L} q_s. \tag{3}$$

Based on the LPA of the phase, the first step of the proposed algorithm computes the maximum likelihood estimate of the absolute phase. As a result, we obtain an unambiguous absolute phase estimates in the interval $[-Q \cdot \pi, Q \cdot \pi)$. Equivalently, we get an $2\pi Q$ periodic estimate. The adaptive window size LPA is a key technical element in the noise suppression and reconstruction of this wrapped $2\pi Q$-phase. The complete unwrapping is achieved by applying an unwrapping algorithm. In our implementation, we use the PUMA algorithm [1], which is able to preserve discontinuities by using graph cut based methods to solve the integer optimization problem associated to the phase unwrapping.

The polynomial modeling is a popular idea for both wrapped phase denoising and noisy phase unwrap. Using the local polynomial fit in terms of the phase tracking for the phase unwrap is proposed in the paper [12]. In the paper [13] the linear local polynomial approximation of height profiles is used for the surface reconstruction from the multifrequency InSAR data. Different modifications of the local polynomial approximation oriented to wrapped phase denoising are introduced in the regularized phase-tracking [14], [15], the multiple-parameter least squares [8], and the windowed Fourier ridges [9]. Compared with these works, the efficiency of the PEARLS algorithm [20] is based on the window size selection adaptiveness introduced by the ICI technique, which locally adapts the amount of smoothing according to the data. In particular, the discontinuities are preserved, what is a sine quo non condition for the success of the posterior unwrapping; in fact, as discussed in [7], it is preferable to unwrap the noisy interferogram than a filtered version in which the discontinuities or the areas of high phase rate have been washed out. In this paper, the PEARLS [20] adaptive filtering is generalized for the multifrequency data. Experiments based on simulations give evidence that the developed unwrapping is very efficient for the continuous as well as discontinuous absolute phase with a range of the phase variation so large that there no alternative algorithms able to unwrap this data.

3 Local Maximum Likelihood Technique

Herein, we adopt the complex-valued (cos/sin) observation model

$$u_s = B_s \exp(j\mu_s\varphi) + n_s, \; s = 1, ..., L, \;\; B_s \geq 0, \tag{4}$$

where B_s are amplitudes of the harmonic phase functions, and n_s is zero-mean independent complex-valued circular Gaussian random noises of variance equal to 1, *i.e.*, $E\{\mathrm{Re}\, n_s\} = 0$, $E\{\mathrm{Im}\, n_s\} = 0$, $E\{\mathrm{Re}\, n_s \cdot \mathrm{Im}\, n_s\} = 0$, $E\{(\mathrm{Re}\, n_s)^2\} = 1/2$, $E\{(\mathrm{Im}\, n_s)^2\} = 1/2$. We assume that the amplitudes B_s are non-negative in order to avoid ambiguities in the phase $\mu_s\varphi$, as the change of the amplitude sign is equivalent to a phase change of $\pm\pi$ in $\mu_s\varphi$. We note that the assumption of equal noise variance for all channel is not limitative as different noise variances can be accounted for by rescaling u_s and A_s in (4) by the corresponding noise standard deviation.

Model (4) accurately describes the acquisition mechanism of many interferometric applications, such as InSAR and magnetic resonance imaging. Furthermore, it retains the principal characteristics of most interferometric applications: it is a 2π-periodic function of $\mu_s\varphi$ and, thus, we have only access to the wrapped phase.

Since we are interested in two-dimensional problems, we assume that the observations are given on a regular $2D$ grid, $X \subset \mathbb{Z}^2$. The unwrapping problem is to reconstruct the absolute phase $\varphi(x, y)$ from the observed wrapped noisy $\psi_s(x, y)$, for $x, y \in X$.

Let us define the parameterized family of first order polynomials

$$\tilde{\varphi}(u, v|\mathbf{c}) = p^T(u, v)\mathbf{c}, \tag{5}$$

where $p = [p_1, p_2, p_3]^T = [1, u, v]^T$ and $\mathbf{c} = [c_1, c_2, c_3]^T$ is a vector of parameters. Assume that in some neighborhood of the point (x, y), the phase φ is well approximated by an element of the family (5); *i.e.*, for (x_l, y_l) in a neighborhood of the origin, there exists a vector $\mathbf{c}$ such that

$$\varphi(x + x_l, y + x_l) \simeq \tilde{\varphi}(x_l, y_l|\mathbf{c}). \tag{6}$$

To infer $\mathbf{c}$ and $\mathbf{B} \equiv \{B_1, \ldots, B_L\}$ (see (4)), we compute

$$\hat{\mathbf{c}} = \arg \min_{\mathbf{c},\mathbf{B}\geq\mathbf{0}} L_h(\mathbf{c}, \mathbf{B}). \tag{7}$$

where $L_h(\mathbf{c}, \mathbf{B})$ is a negative local log-likelihood function given by

$$L_h(\mathbf{c}, \mathbf{B}) = \tag{8}$$
$$\sum_s \frac{1}{\sigma_s^2} \sum_l w_{h,l,s} |u_s(x + x_l, y + y_l) - B_s \exp(j\mu_s \tilde{\varphi}(x_l, y_l|\mathbf{c})|^2.$$

Terms $w_{h,l,s}$ are window weights and can be different for different channels.

The local model $\tilde{\varphi}(u, v|\mathbf{c})$ (5) is the same for all frequency channels. We start by minimization L_h with respect to $\mathbf{B}$, which reduces to decoupled minimizations with respect to $B_s \geq 0$, one for channel. Noting that $\mathrm{Re}[\exp(-j\mu_s c_1)F] = |F| \cos(\mu_s c_1 - \mathrm{angle}(F))$, where F is a complex and $\mathrm{angle}(F) \in [-\pi, \pi[$ is the angle of F, and that $\min_{B\geq 0}\{aB^2 - 2Bc\} = -c_+^2/a$, where $a > 0$ and b are reals and x_+ is the positive part[1] of x, then after some manipulations, we obtain

$$-\tilde{L}_h(\mathbf{c}) = \tag{9}$$
$$\sum_s \frac{1}{\sigma_s^2} \frac{1}{\sum_l w_{h,l,s}} |F_{w,h,s}(\mu_s c_2, \mu_s c_3)|^2 \cos_+^2 \left[\mu_s c_1 - \mathrm{angle}(F_{w,h,s}(\mu_s c_2, \mu_s c_3))\right],$$

where $F_{w,h,s}(\mu c_2, \mu c_3)$ is the windowed/weighted Fourier transform of u_s,

$$F_{w,h,s}(\omega_2, \omega_3) = \sum_l w_{h,l,s} u_s(x + x_l, y + y_l) \exp(-j(\omega_2 x_l + \omega_3 y_l)), \tag{10}$$

calculated for the frequencies $(\omega_2 = \mu_s c_2, \omega_3 = \mu_s c_3)$.

The phase estimate is based on the solution of the optimization of $\tilde{L}_h$ over the three phase variables c_1, c_2, c_3

$$\hat{\mathbf{c}} = \arg \max_{\mathbf{c}} \tilde{L}_h(\mathbf{c}). \tag{11}$$

Let the condition (2) be fulfilled and $Q = \prod q_s$. Given fixed values of c_2 and c_3, the criterion (9) is a periodic function of c_1 with the period $2\pi Q$. Define the main interval for c_1 to be $[-\pi Q, \pi Q)$. Thus the optimization on c_1 is restricted to the interval $[-\pi Q, \pi Q)$. We term this effect *periodization* of the absolute phase φ, given that its estimation is restricted to this interval only. Because $Q \geq \max_s q_s$, this periodization means also a partial unwrapping of the phase from the periods q_s to the larger period Q.

[1] *I.e.*, $x_+ = x$ if $x \geq 0$ and $x_+ = 0$ if $x < 0$.

4 Approximating the ML Estimate

The 3D optimization (11) is quite demanding. Pragmatically, we compute a suboptimal solution based on the assumption

$$F_{w,h,s}(\hat{c}_{2,s},\hat{c}_{3,s}) \simeq F_{w,h,s}(\mu_s\hat{c}_2,\mu_s\hat{c}_3), \tag{12}$$

where $\hat{c}_2$ and $\hat{c}_3$ are the solution of (11) and

$$(\hat{c}_{2,s},\hat{c}_{3,s}) \equiv \arg\max_{c_2,c_3} |F_{w,h,s}(c_2,c_3)|. \tag{13}$$

We note that the assumption (12) holds true at least in two scenarios: a) single channel; b) high signal-to-noise ratio. When the noise power increases, the above assumption is violated and we can not guarantee a performance close to optimal. Nevertheless, we have obtained very good estimates, even in medium to low signal-to-noise ratio scenarios. The comparison between the optimal and suboptimal estimates is, however, beyond the scope of this paper.

Let us introduce the right hand side of (12) into (9). We are then led to the absolute phase estimate $\hat{\varphi} = \hat{c}_1$ calculated by the single-variable optimization

$$\begin{aligned} \hat{c}_1 &= \arg\max_{c_1} \tilde{L}_h(c_1), \\ \tilde{L}_h(c_1) &= \sum_s \frac{1}{\sigma_s^2}\frac{1}{\sum_l w_{h,l,s}} |F_{w,h,s}(\hat{c}_{2,s},\hat{c}_{3,s})|^2 \cos_+^2(\mu_s c_1 - \hat{\psi}_s) \\ \hat{\psi}_s &= \text{angle}(F_{w,h,s}(\hat{c}_{2,s},\hat{c}_{3,s})). \end{aligned} \tag{14}$$

Phases $\hat{\psi}_s$, for $s = 1,\ldots,L$, are the LPA estimates of the corresponding wrapped phases $\psi_s = W(\mu_s\varphi)$. Again note that the criterion $\tilde{L}_h(c_1)$ is periodic with respect to c_1 with period $2\pi Q$. Thus, the optimization can be performed only on the finite interval $[-\pi Q, \pi Q)$:

$$\hat{c}_1 = \arg\max_{c_1\in[-\pi Q,\pi Q)} \tilde{L}_h(c_1). \tag{15}$$

If this interval covers the variation range of the absolute phase φ, $\varphi \in [-\pi Q, \pi Q)$, the estimate (15) gives a solution of the multifrequency phase unwrap problem. If $\varphi \notin [-\pi Q, \pi Q)$, *i.e.*, the range of the absolute phase φ is larger than $2\pi Q$, then $\hat{c}_1$ gives the partial phase unwrapping periodized to the interval $[-\pi Q, \pi Q)$. A complete unwrapping is obtained by applying one of the standard unwrapping algorithms, as these partially unwrapped data can be treated as obtained from a single sensor modulo-$2\pi Q$ wrapped phase. The above formulas define what we call **ML-MF-PEARLS** algorithm as short for **M**aximum **L**ikelihood **M**ulti-**F**requency **P**hase **E**stimation using **A**daptive **R**egularization based on **L**ocal **S**moothing.

5 Experimental Results

Let us we consider a two-frequency scenario with the wavelength $\lambda_1 < \lambda_2$ and compare it versus a single frequency reconstructions with the wavelengths λ_1

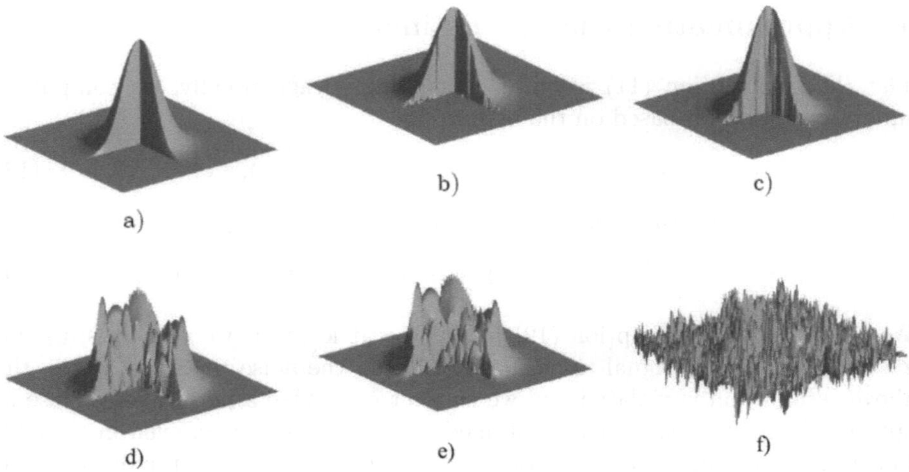

Fig. 1. Discontinuos phase reconstruction: a) true phase surface, b) ML-MF-PEARS reconstruction, ($\mu_1 = 1, \mu_2 = 4/5$), c) ML-MF-PEARS reconstruction, ($\mu_1 = 1, \mu_2 = 9/10$), d) a single frequency PEARLS reconstruction, $\mu_1 = 1$ e) a single frequency PEARLS reconstruction, $\mu_2 = 9/10$, f) a single beat-frequency PEARLS reconstruction, $\mu_{12} = 10$

and λ_2 as well as versus the synthetic wavelength $\Lambda_{1,2} = \lambda_1\lambda_2/(\lambda_2 - \lambda_1)$. The measurement sensitivity is reduced when one considers larger wavelengths. This effect can be modelled by the noise standard deviation proportional to the wavelength. Thus, the noise level in the data corresponding to the wavelength $\Lambda_{1,2}$ is much larger than that for the smaller wavelength λ_1 and λ_2.

The proposed algorithm shows a much better accuracy for the two-frequency data than for the data above mentioned corresponding single frequency scenarios. Another advantage of the multifrequency scenario is its ability to reconstruct the absolute phase for continuous surfaces with huge range and large derivatives. The multifrequency estimation implements an intelligent use of the multichannel data leading to effective phase unwrapping in scenarios in which the unwrapping based on any of the data channels would fail. Moreover, the multifrequency data processing allows to successfully unwrap discontinuous surfaces in situations in which the separate channel processing has no chance for success.

In what follows, we present several experiments illustrating the ML-MF-PEARLS performance for continuous and discontinuous phase surfaces. For the phase unwrap of the filtered wrapped phase, we use the PUMA algorithm [1], which is able to work with discontinuities. LPA is exploited with the uniform square windows w_h defined on the integer symmetric grid $\{(x, y) : |x|, |y| \leq h\}$; thus, the number of pixels of w_h is $(2h+1)$. The ICI parameter was set to $\Gamma = 2.0$ and the window sizes to $H \in \{1, 2, 3, 4\}$. The frequencies (13) were computed via FFT zero-padded to the size 64×64.

As a test function, we use $\varphi(x, y) = A_\varphi \times \exp(-x^2/(2\sigma_x^2) - y^2/(2\sigma_y^2))$, a Gaussian shaped surface, with $\sigma_x = 10$, $\sigma_y = 15$, and $A_\varphi = 40 \times 2\pi$. The surface is defined on a square grid with integer arguments x, y, $-49 \leq x, y \leq 50$. The maximum value of φ is $40 \times 2\pi$ and the maximum values of the first differences are about 15.2 radians. With such high phase differences, any single channel based unwrapping algorithm fail due to many phase differences larger than π. The noisy observations were generated according to (4), for $B_s = 1$.

We produce two groups of experiments assuming that we have two channels observations with ($\mu_1 = 1, \mu_2 = 4/5$) and ($\mu_1 = 1, \mu_2 = 9/10$), respectively. Then for the synthetic wavelength $\Lambda_{1,2}$ we introduce the phase scaling factor as $\mu_{1,2} = 1/\Lambda_{1,2} = \lambda_1 - \lambda_2$. For the selected $\mu_1 = 1$ and $\mu_2 = 4/5$ we have $\mu_{1,2} = 1/5$ or $\Lambda_{1,2} = 5$, and for $\mu_1 = 1$ and $\mu_2 = 9/10$ we have $\mu_{1,2} = 1/10$ or $\Lambda_{1,2} = 10$. Note that for all these cases we have the period Q equal to the corresponding beat wavelength $\Lambda_{1,2} = 5, 10$.

It order to make comparable the accuracy results obtained for the signals of different wavelength, we assume that the noise standard deviation is proportional to the wavelength or inverse proportional to the phase scalling factors μ:

$$\sigma_1 = \sigma/\mu_1,\ \sigma_2 = \sigma/\mu_2,\ \sigma_{1,2} = \sigma/\mu_{1,2}, \tag{16}$$

where σ is a varying parameter. Tables 1 and 2 shows some of the results. The ML-MF-PEARLS shows systematically better accuracy and manage to unwrap the phase when single frequency algorithms fail.

Table 1. RMSE (in rad), $A_\varphi = 40 \times 2\pi$, $\mu_1 = 1, \mu_2 = 4/5$

Algorithm \ σ	.3	.1	.01
PEARLS, $\mu_1 = 1$	fail	fail	fail
PEARLS, $\mu_2 = 4/5$	fail	fail	fail
PEARLS, $\mu_{1,2} = 1/5$	fail	0.722	0.252
ML-MF-PEARLS	0.587	0.206	0.194

Table 2. RMSE (in rad), $A_\varphi = 40 \times 2\pi$, $\mu_1 = 1, \mu_2 = 9/10$

Algorithm \ σ	.3	.1	.01
PEARLS, $\mu_1 = 1$	fail	fail	fail
PEARLS, $\mu_2 = 9/10$	fail	fail	fail
PEARLS, $\mu_{1,2} = 1/10$	fail	3.48	0.496
ML-MF-PEARLS	1.26	0.204	0.194

We now illustrate the potential in handling discontinuities of bringing together the adaptive denoising and the unwrapping. For the test, we use the Gaussian surface with one quarter set to zero. The corresponding results are shown in Fig. 1. The developed algorithm confirms its clear ability to reconstruct a strongly varying discontinuous absolute phase from noisy multifrequency data.

Figure 2 shows results based on a simulated InSAR example supplied in the book [3]. This data set have been generated based on a real digital elevation model of mountainous terrain around Long's Peak using a high-fidelity InSAR

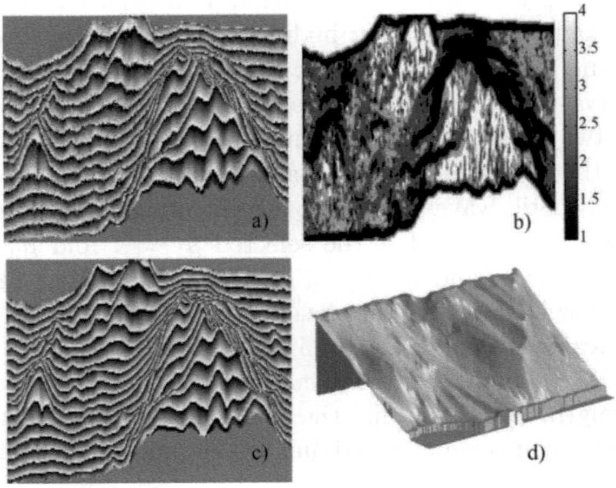

Fig. 2. Simulated SAR based on a real digital elevation model of mountainous terrain around Long's Peak using a high-fidelity InSAR simulator (see [3] for details): a) original interferogram (for $\mu_1 = 1$); b) Window sizes given by ICI; c) LPA phase estimation corresponding to $\psi_1 = W(\mu_1\varphi)$; d) ML-MF-PEARS reconstruction for $\mu_1 = 1$ and $\mu_2 = 4/5$ corresponding to rmse $= 0.3$ rad (see text for details)

simulator that models the SAR point spread function, the InSAR geometry, the speckle noise (4 looks) and the layover and shadow phenomena. To simulate diversity in the acquisition, besides the interferogram supplied with the data, we have generated another interferogram, according to the statistics of a fully developed speckle (see, *e.g.*, [7] for details) with a frequency $\mu_2 = 4/5$.

Figure 2 a) shows the original interferogram corresponding to $\mu_1 = 1$. Due to noise, areas of low coherence, and layover, the estimation of the original phase based on this interferogram is a very hard problem, which does not yield reasonable estimates, unless external information in the form of quality maps is used [3], [7]. Parts b) and c) shows the window sizes given by ICI and the LPA phase estimation corresponding to $\psi_1 = W(\mu_1\varphi)$, respectively. Part d) shows ML-MF-PEARS reconstruction, where the areas of very low coherence were removed and interpolated from the neighbors. We stress that we have not used these quality information in the estimation phase. The estimation error is RMSE $= 0.3$ rad, which, having in mind that the phase range is larger the 120 rad, is a very good figure.

The leading term of the computational complexity of the ML-MF-PEARLS is $O(n^{2.5})$ (n is the number of pixels) due to the PUMA algorithm. This is, however, the worst case figure. The practical complexity is very close to $O(n)$ [1]. In practice, we have observed that a good approximation of the algorithm complexity is given by complexity of nL FFTs, *i.e.*, $(2LP^2 \log_2 P)n$, where L is the number of channels and $P \times P$ is the size of the FFTs. The examples shown is this section took less than 30 seconds in a PC equipped with a dual core CPU running at 3.0GHz

6 Concluding Remarks

We have introduced ML-MF-PEARLS, a new adaptive algorithm to estimate the absolute phase from frequency diverse wrapped observations. The new methodology is based on local maximum likelihood phase estimates. The true phase is approximated by a local polynomial with varying adaptive neighborhood used in reconstruction. This mechanism is critical in preserving the discontinuities of piecewise smooth absolute phase surfaces. The ML-MF-PEARLS, algorithm, besides filtering the noise, yields a $2\pi Q$-periodic solution, where $Q > 1$ is an integer. Depending on the value of Q and of the original phase range, we may obtain complete or partial phase unwrapping. In the latter case, we apply the recently introduced robust (in the sense of discontinuity preserving) PUMA unwrapping algorithm [1]. In a set of experiments, we gave evidence that the ML-MF-PEARLS algorithm is able to produce useful unwrappings, whereas state-of-the art competitors fail.

Acknowledgments

This research was supported by the "Fundação para a Ciência e Tecnologia", under the project PDCTE/CPS/49967/2003, by the European Space Agency, under the project ESA/C1:2422/2003, and by the Academy of Finland, project No. 213462 (Finnish Centre of Excellence program 2006 – 2011).

References

1. Bioucas-Dias, J., Valadão, G.: Phase unwrapping via graph cuts. IEEE Trans. Image Processing 16(3), 684–697 (2007)
2. Graham, L.: Synthetic interferometer radar for topographic mapping. Proceeding of the IEEE 62(2), 763–768 (1974)
3. Ghiglia, D., Pritt, M.: Two-Dimensional Phase Unwrapping. In: Theory, Algorithms, and Software. John Wiley & Sons, New York (1998)
4. Zebker, H., Goldstein, R.: Topographic mapping from interferometric synthetic aperture radar. Journal of Geophysics Research 91(B5), 4993–4999 (1986)
5. Patil, A., Rastogi, P.: Moving ahead with phase. Optics and Lasers in Engineering 45, 253–257 (2007)
6. Goldstein, R., Zebker, H., Werner, C.: Satellite radar interferometry: Two-dimensional phase unwrapping. In: Symposium on the Ionospheric Effects on Communication and Related Systems. Radio Science, vol. 23, pp. 713–720 (1988)
7. Bioucas-Dias, J., Leitao, J.: The $Z\pi M$ algorithm: a method for interferometric image reconstruction in SAR/SAS. IEEE Trans. Image Processing 11(4), 408–422 (2002)
8. Yun, H.Y., Hong, C.K., Chang, S.W.: Least-square phase estimation with multiple parameters in phase-shifting electronic speckle pattern interferometry. J. Opt. Soc. Am. A 20, 240–247 (2003)
9. Kemao, Q.: Two-dimensional windowed Fourier transform for fringe pattern analysis: principles, applications and implementations. Opt. Lasers Eng. 45, 304–317 (2007)

10. Katkovnik, V., Astola, J., Egiazarian, K.: Phase local approximation (PhaseLa) technique for phase unwrap from noisy data. IEEE Trans. on Image Processing 46(6), 833–846 (2008)
11. Katkovnik, V., Egiazarian, K., Astola, J.: Local Approximation Techniques in Signal and Image Processing. SPIE Press, Bellingham (2006)
12. Servin, M., Marroquin, J.L., Malacara, D., Cuevas, F.J.: Phase unwrapping with a regularized phase-tracking system. Applied Optics 37(10), 1917–1923 (1998)
13. Pascazio, V., Schirinzi, G.: Multifrequency InSAR height reconstruction through maximum likelihood estimation of local planes parameters. IEEE Transactions on Image Processing 11(12), 1478–1489 (2002)
14. Servin, M., Cuevas, F.J., Malacara, D., Marroguin, J.L., Rodriguez-Vera, R.: Phase unwrapping through demodulation by use of the regularized phase-tracking technique. Appl. Opt. 38, 1934–1941 (1999)
15. Servin, M., Kujawinska, M.: Modern fringe pattern analysis in interferometry. In: Malacara, D., Thompson, B.J. (eds.) Handbook of Optical Engineering, ch. 12, pp. 373–426, Dekker (2001)
16. Born, M., Wolf, E.: Principles of Optics, 7th edn. Cambridge University Press, Cambridge (2002)
17. Xia, X.-G., Wang, G.: Phase unwrapping and a robust chinese remainder theorem. IEEE Signal Processing Letters 14(4), 247–250 (2007)
18. McClellan, J.H., Rader, C.M.: Number Theory in Digital Signal Processing. Prentice-Hall, Englewood Cliffs (1979)
19. Goldreich, O., Ron, D., Sudan, M.: Chinese remaindering with errors. IEEE Trans. Inf. Theory 46(7), 1330–1338 (2000)
20. Bioucas-Dias, J., Katkovnik, V., Astola, J., Egiazarian, K.: Absolute phase estimation: adaptive local denoising and global unwrapping. Applied Optics 47(29), 5358–5369 (2008)

A New Hybrid DCT and Contourlet Transform Based JPEG Image Steganalysis Technique

Zohaib Khan and Atif Bin Mansoor

College of Aeronautical Engineering,
National University of Sciences & Technology, Pakistan
zohaibkh_27@yahoo.com, atif-cae@nust.edu.pk

Abstract. In this paper, a universal steganalysis scheme for JPEG images based upon hybrid transform features is presented. We first analyzed two different transform domains (Discrete Cosine Transform and Discrete Contourlet Transform) separately, to extract features for steganalysis. Then a combination of these two feature sets is constructed and employed for steganalysis. A Fisher Linear Discriminant classifier is trained on features from both clean and steganographic images using all three feature sets and subsequently used for classification. Experiments performed on images embedded with two variants of F5 and Model based steganographic techniques reveal the effectiveness of proposed steganalysis approach by demonstrating improved detection for hybrid features.

Keywords: Steganography, Steganalysis, Information Hiding, Feature Extraction, Classification.

1 Introduction

The word steganography comes from the Greek words *steganos* and *graphia*, which together mean 'hidden writing' [1]. Steganography is being used to hide information in digital images and later transfer them through the internet without any suspicion. This poses a serious threat to both commercial and military organizations as regards to information security. Steganalysis techniques aim at detecting the presence of hidden messages from inconspicuous stego images.

Steganography is an ancient subject, with its roots lying in ancient Greece and China, where it was already in use thousands of years ago. The prisoners' problem [2] well defines the modern formulation of steganography. Two accomplices Alice and Bob are in a jail. They wish to communicate in order to plan to break the prison. But all communication between the two is being monitored by the warden, Wendy, who will put them in a high security prison if they are suspected of escaping. Specifically, in terms of a steganography model, Alice wishes to send a secret message $\boldsymbol{m}$ to Bob. For this, she hides the secret message $\boldsymbol{m}$ using a shared secret key $\boldsymbol{k}$ into a cover-object $\boldsymbol{c}$ to obtain the stego-object $\boldsymbol{s}$. The stego-object $\boldsymbol{s}$ is then sent by Alice through the public channel to Bob, $\boldsymbol{m}$ unnoticed by Wendy. Once Bob receives the stego-object $\boldsymbol{s}$, he is able to recover the secret message $\boldsymbol{m}$ using the shared secret key $\boldsymbol{k}$.

A.-B. Salberg, J.Y. Hardeberg, and R. Jenssen (Eds.): SCIA 2009, LNCS 5575, pp. 321–330, 2009.

Steganography and cryptography are closely related information hiding techniques. The purpose of cryptography is to scramble a message so that it cannot be understood, while that of steganography is to hide a message so that it cannot be seen. Generally, a message created with cryptographic tools will raise the alarm on a neutral observer while a message created with steganographic tools will not. Sometimes, steganography and cryptography are combined in a way that the message may be encrypted before hiding to provide additional security.

Steganographers who intend to hide communications are countered by steganalysts who intend to reveal it. The specific field to counter steganography is known as *steganalysis*. The goal of a steganalyst is to detect the presence of steganography so that the secret message may be stopped before it is received. Then the further identification of the steganography tool to extract the secret message from the stego file comes under the field of cryptanalysis.

Generally, two approaches are followed for steganalysis; one is to come up with a steganalysis method specific to a particular steganographic algorithm. The other is to develop universal steganalysis techniques which are independent of the steganographic algorithm. Both approaches have their own strengths and weaknesses. A steganalysis technique specific to an embedding method would give very good results when tested only on that embedding method; but might fail on all other steganographic algorithms as in [4], [5], [6] and [7]. On the other hand, a steganalysis technique which is independent of the embedding algorithm might perform less accurately overall but still shows its effectiveness against new and unseen embedding algorithms as in [8], [9], [10] and [11]. Our research work is concentrated on the second approach due to its wide applicability.

In this paper, we propose a steganalysis technique by extracting features from two transform domains; the discrete contourlet transform and the discrete cosine transform. These features are investigated individually and combinatorially. The rest of the paper is organized as follows: In Section 2, we discuss the previous research work related to steganalysis. In Section 3, we present our proposed approach. Experimental results are presented in Section 4. Finally, the paper is concluded in Section 5.

2 Related Work

Due to the increasing availability of new steganography tools over the internet, there has been an increasing interest in the research for new and improved steganalysis techniques which are able to detect both previously seen and unseen embedding algorithms. A good survey of benchmarking of steganography and steganalysis techniques is given by Kharrazi et al. [3].

Fridrich et al. presented a steganalysis method which can reliably detect messages hidden in JPEG images using the steganography algorithm F5, and also estimate their lengths [4]. This method was further improved by Aboalsamh et al. [5] by determining the optimal value of the message length estimation parameter β. Westfeld and Pfitzmann presented visual and statistical attacks on various steganographic systems including EzStego v2.0b3, Jsteg v4, Steganos v1.5 and S-Tools v4.0, by using an embedding filter and the χ^2 statistic [6]. A steganalysis scheme specific to the embedding algorithm Outguess is proposed in [7], by making use of the assumption that the embedding of a message in a stego image will be different than embedding the same into a cover image.

Avcibas et al. proposed that the correlation between the bit planes as well as the binary texture characteristics within the bit planes will differ between a stego image and a cover image, thus facilitating steganalysis [8]. Farid suggested that embedding of a message alters the higher order statistics calculated from a multi-scale wavelet decomposition [9]. Particularly, he calculated the first four statistical moments (mean, variance, skewness and kurtosis) of the distribution of wavelet coefficients at different scales and subbands. These features (moments), calculated from both cover and stego images were then used to train a linear classifier which could distinguish them with a certain success rate. Fridrich showed that a functional obtained from marginal and joint statistics of DCT coefficients will vary between stego and cover images. In particular, a functional such as the global DCT coefficient histogram was calculated for an image and its decompressed, cropped and recompressed versions. Finally the resulting features were obtained as the L_1 norm of the difference between the two. The classifier built with features extracted from both cover and stego images could reliably detect F5, Outguess and Model based steganography techniques [10]. Avcibas et al. used various image quality metrics to compute the distance between a test image and its lowpass filtered versions. Then a classifier built using linear regression showed detection of LSB steganography and various watermarking techniques with a reasonable accuracy [11].

3 Proposed Approach

3.1 Feature Extraction

The addition of a message to a cover image does not affect the visual appearance of the image but may affect some statistics. The features required for the task of steganalysis should be able to catch these minor statistical disorders that are created during the data hiding process. In our approach, we first extract features in the discrete contourlet transform domain, followed by the discrete cosine transform domain and finally combine both extracted features to make a hybrid feature set.

Discrete Contourlet Tranform Features. The contourlet transform is a new two-dimensional extension of the wavelet transform using multiscale and directional filter banks [13]. For extraction of features in the Discrete Contourlet Transform domain, we decomposed image into three pyramidal levels and 2^n directions where $n = 0, 2, 4$. Figure 1 shows the levels and selection of subbands for this decomposition. For the laplacian pyramidal decomposition stage, the *'Haar'* filter was used. For the directional decomposition stage the *'PKVA'* filter was used. In each scale from coarse to fine, the number of directions are 1,4,and 16. By applying the pyramidal directional filter bank decomposition and ignoring the finest lowpass approximation subband, we obtained a total of 23 subbands.

Fig. 1. A three level contourlet decomposition

Various statistical measures are used in our analysis. Particularly, the first three normalized moments of the characteristic function are computed. The K-point discrete Characteristic Function (CF) is defined as

$$\Phi(k) = \sum_{m=0}^{M-1} h(m) e^{\{\frac{j2\pi mk}{K}\}} . \tag{1}$$

where $\{h(m)\}_{m=0}^{M-1}$ is the M bin histogram which is an estimate of the PDF, $p(x)$ of the contourlet coefficients distribution. The nth absolute moment of discrete CF is defined as

$$M_n^A = \sum_{k=0}^{K/2-1} \Phi(k) \sin^n \left(\frac{\pi k}{K}\right) . \tag{2}$$

Finally, the normalized CF moment is defined as

$$\hat{M}_n^A = \frac{M_n^A}{M_0^A} . \tag{3}$$

where M_0^A is the zeroth order moment. We calculated the first three normalized CF moments for each of the 23 subbands, giving a **69-D** feature vector.

DCT Based Features. The DCT based feature set is constructed following the approach of Fridrich [10]. A vector functional $\boldsymbol{F}$ is applied to the JPEG image J_1. This image is then decompressed to the spatial domain, cropped by 4 pixels in each direction and recompressed with the same quantization table as J_1 to obtain J_2. The vector functional $\boldsymbol{F}$ is then applied to J_2. The final feature f is obtained as the L_1 norm of the difference of the functional applied to J_1 and J_2.

$$f = \|\boldsymbol{F}(J_1) - \boldsymbol{F}(J_2)\|_{L_1} . \tag{4}$$

The rational behind this procedure is that the recompression after cropping by 4 pixels does not see the previous JPEG compression's 8×8 block boundary and thus it is not affected by the previous quantization and hence embedding in the DCT domain. So, J_2 can be thought of as an approximation to its cover image.

We calculated the global, individual and dual histograms of the DCT coefficient array $d_{(k)}(i,j)$ as the first order functionals. The symbol $d_{(k)}(i,j)$ denotes the $(i,j)^{th}$ quantized DCT coefficient $(i,j=1,2,...,8)$ in the *k-th* block, $(k=1,2,...,B)$. The global histogram of all 64B DCT coefficients is given as, $H(m)_{m=L}^{R}$, where $L = min_{k,i,j} d_{(k)}(i,j)$ and $R = max_{k,i,j} d_{(k)}(i,j)$. We computed $H/\left\|H\right\|_{L_1}$, the normalized global histogram of DCT coefficients as the first functional.

Steganographic techniques that preserve global DCT coefficients histogram may not necessarily preserve the histogram of individual DCT modes. So, we calculated $h^{ij}/\left\|h^{ij}\right\|_{L_1}$, the normalized individual histograms $h(m)_{m=L}^{R}$ of 5 low frequency DCT modes, $(i,j) = (2,1), (3,1), (1,2), (2,2), (1,3)$ as the next five functionals.

The dual histogram is an 8×8 matrix which indicates the number of how many times the value *'d'* occurs as the $(i,j)^{th}$ DCT coefficient over all blocks B in the image. We computed $g_{ij}^{d}/\left\|g_{ij}^{d}\right\|_{L_1}$, the normalized dual histograms where $g_{ij}^{d} = \sum_{k=1}^{B} \delta(d, d_{(k)}(i,j))$ for 11 values of $d = -5, -4, ..., 4, 5$.

Inter block dependency is captured by the second order features *variation* and *blockiness*. Most steganographic techniques add entropy to the DCT coefficients which is captured by the *variation* (V)

$$V = \frac{\sum_{i,j=1}^{8} \sum_{k=1}^{|I_r|-1} |d_{I_r(k)}(i,j) - d_{I_r(k+1)}(i,j)| + \sum_{i,j=1}^{8} \sum_{k=1}^{|I_c|-1} |d_{I_c(k)}(i,j) - d_{I_c(k+1)}(i,j)|}{|Ir| + |Ic|}. \quad (5)$$

where I_r and I_c denote the vectors of block indices while scanning the image 'by rows' and 'by columns' respectively.

Blockiness is calculated from the decompressed JPEG image and is a measure of discontinuity along the block boundaries over all DCT modes over the whole image. The L_1 and L_2 *blockiness* $(B_\alpha, \alpha = 1, 2)$ is defined as

$$B_\alpha = \frac{\sum_{i=1}^{\lfloor (M-1)/8 \rfloor} \sum_{j=1}^{N} |x_{8i,j} - x_{8i+1,j}|^\alpha + \sum_{j=1}^{\lfloor (N-1)/8 \rfloor} \sum_{i=1}^{M} |x_{i,8j} - x_{i,8j+1}|^\alpha}{N\lfloor (M-1)/8 \rfloor + M\lfloor (N-1)/8 \rfloor} \quad (6)$$

where $x_{i,j}$ are the grayscale intensity values of an image with dimensions $M \times N$.

The final DCT based feature vector is **20-D** (Histograms: 1 global, 5 individual, 11 dual. *Variation*: 1. *Blockiness*: 2).

Hybrid Features. After extracting the features in the discrete cosine transform and the discrete contourlet transform domain, we finally combine the extracted feature sets into one hybrid feature set, giving a **89-D** feature vector, (69 CNT + 20 DCT).

4 Experimental Results

4.1 Image Datasets

Cover Image Dataset. For our experiments, we used 1338 grayscale images of size 512x384 obtained from the Uncompressed Colour Image Database (UCID) constructed by Schaefer and Stich [14], available at [15]. These images contain a wide range of indoor/outdoor, daylight/night scenes, providing a real and challenging environment for a steganalysis problem. All images were converted to JPEG at 80% quality for our experiments.

F5 Stego Image Dataset. Our first stego image dataset is generated by the steganography software F5 [16], proposed by Andreas Westfeld. F5 steganography algorithm embeds information bits by incrementing and decrementing the values of quantized DCT coefficients from compressed JPEG images [17]. F5 also uses an operation known as 'matrix embedding' in which it minimizes the amount of changes made to the DCT coefficients necessary to embed a message of certain length. Matrix embedding has three parameters (c, n, k), where c is the number of changes per group of n coefficients, and k is the number of embedded bits. These parameter values are determined by the embedding algorithm.

F5 algorithm first compresses the input image with a user defined quality factor before embedding the message. We chose a quality factor of 80 for stego images. Messages were successfully embedded at rates of 0.05, 0.10, 0.20, 0.3, 0.40 and 0.60 bpc (bits per non-zero DCT coefficients). We chose F5 because recent results in [8], [9], [12] have shown that F5 is harder to detect than other commercially available steganography algorithms.

MB Stego Image Dataset. Our second stego image dataset is generated by the Model Based steganography method [18], proposed by Phil Sallee [19]. The algorithm first breaks down the quantized DCT coefficients of a JPEG image into two parts and then replaces the perceptually insignificant component

Table 1. The number of images in the stego image datasets given the message length. F5 with matrix embedding turned off $(1, 1, 1)$ and turned on (c, n, k). Model based steganography without deblocking (MB1) and with deblocking (MB2). (U = unachievable rate).

Embedding Rate (bpc)	**F5** $(1, 1, 1)$	**F5** (c, n, k)	**MB1**	**MB2**
0.05	1338	1338	1338	1338
0.10	1338	1338	1338	1338
0.20	1338	1337	1338	1334
0.30	1337	1295	1338	1320
0.40	1332	5	1338	1119
0.60	5	U	1332	117
0.80	U	U	60	U

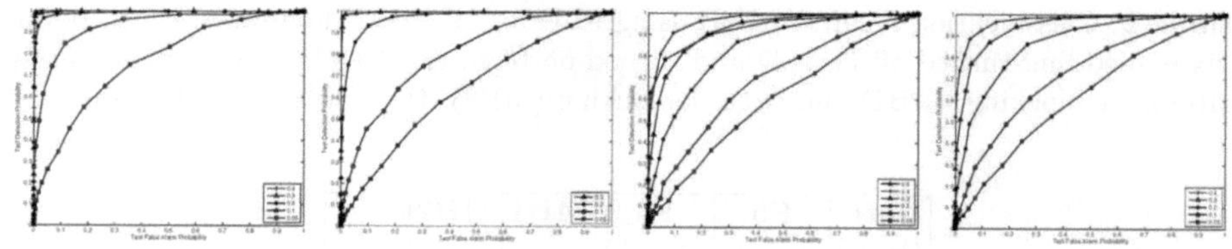

(a) (b) (c) (d)

Fig. 2. ROC curves using DCT based features. (a) F5 (without matrix embedding) (b) F5 (with matrix embedding) (c) MB1 (without deblocking) (d) MB2 (with deblocking).

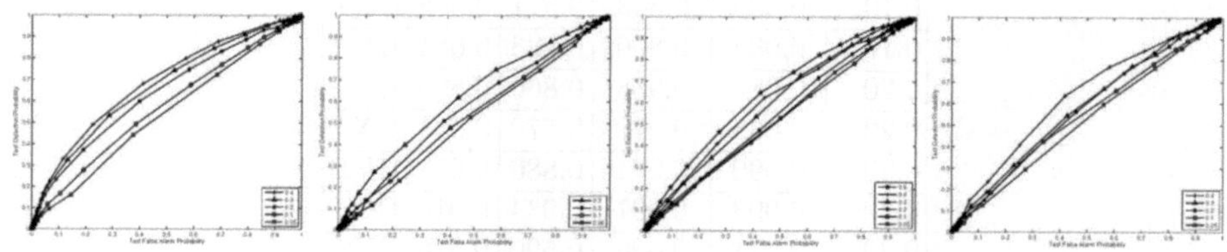

(a) (b) (c) (d)

Fig. 3. ROC curves using CNT based features. (a) F5 (without matrix embedding) (b) F5 (with matrix embedding) (c) MB1 (without deblocking) (d) MB2 (with deblocking).

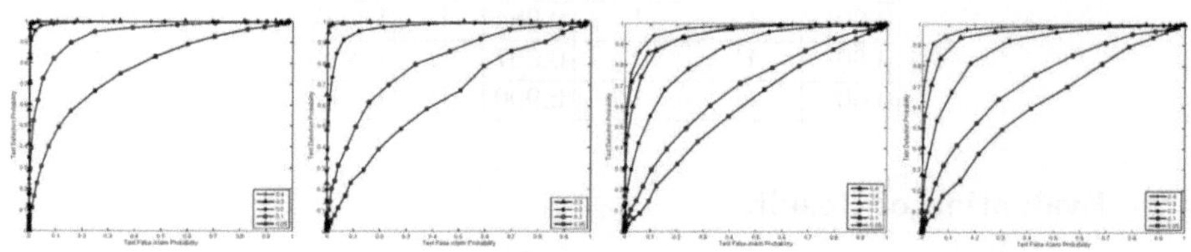

(a) (b) (c) (d)

Fig. 4. ROC curves using Hybrid features. (a) F5 (without matrix embedding) (b) F5 (with matrix embedding) (c) MB1 (without deblocking) (d) MB2 (with deblocking).

with the coded message signal. The algorithm has two types; MB1 is normal steganography and MB2 is steganography with deblocking. The deblocking algorithm adjusts the unused coefficients to reduce the blockiness of the resulting image to the original blockiness. Unlike F5, the Model Based steganography algorithm does not recompress the cover image before embedding. We embed at rates of 0.05, 0.10, 0.20, 0.3, 0.40 0.60 and 0.80 bpc. The model based steganography algorithm has also shown high resistance against steganalysis techniques in [3], [10].

The reason for choosing the message length proportional to the number of non-zero DCT coefficients was to create a stego image database for which the steganalysis is roughly of the same level of difficulty. We further carried out embedding at different rates to observe the steganalysis performance for messages of varying length. It can be seen in Table 1 that the Model based steganography is more efficient in embedding as compared to F5; since longer messages can be accommodated in images using Model based steganography.

Table 2. Classification results (AUC) using FLD for all embedding rates. F5 with matrix embedding turned off (1, 1, 1) and turned on (c, n, k). Model based steganography without deblocking (MB1) and with deblocking (MB2). (U = unachievable rate).

Rate (bpc)	**F5** (1, 1, 1)	**F5** (c, n, k)	**MB1**	**MB2**	
0.05	0.769	0.643	0.611	0.591	DCT
0.05	0.555	0.511	0.529	0.518	CNT
0.05	0.789	0.632	0.624	0.585	HYB
0.10	0.924	0.795	0.721	0.686	DCT
0.10	0.589	0.543	0.511	0.508	CNT
0.10	0.936	0.800	0.723	0.681	HYB
0.20	0.989	0.968	0.860	0.829	DCT
0.20	0.639	0.572	0.570	0.541	CNT
0.20	0.990	0.971	0.886	0.851	HYB
0.30	0.998	0.997	0.934	0.914	DCT
0.30	0.688	0.629	0.590	0.576	CNT
0.30	0.996	0.996	0.953	0.935	HYB
0.40	1.000	U	0.963	0.962	DCT
0.40	0.697	U	0.617	0.619	CNT
0.40	0.997	U	0.978	0.974	HYB
0.60	U	U	0.984	U	DCT
0.60	U	U	0.667	U	CNT
0.60	U	U	0.990	U	HYB

4.2 Evaluation of Results

The Fisher Linear Discriminant classifier [20] was utilized for our experiments. Each steganographic algorithm was analyzed separately for the evaluation of the steganalytic classifier. For a fixed relative message length, we created a database of training images comprising 669 cover and 669 stego images. Both DWT based features (DWT) and DCT based features (DCT) were extracted from the training set and were combined to form a Joint feature set (JNT), according to the procedure explained in Section 3.1. The FLD classifier was then tested on the features extracted from a different database of test images comprising 669 cover and 669 stego images. The Receiver Operating Characteristics (ROC) curves, which give the variation of the Detection Probability (P_d , the fraction of correctly classified stego images) with the False Alarm Probability (P_f , the fraction of stego images wrongly classified as cover image), were computed for each steganographic algorithm and embedding rate. The area under the ROC curve (AUC) was measured to determine the overall classification accuracy.

Figures 2-4 give the obtained ROC curves for the steganographic techniques under test for different embedding rates. Note that due to the space limitation, these figures are displayed in small size. However, readers are encouraged to take a look by using zoom to 400%. We observe that the DCT based features outperform the CNT based features for all embedding rates. As could be expected, the detection of F5 without matrix embedding is better than F5 with matrix embedding since the matrix embedding operation significantly reduces detectability at the expense of message capacity.

Table 2 summarizes the classification results. For F5 without matrix embedding, the proposed Hybrid transform features dominate both DCT and CNT based features for embedding rates till 0.20 bpc. For higher embedding rates the DCT based features perform better. For F5 with matrix embedding, both the proposed hybrid features and the DCT based features are close competitors, though the former performs better at some embedding rates.

For MB1 algorithm (without deblocking), the proposed hybrid features outperform both the DCT and CNT based features for all embedding rates. For MB2 algorithm (with deblocking), the hybrid features perform better compared to both CNT and DCT based features for embedding rates greater than 0.10 bpc. It is observed that the detection of MB1 is better than MB2, as the deblocking algorithm in MB2 reduces the blockiness of the stego image to match the original image.

5 Conclusion

This paper presents a new DCT and CNT based hybrid features approach for universal steganalysis. DCT and CNT based statistical features are investigated individually, followed by research on combined features. The Fisher Linear Discriminant classifier is employed for classification. The experiments were performed on image datasets with different embedding rates for F5 and Model based steganography algorithms. Experiments revealed that for JPEG images the DCT is a better choice for extraction of features as compared to the CNT. The experiments with hybrid transform features reveal that the extraction of features in more than one transform domain improves the steganalysis performance.

References

1. McBride, B.T., Peterson, G.L., Gustafson, S.C.: A new Blind Method for Detecting Novel Steganography. Digital Investigation 2, 50–70 (2005)
2. Simmons, G.J.: 'Prisoners' Problem and the Subliminal Channel. In: CRYPTO 1983-Advances in Cryptology, pp. 51–67 (1984)
3. Kharrazi, M., Sencar, T.H., Memon, N.: Benchmarking Steganographic and Steganalysis Techniques. In: Proc. of SPIE Electronic Imaging, Security, Steganography and Watermarking of Multimedia Contents VII, San Jose, California, USA (2005)
4. Fridrich, J., Goljan, M., Hogea, D.: Steganalysis of JPEG images: Breaking the F5 Algorithm. In: Petitcolas, F.A.P. (ed.) IH 2002. LNCS, vol. 2578, pp. 310–323. Springer, Heidelberg (2003)
5. Aboalsamh, H.A., Dokheekh, S.A., Mathkour, H.I., Assassa, G.M.: Breaking the F5 Algorithm: An Improved Approach. Egyptian Computer Science Journal 29(1), 1–9 (2007)

6. Westfeld, A., Pfitzmann, A.: Attacks on Steganographic Systems. In: Proc. 3rd Information Hiding Workshop, Dresden, Germany, pp. 61–76 (1999)
7. Fridrich, J., Goljan, M., Hogea, D.: Attacking the OutGuess. In: Proc. ACM Workshop on Multimedia and Security 2002. ACM Press, Juan-les-Pins (2002)
8. Avcibas, I., Memon, N., Sankur, B.: Image Steganalysis with Binary Similarity Measures. In: Proc. of the IEEE International Conference on Image Processing, Rochester, New York (September 2002)
9. Farid, H.: Detecting Hidden Messages Using Higher-order Statistical Models. In: Proc. of the IEEE International Conference on Image Processing, vol. 2, pp. 905–908 (2002)
10. Fridrich, J.: Feature-Based Steganalysis for JPEG Images and its Implications for Future Design of Steganographic Schemes. In: Moskowitz, I.S. (ed.) Information Hiding 2004. LNCS, vol. 2137, pp. 67–81. Springer, Heidelberg (2005)
11. Avcibas, I., Memon, N., Sankur, B.: Steganalysis Using Image Quality Metrics. IEEE Transactions on Image Processing 12(2), 221–229 (2003)
12. Wang, Y., Moulin, P.: Optimized Feature Extraction for Learning-Based Image Steganalysis. IEEE Transactions on Information Forensics and Security 2(1) (2007)
13. Po, D.-Y., Do, M.N.: Directional Multiscale Modeling of Images Using the Contourlet Transform. IEEE Transactions on Image Processing 15(6), 1610–1620 (2006)
14. Schaefer, G., Stich, M.: UCID - An Uncompressed Colour Image Database. In: Proc. SPIE, Storage and Retrieval Methods and Applications for Multimedia, San Jose, USA, pp. 472–480 (2004)
15. UCID – Uncompressed Colour Image Database, `http://vision.cs.aston.ac.uk/datasets/UCID/ucid.html` (visited on 02/08/08)
16. Steganography Software F5, `http://wwwrn.inf.tu-dresden.de/~westfeld/f5.html` (visited on 02/08/08)
17. Westfeld, A.: F5 – A Steganographic Algorithm: High capacity despite better steganalysis. In: Moskowitz, I.S. (ed.) IH 2001. LNCS, vol. 2137, pp. 289–302. Springer, Heidelberg (2001)
18. Model Based JPEG Steganography Demo, `http://www.philsallee.com/mbsteg/index.html` (visited on 02/08/08)
19. Sallee, P.: Model-based steganography. In: Kalker, T., Cox, I., Ro, Y.M. (eds.) IWDW 2003. LNCS, vol. 2939, pp. 154–167. Springer, Heidelberg (2004)
20. Duda, R.O., Hart, P.E., Stork, D.G.: Pattern Classification, 2nd edn. John Wiley & Sons, New York (2001)

Improved Statistical Techniques for Multi-part Face Detection and Recognition

Christian Micheloni[1], Enver Sangineto[2], Luigi Cinque[2], and Gian Luca Foresti[1]

[1] Univeristy of Udine
Via delle Scienze 206, 33100 Udine
{michelon,foresti}@dimi.uniud.it
[2] University of Rome "Sapienza"
Via Salaria 113, 00198 Roma
{sangineto,cinque}@di.uniroma1.it

Abstract. In this paper we propose an integrated system for face detection and face recognition based on improved versions of state-of-the-art statistical learning techniques such as Boosting and LDA. Both the detection and the recognition processes are performed on facial features (e.g., the eyes, the nose, the mouth, etc) in order to improve the recognition accuracy and to exploit their statistical independence in the training phase. Experimental results on real images show the superiority of our proposed techniques with respect to the existing ones in both the detection and the recognition phase.

1 Introduction

Face recognition is one of the most studied problems in computer vision, especially w.r.t. security application. Important issues in accurate and robust face recognition is good detection of face patterns and the handling of occlusions. Detecting a face in an image can be solved by applying algorithms developed for pattern recognition tasks. In particular, the goal is to adopt training algorithms like Neural Networks [14], Support Vector Machines [1] etc. that can learn the features that mostly characterize the class of patterns to detect. Within appearance-based method, in the last years boosting algorithms [15,10] have been widely adopted to solve the face detection problem. Although they seemed to have reached a good trade-off between computational complexity and detection efficiency, there are still some considerations that leave room for further improvements in both performance and accuracy. Shapire in [13] proposed the theoretical definition of boosting. A set of *weak* hypotheses $h_1, \ldots, h_T$ is selected and linearly combined to build a more robust *strong* classifier of the form:

$$H(x) = sign\left(\sum_{t=1}^{T} \alpha_t h_t(x)\right) \tag{1}$$

A.-B. Salberg, J.Y. Hardeberg, and R. Jenssen (Eds.): SCIA 2009, LNCS 5575, pp. 331–340, 2009.

On such an idea, the Adabost algorithm [8] proposes an efficient iterative procedure to select at each step the best weak hypothesis from an over complete set of features (e.g. Haar features). Such a result is obtained by maintaining a distribution of weights D over a set of input samples $S = \{x_i, y_i\}$ such that the error ϵ_t introduced by selecting the $t-th$ weak classifier is minimum. The error is defined as:

$$\epsilon_t \equiv Pr_{i \sim D_t}(h_t(x_i) \neq y_i) = \sum_{x_i \in S: h_t(x_i) \neq y_i} D_t(i) \tag{2}$$

where x_i is the sample pattern and y_i its class. Hence, the error introduced by selecting the hypothesis h_t is given by the sum of the current weights associated to those patterns that are misclassified by h_t. To maintain a coherent distribution D_t, that for every step t guarantees the selection of such an optimal weak classifier, the update step is as follows:

$$D_{t+i}(i) = \frac{exp\left(-y_i \sum_t h_t(x_i)\right)}{\prod_t Z_t} \tag{3}$$

where Z_t is a normalization factor that allows to maintain D as a distribution [13]. From this first formulation, new evolutions of AdaBoost have been proposed. RealBoost [9] introduced real values for weak classifiers rather then discrete ones, its development in a cascade of classifiers [16] aims to reduce the computational time for negative samples, while FloatBoost [10] introduces a backtracking mechanism for the rejection of not robust weak classifiers.

Though, all these developments suffer of a high false positive detection rate. The cause can be associated to the high asymmetry of the problem. The number of face patterns into an image is much lower than the number of non-face patterns. To balance the significance of the patterns depending on the belonging classes can be managed only by balancing the cardinality of the positives and negatives training data sets. For such a reason, the training data sets are usually composed of a larger number of negative samples than positives ones. Without this kind of control the so determined classifiers would classify positives and negatives sample in an equal way. Obviously, since we are more interested in detecting face patterns rather than non-face ones we need a mechanism that introduces a degree of asymmetry into the training process regardless the composition of the training set. Viola a Jones in [15], to reproduce the asymmetry of the face detection problem into the training mechanism, introduced a different weighting mechanism for the two classes by modifying the distribution update step. The new updating rule is the following:

$$D_{t+1}(i) = \frac{\exp\left(y_i \log \sqrt{k}\right) \exp\left(-y_i \sum_t h_t(x_i)\right)}{\prod_t Z_t} \tag{4}$$

where k is a user defined parameter that gives a different weight to the samples depending on the belonging class. If $k > 1 (< 1)$ the positive samples are considered more (less) important, if $k = 1$ the algorithm is again the original AdaBoost. Experimentally, the authors noticed that, when determining the asymmetry parameter only at the beginning of the process, the selection of the first classifier absorbs the entire effect of the initial asymmetric weights. The asymmetry is immediately lost and the remaining rounds are entirely symmetric.

For such a reason, in this paper we propose a new learning strategy that tunes the parameter k in order to maintain active the asymmetry for the entire training process. We do that both at strong classifier learning level and at cascade definition. The resulting optimized boosting technique is exploited to train face detectors and to train other classifiers that working on face patterns can detect sub-face patterns (e.g. eyes, nose, mouth, etc.). This important features are used to achieve both a face alignment process (e.g. bringing the eyes axis horizontal) and the block extraction for recognition purposes.

Concerning the face recognition point of view, the existing approaches can be classified in three general categories [19]: *feature-based* , *holistic* and *hybrid* techniques (mixed holistic and feature-based methods). Feature based approaches extract and compares prefixed feature values from some locations on the face. The main drawback of these techniques is their dependence on an exact localization of facial features. In [3], experimental results show the superiority of holistic approaches with respect to feature based ones. On the other hand, holistic approaches consider as input the whole sub-window selected by a previous face detection step. To compress the original space for a reliable estimation of the statistical distribution, statistical "feature extraction techniques" such as Principal Component Analysis (PCA) or Linear Discriminant Analysis (LDA) [5] are usually adopted. Good results have been obtained using Linear Discriminant Analysis (LDA)(e.g., see [18]). The LDA compression technique consists in finding a subspace T of $\mathbb{R}^M$ which maximizes the distances between the points obtained projecting the face clusters into T (where each face class corresponds to a single person). For further details, we refer to [5].

As a consequence of the limited training samples, it is usually hard to reliably learn a correct statistical distribution of the clusters in T, especially when important variability factors are present (e.g., lighting condition changes etc.). In other words, the high variance of the class pattern compared with the limited number of training samples is likely to produce an overfitting phenomenon. Moreover, the necessity of having the whole pattern as input makes it difficult to handle occluded faces. Indeed, face recognition with partial occlusions is an open problem [19] and it is usually not dealt with by holistic approaches.

In this paper we propose a "block-based" holistic technique. Facial feature detection is used to roughly estimate the position of the main facial features such as the eyes, the mouth, the nose, etc. From these positions the face pattern is split in blocks each then separately projected into a dedicated LDA space. At run time a face is partitioned in corresponding blocks and the final recognition is given by the combination of the results separately obtained from each (visible) block.

2 Multi-part Face Detection

To improve the detection rate of a boosting algorithm we considered the Asymboost technique [15] that assigns different weights to the two classes:

$$D_{t+1}(i) = \frac{\exp(y_i \log \sqrt{k}) \exp(-y_i \sum_t h_t(x_i))}{\prod_t Zt} \tag{5}$$

In particular, the idea we propose, instead of considering static the parameter k, aims to tune it on the basis of the current false positives and negatives rate.

2.1 Balancing False Positives Rate

A common way to obtain a cascade classifier with a predetermined False Positives (FP) rate $FP_{cascade}$ is to train the cascade's strong classifiers by equally spreading the FP rate among all the classifiers. This holds to the following equation:

$$FP_{cascade} = \prod_{i=1,\ldots,N} FP_{sc_i} \tag{6}$$

where FP_{sc} is the FP rate that each strong classifier of the cascade has to perform.

However, this method is not enough to allow the strong classifier to automatically control the false positive desired rate in consequence of the history of the false positives rates. In other words, if the previous level obtained a false positive rate that is under the predicted threshold, it is reasonable to suppose that the new strong classifier can consider to have a new "'smoothed"' FP threshold. For this reason, during the training of the classifier at level t we replaced FP_{sc_i} with a dynamic threshold, defined as

$$FP^{*t}_{sc_i} = FP_{sc_i} * \left(\frac{FP^{*t-1}_{sc_i}}{FP^{t-1}_{sc_i}} \right) \tag{7}$$

It is worth noticing how the false positive rate reachable by the classifier is updated at each level to obtain always a reachable rate at the end of the training process. In particular, we can see how such a value increases if at the previous step we added a weak classifier that has reduced it ($FP^{*t-1}_{sc_i} < FP^{t-1}_{sc_i}$) while decreases otherwise.

2.2 Asymmetry Control

As for the false positives rate, we can reduce the total number of false negatives by introducing a constant constraint that at each level forces the training algorithm to keep the false negatives ratio as low as possible (preferable 0). This can be achieved by balancing the asymmetry during the single strong classifier training process. The false positives-false negatives rates represent a trade-off that can be exploited to adopt a tuning strategy in the asymmetry for the two rates.

Supposing that the false negative value at the level i is quite far from the desired threshold FN_{sc_i}; at each step t of the training we can assign a different value to $k_{i,t}$, forcing the false negative ratio to decrease when $k_{i,t}$ is high (greater than one). If we suppose that the magnitude of $k_{i,t}$ directly depends on the variation of false positives obtained at step $t-1$ with respect to the desired value for such a step, we can introduce a tuning equation that increases the weight to positive samples when the false achieved positives rate is low and decreases it otherwise. Hence, for each each step $t = 1, \ldots, T$, $k_{i,t}$ is computed as

$$k_{i,t} = 1 + \frac{FP^{*t-1}_{sc_i} - FP^{t-1}_{sc_i}}{FP^{*t-1}_{sc_i}} \tag{8}$$

This equation returns a value of k that is bigger than 1 when the false positive rate obtained at the previous step has been lower than the desired one.

The Boosting technique described above have been applied both for searching the whole face and for searching some facial features. Specifically, once the face has been located in a new image (producing a candidate window D), we search in D for those candidate sub-windows representing the eyes, the mouth and the nose producing the subwindows D_{le}, D_{re}, D_m, D_n. These are used to completely partition the face pattern and produce subwindows for the forehead, the cheekbones, etc. In the next section we explain how these blocks are used for the face recognition task.

3 Block-Based Face Recognition

At training time each face image ($X^{(j)}$, $j = 1, ..., z$) of the training set is split in h independent blocks $B_i^{(j)}$ ($i = 1, ..., h$; currently $h = 9$: see Figure 1 (a)), each block corresponding to a specific facial feature. For instance, suppose that subwindow $D_m(X^{(j)})$, delimiting the mouth area found in $X^{(j)}$ is composed of the set of pixels $\{p_1, p_2, ...p_o\}$. We first normalize this window by scaling it in order to fit a window of fixed size, used for all the mouth patterns and we obtain $D'_m(X^{(j)}) = \{q_1, ...q_{M_m}\}$, where M_m is the cardinality of the standard mouth window. Block B_m, associated with D'_m is given by the concatenation of the (either gray-level or color) values of all the pixels in D'_m:

$$B_m^{(j)} = ((q_1), ...(q_{M_m}))^T. \tag{9}$$

Using $\{B_i^{(j)}\}$ ($j = 1, ..., z$) we obtain the eigenvectors corresponding to the LDA transformation associated with the i-th block:

$$W_i = (w_1^i, ..., w_{K_i}^i)^T. \tag{10}$$

Each block $B_i^{(j)}$ of each face of the gallery can then be projected by means of W_i into a subspace T_i with K_i dimensions (being $K_i << M_i$):

$$B_i^{(j)} = \mu_i + W_i C_i^{(j)}, \tag{11}$$

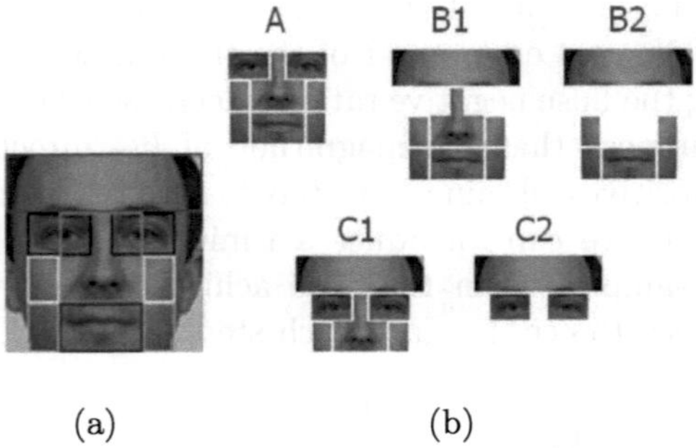

Fig. 1. Examples of missed block tests for occlusion simulation

where μ_i is the mean value of the i-th block and $C_i^{(j)}$ is the vector of coefficients corresponding to the projection values of $B_i^{(j)}$ in T_i. We can now represent each original face $X^{(j)}$ of the gallery by means of the concatenation of the vectors $C_i^{(j)}$:

$$R(X^{(j)}) = (C_1^{(j)} \circ C_2^{(j)} \circ ... \circ C_h^{(j)})^T. \tag{12}$$

$R(X^{(j)})$ is a point in a feature space Q having $K_1 + ... + K_h$ dimensions. Note that, due to the assumed independence of block B_i from block B_j $(i \neq j)$, we can use the same image samples to separately compute both W_i and W_j. The number of necessary training samples is now dependent from the dimension of the largest block $\widehat{K} = \max_{i=1,...,h}\{K_i\}$, being $\widehat{K} < K_1 + ... + K_h$. Splitting the pattern in subpatterns offers us the possibility to deal with lower dimensional feature spaces and then using less training samples. The result is a system more robust to overfitting problems.

At testing time first of all we want to exclude from the recognition process those blocks which are not completely visible (e.g., due to occlusions). One of the problems of holistic techniques, in fact, is the necessity to consider the pattern as a whole, even when only a part of the object to be classified is visible. For this reason, at testing time we use a skin detector in order to estimate the percentage of skin in each face block and we discard from the subsequent recognition process those blocks with insufficient skin pixels. Given a test image X and a set of v visible facial blocks B_{i_l} $(l = 1, ..., v)$ of X we project each B_{i_l} into the corresponding subspace T_{i_l}, obtaining:

$$Z = (C_{i_1} \circ ... \circ C_{i_v})^T. \tag{13}$$

Z represents the *visible* patterns and is a point in the subspace U of Q. The dimensionality of U is $K_{i_1} + ... + K_{i_v}$ and U is obtained projecting Q into the dimensions corresponding to the visible blocks B_{i_l} $(l = 1, ..., v)$. Finally, we use k-Nearest Neighbor (k-NN) to search in U for the points closest to Z which indicate the gallery faces most similar to X that will be ranked and presented to the user.

It is worth noticing that the projection of Q into U is trivial and efficient to compute, since at testing time (when using k-NN) we only have to exclude,

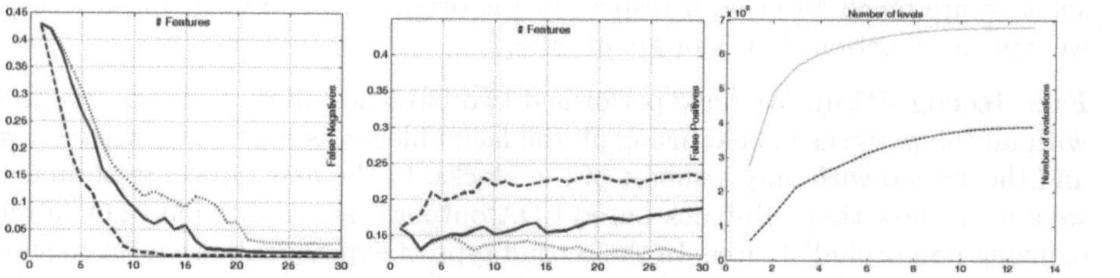

(a) (b) (c)

Fig. 2. False positives (FP) and negatives (FN) obtained while testing small strong classifiers. The continuous, dotted and dashed lines represent performance obtained using respectively AdaBoost, AsymBoost (k=1.1) and the proposed strategy. With the same number of features, the false negatives (a) decrease faster when we apply asymmetry. Even more if we tune the asymmetry. This means our solution has a higher detection rate by using a lower number of features while keeping the false positives low (b). In (c), the lower number of features necessary by the proposed solution (dashed line) to achieve a good detection rate yields to a reduction of about 50% in computation time with respect to Adaboost (continuous line).

in computing the Euclidean distance between Z and an element $R(X^{(j)})$ of the system's database, those coefficients corresponding to the non visible blocks.

4 Experimental Results

Face Detection. The first set of experiments is aimed to compare four small single strong classifiers trained by using the presented algorithm with ones obtained by using standard boosting techniques. The input set consisted on 6500 positive (face) samples and 6500 negative (non–face) samples, collected from different sources and scaled in a standard format 27×27 pixels. In Fig. 2, the false negatives and false positive rates of three considered algorithms are plotted. The compared algorithms are AdaBoost, AsymBoost and the proposed one. Analyzing these plots we can conclude that with the same number of weak classifiers the tuning strategy that we propose achieves a faster reduction of false negatives, while keeping low false positives.

For the second experiment, two cascades of twelve levels have been trained. At each round, while the face set remains the same, a bagging process is applied to negative samples to ensure a better training of the cascade [2]. A first improvement consists in a considerable reduction of the false negatives produced by the proposed solution with respect to AsymBoost. In addition, as showed for single strong classifiers, also for cascades the number of features required by the proposed solution to achieve the same detection rate of AsymBoost is much lower. This means building a cascade with lighter strong classifiers yielding to a faster computation. As matter of fact testing both asymmetric algorithm to a benchmark test set (see Fig. 2(c)), the global evaluation costs for the proposed solution are much lower with respect to the original AsymBoost. In particular, we have a reduction that is of about 50%.

Face Recognition. We have performed two batteries of experiments: the first with all the patterns visible (using all the facial blocks as input, i.e., with $v = h$) and the second with only a subset of the blocks. In the first type of experiments we aim to show that sub-block based LDA outperforms traditional LDA in recognizing non-occluded faces. In the second type of experiments we want to show that the proposed system is effective even with partial information, being able to correctly recognize faces with only few visible blocks.

Both types of experiments have been performed using two different datasets: the gray-scale images of the ORL [12] and (a random subset of) the colour images of the Essex [6] database. Concerning the ORL dataset, for training we have randomly chosen 5 images for each of the 40 individuals this database is composed of and we used the remaining 200 images for testing. Concerning Essex, we have randomly chosen 40 individuals of the dataset, using 5 images each for training and other 582 images of the same individuals for testing.

In the first type of experiments we have used both LDA and PCA techniques in order to provide a comparison between the two most common feature extraction techniques in both block-based and holistic recognition processes. Figure 3 shows the results concerning the top 10 corrected individuals in both the ORL and the Essex dataset. In the (easier) Essex dataset, both holistic and block-based LDA and PCA recognition techniques perform very well, with more than 98% of

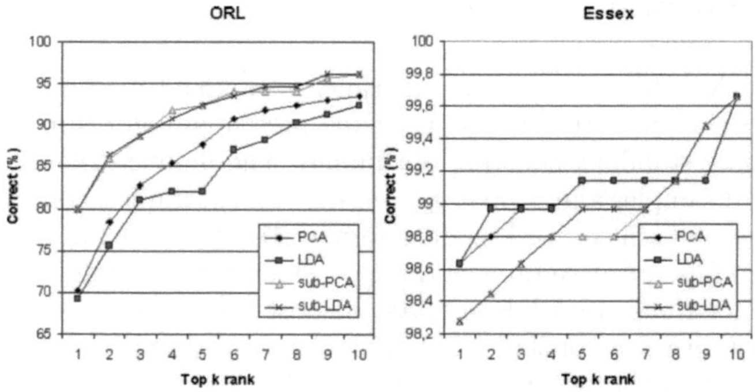

Fig. 3. Comparison between standard and sub-pattern based PCA and LDA with the ORL and the Essex datasets

Table 1. Test results obtained with missed blocks

Occlusion	ORL (%)	Essex (%)
A	71.35	93.47
B1	74.59	98.28
B2	68.11	98.45
C1	69.19	97.42
C2	62.70	96.91

correct individuals retrieved in the very first position. Traditional LDA and PCA as well as their corresponding block based versions (indicated as "sub-LDA" and "sub-PCA" respectively) have comparable results (being the difference among the four tested methods less than 1%). Conversely, in the hardier ORL dataset, sub-PCA and sub-LDA clearly outperform holistic approaches, with a difference in accuracy of about $5 - 10\%$. We think that this result is due to the fact that the lower dimensionality of each block with respect to the whole face window permits the system to more accurately learn the pattern distribution (at training time) with few training data (see Section 3).

Table 1 shows the results obtained using only subsets of the blocks. In details, we have tested the following block combinations (see Figure 1 (b)):

- A: The whole face except the forehead,
- B: The whole face except the eyes-nose zone,
- C: The whole face except the lower part.

Table 1 refers to sub-LDA technique only and to *top 1 ranking* (percentage of correct individuals retrieved in the very first position). As it is evident from the table, even with very incomplete data (e.g., the $C2$ test), block based LDA performs surprisingly well.

5 Conclusions

In this paper we have presented some improvements in state-of-the-art statistical learning techniques for face detection and recognition and we have shown an integrated system performing both tasks. Concerning the detection phase, we propose a method to balance the asymmetry of boosting techniques during the learning phase. In this way the detection performances show a faster detection and a lower FN rate. Moreover, in the recognition step, we propose to combine the results of separate classifications, each one obtained using a particular anatomically significant portion of the face. The resulting system is more robust to overfitting and can better deal with possible face occlusions.

Acknowledgments. This work was partially supported by the Italian Ministry of University and Scientific Research within the framework of the project "Ambient Intelligence: event analysis, sensor reconfiguration and multimodal interfaces"(2006-2008).

References

1. Bassiou, N., Kotropoulos, C., Kosmidis, T., Pitas, I.: Frontal face detection using support vector machines and back-propagation neural networks. In: ICIP (1), Thessaloniki, Greece, October 7–10, 2001, pp. 1026–1029 (2001)
2. Breiman, L.: Bagging predictors. Machine Learning 24(2), 123–140 (1996)
3. Brunelli, R., Poggio, T.: Face recognition: Features versus templates. IEEE Transaction on Pattern Analysis and Machine Intelligence 15(10), 1042–1052 (1993)

4. Cristinacce, D., Cootes, T., Scott, I.: A multi-stage approach to facial feature detection. In: British Machine Vision Conference (BMVC 2004), pp. 277–286 (2004)
5. Duda, R.O., Hart, P.E., Strorck, D.G.: Pattern classification, 2nd edn. Wiley Interscience, Hoboken (2000)
6. University of Essex. The Essex Database (1994), `http://cswww.essex.ac.uk/mv/allfaces/faces94.html`
7. Phillips, P., Wechsler, H., Huang, J., Rauss, P.: The FERET database and evaluation procedure for face recognition algorithms. Image and Vision Computing 16(5), 295–306 (1998)
8. Freund, Y., Schapire, R.E.: Experiments with a new boosting algorithm. In: ICML, Bari, Italy, July 3–6, 1996, pp. 148–156 (1996)
9. Friedman, J., Hastie, T., Tibshirani, R.: Additive logistic regression: A statistical view of boosting. The Annals of Statistics 28, 337–374 (2000)
10. Li, S.Z., Zhang, Z.: Floatboost learning and statistical face detection. IEEE Trans. Pattern Anal. Machine Intell. 26(9), 1112–1123 (2004)
11. Nefian, A., Hayes, M.: Face detection and recognition using hidden markov models. In: ICIP, Chicago, IL, USA, October 4–7, 1998, vol. 1, pp. 141–145 (1998)
12. ATeT Laboratories Cambridge. The ORL Face Database (2004), `http://www.camorl.co.uk/facedatabase.html`
13. Schapire, R.E.: Theoretical views of boosting and applications. In: Watanabe, O., Yokomori, T. (eds.) ALT 1999. LNCS, vol. 1720, pp. 13–25. Springer, Heidelberg (1999)
14. Smach, F., Abid, M., Atri, M., Mitéran, J.: Design of a neural networks classifier for face detection. Journal of Computer Science 2(3), 257–260 (2006)
15. Viola, P.A., Jones, M.J.: Fast and robust classification using asymmetric adaboost and a detector cascade. In: NIPS, Vancouver, British Columbia, Canada, December 3–8, 2001, pp. 1311–1318 (2001)
16. Viola, P.A., Jones, M.J.: Rapid object detection using a boosted cascade of simple features. In: CVPR (1), Kauai, HI, USA, December 8–14, 2001, pp. 511–518 (2001)
17. Wiskott, L., Fellous, J.M., Malsburg, C.V.D.: Face recognition by elastic bunch graph matching. IEEE Trans. Pattern Anal. Machine Intell. 19, 775–779 (1997)
18. Xiang, C., Fan, X.A., Lee, T.H.: Face recognition using recursive fisher linear discriminant. IEEE Transactions on Image Processing 15(8), 2097–2105 (2006)
19. Zhao, W., Chellappa, R., Phillips, P.J., Rosenfeld, A.: Face recognition: A literature survey. CM Computing Surveys 35(4), 399–458 (2003)

Face Recognition under Variant Illumination Using PCA and Wavelets

Mong-Shu Lee*, Mu-Yen Chen, and Fu-Sen Lin

Department of Computer Science and Engineering,
National Taiwan Ocean University, Keelung, Taiwan
Tel.: 886-2-2462-2192; Fax: 886-2-2462-3249
{mslee,chenmy,fslin}@mail.ntou.edu.tw

Abstract. In this paper, an efficient wavelet subband representation method is proposed for face identification under varying illumination. In our presented method, prior to the traditional principal component analysis (PCA), we use wavelet transform to decompose the image into different frequency subbands, and a low-frequency subband with three secondary high-frequency subbands are used for PCA representations. Our aim is to compensate for the traditional wavelet-based methods by only selecting the most discriminating subband and neglecting the scattered characteristic of discriminating features. The proposed algorithm has been evaluated on the Yale Face Database B. Significant performance gains are attained.

Keywords: Face recognition, Principal component analysis, Wavelet transform, Illumination.

1 Introduction

Human face recognition has become a popular area of research in computer vision recently. It is applied to various different fields such as criminal identification, human-machine interaction, and scene surveillance. However, variable illumination is one of the most challenging problems with face recognition, due to variations in light conditions in practical applications. Of the existing face recognition methods, the principal component analysis (PCA) method takes all the pixels in the entire face image as a signal, and proceeds to extract a set of the most representative projection vectors (feature vectors) from the original samples for classification. First, Turk and Pentland [15] extracted noncorrelational features between face objects by PCA, and applied the neighborhood algorithm classification method to face recognition. Yet, the variations between the images of the same face due to illumination and view direction are always larger than the image variations due to a change in face identity [1]. Standard PCA-based methods cannot facilitate division of classes as feature vectors obtained from face image under varying lighting conditions. Hence, if only one upright frontal image per person, which is under severe light variations, is available for training, the performance of PCA will be seriously degraded.

* Corresponding author.

A.-B. Salberg, J.Y. Hardeberg, and R. Jenssen (Eds.): SCIA 2009, LNCS 5575, pp. 341–350, 2009.

Many methods have been presented to deal with the illumination problem. The first approach to handling the effect that results from illumination changes is constructing illumination model from several images acquired under different illumination condition. The representative method, the illumination cone model that can deal with shadow and multiple lighting sources, is introduced by [2, 10]. Although this approach achieved 100% recognition rates, it is not practical to require seven images of each person to obtain the shape and albedo of a face. Zhao and Chellappa [19] developed a shape-based face recognition system by means of an illumination-independent ratio image derived by applying a symmetrical shape-from-shading technique to face images. Shashua and Riklin-Raviv [14] used quotient images to solve the problem of class-based recognition and image synthesis under varying illumination. Xie and Lam [16] adopted a local normalization (LN) technique to images, which can effectively eliminate the effect of uneven illumination. Then the generated images with illumination variation insensitivity property are used for face recognition using different methods, such as PCA, ICA and Gabor wavelets. Discrete Wavelet transform (DWT) has been used successfully in image processing. An advantage of DWT is that with few wavelet coefficients it can capture most of the image energy and the image features. In addition, its ability to characterize local spatial-frequency information of image motivates us to use it for feature extraction. In [9], three-level wavelet transform is performed to decompose the original image into its subbands on which the PCA is applied. The experiments on Yale database show that the third level diagonal details attain the highest correct recognition rate. Later, wavelet face [4] only uses the low-frequency subbands to present the basic figure of an image, and ignore the efficacy of high-frequency subbands. Ekenel and Sankur [7] came up with a fusing scheme by collecting the information coming from the subbands that attain individually high correct recognition rates to improve the classification performance.

Although some studies have been conducted on the discriminatory potential of single frequency subband in DWT, little research has been done on the counterparts of the combination of frequency subbands. In this study, we propose a novel method to handle the problem of face recognition with varying illumination. In our approach, DWT is adopted first to decompose an image into different frequency components. To avoid neglecting the image features resulting from different lighting condition, a low-frequency and three midrange frequency subbands are selected for PCA representation. In the last step of the classification rule, it is the weighting combination of the individual discriminatory potential, applied to the PCA-based face recognition procedure. Experimental results demonstrated that applying PCA on four different DWT subbands, and then merging distinct subbands information with relative weights in classification achieve a rather excellent recognition performance.

2 Wavelet Transform and PCA

2.1 Multi-resolution Property of Wavelet Transform

Over the last decade or so, the wavelet transform (WT) has been successfully adopted to solve various problems of signal and image processing. The wavelet transform is fast, local in the time and the frequency domain, and provides multi-resolution analysis of real-world signals and images. Wavelets are collections of functions in L^2 constructed from a basic wavelet ψ using dilations and translations. Here we will only consider the families of wavelets using dilations by powers of 2 and integer translations:

$$\psi_{j,k}(x) = 2^{\frac{j}{2}}\psi(2^j x - k),\ j,k \in Z.$$

We can see that the time and frequency localization of the wavelet basis functions are adjusted by both scale index j and position index k.

Multi-resolution Analysis is generally an important method for constructing orthonormal wavelet bases for L^2. In multi-resolution schemes, wavelets have corresponding scaling function φ, whose analogously defined dilations and translation $\varphi_{j,k}(x)$ span a nested sequence of multi-resolution space V_j, $j \in$ Z. Wavelets $\{\psi_{j,k}(x) : j,k \in Z\}$ form orthonormal bases for the orthogonal complements $W_j = V_j - V_{j-1}$ and for all of L^2. Therefore, the wavelet transform decomposes a function into a set of orthogonal components describing the signal variations across scales [5]. For one-dimensional wavelet transform, a signal f, is represented by its wavelet expansion as:

$$f(x) = \sum_{k \in Z} c_I(k)\varphi_{I,k}(x) + \sum_{j \geq I}\sum_{k \in Z} d_j(k)\psi_{j,k}(x), \tag{1}$$

where the expansion coefficients $c_I(k)$ and $d_j(k)$ in (1) are obtained by an inner product, for example:

$$d_j(k) = < f, \psi_{j,k} >= \int f(x) 2^{\frac{j}{2}}\psi(2^j x - k)dx.$$

In practice, we usually apply the DWT algorithm corresponding to (1) with finite decomposition levels to obtain the coefficients. Here, the wavelet coefficients of a 1-D signal is calculated by splitting it into two parts, with a low-pass filter (corresponding to the scaling function ϕ) and high-pass filter (corresponding to the wavelet function ψ), respectively. The low frequency part is split again into two parts of high and low frequencies, and the original signal can be reconstructed from the DWT coefficients.

The two-dimensional DWT is performed by consecutively applying one-dimensional DWT to the rows and columns of the two-dimensional data. Two-dimensional DWT decomposes an image into "subbands" that are localized in time and frequency domains. The DWT is created by passing the image through a series of filter bank stages. The high-pass filter and low-pass filter are finite impulse response filters. In other words, the output at each point depends only on a finite portion of the input image. The filtered outputs are then sub-sampled by 2 in the row direction.

These signals are then each filtered by the same filter pair in the column direction. As a result, we have a decomposition of the image into 4 subbands denoted HH, HL, LH, and LL. Each of these subbands can be regarded as a smaller version of the image representing different image contents. The Low-Low (LL) frequency subband preserves the basic content of the image (coarse approximation) and the other three high frequency subbands HH, HL, and LH characterize image variations along diagonal, vertical, and horizontal directions, respectively. Second level decomposition can then be conducted on the LL subband. Such iteration process is continued until the specified number of desired decomposition level is achieved. The multi-resolution decomposition strategy is very useful for the effective feature extraction. Fig. 1 shows the subbands of three-level discrete wavelet decomposition. Fig. 2 displays an example of image Box with its corresponding subbands LL_3, LH_3, HL_3 and HH_3 in Fig. 1.

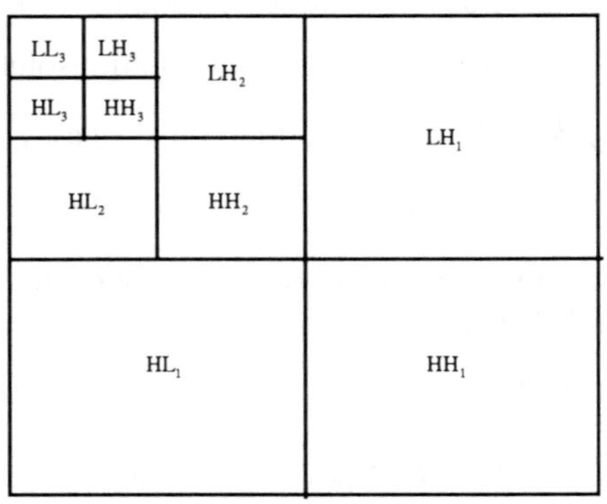

Fig. 1. Different frequency subbands of a three-level DWT

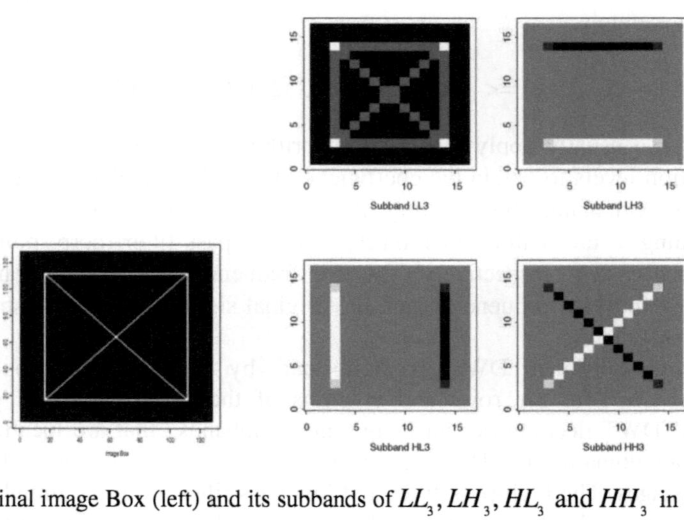

Fig. 2. Original image Box (left) and its subbands of LL_3, LH_3, HL_3 and HH_3 in a three-level DWT

2.2 PCA and Face Eigenspace

Principal component analysis (PCA) is a dimensionality reduction technique based on extracting the desired number of principal components of the multidimensional data. Given an N – dimensional vector representation of each face in a training set of M images, PCA tends to find a t – dimensional subspace whose basis vectors correspond to the maximum variance direction in the original image space. This new subspace is normally a smaller dimension $(t << N)$. These new basis vectors can be calculated in the following way. Let X be the $N \times M$ data matrix whose columns $x_1, x_2, ..., x_M$ are observations of a signal embedded in $\mathbb{R}^N$; in the context of face recognition, M is the available training images, and $N = m \times n$ is the number of pixels in an image. The PCA basis Ψ is obtained by solving the eigenvalue problem $\Lambda = \Psi^T \mathrm{E} \Psi$, where E is the covariance matrix of the data

$$\mathrm{E} = \frac{1}{M} \sum_{i=1}^{M} (x_i - \overline{x})(x_i - \overline{x})^T, \text{ where } \overline{x} \text{ is the mean of } x_i.$$

$\Psi = [\psi_1, ..., \psi_m]^T$ is the eigenvector matrix of E, and Λ is the diagonal matrix with eigenvalues $\lambda_1 \geq .... \geq \lambda_N$ of E on its main diagonal, so ψ_j is the eigenvector corresponding to the jth largest eigenvalue. Thus, to perform PCA and extract t principal components of the data, one must project the data onto Ψ_t, the first t columns of the PCA basis Ψ, which correspond to the t highest eigenvalue of E. This can be regarded as a linear projection $\mathbb{R}^N \rightarrow \mathbb{R}^t$, which retains the maximum energy (i.e., variance) of the signal. This new subspace $\mathbb{R}^t$ defines a subspace of face images called face space. Since the basis vectors constructed by PCA had the same dimension as the input face images, they are named "eigenfaces" by Turk and Pentland [15].

Combined with the effectiveness of capturing image features of DWT and the accuracy of data representation of PCA, we are motivated to develop an efficient scheme for the face recognition in the next section.

3 The Proposed Method

The study is aimed to enhance the recognition rate of the face image under varying lighting conditions by the standard PCA-based methods. In the literature, the DWT was applied in texture classification [3] and image compression [6] due to its powerful capability in multi-resolution decomposition analysis. The wavelet decomposition technique was also used to extract the intrinsic features for face recognition [8]. In [11], a 2D Gabor wavelet representation was sampled on the grid and combined into a labeled graph vector for elastic graph matching of face images. Similar to [9], we apply the multilevel two-dimensional DWT to extract the facial features. In order to reduce the effect of illumination, the pre-processing of training and unknown images may choose to employ histogram equalization before taking DWT.

The whole block diagram of the face recognition system including training stage and recognition stage is as in Fig. 3. A three-level DWT, using the Daubechies' S8 wavelet, is applied to decompose the training image, as illustrated in Fig. 1. Generally, the low frequency subband LL_3 represents and preserves the coarser approximation of an image, and the other three sub-high frequency subbands characterize the details of the image texture in three different directions. Earlier studies concluded that the information in the low spatial frequency bands play a dominant role in face recognition. Naster *et al.* [13] have found that facial expression and small occlusions affect the intensity manifold locally. Under frequency-based representation, only the high frequency spectrum is affected. Moreover, changes in illumination affect the intensity manifold globally, in which only the low frequency spectrum is affected. When there is a change in human face, all frequency components will be affected. Based on these observations, we select the HH_3, LH_3, HL_3 and LL_3 subbands in the third level to employ the PCA procedure in this study. All these frequency components have played their parts with different weights in discriminating face identity.

In the recognition step, distance measurement between the unknown image and the training images in the library is performed to determine whether the input of an

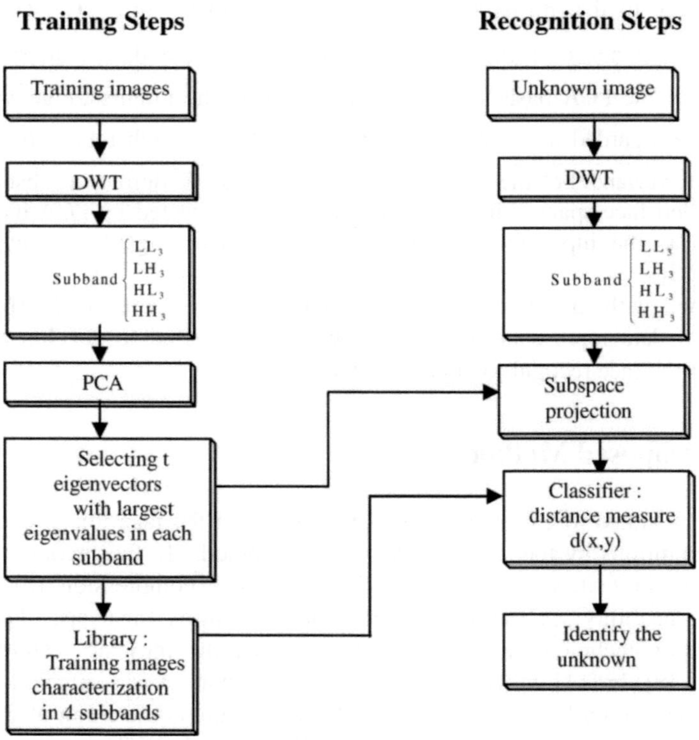

Fig. 3. Block diagram of the proposed recognition system

unknown image matches any of the images in the library. In terms of classifying the criterion, the traditional Euclidean distance cannot measure the similarity very well when there illumination variations on the facial images exist. Yambor [17] reported that a standard PCA classifier performed better when the Mahalanobis distance was used. Therefore, the Mahalanobis distance is also selected as the distance measure in the recognition step of our experiments. The Mahalanobis distance is formally defined in [12], and Yambor [17] gives a simplification, which is adopted here as follows:

$$d_{Mah}(x, y) = -\sum_{i=1}^{t} \frac{1}{\sqrt{\lambda_i}} x_i y_i$$

where x and y are the two face images to be compared and λ_i is the ith eigenvalue corresponding to the ith eigenvector of the covariance matrix E.

Finally, the distance between the unknown image and the training image is a linear combination of their discriminating ability of four wavelet subbands, and is defined as follows:

$$d(x, y) = 0.4d_{Mah}^{HH_3}(x, y) + 0.3d_{Mah}^{LH_3}(x, y) + 0.2d_{Mah}^{HL_3}(x, y) + 0.1d_{Mah}^{LL_3}(x, y) \quad (2)$$

where $d_{Mah}^{HH_3}(x, y)$, $d_{Mah}^{LH_3}(x, y)$, $d_{Mah}^{HL_3}(x, y)$ and $d_{Mah}^{LL_3}(x, y)$ are the Mahalanobis distance measured on the subbands of HH_3, LH_3, HL_3, and LL_3 respectively. The weighting coefficients put in front of each subband in equation (2) were selected on the basis of their recognition performance in the single-band experiment with Subset 3 images of Yale Face Database B. The average recognition accuracy of the four different subbands using Subset 3 images (with and without histogram equalization) is recorded in Table 1. It can be shown that the HH_3 subband gives the best result, and thus the weighting coefficient of subband HH_3 deserves the largest value 0.4 in the classifier equation (2). The weighting coefficients of the other three subbands LH_3, HL_3, and LL_3 are in decreasing order according to their decline in average recognition rate in Table 1.

Table 1. The average recognition performance (with and without histogram equalization) using Subset 3 images of Yale Face Database B on different DWT subbands

DWT Subband	Average recognition accuracy
HH_3	89.2%
LH_3	86.4%
HL_3	81.4%
LL_3	78.6%
Average	83.9%

4 Experimental Results

The performance of our algorithm is evaluated using the popular Yale Face Database B that contains images of 10 persons under 45 different lighting conditions, and the test is performed on all of the 450 images. All the face images are cropped and normalized to a size of 128x128. The images of this database are divided into four subsets according to the lighting angle between the direction of the light source and the camera axis. The first subset (Subset 1) covers the angular range up to 12°, the second subset (Subset 2) covers 12° to 25°, the third subset (Subset 3) covers 25° to 50°, and the fourth subset (Subset 4) covers 50° to 75°. One example images of these four subsets are illustrated in Fig. 4.

For each individual in the Subset 1 and 2, two of their images were used for training (total 20 training images for each set), and the remaining images were used for testing. As a method to overcome left and right face illumination variation that appeared in Subset 3 and Subset 4, we computed the difference between the average pixel value of the left and right face, where the left and right face were divided on the vertical-axis center of the input image. We selected two images with the left and right face difference greater than the threshold value 30 (experimental value) per person from Subset 3 and Subset 4 to form the training image set, and the rest of the images

Fig. 4. Sample images of one individual in the Yale Face Database B under the four subsets of lighting

Table 2. Comparison of recognition methods with Yale Face Database B

(The entries with indicated citation were taken from published papers)

Method	Similarity measure	Size of training sample	The number of eigenfaces	Recognition rate
WT(Fusing six subbands into one-single band) +PCA [7]	Correlation coefficient	2	80	77.1%
WT(subband HH3) + PCA [9]	Correlation coefficient	2	11	84.5%
The proposed method	Mahalanobis distance	2	36	99.3%
LN(local normalization) +HE + PCA [16]	Mahalanobis distance	1	200	99.7%

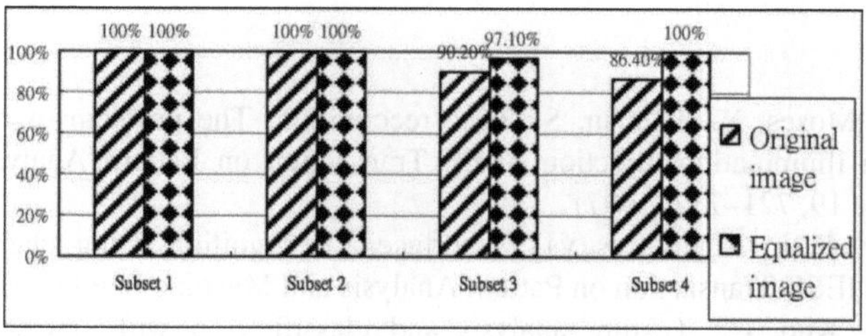

Fig. 5. The recognition performance of the algorithm when applied to the Yale Face Database B

were used as test images. The proposed method was tested on the image database as follows: the existing PCA with the first two eigenvectors excluded, and PCA with histogram equalized images. Fig. 5 tabulates the recognition rates using the images on the database and PCA approaches, where nine eigenvectors in each subbands (total 36 eigenvectors) calculated from the training images were used for face recognition. The result of the PCA application to original images on Subset 1, 2, 3 and 4 with the first two eigenvectors excluded shows high recognition performance of 100%, 100%, 90.2% and 86.4% respectively. Moreover, the result of the PCA application after histogram equalization (HE) on Subset 1, 2, 3 and 4 was recognition performance of 100%, 100%, 97.1% and 100% respectively (with average 99.3%). The PCA-based recognition performance may be influenced by several factors, such as the size of training sample, the number of eigenfaces, and similarity measure. Under similar influence factors, we compare the performance between the proposed method and other PCA-based face recognition methods in Table 2. The local normalization (LN) approach achieved the highest recognition rate 99.7% in Table 2, but they use 200 eigenfaces. Obviously, our recognition rate is comparable to the LN approach and significantly improves the traditional PCA-based face recognition methods.

5 Conclusions

In this study, a novel wavelet-based PCA method for human face recognition under varying lighting condition is proposed. The advantages of our method are summarized as follows:

1. Wavelet PCA offers a method through which we can improve the performance of normal PCA by using low frequency and sub-high frequency components, which lowers the computation cost while keeping the essential feature information needed for face recognition.
2. We carefully design the classification rule, which is a linear combination of four subband contents according to their individual recognition rates in a single-band test. Therefore, the weights for each subband used in the distance function are highly meaningful.

The experimental result shows that the proposed method demonstrates very efficient performance with the histogram-equalized images. The future work includes the evaluation of the other image data with illumination variation, such as CMU PIE database.

References

1. Adini, Y., Moses, Y., Ullman, S.: Face recognition: The problem of compensating for changes in illumination direction. IEEE Transaction on Pattern Analysis and Machine Intelligence 19, 721–732 (1997)
2. Belhumeur, P.N.: Eigenfaces vs. Fisherfaces: Recognition using class specific linear projection. IEEE Transaction on Pattern Analysis and Machine Intelligence 19, 711 (1997)
3. Chang, T., Kuo, C.: Texture analysis and classification with tree-structured wavelet transform. IEEE Tran. on Image Processing 2, 429 (1993)
4. Chien, J.T., Wu, C.C.: Discriminant waveletfaces and nearest feature classifiers for face recognition. IEEE Transaction on Pattern Analysis and Machine Intelligence 24(12), 1644–1649 (2002)
5. Daubechies, I.: Ten Lectures on Wavelets. In: SIAM. CBMB Regional Conference in Applied Mathematics Series, vol. 61 (1993)
6. DeVore, R., Jawerth, B., Lucier, B.: Image compression through wavelet transform coding. IEEE Trans. on Information Theory 38, 719–746 (1992)
7. Ekenel, H.K., Sanker, B.: Multiresolution face recognition. Image and Vision Computing (23), 469–477 (2005)
8. Etemad, K., Chellappa, R.: Face recognition using Discreminant eigenvectors. In: Proceeding IEEE Int'l. Conf. Acoustic, Speech, and Signal Processing, pp. 2148–2151 (1996)
9. Feng, G.C., Yuen, P.C.: Human face recognition using PCA on wavelet subband. Journal of Eectronic Imaging (9), 226–233 (2000)
10. Georghiades, A., Kriegman, D., Belhumeur, P.: Illumination cones for recognition under variable lighting: faces. In: Proceeding IEEE C CVPR SANT B (1998)
11. Lyons, M.J., Budynek, J., Akamatsu, S.: Automatic classification of single facial image. IEEE Transaction on Pattern Analysis and Machine Intelligence 21(12), 1357–1362 (1999)
12. Moon, H., Phillips, J.: Analysis of PCA-based face recognition algorithms. In: Boyer, K., Phillips, J. (eds.) Empirical Evaluation Methods in Computer Vision. World Scientific Press, MD (1998)
13. Nastar, C., Ayach, N.: Frequency-based nongrid motion analysis. IEEE Transaction on Pattern Analysis and Machine Intelligence 18, 1067–1079 (1996)
14. Shashua, A.: The quotient image: Class-based re-rendering and recognition with varying illuminations. IEEE Transaction on Pattern Analysis and Machine Intelligence 23(2), 129–139 (2001)
15. Turk, M., Pentland, A.: Eligenfaces for Recognition. Journal of Cognitive Neuroscience 3, 71 (1991)
16. Xie, X., Lam, K.: An efficient illumination normalization method for face recognition. Pattern Recognition Letters 27(6), 609–617 (2006)
17. Yambor, W., Draper, B., Beveridge, R.: Analyzing PCA-based face recognition algorithms: eigenvector selection and distance measures. In: Christensen, H., Phillips, J. (eds.) Empirical Evaluation Methods in Computer Vision. World Scientific Press, Singapore (2002)
18. Zhao, J., Su, Y., Wang, D., Luo, S.: Illumination ratio image: synthesizing and recognition with varying illuminations. Pattern Recognition Letters (24) (2003)
19. Zhao, J., Chellappa, R.: Illumination-insensitive face recognition using symmetric shape-from-shading. In: Proceeding IEEE conf. CVPR Hilton Head (2000)

On the Spatial Distribution of Local Non-parametric Facial Shape Descriptors

Olli Lahdenoja[1,2], Mika Laiho[1], and Ari Paasio[1]

[1] University of Turku, Department of Information Technology
Joukahaisenkatu 3-5, FIN-20014, Turku, Finland
[2] Turku Centre for Computer Science (TUCS)

Abstract. In this paper we present a method to form pattern specific facial shape descriptors called basis-images for non-parametric LBPs (Local Binary Patterns) and some other similar face descriptors such as Modified Census Transform (MCT) and LGBP (Local Gabor Binary Pattern). We examine the distribution of different local descriptors among the facial area from which some useful observations can be made. In addition, we test the discriminative power of the basis-images in a face detection framework for the basic LBPs. The detector is fast to train and uses only a set of strictly frontal faces as inputs, operating without non-faces and bootstrapping. The face detector performance is tested with the full CMU+MIT database.

1 Introduction

Recently, significant progress in the field of face recognition and analysis has been achieved using partially or fully non-parametric local descriptors which provide invariance against changing illumination conditions. These descriptors include Local Binary Pattern (LBP) [1] which was originally proposed as a texture descriptor in [2] and its extensions such as Local Gabor Binary Pattern (LGBP) [3]. In MCT (Modified Census Transform [4]) the means for forming the descriptor are very similar to LBP, hence it is also called modified LBP. The iLBP method for extending the neighborhood of the MCT for multiple radius was presented in [5].

The above mentioned methods for local feature extraction have been applied also to face detection [6] and facial expression recognition [7] (also using a spatio-temporal approach). In face detection, for MCT [4] a cascade of classifiers was used and in [5] a multiscale strategy for iLBP features in a cascade was proposed. In [6] an SVM approach was adopted using the LBPs as features for face detection.

Although the above mentioned (discrete, i.e. non-continuously valued) local descriptors have become very popular, the individual characteristics of each descriptor has not been intensively studied. In the work of [8], MCT and LBP were compared among some other face normalization methods in face verification performance point of view using the eigenspace approach. In [9] the LBPs

A.-B. Salberg, J.Y. Hardeberg, and R. Jenssen (Eds.): SCIA 2009, LNCS 5575, pp. 351–358, 2009.

were seen as thresholded oriented derivative filters and compared to e.g. Gabor filters.

In this paper we present a systematic procedure for analyzing the local descriptors aiming at finding possible redundancies and improvements as well as deepening the understanding of these descriptors. We also show that the new basis-image concept, which is based on a simple histogram manipulation technique can be applied to face detection based on discrete local descriptors.

2 Background

The fundamental idea of LBP, LGBP, MCT and their extensions is to compare intensity values in a local neighbourhood in a way which produces a representation which is invariant to intensity bias changes and the distribution of the local intensities. In a short period of time after [1] in which a clear improvement in face recognition rates was obtained against many state-of-the-art reference algorithms, very impressive recognition results with the standard FERET database among many other databases have been achieved.

A main characteristic of these methods is that they use histograms to represent a local facial area and classification is performed between the extracted histograms, the bins of which describe discrete micro-textural shapes. The LBP (which is also included in LGBP) is clearly a more commonly used descriptor than MCT, possibly because of reduced dimension of the histogram description (by a factor of two) and further histogram length reduction methods, such as the usage of only uniform patterns [2].

While the main difference between MCT and LBP is that in MCT instead of center pixel the mean of all pixels is used as reference intensity (and that the center pixel is included into resulting pattern), the difference between LGBP and LBP is that in LGBP, Gabor filtering is first applied in different frequencies and orientations, after which the LBPs are extracted for classification. LGBP provide a significant improvement in face recognition accuracy compared to basic LBP, but due to many different Gabor filters (resulting in many histograms) the dimensionality of the LGBP feature vectors becomes extremely high. Therefore dimensionality reduction, e.g. PCA and LDA are applied after feature extraction.

3 Research Methods and Analysis

3.1 Constructing the Facial Shape Descriptors

We used the normalized FERET [10] gallery data set (consisting of 1196 intensity faces) as inputs for histogramming which aimed at constructing a representative set of images (so called basis-images) which describe the response of each individual local pattern (e.g. LBP, MCT, LGBP) to different facial locations (and hence, the shape of these locations). Also, some tests were performed with full 3113 intensity image data containing the fb and dup 1 sets. The construction of the basis-images is described in the following.

In a histogram perspective, a pattern histogram is constructed for each spatial face image location (x-y pixel) through the whole input intensity image set. These histograms are then placed to their corresponding spatial (x-y pixel) locations where they were extracted from, and all the other bins in the histograms, except the bin under investigation are ignored. Thus, the resulting basis-image of a certain pattern consists of a spatial arrangement of bin magnitudes for that pattern. The spatial (x-y) size of a basis image is the same as that of each individual input intensity image. This technique results N basis-images for which N is the total number of patterns (histogram bins). Then each basis image is (separately) normalized according to its total sum of bins. The normalization removes the bias which results from the differences in total number of occurrences of each pattern in facial area and shows the pattern specific shape distribution clearly. These basis-images represent the shape distribution of individual patterns among the facial area on average.

Although the derivation of the basis-images is simple, we consider the existence of these continuously valued images a non-trivial case. This is because, especially LBPs, are usually considered as texture descriptors despite of wide range of applications, instead of descriptors with a certain larger scale shape response.

3.2 Analyzing the Properties of Local Descriptors

We conducted tests on LBP and MCT (and some initial tests with LGBP) in order to find out their responses to facial shapes. Neighborhood with a radius of 1 and sample number of 8 was used in the experiments (i.e. 8-neighborhood), but the method allows for choosing any radius. The basis-images of all uniform LBP descriptors are shown in Figure 1. Also, the four basis-images in the upper right corner represent examples of non-uniform patterns. The uniformity of a LBP refers to the total number of circular 0-1 and 1-0 transitions of the LBP (patterns with uniformity of 0 or 2 are considered as uniform patterns, in general).

It seems that as the uniformity level increases (i.e. non-uniform patterns are considered, see Figure 1) the distribution becomes less spatially detailed. However the patterns that are 'near' to uniform patterns seem to give a more detailed response (e.g. non-uniform pattern 0001011) than patterns far from uniformity criterion (e.g. pattern 00101010). In [11] it was observed, that rounding non-uniform patterns into uniform using a hamming distance measure between them resulted in lower error rates in face recognition. With larger data set (of 3113 input intensity faces) many non-uniform patterns seemed to occur in eye center region. By examining the basis-images it seems that non-uniform patterns can not describe facial shapes in as discriminative manner as uniform patterns (which has previously only empirically been verified). Also, as the uniformity level increases the patterns become more rare, as expected.

When studying the distribution of MCT (Modified Census Transform, also called mLBP), we noticed that with the test set used, uniform patterns formed clear spatial shapes similarly to LBPs, while many non-uniform patterns were very rare (i.e. only distinct occurrences). Hence, we propose using the same

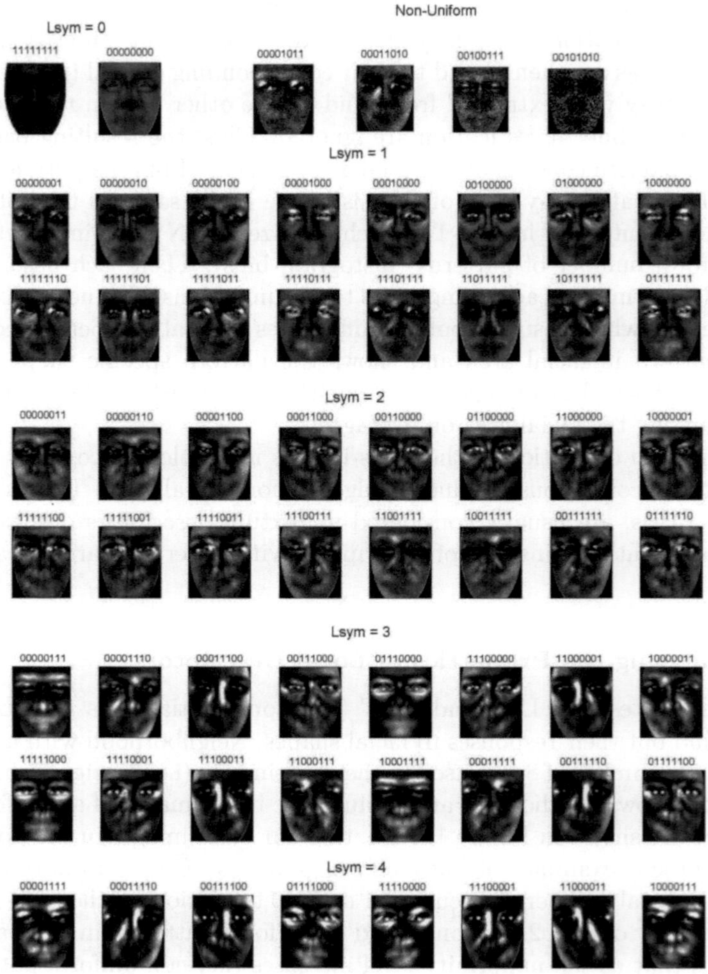

Fig. 1. Selected LBP basis-images

concept of uniform patterns that have been used for LBPs, also with MCTs in face analysis.

In [12] so called symmetry levels for uniform LBPs were presented. Symmetry level L_{sym} of an uniform LBP is defined as the minimum between the total amount of ones and total amount of zeros in a pattern. It was observed in [12] that as the symmetry level of an uniform LBP increases, also the average discriminative efficiency of the LBP increases. This was verified in tests with face recognition using the FERET database. Interestingly, the basis-images of uniform patterns can be divided into classes by their symmetry levels. The spatial distinction between pattern occurrence probabilities gets larger (as occurrence probabilities also mean histogram bin magnitudes, which are now represented as brightness values in Figure 1). Hence, there is a connection between the shape of the basis-images and the discriminative efficiency of the patterns so that as the basis-images become more spatially varied, also the discriminative efficiency of those patterns in face recognition increases [12]. It is also interesting to notice, that the LBPs with a smaller symmetry level seem to give the largest response in the eye regions.

4 Applying Basis-Images for Face Detection

4.1 Motivation

Although the face representation with basis-images is illustrative for examining the response of each pattern to different facial shapes, it can also be used as such in a more quantitative manner. We examined the discriminative power of the basis-image representation in face detection framework since this allows implementing a very compact face detector which requires a negligible time and effort for training or collecting training samples. The training time for the classifier was less than a minute with a P4 processor PC and Matlab.

The simple structure of the classifier and training might be beneficial in certain application environments (e.g. special hardware). However if a state-of-the-art detection rate would be required some of the more complicated procedures (e.g. using also non-faces and bootstrapping) would be necessary. At this point uniform basis-images were used with the basic LBP. However, also MCT and LGBP could be applied in a similar manner for constructing a face detector straightforwardly. The latter methods would lead to a higher dimension of the face description (i.e. more basis-images would have to be used for complete face representation) but might also improve the detection rate and FPR.

4.2 Classification Principle

The face detector implemented operates with a 21x21 search window size. It is slided through all image scales (scaling performed with bilinear subsampling). First the input image is formed for all scales and for each scale the LBP transform is applied. For a certain search window position and scale the LBPs within the search window are replaced by the magnitudes of the corresponding basis-images of these same LBPs in the current spatial locations. For example, if we are in a search window position (x, y) (positions vary between 1 and 21 in x and y directions) we read the LBP of that position (e.g. '00001111') of the input image and use it to find the basis-image of the LBP '00001111', after which the value of that basis-image in the same position (x, y) is summed into accumulator. The 'faceness' measure is then formed by accumulating the magnitudes of the (normalized) basis-image look-ups within the search window area (note that the basis-image concept allows for the normalization procedure). The 'faceness' measure is finally compared against a fixed threshold (determined empirically), which determines whether the sample belongs to class face or non-face. In the current implementation we use 59 basis images, i.e. one for each uniform LBP, and one for describing all the remaining LBPs.

The operations can be performed in cascade, for example, simply by subsampling certain x-y search window positions at a time (possibly first determining which positions belong to the most important ones) and applying a proper threshold for each stage. We tested using two stages to achieve a detection speed of about 4-8 fps with P4 processor and 320x240 resolution in Matlab. However, the detection results reported in this paper were performed without a cascade. In that case the detection speed was approximately 1-2 fps. The search window step in both x and y directions was two in the tests performed. In the experiments a pre-processing step for the input test images (in full scale) and also to basis-images was performed. Both images were low-pass filtered with a 3x3 averaging mask.

4.3 Experimental Results

A detection rate of 78.7% was obtained with 126 false detections with full CMU+MIT database consisting total of 507 faces in cluttered scenes. The total amount of patches searched was about $96.4 * 10^6$ which results in false positive rate in the order of $1.3 * 10^{-6}$. A maximum of 18 scales were used with scale downsampling factor between 1 and 1.2. Many of the faces that were not detected were not fully frontal, hence it explains part of the moderate recognition rate compared to more advanced detectors, which can easily achieve more than 90% detection rates (however, a more versatile set of input samples for classification is provided with them).

We also tested the detection performance with an easier (more frontal faces) subset of the CMU+MIT set which has been used e.g. in [6]. With this subset there were a total of 227 faces in 80 images. We obtained a detection rate of 87.7% (including drawn faces) with 53 false detections. The total amount of patches searched was about $44.4 * 10^6$ which results in false positive rate in the order of $1.2 * 10^{-6}$ with this set. Hence, the discriminative efficiency (FPR, False Positive Rate) shows a relatively good performance considering the simplicity of the detection framework. In the Figures 2 and 3 some detection results are shown.

Fig. 2. Example detection results with the CMU+MIT database

Fig. 3. Example detection results with the CMU+MIT database

5 Discussion

The idea of basis-images could possibly be extended into other face analysis applications. For example, it might be possible to construct person specific basis images if enough face samples would be present. This could be used for increasing the performance of a face recognition system. In facial expression analysis using a proper alignment procedure it could be possible to capture different expressions to different basis-image sets and use these for recognition and illustration. Also, the effect of global illumination on non-parametric local descriptors could be studied using the basis-image framework.

6 Conclusions

In this paper we presented a method for analyzing local non-parametric descriptors in spatial domain, which showed that they can be seen as orientation selective shape descriptors which form a continuously valued holistic facial pattern representation. We established a dependency between the spatial variability of the resulting LBP basis-images and the symmetry level concept presented in [12]. Through the analysis of basis-images we propose that uniform patterns could be beneficial also with MCTs as with LBPs. We also tested the discriminative power of the basis-image representation in face detection, thus resulting in a new kind of face detector implementation, showing a moderate discriminative efficiency (FPR, False Positive Rate).

References

1. Ahonen, T., Hadid, A., Pietikainen, M.: Face Recognition with Local Binary Patterns. In: Pajdla, T., Matas, J. (eds.) ECCV 2004. LNCS, vol. 3021, pp. 469–481. Springer, Heidelberg (2004)
2. Ojala, T., Pietikainen, M., Maenpaa, T.: Multiresolution Gray-scale and Rotation Invariant Texture Classification with Local Binary Patterns. IEEE Transactions on Pattern Analysis and Machine Intelligence 24(7), 971–984 (2002)
3. Zhang, W., Shan, S., Gao, W., Chen, X., Zhang, H.: Local Gabor binary pattern histogram sequence (LGBPHS): a novel non-statistical model for face representation and recognition. In: Tenth IEEE International Conference on Computer Vision, ICCV, October 2005, vol. 1, pp. 786–791 (2005)
4. Froba, B., Ernst, A.: Face detection with the modified census transform. In: Sixth IEEE International Conference on Automatic Face and Gesture Recognition, May 2004, pp. 91–96 (2004)
5. Jin, H., Liu, Q., Tang, X., Lu, H.: Learning Local Descriptors for Face Detection. In: IEEE International Conference on Multimedia and Expo., ICME, July 2005, pp. 928–931 (2005)
6. Hadid, A., Pietikainen, M., Ahonen, T.: A Discriminative Feature Space for Detecting and Recognizing Faces. In: Proceedings of the IEEE Conference on Computer Vision and Pattern Recognition CVPR, Washington, DC, vol. 2, pp. 797–804 (2004)
7. Feng, X., Pietikainen, M., Hadid, A.: Facial expression recognition with local binary patterns and linear programming. Pattern Recognition and Image Analysis 15(2), 546–548 (2005)
8. Ruiz-del-Solar, J., Quinteros, J.: Illumination Compensation and Normalization in Eigenspace-based Face Recognition: A comparative study of different preprocessing approaches. Pattern Recognition Letters 29(14), 1966–1978 (2008)
9. Ahonen, T., Pietikainen, M.: Image description using joint distribution of filter bank responses. Pattern Recognition Letters 30(4), 368–376 (2009)
10. Phillips, P.J., Wechsler, H., Huang, J., Rauss, P.: FERET Database and Evaluation Procedure for Face Recognition Algorithms. Image and Vision Computing 16, 295–306 (1998)
11. Yang, H., Wang, Y.: A LBP-based Face Recognition Method with Hamming Distance Constraint. In: Proceedings of Fourth International Conference on Image and Graphics (ICIG 2007), pp. 645–649 (2007)
12. Lahdenoja, O., Laiho, M., Paasio, A.: Reducing the feature vector length in local binary pattern based face recognition. In: IEEE International Conference on Image Processing, ICIP, September 2005, vol. 2, pp. 914–917 (2005)

Informative Laplacian Projection

Zhirong Yang and Jorma Laaksonen

Department of Information and Computer Science*
Helsinki University of Technology
P.O. Box 5400, FI-02015, TKK, Espoo, Finland
{zhirong.yang,jorma.laaksonen}@tkk.fi

Abstract. A new approach of constructing the similarity matrix for eigendecomposition on graph Laplacians is proposed. We first connect the Locality Preserving Projection method to probability density derivatives, which are then replaced by informative score vectors. This change yields a normalization factor and increases the contribution of the data pairs in low-density regions. The proposed method can be applied to both unsupervised and supervised learning. Empirical study on facial images is provided. The experiment results demonstrate that our method is advantageous for discovering statistical patterns in sparse data areas.

1 Introduction

In image compression and feature extraction, linear expansions are commonly used. An image is projected on the eigenvectors of a certain positive semidefinite matrix, each of which provides one linear feature. One of the classical approaches is the *Principal Component Analysis* (PCA), where the variance of images in the projected space is maximized. However, the projection found by PCA may not always encode locality information properly.

Recently, many dimensionality reduction algorithms using eigendecomposition on a graph-derived matrix have been proposed to address this problem. This stream of research has been stimulated by the methods *Isomap* [1] and *Local Linear Embedding* [2], which have later been unified as special cases of the *Laplacian Eigenmap* [3]. The latter minimizes the local variance while maximizing the weighted global variance. The Laplacian Eigenmap has also shown to be a good approximation of both the Laplace-Beltrami operator for a Riemannian manifold [3] and the *Normalized Cut* for finding data clusters [4]. A linear version of the Laplacian Eigenmap algorithm, the *Locality Preserving Projection* (LPP) [5], as well as many other locality-sensitive transformation methods such as the *Hessian Eigenmap* [6] and the *Local Tangent Space Alignment* [7], have also been developed.

However, little research effort has been devoted to graph construction. Locality in the above methods is commonly defined as a spherical neighborhood

* Supported by the Academy of Finland in the project *Finnish Centre of Excellence in Adaptive Informatics Research.*

A.-B. Salberg, J.Y. Hardeberg, and R. Jenssen (Eds.): SCIA 2009, LNCS 5575, pp. 359–368, 2009.

around a vertex (e.g. [1,8]). Two data points are linked with a large weight if and only if they are close, regardless of their relationship to other points. A Laplacian Eigenmap based on such a graph tends to overly emphasize the data pairs in dense areas and is therefore unable to discover the patterns in sparse areas. A widely used alternative to define the locality (e.g. [1,9]) is by k-nearest neighbors (k-NN, $k \geq 1$). Such definition however assumes that relations in each neighborhood are uniform, which may not hold for most real-world data analysis problems. Combination of a spherical neighborhood and the k-NN threshold has also been used (e.g. [5]), but how to choose a suitable k remains unknown. In addition, it is difficult to connect the k-NN locality to the probability theory.

Sparse patterns, which refer to the rare but characteristic properties of samples, play essential roles in pattern recognition. For example, moles or scars often help people identify a person by appearance. Therefore, facial images with such features should be more precious than those with an average face when for example training a face recognition system. A good dimensionality reduction method ought to make the most use of the former kind of samples while associating relatively low weights to the latter.

We propose a new approach to construct a graph similarity matrix. First we express the LPP objective in terms of Parzen estimation, after which the derivatives of the density function with respect to difference vectors are replaced by the informative score vectors. The proposed normalization principle penalizes the data pairs in dense areas and thus helps discover useful patterns in sparse areas for exploratory analysis. The proposed *Informative Laplacian Projection* (ILP) method can then reuse the LPP optimization algorithm. ILP can be further adapted to the supervised case with predictive densities. Moreover, empirical results of the proposed method on facial images are provided for both unsupervised and supervised learning tasks.

The remaining of the paper is organized as follows. The next section briefly reviews the Laplacian Eigenmap and its linear version. In Section 3 we connect LPP to the probability theory and present the Informative Laplacian Projection method. The supervised version of ILP is described in Section 4. Section 5 provides the experiment results on unsupervised and supervised learning. Conclusions as well as future work is finally discussed in Section 6.

2 Laplacian Eigenmap

Given a collection of zero-mean samples $\mathbf{x}^{(i)} \in \mathbb{R}^M$, $i = 1, \ldots, N$, the *Laplacian Eigenmap* [3] computes an implicit mapping $f : \mathbb{R}^M \mapsto \mathbb{R}$ such that $y^{(i)} = f(\mathbf{x}^{(i)})$. The mapped result $\mathbf{y} = \left[y^{(1)}, \ldots, y^{(N)}\right]^T$ minimizes

$$\mathcal{J}(\mathbf{y}) = \sum_{i=1}^{N} \sum_{j=1}^{N} S_{ij} \left(y^{(i)} - y^{(j)}\right)^2 \quad (1)$$

subject to $\mathbf{y}^T \mathbf{D} \mathbf{y} = 1$, where $\mathbf{S}$ is a similarity matrix and $\mathbf{D}$ a diagonal matrix with $D_{ii} = \sum_{j=1}^{N} S_{ij}$. A popular choice of $\mathbf{S}$ is the *radial Gaussian kernel*:

$$S_{ij} = \exp\left(-\frac{\|\mathbf{x}^{(i)} - \mathbf{x}^{(j)}\|^2}{2\sigma^2}\right), \tag{2}$$

with a positive kernel parameter σ. $\{S_{ij}\}_{i,j=1}^{N}$ can also be regarded as the edge weights of a graph where the data points serve as vertices.

The solution of the Laplacian Eigenmap (1) can be found by solving the generalized eigenproblem

$$(\mathbf{D} - \mathbf{S})\mathbf{y} = \lambda \mathbf{D}\mathbf{y}. \tag{3}$$

An R-dimensional ($R \ll M$) compact representation of the data set is then given by the eigenvectors associated with the second least to $(R+1)$-th least eigenvalues.

The Laplacian Eigenmap outputs only the transformed results of the training data points without an explicit mapping function. One has to rerun the whole algorithm for newly coming data. This drawback can be overcome by using parameterized transformations, among which the simplest way is to restrict the mapping to be linear: $y = \mathbf{w}^T\mathbf{x}$ for any input vector $\mathbf{x}$ with $\mathbf{w} \in \mathbb{R}^M$. Let $\mathbf{X} = \left[\mathbf{x}^{(1)}, \ldots, \mathbf{x}^{(N)}\right]$. The linearization leads to the *Locality Preserving Projection* (LPP) [5] whose optimization problem is

$$\text{minimize} \quad \mathcal{J}_{\text{LPP}}(\mathbf{w}) = \mathbf{w}^T\mathbf{X}(\mathbf{D} - \mathbf{S})\mathbf{X}^T\mathbf{w} \tag{4}$$

$$\text{subject to} \quad \mathbf{w}^T\mathbf{X}\mathbf{D}\mathbf{X}^T\mathbf{w} = 1, \tag{5}$$

with the corresponding eigenvalue solution:

$$\mathbf{X}(\mathbf{D} - \mathbf{S})\mathbf{X}^T\mathbf{w} = \lambda\mathbf{X}\mathbf{D}\mathbf{X}^T\mathbf{w}. \tag{6}$$

Then the eigenvectors with the second least to $(R+1)$-th least eigenvalues form the columns of the R-dimensional transformation matrix $\mathbf{W}$.

3 Informative Laplacian Projection

With the radial Gaussian kernel, the Laplacian Eigenmap or LPP objective (1) weights the data pairs only according to their distance without considering the relationship between their vertices and other data points. Moreover, it is not difficult to see that the $\mathbf{D}$ matrix actually measures the "importance" of data points by their densities, which overly emphasizes some almost identical samples. Consequently, the Laplacian Eigenmap and LPP might fail to preserve the statistical information of the manifold in some sparse areas even though a vast amount of training samples were available. Instead, they could encode some tiny details which are difficult to interpret (see e.g. Fig. 4 in [5] and Fig. 2 in [10]).

To attack this problem, let us first rewrite the objective (1) with the density estimation theory:

$$\mathcal{J}_{\text{LPP}}(\mathbf{w}) = \mathbf{w}^T \left[\sum_{i=1}^{N} \sum_{j=1}^{N} S_{ij} \left(\mathbf{x}^{(i)} - \mathbf{x}^{(j)} \right) \left(\mathbf{x}^{(i)} - \mathbf{x}^{(j)} \right)^T \right] \mathbf{w} \tag{7}$$

$$= -\mathbf{w}^T \left[\sum_{i=1}^{N} \sum_{j=1}^{N} \frac{N}{2\sigma^2} \frac{\partial \left(\frac{1}{N} \sum_{k=1}^{N} S_{ik} \right)}{\partial \left(\mathbf{x}^{(i)} - \mathbf{x}^{(j)} \right)} \left(\mathbf{x}^{(i)} - \mathbf{x}^{(j)} \right)^T \right] \mathbf{w} \tag{8}$$

$$= \text{const} \cdot \mathbf{w}^T \left[\sum_{i=1}^{N} \sum_{j=1}^{N} \frac{\partial \hat{p} \left(\mathbf{x}^{(i)} \right)}{\partial \boldsymbol{\Delta}^{(ij)}} \left(\boldsymbol{\Delta}^{(ij)} \right)^T \right] \mathbf{w}, \tag{9}$$

where $\boldsymbol{\Delta}^{(ij)}$ denotes $\mathbf{x}^{(i)} - \mathbf{x}^{(j)}$ and $\hat{p}\left(\mathbf{x}^{(i)}\right) = \sum_{k=1}^{N} S_{ik}/N$ is recognized as a Parzen window estimation of $p\left(\mathbf{x}^{(i)}\right)$.

Next, we propose the *Informative Laplacian Projection* (ILP) method by using the information function $\log \hat{p}$ instead of raw densities $\hat{p}$:

$$\text{minimize} \quad \mathcal{J}_{\text{ILP}}(\mathbf{w}) = -\mathbf{w}^T \left[\sum_{i=1}^{N} \sum_{j=1}^{N} \frac{\partial \log \hat{p} \left(\mathbf{x}^{(i)} \right)}{\partial \boldsymbol{\Delta}^{(ij)}} \left(\boldsymbol{\Delta}^{(ij)} \right)^T \right] \mathbf{w} \tag{10}$$

$$\text{subject to} \quad \mathbf{w}^T \mathbf{X} \mathbf{X}^T \mathbf{w} = 1. \tag{11}$$

The use of the log function arises from the fact that partial derivatives on the log-density can yield a normalization factor:

$$\mathcal{J}_{\text{ILP}}(\mathbf{w}) = \mathbf{w}^T \left[\sum_{i=1}^{N} \sum_{j=1}^{N} \frac{S_{ij}}{\sum_{k=1}^{N} S_{ik}} \boldsymbol{\Delta}^{(ij)} \left(\boldsymbol{\Delta}^{(ij)} \right)^T \right] \mathbf{w} \tag{12}$$

$$= \sum_{i=1}^{N} \sum_{j=1}^{N} E_{ij} \left(y^{(i)} - y^{(j)} \right)^2, \tag{13}$$

where $E_{ij} = S_{ij} / \sum_{k=1}^{N} S_{ik}$. We can then employ the symmetrized version $\mathbf{G} = (\mathbf{E} + \mathbf{E}^T)/2$ to replace $\mathbf{S}$ in (6) and reuse the optimization algorithm of LPP except that the weighting in the constraint of LPP is omitted, i.e. $\mathbf{D} = \mathbf{I}$, because such weighting excessively stresses the samples in dense areas.

The projection found by our method is also locality preserving. Actually the ILP is identical to LPP for the manifolds such as the "Swiss roll" [1,2] or S-manifold [11] where the data points are uniformly distributed. However, ILP behaves very differently from LPP otherwise. The above normalization, as well as omitting the sample weights, penalizes the pairs in dense regions while increases the contribution of those in areas of lower-density, which is conducive to discovering sparse patterns.

4 Supervised Informative Laplacian Projection

The Informative Laplacian Projection can be extended to the supervised case where each sample $\mathbf{x}^{(i)}$ is associated with a class label c_i in $\{1, \ldots, Q\}$. The discriminative version just replaces $\log p(\mathbf{x}^{(i)})$ in (10) with $\log p(c_i|\mathbf{x}^{(i)})$. The resulting *Supervised Informative Laplacian Projection* (SILP) minimizes

$$\mathcal{J}_{\text{SILP}}(\mathbf{w}) = -\mathbf{w}^T \left[\sum_{i=1}^{N} \sum_{j=1}^{N} \frac{\partial \log \hat{p}\left(c_i|\mathbf{x}^{(i)}\right)}{\partial \boldsymbol{\Delta}^{(ij)}} \left(\boldsymbol{\Delta}^{(ij)}\right)^T \right] \mathbf{w} \tag{14}$$

subject to $\mathbf{w}^T \mathbf{X}\mathbf{X}^T \mathbf{w} = 1$. According to the Bayes theorem, we can write out the partial derivative with Parzen density estimations:

$$-\frac{\partial \log p\left(c_i|\mathbf{x}^{(i)}\right)}{\partial \boldsymbol{\Delta}^{(ij)}} = S_{ij} \cdot \frac{1}{\sum_{k=1}^{N} S_{ik}} \cdot \left(\phi_{ij} \frac{\sum_{k=1}^{N} S_{ik}}{\sum_{k=1}^{N} S_{ik}\phi_{ik}} - 1 \right) \boldsymbol{\Delta}^{(ij)}, \tag{15}$$

where $\phi_{ij} = 1$ if $c_i = c_j$ and 0 otherwise. The optimization of SILP is analogous to the unsupervised algorithm except

$$E_{ij} = S_{ij} \cdot \frac{1}{\sum_{k=1}^{N} S_{ik}} \cdot \left(\phi_{ij} \frac{\sum_{k=1}^{N} S_{ik}}{\sum_{k=1}^{N} S_{ik}\phi_{ik}} - 1 \right). \tag{16}$$

The first two factors in (16) are identical to the unsupervised case, favoring local pairs, but penalizing those in dense areas. The third factor in parentheses, denoted by ρ_{ij}, takes the class information into account. It approaches zero when $\phi_{ij} = 1$ and the class label remains almost unchanged in the neighborhood of $\mathbf{x}^{(i)}$. This neglects pairs that are far apart from the classification boundary. For other equi-class pairs, ρ_{ij} takes a positive value if different class labels are mixed in the neighborhood, i.e. the pair is near the classification boundary. In this case SILP minimizes the variance of their difference vectors, which reflects the idea of increasing class cohesion. Finally, $\rho_{ij} = -1$ if $\phi_{ij} = 0$, i.e. the vertices belong to different classes. SILP actually maximizes the norm of such edges in the projected space. This results in dilation around the classification boundary in the projected space, which is desired for discriminative purposes.

Unlike the conventional Fisher's *Linear Discriminant Analysis* (LDA) [12], our method does not rely on the between-class scatter matrix, which is often of low-rank and restricts the number of discriminants. Instead, SILP can produce discriminative components as many as the dimensionality of the original data. The additional dimensions can be beneficial for classification accuracy, as will be shown in Section 5.2.

5 Experiments

5.1 Learning of Turning Angles of Facial Images

This section demonstrates the application of ILP on facial images. We have used 2,662 facial images from the FERET collection [13], in which 2409 are of pose

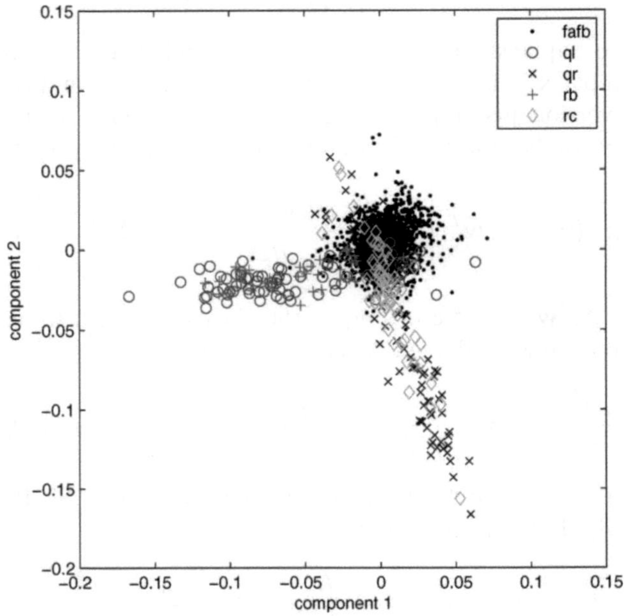

Fig. 1. FERET faces in the subspace found by ILP

fa or *fb*, 81 of *ql*, 78 of *qr*, 32 of *rb*, and 62 of *rc*. The meanings of the FERET pose abbreviations are:

- *fa*: regular frontal image;
- *fb*: alternative frontal image, taken shortly after the corresponding *fa* image;
- *ql*: quarter left – head turned about 22.5 degrees left;
- *qr*: quarter right – head turned about 22.5 degrees right;
- *rb*: random image – head turned about 15 degree left;
- *rc*: random image – head turned about 15 degree right.

In summary, most images are of frontal pose except about 10 percent turning to the left or to the right. The unsupervised learning goal is to find the components that correspond to the left- and right-turning directions. In this work we obtained the coordinates of the eyes from the ground truth data of the collection. Afterwards, all face boxes were normalized to the size of 64×64, with fixed locations for the left eye (53,17) and the right eye (13,17).

We have tested three methods that use the eigenvalue decomposition on a graph: ILP (10)–(11), LPP (4)–(5), and the linearized *Modularity* [14] method. The original facial images were first preprocessed by Principal Component Analysis and reduced to feature vectors of 100 dimensions. The neighborhood parameter for the similarity matrix was empirically set to $\sigma = 3.5$ in (2) for all the compared algorithms.

The data points in the subspace learned by ILP are shown in Figure 1. It can be seen that the faces with left-turning poses (*ql* and *rb*) mainly distribute along

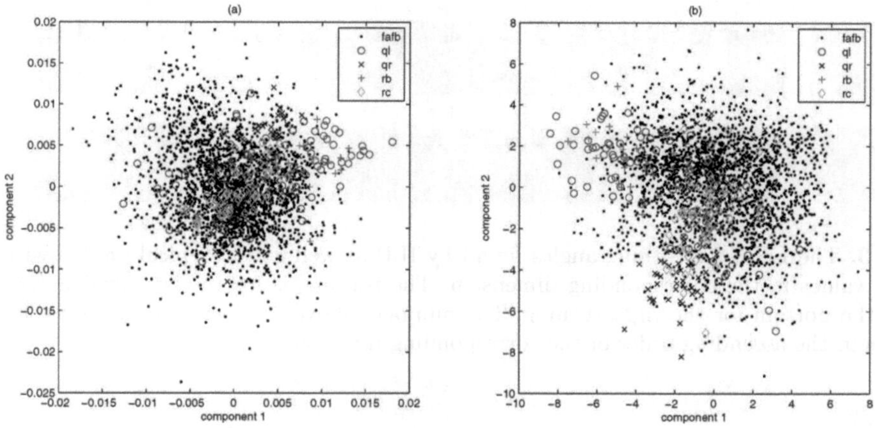

Fig. 2. FERET faces in the subspaces found by (a) LPP and (b) Modularity

the horizontal dimension while the right-turning faces (*qr* and *rc*) roughly along the vertical. The projected results of LPP and Modularity are shown in Figure 2. As one can see, it is almost impossible to distinguish any direction related to a facial pose in the subspace learned by LPP. For the Modularity method, one can barely perceive the left-turning direction is associated with the horizontal dimension while the right-turning with the vertical. All in all, the faces with turning poses are heavily mixed with the frontal ones.

The resulting **W** contains three columns, each of which has the same dimensionality as the input feature vector and can thus be reconstructed to a *filtering image* via the inverse PCA transformation. If a transformation matrix works well for a given learning problem, it is expected to find some semantic connections between its filtering images and our common prior knowledge of the discrimination goal. The filtering images of ILP are displayed in the left-most column of Figure 3, from which one can easily connect the contrastive parts in these filtering images with the turning directions. The facial images on the right of the filtering images are the every sixth images with the least 55 projected values in the corresponding projected dimension.

5.2 Discriminant Analysis on Eyeglasses

Next we performed experiments for discriminative purposes on a larger facial image data set in the University of Notre Dame biometrics database distribution, collection B [15]. The preprocessing was similar to that for the FERET database. We segmented the inner part from 7,200 facial images, among which 2,601 are labeled as the subject in the image is wearing eyeglasses. We randomly selected 2,000 eyeglasses and 4,000 non-eyeglasses images for training and the rest for testing. The images of a same subject were assigned to either the training set or the testing set, never to both. The supervised learning task here is to analyze the discriminative components for recognizing eyeglasses.

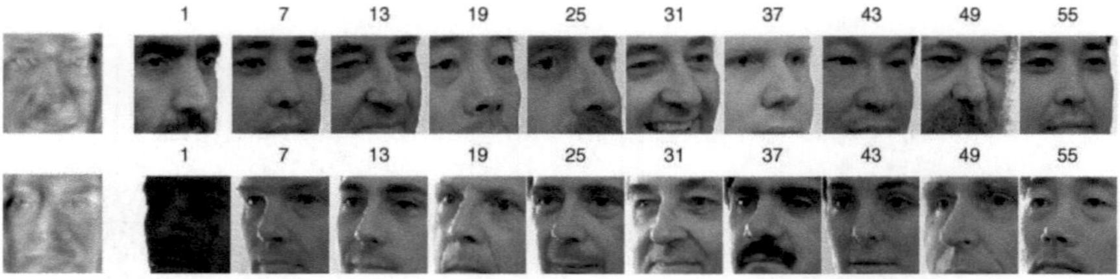

Fig. 3. The bases for turning angles found by ILP as well as the typical images with least values in the corresponding dimension. The top line is for the left-turning pose and the bottom for the right-turning. The numbers above the facial images are their ranks in the ascending order of the corresponding dimension.

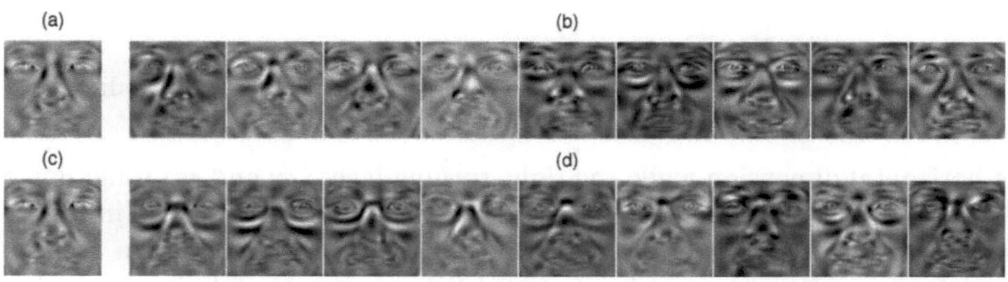

Fig. 4. Filtering images of four discriminant analysis methods: (a) LDA, (b) LSDA, (c) LSVM, and (d) SILP

We have compared four discriminant analysis methods: LDA [12], the *Linear Support Vector Machine* (LSVM) [16], the *Locality Sensitive Discriminant Analysis* (LSDA) [9], and SILP (14). The neighborhood width parameter σ in (2) was empirically set to 300 for LSDA and SILP. The tradeoff parameters in LSVM and LSDA were determined by five-fold cross-validations. The filtering images learned by the above methods are displayed in Figure 4. LDA and LSVM can produce only one discriminative component for two-class problems. In this experiment, their resulting filtering images are very similar except some tiny differences, where the major effective filtering part appears in and between the eyes. The number of discriminants learned by LSDA or SILP is not restricted to one. One can see different contrastive parts in the filtering images of these two methods. In comparison, the top SILP filters are more Gabor-like and the wave packets are mostly related with the bottom rim of the glasses.

After transforming the data, we predicted the class label of each test sample by its nearest neighbor in the training set using the Euclidean distance. Figure 5 illustrates the classification error rates versus the number of discriminative components used. The performance of LDA and LSVM only depends on the first component, with classification error rates 16.98% and 15.51%, respectively. Although the first discriminant of LSDA and SILP work not as well as the one of LDA, they both supersede LDA or even outperform LSVM with subsequent components added. With the first 11 projected dimensions, LSDA achieves its

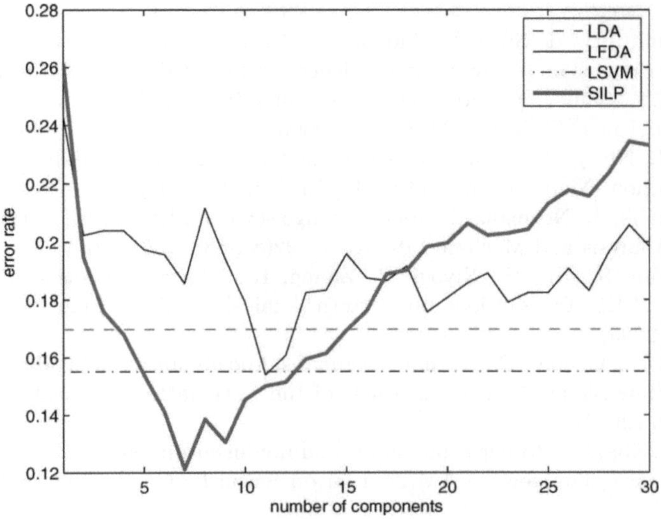

Fig. 5. Nearest neighbor classification error rates with different number of discriminative components used

least error rate 15.37%. SILP is more promising in the sense that the error rate keeps decreasing with its first seven components, attaining the least classification error rate 12.29%.

6 Conclusions

In this paper, we have incorporated the information theory into the Locality Preserving Projection and developed a new dimensionality reduction technique named Informative Laplacian Projection. Our method defines the neighborhood of a data point with its density considered. The resulting normalization factor enables the projection to encode patterns with high fidelity in sparse data areas. The proposed algorithm has been extended for extracting relevant components in supervised learning problems. The advantages of the new method have been demonstrated by empirical results on facial images.

The approach described in this paper sheds light on discovering statistical patterns for non-uniform distributions. The normalization technique may be applied to other graph-based data analysis algorithms. Yet, the challenging work is still ongoing. Adaptive neighborhood functions could be defined using advanced Bayesian learning, as spherical Gaussian kernels calculated in the input space might not work well for all kinds of data manifolds. Moreover, the transformation matrix learned by the LPP algorithm is not necessarily orthogonal. One could employ the orthogonalization techniques in [10] to enforce this constraint. Furthermore, the linear projection methods are readily extended to their nonlinear version by using the kernel technique (see e.g. [9]).

References

1. Tenenbaum, J.B., de Silva, V., Langford, J.C.: A global geometric framework for nonlinear dimensionality reduction. Science. Science 290(5500), 2319–2323 (2000)
2. Roweis, S.T., Saul, L.K.: Nonlinear dimensionality reduction by locally linear embedding. Science 290(5500), 2323–2326 (2000)
3. Belkin, M., Niyogi, P.: Laplacian eigenmaps for dimensionality reduction and data representation. Neural Computation 15, 1373–1396 (2003)
4. Shi, J., Malik, J.: Normalized cuts and image segmentation. IEEE Transactions on Pattern Analysis and Machine Intelligence 22(8), 888–905 (2000)
5. He, X., Yan, S., Hu, Y., Niyogi, P., Zhang, H.J.: Face recognition using Laplacianfaces. IEEE Transactions on Pattern Analysis And Machine Intelligence 27, 328–340 (2005)
6. Donoho, D.L., Grimes, C.: Hessian eigenmaps: Locally linear embedding techniques for high-dimensional data. Proceedings of the National Academy of Sciences 100, 5591–5596 (2003)
7. Zhang, Z., Zha, H.: Principal manifolds and nonlinear dimensionality reduction via tangent space alignment. SIAM Journal on Scientific Computing 26(1), 318–338 (2005)
8. Belkin, M., Niyogi, P.: Laplacian eigenmaps and spectral techniques for embedding and clustering. In: Advances in Neural Information Processing Systems, vol. 14, pp. 585–591 (2002)
9. Cai, D., He, X., Zhou, K., Han, J., Bao, H.: Locality sensitive discriminant analysis. In: Proceedings of the 20th International Joint Conference on Artificial Intelligence, Hyderabad, India, January 2007, pp. 708–713 (2007)
10. Cai, D., He, X., Han, J., Zhang, H.J.: Orthogonal laplacianfaces for face recognition. IEEE Transactions on Image Processing 15(11), 3608–3614 (2006)
11. Saul, L.K., Roweis, S.: Think globally, fit locally: Unsupervised learning of low dimensional manifolds. Journal of Machine Learning Research 4, 119–155 (2003)
12. Fisher, R.A.: The use of multiple measurements in taxonomic problems. Annals of Eugenics 7, 179–188 (1963)
13. Phillips, P.J., Moon, H., Rizvi, S.A., Rauss, P.J.: The FERET evaluation methodology for face recognition algorithms. IEEE Trans. Pattern Analysis and Machine Intelligence 22, 1090–1104 (2000)
14. Newman, M.E.J.: Finding community structure in networks using the eigenvectors of matrices. Phys. Rev. 74(036104) (2006)
15. Flynn, P.J., Bowyer, K.W., Phillips, P.J.: Assessment of time dependency in face recognition: An initial study. In: Audio- and Video-Based Biometric Person Authentication, pp. 44–51 (2003)
16. Cristianini, N., Shawe-Taylor, J.: An Introduction to Support Vector Machines. Cambridge University Press, Cambridge (2000)

Segmentation of Highly Lignified Zones in Wood Fiber Cross-Sections

Bettina Selig[1], Cris L. Luengo Hendriks[1], Stig Bardage[2], and Gunilla Borgefors[1]

[1] Centre for Image Analysis,
Swedish University of Agricultural Sciences, Box 337,
SE-751 05 Uppsala, Sweden
{bettina,cris,gunilla}@cb.uu.se
[2] Department of Forest Products,
Swedish University of Agricultural Sciences, Vallvägen 9 A-D,
SE-750 07 Uppsala, Sweden
Stig.Bardage@sprod.slu.se

Abstract. Lignification of wood fibers has important consequences to the paper production, but its exact effects are not well understood. To correlate exact levels of lignin in wood fibers to their mechanical properties, lignin autofluorescence is imaged in wood fiber cross-sections. Highly lignified areas can be detected and related to the area of the whole cell wall. Presently these measurements are performed manually, which is tedious and expensive. In this paper a method is proposed to estimate the degree of lignification automatically. A multi-stage snake-based segmentation is applied on each cell separately. To make a preliminary evaluation we used an image which contained 17 complete cell cross-sections. This image was segmented both automatically and manually by an expert. There was a highly significant correlation between the two methods, although a systematic difference indicates a disagreement in the definition of the edges between the expert and the algorithm.

1 Introduction

1.1 Background

Wood is composed of cells that are not visible to the naked eye. The majority of wood cells are hollow fibers. They are up to 2 mm long and 30 μm in diameter and mainly consist of cellulose, hemicellulose and lignin [1]. Wood fibers are composed of a cell wall and a empty space in the center which is called lumen (see Fig. 1). The middle lamellae occupies the space between the fibers and contains lignin, which binds the cells together. Lignin also occurs within the cell walls and gives them rigidity [1,2].

The process of lignin diffusion into the cell is called lignification: Lignin precursors diffuse from the lumen to the cell wall and middle lamellae. They condensate (lignificate) starting at the middle lamellae into the cell wall. A so-called condensation front arises (see Fig. 2) that separates the highly lignified zone from the normally lignified zone [2].

A.-B. Salberg, J.Y. Hardeberg, and R. Jenssen (Eds.): SCIA 2009, LNCS 5575, pp. 369–378, 2009.

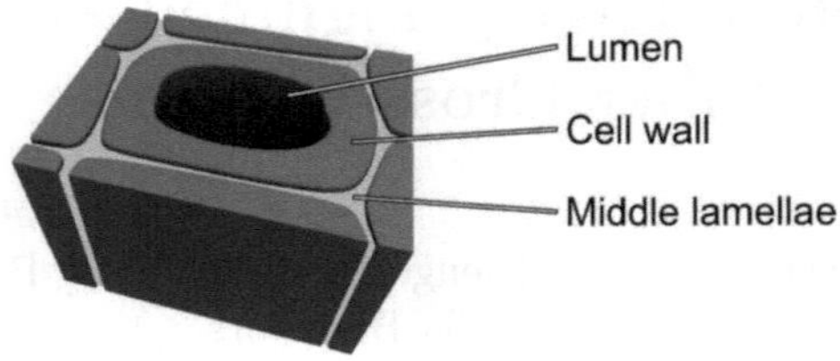

Fig. 1. A wood cell consists of two main structures: lumen and cell wall. The middle lamellae fills the space between the fibres.

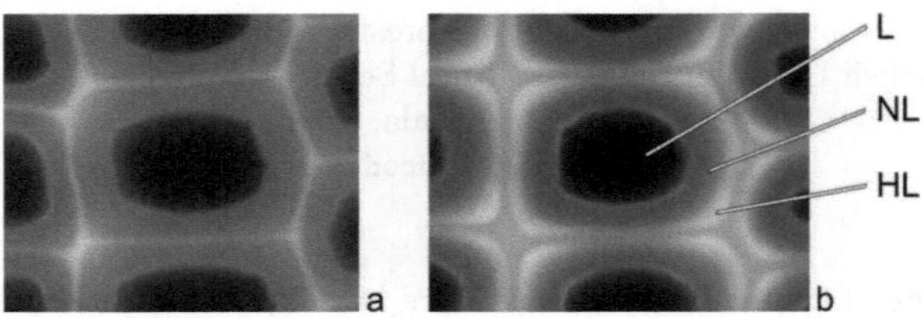

Fig. 2. Cross-section of a normal lignified wood cell (a) and a wood cell with highly lignified zone (b). The area of the lumen (L), the normally lignified zone (NL) and the highly lignified zone (HL) are well-defined in the autofluorescence microscope images. The boundary between NL and HL is called condensation front.

The effects of high lignification on the mechanical properties of wood fibers are especially important in paper production, but are not well understood. A high amount of lignin in the fibers causes bad paper quality. To study these effects it is necessary to measure the distribution of lignin throughout the fiber.

Because lignin is autofluorescent [3], it is possible to image a wood section in a fluorescence microscope with little preparation. The areas of lumen (L), normal lignified cell wall (NL) and highly lignified cell wall (HL) have to be identified so that they can be measured individually. The aim is to relate HL to the area of the whole cell wall. Presently this is done manually, but manual measurements are tedious, expensive and non-reproducible. To our knowledge there exists no automatic method to determine the size of HL in the cell wall. Therefore we are developing a proceeding to analyze large amounts of wood fiber cross-sections automatically. The resulting program will be used by wood scientists.

In fluorescence images edges are in general not sharp. This complicates the boundary detection seriously. Additionally, the condensation front is fuzzy nature and the boundaries around the cell walls have very low contrast at some points, which makes the detection by thresholding impossible.

1.2 Active Contour Models

Active contour models [4,5], known as snakes, are often used to detect the boundary of an object in an image especially when the boundary is partly missing or partly difficult to detect. After an initial guess, the snake $v(s)$ is deformed according to an energy function and converges to local minima which correspond mainly to edges in the image. This energy function is calculated from so called internal and external forces.

$$E_{snake} = E_{int} + E_{ext} \tag{1}$$

The internal force defines the most likely shape of the contour to be found. Its parameters, elasticity α and rigidity β, have to be well chosen to achieve a good result.

$$E_{int} = \alpha|\frac{dv}{ds}|^2 + \beta|\frac{d^2v}{ds^2}|^2 \tag{2}$$

The external force moves the snake towards the most probable position in the image. There exist many ways to calculate the external force. In this paper we use traditional snakes, in which the external force is based on the gradient magnitude of the image I. Therefore, regions with a large gradient attract the snakes.

$$E_{ext} = -|\bigtriangledown I(x,y)|^2 \tag{3}$$

A balloon force is added that forces the active contour to grow outwards (towards the normal direction $\vec{n}(s)$) like a balloon [6]. This enables the snake to overcome regions where the gradient magnitude is too small to move it.

$$F_{ext} = -\kappa\frac{\bigtriangledown E_{ext}}{\| \bigtriangledown E_{ext}\|} + \kappa_p \vec{n}(s) \tag{4}$$

The difficulty with using active contour models lies in finding suitable weights for the different forces. The snake can get stuck in an area with a low gradient if the balloon force is too weak, or the active contour will overshoot the desired boundary if the balloon force is too strong compared to the traditional external force.

In section 2.2 a method is proposed, that considers the mentioned difficulties and expands the snake-based detection in order to find and segment the different regions of highly lignified wood cells in fluorescence light microscopy images.

2 Materials and Methods

2.1 Material

A sample of approximately 2×1×1 cm^3 was cut from a wood disk of Scots pine, *Pinus sylvestris L.*, and sectioned using a sledge microtome. Afterwards, transverse sections 20 µm thick were cut and transferred onto an objective glass, together with some drops of distilled water, and covered with a cover glass. The images were acquired using a Leica DFC490 CCD camera attached to an epifluorescence microscope.

The acquired images have 1300×1030 pixels and a pixel size of 0.3 µm. Only the green channel was used for further processing, as the red and blue channels contain no additional information.

In this paper we illustrate the proposed algorithm using an image section of 340×500 pixels shown in Fig. 3. This section contains representative cells with high lignification. An expert on wood fiber properties segmented manually 17 cells from this image for comparison.

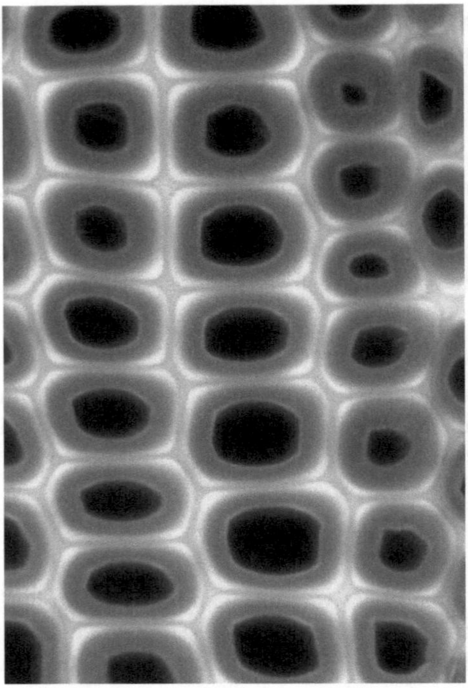

Fig. 3. Sample image with 17 representative cells used to illustrate the proposed algorithm

2.2 Method

The segmentation of the different regions is performed individually for each cell. The lumen is used as seed area for the snake-based segmentation. By expanding the active contour the relevant regions can be found and measured.

Finding Lumen. The lumen of a cell is significantly darker than the cell wall and the middle lamellae. This makes it possible to detect the lumens using a suitable global threshold. However, the histogram gives little help in determining an appropriate threshold level. Therefore we used a more complex automatic method based on a rough segmentation of the image by edges, yielding a few sample lumens and cell walls. These sample regions were then used to determine the correct global threshold level. The rough segmentation was accomplished as follows.

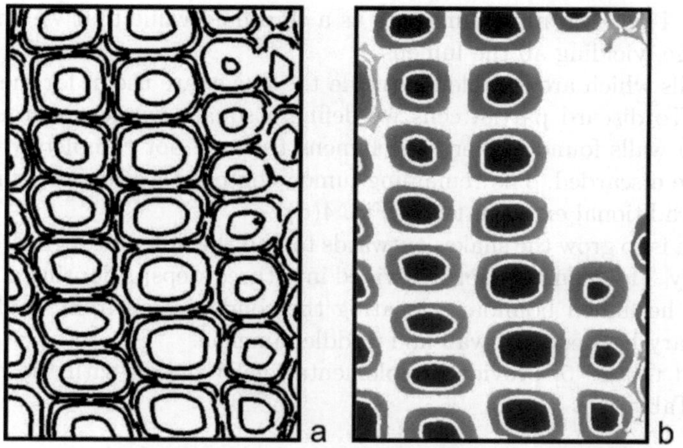

(a) Edge map (b) Set of regions surrounded by another region

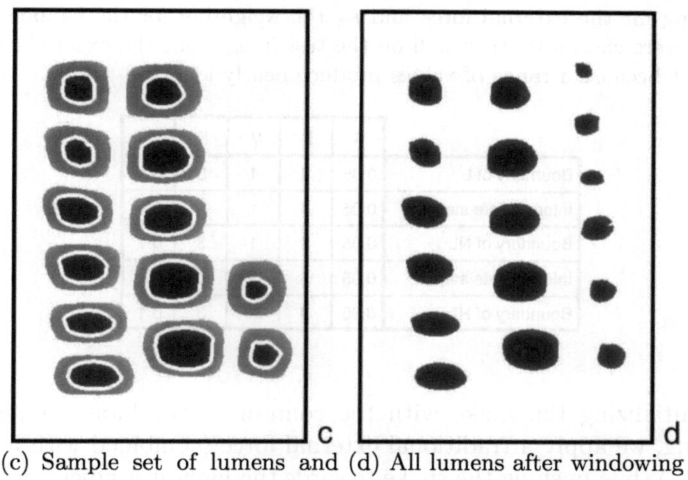

(c) Sample set of lumens and cell walls (d) All lumens after windowing

Fig. 4. Steps followed to find the fiber lumens in the image of Fig. 3

The Canny edge detector [7] followed by a small dilation yields a continuous boundary for most lumens and many of the cell walls (Fig. 4(a)). Each of these closed regions is individually labeled. Because a lumen is always surrounded by a cell wall, we now look for regions that are completely surrounded by another region (Fig. 4(b)). To avoid misclassification, we further constrain this selection to outer regions that are convex (the cross-section of a wood fiber is expected to be convex).

We now have a set of sample lumens and corresponding cell wall regions (Fig. 4(c)). The gray values from only these regions are compiled into a histogram, which typically is nicely bimodal with a strong minimum between the two peaks. This local minimum gives us a threshold value that we apply to the whole image, yielding all the lumens.

Only cells which are completely inside the image are useful for measurement purposes. To discard partial cells we define a square window surrounding the sample cell walls found earlier. The lumens that are not completely inside this window are discarded. The remaining lumen contours are refined using a snake with the traditional external force (Fig. 4(d)).

The idea is to grow the snakes outwards to find the different regions of the cells successively. The segmentation is divided into three steps: Adapting a reasonable shape for the lumen boundary, locating the condensation front, and detecting the boundary between cell wall and middle lamellae.

We used the in [5] provided implementation of snakes with the parameters shown in Table 1.

Table 1. Parameters used for the implementation of the algorithm, where α is elasticity and β rigidity for the internal force, γ viscosity (weighting of the original position), κ the weighting for the external force and κ_p the weighting for the balloon force. The parameters were chosen to work well on the test image, but the exact choices are not so important because a range of values produce nearly identical results.

	α	β	γ	κ	κ_p
Boundary of L	0.05	3	1	3	0.1
Intermediate step 1	0.05	3	1	3	0.1
Boundary of NL	0.05	3	1	3	0.1
Intermediate step 2	0.05	3	1	3	0
Boundary of HL	0.05	3	20	3	0.1

After initializing the snake with the contour of the lumen found through thresholding, we apply a traditional external force (combined with a small balloon force). While pushing the snake towards the highest gradient, we refine the position of the lumen boundary.

Finding condensation front. The result from the first step is used as a starting point for the second step. Since the lumen boundary and the condensation front are very similar (both edges have the same polarity) it is impossible to push the snake away from the first edge and at the same time make sure it settles at the second edge. To solve this problem we use an intermediate step with a new external energy, which has its minima in regions with a small gradient magnitude.

$$E_1 = +|\nabla I(x, y)|^2 \tag{5}$$

Combined with a small balloon force, the snake converges to the region with the lowest gradient between the two edges. From this point, the condensation front can be found with a snake using a small balloon force and the traditional external force.

Finding cell wall boundary. To locate the boundary between the cell wall and the middle lamellae a similar two-stage snake is applied. This time an external energy is used which has its minima in the areas with high gray values.

$$E_2 = -I(x, y) \tag{6}$$

Since the highly lignified zones are very bright, the snake will converge in the middle of these regions. Afterwards, a traditional external force is used to push the snake outwards to detect the boundary between cell wall and middle lamellae.

Typically traditional snakes do not terminate. However, due to the combination of the chosen forces all snakes described in this paper converge to their final position after 10-20 steps. Afterwards only little changes occur and the algorithm is stopped after 30 steps.

3 Results

To make a preliminary evaluation of the proposed method we used an image which contained 17 detectable wood cells. This image was segmented independently by the proposed algorithm and manually by an expert. This delineation was performed after the algorithm was finished, and it was not used to define the algorithm. The regions L, NL and HL were measured and compared.

The results from the two analyses and the area HL related to the whole cell wall are compared in Fig. 6. Here, the horizontal axis represent the results from the automatic method and the vertical axis from manual measurements. The solid line in the figure is the identity. Measurements on this line had the same result in the manual and automatic method. Values left of this line were underestimated by the proposed algorithm and values right of this line were overestimated.

The area of the lumen was measured well, whereas NL was a bit overestimated and HL generally underestimated.

The relative area of the highly lignified zone was computed as $p = \frac{HL}{NL+HL}$. These results reflect the overestimated measurements of HL.

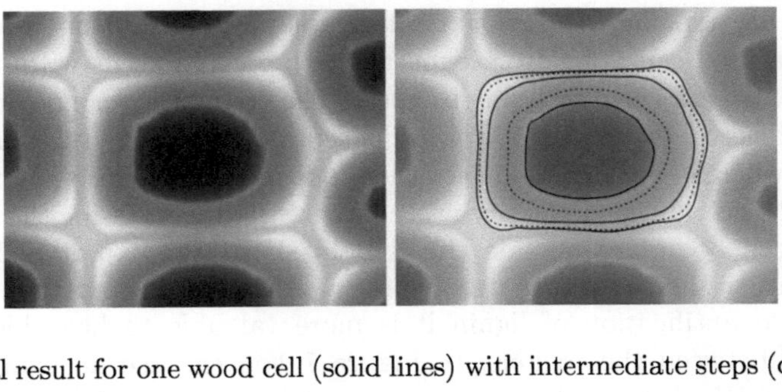

Fig. 5. Final result for one wood cell (solid lines) with intermediate steps (dotted lines)

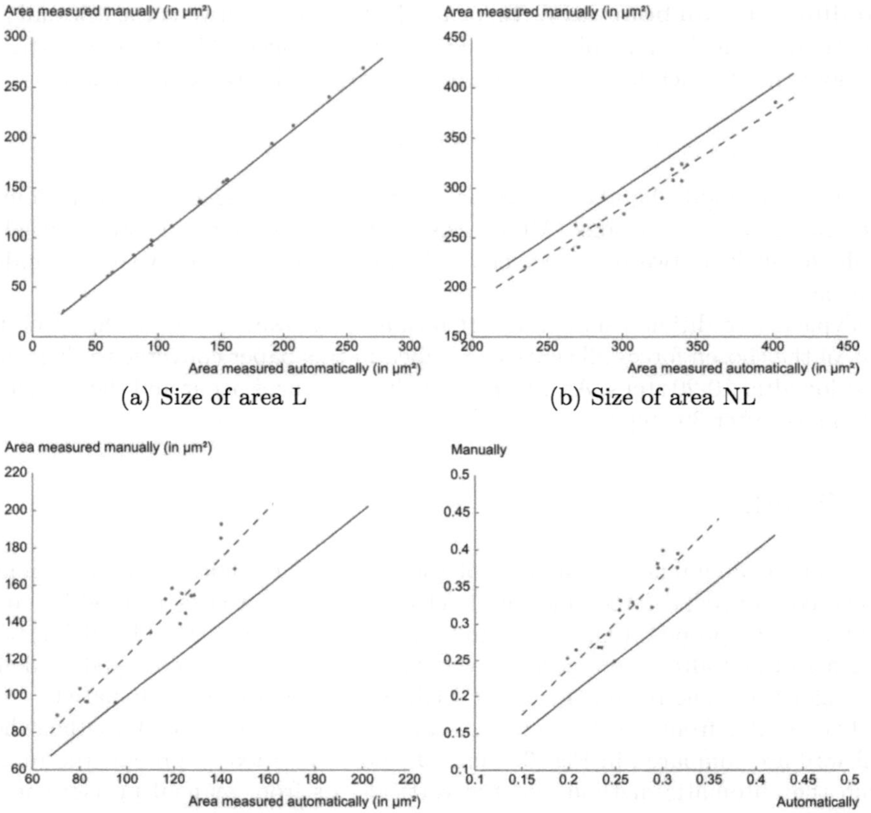

(a) Size of area L

(b) Size of area NL

(c) Size of area HL

(d) Size of HL in relation to size of the whole cell wall.

Fig. 6. Comparison between manual and automatic method

4 Discussion and Conclusions

The automatic labeling and the expert agreed to a different degree for each of the boundaries. These disparities have various reasons.

First of all, manual measuring is always subjective and not deterministic. The criteria used can differ from expert to expert, as well as within a series of measurements performed by a single expert. The boundaries can be drawn inside, outside or directly on the edge. The proposed algorithm sets the boundaries on the edges, whereas our expert places them depending on the type of boundary. For example, the lumen boundary was consistently drawn inside the lumen, and the outer cell boundary outside the cell. In short, the expert delineated the cell wall rather than marking its boundary. It can be argued that for further automated quantification of lignin it is more valuable to have identified the boundaries between the regions. In Figure 7 you can see an example of the boundary of HL done both automatically and manually. Here it is apparent

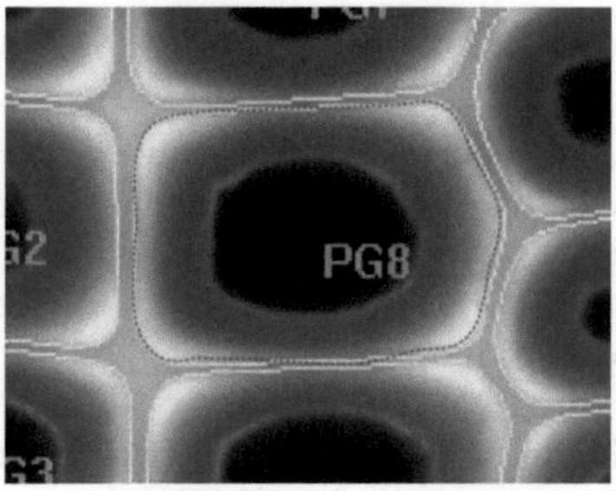

Fig. 7. Manual (solid line) and automatic (dotted line) segmentation of the outer boundary of a cell

that the manually placed boundary lies outside the one created by the proposed algorithm.

Although the results of HL do not follow the identity line, they are scattered around a (virtual) second line which is slightly tilted and shifted relative to the identity. This systematic error shows that even though the measurements followed slightly different criteria a close relation exists.

Another characteristic of the edges can be detected in the result graphs. The region NL has blurry and fuzzy boundaries and the edges around HL have very low contrast at some points. Both are difficult to detect either manually or automatically. Therefore, the plots for these boundaries show a larger degree of scatter then the highly correlated plot of L. The lumen has a sharp and well defined boundary that allows for a more precise measurement. But in spite of everything, the calculated correlation is high for all the regions (see Table 2).

Table 2. Correlation between manual and automatic measurements of the areas L, NL and HL. (All the p-values are less than 10^{-8}).

	L	NL	HL
Correlation	0.9998	0.9664	0.9495

We tested the algorithm on other images and obtained similar results. In this paper we show the algorithm applied to this one particular image because that is the one we have a manual segmentation for and therefore it's the only data we have we can do comparisons on.

Currently the algorithm is applied on each cell separately. An improvement will be to grow regions simultaneously, allowing them to compete for space (e.g. [8]). This would be particularly useful when segmenting not highly lignified cells, because for these cells the current algorithm is not able to distinguish the edges, producing overlapping regions.

References

1. Haygreen, J.G., Bowyer, J.L.: Forest Products and Wood Science: An Introduction, 3rd edn. Iowa State University Press, Ames (1996)
2. Barnett, J.R., Jeronimidis, G.: Wood Quality and its Biological Basis, 1st edn. Blackwell Publishing Ltd., Malden (2003)
3. Ruzin, S.E.: Plant Microtechnique and Microscopy, 1st edn. Oxford University Press, Oxford (1999)
4. Sonka, M., Hlavac, V., Boyle, R.: Ch. 7.2. In: Image Processing, Analysis, and Machine Vision, 3rd edn. Thomson Learning (2008)
5. Xu, C., Prince, J.L.: Snakes, shapes, and gradient vector flow. IEEE Transaction on Image Processing 7(3), 359–369 (1998)
6. Cohen, L.D.: On active contour models and balloons. CVGIP: Image Understanding 53(2), 211–218 (1991)
7. Sonka, M., Hlavac, V., Boyle, R.: Ch. 5.3.5. In: Image Processing, Analysis, and Machine Vision, 3rd edn. Thomson Learning (2008)
8. Kerschner, M.: Homologous twin snakes integrated in a bundle block adjustment. In: International Archives of Photogrammetry and Remote Sensing, vol. XXXII, Part 3/1, pp. 244–249 (1998)

Dense and Deformable Motion Segmentation for Wide Baseline Images

Juho Kannala, Esa Rahtu, Sami S. Brandt, and Janne Heikkilä

Machine Vision Group, University of Oulu, Finland
{jkannala,erahtu,sbrandt,jth}@ee.oulu.fi

Abstract. In this paper we describe a dense motion segmentation method for wide baseline image pairs. Unlike many previous methods our approach is able to deal with deforming motions and large illumination changes by using a bottom-up segmentation strategy. The method starts from a sparse set of seed matches between the two images and then proceeds to quasi-dense matching which expands the initial seed regions by using local propagation. Then, the quasi-dense matches are grouped into coherently moving segments by using local bending energy as the grouping criterion. The resulting segments are used to initialize the motion layers for the final dense segmentation stage, where the geometric and photometric transformations of the layers are iteratively refined together with the segmentation, which is based on graph cuts. Our approach provides a wider range of applicability than the previous approaches which typically require a rigid planar motion model or motion with small disparity. In addition, we model the photometric transformations in a spatially varying manner. Our experiments demonstrate the performance of the method with real images involving deforming motion and large changes in viewpoint, scale and illumination.

1 Introduction

The problem of motion segmentation typically arises in a situation where one has a sequence of images containing differently moving objects and the task is to extract the objects from the images using the motion information. In this context the motion segmentation problem consists of the following two subproblems: (1) determination of groups of pixels in two or more images that move together, and (2) estimation of the motion fields associated with each group [1].

Motion segmentation has a wide variety of applications. For example, representing the moving images with a set of overlapping motion layers may be useful for video coding and compression as well as for video mosaicking [2,1]. Furthermore, the object-level segmentation and registration could be directly used in recognition and reconstruction tasks [3,1].

Many early approaches to motion segmentation assume small motion between consecutive images and use dense optical flow techniques for motion estimation [2,4]. The main limitation of optical flow based methods is that they are not suitable for large motions. Some approaches try to alleviate this problem by using feature point correspondences for initializing the motion models [5,6,1]. However, the implementations described in [5] and [6] still require that the motion is relatively small and approximately planar. The approach in [1] can deal with large planar motions.

A.-B. Salberg, J.Y. Hardeberg, and R. Jenssen (Eds.): SCIA 2009, LNCS 5575, pp. 379–389, 2009.

Fig. 1. An example image pair, courtesy of [3], and the extracted motion components (middle) with the associated geometric and photometric transformations (right)

In this work, we address the motion segmentation problem in the context of wide baseline image pairs. This means that we consider cases where the motion of the objects between the two images may be very large due to non-rigid deformations and viewpoint variations. Another challenge in the wide baseline case is that the appearance of objects usually changes with illumination. For example, spatially varying illumination changes, such as shadows, occur frequently in wide baseline imagery and may further complicate object detection and segmentation. In order to address these challenges we propose a bottom-up motion segmentation approach which gradually expands and merges the initial matching regions into smooth motion layers and finally provides a dense assignment of pixels into these layers. Besides segmentation, the proposed method provides the geometric and photometric transformations for each layer.

The previous works closest to ours are [1,7,8]. In [1] the problem statement is the same as here, i.e., two-view motion segmentation for large motions. However, the solution proposed there requires approximately planar motion and does not model varying lighting conditions. The problem setting in [7] and [8] is slightly different than here since there the main focus is on object recognition. Nevertheless, the ideas of [7] and [8] can be utilized in motion segmentation and we develop them further towards a dense and deformable two-view motion segmentation method. In particular, we use the quasi-dense matching technique of [8] for initializing the motion layers. This allows us to avoid the planar motion assumption and makes the initialization more robust to extensive background clutter. In order to get the pixel level segmentation, we use graph cut based optimization together with a somewhat similar probabilistic model as in [7]. However, unlike in [7], we do not use any presegmented reference images but detect and segment the common regions automatically from both images. Furthermore, we propose a spatially varying photometric transformation model which is more expressive than the global model in [7].

In addition to the aforementioned publications, there are also other recent works related to the topic. For example, [9] describes an approach for computing layered motion segmentations of video. However, that work uses continuous video sequences and hence avoids the problems of large geometric and photometric transformations which make the wide baseline case difficult. Another related work is [10] which describes a layered image formation model for motion segmentation. Nevertheless, [10] does not address the problem of model initialization which is essential for large motions.

Algorithm 1. Outline of the method	**Algorithm 2.** Dense motion segmentation
Input: two images I and I' and a set of seed matches **Algorithm:** 1. Grow and group the seed matches [8] 2. Verify the grown groups of matches 3. Initialize motion layers 4. Perform dense segmentation of both images 5. Enforce the consistency of segmentations **Output:** a dense assignment of pixels to layers which define the motion for each pixel	**Input:** • the image to be segmented (I) and the other image (I') • a set of motion layers ($\mathcal{L}_j$) with geometric and photometric transformations ($\mathcal{G}_j$ and $\mathcal{F}_j$) • initial segmentation S **Algorithm:** 1. Update the photometric transformations $\mathcal{F}_j$ 2. Update the geometric transformations $\mathcal{G}_j$ 3. Update the segmentation S 4. Repeat steps 1-3 until S does not change

2 Overview

This section gives a brief overview of our approach whose main stages are summarized in Algorithm 1. The particular focus of this paper is on the dense segmentation method which is described in Algorithm 2 and detailed in Section 3.

2.1 Hypothesis Generation and Verification

First, given a pair of images and a sparse set of seed matches between them, we compute our motion hypotheses by region growing and grouping. That is, we first use the match propagation technique [8] to obtain more matching pixels in the spatial neighborhoods of the seed matches, which are acquired using standard region detectors and SIFT-based matching [11]. After the propagation, the coherently moving matches are grouped together by using a similar approach as in [8], where the neighboring quasi-dense matches, connected by Delaunay triangulation, are merged to the same group if the triangulation is consistent with the local affine motions estimated during the propagation. However, instead of the heuristic criterion in [8], we use the bending energy of locally fitted thin plate splines [12] to measure the consistency of triangulations.

Then, the grouped correspondences are verified in order to reject incorrect matches. The idea is to improve the precision of keypoint based matching by examining the grown regions, as in [3,8,13,14]. In our current implementation the verification is based on the size of the matching regions [8] but also other decision criteria could be used in the proposed framework (cf. [14]). Finally, the verified groups of correspondences are used to initialize the tentative motion layers illustrated in Fig. 2.

2.2 Motion Segmentation

The tentative motion layers are refined in the dense segmentation stage (Step 4, Alg. 1) where the assignment of pixels to layers is first done separately for each image whereafter the final layers are obtained by checking the inverse consistency of the two assignments as in [1] (Step 5, Alg. 1). The segmentation procedure (Alg. 2) iterates the

Fig. 2. Left: the seed regions (yellow ellipses) and the propagated quasi-dense matches. Middle: the grouped matches (each group has own color, the yellow lines are the Delaunay edges joining initial groups [8]). Right: the six largest groups and their support regions.

following steps: (1) estimation of photometric transformations for each color channel, (2) estimation of geometric transformations, and (3) graph cut based segmentation of pixels to layers. The details of the iteration are described in Sect. 3 but the core idea is the following: when the segmentation is updated some pixels change their layer to a better one and this allows to improve the estimates for the geometric and photometric transformations of the layers (which then again improves the segmentation and so on).

The final motion layers for the example image pair of Fig. 2 are illustrated in the last column of Fig. 1 where the meshes illustrate the geometric transformations and the colors visualize the photometric transformations. The colors show how the gray color, shown on the background layer, would be transformed from the other image to the colored image. The result indicates that the white balance is different in the two images. Note also the shadow on the corner of the foremost magazine in the first image.

3 Dense and Deformable Motion Segmentation

3.1 Layered Model

Our layer-based model describes each one of the two images as a composition of layers which are related to the other image by different geometric and photometric transformations. In the following, we assume that image I is the image to be segmented and I' is the reference image. The other case is obtained by changing the roles of I and I'.

The model consists of a set of motion layers, denoted by $\mathcal{L}_j$, $j = 0, \ldots, L$. The segmentation of image I is defined by the label matrix S which has the same size as I (i.e. $m \times n$). So, $S(\mathbf{p}) = j$ means that the pixel $\mathbf{p}$ in I is labeled to layer j. The layer $j = 0$ is the background layer reserved for those pixels which are not visible in I'. The label matrix S is sufficient for representing the final assignment of pixels to layers. However, it is not sufficient for the initialization of our iterative segmentation method since some of the tentative layers may overlap as shown in Fig. 2. Therefore, for later use, we introduce additional label matrices S_j so that $S_j(\mathbf{p}) = 1$ if $\mathbf{p}$ belongs to layer j and $S_j(\mathbf{p}) = 0$ otherwise.

The geometric transformation associated to layer j ($j \neq 0$) is denoted by $\mathcal{G}_j$. In detail, the motion field $\mathcal{G}_j$ transforms the pixels in I to the other image and is represented by two matrices of size $m \times n$ (one for each coordinate). Thus, $\mathcal{G}_j(\mathbf{p}) = \mathbf{p}'$ means that pixel $\mathbf{p}$ is mapped to position $\mathbf{p}'$ in the other image if it belongs to layer j.

The photometric transformation of layer j ($j \neq 0$) is denoted by $\mathcal{F}_j$ and its parameters define an affine intensity transformation for each color channel at every pixel. Hence, if the number of color channels is K, then $\mathcal{F}_j$ is represented by a set of $2K$ matrices each of which has size $m \times n$. So, the modeled intensity for color channel k at pixel $\mathbf{p}$ is defined by

$$\hat{I}_j^k(\mathbf{p}) = \mathcal{F}_j^k(\mathbf{p}) \cdot I'^k(\mathcal{G}_j(\mathbf{p})) + \mathcal{F}_j^{(K+k)}(\mathbf{p}), \tag{1}$$

where the superscript of $\mathcal{F}_j$ indicates which ones of the $2K$ transformation parameters correspond to channel k.

Given the latent variables S, $\mathcal{G}_j$, $\mathcal{F}_j$ and the reference image I', the relation (1) provides a generative model for I. In fact, the goal in the dense segmentation stage is to determine the latent variables so that the resulting layered model would explain well the observed intensities in I. This is acquired by minimizing an energy function which is introduced in Sect. 3.3. However, first, we describe how the layered model is initialized.

3.2 Model Initialization

The motion hypotheses which pass the verification stage are represented as groups of two-view point correspondences and each of them is used to initialize a motion layer.

First, the initialization of the label matrices S_j is obtained directly from the support regions of the grouped correspondences. That is, we give a label $j > 0$ for each group and assign $S_j(\mathbf{p}) = 1$ for those pixels $\mathbf{p}$ that are inside the support region of group j. At this stage there may be pixels which are assigned to several layers. However, these conflicting assignments are eventually solved when the final segmentation S is produced (see Sect. 3.4).

Second, the initialization of the motion fields $\mathcal{G}_j$ is done by fitting a regularized thin-plate spline to the point correspondences of each group [12]. The thin-plate spline is a parametrized mapping which allows extrapolation, i.e., it defines the motion also for those pixels that are outside the particular layer. Thus, each motion field $\mathcal{G}_j$ is initialized by evaluating the thin-plate spline for all pixels $\mathbf{p}$.

Third, the coefficients of the photometric transformations $\mathcal{F}_j$ are initialized with constant values determined from the intensity histograms of the corresponding regions in I and I'. In fact, when $\mathcal{F}_j^k(\mathbf{p})$ and $\mathcal{F}_j^{K+k}(\mathbf{p})$ are the same for all $\mathbf{p}$, (1) gives simple relations for the standard deviations and means of the two histograms for each color channel k. Hence, one may estimate $\mathcal{F}_j^k$ and $\mathcal{F}_j^{K+k}$ by computing robust estimates for the standard deviations and means of the histograms. The estimates are later refined in a spatially varying manner as described in Sect. 3.5.

3.3 Energy Function

The aim is to determine the latent variables $\boldsymbol{\theta} = \{S, \mathcal{G}_j, \mathcal{F}_j\}$ so that the resulting layered model explains the observed data $\mathcal{D} = \{I, I'\}$ well. This is done by maximizing the posterior probability $P(\boldsymbol{\theta}|\mathcal{D})$, which is modeled in the form $P(\boldsymbol{\theta}|\mathcal{D}) = \psi \exp\left(-E(\boldsymbol{\theta}, \mathcal{D})\right)$, where the normalizing factor ψ is independent of $\boldsymbol{\theta}$ [9]. Maximizing $P(\boldsymbol{\theta}|\mathcal{D})$ is equivalent to minimizing the energy

$$E(\boldsymbol{\theta},\mathcal{D}) = \sum_{\mathbf{p}\in\mathcal{P}} U_{\mathbf{p}}(\boldsymbol{\theta},\mathcal{D}) + \sum_{(\mathbf{p},\mathbf{q})\in\mathcal{N}} V_{\mathbf{p},\mathbf{q}}(\boldsymbol{\theta},\mathcal{D}), \tag{2}$$

where $U_{\mathbf{p}}$ is the unary energy for pixel $\mathbf{p}$ and $V_{\mathbf{p},\mathbf{q}}$ is the pairwise energy for pixels $\mathbf{p}$ and $\mathbf{q}$, $\mathcal{P}$ is the set of pixels in image I and $\mathcal{N}$ is the set of adjacent pairs of pixels in I.

The unary energy in (2) consists of two terms,

$$\sum_{\mathbf{p}\in\mathcal{P}} U_{\mathbf{p}}(\boldsymbol{\theta},\mathcal{D}) = \sum_{\mathbf{p}\in\mathcal{P}} -\log P_{\mathbf{p}}(I|\boldsymbol{\theta},I') - \log P_{\mathbf{p}}(\boldsymbol{\theta}) =$$

$$\sum_{j=0}^{L} \sum_{\mathbf{p}|S(\mathbf{p})=j} -\log P_{\mathrm{l}}(I(\mathbf{p})|\mathcal{L}_j,I') - \log P(S(\mathbf{p})=j), \tag{3}$$

where the first one is the likelihood term defined by P_{l} and the second one is the pixel-wise prior for $\boldsymbol{\theta}$. The pairwise energy in (2) is defined by

$$V_{\mathbf{p},\mathbf{q}}(\boldsymbol{\theta},\mathcal{D}) = \gamma(1-\delta_{S(\mathbf{p}),S(\mathbf{q})}) \exp\left(\frac{-\max_k |\nabla I^k(\mathbf{p}) \cdot \frac{\mathbf{p}-\mathbf{q}}{\|\mathbf{p}-\mathbf{q}\|}|^2}{\beta}\right), \tag{4}$$

where $\delta_{\cdot,\cdot}$ is the Kronecker delta function and γ and β are positive scalars. In the following, we describe the details behind the expressions in (3) and (4).

Likelihood term. The term $P_{\mathbf{p}}(I|\boldsymbol{\theta},I')$ measures the likelihood that the pixel $\mathbf{p}$ in I is generated by the layered model $\boldsymbol{\theta}$. This likelihood depends on the parameters of the particular layer $\mathcal{L}_j$ to which $\mathbf{p}$ is assigned and it is modeled by

$$P_{\mathrm{l}}(I(\mathbf{p})|\mathcal{L}_j,I') = \begin{cases} \kappa & j=0 \\ P_{\mathrm{c}}(I(\mathbf{p})|\hat{I}_j)P_{\mathrm{t}}(I(\mathbf{p})|\hat{I}_j) & j\neq 0 \end{cases} \tag{5}$$

Thus, the likelihood of the background layer ($j=0$) is κ for all pixels. On the other hand, the likelihood of the other layers is modeled by a product of two terms, P_{c} and P_{t}, which measure the consistency of color and texture between the images I and $\hat{I}_j$, where $\hat{I}_j$ is defined by $\mathcal{G}_j$, $\mathcal{F}_j$, and I' according to (1). In other words, $\hat{I}_j$ is the image generated from I' by $\mathcal{L}_j$ and $P_{\mathrm{l}}(I(\mathbf{p})|\mathcal{L}_j,I')$ measures the consistency of appearance of I and $\hat{I}_j$ at $\mathbf{p}$.

The color likelihood $P_{\mathrm{c}}(I(\mathbf{p})|\hat{I}_j)$ is a Gaussian density function whose mean is defined by $\hat{I}_j(\mathbf{p})$ and whose covariance is a diagonal matrix with predetermined variance parameters. For example, if the RGB color space is used then the density is three-dimensional and the likelihood is large when $I(\mathbf{p})$ is close to $\hat{I}_j(\mathbf{p})$.

Here the texture likelihood $P_{\mathrm{t}}(I(\mathbf{p})|\hat{I}_j)$ is also modeled with a Gaussian density. That is, we compute the normalized grayscale cross-correlation between two small image patches extracted from I and $\hat{I}_j$ around $\mathbf{p}$ and denote it by $t_j(\mathbf{p})$. Thereafter the likelihood is obtained by setting $P_{\mathrm{t}}(I(\mathbf{p})|\hat{I}_j) = N(t_j(\mathbf{p})|1,\nu)$, where $N(\cdot|1,\nu)$ is a one-dimensional Gaussian density with mean 1 and variance ν.

Prior term. The term $P_{\mathbf{p}}(\boldsymbol{\theta})$ in (3) denotes the pixelwise prior for $\boldsymbol{\theta}$ and it is defined by the probability $P(S(\mathbf{p}) = j)$ with which $\mathbf{p}$ is labeled with j. If there is no prior information available one may here use the uniform distribution which gives equal probability for all labels. However, in our iterative approach, we always have an initial estimate $\boldsymbol{\theta}_0$ for the parameters $\boldsymbol{\theta}$ while minimizing (2), and hence, we may use the initial estimate S_0 to define a prior for the label matrix S. In fact, we model the spatial distribution of labels with a mixture of two-dimensional Gaussian densities, where each label j is represented by one mixture component, whose portion of the total density is proportional to the number of pixels with the label j. The mean and covariance of each component are estimated from the correspondingly labeled pixels in S_0.

The spatially varying prior term is particularly useful in such cases where the colors of some uniform background regions accidentally match for some layer. (This is actually quite common when both images contain a lot of background clutter.) If these regions are distant from the objects associated to that particular layer, as they usually are, the non-uniform prior may help to prevent incorrect layer assignments.

Pairwise term. The purpose of the term $V_{\mathbf{p},\mathbf{q}}(\boldsymbol{\theta}, \mathcal{D})$ in (2) is to encourage piecewise constant labelings where the layer boundaries lie on the intensity edges. The expression (4) has the form of a generalized Potts model [15], which is commonly used in segmentation approaches based on Markov Random Fields [1,7,9]. The pairwise term (4) is zero for such neighboring pairs of pixels which have the same label and greater than zero otherwise. The cost is highest for differently labeled pixels in uniform image regions where ∇I^k is zero for all color channels k. Hence, the layer boundaries are encouraged to lie on the edges, where the directed gradient is non-zero. The parameter γ determines the weighting between the unary term and the pairwise term in (2).

3.4 Algorithm

The minimization of (2) is performed by iteratively updating each of the variables S, $\mathcal{G}_j$ and $\mathcal{F}_j$ in turn so that the smoothness of the geometric and photometric transformation fields, $\mathcal{G}_j$ and $\mathcal{F}_j$, is preserved during the updates. The approach is summarized in Alg. 2 and the update steps are detailed in the following sections.

In general, the approach of Alg. 2 can be used for any number of layers. However, after the initialization (Sect. 3.2), we do not directly proceed to the multi-layer case but first verify the initial layers individually against the background layer. In detail, for each initial layer j, we run one iteration of Alg. 2 by using uniform prior for the two labels in S_j and a relatively high value of γ. Here the idea is that those layers j, which do not generate high likelihoods $P_l(I(\mathbf{p})|\mathcal{L}_j, I')$ for a sufficiently large cluster of pixels, are completely replaced by the background. For example, the four incorrect initial layers in Fig. 2 were discarded at this stage. Then, after the verification, the multi-label matrix S is initialized (by assigning the label with the highest likelihood $P_l(I(\mathbf{p})|\mathcal{L}_j, I')$ for ambiguous pixels) and the layers are finally refined by running Alg. 2 in the multi-label case, where the spatially varying prior is used for the labels.

3.5 Updating the Photometric Transformations

The spatially varying photometric transformation model is an important element of our approach. Given the segmentation S and the geometric transformation $\mathcal{G}_j$, the coefficients of the photometric transformation $\mathcal{F}_j$ are estimated from linear equations by using Tikhonov regularization [16] to ensure the smoothness of solution.

In detail, according to (1), each pixel $\mathbf{p}$ assigned to layer j provides a linear constraint for the unknowns $\mathcal{F}_j^k(\mathbf{p})$ and $\mathcal{F}_j^{(K+k)}(\mathbf{p})$. By stacking the elements of $\mathcal{F}_j^k$ and $\mathcal{F}_j^{(K+k)}$ into a vector, denoted by $\mathbf{f}_j^k$, we may represent all these constraints, generated by the pixels in layer j, in matrix form $\mathbf{M}\mathbf{f}_j^k = \mathbf{b}$, where the number of unknowns in $\mathbf{f}_j^k$ is larger than the number of equations. Then, we use Tikhonov regularization and solve

$$\min_{\mathbf{f}_j^k} ||\mathbf{M}\mathbf{f}_j^k - \mathbf{b}||^2 + \lambda||\mathbf{L}\mathbf{f}_j^k||^2, \tag{6}$$

where λ is the regularization parameter and the difference operator $\mathbf{L}$ is here defined so that $||\mathbf{L}\mathbf{f}_j^k||^2$ is a discrete approximation to

$$\int ||\nabla\mathcal{F}_j^k(\mathbf{p})||^2 + ||\nabla\mathcal{F}_j^{(K+k)}(\mathbf{p})||^2 d\mathbf{p}. \tag{7}$$

Since the number of unknowns is large in (6) (i.e. two times the number of pixels in I) we use conjugate gradient iterations to solve the related normal equations [16]. The initial guess for the iterative solver is obtained from the current estimate of $\mathcal{F}_j$. Since we initially start from a constant photometric transformation field (Sect. 3.2) and our update step aims at minimizing (6), thereby increasing the likelihood $P_1(\mathbf{p}|\hat{I}_j)$ in (3), it is clear that the energy (2) is decreased in the update process.

3.6 Updating the Geometric Transformations

The geometric transformations $\mathcal{G}_j$ are updated by optical flow [17]. Given S and $\mathcal{F}_j$ and the current estimate of $\mathcal{G}_j$, we generate the modeled image $\hat{I}_j$ by (1) and determine the optical flow from I to $\hat{I}_j$ in a domain which encloses the regions currently labeled to layer j [17] (color images are transformed to grayscale before computation). Then, the determined optical flow is used for updating $\mathcal{G}_j$. However, the update is finally accepted only if it decreases the energy (2).

3.7 Updating the Segmentation

The segmentation is performed by minimizing the energy function (2) over different labelings S using graph cut techniques [15]. The exact global minimum is found only in the two-label case and in the multi-label case efficient approximate minimization is produced by the α-expansion algorithm of [15]. Here the computations were performed using the implementations provided by the authors of [15,18,19,20].

4 Experiments

Experimental results are illustrated in Figs. 3 and 4. The example in Fig. 3 shows the first and last frame from a classical benchmark sequence [2,4], which contains three different planar motion layers. Good motion segmentation results have been obtained

Fig. 3. Left: two images and the final three-layer segmentation. Middle: the grouped matches generating 12 tentative layers. Right: the layers of the first image mapped to the second.

Fig. 4. Five examples. The bottom row illustrates the geometric and photometric registrations.

for this sequence by using all the frames [2,6,9]. However, if the intermediate frames are not available the problem is harder and it has been studied in [1]. Our results in Fig. 3 are comparable to [1]. Nevertheless, compared to [1], our approach has better applicability in cases where (a) only a very small fraction of keypoint matches is correct, and (b) the motion can not be described with a low-parametric model. Such cases are illustrated in Figs. 1 and 4.

The five examples in Fig. 4 show motion segmentation results for scenes containing non-planar objects, non-uniform illumination variations, multiple objects, and deforming surfaces. For example, the recovered geometric registrations illustrate the 3D shape of the toy lion and the car as well as the bending of the magazines. In addition, the varying illumination of the toy lion is correctly recovered (the shadow on the backside of the lion is not as strong as elsewhere). On the other hand, if the changes of illumination are too abrupt or if some primary colors are not present in the initial layer (implying that the estimated transformation may not be accurate for all colors), it is difficult to achieve perfect segmentation. For example, in the last column of Fig. 4, the letter "F" on the car, where the intensity is partly saturated, is not included in the car layer.

Besides illustrating the capabilities and limitations of the proposed method, the results in Fig. 4 also suggest some topics for future improvements. Firstly, improving the initial verification stage might give a better discrimination between the correct and incorrect correspondences (the magenta region in the last example is incorrect). Secondly, some postprocessing method could be used to join distant coherently moving segments if desired (the green and cyan region in the fourth example belong to the same rigid object). Thirdly, if the change in scale is very large, more careful modeling of the sampling rate effects might improve the accuracy of registration and segmentation (magazines).

5 Conclusion

This paper describes a dense layer-based two-view motion segmentation method, which automatically detects and segments the common regions from the two images and provides the related geometric and photometric registrations. The method is robust to extensive background clutter and is able to recover the correct segmentation and registration of the imaged surfaces in challenging viewing conditions (including uniform image regions where mere match propagation can not provide accurate segmentation). Importantly, in the proposed approach both the initialization stage and the dense segmentation stage can deal with deforming surfaces and spatially varying lighting conditions, unlike in the previous approaches. Hence, in the future, it might be interesting to study whether the techniques can be extended to multi-frame image sequences.

References

1. Wills, J., Agarwal, S., Belongie, S.: A feature-based approach for dense segmentation and estimation of large disparity motion. IJCV 68, 125–143 (2006)
2. Wang, J.Y.A., Adelson, E.H.: Representing moving images with layers. IEEE Transactions on Image Processing 3(5), 625–638 (1994)
3. Ferrari, V., Tuytelaars, T., Van Gool, L.: Simultaneous object recognition and segmentation from single or multiple model views. IJCV 67, 159–188 (2006)
4. Weiss, Y.: Smoothness in layers: Motion segmentation using nonparametric mixture estimation. In: CVPR (1997)
5. Torr, P.H.S., Szeliski, R., Anandan, P.: An integrated bayesian approach to layer extraction from image sequences. TPAMI 23(3), 297–303 (2001)
6. Xiao, J., Shah, M.: Motion layer extraction in the presence of occlusion using graph cuts. TPAMI 27, 1644–1659 (2005)
7. Simon, I., Seitz, S.M.: A probabilistic model for object recognition, segmentation, and non-rigid correspondence. In: CVPR (2007)
8. Kannala, J., Rahtu, E., Brandt, S.S., Heikkilä, J.: Object recognition and segmentation by non-rigid quasi-dense matching. In: CVPR (2008)
9. Kumar, M.P., Torr, P.H.S., Zisserman, A.: Learning layered motion segmentations of video. IJCV 76, 301–319 (2008)
10. Jackson, J.D., Yezzi, A.J., Soatto, S.: Dynamic shape and appearance modeling via moving and deforming layers. IJCV 79, 71–84 (2008)
11. Lowe, D.: Distinctive image features from scale invariant keypoints. IJCV 60, 91–110 (2004)
12. Donato, G., Belongie, S.: Approximate thin plate spline mappings. In: Heyden, A., Sparr, G., Nielsen, M., Johansen, P. (eds.) ECCV 2002. LNCS, vol. 2352, pp. 21–31. Springer, Heidelberg (2002)

13. Vedaldi, A., Soatto, S.: Local features, all grown up. In: CVPR (2006)
14. Čech, J., Matas, J., Perd'och, M.: Efficient sequential correspondence selection by cosegmentation. In: CVPR (2008)
15. Boykov, Y., Veksler, O., Zabih, R.: Fast approximate energy minimization via graph cuts. TPAMI 23(11), 1222–1239 (2001)
16. Hansen, P.C.: Rank-Deficient and Discrete Ill-Posed Problems. SIAM, Philadelphia (1998)
17. Horn, B.K.P., Schunk, B.G.: Determining optical flow. Artificial Intelligence (1981)
18. Boykov, Y., Kolmogorov, V.: An experimental comparison of min-cut/max-flow algorithms for energy minimization in vision. TPAMI 26(9), 1124–1137 (2004)
19. Kolmogorov, V., Zabih, R.: What energy functions can be minimized via graph cuts? TPAMI 26(2), 147–159 (2004)
20. Bagon, S.: Matlab wrapper for graph cut (2006), http://www.wisdom.weizmann.ac.il/~bagon

A Two-Phase Segmentation of Cell Nuclei Using Fast Level Set-Like Algorithms

Martin Maška[1], Ondřej Daněk[1], Carlos Ortiz-de-Solórzano[2], Arrate Muñoz-Barrutia[2], Michal Kozubek[1], and Ignacio Fernández García[2]

[1] Centre for Biomedical Image Analysis, Faculty of Informatics
Masaryk University, Brno, Czech Republic
xmaska@fi.muni.cz
[2] Center for Applied Medical Research (CIMA)
University of Navarra, Pamplona, Spain

Abstract. An accurate localization of a cell nucleus boundary is inevitable for any further quantitative analysis of various subnuclear structures within the cell nucleus. In this paper, we present a novel approach to the cell nucleus segmentation in fluorescence microscope images exploiting the level set framework. The proposed method works in two phases. In the first phase, the image foreground is separated from the background using a fast level set-like algorithm by Nilsson and Heyden [1]. A binary mask of isolated cell nuclei as well as their clusters is obtained as a result of the first phase. A fast topology-preserving level set-like algorithm by Maška and Matula [2] is applied in the second phase to delineate individual cell nuclei within the clusters. The potential of the new method is demonstrated on images of DAPI-stained nuclei of a lung cancer cell line A549 and promyelocytic leukemia cell line HL60.

1 Introduction

Accurate segmentation of cells and cell nuclei is crucial for the quantitative analyses of microscopic images. Measurements related to counting of cells and nuclei, their morphology and spatial organization, and also a distribution of various subcellular and subnuclear components can be performed, provided the boundary of individual cells and nuclei is known. The complexity of the segmentation task depends on several factors. In particular, the procedure of specimen preparation, the acquisition system setup, and the type of cells and their spatial arrangement influence the choice of the segmentation method to be applied.

The most commonly used cell nucleus segmentation algorithms are based on thresholding [3,4] and region-growing [5,6] approaches. Their main advantage consists in the automation of the entire segmentation process. However, these methods suffer from oversegmentation and undersegmentation, especially when the intensities of the nuclei vary spatially or when the boundaries contain weak edges.

Ortiz de Solórzano et al. [7] proposed a more robust approach exploiting the geodesic active contour model [8] for the segmentation of fluorescently labeled

A.-B. Salberg, J.Y. Hardeberg, and R. Jenssen (Eds.): SCIA 2009, LNCS 5575, pp. 390–399, 2009.

cell nuclei and membranes in two-dimensional images. The method needs one initial seed to be defined in each nucleus. The sensitivity to proper initialization and, in particular, the computational demands of the narrow band algorithm [9] severely limit the use of this method in unsupervised real-time applications. However, the research addressed to the application of partial differential equations (PDEs) to image segmentation has been extensive, popular, and rather successful in recent years. Several fast algorithms [10,1,11] for the contour evolution were developed recently and might serve as an alternative to common cell nucleus segmentation algorithms.

The main motivation of this work is the need for a robust, as automatic as possible, and fast method for the segmentation of cell nuclei. Our input image data typically contains both isolated as well as touching nuclei with different average fluorescent intensities in a variable but often bright background. Furthermore, the intensities within the nuclei are significantly varying and their boundaries often contain holes and weak edges due to the non-uniformity of chromatin organization as well as abundant occurrence of nucleoli within the nuclei. Since the basic techniques, such as thresholding or region-growing, produce inaccurate results on this type of data, we present a novel approach to the cell nucleus segmentation in 2D fluorescence microscope images exploting the level set framework. The proposed method works in two phases. In the first phase, the image foreground is separated from the background using a fast level set-like algorithm by Nilsson and Heyden [1]. A binary mask of isolated cell nuclei as well as their clusters is obtained as a result of the first phase. A fast topology-preserving level set-like algorithm by Maška and Matula [2] is applied in the second phase to delineate individual cell nuclei within the clusters. We demonstrate the potential of the new method on images of DAPI-stained nuclei of a lung cancer cell line A549 and promyelocytic leukemia cell line HL60.

The organization of the paper is as follows. Section 2 shortly reviews the basic principle of the level set framework. The properties of input image data are presented in Section 3. Section 4 describes our two-phase approach to the cell nucleus segmentation. Section 5 is devoted to experimental results of the proposed method. We conclude the paper with discussion and suggestions for future work in Section 6.

2 Level Set Framework

This section is devoted to the level set framework. First, we briefly describe its basic principle, advantages, and also disadvantages. Second, a short review of fast approximations aimed at speeding up the basic framework is presented. Finally, we briefly discuss the topological flexibility of this framework.

Implicit active contours [12,8] have been developed as an alternative to parametric snakes [13]. Their solution is usually carried out using the level set framework [14], where the contour is represented implicitly as the zero level set (also called *interface*) of a scalar, higher-dimensional function ϕ. This representation has several advantages over the parametric one. In particular, it avoids parametrization problems, the topology of the contour is handled inherently, and the extension into higher dimensions is straightforward.

The contour evolution is governed by the following PDE:

$$\phi_t + F|\nabla\phi| = 0\,, \tag{1}$$

where F is an appropriately chosen speed function that describes the motion of the interface in the normal direction. A basic PDE-based solution using an explicit finite difference scheme results in a significant computational burden limiting the use of this approach in near real-time applications.

Many approximations, aimed at speeding up the basic level set framework, have been proposed in last two decades. They can be divided into two groups. First, methods based on the additive operator splittings scheme [15,16] have emerged to decrease the time step restriction. Therefore, a considerable lower number of iterations has to be performed to obtain the final contour in contrast to standard explicit scheme. However, these methods require maintaining the level set function in the form of signed distance function that is computationally expensive. Second, since one is usually interested in the single isocontour – the interface – in the context of image segmentation, other methods have been suggested to minimize the number of updates of the level set function ϕ in each iteration, or even to approximate the contour evolution in a different way. These include the narrow band [9], sparse-field [17], or fast marching method [10]. Another interesting approaches based on a pointwise scheduled propagation of the implicit contour can be found in the work by Deng and Tsui [18] or Nilsson and Heyden [1]. We also refer the reader to the work by Shi and Karl [11].

The topological flexibility of the evolving implicit contour is a great benefit since it allows to detect several objects simultaneously without any a priori knowledge. However, in some applications this flexibility is not desirable. For instance, when the topology of the final contour has to coincide with the known topology of the desired object (e.g. brain segmentation), or when the final shape must be homeomorphic to the initial one (e.g. segmentation of two touching nuclei starting with two separated contours, each labeling exactly one nucleus). Therefore, imposing topology-preserving constraints on evolving implicit contours is often more convenient than including additional postprocessing steps. We refer the reader to the work by Maška and Matula [2], and references therein for further details on this topic.

3 Input Data

The description and properties of two different image data sets that have been used for our experiments (see Sect. 5) are outlined in this section.

The first set consists of 10 images (16-bit grayscale, $1392 \times 1040 \times 40$ voxels) of DAPI-stained nuclei of a lung cancer cell line A549. The images were acquired using a conventional fluorescence microscope and deconvolved using the Maximum Likelihood Estimation algorithm provided by the Huygens software (Scientific Volume Imaging BV, Hilversum, The Netherlands). They typically contain both

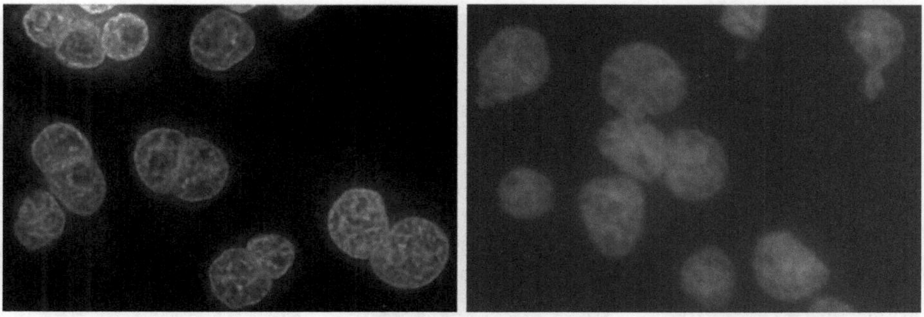

Fig. 1. Input image data. Left: An example of DAPI-stained nuclei of a lung cancer cell line A549. Right: An example of DAPI-stained nuclei of a promyelocytic leukemia cell line HL60.

isolated as well as touching, bright and dark nuclei with bright background in their surroundings originating from fluorescence coming from non-focal planes and from reflections of the light coming from the microscope glass slide surface. Furthermore, the intensities within the nuclei are significantly varying and their boundaries often contain holes and weak edges due to the non-uniformity of chromatin organization and abundant occurrence of nucleoli within the nuclei.

To demonstrate the potential of the proposed method (at least its second phase) on a different type of data, the second set consists of 40 images (8-bit grayscale, $1300 \times 1030 \times 60$ voxels) of DAPI-stained nuclei of a promyelocytic leukemia cell line HL60. The images were acquired using a confocal fluorescence microscope and typically contain isolated as well as clustered nuclei with just slightly varying intensities within them.

Since we presently focus only on the 2D case, 2D images (Fig. 1) were obtained as maximal projections of the 3D ones to the xy plane.

4 Proposed Approach

In this section, we describe the principle of our novel approach to cell nucleus segmentation. In order to cope better with the quality of input image data (see Sect. 3), the segmentation process is performed in two phases. In the first phase, the image foreground is separated from the background to obtain a binary mask of isolated nuclei and their clusters. The boundary of each nucleus within the previously identified clusters is found in the second phase.

4.1 Background Segmentation

The first phase is focused on separating the image foreground from the background. To achieve high-quality results during further analysis, we start with preprocessing of input image data. A white top-hat filter with a large circular structuring element is applied to eliminate bright background (Fig. 2a) in the

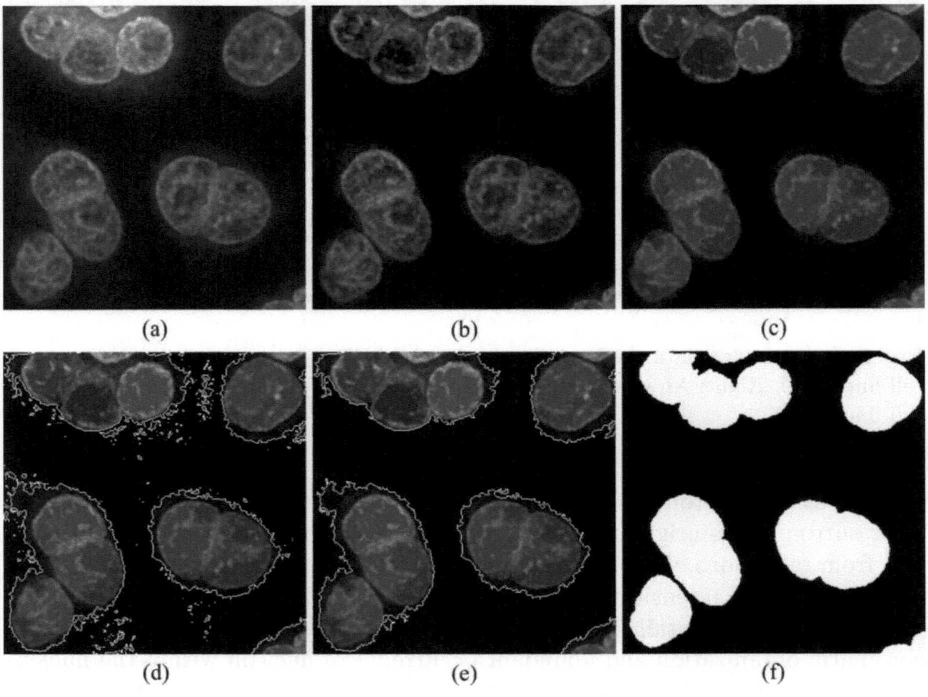

Fig. 2. Background segmentation. (a) An original image. (b) The result of a white top-hat filtering. (c) The result of a hole filling algorithm. (d) The initial interface defined as the boundary of foreground components obtained by applying the unimodal thresholding. (e) The initial interface when the small components are filtered out. (f) The final binary mask of the image foreground.

nucleus surroundings, as illustrated in Fig. 2b. Due to frequent inhomogeneity in the nucleus intensities, the white top-hat filtering might result in dark holes within the nuclei. This undesirable effect is reduced (Fig. 2c) by applying a hole filling algorithm based on a morphological reconstruction by erosion.

Segmentation of a preprocessed image I is carried out using the level set framework. A solution of a PDE related to the geodesic active contour model [8] is exploited for this purpose. The speed function F is defined as

$$F = g_I(c + \varepsilon\kappa) + \beta \cdot \nabla P \cdot n \ . \quad (2)$$

The function $g_I = \frac{1}{1+|\nabla G_\sigma * I|}$ is a strictly decreasing function that slows down the interface speed as it approaches edges in a smoothed version of I. The smoothing is performed by convolving the image I with a Gaussian filter G_σ ($\sigma = 1.3$, radius $r = 3.0$). The constant c corresponds to the inflation (deflation) force. The symbol κ denotes the mean curvature that affects the interface smoothness. Its relative impact is determined by the constant ε. The last term $\beta \cdot \nabla P \cdot n$, where $P = |\nabla G_\sigma * I|$, β is a constant, and the symbol n denotes the normal to the interface, attracts the interface towards the edges in the smoothed version of I. We exploit the Nilsson and Heyden's algorithm [1], a fast approximation of the level set framework, for tracking the interface evolution.

To define an initial interface automatically, the boundary of foreground components, obtained by the unimodal thresholding, is used (Fig. 2d). It is important to notice that not every component has to be taken into account. The small components enclosing foreign particles like dust or other inpurities can be filtered out (Fig. 2e). The threshold

$$size_{min} = k \cdot size_{avg} \ , \qquad (3)$$

where $k \geq 1$ is a constant and $size_{avg}$ is an average component size (in pixels), ensures that only the largest components (denote them S) enclosing desired cell nuclei remain.

To prevent the interface from propagating inside a nucleus due to discontinuity of its boundary (see Fig. 3), we ommit the deflation force ($c = 0$) from (2).

Since the image data contains bright nuclei as well as dark ones, it is difficult to segment all the images accurately with the same value of β and ε. Instead of setting these parameters properly for each particular image, we perform two runs of the Nilsson and Heyden's algorithm that differ only in the parameter ε. The parameter β remains unchanged. In the first run, a low value of ε is applied to detect dark nuclei. In the case of bright ones, the evolving interface might be attracted to a brighter background in their surroundings as its intensity is often similar to the intensity of dark nuclei. To overcome such problem, a high value of ε is used in the second run to enforce the interface to pass through the brighter background (and obviously also through the dark nuclei) and detects the bright nuclei correctly. Finally, the results of both runs are combined together to obtain a binary mask M of the image foreground, as illustrated in Fig. 2f.

The number of performed iterations is considered as a stopping criterion. In each run, we conduct the same number of iterations determined as

$$N_1 = k_1 \cdot \sum_{s \in S} size(s) \ , \qquad (4)$$

where k_1 is a positive constant and $size(s)$ corresponds to the size (in pixels) of the component s.

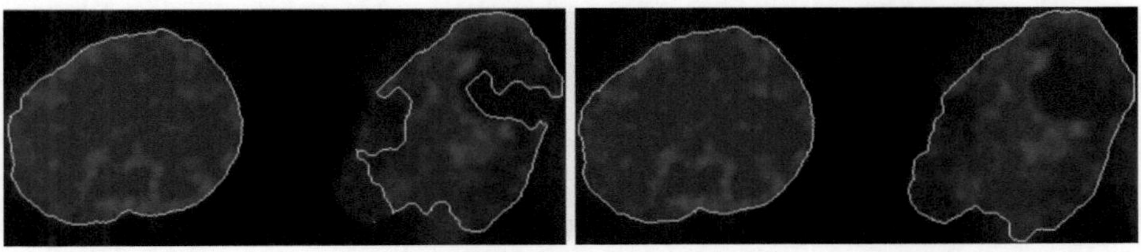

Fig. 3. The influence of the deflation force in (2). Left: The deflation force is applied ($c = -0.01$). Right: The deflation force is omitted ($c = 0$).

4.2 Cluster Separation

The second phase addresses the separation of touching nuclei detected in the first phase. The binary mask M is considered as the computational domain in this phase. Each component m of M is considered as a cluster and processed separately. Since the image preprocessing step degrades significantly the information within the nuclei, the original image data is processed in this phase.

The number of nuclei within the cluster m is computed first. A common approach based on finding peaks in a distance transform of m using an extended maxima transformation is exploited for this purpose. The number of peaks is established as the number of nuclei within the cluster m. If m contains just one peak (i.e. m corresponds to an isolated nucleus), its processing is over. Otherwise, the cluster separation is performed.

The peaks are considered as an initial interface that is evolved using a fast topology-preserving level set-like algorithm [2]. This algorithm integrates the Nilsson and Heyden's one [1] with the simple point concept from digital geometry to prevent the initial interface from changing its topology. Starting with separated contours (each labeling a different nucleus within the cluster m), the topology-preserving constraint ensures that the number of contours remains unchanged during the deformation. Furthermore, the final shape of each contour corresponds to the boundary of the nucleus that it labeled at the beginning.

Similarly to the first phase, (1) with the speed function (2) governs the contour evolution. In order to propagate the interface over the high gradients within the nuclei, a low value of β (approximately two orders of the magnitude lower than the value used in the first phase) has to be applied. As a consequence, the contour is stopped at the boundary of touching nuclei mainly due to the topology-preserving constraint. The use of a constant inflation force might, therefore, result in inaccurate segmentation results in the case of complex nucleus shape or when a smaller nucleus touches a larger one, as illustrated in Fig. 4. To overcome such complication, a position-dependent inflation force defined as a magnitude of the distance transform of m is applied. This ensures that the closer to the nucleus boundary the interface is, the lower is the inflation force.

The number of performed iterations reflecting the size of the cluster m:

$$N_2 = k_2 \cdot size(m) \ , \tag{5}$$

where k_2 is a positive constant, is considered again as a stopping criterion.

5 Results and Discussion

In this section, we present the results of the proposed method on both image data sets and discuss briefly the choice of parameters as well as its limitations. The experiments have been performed on a common workstation (Intel Core2 Duo 2.0 GHz, 2 GB RAM, Windows XP Professional).

The parameters k, k_1, k_2, β, and ε were empirically set. Their values used in each phase are listed in Table 1. As expected, only β, which defines the sensitivity of the interface attraction force on the image gradient, had to be carefully set

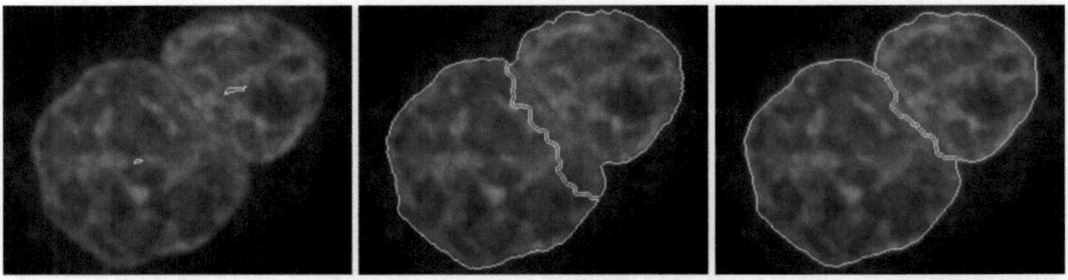

Fig. 4. Cluster separation. Left: The original image containing initial interface. Centre: The result when a constant inflation force $c = 1.0$ is applied. Right: The result when a position-dependent inflation force is applied.

according to the properties of specific image data. It is also important to notice that the computational time of the second phase mainly depends on the number and shape of clusters in the image, since the isolated nuclei are not further processed in this phase. Regarding the images of HL60 cell nuclei, the first phase of our approach was not used due to a good quality of image data. Instead, a low-pass filtering followed by the Otsu thresholding were applied to obtain the foreground mask. Subsequently, the cluster separation was performed using the second phase of our method. Some examples of the final segmentation are illustrated in Fig. 5. To evaluate an accuracy of the proposed method, a measure *Acc* defined as a product of sensitivity (*Sens*) and specificity (*Spec*) was applied. A manual segmentation done by an expert was considered as a ground truth. The product was computed for each nucleus and averaged over all images of a cell line. The results are listed in Table 1.

Our method, as it was described in Sect. 4, is directly applicable to the segmentation of 3D images. However, its main limitation stems from the computation of the number of nuclei within a cluster and initialization of the second phase. The approach based on finding the peaks of the distance transform is not well applicable to more complex clusters that appear, for instance, in thick tissue sections. A possible solution might consist in defining the initial interface either interactively by a user or as a skeleton of each particular nucleus. The former is computationally expensive in the case of processing a huge amount of data. On the other hand, finding the skeleton of each particular nucleus is not trivial in more complex clusters. This problem will be addressed in future work.

Table 1. The parameters, average computation times and accuracy of our method. The parameter that is not applicable in a specific phase is denoted by the symbol –.

Cell line	Phase	k	k_1	k_2	ε		β	Time	*Sens*	*Spec*	*Acc*
A549	1	2	1.8	–	0.15	0.6	$0.16 \cdot 10^{-5}$	$5.8\,s$			
	2	–	–	1.5	0.3		$0.18 \cdot 10^{-7}$	$3.2\,s$	96.37%	99.97%	96.34%
HL60	1	–	–	–	–		–	$< 1\,s$			
	2	–	–	1.5	0.3		$0.08 \cdot 10^{-4}$	$2.9\,s$	95.91%	99.95%	95.86%

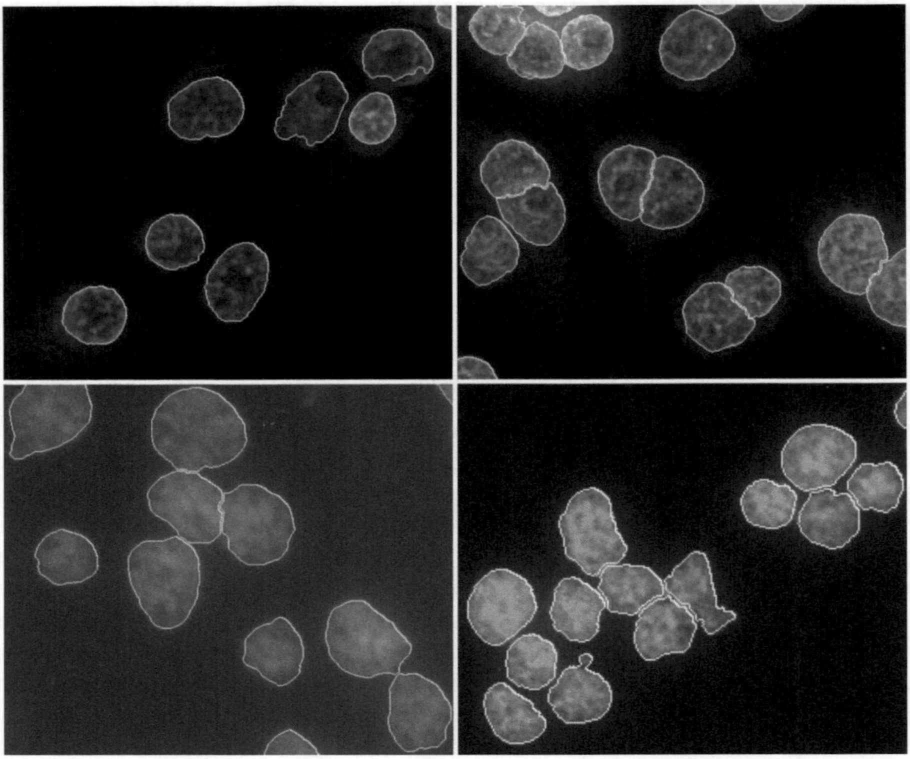

Fig. 5. Segmentation results. Upper row: The final segmentation of the A549 cell nuclei. Lower row: The final segmentation of the HL60 cell nuclei.

6 Conclusion

In this paper, we have presented a novel approach to the cell nucleus segmentation in fluorescence microscopy demonstrated on examples of images of a lung cancer cell line A549 as well as promyelocytic leukemia cell line HL60. The proposed method exploits the level set framework and works in two phases. In the first phase, the image foreground is separated from the background using a fast level set-like algorithm by Nilsson and Heyden. A binary mask of isolated cell nuclei as well as their clusters is obtained as a result of the first phase. A fast topology-preserving level set-like algorithm by Maška and Matula is applied in the second phase to delineate individual cell nuclei within the clusters. Our results show that the method succeeds in delineating each cell nucleus correctly in almost all cases. Furthermore, the proposed method can be reasonably used in near real-time applications due to its low computational time demands. A formal quantitative evaluation involving, in particular, the comparison of our approach with watershed-based as well as graph-cut-based methods on both real and simulated image data will be addressed in future work. We also intend to adapt the method to more complex clusters that appear in thick tissue sections.

Acknowledgments. This work has been supported by the Ministry of Education of the Czech Republic (Projects No. MSM-0021622419, No. LC535 and No. 2B06052). COS, AMB, and IFG were supported by the Marie Curie IRG Program (grant number MIRG CT-2005-028342), and by the Spanish Ministry of Science and Education, under grant MCYT TEC 2005-04732 and the Ramon y Cajal Fellowship Program.

References

1. Nilsson, B., Heyden, A.: A fast algorithm for level set-like active contours. Pattern Recognition Letters 24(9-10), 1331–1337 (2003)
2. Maška, M., Matula, P.: A fast level set-like algorithm with topology preserving constraint. In: CAIP 2009 (March 2009) (submitted)
3. Netten, H., Young, I.T., van Vliet, L.J., Tanke, H.J., Vrolijk, H., Sloos, W.C.R.: Fish and chips: Automation of fluorescent dot counting in interphase cell nuclei. Cytometry 28(1), 1–10 (1997)
4. Gué, M., Messaoudi, C., Sun, J.S., Boudier, T.: Smart 3D-fish: Automation of distance analysis in nuclei of interphase cells by image processing. Cytometry 67(1), 18–26 (2005)
5. Malpica, N., Ortiz de Solórzano, C., Vaquero, J.J., Santos, A., Vallcorba, I., García-Sagredo, J.M., del Pozo, F.: Applying watershed algorithms to the segmentation of clustered nuclei. Cytometry 28(4), 289–297 (1997)
6. Wählby, C., Sintorn, I.M., Erlandsson, F., Borgefors, G., Bengtsson, E.: Combining intensity, edge and shape information for 2D and 3D segmentation of cell nuclei in tissue sections. Journal of Microscopy 215(1), 67–76 (2004)
7. Ortiz de Solórzano, C., Malladi, R., Leliévre, S.A., Lockett, S.J.: Segmentation of nuclei and cells using membrane related protein markers. Journal of Microscopy 201(3), 404–415 (2001)
8. Caselles, V., Kimmel, R., Sapiro, G.: Geodesic active contours. International Journal of Computer Vision 22(1), 61–79 (1997)
9. Chopp, D.: Computing minimal surfaces via level set curvature flow. Journal of Computational Physics 106(1), 77–91 (1993)
10. Sethian, J.A.: A fast marching level set method for monotonically advancing fronts. Proceedings of the National Academy of Sciences 93(4), 1591–1595 (1996)
11. Shi, Y., Karl, W.C.: A real-time algorithm for the approximation of level-set-based curve evolution. IEEE Transactions on Image Processing 17(5), 645–656 (2008)
12. Caselles, V., Catté, F., Coll, T., Dibos, F.: A geometric model for active contours in image processing. Numerische Mathematik 66(1), 1–31 (1993)
13. Kass, M., Witkin, A., Terzopoulos, D.: Snakes: Active contour models. International Journal of Computer Vision 1(4), 321–331 (1987)
14. Osher, S., Fedkiw, R.: Level Set Methods and Dynamic Implicit Surfaces. Springer, New York (2003)
15. Goldenberg, R., Kimmel, R., Rivlin, E., Rudzsky, M.: Fast geodesic active contours. IEEE Transactions on Image Processing 10(10), 1467–1475 (2001)
16. Kühne, G., Weickert, J., Beier, M., Effelsberg, W.: Fast implicit active contour models. In: Van Gool, L. (ed.) DAGM 2002. LNCS, vol. 2449, pp. 133–140. Springer, Heidelberg (2002)
17. Whitaker, R.T.: A level-set approach to 3D reconstruction from range data. International Journal of Computer Vision 29(3), 203–231 (1998)
18. Deng, J., Tsui, H.T.: A fast level set method for segmentation of low contrast noisy biomedical images. Pattern Recognition Letters 23(1-3), 161–169 (2002)

A Fast Optimization Method for Level Set Segmentation

Thord Andersson[1,3], Gunnar Läthén[2,3], Reiner Lenz[2,3], and Magnus Borga[1,3]

[1] Department of Biomedical Engineering, Linköping University
[2] Department of Science and Technology, Linköping University
[3] Center for Medical Image Science and Visualization (CMIV), Linköping University

Abstract. Level set methods are a popular way to solve the image segmentation problem in computer image analysis. A contour is implicitly represented by the zero level of a signed distance function, and evolved according to a motion equation in order to minimize a cost function. This function defines the objective of the segmentation problem and also includes regularization constraints. Gradient descent search is the de facto method used to solve this optimization problem. Basic gradient descent methods, however, are sensitive for local optima and often display slow convergence. Traditionally, the cost functions have been modified to avoid these problems. In this work, we instead propose using a modified gradient descent search based on resilient propagation (Rprop), a method commonly used in the machine learning community. Our results show faster convergence and less sensitivity to local optima, compared to traditional gradient descent.

Keywords: Image segmentation, level set method, optimization, gradient descent, Rprop, variational problems, active contours.

1 Introduction

In order to find objects such as tumors in medical images or roads in satellite images, an image segmentation problem has to be solved. One approach is to use calculus of variations. In this context, a contour parameterizes an energy functional defining the objective of the segmentation problem. The functional depends on properties of the image such as gradients, curvatures and intensities, as well as regularization terms, e.g. smoothing constraints. The goal is to find the contour which, depending on the formulation, maximizes or minimizes the energy functional. In order to solve this optimization problem, the gradient descent method is the de facto standard. It deforms an initial contour in the steepest (gradient) descent of the energy. The equations of motion for the contour, and the corresponding energy gradients, are derived using the Euler-Lagrange equation and the condition that the first variation of the energy functional should vanish at a (local) optimum. Then, the contour is evolved to convergence using these equations. The use of a gradient descent search commonly leads to problems with convergence to small local optima and slow/poor convergence in

A.-B. Salberg, J.Y. Hardeberg, and R. Jenssen (Eds.): SCIA 2009, LNCS 5575, pp. 400–409, 2009.

general. The problems are accentuated with noisy data or with a non-stationary imaging process, which may lead to varying contrasts for example. The problems may also be induced by bad initial conditions for certain applications. Traditionally, the energy functionals have been modified to avoid these problems by, for example, adding regularizing terms to handle noise, rather than to analyze the performance of the applied optimization method. This is however discussed in [1,2], where the metric defining the notion of steepest descent (gradient) has been studied. By changing the metric in the solution space, local optima due to noise are avoided in the search path.

In contrast, we propose using a modified gradient descent search based on *resilient propagation (Rprop)* [3][4], a method commonly used in the machine learning community. In order to avoid the typical problems of gradient descent search, *Rprop* provides a simple but effective modification which uses individual (one per parameter) adaptive step sizes and considers only the sign of the gradient. This modification makes Rprop more robust to local optima and avoids the harmful influence of the size of the gradient on the step size. The individual adaptive step sizes also allow for cost functions with very different behaviors along different dimensions because there is no longer a single step size that should fit them all. In this paper, we show how Rprop can be used for image segmentation using level set methods. The results show faster convergence and less sensitivity to local optima.

The paper will proceed as follows. In Section 2, we will describe gradient descent with Rprop and give an example of a representative behavior. Then, Section 3 will discuss the level set framework and how Rprop can be used to solve segmentation problems. Experiments, where segmentations are made using Rprop for gradient descent, are presented in Section 4 together with implementation details. In Section 5 we discuss the results of the experiments and Section 6 concludes the paper and presents ideas for future work.

2 Gradient Descent with Rprop

Gradient descent is a very common optimization method which appeal lies in the combination of its generality and simplicity. It can handle many types of cost functions and the intuitive approach of the method makes it easy to implement. The method always moves in the negative direction of the gradient, locally minimizing the cost function. The steps of gradient descent are also easy and fast to calculate since they only involve the first order derivatives of the cost function. Unfortunately, gradient descent is known to exhibit slow convergence and to be sensitive to local optima for many practical problems. Other, more advanced, methods have been invented to deal with the weaknesses of gradient descent, e.g. the methods of conjugate gradient, Newton, Quasi-Newton etc, see [5]. Rprop, proposed by the machine learning community [3], provides an intermediate level between the simplicity of gradient descent and the complexity of these more theoretically sophisticated variants.

Gradient descent may be expressed using a standard line search optimization:

$$\boldsymbol{x}_{k+1} = \boldsymbol{x}_k + \boldsymbol{s}_k \tag{1}$$

$$\boldsymbol{s}_k = \alpha_k \boldsymbol{p}_k \tag{2}$$

where $\boldsymbol{x}_k$ is the current iterate, $\boldsymbol{s}_k$ is the next step consisting of length α_k and direction $\boldsymbol{p}_k$. To guarantee convergence, it is often required that $\boldsymbol{p}_k$ be a descent direction while α_k gives a *sufficient decrease* in the cost function. A simple realization of this is gradient descent which moves in the steepest descent direction according to $\boldsymbol{p}_k = -\boldsymbol{\nabla} f_k$, where f is the cost function, while α_k satisfies the *Wolfe conditions* [5].

In standard implementations of steepest descent search, $\alpha_k = \alpha$ is a constant not adapting to the shape of the cost-surface. Therefore if we set it too small, the number of iterations needed to converge to a local optima may be prohibitive. On the other hand, a too large value of α may lead to oscillations causing the search to fail. The optimal α does not only depend on the problem at hand, but varies along the cost-surface. In shallow regions of the surface a large α may be needed to obtain an acceptable convergence rate, but the same value may lead to disastrous oscillations in neighboring regions with larger gradients or in the presence of noise. In regions with very different behaviors along different dimensions it may be hard to find an α that gives acceptable convergence performance.

The Resilient Propagation (Rprop) algorithm was developed [3] to overcome these inherent disadvantages of standard gradient descent using adaptive step-sizes $\boldsymbol{\Delta}_k$ called *update-values*. There is one update-value per dimension in $\boldsymbol{x}$, i.e. $\dim(\boldsymbol{x}_k) = \dim(\boldsymbol{\Delta}_k)$. However, the defining feature of Rprop is that the size of the gradient is never used, only the signs of the partial derivatives are considered in the update rule. There are other methods using both adaptive step-sizes and the size of the gradient, but the unpredictable behavior of the derivatives often counter the careful adaption of the step-sizes. Another advantage of Rprop, very important in practical use, is the robustness of its parameters; Rprop will work out-of-the-box in many applications using only the standard values of its parameters [6].

We will now describe the Rprop algorithm briefly, but for implementation details of Rprop we refer to [4]. For Rprop, we choose a search direction $\boldsymbol{s}_k$ according to:

$$\boldsymbol{s}_k = -\text{sign}\left(\boldsymbol{\nabla} f_k\right) * \boldsymbol{\Delta}_k \tag{3}$$

where $\boldsymbol{\Delta}_k$ is a vector containing the current update-values, a.k.a. *learning rates*, $*$ denotes elementwise multiplication and $\text{sign}(\cdot)$ the elementwise sign function. The individual update-value $\boldsymbol{\Delta}_k^i$ for dimension i is calculated according to the rule:

$$\boldsymbol{\Delta}_k^i = \begin{cases} \min\left(\boldsymbol{\Delta}_{k-1}^i \cdot \eta^+, \Delta_{\max}\right) & , \boldsymbol{\nabla}^i f_k \cdot \boldsymbol{\nabla}^i f_{k-1} > 0 \\ \max\left(\boldsymbol{\Delta}_{k-1}^i \cdot \eta^-, \Delta_{\min}\right) & , \boldsymbol{\nabla}^i f_k \cdot \boldsymbol{\nabla}^i f_{k-1} < 0 \\ \boldsymbol{\Delta}_{k-1}^i & , \boldsymbol{\nabla}^i f_k \cdot \boldsymbol{\nabla}^i f_{k-1} = 0 \end{cases} \tag{4}$$

where $\boldsymbol{\nabla}^i f_k$ denotes the partial derivative i in the gradient. Note that this is Rprop without backtracking as described in [4]. The update rule will accelerate the update-value with a factor η^+ when consecutive partial derivatives have the same sign, decelerate with the factor η^- if not. This will allow for greater steps in favorable directions, causing the rate of convergence to be increased while overstepping eventual local optima.

3 Energy Optimization for Segmentation

As discussed in the introduction, segmentation problems can be approached by using the calculus of variations. Typically, an energy functional is defined representing the objective of the segmentation problem. The functional is described in terms of the contour and the relevant image properties. The goal is to find a contour that represents a solution which, depending on the formulation, maximizes or minimizes the energy functional. These extrema are found using the Euler-Lagrange equation which is used to derive equations of motion, and the corresponding energy gradients, for the contour [7]. Using these gradients, a gradient descent search in contour space is commonly used to find a solution to the segmentation problem. Consider, for instance, the derivation of the *weighted region* (see [7]) described by the following functional:

$$E(C) = \iint_{\Omega_C} f(x,y)dxdy \tag{5}$$

where C is a 1D curve embedded in a 2D space, Ω_C is the region inside of C, and $f(x,y)$ is a scalar function. This functional is used to maximize some quantity given by $f(x,y)$ inside C. If $f(x,y) = 1$ for instance, the area will be maximized. Calculating the first variation of Eq. 5 yields the evolution equation:

$$\frac{\partial C}{\partial t} = -f(x,y)\boldsymbol{n} \tag{6}$$

where $\boldsymbol{n}$ is the curve normal. If we anew set $f(x,y) = 1$, this will give a constant flow in the normal direction, commonly known as the "balloon force".

The contour is often implicitly represented by the zero level of a time dependent signed distance function, known as the level set function. The *level set method* was introduced by Osher and Sethian [8] and includes the advantages of being parameter free, implicit and topologically adaptive. Formally, a contour C is described by $C = \{\boldsymbol{x} : \phi(\boldsymbol{x},t) = 0\}$. The contour C is evolved in time using a set of partial differential equations (PDEs). A motion equation for a parameterized curve $\frac{\partial C}{\partial t} = \gamma\boldsymbol{n}$ is in general translated into the level set equation $\frac{\partial \phi}{\partial t} = \gamma\,|\boldsymbol{\nabla}\phi|$, see [7]. Consequently, Eq. 6 gives the familiar level set equation:

$$\frac{\partial \phi}{\partial t} = -f(x,y)\,|\boldsymbol{\nabla}\phi| \tag{7}$$

3.1 Rprop for Energy Optimization Using Level Set Flow

When solving an image segmentation problem, we can represent the entire level set function (corresponding to the image) as one vector, $\phi(t_n)$. In order to perform a gradient descent search as discussed earlier, we can approximate the gradient as the finite difference between two time instances:

$$\nabla f(t_n) \approx \frac{\phi(t_n) - \phi(t_{n-1})}{\Delta t} \tag{8}$$

where $\Delta t = t_n - t_{n-1}$ and ∇f is the gradient of a cost function f as discussed in Section 2. Using the update values estimated by Rprop (as in Section 2), we can update the level set function:

$$s(t_n) = -\text{sign}\left(\frac{\widetilde{\phi}(t_n) - \phi(t_{n-1})}{\Delta t}\right) * \boldsymbol{\Delta}(t_n) \tag{9}$$

$$\phi(t_n) = \phi(t_{n-1}) + s(t_n) \tag{10}$$

where $*$ as before denotes elementwise multiplication. The complete procedure works as follows:

Procedure `UpdateLevelset`

1 Given the level set function $\phi(t_{n-1})$, compute the next (intermediate) time step $\widetilde{\phi}(t_n)$. This is performed by evolving ϕ according to a PDE (such as Eq. 7) using standard techniques (e.g. Euler integration).

2 Compute the approximate gradient by Eq. 8.

3 Compute a step $s(t_n)$ according to Eq. 9. This step effectively modifies the gradient direction by using the Rprop derived update values.

4 Compute the next time step $\phi(t_n)$ by Eq. 10. Note that this replaces the intermediate level set function computed in Step 1.

The procedure is very simple and can be used directly with any type of level set implementation.

4 Experiments

We will now evaluate our idea by solving two example segmentation tasks using a simple energy functional. Both examples use 1D curves in 2D images but our approach also supports higher dimensional contours, e.g. 2D surfaces in 3D volumes.

4.1 Implementation Details

We have implemented Rprop in Matlab as described in [4]. The level set algorithm has also been implemented in Matlab based on [9,10]. Some notable implementation details are:

- Any explicit or implicit time integration scheme can be used in Step 1. Due to its simplicity, we have used explicit Euler integration which might require several inner iterations in Step 1 to advance the level set function by Δt time units.

- The level set function is reinitialized (reset to a signed distance function) after Step 1 and Step 4. This is typically performed using the fast marching [11] or fast sweeping algorithms [12]. This is required for stable evolution in time due to the use of explicit Euler integration in Step 1.
- The reinitializations of the level set function can disturb the adaptation of the individual step sizes outside the contour, causing spurious "islands" close to the contour. In order to avoid them we set the maximum step size to a low value once the target function integral has converged:

$$\left(\iint_{\Omega_{C(t)}} f(x,y)dxdy - \iint_{\Omega_{C(t-k)}} f(x,y)dxdy\right) < 0 \tag{11}$$

where k denotes the time under which the target function integral should not have increased.

4.2 Weighted Region Based Flow

In order to test and evaluate our idea, we have used a simple energy functional to control the segmentation. It is based on a weighted region term (Eq. 5) combined with a penalty on curve length for regularization. The goal is to maximize:

$$E(C) = \iint_{\Omega_C} f(x,y)dxdy - \alpha \oint_C ds \tag{12}$$

where α is a regularization parameter adjusting the penalty of the curve length. The target function $f(x,y)$ is here the real part of a global phase image, derived from the original image using the method in [13]. This method uses quadrature filters [14] across multiple scales to generate a global phase image that represents line structures. The function $f(x,y)$ will have positive values on the inside of linear structures, negative on the outside, and zero on the edges. A level set PDE can be derived from Eq. 12 (see [7]) just as in section Section 3:

$$\frac{\partial \phi}{\partial t} = -f(x,y)\left|\nabla\phi\right| + \alpha\kappa\left|\nabla\phi\right| \tag{13}$$

where κ is the curvature of the contour.

We will now evaluate gradient descent with and without Rprop using Eq. 13 on a synthetic test image shown in Figure 1(a). The image illustrates a line-like structure with a local dip in contrast. This dip results in a local optimum in the contour space, see Figure 2, and will help us test the robustness of our method. We let the target function $f(x,y)$, see Figure 1(b), be the real part of the global phase image as discussed above. The bright and dark colors indicate positive and negative values respectively. Figure 2 shows the results after an ordinary gradient search has converged. We define convergence as $\left|\nabla f\right|_\infty < 0.03$ (using the L^∞-norm), with ∇f given in Eq. 8. For this experiment we used

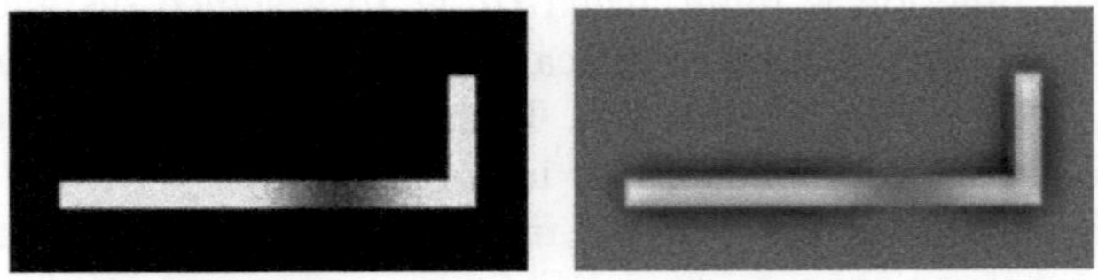

(a) Synthetic test image (b) Target function $f(x, y)$

Fig. 1. Synthetic test image spawning a local optima in the contour space

(a) $t = 0$ (b) $t = 40$ (c) $t = 100$ (d) $t = 170$ (e) $t = 300$ (f) $t = 870$

Fig. 2. Iterations without Rprop (Time units per iteration: $\Delta t = 5$)

(a) $t = 0$ (b) $t = 60$ (c) $t = 75$ (d) $t = 160$ (e) $t = 170$ (f) $t = 245$

Fig. 3. Iterations using Rprop (Time units per iteration: $\Delta t = 5$)

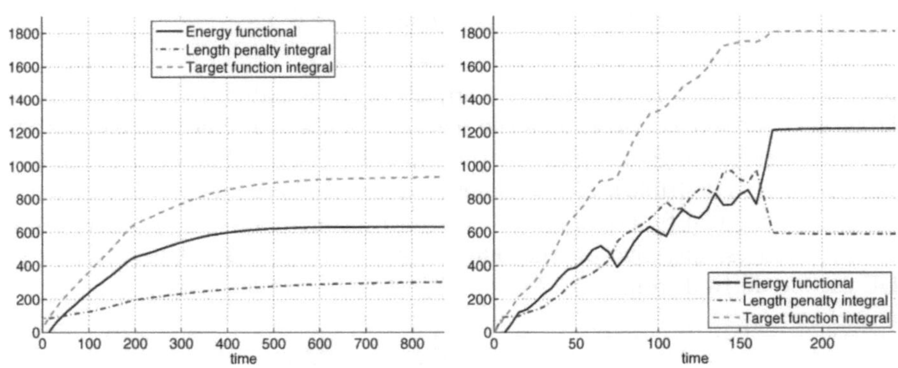

(a) Without Rprop (b) With Rprop

Fig. 4. Plots of energy functionals for synthetic test image in Figure 1(a)

parameters $\alpha = 0.7$ and we reinitialized the level set function every fifth iteration. For comparison, Figure 3 shows the results after running our method using default Rprop parameters $\eta^+ = 1.2, \eta^- = 0.5$, and other parameters set to $\Delta_0 = 2.5, s_{max} = 30$ and $\Delta t = 5$. Plots of the energy functional for both experiments are shown in Figure 4. Here, we plot the weighted area term and the length penalty term separately, to illustrate the balance between the two. Note that the functional without Rprop in Figure 4(a) is monotonically increasing as would be expected of gradient descent, while the functional with Rprop visits a number of local maxima during the search. The effect of setting the maximum

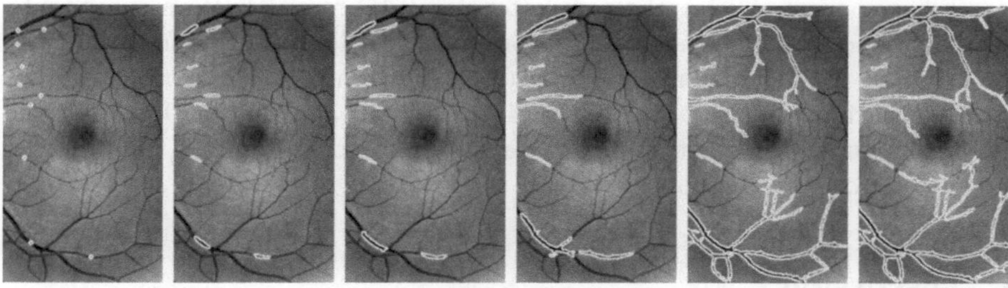

(a) $t = 0$ (b) $t = 20$ (c) $t = 40$ (d) $t = 100$ (e) $t = 500$ (f) $t = 970$

Fig. 5. Iterations without Rprop (Time units per iteration: $\Delta t = 10$)

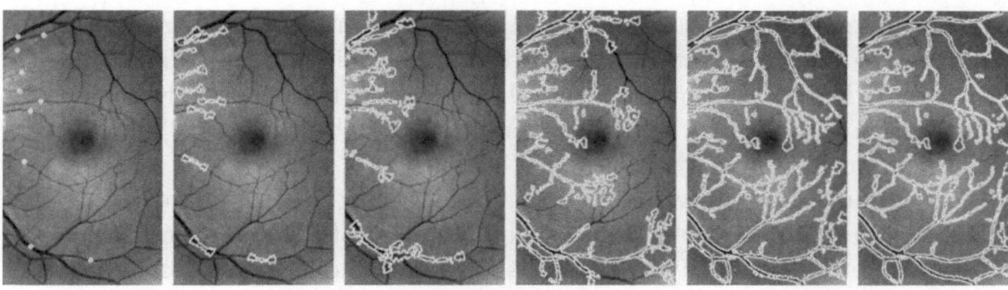

(a) $t = 0$ (b) $t = 40$ (c) $t = 80$ (d) $t = 200$ (e) $t = 600$ (f) $t = 990$

Fig. 6. Iterations using Rprop (Time units per iteration: $\Delta t = 10$)

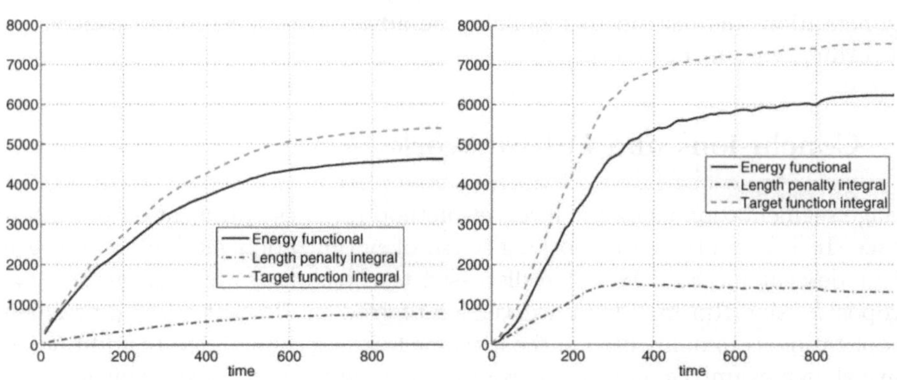

(a) Without Rprop (b) With Rprop

Fig. 7. Plots of energy functionals for the retinal image as seen in Figure 5

step size to a low value at $t = 160$, as discussed above (Eq. 11), effectively cancels the issue of spurious "islands" close to the contour in only two iterations. As a second test image we used a 458×265 retinal image from the DRIVE database [15], as seen in Figure 5. The target function $f(x, y)$ is, as before, the real part of the global phase image. Figure 5 shows the results after an ordinary gradient search has converged using the parameter $\alpha = 0.15$, reinitialization every tenth time unit and with the initial condition given in Figure 5(a). We have again used $|\nabla f|_\infty < 0.03$ as convergence criteria. If we instead use Rprop together with the parameters $\alpha = 0.15, \Delta_0 = 4, s_{max} = 10$ and $\Delta t = 10$, we get the result in Figure 6. The energy functionals are plotted in Figure 7, showing the convergence of both methods.

5 Discussion

The synthetic test image in Figure 1(a) spawns a local optimum in the contour space when we apply the set of parameters used in our first experiment. The standard gradient descent method converges as expected, see Figure 2, to this local optimum. Gradient descent with Rprop, however, accelerates along the linear structure due to the stable sign of the gradient in this area. The adaptive step-sizes of Rprop consequently grow large enough to overstep the local optimum. This is followed by a fast convergence to the global optimum. The progress of the method is shown in Figure 3.

Our second example evaluates our method using real data from a retinal image. The standard gradient descent method does not succeed to segment blood vessels where the signal to noise ratio is low. This is due to the local optima in these areas, induced by noise and blood vessels with low contrast. Gradient descent using Rprop, however, succeeds to segment practically all visible vessels, see Figure 6. Observe that the quality and accuracy of the segmentation have not been verified and is out of scope of this paper. The point of this experimental segmentation was instead to highlight the advantages of Rprop in contrast to the ordinary gradient descent.

6 Conclusions and Future Work

Image segmentation using the level set method involves optimization in contour space. In this context, the working horse of optimization methods is the gradient descent method. We have discussed the weaknesses of this method and proposed using Rprop, a modified version of gradient descent based on resilient propagation, commonly used in the machine learning community. In addition, we have shown examples on how the solution is improved by Rprop, which adapts its individual update values to the behavior of the cost surface. Using Rprop, the optimization gets less sensitive to local optima and the convergence rate is improved. In contrast to much of the previous work, we have improved the solution by changing the method of solving the optimization problem rather than modifying the energy functional.

Future work includes further study of the general optimization problem of image segmentation and verification of the segmentation quality in real applications. The issue of why the reinitializations disturb the adaptation of the step sizes also has to be studied further.

References

1. Charpiat, G., Keriven, R., Pons, J.P., Faugeras, O.: Designing spatially coherent minimizing flows for variational problems based on active contours. In: Tenth IEEE International Conference on Computer Vision, ICCV 2005, October 2005, vol. 2, pp. 1403–1408 (2005)
2. Sundaramoorthi, G., Yezzi, A., Mennucci, A.: Sobolev active contours. International Journal of Computer Vision 73(3), 345–366 (2007)
3. Riedmiller, M., Braun, H.: A direct adaptive method for faster backpropagation learning: The rprop algorithm. In: Proceedings of the IEEE International Conference on Neural Networks, pp. 586–591 (1993)
4. Riedmiller, M., Braun, H.: Rprop – description and implementation details. Technical report, Universitat Karlsruhe (1994)
5. Nocedal, J., Wright, S.J.: Numerical Optimization, 2nd edn. Springer, Heidelberg (2006)
6. Schiffmann, W., Joost, M., Werner, R.: Comparison of optimized backpropagation algorithms. In: Proc. of ESANN 1993, Brussels, pp. 97–104 (1993)
7. Kimmel, R.: Fast edge integration. In: Geometric Level Set Methods in Imaging, Vision and Graphics. Springer, Heidelberg (2003)
8. Osher, S., Sethian, J.A.: Fronts propagating with curvature-dependent speed: Algorithms based on Hamilton-Jacobi formulations. Journal of Computational Physics 79, 12–49 (1988)
9. Osher, S., Fedkiw, R.: Level Set and Dynamic Implicit Surfaces. Springer, New York (2003)
10. Peng, D., Merriman, B., Osher, S., Zhao, H.K., Kang, M.: A pde-based fast local level set method. Journal of Computational Physics 155(2), 410–438 (1999)
11. Sethian, J.: A fast marching level set method for monotonically advancing fronts. Proceedings of the National Academy of Science 93, 1591–1595 (1996)
12. Zhao, H.K.: A fast sweeping method for eikonal equations. Mathematics of Computation (74), 603–627 (2005)
13. Läthén, G., Jonasson, J., Borga, M.: Phase based level set segmentation of blood vessels. In: Proceedings of 19th International Conference on Pattern Recognition, IAPR, Tampa, FL, USA (December 2008)
14. Granlund, G.H., Knutsson, H.: Signal Processing for Computer Vision. Kluwer Academic Publishers, Netherlands (1995)
15. Staal, J., Abramoff, M., Niemeijer, M., Viergever, M., van Ginneken, B.: Ridge based vessel segmentation in color images of the retina. IEEE Transactions on Medical Imaging 23(4), 501–509 (2004)

Segmentation of Touching Cell Nuclei Using a Two-Stage Graph Cut Model

Ondřej Daněk[1], Pavel Matula[1], Carlos Ortiz-de-Solórzano[2], Arrate Muñoz-Barrutia[2], Martin Maška[1], and Michal Kozubek[1]

[1] Centre for Biomedical Image Analysis, Faculty of Informatics
Masaryk University, Brno, Czech Republic
xdanek2@fi.muni.cz
[2] Center for Applied Medical Research (CIMA)
University of Navarra, Pamplona, Spain

Abstract. Methods based on combinatorial graph cut algorithms received a lot of attention in the recent years for their robustness as well as reasonable computational demands. These methods are built upon an underlying Maximum a Posteriori estimation of Markov Random Fields and are suitable to solve accurately many different problems in image analysis, including image segmentation. In this paper we present a two-stage graph cut based model for segmentation of touching cell nuclei in fluorescence microscopy images. In the first stage voxels with very high probability of being foreground or background are found and separated by a boundary with a minimal geodesic length. In the second stage the obtained clusters are split into isolated cells by combining image gradient information and incorporated a priori knowledge about the shape of the nuclei. Moreover, these two qualities can be easily balanced using a single user parameter. Preliminary tests on real data show promising results of the method.

1 Introduction

Image segmentation is one of the most crucial tasks in fluorescence microscopy and image cytometry. Due to its importance many methods were proposed for solving this problem in the past. For simple cases basic techniques like thresholding [1], region growing [2] or watershed algorithm [2] are the most popular. However, when the data is severely degraded or contains complex structures requiring isolation of touching objects these simple methods are not powerful enough. Unfortunately, these scenarios are quite frequent. For this type of images more sophisticated methods have been designed in the past [3,4,5]. Their results although quite satisfactory, have some limitations: 1) in some cases suffer from over/undersegmentation, 2) need for human input, 3) require specific preparation of the biological samples.

The graph cut segmentation framework, first outlined by Boykov and Jolly [6,7], received a lot of attention in the recent years due to its robustness, reasonable computational demands and the ability to integrate visual cues, contextual information and topological constraints while offering several favourable characteristics

A.-B. Salberg, J.Y. Hardeberg, and R. Jenssen (Eds.): SCIA 2009, LNCS 5575, pp. 410–419, 2009.

like global optima [8], unrestricted topological properties and applicability to N-D problems. The core of their solution relies on modeling the segmentation process as a labelling problem with an associated energy function. This function is then optimized by finding a minimal cut in a specially designed graph. The method can be also formulated in terms of Maximum a Posteriori estimate of a Markov Random Field (MAP-MRF) [9,10].

In this paper we present a two-stage fully automated graph cut based model for segmentation of touching cell nuclei addressing most of the problems associated with the segmentation of fluorescence microscopy images. In the first stage background segmentation is performed. Voxels with very high probability of being foreground or background are located and separated by a boundary with a minimal geodesic length. In the second stage the obtained clusters are split into isolated cells by combining image gradient information and incorporated a priori knowledge about the shape of the nuclei. Moreover, these two qualities can be easily balanced using a single user parameter, allowing to control the placement of the dividing line in a desired way. This is a great advantage over the standard methods. Our algorithm can work on both 2-D and 3-D data sets. We demonstrate its potential on segmentation of 2-D cancer cell line images.

The organization of the paper is as follows. Graph cut segmentation framework is briefly reviewed in Section 2. A detailed description of our two-stage model is presented in Section 3 with experimental results in Section 4. In Section 5 we discuss the benefits and limitations of our method. Finally, we conclude our work in Section 6.

2 Graph Cut Segmentation Framework

In this section we briefly revisit the graph cut segmentation framework and related terms [6,7,11,10]. Because our method exploits both two-terminal and multi-terminal graph cuts we are going to describe the latter case which is a generalization of the former.

Consider an N-D image $\mathcal{I}$ consisting of set of voxels $\mathcal{P}$ and some neighbourhood system, denoted $\mathcal{N}$, containing all unordered pairs $\{p, q\}$ of neighbouring elements in $\mathcal{P}$. Further, let us consider a set of labels $\mathcal{L} = \{l_1, l_2, \ldots, l_n\}$ that should be assigned to each voxel in the image. Now, let $A = (A_1, \ldots, A_{|\mathcal{P}|})$ be a vector, where $A_i \in \{1, \ldots, n\}$ specifies the assignment of labels $\mathcal{L}$ to voxels $\mathcal{P}$.

The energy corresponding to a given labelling A is constructed as a linear combination of regional (data dependent) and boundary (smoothness) term and takes the form of:

$$E(A) = \lambda \cdot \sum_{p \in \mathcal{P}} R_p(A_p) + \sum_{(p,q) \in \mathcal{N}} B_{(p,q)} \cdot \delta_{A_p \neq A_q}, \qquad (1)$$

where $R_p(l)$ is the regional term evaluating the penalty for assigning voxel p to label l, $B_{(p,q)}$ is the boundary term evaluating the penalty for assigning neighbouring voxels p and q to different labels, δ is the Kronecker delta and λ is a weighting factor. The choice of the two evaluating functions R_p and $B_{(p,q)}$ is crucial for the segmentation. Based on the underlying MAP-MRF, the values of R_p are usually calculated as follows:

$$R_p(l) = -\log \Pr(p|l), \tag{2}$$

where $\Pr(p|l)$ is the probability that voxel p matches the label l. It is assumed that these probabilities are known a priori. However, in practice it is often hard to estimate them. The boundary term function can be naturaly expressed using the image contrast information [6,7] and can also approximate any Euclidean or Riemmannian metric [12]. The choice of $B_{(p,q)}$ for cell nuclei segmentation is discussed in Sect. 3.1.

Equation 1 can be minimized by finding a minimal cut in a specially designed graph (network). Construction of such graph is depicted in Fig. 1. In the first step a node is added for each voxel and these nodes are connected according to the neighbourhood $\mathcal{N}$. The edges connecting these nodes are denoted n-links and their weights (capacities) are determined by the function $B_{(p,q)}$. In the next step terminal nodes $\{t_1, t_2, \ldots, t_n\}$ corresponding to labels in $\mathcal{L}$ are added and each of them is connected with all nodes created in the first step. The resulting edges are called t-links and their capacities are given by the function R_p [10].

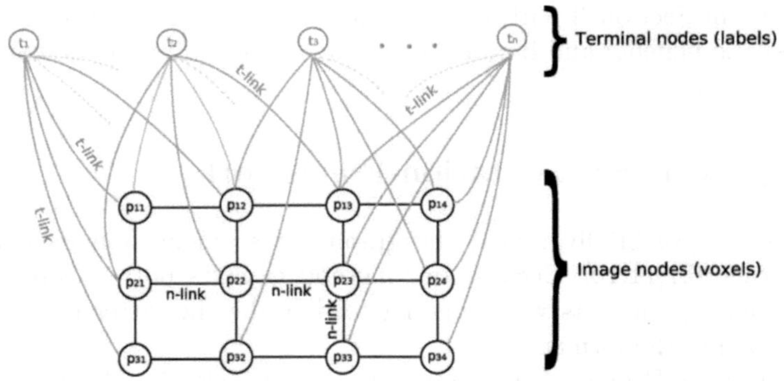

Fig. 1. Graph construction for given 2-D image, $\mathcal{N}_4$ neighbourhood system and set of terminals $\{t_1, \ldots, t_n\}$ (not all t-links are included for the sake of lucidity)

The minimal cut splits the graph into disjoint components $C_1, \ldots, C_n$, such that t_i lies in C_i for all $i \in \{1, \ldots, n\}$ and the sum of capacities of the removed edges is minimal. Consequently, every voxel receives the label of the terminal node in its component. In case of only two labels (terminals) the minimal cut can be found effectively in polynomial time using one of the well-known max-flow algorithms [11]. Unfortunately, for more than two terminals the problem is NP-complete [13] and an approximation of the minimal cut is calculated [10]. In this framework it is also possible to set up hard constraints in an elegant way. A binding of voxel p to a chosen label $\hat{l}$ is done by setting $R_p(l \neq \hat{l}) = \infty$ (refer to [7] for implementation details).

3 Cell Nuclei Segmentation

In this section we are going to give a detailed description of our fully automated two-stage graph cut model for segmentation of touching cell nuclei. The images that we cope with are acquired using fluorescence microscopy, meaning they are blurred, noisy and low contrast. They contain bright objects of mostly spherical shape on a dark background. Also the nuclei are often tightly packed and form clusters with indistinctive frontiers. Moreover, the interior of the nuclei can be greatly non-homogeneous and can contain dark holes incised into the nucleus boundary (caused by nucleoli, non-uniformity of chromatin organization or imperfect staining). See Sect. 4 for examples of such data.

In the first stage of our method foreground/background segmentation is performed, while in the second stage individual cells are identified in the obtained cell clusters and separated. The algorithm can work on both 2-D and 3-D data sets.

3.1 Background Segmentation

In this stage we are interested in binary labelling of the voxels with either a foreground or background label. The voxels that receive the foreground label are then treated as cluster masks and are separated into individual nuclei in the second stage. Because we deal with binary labelling only, the standard two-terminal graph cut algorithm [7] together with fast optimization methods [11] can be used. To obtain correct segmentation of the background, functions $B_{(p,q)}$ and R_p in (1) have to be set properly.

As the choice for $B_{(p,q)}$ we suggest the Riemmanian metric based edge capacities proposed in [12]. The equations in [12] can be simplified to the following form (assumming p and q are voxel coordinates):

$$B_{(p,q)} = \frac{\|q-p\|^2 \cdot \Delta\Phi \cdot g(p)}{2 \cdot \left[g(p) \cdot \|q-p\|^2 + (1-g(p)) \cdot \left\langle q-p, \frac{\nabla I_p}{|\nabla I_p|} \right\rangle^2 \right]^{\frac{3}{2}}}, \tag{3}$$

where $\Delta\Phi$ is $\frac{\pi}{4}$ for 8-neighbourhood and $\frac{\pi}{2}$ for 4-neighbourhood system respectively, $\langle\cdot\rangle$ denotes the dot product, ∇I_p is image gradient in voxel p and

$$g(p) = \exp\left(-\frac{|\nabla I_p|^2}{2\sigma^2}\right), \tag{4}$$

with σ being estimated as the average gradient magnitude in the image. Note that this equation applies to the 2-D case and that it is slightly different for 3-D [12]. It is also advisable to smooth the input image (e.g. using a Gaussian filter) before calculating the capacities.

Setting the capacities of t-links is the tricky part of this stage. In most approaches [5] homogeneous interior of the nuclei is assumed, allowing some simplifications of the algorithms. While this may be true in some situations, often it is not, as mentioned before. Hence, it is really hard to estimate the probability of the voxel being foreground or background based solely on its intensity. For example, the bright voxels among the cell nuclei in the top cluster in Fig. 2 are part of the background. To avoid introduction of false information into the model we suggest to stick to hard constraints only. We place them into voxels with very high probability of being background or foreground and ignore the intensity information elsewhere.[1] To find such voxels in the image we perform bilevel histogram analysis, find the two peaks corresponding to background and foreground and take the centres of these two peaks as our background/foreground thresholds. For voxels with intensity below the background threshold (black pixels in Fig. 2b) the corresponding capacity of the t-link going to background terminal is set to ∞ and analogously for voxels with intensity above the foreground threshold (white pixels in Fig. 2b). Remaining voxels (grey pixels in Fig. 2b) are left without any affiliation and both their t-link capacities are set to zero. As a consequence, λ value in (1) is irrelevant in this situation.

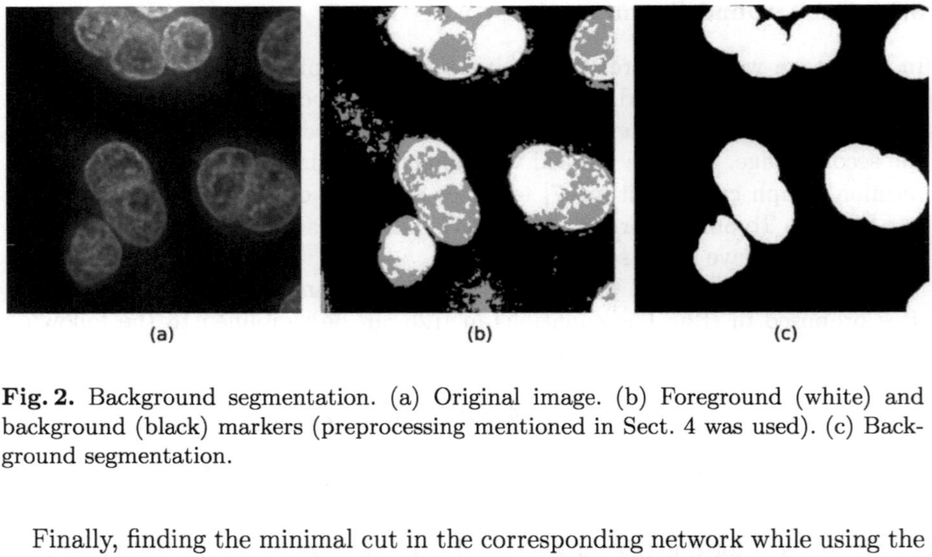

Fig. 2. Background segmentation. (a) Original image. (b) Foreground (white) and background (black) markers (preprocessing mentioned in Sect. 4 was used). (c) Background segmentation.

Finally, finding the minimal cut in the corresponding network while using the capacities described in this subsection gives us the background segmentation, that is shown in Fig. 2c. The result is a segmentation separating the background and foreground hard constraints with a minimal geodesic boundary length with respect to chosen metric. It is worth mentioning, that due to the nature of graph cuts, effective interactive correction of the segmentation could be involved at this stage of the method whenever required.

3.2 Cluster Separation

Whereas in the first stage of our method the segmentation is driven largely by the image gradient (n-links), trying to satisfy the hard constraints at the same

[1] Note that the intensity gradient information is included in n-link weights.

time, in the second stage we employ a different approach and stick to the cluster morphology. That is motivated by the fact, that the image gradient inside of the nuclei does not provide us with reliable information. The interior of the nuclei can be greatly non-homogeneous and the dividing line between the touching nuclei not distinct enough, while some other parts of the nuclei can contain very sharp gradients. However, our solution allows us to tune the algorithm to different scenarios by simply changing the value of the parameter λ in (1). The clusters obtained in the first stage are treated separately in the second stage, so the following procedures refer to the process of division of one particular cluster.

First of all, the number of cell nuclei in the cluster is estabilished. To do this we calculate a distance transform of the cluster interior and find peaks in the resulting image using a morphological extended maxima transformation [2] with the maxima height chosen as 5% of the maximum value. The number of peaks in the distance transform is then taken as the number of cell nuclei in the cluster. If the cluster contains only one cell nucleus the second stage is over, otherwise we proceed to the separation of the touching nuclei. In the following text we will denote M_l the connected set of voxels corresponding to one peak in the distance transform, where $l \in \{1, \ldots, n\}$ and n is the number of nuclei in the cluster. An estimation of the nucleus radius σ_l is calculated as the mean value of the distance transform across voxels in M_l for each nucleus.

To find the dividing line among the cell nuclei a graph cut in a network with n terminals is used. The n-link capacities are set up in exactly the same way as in the first stage. The t-link weights are assigned as follows. For each label l and each voxel p in the cluster mask we define $d_l(p)$ to be the Euclidean distance of the voxel p to the nearest voxel in M_l. The values of d_l for all voxels and labels can be effectively calculated using n distance transforms. Further, we estimate the probability of voxel p matching label l as:

$$\Pr(p|l) = \exp\left(-\frac{d_l(p)^2}{2\sigma_l}\right), \tag{5}$$

which corresponds to a normal distribution with the probability inversely proportional to the distance of the voxel p from the set M_l and standard deviation $\sqrt{\sigma_l}$. The normalizing factor is omitted to ensure uniform amplitude of the probabilities. As a consequence of (2) the regional penalties are calculated as:

$$R_p(l) = -\log \Pr(p|l) = \frac{d_l(p)^2}{2\sigma_l} \ . \tag{6}$$

Indeed, hard constraints are set up for voxels in M_l. Such regional penalties (proportional to the distance from the M_l sets) incorporate an a priori shape information into the model and help us to push the dividing line of the neighbouring nuclei to its expected position and ignore the possibly strong gradients near the nucleus center. How much it will be pushed depends on the parameter λ in (1). The influence of this parameter is illustrated in Fig. 3. Generally, the smaller λ is, the higher importance will be given to the image gradient.

If the given cluster contains more than two cell nuclei (and more than two terminals in consequence) standard max-flow algorithms can not be used to find

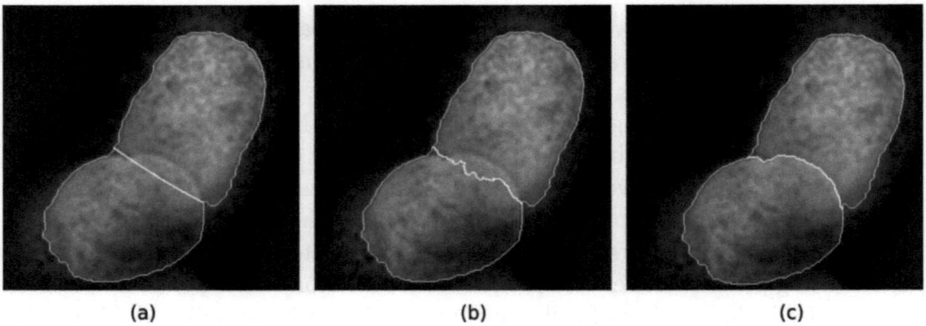

Fig. 3. Influence of the λ parameter on data with distinct frontier between the nuclei. (a) $\lambda = 1000$ (b) $\lambda = 0.15$ (c) $\lambda = 0$.

the minimal cut. Due to the NP-completeness of the problem [13], it is necessary to use approximations. We use the α-β-swap iterative algorithm proposed in [10], that is based on repeated calculations of standard minimal cut for all pairs of labels.[2] According to our tests this approximation converges very fast and three or four iterations are usually enough to reach the minimum. To obtain an initial labelling we assign a label l' to voxel p such as $l' = \arg\min_{l \in \mathcal{L}} R_p(l)$.

4 Experimental Results

Results obtained using an implementation of our model for 2-D images are presented in this section. We have tested our method on two different data sets. The first one consisted of 40 images (16-bit grayscale, 1300 × 1030 pixels) of DAPI stained HL60 (human promyelocytic leukemia cells) cell nuclei. The second one consisted of 10 images (16-bit grayscale, 1392 × 1040 pixels) of DAPI stained A549 (lung epithelial cells) cell nuclei deconvolved using the Maximum Likelihood Estimation algorithm, provided by the Huygens software (Scientific Volume Imaging BV, Hilversum, The Netherlands). In both cases the 2-D images were obtained as maximum intensity projections of 3-D images to the xy plane. Samples of the final segmentation are depicted in Fig. 4.

Each of the images in the data sets consisted of 10 to 20 clustered cell nuclei. Even though the clusters are quite complicated (particularly in the HL60 case) and the image quality is low, all of the nuclei are reliably identified, as can be seen in the figure. To quantitatively measure the accuracy of the segmentation, we have used the following sensitivity and specificity measures with respect to an expert provided ground truth:

$$Sens_i(f) = \frac{TP_i}{TP_i + FN_i} \qquad Spec_i(f) = \frac{TN_i}{TN_i + FP_i}, \tag{7}$$

[2] It is also possible to use the stronger α-expansion algorithm described in the same paper, because our $B_{(p,q)}$ is a metric.

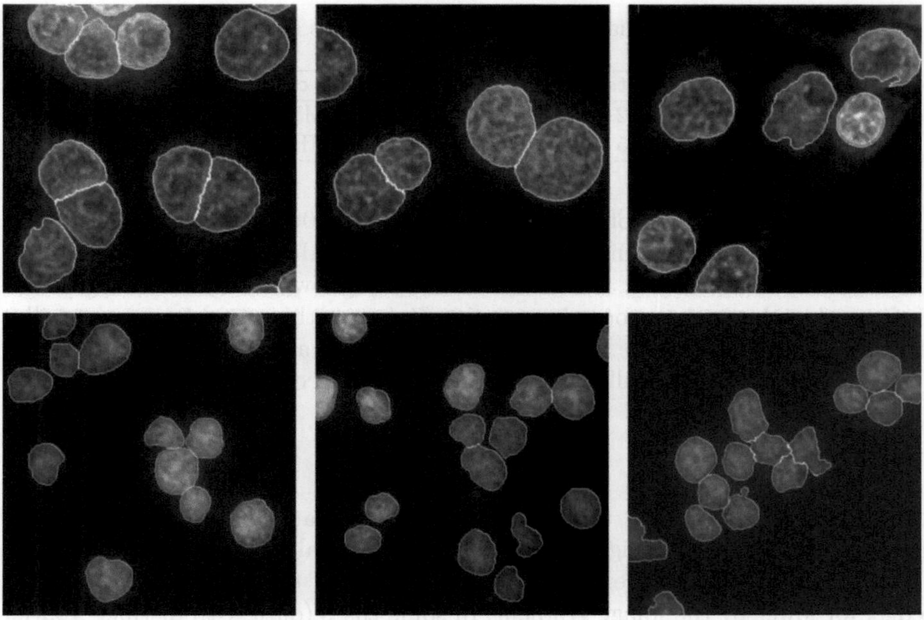

Fig. 4. Samples of the final segmentation. Top row: A549 cell nuclei. Bottom row: HL60 cell nuclei.

where i is a particular cell nucleus, f is the final segmentation and TP_i (true positive), TN_i (true negative), FP_i (false positive) and FN_i (false negative) denote the number of voxels correctly (true) and incorrectly (false) segmented as nucleus i (positive) and background or another nucleus (negative), respectively. Average and worst case values of both measures are listed in Table 1.

Table 1. Quantitative evaluation of the segmentation. Average and worst case values of *sensitivity* and *specificity* measures calculated against expert provided ground truth.

Cell line	$Sens_{worst}(f)$	$Spec_{worst}(f)$	$Sens_{avg}(f)$	$Spec_{avg}(f)$
A549	91.42%	92.98%	98.38%	97.00%
HL60	88.60%	95.68%	97.43%	98.12%

The computational time demands and memory consumption of our algorithm are listed in Table 2, they were approximately the same for both data sets (measured on a PC equipped with an Intel Q6600 processor and 2 GB RAM). The standard max-flow algorithm [7] was used to find the minimal cut in two-terminal networks. The memory footprint is smaller in the second stage, that is due to the fact that only parts of the image are processed. Also the computational time of the second stage depends on the number of nuclei clusters and on their complexity.

Table 2. Computational demands on tested images (≈ 1300 × 1000 pixels)

Stage	Total time	Peak memory consumption
1	2 sec	150 MB
2	5 sec	30 MB

For the segmentation of HL60 cell nuclei $\lambda = 0.001$ was used, because the interior of the nuclei is quite homegeneous and the dividing lines are perceptible. In the second case, $\lambda = 0.15$ was used, giving lower weight to the gradient information. Image preprocessing consisted of smoothing and background illumination correction in the first case and white top hat transformation followed by a morphological hole filling algorithm [2] in the second.

5 Discussion

The method described in this paper is fully automatic with the only tunable parameter being the λ weighting factor. For higher values of λ the segmentation is driven mostly by the regional term incorporating the a priori shape knowledge, for lower by the image gradient. In some cases (data with distinct frontier between the nuclei, such as the one in Fig. 3) it is even possible to use $\lambda = 0$. Such simple tuning of the algorithm is not possible with standard methods.

An important aspect of the second stage of our method is the incorporation of a priori shape information into the model. The proposed approach is well suited to a wide range of shapes, not only circular, provided that the M_l sets mentioned in Sect. 3.2 approximate the skeletons of the objects being sought. It is obvious that in case of mostly circular nuclei the skeletons correspond to centres and our method looking for peaks in the distance transform of the cluster is applicable. However, in case of more complex shapes it might be harder to find the initial M_l sets and the number of objects.

The implementation of our method in 3-D is straightforward. However, some complications may arise, which include a slower computation due to the huge size of the graphs and those related to low resolution and significant blur of the fluorescence microscope images in the axial direction.

6 Conclusion

A fully automated two-stage segmentation method based on the graph cut framework for the segmentation of touching cell nuclei in fluorescence microscopy has been presented in this paper. Our main contribution was to show how to cope with low image quality that is unfortunately common in optical microscopy. This is achieved particularly by combining image gradient information and incorporated a priori knowledge about the shape of the nuclei. Moreover, these two qualities can be easily balanced using a single user parameter.

We plan to compare the proposed approach with other segmentation methods, in particular, level-sets and the watershed transform. The quantitative evaluation in terms of computational time and accuracy will be done on both synthetic data with a ground truth and real images. Our goal is also to implement the method in 3-D and improve its robustness for more complex types of clusters, that appear in thick tissue sections.

Acknowledgments. This work has been supported by the Ministry of Education of the Czech Republic (Projects No. MSM-0021622419, No. LC535 and No. 2B06052). COS and AMB were supported by the Marie Curie IRG Program (grant number MIRG CT-2005-028342), and by the Spanish Ministry of Science and Education, under grant MCYT TEC 2005-04732 and the Ramon y Cajal Fellowship Program.

References

1. Pratt, W.K.: Digital Image Processing. Wiley, Chichester (1991)
2. Soille, P.: Morphological Image Analysis, 2nd edn. Springer, Heidelberg (2004)
3. Ortiz de Solórzano, C., Malladi, R., Leliévre, S.A., Lockett, S.J.: Segmentation of nuclei and cells using membrane related protein markers. Journal of Microscopy 201, 404–415 (2001)
4. Malpica, N., Ortiz de Solórzano, C., Vaquero, J.J., Santos, A., Lockett, S.J., Vallcorba, I., García-Sagredo, J.M., Pozo, F.d.: Applying watershed algorithms to the segmentation of clustered nuclei. Cytometry 28, 289–297 (1997)
5. Nilsson, B., Heyden, A.: Segmentation of dense leukocyte clusters. In: Proceedings of the IEEE Workshop on Mathematical Methods in Biomedical Image Analysis, pp. 221–227 (2001)
6. Boykov, Y., Jolly, M.P.: Interactive graph cuts for optimal boundary & region segmentation of objects in n-d images. In: IEEE International Conference on Computer Vision, July 2001, vol. 1, pp. 105–112 (2001)
7. Boykov, Y., Funka-Lea, G.: Graph cuts and efficient n-d image segmentation. International Journal of Computer Vision 70(2), 109–131 (2006)
8. Kolmogorov, V., Zabih, R.: What energy functions can be minimized via graph cuts? IEEE Transactions on Pattern Analysis and Machine Intelligence 26(2), 147–159 (2004)
9. Boykov, Y., Veksler, O., Zabih, R.: Markov random fields with efficient approximations. In: Proceedings of the IEEE Computer Society Conference on Computer Vision and Pattern Recognition, pp. 648–655. IEEE Computer Society, Los Alamitos (1998)
10. Boykov, Y., Veksler, O., Zabih, R.: Fast approximate energy minimization via graph cuts. IEEE Transactions on Pattern Analysis and Machine Intelligence 23, 1222–1239 (2001)
11. Boykov, Y., Kolmogorov, V.: An experimental comparison of min-cut/max-flow algorithms for energy minimization in vision. IEEE Transactions on Pattern Analysis and Machine Intelligence 26(9), 1124–1137 (2004)
12. Boykov, Y., Kolmogorov, V.: Computing geodesics and minimal surfaces via graph cuts. In: IEEE International Conference on Computer Vision, vol. 1, pp. 26–33 (2003)
13. Dahlhaus, E., Johnson, D.S., Papadimitriou, C.H., Seymour, P.D., Yannakakis, M.: The complexity of multiterminal cuts. SIAM J. Comput. 23(4), 864–894 (1994)

Parallel Volume Image Segmentation with Watershed Transformation

Björn Wagner[1], Andreas Dinges[2], Paul Müller[3], and Gundolf Haase[4]

[1] Fraunhofer ITWM, 67663 Kaiserslautern, Germany
bjoern.wagner@itwm.fraunhofer.de
[2] Fraunhofer ITWM, 67663 Kaiserslautern, Germany
[3] University Kaiserslautern, 67663 Kaiserslautern, Germany
[4] Karl-Franzens University Graz, A-8010 Graz, Austria

Abstract. We present a novel approach to parallel image segmentation of volume images on shared memory computer systems with watershed transformation by immersion. We use the domain decomposition method to break the sequential algorithm in multiple threads for parallel computation. The use of a chromatic ordering allows us to gain a correct segmentation without an examination of adjacent domains or a final relabeling. We will briefly discuss our approach and display results and speedup measurements of our implementation.

1 Introduction

The watershed transformation is a powerful region-based method for greyscale image segmentation introduced by H. Digabel and C. Lantuéjoul [2]. The grey-values of an image are considered as the altitude of a topological relief. The segmentation is computed by a simulated immersion of this greyscale range. Each local minimum induces a new basin which grows during the flooding by iterative assigning adjacent pixels. If two basins clash the contact pixels are marked as watershed lines.

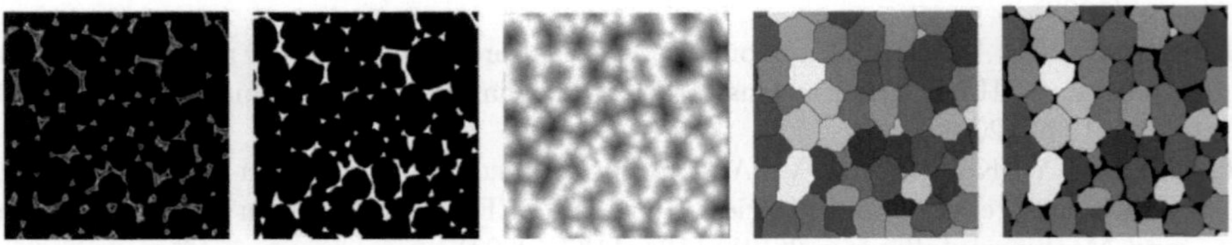

(a) original scan (b) segmented and closed edge system (c) inverted distance map of the background (d) watershed transformation of the distance map (e) reconstructed cells

Fig. 1. Cell reconstruction sequence of a metal foam

A.-B. Salberg, J.Y. Hardeberg, and R. Jenssen (Eds.): SCIA 2009, LNCS 5575, pp. 420–429, 2009.

In 3d image processing the watershed transformation can be used for object reconstruction. This is shown in figure 1 for the reconstruction of the cells of a metal foam[1] from a computer tomographic image. Due to the huge size of volume datasets the watershed transformation is a very computation intense task and the parallelization pays off.

The paper is organized as follows. Section 2 describes the sequential algorithm we used as a base for our parallel implementation. Section 3 gives a detailed description of our parallel approach and in section 4 we present some benchmarks and discuss the results.

2 The Sequential Watershed Algorithm

2.1 Preliminary Definitions

This section outlines some basic definitions, detailed in [6], [4] and [3].

A graph $G = (V, E)$ consists of a set V of vertices and a finite set $E \subseteq V \times V$ of pairs defining the connectivity. If there is a pair $e = (p, q) \in E$ we call p and q neighbors, or we say p and q are adjacent. The set of neighbors $N(p)$ of a vertex p is called the neighborhood of p.

A path $\pi = (v_0, v_1, \ldots, v_l)$ on a graph G from vertex p to vertex q is a sequence of vertices where $v_0 = p, v_l = q$ and $(v_i, v_{i+1}) \in E$ with $i \in [0, \ldots, l)$. The length of a path is denoted with $length(\pi) = l + 1$.

The geodesic distance $d_G(p, q)$ is defined as the length of the shortest path among two vertices p and q. The geodesic distance between a vertex p and a subset of vertices Q is defined by $d_G(p, Q) = \min_{q \in Q}(d_G(p, q))$.

A digital grid is a special kind of graph. For volume images usually the domain is defined by a cubic grid $D \subseteq \mathbb{Z}^3$, which is arranged as graph structure $G = (D, E)$. For E a subset of $\mathbb{Z}^3 \times \mathbb{Z}^3$ defining the connectivity is chosen. Usual choices are the 6-Connectivity, where each vertex has edges to its horizontal, vertical, front and back neighbors, or the 26-Connectivity, where a point is connected to all its immediate neighbors. The vertices of a cubic digital grid are called voxels.

A greyscale volume image is a digital grid where the vertices are valued by a function $g : D \to [h_{min}..h_{max}]$ with $D \subseteq \mathbb{Z}^3$ the domain of the image and h_{min} and h_{max} the minimum and the maximum greyvalue.

A label volume image is a digital grid where the vertices are valued by a function $l : D \to \mathbb{N}$ with $D \subseteq \mathbb{Z}^3$ the domain of the image.

2.2 Overview of the Algorithm

Vicent and Soille [7] gave an algorithmic definition of a watershed transformation by simulated immersion. The sequential procedure our parallel algorithm is derived from is based on a modified version of their method.

[1] Chrome-nickel foam provided by Recemat International (RCM-NC-2733).

The input image is a greyvalue image $g : D \rightarrow [h_{min}..h_{max}]$, with D the domain of the image and h_{min} and h_{max} are the minimum and maximum greyvalues respectively, and the output image $l : D \rightarrow \mathbb{N}$ is a label image containing the segmentation result.

The algorithm is performed in two parts. In the first part an ordered sequence $(L_{h_{min}}, \ldots, L_{h_{max}})$ of voxel lists is created, one list L_h for each greylevel $h \in [h_{min}, \ldots, h_{max}]$ of the input image g. The lists are filled with voxels p of the image domain D so that L_h contains all voxels $p \in D$ with $g(p) = h$. Moreover each voxel is tagged with the special label λ_{INIT}, indicating that this voxel has not been processed.

We have to use several particular labels to denote special states of a voxel. To distinguish them easily from the labels of the basins their value is always below the first basin label λ_0.

To assign a label λ to a voxel p the label image at coord p is set to λ, $l(p) = \lambda$.

The second part of the sequence of lists is processed in iterative steps starting at the lowest greylevel of the input image h_{min}. For each greylevel h new basins are created respectively to local minimas of the current level h and get a distinct label λ_i assigned. Further already existing basins, from former iteration steps, are expanded if they have adjoining pixels of the greyvalue h.

The expansion of the basins at greylevel h is done before the initiation of new basins by using a breadth-first algorithm [1]. Therefore each voxel of L_h is tagged with the special label λ_{MASK}, to denote it belongs to the current greylevel and has to be processed in this iteration step. This is also called masking level h. The set M_h contains all voxels p of level h with $l(p) = \lambda_{MASK}$.

Each voxel p which has at least one immediate neighbor q that is already assigned to a basin, so $l(q) \geq \lambda_0$ is appended to a FIFO queue Q_{ACTIVE}. Further p is tagged with the special label λ_{QUEUE}, indicating that it is in a queue.

Starting from these pixels the adjacent basins are propagated into the set of the of the masked pixels M_h. Each pixel of the active queue is processed sequential as follows:

- If a pixel has only one adjacent basin, it is labeled with the same label as the neighboring basin.
- If it is adjoining at least two different basins, it is labeled with the special label for Watersheds $\lambda_{WATERSHED}$.

All neighboring pixels which are marked with the label λ_{MASK} are appended to a FIFO queue $Q_{NOMINEE}$ and are labeled with the label λ_{QUEUE}.

When the queue Q_{ACTIVE} is empty the queue $Q_{NOMINEE}$ becomes the new Q_{ACTIVE} and a new queue $Q_{NOMINEE}$ is created. The propagation of the basins stops when there are no more pixels in one of the queues.

For each pixel $p \in Q_{ACTIVE}$ the distance $d_G(p, q)$ to the next pixel q with a lower greyvalue is the same. That condition also holds for $Q_{NOMINEE}$. Further for all voxel $q \in Q_{NOMINEE}$ it is true $d(q) = d(p) + 1$ for all $p \in Q_{ACTIVE}$.

After the expansion the pixels of the current greylevel are scanned sequential a second time. If a voxel is still tagged with the label λ_{MASK} a new basin is created starting at this voxel. Therefore the pixel is labeled with a new distinct label and this label is propageted to all adjacent masked voxels, using a breadth-first algorithm [1] as in the flooding process. The propagation stops when no more pixels can be associated to the new basin. When there are still voxels with $l(p) = \lambda_{MASK}$ left, further basins are created in the same way until no more voxels with λ_{MASK} label exist.

When all pixels of a greylevel are processed the algorithm continues with the following greylevel until the maximum greylevel h_{max} has been processed.

Figure 3 shows a simplified example of a watershed transformation sequence on a two dimensional image.

3 The Parallel Watershed Algorithm

For the parallel watershed transformation we apply the divide and conquer principle. The image domain D is divided into several non-overlapping subdomains $S \subseteq D$, usually into slices or blocks of a fixed size, on which the iterative steps of the transformation are performed concurrently. For each subdomain S an own ordered sequence of pixel lists $(L^S_{h_{min}}, \ldots, L^S_{h_{max}})$ is created and initialized with the voxels of S in the same way as for the sequential procedure. Further separate FIFO queues Q^S_{ACTIVE} and $Q^S_{NOMINEE}$ are created for each S.

As in the serial case, the sequences are processed in iterative steps starting at the lowest greylevel of the image. For each greylevel the parallel algorithm expands existing basins and creates new basins for each subdomain concurrently. Due to the recursive nature of the algorithm we have to synchronize the performance between the subdomains to get correct results. The masking step, in which each voxel of the current greylevel is marked with the label λ_{MASK} and the starting voxels for the label propagation are collected can be performed concurrently. The masking itself does not interact with any other subdomain. Further if a voxel of an adjacent subdomain must be checked whether it is already labeled there is also no problem with synchronization, because the relevant labels do not change during this step.

When all subdomains are masked, the algorithm can continue with the expansion of already detected basins. The algorithm implies a sequence of labeling events τ_p (read as labeling of pixel p), which is given by the greyvalue gradient of the input image, the ordering of the voxel lists L^S_h and the scanning order of the used breadth-first algorithm. The order of labeling events was defined by sequential appending of the pixels to the queues. It can be said that if q is appended to the queue after p then follows $\tau_p \prec \tau_q$ (to be read p is labeled before q). Further for all $p \in Q^S_{ACTIVE}, \forall S$ and for all $q \in Q^S_{NOMINEE}, \forall S$ follows $\tau_p \prec \tau_q$. The label assigned to a voxel p during the expansion depends on the labels of the already labeled voxels. The expansion relation can be formulated as follows:

$$l(p) = \begin{cases} c & \text{if } l(q) = c \; \forall q \in N^{\prec}(p) \\ \lambda_{WATERSHED} & \text{else} \end{cases} \tag{1}$$

where $N^{\prec}(p) = \{q \in N(p) : q \prec p \wedge l(q) \neq \lambda_{WATERSHED}$. If the sequence changes, for e.g. when the scanorder of the breadth-first algorithm is changed, the segmentation results also differ occasionally. Figure 2 shows an example for such a case for a simple example in one dimension. The pixels 1 and 2 are marked for labeling and are already appended to the queue Q_{ACTIVE}. In figure 2(a) pixel 1 will be labeled before pixel 2 and in figure 2(a) pixel 2 will be labeled before pixel 1.

As it can be seen the results of both sequences differ, because the labeling of the second pixel was influenced by the result of the first labeling. Thus it appears that we have to take care of the sequence of labeling events when performing a parallel expansion.

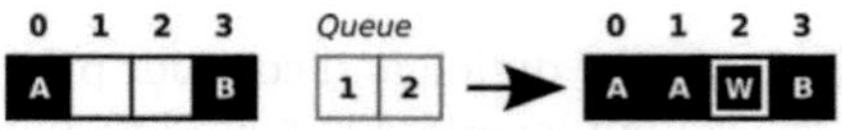

(a) sequence a

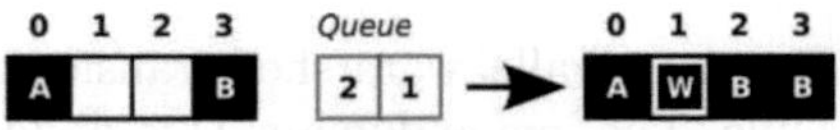

(b) sequence b

Fig. 2. Sequence dependend labeling

So if the concurrent performance does not follow the same sequence for each execution the results may be unpredictable. Therefore we introduce a further level of ordering of the labeling events.

Let $\mathcal{S}$ be the set of all subdomains of the image domain D. Further $E : \mathcal{S} \rightarrow \mathfrak{P}(\mathcal{S}) = \{X | X \subseteq \mathcal{S}\}$ defines the environment of a subdomain with

$$E(S) = \{T | \exists p \in S \text{ with } \exists q \in N(p) \wedge q \in T\} \tag{2}$$

We define a coloring function $\Gamma : \mathcal{S} \rightarrow \mathcal{C}$ for the subdomains, with $\mathcal{C}$ an ordered set of colors, so that for a subdomain S the condition

$$\forall U, V \in \mathcal{E}(S) \cup S : \Gamma(U) \neq \Gamma(V) \tag{3}$$

holds.

Further we define a coloring for the pixels $\gamma : D \rightarrow \mathcal{C}$ so that the condition

$$\forall p \in S : \gamma(p) = \Gamma(S) \tag{4}$$

holds.

The parallel expansion of the basins works as follows. For each color $c \in \mathcal{C}$ the propagation is performed for all voxels in the Q^S_{ACTIVE} queues of all subdomains S with $\Gamma(S) = c$. This is done in the sequence defined by the ordering of the colors. For two subdomains U, V with $\Gamma(U) < \Gamma(V)$, U is processed before V.

Inside of a subdomain the propagation still performs sequential as depicted in section 2.2, but subdomains S, T with $\Gamma(S) = \Gamma(T)$ can be performed concurrently.

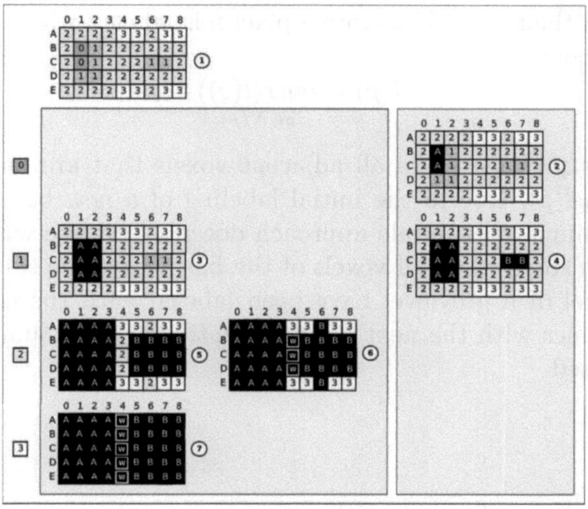

Fig. 3. Watershed transformation sequence

All neighboring pixels which are marked with the label λ_{MASK} are appended to the FIFO queue $Q^S_{NOMINEE}$ of the subdomain they are element of and are labeled with the label λ_{QUEUE}.

After all colors have been processed the $Q^S_{NOMINEE}$ queues become the new Q^S_{ACTIVE} queues and the propagation is continued until none of the queues of any subdomain contains any more voxels.

Due to the color depended performance of the expansion, it never happens that two voxels of adjacent subdomains are processed concurrently. So if voxel of adjacent subdomain have to be checked this can be performed without additional synchronization. Further for all pixels of any Q^S_{ACTIVE} queue follows:

$$\forall p \in Q^S_{ACTIVE}, q \in Q^T_{ACTIVE}, S \neq T, \gamma(p) < \gamma(q) \rightsquigarrow p \prec q \tag{5}$$

So the results only depend on the domain decomposition of the image and the order of the assigned colors.

When the expansion has finished in all subdomains, the creation of new basins is performed. This can also be done concurrently in a similar way as by the expansion step. For each subdomain S we create an own label counter $nextlabel^S$ which is initialized with the value $\lambda_{WATERSHED} + \mathcal{I}(S)$, where $\mathcal{I} : \mathcal{S} \rightarrow [1..\|\mathcal{S}\|]$ is a function assigning a distinct identifier to each subdomain. When a minimum is detected in a subdomain S, a new basin with the label $nextlabel^S$ is created and the counter is increased by $\|\mathcal{S}\|$. The increasing by $\|\mathcal{S}\|$ avoids duplicate labels in the subdomains.

Inside of a subdomain the propagation of a new label still performs sequential as depicted in section 2.2, but subdomains S, T with $\Gamma(S) = \Gamma(T)$ can be performed concurrently, as in the expansion step. It may happen that a local minimum spreads over several subdomains and gets different labels in each subdomain. To merge the different labels the propagation overrides all labels with a value lesser than their own. Therefore a pixel p is labeled with the highest label of its neighborhood:

$$l(p) = \max_{q \in N(p)}(l(q)) \tag{6}$$

and this label is propageted to all adjacent voxels that are masked of have a label lower than $l(p)$. Due to the initial labeling of a new basin only affecting the pixels of minima, this simple approach doesn't interfere with other basins. The propagation stops when all voxels of the basin have the same label.

When all voxel of a greylevel have been labeled with the correct label the algorithm continues with the next greylevel until the maximum greylevel h_{max} has been processed.

4 Results

To verify the efficiency of our algorithm we measured the speedup for datasets of different sizes[2], ranging from 100^3 pixels to 1000^3 pixels with cubic subdomains of a size of 32^3 pixels on a usual shared memory machine[3]. We have chosen simulated data to be able to compare datasets of different sizes without clipping scanned datasets and influencing the results. As it is shown in figure 4(b) our algorithm scales well for image sizes above 200^3 pixels. For images with 100^3 and 200^3 pixels there are not enough subdomains available for simultaneous computation to utilize the machine.

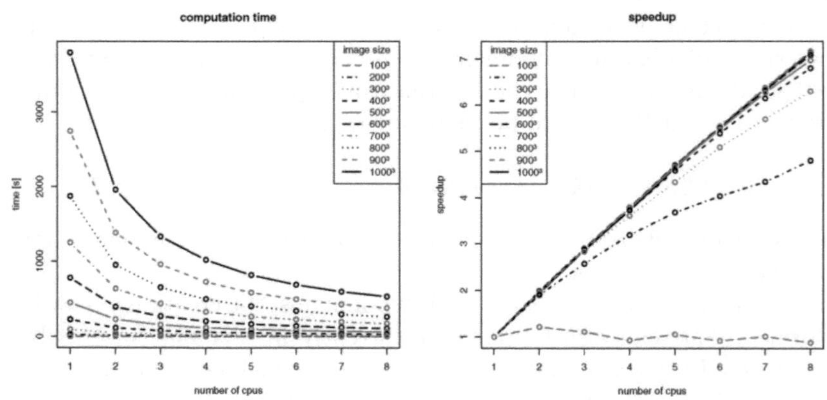

(a) computation time (b) speedup

Fig. 4. Computation time and speedup or different image sizes

To prove the efficiency of our algorithm also for real volume datasets, we measured the speedup and the timing for the watershed transformation of the reconstruction pipeline mentioned in the introduction (see figure 1) for different

[2] Simulated foam structures.

[3] Dual Intel Xeon X5450@3.00GHz Quadcore.

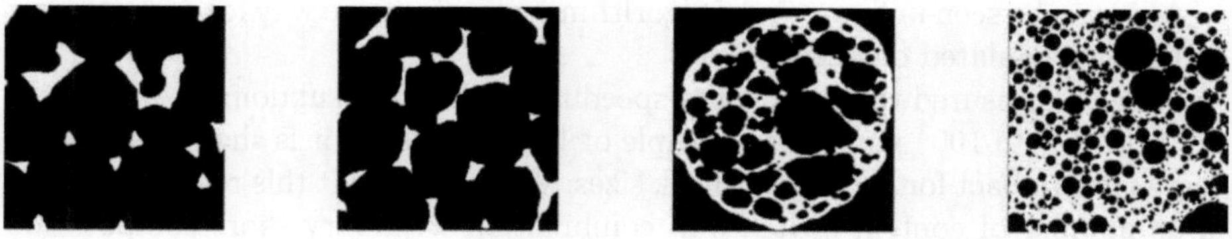

(a) recemat2733 (b) recemat4573 (c) ceramic grain (d) gas concrete

Fig. 5. Segmented datasets

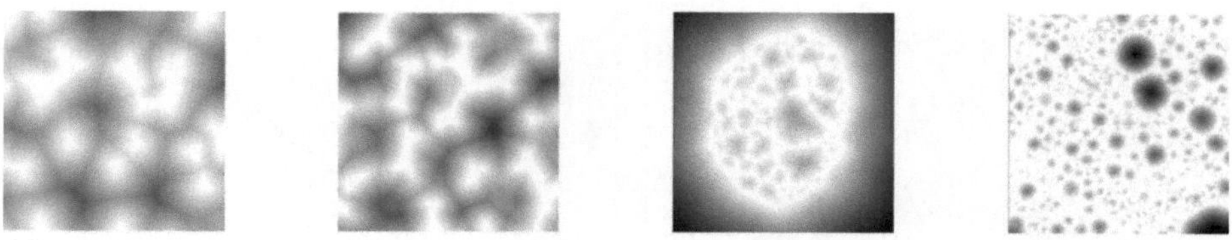

(a) recemat2733 (b) recemat4573 (c) ceramic grain (d) gas concrete

Fig. 6. Distance maps

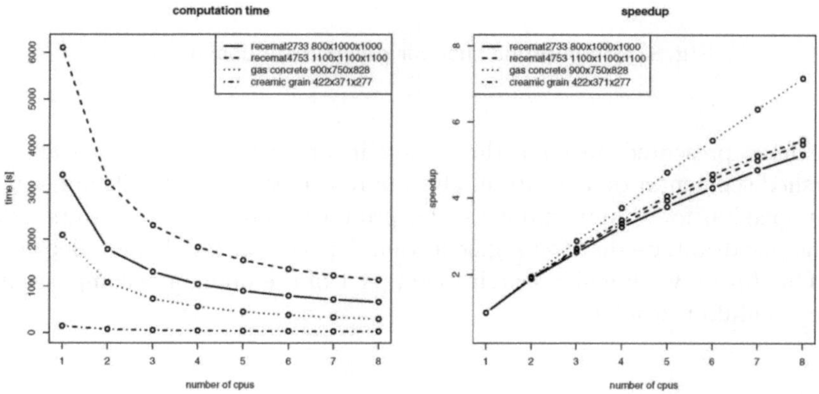

(a) computation time (b) speedup

Fig. 7. Computation time and speedup for different volume datasets

datasets. Figure 5 shows crosssections of the used datasets. In figure 5(a) and figure 5(b) segmentations of two different chrome-nickel foam provided by Recemat International are depicted, figure 5(c) shows a segmented ceramic grain and figure 5(d) displays the pores of a gas concrete sample. The corresponding distance maps are shown in figure 6.

As it can be seen in figure 7 our algorithm scales the same way for real datasets as for the simulated datasets.

We also measured the timing and speedup for different subdomain sizes ranging from 10^3 to 100^3 pixels for a sample of 1000^3 pixel. As it is shown in figure 8 there is an impact for very small block sizes. We assume that this results from the large number of context switches in combination with very short computation times for one subdomain.

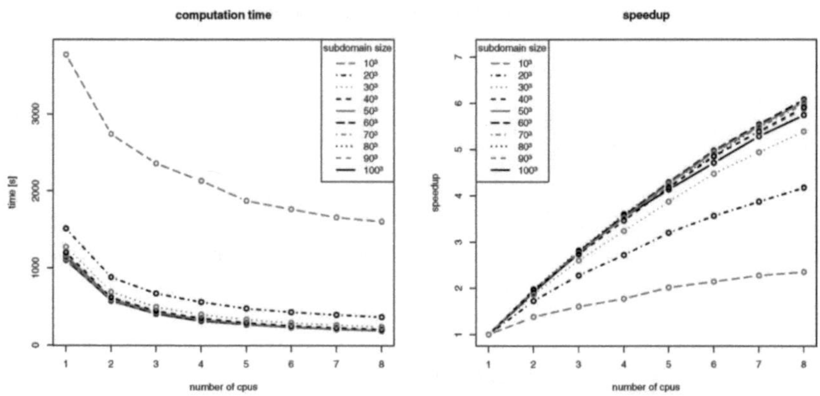

(a) computation time (b) speedup

Fig. 8. Computation time for different subdomain sizes

We have presented an algorithm study in order to efficiently parallelize a watershed segmentation algorithm. Our approach leads to a significant segmentation speedup for volume datasets and produces deterministic results. It still has the disadvantage that the segmentation depends on the domain decomposition. Our future work will research the impact of the domain decomposition on the segmentation results.

References

1. Cormen, T.H., Leiserson, C.E., Rivest, R.L., Stein, C.: Introduction to Algorithms, 2nd edn. MIT Press, Cambridge (2001)
2. Digabel, H., Lantuejoul, C.: Iterative algorithms. In: Actes du second symposium europeen d'analyse quantitative des microstructures en sciences des materiaux, biologie et medecine (1977)
3. Klette, R., Rosenfeld, A.: Digital Geometry: Geometric Methods for Digital Image Analysis. The Morgan Kaufmann Series in Computer Graphics. Morgan Kaufmann, San Francisco (2004)
4. Lohmann, G.: Volumetric Image Processing. John Wiley & Sons, B.G. Teubner Publishers, Chichester (1998)

5. Moga, A.N., Gabbouj, M.: Parallel image component labeling with watershed transformation. IEEE Transactions on Pattern Analysis and Machine Intelligence 19, 441–450 (1997)
6. Roerdink, J.B.T.M., Meijster, A.: Ios press the watershed transform: Definitions, algorithms and parallelization strategies
7. Vincent, L., Soille, P.: Watersheds in digital spaces: An efficient algorithm based on immersion simulations. IEEE Trans. Pattern Anal. Mach. Intell. 13(6), 583–598 (1991)

Fast-Robust PCA

Markus Storer, Peter M. Roth, Martin Urschler, and Horst Bischof

Institute for Computer Graphics and Vision,
Graz University of Technology,
Inffeldgasse 16/II, 8010 Graz, Austria
{storer,pmroth,urschler,bischof}@icg.tugraz.at

Abstract. Principal Component Analysis (PCA) is a powerful and widely used tool in Computer Vision and is applied, e.g., for dimensionality reduction. But as a drawback, it is not robust to outliers. Hence, if the input data is corrupted, an arbitrarily wrong representation is obtained. To overcome this problem, various methods have been proposed to robustly estimate the PCA coefficients, but these methods are computationally too expensive for practical applications. Thus, in this paper we propose a novel fast and robust PCA (FR-PCA), which drastically reduces the computational effort. Moreover, more accurate representations are obtained. In particular, we propose a two-stage outlier detection procedure, where in the first stage outliers are detected by analyzing a large number of smaller subspaces. In the second stage, remaining outliers are detected by a robust least-square fitting. To show these benefits, in the experiments we evaluate the FR-PCA method for the task of robust image reconstruction on the publicly available ALOI database. The results clearly show that our approach outperforms existing methods in terms of accuracy and speed when processing corrupted data.

1 Introduction

Principal Component Analysis (PCA) [1] also known as Karhunen-Loève transformation (KLT) is a well known and widely used technique in statistics. The main idea is to reduce the dimensionality of data while retaining as much information as possible. This is assured by a projection that maximizes the variance but minimizes the mean squared reconstruction error at the same time. Murase and Nayar [2] showed that high dimensional image data can be projected onto a subspace such that the data lies on a lower dimensional manifold. Thus, starting from face recognition (e.g., [3,4]) PCA has become quite popular in computer vision[1], where the main application of PCA is dimensionality reduction. For instance, a number of powerful model-based segmentation algorithms such as Active Shape Models [8] or Active Appearance Models [9] incorporate PCA as a fundamental building block.

In general, when analyzing real-world image data, one is confronted with unreliable data, which leads to the need for robust methods (e.g., [10,11]). Due to

[1] For instance, at CVPR 2007 approximative 30% of all papers used PCA at some point (e.g., [5,6,7]).

A.-B. Salberg, J.Y. Hardeberg, and R. Jenssen (Eds.): SCIA 2009, LNCS 5575, pp. 430–439, 2009.

its least squares formulation, PCA is highly sensitive to outliers. Thus, several methods for robustly learning PCA subspaces (e.g., [12,13,14,15,16]) as well as for robustly estimating the PCA coefficients (e.g., [17,18,19,20]) have been proposed. In this paper, we are focusing on the latter case. Thus, in the learning stage a reliable model is estimated from undisturbed data, which is then applied to robustly reconstruct unreliable values from the unseen corrupted data.

To robustly estimate the PCA coefficients Black and Jepson [18] applied an M-estimator technique. In particular, they replaced the quadratic error norm with a robust one. Similarly, Rao [17] introduced a new robust objective function based on the MDL principle. But as a disadvantage, an iterative scheme (i.e., EM algorithm) has to be applied to estimate the coefficients. In contrast, Leonardis and Bischof [19] proposed an approach that is based on sub-sampling. In this way, outlying values are discarded iteratively and the coefficients are estimated from inliers only. Similarly, Edwards and Murase introduced adaptive masks to eliminate corrupted values when computing the sum-squared errors.

A drawback of these methods is their computational complexity (i.e., iterative algorithms, multiple hypotheses, etc.), which limits their practical applicability. Thus, we develop a more efficient robust PCA method that overcomes this limitation. In particular, we propose a two-stage outlier detection procedure. In the first stage, we estimate a large number of smaller subspaces sub-sampled from the whole dataset and discard those values that are not consistent with the subspace models. In the second stage, the data vector is robustly reconstructed from the thus obtained subset. Since the subspaces estimated in the first step are quite small and only a few iterations of the computationally more complex second step are required (i.e., most outliers are already discarded by the first step), the whole method is computationally very efficient. This is confirmed by the experiments, where we show that the proposed method outperforms existing methods in terms of speed and accuracy.

This paper is structured as follows. In Section 2, we introduce and discuss the novel fast-robust PCA (FR-PCA) approach. Experimental results for the publicly available ALOI database are given in Section 3. Finally, we discuss our findings and conclude our work in Section 4.

2 Fast-Robust PCA

Given a set of n high-dimensional data points $\mathbf{x}_j \in \mathbb{R}^m$ organized in a matrix $\mathbf{X} = [\mathbf{x}_1, \ldots, \mathbf{x}_n] \in \mathbb{R}^{m \times n}$, then the PCA basis vectors $\mathbf{u}_1, \ldots, \mathbf{u}_{n-1}$ correspond to the eigenvectors of the sample covariance matrix

$$\mathbf{C} = \frac{1}{n-1} \hat{\mathbf{X}} \hat{\mathbf{X}}^\top , \tag{1}$$

where $\hat{\mathbf{X}} = [\hat{\mathbf{x}}_1, \ldots, \hat{\mathbf{x}}_n]$ is the mean normalized data with $\hat{\mathbf{x}}_j = \mathbf{x}_j - \bar{\mathbf{x}}$. The sample mean $\bar{\mathbf{x}}$ is calculated by

$$\bar{\mathbf{x}} = \frac{1}{n} \sum_{j=1}^{n} \mathbf{x}_j . \tag{2}$$

Given the PCA subspace $\mathbf{U}_p = [\mathbf{u}_1, \ldots, \mathbf{u}_p]$ (usually only p, $p < n$, eigenvectors are sufficient), an unknown sample $\mathbf{x} \in \mathbb{R}^m$ can be reconstructed by

$$\tilde{\mathbf{x}} = \mathbf{U}_p\mathbf{a} + \bar{\mathbf{x}} = \sum_{j=1}^{p} a_j\mathbf{u}_j + \bar{\mathbf{x}}\,, \tag{3}$$

where $\tilde{\mathbf{x}}$ denotes the reconstruction and $\mathbf{a} = [a_1, \ldots, a_p]$ are the PCA coefficients obtained by projecting $\mathbf{x}$ onto the subspace $\mathbf{U}_p$.

If the sample $\mathbf{x}$ contains outliers, Eq. (3) does not yield a reliable reconstruction; a robust method is required (e.g., [17,18,19,20]). But since these methods are computationally very expensive (i.e., they are based on iterative algorithms) or can handle only a small amount of noise, they are often not applicable in practice. Thus, in the following we propose a new fast robust PCA approach (FR-PCA), which overcomes these problems.

2.1 FR-PCA Training

The training procedure, which is sub-divided into two major parts, is illustrated in Figure 1. First, a standard PCA subspace $\mathbf{U}$ is generated using the full available training data. Second, N sub-samplings $\mathbf{s}_n$ are established from randomly selected values from each data point (illustrated by the red points and the green crosses). For each sub-sampling $\mathbf{s}_n$, a smaller subspace (sub-subspace) $\mathbf{U}^n$ is estimated, in addition to the full subspace.

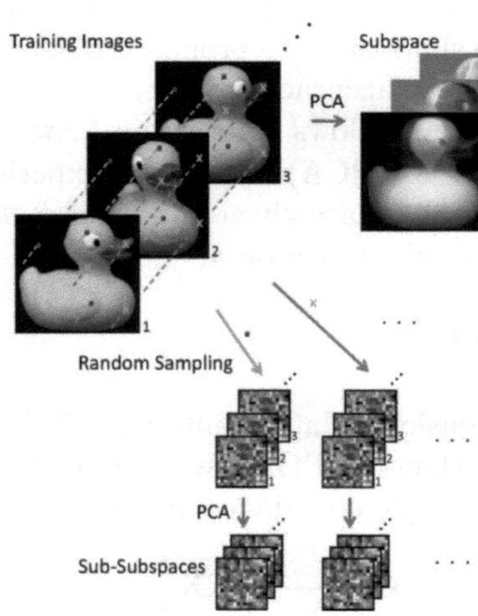

Fig. 1. FR-PCA training: A global PCA subspace and a large number of smaller PCA sub-subspaces are estimated in parallel. Sub-subspaces are derived by randomly sub-sampling the input data.

2.2 FR-PCA Reconstruction

Given a new unseen test sample $\mathbf{x}$, the robust reconstruction $\tilde{\mathbf{x}}$ is estimated in two stages. In the first stage (*gross outlier detection*), the outliers are detected based on the reconstruction errors of the sub-subspaces. In the second stage (*refinement*), using the thus estimated inliers, a robust reconstruction $\tilde{\mathbf{x}}$ of the whole sample is generated.

In the *gross outlier detection*, first, N sub-samplings $\mathbf{s}_n$ are generated according to the corresponding sub-subspaces $\mathbf{U}^n$, which were estimated as described in Section 2.1. In addition, we define the set of "inliers" $\mathbf{r}$ as the union of all selected pixels: $\mathbf{r} = \mathbf{s}_1 \cup \ldots \cup \mathbf{s}_N$, which is illustrated in Figure 2(a) (green points). Next, for each sub-sampling $\mathbf{s}_n$ a reconstruction $\tilde{\mathbf{s}}_n$ is estimated by Eq. (3), which allows to estimate the error-maps

$$\mathbf{e}_n = |\mathbf{s}_n - \tilde{\mathbf{s}}_n| \;, \tag{4}$$

the mean reconstruction error $\bar{e}$ over all sub-samplings, and the mean reconstruction errors $\bar{e}_n$ for each of the N sub-samplings.

Based on these errors, we can detect the outliers by local and global thresholding. The local thresholds (one for each sub-sampling) are defined by $\theta_n = \bar{e}_n w_n$, where w_n is a weighting parameter and the global threshold θ is set to the mean error $\bar{e}$. Then, all points $s_{n,(i,j)}$ for which

$$e_{n,(i,j)} > \theta_n \quad \text{or} \quad e_{n,(i,j)} > \theta \tag{5}$$

are discarded from the sub-samplings $\mathbf{s}_n$ obtaining $\hat{\mathbf{s}}_n$. Finally, we re-define the set of "inliers" by

$$\mathbf{r} = \hat{\mathbf{s}}_1 \cup \ldots \cup \hat{\mathbf{s}}_q \;, \tag{6}$$

where $\hat{\mathbf{s}}_1, \ldots, \hat{\mathbf{s}}_q$ indicate the first q sub-samplings (sorted by $\bar{e}_n$) such that $|\mathbf{r}| \leq k$; k is the pre-defined maximum number of points. The thus obtained "inliers" are shown in Figure 2(b).

The *gross outlier detection* procedure allows to remove most outliers. Thus, the obtained set $\mathbf{r}$ contains almost only inliers. To further improve the final result in the *refinement* step, the final robust reconstruction is estimated similar to [19]. Starting from the point set $\mathbf{r} = [r_1, \ldots, r_k]$, $k > p$, obtained from the *gross outlier detection*, repeatedly reconstructions $\tilde{\mathbf{x}}$ are computed by solving an over-determined system of equations minimizing the least squares reconstruction error

$$E(\mathbf{r}) = \sum_{i=1}^{k} \left(x_{r_i} - \sum_{j=1}^{p} a_j \mathbf{u}_{j,r_i} \right)^2 . \tag{7}$$

Thus, in each iteration those points with the largest reconstruction errors can be discarded from $\mathbf{r}$ (selected by a reduction factor α). These steps are iterated until a pre-defined number of remaining points is reached. Finally, an outlier-free subset is obtained, which is illustrated in Figure 2(c).

A robust reconstruction result obtained by the proposed approach compared to a non-robust method is shown in Figure 3. One can clearly see that the robust

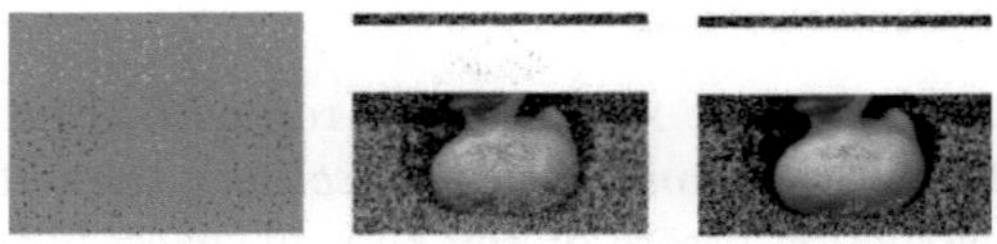

Fig. 2. Data point selection process: (a) data points sampled by all sub-subspaces, (b) occluded image showing the remaining data points after applying the sub-subspace procedure, and (c) resulting data points after the iterative refinement process for the calculation of the PCA coefficients. This figure is best viewed in color.

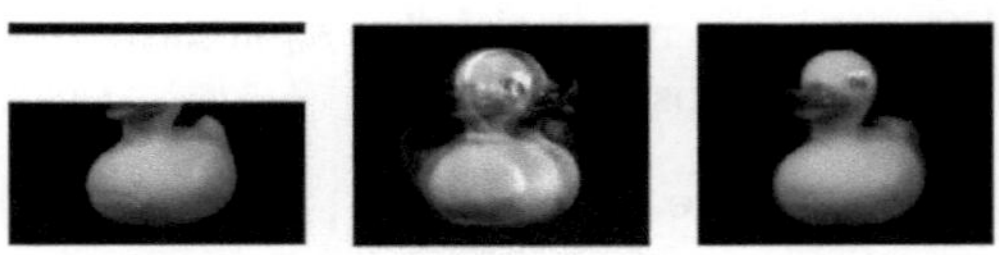

Fig. 3. Demonstration of the insensitivity of the robust PCA to noise (i.e., occlusions): (a) occluded image, (b) reconstruction using standard PCA, and (c) reconstruction using the FR-PCA

method considerably outperforms the standard PCA. Note, the blur visible in the reconstruction of the FR-PCA is the consequence of taking into account only a limited number of eigenvectors.

In general, the robust estimation of the coefficients is computationally very efficient. In the *gross outlier detection* procedure, only simple matrix operations have to be performed, which are very fast; even if hundreds of sub-subspace reconstructions have to be computed. The computationally more expensive part is the *refinement* step, where repeatedly an overdetermined linear system of equations has to be solved. Since only very few refinement iterations have to be performed due to the preceding *gross outlier detection*, the total runtime is kept low.

3 Experimental Results

To show the benefits of the proposed fast robust PCA method (FR-PCA), we compare it to the standard PCA (PCA) and the robust PCA approach presented in [19] (R-PCA). We choose the latter one, since it yields superior results among the presented methods in the literature and our refinement process is similar to theirs.

In particular, the experiments are evaluated for the task of robust image reconstruction on the "Amsterdam Library of Object Images (ALOI)" database [21]. The ALOI database consists of 1000 different objects. Over hundred images of each object are recorded under different viewing angles, illumination angles and illumination colors, yielding a total of 110,250 images. For our experiments we arbitrarily choose 30 categories (009, 018, 024, 032, 043, 074, 090, 093, 125, 127,

Fig. 4. Illustrative examples of ALOI database objects [21] used in the experiments

135, 138, 151, 156, 171, 174, 181, 200, 299, 306, 323, 354, 368, 376, 409, 442, 602, 809, 911, 926), where an illustrative subset of objects is shown in Figure 4.

In our experimental setup, each object is represented in a separate subspace and a set of 1000 sub-subspaces, where each sub-subspace contains 1% of data points of the whole image. The variance retained for the sub-subspaces is 95% and 98% for the whole subspace, which is also used for the standard PCA and the R-PCA. Unless otherwise noted, all experiments are performed with the parameter settings given in Table 1.

Table 1. Parameters for the FR-PCA (a) and the R-PCA (b) used for the experiments

(a)

FR-PCA	
Number of initial points k	130*p*
Reduction factor α	0.9

(b)

R-PCA	
Number of initial hypotheses H	30
Number of initial points k	48*p*
Reduction factor α	0.85
K_2	0.01
Compatibility threshold	100

A 5-fold cross-validation is performed for each object category, resulting in 80% training- and 20% test data, corresponding to 21 test images per iteration. The experiments are accomplished for several levels of spatially coherent occlusions and several levels of salt & pepper noise. Quantitative results for the root-mean-squared (RMS) reconstruction-error per pixel for several levels of occlusions are given in Table 2. In addition, in Figure 5 we show box-plots of the RMS reconstruction-error per pixel for different levels of occlusions. Analogously, the RMS reconstruction-error per pixel for several levels of salt & pepper noise is presented in Table 3 and the corresponding box-plots are shown in Figure 6.

From Table 2 and Figure 5 it can be seen – starting from an occlusion level of 0% – that all subspace methods exhibit nearly the same RMS reconstruction-error. Increasing the portion of occlusion, the standard PCA shows large errors

Table 2. Comparison of the reconstruction errors of the standard PCA, the R-PCA and the FR-PCA for several levels of occlusion showing RMS reconstruction-error per pixel given by mean and standard deviation

Occlusion	Error per Pixel											
	0%		10%		20%		30%		50%		70%	
	mean	std	mean	std	mean	std	mean	std	mean	std	mean	std
PCA	10.06	6.20	21.82	8.18	35.01	12.29	48.18	15.71	71.31	18.57	92.48	18.73
R-PCA	11.47	7.29	11.52	7.31	12.43	9.24	22.32	21.63	59.20	32.51	94.75	43.13
FR-PCA	10.93	6.61	11.66	6.92	11.71	6.95	11.83	7.21	26.03	23.05	83.80	79.86

Table 3. Comparison of the reconstruction errors of the standard PCA, the R-PCA and the FR-PCA for several levels of salt & pepper noise showing RMS reconstruction-error per pixel given by mean and standard deviation

Salt&Pepper Noise	Error per Pixel									
	10%		20%		30%		50%		70%	
	mean	std	mean	std	mean	std	mean	std	mean	std
PCA	11.77	5.36	14.80	4.79	18.58	4.80	27.04	5.82	36.08	7.48
R-PCA	11.53	7.18	11.42	7.17	11.56	7.33	11.63	7.48	15.54	10.15
FR-PCA	11.48	6.86	11.30	6.73	11.34	6.72	11.13	6.68	14.82	7.16

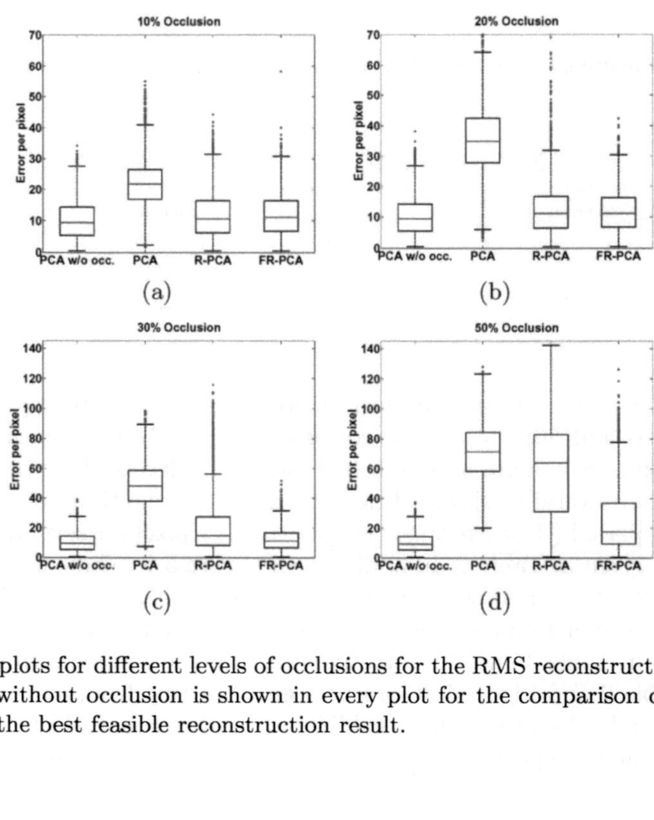

Fig. 5. Box-plots for different levels of occlusions for the RMS reconstruction-error per pixel. PCA without occlusion is shown in every plot for the comparison of the robust methods to the best feasible reconstruction result.

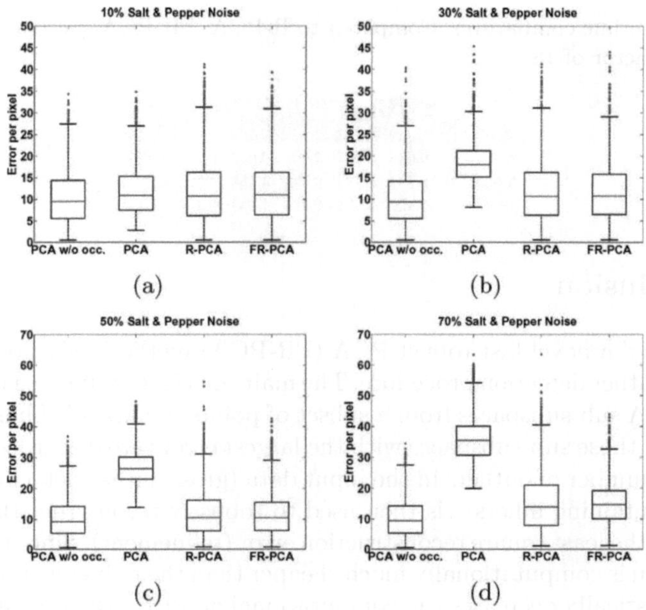

Fig. 6. Box-plots for different levels of salt & pepper noise for the RMS reconstruction-error per pixel. PCA without occlusion is shown in every plot for the comparison of the robust methods to the best feasible reconstruction result.

whereas the robust methods are still comparable to the non-disturbed (best feasible) case, where our novel FR-PCA presents the best performance. In contrast, as can be seen from Table 3 and Figure 6, all methods can generally cope better with salt & pepper noise. However, also for this experiment FR-PCA yields the best results.

Finally, we evaluated the runtime[1] for the applied different PCA reconstruction methods, which are summarized in Table 4. It can be seen that for the given setup compared to R-PCA for a comparable reconstruction quality the robust reconstruction can be speeded up by factor of 18! This drastic speed-up can be explained by the fact that the refinement process is started from a set of data points mainly consisting of inliers. In contrast, in [19] several point sets (hypotheses) have to be created and the iterative procedure has to be run for every set resulting in a poor runtime performance. Reducing the number of hypotheses or the number of initial points would decrease the runtime, but, however, the reconstruction accuracy gets worse. In particular, the runtime of our approach only depends slightly on the number of starting points, thus having nearly constant execution times. Clearly, the runtime depends on the number and size of used eigenvectors. Increasing one of those values, the gap between the runtime for both methods is even getting larger.

---

[1] The runtime is measured in MATLAB using an Intel Xeon processor running at 3GHz. The resolution of the images is 192x144 pixels.

Table 4. Runtime comparison. Compared to R-PCA, FR-PCA speeds-up the computation by a factor of 18.

	Mean Runtime [s]					
Occlusion	0%	10%	20%	30%	50%	70%
PCA	0.006	0.007	0.007	0.007	0.008	0.009
R-PCA	6.333	6.172	5.435	4.945	3.193	2.580
FR-PCA	0.429	0.338	0.329	0.334	0.297	0.307

4 Conclusion

We developed a novel fast robust PCA (FR-PCA) method based on an efficient two-stage outlier detection procedure. The main idea is to estimate a large number of small PCA sub-subspaces from a subset of points in parallel. Thus, for a given test sample, those sub-subspaces with the largest errors are discarded first, which reduce the number of outliers in the input data (gross outlier detection). This set – almost containing inliers – is then used to robustly reconstruct the sample by minimizing the least square reconstruction error (refinement). Since the gross outlier detection is computationally much cheaper than the refinement, the proposed method drastically decreases the computational effort for the robust reconstruction. In the experiments, we show that our new fast robust PCA approach outperforms existing methods in terms of speed and accuracy. Thus, our algorithm is applicable in practice and can be applied for real-time applications such as robust Active Appearance Model (AAM) fitting [22]. Since our approach is quite general, FR-PCA is not restricted to robust image reconstruction.

Acknowledgments

This work has been funded by the Biometrics Center of Siemens IT Solutions and Services, Siemens Austria. In addition, this work was supported by the FFG project AUTOVISTA (813395) under the FIT-IT programme, and the Austrian Joint Research Project Cognitive Vision under projects S9103-N04 and S9104-N04.

References

1. Jolliffe, I.T.: Principal Component Analysis. Springer, Heidelberg (2002)
2. Murase, H., Nayar, S.K.: Visual learning and recognition of 3-d objects from appearance. Intern. Journal of Computer Vision 14(1), 5–24 (1995)
3. Kirby, M., Sirovich, L.: Application of the karhunen-loeve procedure for the characterization of human faces. IEEE Trans. on Pattern Analysis and Machine Intelligence 12(1), 103–108 (1990)
4. Turk, M., Pentland, A.: Eigenfaces for recognition. Journal of Cognitive Neuroscience 3(1), 71–86 (1991)
5. Wang, Y., Huang, K., Tan, T.: Human activity recognition based on r transform. In: Proc. CVPR (2008)

6. Tai, Y.W., Brown, M.S., Tang, C.K.: Robust estimation of texture flow via dense feature sampling. In: Proc. CVPR (2007)
7. Lee, S.M., Abbott, A.L., Araman, P.A.: Dimensionality reduction and clustering on statistical manifolds. In: Proc. CVPR (2007)
8. Cootes, T.F., Taylor, C.J., Cooper, D.H., Graham, J.: Active shape models - their training and application. Computer Vision and Image Understanding 61, 38–59 (1995)
9. Cootes, T.F., Edwards, G.J., Taylor, C.J.: Active appearance models. IEEE Trans. on Pattern Analysis and Machine Intelligence 23(6), 681–685 (2001)
10. Huber, P.J.: Robust Statistics. John Wiley & Sons, Chichester (2004)
11. Hampel, F.R., Ronchetti, E.M., Rousseeuw, P.J., Stahel, W.A.: Robust Statistics: The Approach Based on Influence Functions. John Wiley & Sons, Chichester (1986)
12. Xu, L., Yuille, A.L.: Robust principal component analysis by self-organizing rules based on statistical physics approach. IEEE Trans. on Neural Networks 6(1), 131–143 (1995)
13. Torre, F.d., Black, M.J.: A framework for robust subspace learning. Intern. Journal of Computer Vision 54(1), 117–142 (2003)
14. Roweis, S.: EM algorithms for PCA and SPCA. In: Advances in Neural Information Processing Systems, pp. 626–632 (1997)
15. Tipping, M.E., Bishop, C.M.: Probabilistic principal component analysis. Journal of the Royal Statistical Society B 61, 611–622 (1999)
16. Skočaj, D., Bischof, H., Leonardis, A.: A robust PCA algorithm for building representations from panoramic images. In: Heyden, A., Sparr, G., Nielsen, M., Johansen, P. (eds.) ECCV 2002. LNCS, vol. 2353, pp. 761–775. Springer, Heidelberg (2002)
17. Rao, R.: Dynamic appearance-based recognition. In: Proc. CVPR, pp. 540–546 (1997)
18. Black, M.J., Jepson, A.D.: Eigentracking: Robust matching and tracking of articulated objects using a view-based representation. In: Proc. European Conf. on Computer Vision, pp. 329–342 (1996)
19. Leonardis, A., Bischof, H.: Robust recognition using eigenimages. Computer Vision and Image Understanding 78(1), 99–118 (2000)
20. Edwards, J.L., Murase, J.: Coarse-to-fine adaptive masks for appearance matching of occluded scenes. Machine Vision and Applications 10(5–6), 232–242 (1998)
21. Geusebroek, J.M., Burghouts, G.J., Smeulders, A.W.M.: The Amsterdam Library of Object Images. International Journal of Computer Vision 61(1), 103–112 (2005)
22. Storer, M., Roth, P.M., Urschler, M., Bischof, H., Birchbauer, J.A.: Active appearance model fitting under occlusion using fast-robust PCA. In: Proc. International Conference on Computer Vision Theory and Applications (VISAPP), February 2009, vol. 1, pp. 130–137 (2009)

Efficient K-Means VLSI Architecture for Vector Quantization

Hui-Ya Li, Wen-Jyi Hwang*, Chih-Chieh Hsu, and Chia-Lung Hung

Department of Computer Science and Information Engineering,
National Taiwan Normal University, Taipei, 117, Taiwan
royalfay@gmail.com, whwang@ntnu.edu.tw,
andy730215@msn.com, nicky730216@gmail.com

Abstract. A novel hardware architecture for k-means clustering is presented in this paper. Our architecture is fully pipelined for both the partitioning and centroid computation operations so that multiple training vectors can be concurrently processed. The proposed architecture is used as a hardware accelerator for a softcore NIOS CPU implemented on a FPGA device for physical performance measurement. Numerical results reveal that our design is an effective solution with low area cost and high computation performance for k-means design.

1 Introduction

Cluster analysis is a method for partitioning a data set into classes of similar individuals. The clustering applications in various areas such as signal compression, data mining and pattern recognition, etc., are well documented. In these clustering methods the k-means [9] algorithm is the most well-known clustering approach which restricts each point of the data set to exactly one cluster.

One drawback of the k-means algorithm is the high computational complexity for large data set and/or large number of clusters. A number of fast algorithms [2,6] has been proposed for reducing the computational time of the k-means algorithm. Nevertheless, only moderate acceleration can be achieved in these software approaches.

Other alternatives for expediting the k-means algorithm are based on hardware. As compared with the software counterparts, the hardware implementations may provide higher throughput for distance computation. Efficient architectures for distance calculation and data set partitioning process have been proposed in [3,5,10]. Nevertheless, the centroid computation is still conducted by software in some architectures. This may limit the speed of the systems. Although hardware dividers can be employed for centroid computation, the hardware cost of the circuit may be high because of the high hardware complexity for the divider design. In addition, when the usual multi-cycle sequential divider architecture is employed, the implementation of pipeline architecture for both clustering and partitioning process may be difficult.

* To whom all correspondence should be sent.

A.-B. Salberg, J.Y. Hardeberg, and R. Jenssen (Eds.): SCIA 2009, LNCS 5575, pp. 440–449, 2009.

The goal of this paper is to present a novel pipeline architecture for the k-means algorithm. The architecture adopts a low-cost and fast hardware divider for centroid computation. The divider is based on simple table lookup, multiplication and shift operations so that the division can be completed in one clock cycle. The centroid computation therefore can be implemented as a pipeline. In our design, the data partitioning process can also be implemented as a c-stages pipeline for clustering a data set into c clusters. Therefore, our complete k-means architecture contains $c+2$ pipeline stages, where the first c stages are used for the data set partitioning, and the final two stages are adopted for the centroid computation.

The proposed architecture has been implemented on field programmable gate array (FPGA) devices [8] so that it can operate in conjunction with a softcore CPU [12]. Using the reconfigurable hardware, we are then able to construct a system on programmable chip (SOPC) system for the k-means clustering. The applications considered in our experiments are the vector quantization (VQ) for signal compression [4]. Although some VLSI architectures [1,7,11] have been proposed for VQ applications, these architectures are used only for VQ encoding. The proposed architecture is used for the training of VQ codewords. As compared with its software counterpart running on Pentium IV CPU, our system has significantly lower computational time for large training set. All these facts demonstrate the effectiveness of the proposed architecture.

2 Preliminaries

We first give a brief review of the k-means algorithm for the VQ design. Consider a full-search VQ with c codewords $\{y_1, ..., y_c\}$. Given a set of training vectors $T = \{x_1, ..., x_t\}$, the average distortion of the VQ is given by

$$D = \frac{1}{wt}\sum_{j=1}^{t} d(x_j, y_{\alpha(x_j)}), \tag{1}$$

where w is the vector dimension, t is the number of training vectors, $\alpha()$ is the source encoder, and $d(u, v)$ is the squared distance between vectors u and v. The k-means algorithm is an iterative approach finding the solution of $\{y_1, ..., y_c\}$ locally minimizing the average distortion D given in eq.(1). It starts with a set of initial codewords. Given the set of codewords, an optimal partition $T_1, T_2, ..., T_c$ is obtained by

$$T_i = \{x : x \in T, \alpha(x) = i\}, \tag{2}$$

where

$$\alpha(x) = \arg\min_{1\le j\le c} d(x, y_j). \tag{3}$$

After that, given the optimal partition obtained from the previous step, a set of optimal codewords is computed by

$$y_i = \frac{1}{Card(T_i)}\sum_{x\in T_i} x. \tag{4}$$

The same process will be repeated until convergence of the average distortion D of the VQ is observed.

3 The Proposed Architecture

As shown in Fig. 1, the proposed k-means architecture can be decomposed into two units: the partitioning unit and the centroid computation unit. These two units will operate concurrently for the clustering process. The partitioning unit uses the codewords stored in the register to partition the training vectors into c clusters. The centroid computation unit concurrently updates the centroid of clusters. Note that, both the partitioning process and centroid computation process should operate iteratively in software. However, by adopting a novel pipeline architecture, our hardware design allows these two processes to operate in parallel for reducing the computational time. In fact, our design allows the concurrent computation of c+2 training vectors for the clustering operations.

Fig. 2 shows the architecture of the partitioning unit, which is a c-stage pipeline, where c is the number of codewords (i.e., clusters). The pipeline fetch one training vector per clock from the input port. The i-th stage of the pipeline compute the squared distance between the training vector at that stage and the i-th codeword of the codebook. The squared distance is then compared with the current minimum distance up to the i-th stage. If distance is smaller than the current minimum, then the i-th codeword becomes the new current optimal codeword, and the corresponding distance becomes the new current minimum distance. After the computation at the c-th stage is completed, the current optimal codeword and current minimum distance are the actual optimal codeword and the actual minimum distance, respectively. The index of the actual optimal codeword and its distance will be delivered to the centroid computation unit for computing the centroid and overall distortion.

As shown in Fig. 2, each pipeline stage i has input ports *training_vector_in, codeword_in, D_in, index_in,* and output ports *training_vector_out, D_out, index_out.* The *training_vector_in* is the input training vector. The *codeword_in* is the i-th codeword. The *index_in* contains index of the current optimal codeword up to the stage i. The *D_in* is the current minimum distance. Each stage i first computes the squared distance between the input training vector and the i-th codeword (denoted by D_i), and then compared it with the *D_in.* When

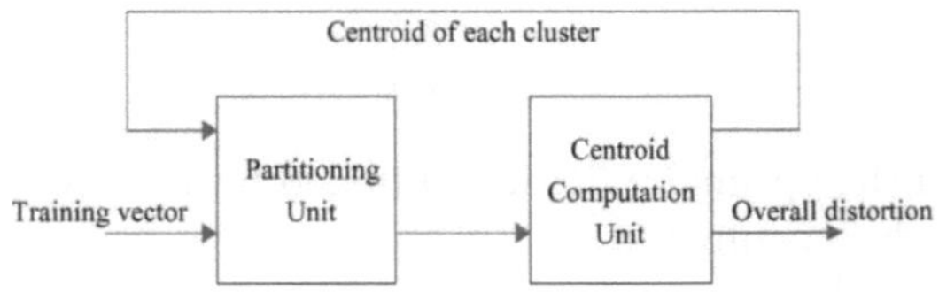

Fig. 1. The proposed k-means architecture

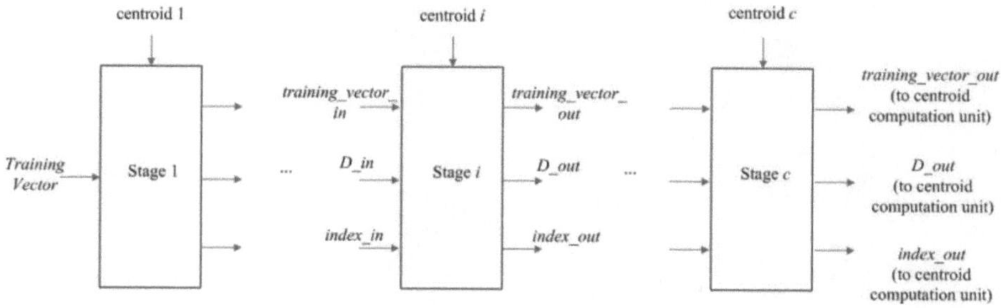

Fig. 2. The architecture of the partitioning unit

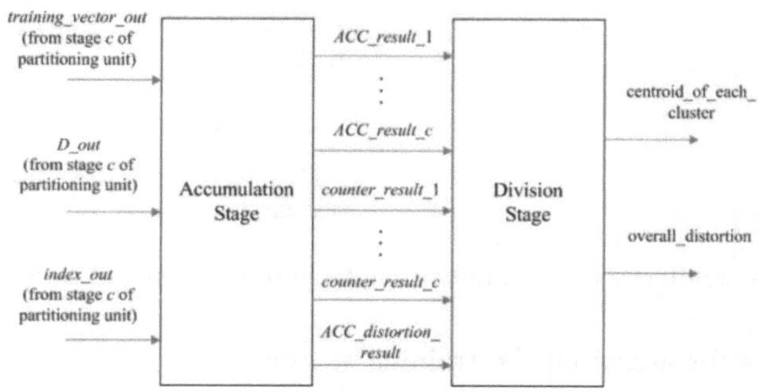

Fig. 3. The architecture of the centroid computation unit

the squared distance is greater than *D_in*, we have *index_out* $\leftarrow$ *index_in* and *D_out* $\leftarrow$ *D_in*. Otherwise, *index_out* $\leftarrow i$, and the *D_out* $\leftarrow D_i$. Note that the output ports *training_vector_out, D_out* and *index_out* at stage i are connected to the input ports *training_vector_in, D_in,* and *index_in* at the stage i+1, respectively. Consequently, the computational results at stage i at the current clock cycle will propagate to stage i+1 at the next clock cycle. When the training vector reaches the c-th stage, the final *index_out* indicates the index of the actual optimal codeword, and the *D_out* contains the corresponding distance.

Fig. 3 depicts the architecture of the centroid computation unit, which can be viewed as a two-stage pipeline. In this paper, we call these two stages, the accumulation stage and division stage, respectively. Therefore, there are $c + 2$ pipeline stages in the k-means unit. The concurrent computation of c+2 training vectors therefore is allowed for the clustering operations.

As shown in Fig. 4, there are c accumulators (denoted by $ACCi, i = 1, .., c$) and c counters for the centroid computation in the accumulation stage. The i-th accumulator records the *current* sum of the training vectors assigned to cluster i. The i-th counter contains the current number of training vectors mapped to cluster i. The *training_vector_out, D_out* and *index_out* in Fig. 4 are actually the outputs of the c-th pipeline stage of the partitioning unit. The *index_out* is used

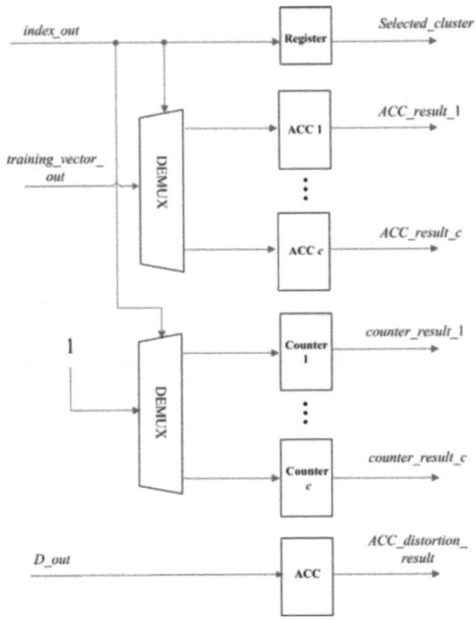

Fig. 4. The architecture of accumulation stage of the centroid computation unit

as control line for assigning the training vector (i.e. *training_vector_out*) to the optimal cluster found by the partitioning unit.

The circuit of division stage is shown in Fig. 5. There is only one divider in the unit because only one centroid computation is necessary at a time. Suppose the final *index_out* is i for the j-th vector in the training set. The centroid of the i-th cluster then need to be updated. The divider and the i-th accumulator and counter are responsible for the computation of the centroid of the i-th cluster. Upon the completion of the j-th training vector at the centroid computation unit, the i-th counter records the number of training vectors (up to j-th vector in the training set) which are assigned to the i-th cluster. The i-th accumulator contains the sum of these training vectors in the i-th cluster. The output of the divider is then the mean value of the training vectors in the i-th cluster.

The architecture of the divider is shown in Fig. 6, which contains w units (w is the vector dimension). Each unit is a scalar divider consisting of an encoder, a ROM, a multiplier and a shift unit. Recall that the goal of the divider is to find the mean value as shown in eq.(4). Because the vector dimension is w, the sum of vectors $\sum_{x \in T_i} x$ has w elements, which are denoted by $S_1, ..., S_w$ in the Fig. 6.(a). For the sake of simplicity, we let S be an element of $\sum_{x \in T_i} x$, and $Card(T_i) = M$. Note that both S and M are integers. It can then be easily observed that

$$\frac{S}{M} = S \times \frac{2^k}{M} \times 2^{-k}, \tag{5}$$

for any integer $k > 0$. Given a positive integer k, the ROM in Fig. 6.(b) in its simplest form have 2^k entries. The m-th, $m = 1, ..., 2^k$, entry of the ROM

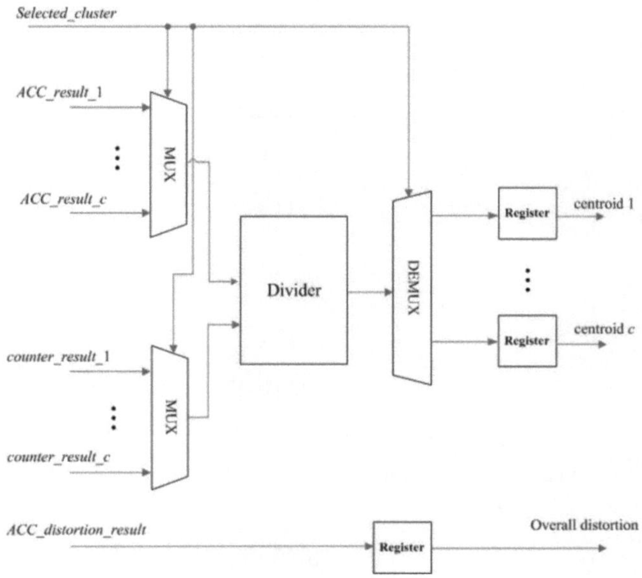

Fig. 5. The architecture of division stage of the centroid computation unit

contains the value $\frac{2^k}{m}$. Consequently, for any positive $M \leq 2^k$, $\frac{2^k}{M}$ can be found by a simple table lookup process from the ROM. The output of the ROM is then multiplied by S, as shown in the Fig. 6.(b). The multiplication result is then shifted right by k bits for the completion of the division operation $\frac{S}{M}$.

In our implementation, each $\frac{2^k}{m}, m = 1, ..., 2^k$, has only finite precision with fixed-point format. Since the maximum value of $\frac{2^k}{m}$ is 2^k, the integer part of $\frac{2^k}{m}$ has k bits. Moreover, the fractional part of $\frac{2^k}{m}$ contains b bits. Each $\frac{2^k}{m}$ therefore is represented by $(k + b)$ bits. There are 2^k entries in the ROM. The ROM size therefore is $(k + b) \times 2^k$ bits.

It can be observed from the Fig. 6 that the division unit also evaluates the overall distortion of the codebook. This can be accomplished by simply accumulating the minimum distortion associated with each training vector after the completion of the partitioning process. The overall distortion is used for both the performance evaluation and the convergence test of the k-means algorithm.

The proposed architecture is used as a custom user logic in a SOPC system consisting of softcore NIOS CPU, DMA controller and SDRAM, as depicted in Fig. 7. The set of training vectors is stored in the SDRAM. The training vectors are then delivered to the proposed circuit one at a time by the DMA controller for k-means clustering. The softcore NIOS CPU only has to activate the DMA controller for the training vector delivery, and then collects the clustering results after the DMA operations are completed. It does not participate in the partitioning and centroid computation processes of the k-means algorithm. The computational time for k-means clustering can then be lowered effectively.

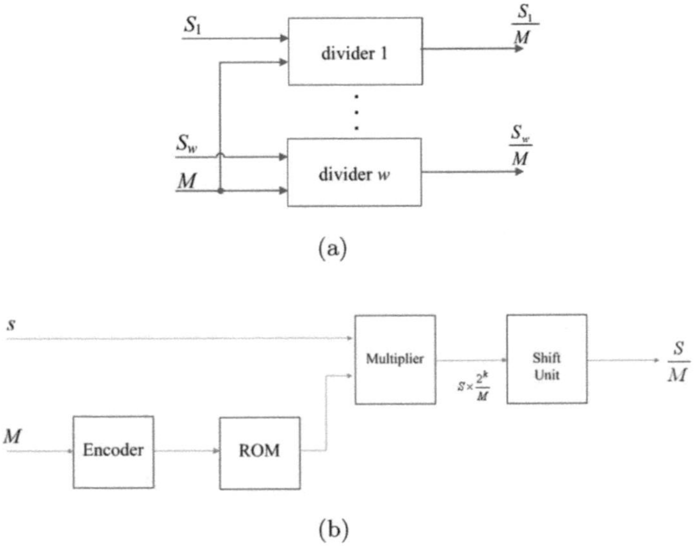

Fig. 6. The architecture of divider: (a) The divider contains w units; (b) Each unit is a scalar divider consisting of an encoder, a ROM, a multiplier, and a shift unit

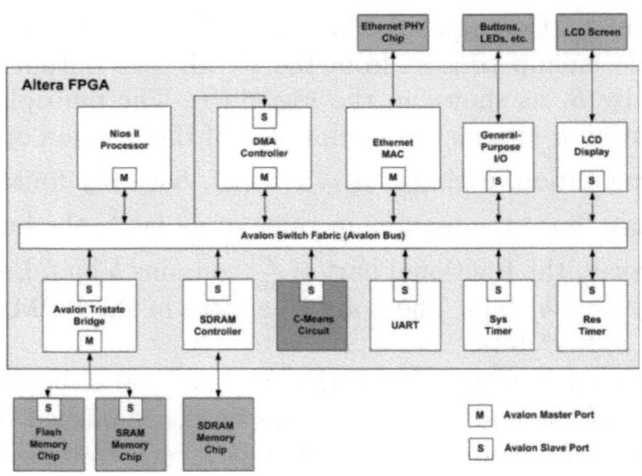

Fig. 7. The architecture of the SOPC using the proposed k-means circuit as custom user logic

4 Experimental Results

This section presents some experimental results of the proposed architecture. The k-means algorithm is used for VQ design for image coding in the experiments. The vector dimension is $w = 2 \times 2$. There are 64 codewords in the VQ. The target FPGA device for the hardware design is Altera Stratix II 2S60.

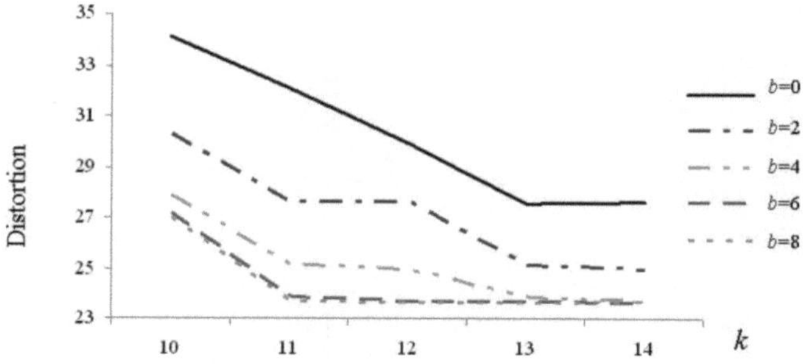

Fig. 8. The performance of the proposed k-means circuit for various sets of parameters k and b

We first consider the performance of the divider for the centroid computation of the k-means algorithm. Recall that our design adopts a novel divider based on table lookup, multiplication and shift operations, as shown in eq.(5). The ROM size of the divider for table lookup is dependent on the parameters k and b. Higher k and b values may improve the k-means performance at the expense of larger ROM size.

Fig. 8 shows the performance of the proposed circuit for various sets of parameters k and b. The training set for VQ design contains 30000 training vectors drawn from the image "Lena" [13]. The performance is defined as the average distortion of the VQ defined in eq.(1). All the VQs in the figure starts with the same set of initial codewords. It can be observed from the figure that the average distortion is effectively lowered as k increases for fixed b. This is because the parameter k set an upper bound on the number of vectors (i.e., M in eq.(5)) in each cluster. In fact, the upper bound of M is 2^k. Higher k values reduce the possibility that actual M is larger than 2^k. This may enhance the accuracy for centroid computation. We can also see from Fig. 8 that larger b can reduce the average distortion as well. Larger b values increase the precision for the representation of $\frac{2^k}{m}$; thereby improve the division accuracy.

The area cost of the proposed k-means circuit for various sets of parameters k and b is depicted in Fig. 9. The area cost is measured by the number of adaptive logic modules (ALMs) consumed by the circuit. It can be observed from the figure that the area cost of our circuit reduces significantly when k and/or b becomes small. However, improper selection of k and b for area cost reduction may increase the average distortion of the VQ. We can see from Fig. 8 that the division circuit with $b = 8$ has performance less susceptible to k. It can be observed from Fig. 8 and 9 that the average distortion of the circuit with $(b = 8, k = 11)$ is almost identical to that of the circuit with $(b = 8, k = 14)$. Moreover, the area cost of the centroid computation unit with $(b = 8, k = 11)$ is significantly lower than that of the circuit with $(b = 8, k = 14)$. Consequently, in our design, we select $b = 8$ and $k = 11$ for the divider design.

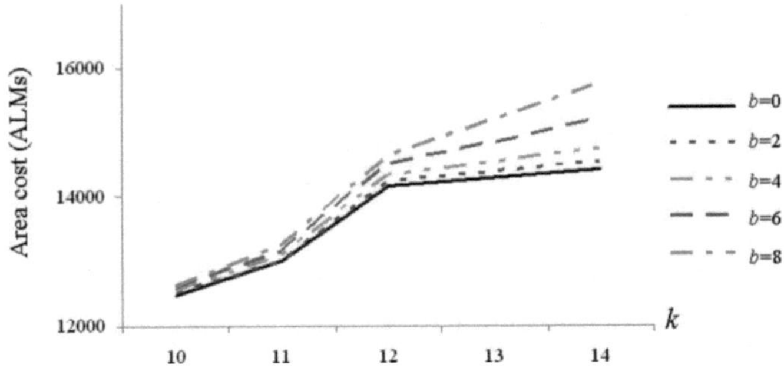

Fig. 9. The area cost of the k-means circuit for various sets of parameters k and b

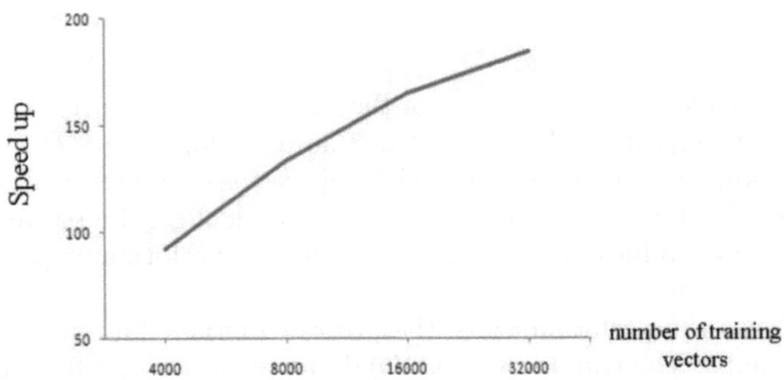

Fig. 10. Speedup of the proposed system over its software counterpart

Our SOPC system consists of softcore NIOS CPU, DMA controller, 10 M bytes SDRAM and the proposed k-means circuit. The k-means circuit consumes 13253 ALMs, 8192 embedded memory bits and 288 DSP elements. The NIOS softcore CPU of our system also consumes hardware resources. The entire SOPC system uses 17427 ALMs and 604928 memory bits.

Fig. 10 compares the CPU time of our system with its software counterpart running on 3 GHz Pentium IV CPU for various sizes of training data set. It can be observed from the figure that the execution time of our system is significantly lower than that of its software counterpart. In addition, gap in CPU time enlarges as the the training set size increases. This is because our system is based on efficient pipelined computation for partitioning and centroid operations. When the training set size is 32000 training vectors, the CPU time of our system is only 3.95 mini seconds, which is only 0.54% of the CPU time of its software counterpart. The speedup of our system over software implementation is 185.18.

5 Concluding Remarks

The proposed architecture has been found to be effective for k-means design. It is fully pipelined with simple divider for centroid computation. It has high throughput, allowing concurrent partitioning and centroid operations for $c+2$ training vectors. The architecture can be efficiently used as an hardware accelerator for a general processor. As compared with the software k-means running on Pentium IV, the NIOS-based SOPC system incorporating our architecture has significantly lower execution time. The proposed architecture therefore is beneficial for reducing computational complexity for clustering analysis.

References

1. Bracco, M., Ridella, S., Zunino, R.: Digital implementation of hierarchical vector quantization. IEEE Trans. Neural Networks, 1072–1084 (2003)
2. Elkan, C.: Using the triangle inequality to accelerate K-Means. In: Proc. International Conference on Machine Learning (2003)
3. Estlick, M., Leeser, M., Theiler, J., Szymanski, J.J.: Algorithmic transformations in the implementation of K- means clustering on reconfigurable hardware. In: Proc. of ACM/SIGDA 9th International Symposium on Field Programmable Gate Arrays (2001)
4. Gersho, A., Gray, R.M.: Vector Quantization and Signal Compression. Kluwer, Norwood (1992)
5. Gokhale, M., Frigo, J., Mccabe, K., Theiler, J., Wolinski, C., Lavenier, D.: Experience with a Hybrid Processor: K-Means Clustering. The Journal of Supercomputing, 131–148 (2003)
6. Hwang, W.J., Jeng, S.S., Chen, B.Y.: Fast Codeword Search Algorithm Using Wavelet Transform and Partial Distance Search Techniques. Electronic Letters 33, 365–366 (1997)
7. Hwang, W.J., Wei, W.K., Yeh, Y.J.: FPGA Implementation of Full-Search Vector Quantization Based on Partial Distance Search. Microprocessors and Microsystems, 516–528 (2007)
8. Hauck, S., Dehon, A.: Reconfigurable Computing. Morgan Kaufmann, San Francisco (2008)
9. MacQueen, J.: Some Methods for Classi cation and Analysis of Multivariate Observations. In: Proc. of the 5th Berkeley Symposium on Mathematical Statistics and Probability, pp. 281–297 (1967)
10. Maruyama, T.: Real-time K-Means Clustering for Color Images on Reconfigurable Hardware. In: Proc. 18th International Conference on Pattern Recognition (2006)
11. Wang, C.L., Chen, L.M.: A New VLSI Architecture for Full-Search Vector Quantization. IEEE Trans. Circuits and Sys. for Video Technol., 389–398 (1996)
12. NIOS II Processor Reference Handbook, Altera Corporation (2007), `http://www.altera.com/literature/lit-nio2.jsp`
13. USC-SIPI Lab, `http://sipi.usc.edu/database/misc/4.2.04.tiff`

Joint Random Sample Consensus and Multiple Motion Models for Robust Video Tracking

Petter Strandmark[1,2] and Irene Y.H. Gu[1]

[1] Dept. of Signals and Systems, Chalmers Univ. of Technology, Sweden
{irenegu,petters}@chalmers.se

[2] Centre for Mathematical Sciences, Lund University, Sweden
petter@maths.lth.se

Abstract. We present a novel method for tracking multiple objects in video captured by a non-stationary camera. For low quality video, RANSAC estimation fails when the number of good matches shrinks below the minimum required to estimate the motion model. This paper extends RANSAC in the following ways: (a) Allowing multiple models of different complexity to be chosen at random; (b) Introducing a conditional probability to measure the suitability of each transformation candidate, given the object locations in previous frames; (c) Determining the best suitable transformation by the number of consensus points, the probability and the model complexity. Our experimental results have shown that the proposed estimation method better handles video of low quality and that it is able to track deformable objects with pose changes, occlusions, motion blur and overlap. We also show that using multiple models of increasing complexity is more effective than just using RANSAC with the complex model only.

1 Introduction

Multiple object tracking in video has been intensively studied in recent years, largely driven by an increasing number of applications ranging from video surveillance, security and traffic control, behavioral studies, to database movie retrievals and many more. Despite the enormous research efforts, many challenges and open issues still remain, especially for multiple non-rigid moving objects in complex and dynamic backgrounds with non-stationary cameras. Despite that human eyes may easily track objects with changing poses, shape, appearances, illuminations and occlusions, robust machine tracking remains a challenging issue.

Blob-tracking is one of the most commonly used approaches, where a bounding box is used for a target object region of interest [6]. Another family of approaches is through exploiting local point features of objects and finding correspondences between points in different image frames. Scale-Invariant Feature Transform (SIFT) [7] is a common local feature extraction and matching method that can be used for tracking. Speeded-Up Robust Features (SURF) [1], has been proposed for speeding up the SIFT through the use of integral images. Both methods provide high-dimensional (e.g. 128) feature descriptors that are invariant to object rotation and scaling, and affine changes in image intensities.

A.-B. Salberg, J.Y. Hardeberg, and R. Jenssen (Eds.): SCIA 2009, LNCS 5575, pp. 450–459, 2009.

Typically, not all correspondences are correct. Often, a number of erroneous matches far away from the correct position are returned. To alleviate this problem, RANSAC [3] is used to estimate the inter-frame transformations [2,4,5,8,10,11]. It estimates a transformation by choosing a random sample of point correspondences, fitting a motion model and counting the number of agreeing points. The transformation candidate with the highest number of agreeing points is chosen (consensus). However, the number of good matches obtained by SIFT or SURF may often momentarily be very low. This is caused by motion blur and compression artifacts for video of low quality, or by object deformations, pose changes or occlusion. If the number of good matches shrinks below the minimum required number needed to estimate the prior transformation model, RANSAC will fail. A key observation is that it is difficult to predict whether a sufficient number of good matches is available for transformation estimation, since the ratio of good matches to the number of outliers is unknown.

There are other methods for removing outliers from a set of matches. [12] recently proposed a method with no prior motion model. However, just like RANSAC the methods assumes that several correct matches are available, which is not always the case for the fast-moving video sequences considered in this work.

Motivated by the above, we propose a robust estimation method by allowing multiple models of different complexity to be considered when estimating the inter-frame transformation. The idea is that when many good matches are available, a complex model should be employed. Conversely, when few good matches are available, a simple model should be used. To determine which model to choose, a probabilistic method is introduced that evaluates each transformation candidate using a prior from previous frames.

2 Tracking System Description

To give a big picture, Fig. 1 shows a block diagram of the proposed method. For a given image $\mathbf{I}_t(n,m)$ at the current frame t, a set of candidate feature points $\mathbf{F}_t^c$ are extracted from the entire image area (block 1). These features are then matched against the feature set of the tracked object $\mathbf{F}_{t-1}^{\text{obj}}$, resulting in a matched feature subset $\mathbf{F}_t' \subset \mathbf{F}_t^c$ (block 2). The best transformation is estimated by evaluating different candidates with respect to the number of consensus points and an estimated probability (block 3). The feature subset $\mathbf{F}_t'$ is then updated by

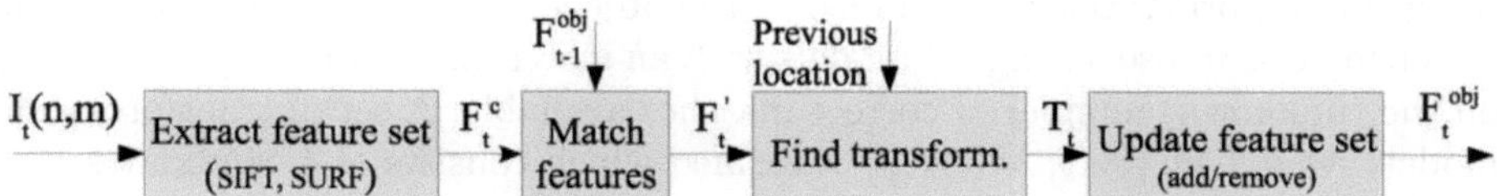

Fig. 1. Block diagram for the proposed tracking method

allowing adding new features within the new object location (block 4). Within object intersections or overlaps updating is not performed. This yields the final feature set for the tracked object $\mathbf{F}_t^{\text{obj}}$ in the current frame t. Block 3 and 4 are described in section 3 and 4, respectively.

3 Random Model and Sample Consensus

To make the motion estimation method robust when the number of good matches becomes very low, our proposed method, RAMOSAC, chooses both the model used for estimation and the sample of point correspondences randomly. The main novelties are: **(a)** Using four types of transformations (see section 3.1), we allow the model itself to be chosen at random from a set of models of different complexity. **(b)** A probability is defined to measure the suitability of each transformation candidate, given the object locations in previous frames. **(c)** The best suitable transformation is determined by the maximum score, defined as the combination of the number of consensus points, the probability of the given candidate transformation, and the complexity of the model. It is worth mentioning that while RANSAC uses only the number of consensus points as the measure of a model, our method differs by using a combination of the number of consensus points and a conditional probability to choose a suitable transformation. Briefly, the proposed RAMOSAC operates in an iterative fashion similar to RANSAC in the following manner:

1. Choose a model at random;
2. Choose a random subset of feature points;
3. Estimate the model using this subset;
4. Evaluate the resulting transformation based on number of agreeing points and the probability given the previous movement.
5. Repeat 1–4 several times and choose the candidate T with the highest score.

Alternatively, each of the possible motion models could be evaluated a fixed number of times. However, because the algorithm is typically iterated until the next frame arrives, the total number of iterations is not known. Choosing a model at random every iteration ensures that no motion model is unduly favored over another. Detailed description of RAMOSAC will be given in the remaining of this section.

3.1 Multiple Transformation Models

Several transformations are included in the object motion model set. The basic idea behind is to use a range of models with an increasing complexity, depending on the (unknown) number of correct matches available. A set of transformation models $\mathcal{M} = \{\mathcal{M}_a, \mathcal{M}_s, \mathcal{M}_t, \mathcal{M}_p\}$ is formed which consists of 4 candidates:

1. Pure translation $\mathcal{M}_t$, with 2 unknown parameters;
2. Similarity transformation $\mathcal{M}_s$, with 4 unknown parameters: rotation, scaling and translation;

3. Affine transformation $\mathcal{M}_a$, with 6 unknown parameters;
4. Projective transformation (described by a 3×3 matrix) $\mathcal{M}_p$, with 8 unknown parameters (since the matrix is indifferent to scale).

The minimum required number of correspondence points for estimating the parameters for the models $\mathcal{M}_t$, $\mathcal{M}_s$, $\mathcal{M}_a$ and $\mathcal{M}_p$ are $n_{\min}$=1, 2, 3 and 4, respectively. If the number of correspondence points available is larger than the minimum required number, least-squares (LS) estimation should be used to solve the over-determined set of equations.

One can see that a range of complexity is involved in these four types of transformations: The simplest motion model is translation, which can be described by a single point correspondence, or by the mean displacement if more points are available. If more matched correspondence points are available, a more detailed motion model can be considered: with a minimum of 2 matched correspondences, the motion can be descried in terms of scaling, rotation and translation by $\mathcal{M}_s$. With 3 matched correspondences, affine motion can be described by adding more parameters such as skew and separate scales in two directions using $\mathcal{M}_a$. With 4 matched correspondences, projective motion can be described by the transformation $\mathcal{M}_p$, which completely describes the image transformation of a surface moving freely in 3 dimensions.

3.2 Probability for Choosing a Transformation

To assess whether a candidate transformation T estimated from a model $\mathcal{M} \in \{\mathcal{M}_t, \mathcal{M}_s, \mathcal{M}_a, \mathcal{M}_p\}$ is suitable for describing the motion of the tracked object, a distance measure and a conditional probability are defined by using the position of the object from the previous frame $t-1$. We assume that the object movement follows the same distribution in two consecutive image frames. Let the normalized boundary of the tracked object be $\boldsymbol{\gamma} : [0,1] \mapsto \mathbf{R}^2$, and the normalized boundary of the tracked object under a candidate transformation be $T(\boldsymbol{\gamma})$. A distance measure is defined as the movement of the boundary under the transformation T:

$$\text{dist}(T|\boldsymbol{\gamma}) = \int_0^1 ||\boldsymbol{\gamma}(t) - T(\boldsymbol{\gamma}(t))||dt. \tag{1}$$

When the boundary can be described by a polygon $\boldsymbol{p}_t = \{p_t^k\}_{k=1}^n$, only the distances moved by the points are considered:

$$\text{dist}(T|\boldsymbol{p}_{t-1}) = \sum_{k=1}^{n} ||p_{t-1}^k - T(p_{t-1}^k)||. \tag{2}$$

A distribution that have been empirically proven to approximate the inter-frame movement is the exponential distribution (density function $\lambda e^{\lambda x}$). The parameter λ is estimated from the movements measured in previous frames. The probability of a candidate transformation T is the probability of a movement with greater or equal magnitude. Given the previous object boundary and the decay rate λ this probability is:

$$P(T|\lambda, p_{t-1}) = e^{-\lambda \operatorname{dist}(T|\boldsymbol{p}_{t-1})} \tag{3}$$

This way, transformations resulting in big movements are penalized, while transformations resulting in small movements are favored. In addition to the number of consensus points, this is the criterion used to select the correct transformation.

3.3 Criterion for Selecting a Transformation Model

A score is defined for choosing the best transformation and is computed for every transformation candidate T, which are estimated using a random model and a random choice of point correspondences:

$$\operatorname{score}(T) = \#(C) + \log_{10} P(T|\lambda, \boldsymbol{p}^{t-1}) + \varepsilon n_{\min}, \tag{4}$$

where $\#(C)$ is the number of consensus points, and $n_{\min}$ is the minimum number of points needed to estimate the model correctly. The last term $\varepsilon n_{\min}$ is introduced to slightly favor a more complicated model. Otherwise, if the movement is small, both a simple and a complex model might have the same number of consensus points and approximately the same probability, resulting in the selection of a simple model. This would ignore the increased accuracy of the advanced model, and could lead to unnecessary error accumulation over time. Adding the last term hence enable, if all other terms are equal, the choice of a more advanced model. $\varepsilon = 0.1$ was used in our experiments.

The score is computed for every candidate transformation. The transformation T having the highest score is then chosen as the correct transformation model for the current video frame, after LS re-estimation over the consensus set. It is worth noting that the score in the RANSAC is $\operatorname{score}(T) = \#(C)$ with only one model. Table 1 summarizes the proposed algorithm.

4 Updating Point Feature Set

It is essential that a good feature set of the tracked object $\mathbf{F}_t^{\text{obj}}$ is maintained and updated. A simple method is proposed here for updating the feature set of the tracked object, through dynamically adding and pruning feature points. To achieve this, a score S_t is assigned to each object feature point. All feature points are then sorted according to their score values. Only the top M feature points are used for matching the object. The score for each feature point is then updated based on the matching result and motion estimation:

$$S_t = \begin{cases} S_{t-1} + 2 & \text{matched, consensus point} \\ S_{t-1} - 1 & \text{matched, outlier} \\ S_{t-1} & \text{not matched} \end{cases} \tag{5}$$

Table 1. The RAMOSAC algorithm in pseudo-code

```
Input: Models M_i, i = 1,...,m, Point correspondences (x_k^(t-1), x'_k^(t)),
       x_k^(t-1) ∈ F^obj_(t-1), x'_k^(t) ∈ F'_t, λ, p_(t-1)
Parameters: i_max = 30, d_thresh = 3
s_best ← -∞
for i ← 1...i_max do
    Randomly pick M from M_1...M_m
    n_min ← number of points to estimate M
    Randomly choose a subset of n_min index points
    Using M, estimate T from this subset
    C ← {}
    foreach (x_k, x'_k) do
        if ||x'_k - T(x_k)||^2 < d_thresh then Add k to C
    end
    s ← #(C) + log_10 P(T|λ, p_(t-1)) + ε n_min
    if s > s_best then
        M_best ← M
        C_best ← C
        s_best ← s
    end
end
Using M_best, estimate T from C_best
return T
```

Initially, the score of a feature point is set to be the median of the feature points currently used for matching. In that way, all new feature points will be tested in the next frame without interfering with the important feature points that have the highest scores. For low-quality video with significant motion blur, this simple method was proven successful. It allows the inclusion of new features while maintaining stable feature points.

Pruning of feature points: In practice, only a small portion of the candidate points with high score are kept in the memory. The remaining feature points are pruned for maintaining a manageable size of feature list. Since these pruned feature points have low scores, they are unlikely to be used as the key feature points for tracking the target objects. Figure 2 shows the final score distribution of the 3568 features collected throughout the test video "Picasso", with $M = 100$.

Updating of feature points when two objects intersect or overlap: When multiple objects intersect or overlap, feature points located in the intersection need special care in order to be assigned to the correct object. This is solved by examining the matches within the intersection. The object having consensus points within the intersection area is considered the foreground object and any new features within that area are assigned to it. No other special treatment is required for tracking multiple objects. Figure 5 shows an example of tracking results with two moving objects (walking persons) using the proposed method.

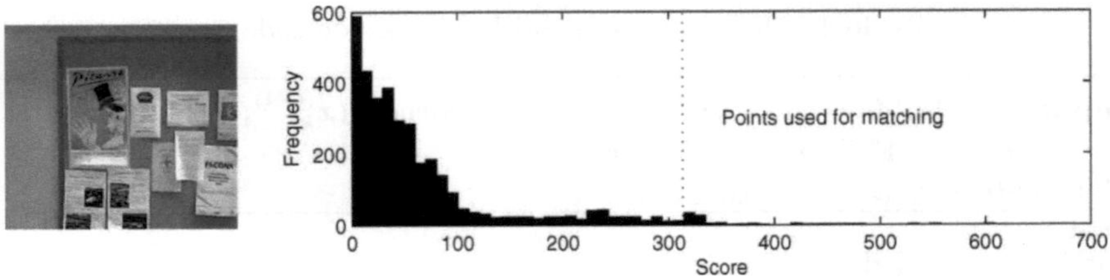

Fig. 2. Final score distribution for the "Picasso" video. The $M = 100$ highest scoring features were used for matching.

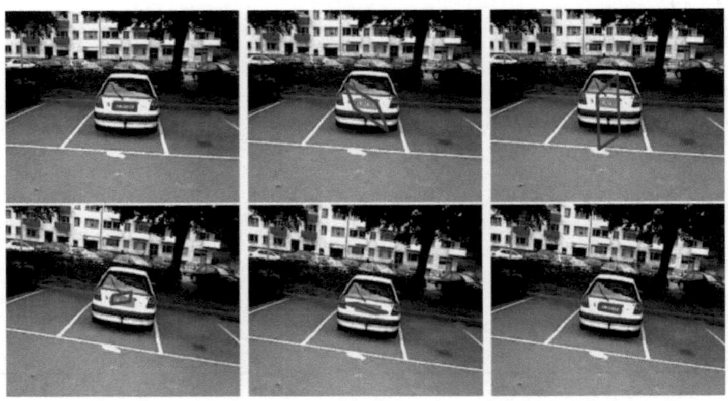

Fig. 3. RANSAC (red) compared to proposed method RAMOSAC (green) for frames #68–#70, #75–#77 of the "'Car" sequence. See also Fig. 6 for comparison. For some frames in this sequence, there is a single correct match with several outliers, making RANSAC estimation impossible.

Fig. 4. Tracking results from the proposed method RAMOSAC for the video "David" [9], showing matched points (green), outliers (red) and newly added points (yellow)

Fig. 5. Tracking two overlapping pedestrians (marked by red and green) using the proposed method

5 Experiments and Results

The proposed method RAMOSAC have been tested for a range of scenarios, including tracking rigid objects, deformable objects, objects with pose changes and multiple overlapping objects. The video used for our tests were recorded by using a cell phone camera with a resolution of 320 × 200 pixels. Three examples are included: In Fig. 3 we show an example of tracking a rigid license plate in video with a very high amount of motion blur, resulting in a low number of good matches. Results from the proposed method and from RANSAC are included for comparison. In the 2nd example, shown in the first row of Fig. 4, a face (with pose changes) was captured with a non-stationary camera. The 3rd example, shown in the 2nd row of Fig. 5, simultaneously tracks two walking persons (containing overlap). By observing the results from these videos in our tests, and from the results shown in these figures, one can see that the proposed method is robust for tracking moving objects with a range of complex scenarios.

The algorithm (implemented in MATLAB) runs in real-time on a modern desktop computer for 320 × 200 video if the faster SURF features are used. It should be noted that over 90% of the processing time is nevertheless spent calculating features. Therefore, any additional processing required by our algorithm is not an issue. Also, both the extraction of features and the estimation of the transformation is amenable to parallelization over multiple CPU cores.

All video files used in this paper are available for download at `http://www.maths.lth.se/matematiklth/personal/petter/video.php`

5.1 Performance Evaluation

To evaluate the performance, and compare the proposed RAMOSAC estimation with RANSAC estimation, the "ground truth" rectangle for each frame of the "Car" sequence (see Fig. 3) was manually marked. The Euclidean distance between the four corners of the tracked object (i.e. car license plate) and the ground truth

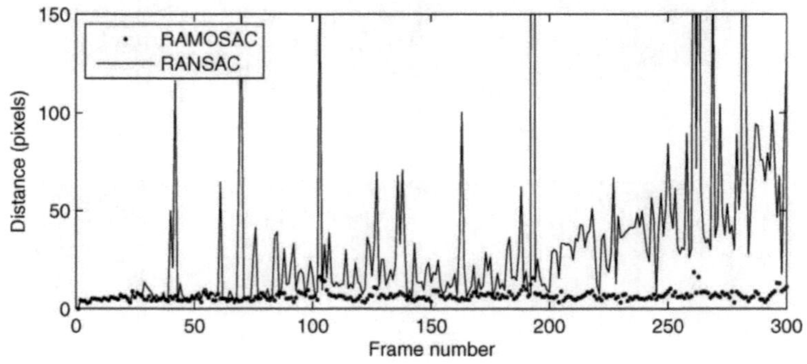

Fig. 6. Euclidean distance between the four corners of the tracked license plate and the ground truth license plate vs. frame numbers, for the "Car" video. Dotted blue line: the proposed RAMOSAC. Solid line: RANSAC.

was then calculated over all frames. Figure 6 shows the distance as a function of image frame for the "Car" sequence. In this comparison, RANSAC always used an affine transformation, whereas RAMOSAC chose from translation, similarity and an affine transformation. The increased robustness obtained from allowing models of lower complexity during difficult passages is clearly seen in Fig. 6.

6 Conclusion

Motion estimation based on RANSAC and (e.g.) an affine motion model requires that at least three correct point correspondences are available. This is not always the case. If less than the minimum number of correct correspondences are available, the resulting motion estimation will always be erroneous.

The proposed method, based on using multiple motion transformation models and finding the maximum number of consensus feature points, as well as a dynamic updating procedure for maintaining feature sets of tracked objects, has been tested for tracking moving objects in videos. Experiments have been conducted on tracking moving objects over a range of video scenarios, including rigid or deformable objects with pose changes, occlusions and two objects with intersect and overlap. Results have shown that the proposed method is capable of, and relatively robust in handling such scenarios.

The method has shown especially effective for tracking in low quality videos (e.g. captured by mobile phone, or videos with large motion blur) where motion estimation using RANSAC runs into some problems. We have shown that using multiple models of increasing complexity is more effective than RANSAC with the complex model only.

Acknowledgments

This project was sponsored by the Signal Processing Group at Chalmers University of Technology and in part by the European Research Council (GlobalVision grant no. 209480), the Swedish Research Council (grant no. 2007-6476) and the Swedish Foundation for Strategic Research (SSF) through the programme Future Research Leaders.

References

1. Bay, H., Ess, A., Tuytelaars, T., Van Gool, L.: SURF: Speeded up robust features. Computer Vision and Image Understanding (CVIU) 110(3), 346–359 (2008)
2. Clarke, J.C., Zisserman, A.: Detection and tracking of independent motion. Image and Vision Computing 14, 565–572 (1996)
3. Fischler, M.A., Bolles, R.C.: Random sample consensus: a paradigm for model fitting with applications to image analysis and automated cartography. Commun. ACM 24(6), 381–395 (1981)
4. Gee, A.H., Cipolla, R., Gee, A., Cipolla, R.: Fast visual tracking by temporal consensus. Image and Vision Computing 14, 105–114 (1996)
5. Grabner, M., Grabner, H., Bischof, H.: Learning features for tracking. In: IEEE Conference on Computer Vision and Pattern Recognition, CVPR 2007, June 2007, pp. 1–8 (2007)
6. Li, L., Huang, W., Gu, I.Y.-H., Luo, R., Tian, Q.: An efficient sequential approach to tracking multiple objects through crowds for real-time intelligent cctv systems. IEEE Trans. on Systems, Man, and Cybernetics 38(5), 1254–1269 (2008)
7. Lowe, D.: Distinctive image features from scale-invariant keypoints. International Journal of Computer Vision 20, 91–110 (2004)
8. Malik, S., Roth, G., McDonald, C.: Robust corner tracking for real-time augmented reality. In: VI 2002, p. 399 (2002)
9. Ross, D., Lim, J., Lin, R.-S., Yang, M.-H.: Incremental learning for robust visual tracking. International Journal of Computer Vision 77(1), 125–141 (2008)
10. Simon, G., Fitzgibbon, A.W., Zisserman, A.: Markerless tracking using planar structures in the scene. In: IEEE and ACM International Symposium on Augmented Reality (ISAR 2000). Proceedings (2000)
11. Skrypnyk, I., Lowe, D.G.: Scene modelling, recognition and tracking with invariant image features. In: ISMAR 2004, Washington, DC, USA, pp. 110–119. IEEE Comp. Society, Los Alamitos (2004)
12. Li, X.-R., Li, X.-M., Li, H.-L., Cao, M.-Y.: Rejecting outliers based on correspondence manifold. Acta Automatica Sinica (2008)

Extending GKLT Tracking—Feature Tracking for Controlled Environments with Integrated Uncertainty Estimation

Michael Trummer[1], Christoph Munkelt[2], and Joachim Denzler[1]

[1] Friedrich-Schiller University of Jena, Chair for Computer Vision
Ernst-Abbe-Platz 2, 07743 Jena, Germany
{michael.trummer,joachim.denzler}@uni-jena.de
[2] Fraunhofer Society, Optical Systems
Albert-Einstein-Straße 7, 07745 Jena, Germany
christoph.munkelt@iof.fraunhofer.de

Abstract. Guided Kanade-Lucas-Tomasi (GKLT) feature tracking offers a way to perform KLT tracking for rigid scenes using known camera parameters as prior knowledge, but requires manual control of uncertainty. The uncertainty of prior knowledge is unknown in general. We present an extended modeling of GKLT that overcomes the need of manual adjustment of the uncertainty parameter. We establish an extended optimization error function for GKLT feature tracking, from which we derive extended parameter update rules and a new optimization algorithm in the context of KLT tracking. By this means we give a new formulation of KLT tracking using known camera parameters originating, for instance, from a controlled environment. We compare the extended GKLT tracking method with the original GKLT and the standard KLT tracking using real data. The experiments show that the extended GKLT tracking performs better than the standard KLT and reaches an accuracy up to several times better than the original GKLT with an improperly chosen value of the uncertainty parameter.

1 Introduction

Three-dimensional (3D) reconstruction from digital images requires, more or less explicitly, a solution to the correspondence problem. A solution can be found by matching and tracking algorithms. The choice between matching and tracking depends on the problem setup, in particular on the camera baseline, available prior knowledge, scene constraints and requirements in the result.

Recent research [1,2] deals with the special problem of active, purposive 3D reconstruction inside a controlled environment, like the robotic arm in Fig. 1, with active adjustment of sensor parameters. These methods, also known as next-best-view (NBV) planning methods, use the controllable sensor and the additional information about camera parameters endowed by the controlled environment to meet the reconstruction goals (e.g. no more than n views, defined reconstruction accuracy) in an optimal manner.

A.-B. Salberg, J.Y. Hardeberg, and R. Jenssen (Eds.): SCIA 2009, LNCS 5575, pp. 460–469, 2009.

Matching algorithms suffer from ambiguities. On the other hand, feature tracking methods are favored by small baselines that can be generated in the context of NBV planning methods. Thus, KLT tracking turns into the method of choice for solving the correspondence problem within NBV procedures. Previous work has shown that it is worth to look for possible improvements of the KLT tracking method by incorporating prior knowledge about camera parameters. This additional knowledge may originate from a controlled environment or from an estimation step within the reconstruction process. Using an estimation of the camera parameters implicates the need to address the uncertainty of this information explicitly.

Originally, the formulation of feature tracking based on an iterative optimization process is the work of Lucas and Kanade [3]. Since then a rich variety of extensions to the original formulation has been published, as surveyed by Baker and Matthews [4]. These extensions may be used independently from the incorporation of camera parameters. For example, Fusiello et al. [5] deal with the removal of spurious correspondences by using robust statistics. Zinsser et al. [6] propose a separated tracking process by inter-frame translation estimation using block matching followed by estimating the affine motion with respect to the template image. Heigl [7] uses an estimation of camera parameters to move features along their epipolar line, but he does not consider the uncertainty of the estimation. Trummer et al. [8,9] give a formulation of KLT tracking, called *Guided KLT tracking (GKLT)*, with known camera parameters regarding uncertainty, using the traditional optimization error function. They adjust uncertainty manually and do not estimate it within the optimization process.

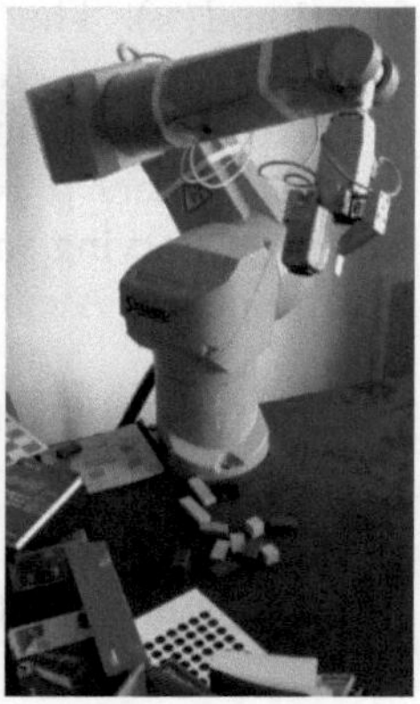

Fig. 1. Robotic arm *Stäubli RX90L* as an example of a controlled environment

This paper contributes to the solution of the correspondence problem by incorporating known camera parameters into the model of KLT tracking under explicit treatment of uncertainty. The resulting extension of GKLT tracking estimates the feature warping together with the amount of uncertainty during the optimization process. Inspired by the EM approach [10], the extended GKLT tracking algorithm uses alternating iterative estimation of hidden information and result values.

The remainder of the paper is organized as follows. Section 2 gives a repetition of KLT tracking basics and defines the notation. It also views the adaptations of GKLT tracking. The incorporation of known camera parameters into the KLT framework with uncertainty estimation is presented in Sect. 3. Section 4 lists experimental results that allow the comparison between the standard KLT, GKLT and the extended GKLT tracking presented in Sect. 3. The paper is concluded in Sect. 5 by summary and outlook to future work.

2 KLT and GKLT Tracking

For the sake of clarity of the explanations in the following sections, we first review the basic KLT tracking and the adaptations for GKLT tracking. The complete derivations can be found in [3,4] (KLT) and [8] (GKLT).

2.1 KLT Tracking

Given a feature position in the initial frame, KLT feature tracking aims at finding the corresponding feature position in the consecutive input frame with intensity function $I(\mathbf{x})$. The initial frame is the template image with intensity function $T(\mathbf{x}), \mathbf{x} = (x, y)^T$. A small image region and the intensity values inside describe a feature. This descriptor is called feature patch P. Tracking a feature means that the parameters $\mathbf{p} = (p_1, ..., p_n)^T$ of a warping function $W(\mathbf{x}, \mathbf{p})$ are estimated iteratively, trying to minimize the squared intensity error over all pixels in the feature patch. A common choice is affine warping by

$$W^a(\mathbf{x}, \mathbf{p}^a) = \begin{pmatrix} a_{11} & a_{12} \\ a_{21} & a_{22} \end{pmatrix} \begin{pmatrix} x \\ y \end{pmatrix} + \begin{pmatrix} \Delta x \\ \Delta y \end{pmatrix} \tag{1}$$

with $\mathbf{p}^a = (\Delta x, \Delta y, a_{11}, a_{12}, a_{21}, a_{22})^T$. The error function of the optimization problem can be written as

$$\epsilon(\mathbf{p}) = \sum_{\mathbf{x} \in P} (I(W(\mathbf{x}, \mathbf{p})) - T(\mathbf{x}))^2, \tag{2}$$

where the goal is to find $\arg\min_{\mathbf{p}} \epsilon(\mathbf{p})$. Following the additional approach (cf. [4]), the error function is reformulated yielding

$$\epsilon(\Delta\mathbf{p}) = \sum_{\mathbf{x} \in P} (I(W(\mathbf{x}, \mathbf{p} + \Delta\mathbf{p})) - T(\mathbf{x}))^2. \tag{3}$$

To resolve for $\Delta\mathbf{p}$ in the end, first-order Taylor approximations are applied to clear the functional dependencies of $\Delta\mathbf{p}$. Two approximation steps give

$$\epsilon'(\Delta\mathbf{p}) = \sum_{\mathbf{x} \in P} (I(W(\mathbf{x}, \mathbf{p})) + \nabla I \nabla_{\mathbf{p}} W(\mathbf{x}, \mathbf{p}) \Delta\mathbf{p} - T(\mathbf{x}))^2 \tag{4}$$

with $\epsilon(\Delta\mathbf{p}) \approx \epsilon'(\Delta\mathbf{p})$ for small $\Delta\mathbf{p}$. The expression in (4) is differentiated with respect to $\Delta\mathbf{p}$ and set to zero. After rearranging the terms it follows that

$$\Delta\mathbf{p} = \mathbf{H}^{-1} \sum_{\mathbf{x} \in P} (\nabla I \nabla_{\mathbf{p}} W(\mathbf{x}, \mathbf{p}))^T (T(\mathbf{x}) - I(W(\mathbf{x}, \mathbf{p}))) \tag{5}$$

using the first-order approximation $\mathbf{H}$ of the Hessian,

$$\mathbf{H} = \sum_{\mathbf{x} \in P} (\nabla I \nabla_{\mathbf{p}} W(\mathbf{x}, \mathbf{p}))^T (\nabla I \nabla_{\mathbf{p}} W(\mathbf{x}, \mathbf{p})). \tag{6}$$

Equation (5) delivers the iterative update rule for the warping parameter vector.

2.2 GKLT Tracking

In comparison to standard KLT tracking, GKLT [8] uses knowledge about intrinsic and extrinsic camera parameters to alter the translational part of the warping function. Features are moved along their respective epipolar line, but allowing for translations perpendicular to the epipolar line caused by the uncertainty in the estimate of the epipolar geometry. The affine warping function from (1) is changed to

$$W^a_{EU}(\mathbf{x}, \mathbf{p}^a_{EU}, \mathbf{m}) = \begin{pmatrix} a_{11} & a_{12} \\ a_{21} & a_{22} \end{pmatrix} \begin{pmatrix} x \\ y \end{pmatrix} + \begin{pmatrix} \frac{-l_3}{l_1} - \lambda_1 l_2 + \lambda_2 l_1 \\ \lambda_1 l_1 + \lambda_2 l_2 \end{pmatrix} \tag{7}$$

with $\mathbf{p}^a_{EU} = (\lambda_1, \lambda_2, a_{11}, a_{12}, a_{21}, a_{22})^T$; the respective epipolar line $\mathbf{l} = (l_1, l_2, l_3)^T = \mathbf{F}\tilde{\mathbf{m}}$ is computed using the fundamental matrix $\mathbf{F}$ and the feature position (center of feature patch) $\tilde{\mathbf{m}} = (x_m, y_m, 1)^T$. In general, the warping parameter vector is $\mathbf{p}_{EU} = (\lambda_1, \lambda_2, p_3, ..., p_n)^T$. The parameter λ_1 is responsible for movements along the respective epipolar line, λ_2 for the perpendicular direction. The optimization error function of GKLT is the same as the one from KLT (2), but using substitutions for the warping parameters and the warping function. The parameter update rule of GKLT derived from the error function,

$$\Delta\mathbf{p}_{EU} = \mathbf{A}_w \mathbf{H}^{-1}_{EU} \sum_{\mathbf{x}\in P} (\nabla I \nabla_{\mathbf{p}_{EU}} W_{EU}(\mathbf{x}, \mathbf{p}_{EU}, \mathbf{m}))^T (T(\mathbf{x}) - I(W_{EU}(\mathbf{x}, \mathbf{p}_{EU}, \mathbf{m}))), \tag{8}$$

also looks very similar to the one of KLT (5). The difference is the weighting matrix

$$\mathbf{A}_w = \begin{pmatrix} w & 0 & 0 & \cdots & 0 \\ 0 & 1-w & 0 & & \\ 0 & 0 & 1 & & \vdots \\ \vdots & & & \ddots & 0 \\ 0 & & \cdots & 0 & 1 \end{pmatrix}, \tag{9}$$

which enables the user to weight the translational changes (along/perpendicular to the epipolar line) by the parameter $w \in [0, 1]$ called *epipolar weight.* In [8] the authors associate $w = 1$ with the case of a perfectly accurate estimate of the epipolar geometry, since only feature translations along the respective epipolar line are realized. The more uncertain the epipolar estimate the smaller is w said to be. The case of no knowledge about the epipolar geometry is linked with $w = 0.5$, when translations along and perpendicular to the respective epipolar line are realized equally weighted.

3 GKLT Tracking with Uncertainty Estimation

The previous section briefly reviewed a way to incorporate knowledge about camera parameters into the KLT tracking model. The resulting GKLT tracking requires manual adjustment of the weighting factor w that controls the translational parts of the warping function and thereby handles an uncertain epipolar geometry. For practical application, it is questionable how to find an optimal w and whether one allocation of w holds for all features in all sequences produced within the respective controlled environment. Hence, we propose to estimate the uncertainty parameter w for each feature during the feature tracking process.

In the following we present a new approach for GKLT where the warping parameters and the epipolar weight are optimally computed in a combined estimation step. Like the EM algorithm [10], our approach uses an alternating iterative estimation of hidden information and result values. The first step in deriving the extended iterative optimization procedure is the specification of the optimization error function of GKLT tracking with respect to the uncertainty parameter.

3.1 Modifying the Optimization Error Function

In the derivation of GKLT from [8], the warping parameter update rule is constructed from the standard error function and in the last step augmented by the weighting matrix $\mathbf{A}_w$ to yield (8). Instead, we suggest to directly include the weighting matrix in the optimization error function. Thus, we reparameterize the standard error function to get the new optimization error function

$$\epsilon(\Delta\mathbf{p}_{EU}, \Delta w) = \sum_{\mathbf{x}\in P} (I(W_{EU}(\mathbf{x}, \mathbf{p}_{EU} + \mathbf{A}_{w,\Delta w}\Delta\mathbf{p}_{EU}, \mathbf{m})) - T(\mathbf{x}))^2. \tag{10}$$

Following the additional approach for the matrix $\mathbf{A}_w$ from (9), we substitute $w + \Delta w$ instead of w to reach the weighting matrix $\mathbf{A}_{w,\Delta w}$ used in (10). We achieve an approximation of this error function by first-order Taylor approximation applied twice,

$$\epsilon'(\Delta\mathbf{p}_{EU},\Delta w) = \sum_{\mathbf{x}\in P} (I(W_{EU}(\mathbf{x},\mathbf{p}_{EU},\mathbf{m})) + \nabla I \nabla_{\mathbf{p}_{EU}} W_{EU}(\mathbf{x},\mathbf{p}_{EU},\mathbf{m})\mathbf{A}_{w,\Delta w}\Delta\mathbf{p}_{EU} - T(\mathbf{x}))^2 \tag{11}$$

with $\epsilon(\Delta\mathbf{p}_{EU}, \Delta w) \approx \epsilon'(\Delta\mathbf{p}_{EU}, \Delta w)$ for small $\mathbf{A}_{w,\Delta w}\Delta\mathbf{p}_{EU}$. This allows for direct access to the warping and uncertainty parameters.

3.2 The Modified Update Rule for the Warping Parameters

We calculate the warping parameter change $\Delta\mathbf{p}_{EU}$ by minimization of the approximated error term (11) with respect to $\Delta\mathbf{p}_{EU}$ in the sense of steepest descent, $\frac{\partial\epsilon'(\Delta\mathbf{p}_{EU},\Delta w)}{\partial\Delta\mathbf{p}_{EU}} \stackrel{!}{=} \mathbf{0}$. We get as the update rule for the warping parameters

$$\Delta\mathbf{p}_{EU} = \mathbf{H}^{-1}_{\Delta\mathbf{p}_{EU}} \sum_{\mathbf{x}\in P} (\nabla I \nabla_{\mathbf{p}_{EU}} W_{EU}(\mathbf{x},\mathbf{p}_{EU},\mathbf{m})\mathbf{A}_{w,\Delta w})^T (T(\mathbf{x}) - I(W_{EU}(\mathbf{x},\mathbf{p}_{EU},\mathbf{m}))) \tag{12}$$

with the approximated Hessian

$$\mathbf{H}_{\Delta\mathbf{p}_{EU}} = \sum_{\mathbf{x}\in P} (\nabla I \nabla_{\mathbf{p}_{EU}} W_{EU}(\mathbf{x},\mathbf{p}_{EU},\mathbf{m})\mathbf{A}_{w,\Delta w})^T (\nabla I \nabla_{\mathbf{p}_{EU}} W_{EU}(\mathbf{x},\mathbf{p}_{EU},\mathbf{m})\mathbf{A}_{w,\Delta w}). \tag{13}$$

3.3 The Modified Update Rule for the Uncertainty Estimate

For calculating the change Δw of the uncertainty estimate we again perform minimization of (11), but with respect to Δw, $\frac{\partial \epsilon'(\Delta \mathbf{p}_{EU}, \Delta w)}{\partial \Delta w} \stackrel{!}{=} \mathbf{0}$. This claim yields

$$\sum_{\mathbf{x}\in P} \left(\frac{\partial}{\partial \Delta w}(\nabla I \nabla_{\mathbf{p}_{EU}} W_{EU}(\mathbf{x}, \mathbf{p}_{EU}, \mathbf{m})\mathbf{A}_{w,\Delta w}\Delta \mathbf{p}_{EU})\right)\cdot$$

$$(I(W_{EU}(\mathbf{x}, \mathbf{p}_{EU}, \mathbf{m})) + \nabla I \nabla_{\mathbf{p}_{EU}} W_{EU}(\mathbf{x}, \mathbf{p}_{EU}, \mathbf{m})\mathbf{A}_{w,\Delta w}\Delta \mathbf{p}_{EU} - T(\mathbf{x})) \stackrel{!}{=} \mathbf{0}. \tag{14}$$

We specify

$$\frac{\partial}{\partial \Delta w}(\nabla I \nabla_{\mathbf{p}_{EU}} W_{EU}(\mathbf{x},\mathbf{p}_{EU},\mathbf{m})\mathbf{A}_{w,\Delta w}\Delta \mathbf{p}_{EU}) = \nabla I \nabla_{\mathbf{p}_{EU}} W_{EU}(\mathbf{x},\mathbf{p}_{EU},\mathbf{m})\frac{\partial \mathbf{A}_{w,\Delta w}}{\partial \Delta w}\Delta \mathbf{p}_{EU}. \tag{15}$$

By rearrangement of (14) and using (15) we get

$$\overbrace{\sum_{\mathbf{x}\in P}(\nabla I \nabla_{\mathbf{p}_{EU}} W_{EU}(\mathbf{x},\mathbf{p}_{EU},\mathbf{m})\frac{\partial A_{w,\Delta w}}{\partial \Delta w}\Delta \mathbf{p}_{EU})(\nabla I \nabla_{\mathbf{p}_{EU}} W_{EU}(\mathbf{x},\mathbf{p}_{EU},\mathbf{m}))}^{\mathbf{h}_{\Delta w}} \mathbf{A}_{w,\Delta w}\Delta \mathbf{p}_{EU}$$

$$= \underbrace{\sum_{\mathbf{x}\in P}(\nabla I \nabla_{\mathbf{p}_{EU}} W_{EU}(\mathbf{x},\mathbf{p}_{EU},\mathbf{m})\frac{\partial A_{w,\Delta w}}{\partial \Delta w}\Delta \mathbf{p}_{EU})(T(\mathbf{x}) - I(W_{EU}(\mathbf{x},\mathbf{p}_{EU},\mathbf{m})))}_{e},$$

i.e.

$$\mathbf{h}_{\Delta w}\mathbf{A}_{w,\Delta w}\Delta \mathbf{p}_{EU} = e. \tag{16}$$

Since e is real-valued, (16) provides one linear equation in Δw. With $\mathbf{h}_{\Delta w} = (h_1, ..., h_n)^T$ and $\Delta \mathbf{p}_{EU} = (\Delta\lambda_1, \Delta\lambda_2, \Delta p_3, ..., \Delta p_n)^T$ we reach the update rule for the uncertainty estimate,

$$\Delta w = \frac{e - h_2\Delta\lambda_2 - h_3\Delta p_3 - ... - h_n\Delta p_n}{h_1\Delta\lambda_1 - h_2\Delta\lambda_2} - w. \tag{17}$$

3.4 The Modified Optimization Algorithm

In comparison to the KLT and GKLT tracking, we now have two update rules: one for $\mathbf{p}_{EU}$ and one for w. These update rules, just as in the previous KLT versions, compute optimal parameter changes in the sense of least-squares estimation found by steepest descent of an approximated error function. We combine the two update rules in an EM-like approach. For one iteration of the optimization algorithm, we calculate $\Delta \mathbf{p}_{EU}$ (using $\Delta w = 0$) followed by the computation of Δw with respect to the $\Delta \mathbf{p}_{EU}$ just computed in this step. Then we apply the change to the warping parameter using the actual w.

The modified optimization algorithm as a whole is:

1. initialize $\mathbf{p}_{EU}$ and w
2. compute $\Delta \mathbf{p}_{EU}$ by (12)
3. compute Δw by (17) using $\Delta \mathbf{p}_{EU}$

4. update $\mathbf{p}_{EU}$: $\mathbf{p}_{EU} \leftarrow \mathbf{p}_{EU} + \mathbf{A}_{w,\Delta w}\Delta\mathbf{p}_{EU}$
5. update w: $w \leftarrow w + \Delta w$
6. if changes are small, stop; else go to step 2.

This new optimization algorithm for feature tracking with known camera parameters uses the update rules derived from the extended optimization error function (12) for GKLT tracking. Most importantly, these steps provide a combined estimation of the warping and the uncertainty parameters. Hence, there is no more need to adjust the uncertainty parameter manually as in [8].

4 Experimental Evaluation

Let us denote the extended GKLT tracking method shown in the previous section by GKLT_2, the original formulation [8] by GKLT_1. In this section we quantitatively compare the performances of the KLT, GKLT_1 and GKLT_2 feature tracking methods with and without the presence of noise in the prior knowledge about camera parameters. For GKLT_1, we measure its performance with respect to different values of the uncertainty parameter w.

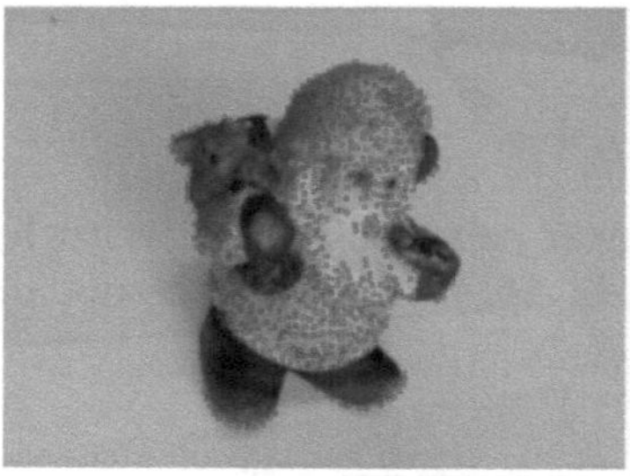

(a) Initial frame of the test sequence with 746 features selected.

(b) View of the set of 3D reference points. Surface mesh for illustration only.

Fig. 2. Test and reference data

As performance measure we use *tracking accuracy*. Assuming that accurately tracked features lead to an accurate 3D reconstruction, we visualize the tracking accuracy by plotting the mean error distances μ_E and standard deviations σ_E of the resulting set of 3D points, reconstructed by plain triangulation, compared to a 3D reference. We also note mean trail lengths.

Figure 2 shows a part of the data we used for our experiments. The image in Fig. 2(a) is the first frame of our test sequence of 26 frames taken from a Santa Claus figurine. The little squares indicate the positions of 746 features initialized for the tracking procedure. Each of the trackers (KLT, GKLT_1 with $w = 0$, ..., GKLT_1 with $w = 1$, GKLT_2) has to track these features through the following frames of the test sequence. We store the resulting trails and calculate the mean trail length for each tracker. Using the feature trails and the camera parameters, we do a 3D reconstruction by plain triangulation for each feature that has a trail length of at least five frames. The resulting set of 3D points is rated by comparison with the reference set shown in Fig. 2(b). This yields μ_E, σ_E of the error distances between each reconstructed point and the actual closest point of the reference set for each tracker. The 3D reference points are provided by a highly accurate (measurement error below 70μm) fringe-projection measurement system [11]. We register these reference points into our measurement coordinate frame by manual registration of distinctive points and an optimal estimation of a 3D Euclidean transformation using dual number quaternions [12]. The camera parameters we apply are provided by our robot arm *Stäubli RX90L* illustrated in Fig. 1. Throughout the experiments, we initialize GKLT_2 with $w = 0.5$.

The extensions of GKLT_1 and GKLT_2 affect the translational part of the feature warping function only. Therefore, we assume and estimate pure translation of the feature positions in the test sequence.

Table 1. Accuracy evaluation by mean error distance μ_E(mm) and standard deviation σ_E(mm) for each tracker. GKLT_1 showed accuracy from 9% better to 269% worse than KLT, depending on choice of w relative to respective uncertainty of camera parameters. GKLT_2 performed better than standard KLT in any case tested. Without additional noise accuracy of GKLT_2 was 5% better than KLT.

	KLT	**GKLT_1, w equals:**											**GKLT_2**
		0.0	0.1	0.2	0.3	0.4	0.5	0.6	0.7	0.8	0.9	1.0	
Using camera parameters without additional noise:													
μ_E(mm)	2.68	9.90	3.52	3.15	2.93	2.77	2.77	2.65	2.62	2.51	2.45	3.90	2.56
σ_E(mm)	3.70	6.99	4.65	4.08	3.63	3.38	3.63	3.55	3.41	3.17	2.77	5.12	3.36
Using disturbed camera parameters:													
μ_E(mm)	2.68	5.09	2.76	2.68	2.75	2.76	2.77	2.78	2.88	3.05	3.35	7.98	2.66
σ_E(mm)	3.70	5.60	3.40	3.37	3.60	3.71	3.63	3.50	4.05	4.08	4.30	6.90	3.61

Throughout the experiments GKLT_2 produced trail lengths that are comparable to standard KLT. The mean runtimes (Intel Core2 Duo, 2.4 GHz, 4 GB RAM) per feature and frame were 0.03 ms for standard KLT, 0.14 ms for GKLT_1 with $w = 0.9$ and 0.29 ms for GKLT_2.

The modified optimization algorithm presented in the last section performs two non-linear optimizations in each step. This results in larger runtimes compared to KLT and GKLT_1 which use one non-linear optimization in each step. The quantitative results of the tracking accuracy are printed in Table 1.

Results using camera parameters without additional noise. GKLT_2 showed a mean error 5% less than KLT, standard deviation was reduced by 9%. The results of $GKLT_1$ were scattered for different values of w. The mean error reached from 9% less at $w = 0.9$ to 269% larger at $w = 0$ than with KLT. The mean trail length of $GKLT_1$ was comparable to KLT at $w = 0.9$, but up to 50% less for all other values of w. An optimal allocation of $w \in [0, 1]$ for the image sequence used is likely to be in $]0.8, 1.0[$, but it is unknown.

Results using disturbed camera parameters. To simulate serious disturbance of the prior knowledge used for tracking, the camera parameters were selected completely random for this test. In the case of fully random prior information, $GKLT_2$ could adapt the uncertainty parameter for each feature in each frame to reduce the mean error by 1% and the standard deviation by 2% relative to KLT. Instead, $GKLT_1$ uses a global value of w for all features in all frames. Again it showed strongly differing performance with respect to the value of w. In the case tested $GKLT_1$ reached the result of KLT at $w = 0.2$ considering mean error and mean trail length. For any other allocation of the uncertainty parameter the mean reconstruction error was up to 198% larger and the mean trail length up to 56% less than with KLT.

5 Summary and Outlook

In this paper we presented a way to extend the GKLT tracking model for integrated uncertainty estimation. For this, we incorporated the uncertainty parameter into the optimization error function resulting in modified parameter update rules. We established a new EM-like optimization algorithm for combined estimation of the tracking and the uncertainty parameters.

The experimental evaluation showed that our extended GKLT performed better than standard KLT tracking in each case tested, even in the case of completely random camera parameters. In contrast the results of the original GKLT varied seriously. An improper choice of the uncertainty parameter caused errors several times larger than with standard KLT. The fitness of the respectively chosen value of the uncertainty parameter was shown to depend on the uncertainty of prior knowledge, which is unknown in general.

Considering the experiments conducted, there are few configurations of the original GKLT that yield better results than KLT and the extended GKLT. Future work is necessary to examine these cases of properly chosen values of the uncertainty parameter. This is a precondition for improving the extended GKLT to reach results closer to the best ones of the original GKLT tracking method.

References

1. Wenhardt, S., Deutsch, B., Angelopoulou, E., Niemann, H.: Active Visual Object Reconstruction using D-, E-, and T-Optimal Next Best Views. In: Computer Vision and Pattern Recognition, CVPR 2007, June 2007, pp. 1–7 (2007)
2. Chen, S.Y., Li, Y.F.: Vision Sensor Planning for 3D Model Acquisition. IEEE Transactions on Systems, Man and Cybernetics – B 35(4), 1–12 (2005)

3. Lucas, B., Kanade, T.: An iterative image registration technique with an application to stereo vision. In: Proceedings of 7th International Joint Conference on Artificial Intelligence, pp. 674–679 (1981)
4. Baker, S., Matthews, I.: Lucas-Kanade 20 Years On: A Unifying Framework. International Journal of Computer Vision 56, 221–255 (2004)
5. Fusiello, A., Trucco, E., Tommasini, T., Roberto, V.: Improving feature tracking with robust statistics. Pattern Analysis and Applications 2, 312–320 (1999)
6. Zinsser, T., Graessl, C., Niemann, H.: High-speed feature point tracking. In: Proceedings of Conference on Vision, Modeling and Visualization (2005)
7. Heigl, B.: Plenoptic Scene Modelling from Uncalibrated Image Sequences. PhD thesis, Friedrich-Alexander-Universität Erlangen-Nürnberg (2003)
8. Trummer, M., Denzler, J., Munkelt, C.: KLT Tracking Using Intrinsic and Extrinsic Camera Parameters in Consideration of Uncertainty. In: Proceedings of 3rd International Conference on Computer Vision Theory and Applications (VISAPP), vol. 2, pp. 346–351 (2008)
9. Trummer, M., Denzler, J., Munkelt, C.: Guided KLT Tracking Using Camera Parameters in Consideration of Uncertainty. Lecture Notes in Communications in Computer and Information Science (CCIS). Springer, Heidelberg (to appear)
10. Dempster, A., Laird, N., Rubin, D.: Maximum likelihood from incomplete data. Journal of the Royal Statistical Society 39, 1–38 (1977)
11. Kuehmstedt, P., Munkelt, C., Matthins, H., Braeuer-Burchardt, C., Notni, G.: 3D shape measurement with phase correlation based fringe projection. In: Osten, W., Gorecki, C., Novak, E.L. (eds.) Optical Measurement Systems for Industrial Inspection V, vol. 6616, p. 66160B. SPIE (2007)
12. Walker, M.W., Shao, L., Volz, R.A.: Estimating 3-D location parameters using dual number quaternions. CVGIP: Image Understanding 54(3), 358–367 (1991)

Image Based Quantitative Mosaic Evaluation with Artificial Video

Pekka Paalanen, Joni-Kristian Kämäräinen*, and Heikki Kälviäinen

Machine Vision and Pattern Recognition Research Group (MVPR)
*MVPR/Computational Vision Group, Kouvola
Lappeenranta University of Technology

Abstract. Interest towards image mosaicing has existed since the dawn of photography. Many automatic digital mosaicing methods have been developed, but unfortunately their evaluation has been only qualitative. Lack of generally approved measures and standard test data sets impedes comparison of the works by different research groups. For scientific evaluation, mosaic quality should be quantitatively measured, and standard protocols established. In this paper the authors propose a method for creating artificial video images with virtual camera parameters and properties for testing mosaicing performance. Important evaluation issues are addressed, especially mosaic coverage. The authors present a measuring method for evaluating mosaicing performance of different algorithms, and showcase it with the root-mean-squared error. Three artificial test videos are presented, ran through real-time mosaicing method as an example, and published in the Web to facilitate future performance comparisons.

1 Introduction

Many automatic digital mosaicing (stitching, panorama) methods have been developed [1,2,3,4,5], but unfortunately their evaluation has been only qualitative. There seems to exist some generally used image sets for mosaicing, for instance the "S. Zeno" (e.g. in [4]), but being real world data, they lack proper ground truth information for basis of objective evaluation, especially intensity and color ground truth. Evaluations have been mostly based on human judgment, while others use ad hoc computational measures such as image blurriness [4]. The ad hoc measures are usually tailored for specific image registration and blending algorithms, possibly giving meaningless results for other mosaicing methods and failing in many simple cases. On the other hand, comparison to any reference mosaic is misleading, if the reference method does not generate an ideal reference mosaic. The very definition of ideal mosaic is ill-posed in most real world scenarios. Ground truth information is crucial for evaluating mosaicing methods on an absolute level and an important research question remains how the ground truth can be formed.

In this paper we propose a method for creating artificial video images for testing mosaicing performance. The problem with real world data is that ground truth information is nearly impossible to gather at sufficient accuracy. Yet ground

A.-B. Salberg, J.Y. Hardeberg, and R. Jenssen (Eds.): SCIA 2009, LNCS 5575, pp. 470–479, 2009.

truth must be the foundation for quantitative analysis. Defining the ground truth ourselves and from it generating the video images (frames) allows to use whatever error measures required. Issues with mosaic coverage are addressed, what to do when a mosaic covers areas it should not cover and vice versa. Finally, we propose an evaluation method, or more precisely, a visualization method which can be used with different error metrics (e.g. root-mean-squared error).

The terminology is used as follows. Base image is the large high resolution image that is decided to be the ground truth. Video frames, small sub-images that represent (virtual) camera output, are generated from the base image. An intermediate step between the base image and the video frame is an optical image, which covers the area the camera sees at a time, and has a higher resolution than the base image. Sequence of video frames, or the video, is fed to a mosaicing algorithm producing a mosaic image. Depending on the camera scanning path (location and orientation of the visible area at each video frame), even the ideal mosaic would not cover the whole base image. The area of the base image, that would be covered by the ideal mosaic, is called the base area.

The main contributions of this work are 1) a method for generating artificial video sequences, as seen by a virtual camera with the most significant camera parameters implemented, and photometric and geometric ground truth, 2) a method for evaluating mosaicing performance (photometric error representation) and 3) publicly available video sequences and ground truth facilitating future comparisons for other research groups.

1.1 Related Work

The work by Boutellier et al. [6] is in essence very similar to ours. They also have the basic idea of creating artificial image sequences and then comparing generated mosaics to the base image. The generator applies perspective and radial geometric distortions, vignetting, changes in exposure, and motion blur. Apparently they assume that a camera mainly rotates when imaging different parts of a scene. Boutellier uses an interest point based registration and a warping method to align the mosaic to the base image for pixel-wise comparison. Due to additional registration steps this evaluation scheme will likely be too inaccurate for superresolution methods. It also presents mosaic quality as a single number, which cannot provide sufficient information.

Möller et al. [7] present a taxonomy of image differences and classify error types into registration errors and visual errors. Registration errors are due to incorrect geometric registration and visual errors appear because of vignetting, illumination and small moving objects in images. Based on pixel-wise intensity and gradient magnitude differences and edge preservation score, they have composed a voting scheme for assigning small image blocks labels depicting present error types. Another voting scheme then suggests what kind of errors an image pair as a whole has, including radial lens distortion and vignetting. Möller's evaluation method is aimed to evaluate mosaics as such, but ranking mosaicing algorithms by performance is more difficult.

Image fusion is basically very different from mosaicing. Image fusion combines images from different sensors to provide a sum of information in the images. One sensor can see something another cannot, and vice versa, the fused image should contain both modes of information. In mosaicing all images come from the same sensor and all images should provide the same information from a same physical target. It is still interesting to view the paper by Petrović and Xydeas [8]. They propose an objective image fusion performance metric. Based on gradient information they provide models for information conservation and loss, and artificial information (fusion artifacts) due to image fusion.

ISET vCamera [9] is a Matlab software that simulates imaging with a camera to utmost realism and processes spectral data. We did not use this software, because we could not find a direct way to image only a portion of a source image with rotation. Furthermore, the level of realism and spectral processing was mostly unnecessary in our case contributing only excessive computations.

2 Generating Video

The high resolution base image is considered as the ground truth, an exact representation of the world. All image discontinuities (pixel borders) belong to the exact representation, i.e. the pixel values are not just samples from the world in the middle of logical pixels but the whole finite pixel area is of that uniform color. This decision makes the base image solid, i.e., there are no gaps in the data and nothing to interpolate. It also means that the source image can be sampled using the nearest pixel method. For simplicity, the mosaic image plane is assumed to be parallel to the base image. To avoid registering the future mosaic to the base image, the pose of the first frame in a video is fixed and provides the coordinate reference. This aligns the mosaic and the base image at sub-pixel accuracy and allows to evaluate also superresolution methods.

The base image is sampled to create an optical image, that spans a virtual sensor array exactly. Resolution of the optical image is k_{interp} times the base image resolution, and it must be considerably higher than the array resolution. Note, that resolution here means the number of pixels per physical length unit, not the image size. The optical image is formed by accounting the virtual camera location and orientation. The area of view is determined by a magnification factor k_{magn} and the sensor array size $w_{\mathrm{s}}, h_{\mathrm{s}}$ such that the optical image in terms of base image pixels is of the size $\frac{w_{\mathrm{s}}}{k_{\mathrm{magn}}}, \frac{h_{\mathrm{s}}}{k_{\mathrm{magn}}}$. All pixels are square.

The optical image is integrated to form the sensor output image. Figure 1(a) presents the structure coordinate system of the virtual sensor array element. A "light sensitive" area inside each logical pixel is defined by its location $(x, y) \in ([0,1],[0,1])$ and size w, h such that $x + w \leq 1$ and $y + h \leq 1$. The pixel fill ratio, as related to true camera sensor arrays, is wh. The value of a pixel in the output image is calculated by averaging the optical image over the light sensitive area. Most color cameras currently use a Bayer mask to reproduce the three color values R, G and B. The Bayer-mask is a per-pixel color mask which transmits only one of the color components. This is simulated by discarding the other two color components for each pixel.

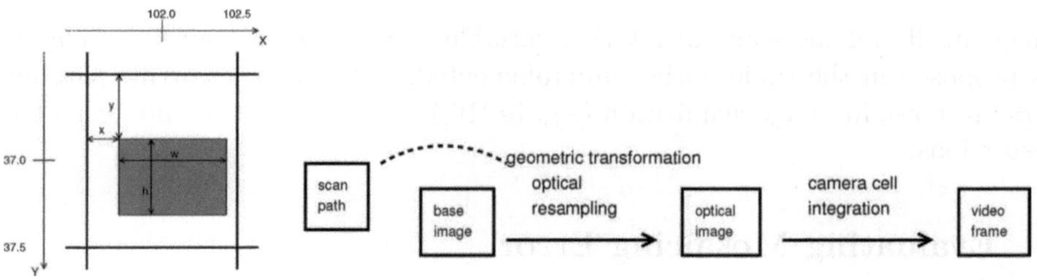

Fig. 1. (a) The structure of a logical pixel in the artificial sensor array. Each logical pixel contains a rectangular "light sensitive" area (the gray box) which determines the value of the pixel. (b) Flow of the artificial video frame generation from a base image and a scan path.

Table 1. Parameters and features used in the video generator

Base image.	The selected ground truth image. Its contents are critical for automatic mosaicing and photometric error scores.
Scan path.	The locations and orientations of the snapshots from a base image. Determines motion velocities, accelerations, mosaic coverage and video length. Video frames must not cross base image borders.
Optical magnification, $k_{\text{magn}} = 0.5$.	Pixel size relationship between base image and video frames. Must be less than one when evaluating superresolution.
Optical interpolation factor, $k_{\text{interp}} = 5$.	Additional resolution multiplier for producing more accurate projections of the base image, defines the resolution of the optical image.
Camera cell array size, 400×300 *pix*.	Affects directly the visible area per frame in the base image. The video frame size.
Camera cell structure, $x = 0.1, y = 0.1, w = 0.8, h = 0.8$.	The size and position of the rectangular light sensitive area inside each camera pixel (Figure 1(a)). In reality this approximation is also related to the point spread function (PSF), as we do not handle PSF explicitly.
Camera color filter.	Either 3CCD (every color channel for each pixel) or Bayer mask. We use 3CCD model.
Video frame color depth.	The same as we use for the base image: 8 bits per color channel per pixel.
Interpolation method in image trans.	Due to the definition of the base image we can use nearest pixel interpolation in forming the optical image.
Photometric error measure.	A pixel-wise error measure scaled to the range $[0, 1]$. Two options: i) root-mean-squared error in RGB space, and ii) root-mean-squared error in L*u*v* space assuming the pixels are in sRGB color space.
Spatial resolution of photometric error.	The finer resolution of the base image and the mosaic resolutions.

An artificial video is composed of output images defined by a scan path. The scan path can be manually created by a user plotting ground truth locations with orientation on the base image. For narrow baseline videos cubic interpolation is used to create a denser path. A diagram of the artificial video generation is presented in Figure 1(b).

Instead of describing the artificial video generator in detail we list the parameters which are included in our implementation and summarize their values and meaning in Table 1. The most important parameters we use are the base image itself and the scan path. Other variables can be fixed to sensible defaults as proposed in the table. Other unimplemented, but still noteworthy, parameters are noise in image acquisition (e.g. in [10]) and photometric and geometric distortions.

3 Evaluating Mosaicing Error

Next we formulate a mosaic image quality representation or visualization, referenced to as *coverage–cumulative error score graph*, for comparing mosaicing methods. First we justify the use of solely photometric information in the representation and second we introduce the importance of coverage information.

3.1 Geometric vs. Photometric Error

Mosaicing, in principle, is based on two rather separate processing steps: registration of video frames, in which the spatial relations between frames is estimated, and blending the frames into a mosaic image, that is deriving mosaic pixel values from the frame pixel values. Since the blending requires accurate registration of frames, especially in superresolution methods, it sounds reasonable to measure the registration accuracy or the geometric error. However, in the following we describe why measuring the success of a blending result (photometric error) is the correct approach.

Geometric error occurs, and typically also cumulates, due to image registration inaccuracy or failure. The geometric error can be considered as errors in geometric transformation parameters, assuming that the transformation model is sufficient. In the simplest case this is the error in frame pose in reference coordinates. Geometric error is the error in pixel (measurement) location.

Two distinct sources for photometric error exist. The first is due to geometric error, e.g., points detected to overlap are not the same point in reality. The second is due to the imaging process itself. Measurements from the same point are likely to differ because of noise, changing illumination, exposure or other imaging parameters, vignetting, and spatially varying response characteristics of the camera. Photometric error is the error in pixel (measurement) value.

Usually a reasonable assumption is that geometric and photometric errors correlate. This is true for natural, diverse scenes, and constant imaging process. It is easy, however, to show pathological cases, where the correlation does not hold. For example, if all frames (and the world) are of uniform color, the photometric error can be zero, but geometric error can be arbitrarily high. On the other hand, if geometric error is zero, the photometric error can be arbitrary by radically changing the imaging parameters. Moreover, even if the geometric error is zero and photometric information in frames is correct, non-ideal blending process may introduce errors. This is the case especially in superresolution methods (the same world location is swiped several times) and the error certainly belongs to the category of photometric error.

From the practical point of view, common for all mosaicing systems is that they take a set of images as input and the mosaic is the output. Without any further insight into a mosaicing system only the output is measurable and, therefore, a general evaluation framework should be based on photometric error. Geometric error cannot be computed if it is not available. For this reason we concentrate on photometric error, which allows to take any mosaicing system as a black box (including proprietary commercial systems).

3.2 Quality Computation and Representation

Seemingly straightforward measure is to compute the mean squared error (MSE) between a base image and a corresponding aligned mosaic. However, in many cases the mosaic and the base image are in different resolutions, having different pixel sizes. The mosaic may not cover all of the base area of the base image, and it may cover areas outside the base area. For these reasons it is not trivial to define as what should be computed for MSE. Furthermore, MSE as such does not really tell the "quality" of a mosaic image. If the average pixel-wise error is constant, MSE is unaffected by coverage. The sum of squared error (SSE) suffers from similar problems.

Interpretation of the base image is simple compared to the mosaic. The base image, and also the base area, is defined as a two-dimensional function with complete support. The pixels in a base image are not just point samples but really cover the whole pixel area. How should the mosaic image be interpreted; as point samples, full pixels, or maybe even with a point spread function (PSF)? Using a PSF would imply that the mosaic image is taken with a virtual camera having the PSF. What should the PSF be? Point sample covers an infinitely small area, which is not realistic. Interpreting the mosaic image the same way as the base image seems the only feasible solution, and is justified by the graphical interpretation of an image pixel (a solid rectangle).

Combing the information about SSE and coverage in a graph can better visualize the quality differences between mosaic images. We borrow from the idea of Receiver Operating Characteristic curve and propose to draw the SSE as a function of coverage. SSE here is the smallest possible SSE when selecting n determined pixels from the mosaic image. This makes all graphs monotonically increasing and thus easily comparable. Define N as the number of mosaic image pixels required to cover exactly the base area. Then coverage $a = n/N$. Note that n must be integer to correspond to binary decision on each mosaic pixel whether to include that pixel. Section 4 contains many graphs as examples.

How to account for differences in resolution, i.e., pixel size? Both the base image and the mosaic have been defined as functions having complete support and composing of rectangular or preferably square constant value areas. For error computation each mosaic pixel is always considered as a whole. The error value for the pixel is the squared error integrated over the pixel area. Whether the resolution of the base image is coarser or finer does not make a difference.

How to deal with undetermined or excessive pixels? Undetermined pixels are areas the mosaic should have covered according to the base area but are not determined. Excessive pixels are pixels in the mosaic covering areas outside the base area. Undetermined pixels do not contribute to the mosaic coverage or error score. If a mosaicing method leaves undetermined pixels, the error curve does not reach 100% coverage. Excessive pixels contribute the theoretical maximum error to the error score, but the effect on coverage is zero. This is justified by the fact that in this case the mosaicing method is giving measurements from an area that is not measured, creating false information.

4 Example Cases

As example methods two different mosaicing algorithms are used. The first one, referenced to as the ground truth mosaic, is a mosaic constructed based on the ground truth geometric transformations (no estimated registration), using nearest pixel interpolation in blending video frames into a mosaic one by one. There is also an option to use linear interpolation for resampling. The second mosaicing algorithm is our real-time mosaicing system that estimates geometric transformations from video images using point trackers and random sample consensus, and uses OpenGL for real-time blending of frames into a mosaic. Neither of these algorithms uses a superresolution approach.

Three artificial videos have been created, each from a different base image. The base images are in Figure 2. The bunker image (2048 × 3072 px) contains a natural random texture. The device image (2430 × 1936 px) is a photograph with strong edges and smooth surfaces. The face image (3797 × 2762 px) is scanned from a print at such resolution that the print raster is almost visible and produces interference patterns when further subsampled (we have experienced this situation with our real-time mosaicing system's imaging hardware). As noted in Table 1, $k_{\text{magn}} = 0.5$ so the resulting ground truth mosaic is in half the resolution, and is scaled up by repeating pixel rows and columns. The real-time mosaicing system uses a scale factor 2 in blending to compensate.

Figure 3 contains coverage–cumulative error score curves of four mosaics created from the same video of the bunker image. In Figure 3(a) it is clear that the real-time methods getting larger error and slightly less coverage are inferior to the ground truth mosaics. The real-time method with sub-pixel accuracy point

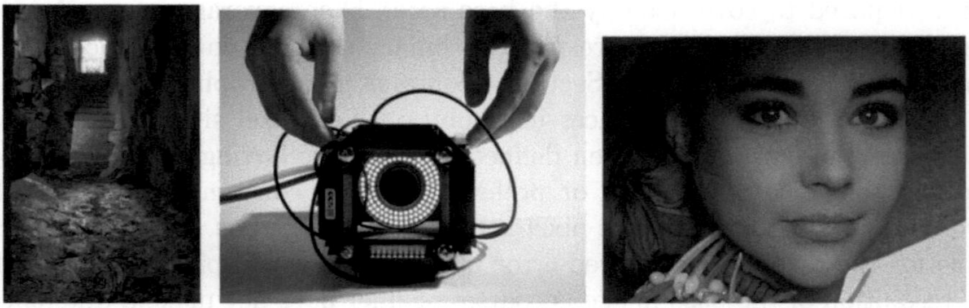

Fig. 2. The base images. (a) Bunker. (b) Device. (c) Face.

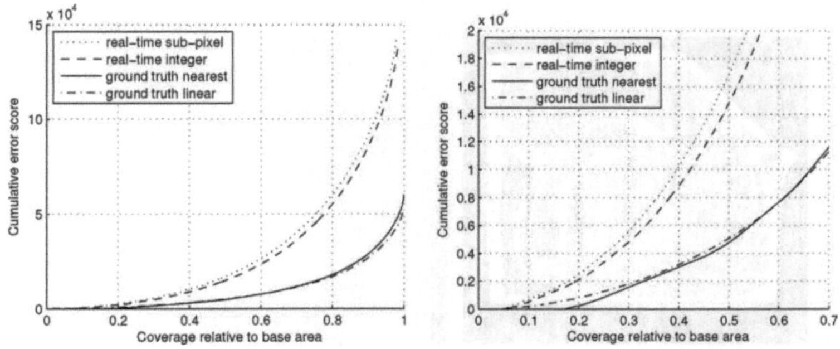

Fig. 3. Quality curves for the Bunker mosaics. (a) Full curves. (b) Zoomed in curves.

Table 2. Coverage–cumulative error score curve end values for the bunker video

mosaicing method	max coverage	error at max coverage	total error
real-time sub-pixel	0.980	143282	143282
real-time integer	0.982	137119	137119
ground truth nearest	1.000	58113	60141
ground truth linear	0.997	50941	50941

tracking is noticeably worse than integer accuracy point tracking, suggesting that the sub-pixel estimates are erroneous. The ground truth mosaic with linear interpolation of frames in blending phase seems to be a little better than using nearest pixel method. However, when looking at the magnified graph in Figure 3(b) the case is not so simple anymore. The nearest pixel method gets some pixel values more correct than linear interpolation, which appears to always make some error. But, when more and more pixels of the mosaics are considered, the nearest pixel method starts to accumulate error faster. If there would be way to select the 50% of the most correct pixels of a mosaic, then in this case the nearest pixel method would be better. A single image quality number, or even coverage and quality together, cannot express this situation. Table 2 shows the maximum coverage values and cumulative error scores without (at max coverage) and with (total) excessive pixels.

To more clearly demonstrate the effect of coverage and excessive pixels, an artificial case is shown in Figure 4. Here the video from the device image is processed with the real-time mosaicing system (integer version). An additional mosaic scale factor was set to 0.85, 1.0 and 1.1. Figure 4(b) presents the resulting graphs along with the ground truth mosaic. When the mosaic scale is too small by factor 0.85, the curve reaches only 0.708 coverage and due to a particular scan path there are no excessive pixels. Too large scale by factor 1.1 introduces a great amount of excessive pixels, which are seen in the coverage–cumulative error score curve as a vertical spike at the end.

The face video is the most controversial because it should have been low-pass filtered to smooth interferences. The non-zero pixel fill ratio in creating the video

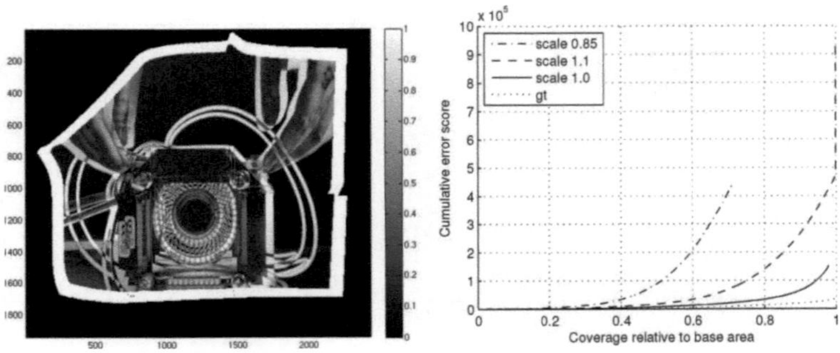

Fig. 4. Effect of mosaic coverage. (a) error image with mosaic scale 1.1. (b) Quality curves for different scales in the real-time mosaicing, and the ground truth mosaic **gt**.

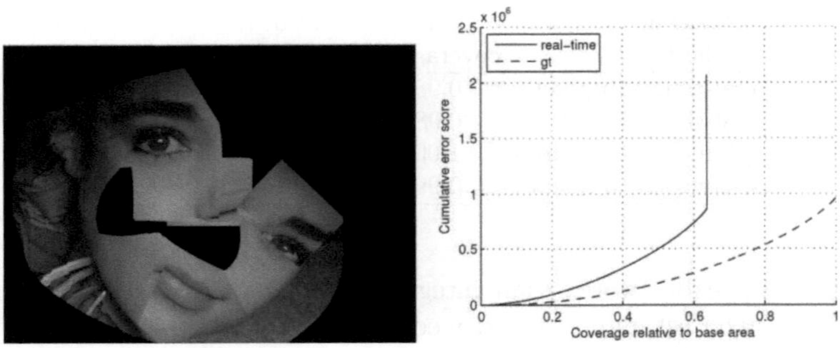

Fig. 5. The real-time mosaicing fails. (a) Produced mosaic image. (b) Quality curves for the real-time mosaicing, and the ground truth mosaic **gt**.

removed the worst interference patterns. This is still a usable example, for the real-time mosaicing system fails to properly track the motion. This results in excessive and undetermined pixels as seen in Figure 5, where the curve does not reach full coverage and exhibits the spike at the end. The relatively high error score of ground truth mosaic compared to the failed mosaic is explained by the difficult nature of the source image.

5 Discussion

In this paper we have proposed the idea of creating artificial videos from a high resolution ground truth image (base image). The idea of artificial video is not new, but combined with our novel way of representing the errors between a base image and a mosaic image it unfolds new views into comparing the performance of different mosaicing methods. Instead of inspecting the registration errors we consider the photometric or intensity and color value error. Using well-chosen base images the photometric error cannot be small if registration accuracy is lacking. Photometric error also takes into account the effect of blending video frames into a mosaic, giving a full view of the final product quality.

The novel representation is the coverage–cumulative error score graph, which connects the area covered by a mosaic to the photometric error. It must be noted, that the graphs are only comparable when they are based on the same artificial video. To demonstrate the graph, we used a real-time mosaicing method and a ground truth transformations based mosaicing method to create different mosaics. The pixel-wise error metric for computing photometric error was selected to be the simplest possible: length of the normalized error vector in RGB color space. This is likely not the best metric and for instance Structural Similarity Index [11] could be considered.

The base images and artificial videos used in this paper are available at `http://www.it.lut.fi/project/rtmosaic` along with additional related images. Ground truth transformations are provided as Matlab data files and text files.

References

1. Brown, M., Lowe, D.: Recognizing panoramas. In: ICCV, vol. 2 (2003)
2. Heikkilä, M., Pietikäinen, M.: An image mosaicing module for wide-area surveillance. In: ACM international workshop on Video Surveillance & Sensor Networks (2005)
3. Jia, J., Tang, C.K.: Image registration with global and local luminance alignment. In: ICCV, vol. 1, pp. 156–163 (2003)
4. Marzotto, R., Fusiello, A., Murino, V.: High resolution video mosaicing with global alignment. In: CVPR, vol. 1, pp. I–692–I–698 (2004)
5. Tian, G., Gledhill, D., Taylor, D.: Comprehensive interest points based imaging mosaic. Pattern Recognition Letters 24(9–10), 1171–1179 (2003)
6. Boutellier, J., Silvén, O., Korhonen, L., Tico, M.: Evaluating stitching quality. In: VISAPP (March 2007)
7. Möller, B., Garcia, R., Posch, S.: Towards objective quality assessment of image registration results. In: VISAPP (March 2007)
8. Petrović, V., Xydeas, C.: Objective image fusion performance characterisation. In: ICCV, vol. 2, pp. 1866–1871 (2005)
9. ISET vcamera, `http://www.imageval.com/public/Products/ISET/ISET_vCamera/vCamera_main.htm`
10. Ortiz, A., Oliver, G.: Radiometric calibration of CCD sensors: Dark current and fixed pattern noise estimation. In: ICRA, vol. 5, pp. 4730–4735 (2004)
11. Wang, Z., Bovik, A., Sheikh, H., Simoncelli, E.: Image quality assessment: From error visibility to structural similarity. Image Processing 13(4), 600–612 (2004)

Improving Automatic Video Retrieval with Semantic Concept Detection*

Markus Koskela, Mats Sjöberg, and Jorma Laaksonen

Department of Information and Computer Science,
Helsinki University of Technology (TKK), Espoo, Finland
{markus.koskela,mats.sjoberg,jorma.laaksonen}@tkk.fi
http://www.cis.hut.fi/projects/cbir/

Abstract. We study the usefulness of intermediate semantic concepts in bridging the semantic gap in automatic video retrieval. The results of a series of large-scale retrieval experiments, which combine text-based search, content-based retrieval, and concept-based retrieval, is presented. The experiments use the common video data and sets of queries from three successive TRECVID evaluations. By including concept detectors, we observe a consistent improvement on the search performance, despite the fact that the performance of the individual detectors is still often quite modest.

1 Introduction

Extracting semantic concepts from visual data has attracted a lot of attention recently in the field of multimedia analysis and retrieval. The aim of the research has been to facilitate semantic indexing of and concept-based retrieval from visual content. The leading principle has been to build semantic representations by extracting intermediate semantic levels (events, objects, locations, people, etc.) from low-level visual and aural features using machine learning techniques.

In early content-based image and video retrieval systems, the retrieval was usually based solely on querying by examples and measuring the similarity of the database objects (images, video shots) with *low-level features* automatically extracted from the objects. Generic low-level features are often, however, insufficient to discriminate content well on a conceptual level. This "semantic gap" is the fundamental problem in multimedia retrieval. The modeling of *mid-level semantic concepts* can be seen as an attempt to fill, or at least reduce, the semantic gap. Indeed, in recent studies it has been observed that, despite the fact that the accuracy of the concept detectors is far from perfect, they can be useful in supporting *high-level indexing and querying* on multimedia data [1]. This is mainly because such semantic concept detectors can be trained off-line with computationally more demanding algorithms and considerably more positive and negative examples than what are typically available at query time.

* Supported by the Academy of Finland in the *Finnish Centre of Excellence in Adaptive Informatics Research* project and by the TKK MIDE programme project UI-ART.

A.-B. Salberg, J.Y. Hardeberg, and R. Jenssen (Eds.): SCIA 2009, LNCS 5575, pp. 480–489, 2009.

In recent years, the TRECVID[1] [2] evaluations have emerged arguably as the leading venue for research on content-based video analysis and retrieval. TRECVID is an annual workshop series which encourages research in multimedia information retrieval by providing large test collections, uniform scoring procedures, and a forum for comparing results for participating organizations.

In this paper, we present a systematic study of the usefulness of semantic concept detectors in automatic video retrieval based on our experiments in three successive TRECVID workshops in the years 2006–2008. Overall, the experiments consist of 96 search topics with associated ground truth in test video corpora of 50–150 hours in duration. A portion of these experiments have been submitted to the official TRECVID evaluations, but due to the submission limitations in TRECVID, some of the presented experiments have been evaluated afterwards using the ground-truth provided by the TRECVID organizers.

The rest of the paper is organized as follows. Section 2 provides an overview of semantic concept detection and the method employed in our experiments. Section 3 discusses briefly the use of semantic concepts in automatic and interactive video retrieval. In Section 4, we present a series of large-scale experiments in automatic video retrieval, which combine text-based search, content-based retrieval, and concept-based retrieval. Conclusions are then given in Section 5.

2 Semantic Concept Detection

The detection and modeling of semantic mid-level concepts has emerged as a prevalent method to improve the accuracy of content-based multimedia retrieval. Recently published large-scale multimedia ontologies such as the Large Scale Concept Ontology for Multimedia (LSCOM) [3] as well as large annotated datasets (e.g. TRECVID, PASCAL Visual Object Classes[2], MIRFLICKR Image Collection[3]) have allowed an increase in multimedia concept lexicon sizes by orders of magnitude. As an example, Figure 1 lists and exemplifies the 36 semantic concepts detected for the TRECVID 2007 high-level feature extraction task. It should be elaborated that high-level feature extraction in TRECVID terminology corresponds to mid-level semantic concept detection.

Disregarding certain specific concepts for which specialized detectors exist (e.g. *human faces*, *speech*), the predominant approach to producing semantic concept detectors is to treat the problem as a generic learning problem, which makes it scalable to large ontologies. The concept-wise training data is used to learn independent detectors for the concepts over selected low-level feature distributions. For building such detectors, a popular approach is to use discriminative methods, such as SVMs, k-nearest neighbor classifiers, or decision trees, to classify between the positive and negative examples of a certain concept. In particular, SVM-based concept detection can be considered as the current *de facto* standard. The SVM detectors require, however, considerable computational resources for training the classifiers. Furthermore, the effect of varying background

[1] http://www-nlpir.nist.gov/projects/trecvid/

[2] http://pascallin.ecs.soton.ac.uk/challenges/VOC/

[3] http://press.liacs.nl/mirflickr/

Fig. 1. The set of 36 semantic concepts detected in TRECVID 2007

is often reduced by using local features such as the SIFT descriptors [4] extracted from a set of interest or corner points. Still, the current concept detectors tend to overfit to the idiosyncrasies of the training data, and their performance often drops considerably when applied to test data from a different source.

2.1 Concept Detection with Self-Organizing Maps

In the experiments reported in this paper, we take a generative approach in which the probability density function of a semantic concept is estimated based on existing training data using kernel density estimation. Only a brief overview is provided here; the proposed method is described in detail in [5].

A large set of low-level features is extracted from the video shots, keyframes extracted from the shots, and the audio track. Separate Self-Organizing Maps (SOMs) are first trained on each of these features to provide a common indexing structure across the different modalities. The positive examples in the training data for each concept are then mapped into the SOMs by finding the best matching unit for each example and inserting a local kernel function. These class-conditional distributions can then be considered as estimates of the true distributions of the semantic concepts in question—not on the original high-dimensional feature spaces, but on the discrete two-dimensional grids defined by the used SOMs. This reduction of dimensionality drastically reduces the computational requirements for building new concept models.

The particular feature-wise SOMs used for each concept detector are obtained by using some feature selection algorithm, e.g. sequential forward selection.

In the TRECVID high-level feature extraction experiments, the used approach has reached relatively good performance, although admittedly failing to reach the level of the current state-of-the-art detectors, which are usually based on SVM classifiers and thus require substantial computational resources for parameter optimization. Our method has, however, proven to be readily scalable to a large number of concepts, which has enabled us to model e.g. a total of 294 concepts from the LSCOM ontology and utilize these concept detectors in various TRECVID experiments without excessive computational requirements.

3 Concept-Based Video Retrieval

The objective of video retrieval is to find relevant video content for a specific information need of the user. The conventional approach has been to rely on textual descriptions, keywords, and other meta-data to achieve this functionality, but this requires manual annotation and does not usually scale well to large and dynamic video collections. In some applications, such as YouTube, the text-based approach works reasonably well, but it fails when there is no meta-data available or when the meta-data cannot adequately capture the essential content of the video material.

Content-based video retrieval, on the other hand, utilizes techniques from related research fields such as image and audio processing, computer vision, and machine learning, to automatically index the video material with low-level features (color layout, edge histogram, Gabor texture, SIFT features, etc.). Content-based queries are typically based on a small number of provided examples (i.e. *query-by-example*) and the database objects are rated based on their similarity to the examples according to the low-level features.

In recent works, the content-based techniques are commonly combined with separately pre-trained detectors for various semantic concepts (*query-by-concepts*) [6,1]. However, the use of concept detectors brings out a number of important research questions, including how to select the concepts to be detected, which methods to use when training the detectors, how to deal with the mixed performance of the detectors, how to combine and weight multiple concept detectors, and how to select the concepts used for a particular query instance.

Automatic Retrieval. In automatic concept-based video retrieval, the fundamental problem is how to map the user's information need into the space of available concepts in the used concept ontology [7]. The basic approach is to select a small number of concept detectors as active and weight them based either on the performance of the detectors or their estimated suitability for the current query. Negative or complementary concepts are not typically used.

In [7], Natsev et al. divide the methods for automatic selection of concepts into three categories: *text-based*, *visual-example-based*, and *results-based methods*. Text-based methods use lexical analysis of the textual query and resources such as WordNet [8] to map query words into concepts. Methods based on visual examples measure the similarity between the provided example objects and the concept detectors to identify suitable concepts. Results-based methods perform an initial retrieval step and analyze the results to determine the concepts that are then incorporated into the actual retrieval algorithm.

The second problem is how to fuse the output of the concept detectors with the other modalities such as text search and content-based retrieval. It has been observed that the relative performances of the modalities significantly depend on the types of queries [9,7]. For this reason, a common approach is to use *query-dependent fusion* where the queries are classified into one of a set of predetermined query classes (e.g. *named entity*, *scene query*, *event query*, *sports query*, etc.) and the weights for the modalities are set accordingly.

Interactive Retrieval. In addition to automatic retrieval, interactive methods constitute a parallel retrieval paradigm. Interactive video retrieval systems include the user in the loop at all stages of the retrieval session and therefore require sophisticated and flexible user interfaces. A global database visualization tool providing an overview of the database as well as a localized point-of-interest with increased level of detail are typically needed. Relevance feedback can also be used to manipulate the system toward video material the user considers relevant.

In recent works, semantic concept detection has been recognized as an important component also in interactive video retrieval [1], and current state-of-the-art interactive video retrieval systems (e.g. [10]) typically use concept detectors as a starting point for the interactive search functionality. A specific problem in concept-based interactive retrieval is how to present to a non-expert user the list of available concepts from a large and unfamiliar concept ontology.

4 Experiments

In this section, we present the results of our experiments in fully-automatic video search in the TRECVID evaluations of 2006–2008. The setup combines text-based search, content-based retrieval, and concept-based retrieval, in order to study the usefulness of existing semantic concept detectors in improving video retrieval performance.

4.1 TRECVID

The video material and the search topics used in these experiments are from the TRECVID evaluations [2] in 2006–2008. TRECVID is an annual workshop series organized by the National Institute of Standards and Technology (NIST), which provides the participating organizations large test collections, uniform scoring procedures, and a forum for comparing the results. Each year TRECVID contains a variable set of video analysis tasks such as high-level feature (i.e. concept) detection, video search, video summarization, and content-based copy detection. For video search, TRECVID specifies three modes of operation: fully-automatic, manual, and interactive search. Manual search refers to the situation where the user specifies the query and optionally sets some retrieval parameters based on the search topic before submitting the query to the retrieval system.

In 2006 the type of used video material was recorded broadcast TV news in English, Arabic, and Chinese, and in 2007 and 2008 the material consisted of documentaries, news reports, and educational programming from Dutch TV. The video data is always divided into separate development and test sets, with the amount of test data being approximately 150, 50, and 100 hours in 2006, 2007 and 2008, respectively. NIST also defines sets of standard search topics for the video search tasks and then evaluates the results submitted by the participants. The search topics contain a textual description along with a small number of both image and video examples of an information need. Figure 2 shows an example of a search topic, including a possible mapping of concept detectors from a concept

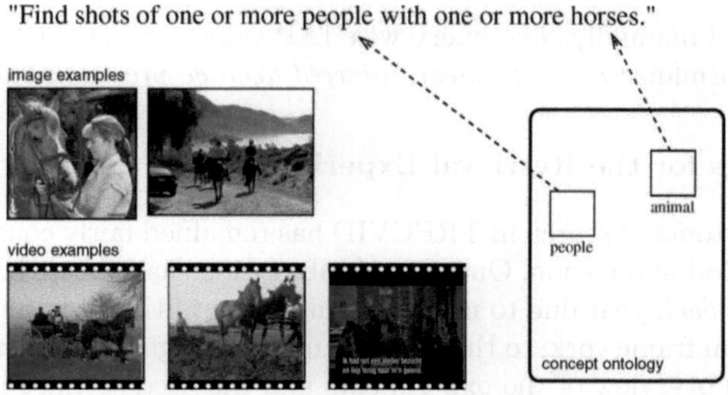

Fig. 2. An example TRECVID search topic, with one possible lexical concept mapping from a concept ontology

ontology based on the textual description. The number of topics evaluated for automatic search was 24 for both 2006 and 2007 and 48 for the year 2008. Due to the limited space, the search topics are not listed here, but are available in the TRECVID guidelines documents at *http://www-nlpir.nist.gov/projects/trecvid/*

The video material used in the search tasks is divided into shots in advance and these reference shots are used as the unit of retrieval. The output from an automatic speech recognition (ASR) software is provided to all participants. In addition, the ASR result from all non-English material is translated into English by using automatic machine translation.

Due to the size of the test corpora, it is infeasible within the resources of the TRECVID initiative to perform an exhaustive examination in order to determine the topic-wise ground truth. Therefore, the following pooling technique is used instead. First, a pool of possibly relevant shots is obtained by gathering the sets of shots returned by the participating teams. These sets are then merged, duplicate shots are removed, and the relevance of only this subset of shots is assessed manually. It should be noted that the pooling technique can result in the underestimation of the performance of new algorithms and, to a lesser degree, new runs, which were not part of the official evaluation, as all unique relevant shots retrieved by them will be missing from the ground truth.

The basic performance measure in TRECVID is *average precision* (AP):

$$\mathrm{AP} = \frac{\sum_{r=1}^{N}(P(r) \times R(r))}{N_{rel}} \tag{1}$$

where r is the rank, N is the number of retrieved shots, $R(r)$ is a binary function stating the relevance of the shot retrieved with rank r, $P(r)$ is the precision at the rank r, and N_{rel} is the total number of relevant shots in the test set. In TRECVID search tasks, N is set to 1000. The mean of the average precision values over a set of queries, *mean average precision* (MAP) has been the standard evaluation measure in TRECVID. In recent years, however, average precision has been gradually replaced by *inferred average precision* (IAP) [11], which approximates the AP measure very closely but requires only a subset of the pooled results to be evaluated manually. The query-wise IAP values are similarly combined to form the performance measure *mean inferred average precision* (MIAP).

4.2 Settings for the Retrieval Experiments

The task of automatic search in TRECVID has remained fairly constant over the three year period in question. Our annual submissions have been, however, somewhat different each year due to modifications and additions to our PicSOM [12] retrieval system framework, to the used features and algorithms, etc. For brevity, only a general overview of the experiments and the used settings is provided in this paper. More detailed descriptions can be found in our annual TRECVID workshop papers [13,14,15]. In all experiments, we combine content-based retrieval based on the topic-wise image and video examples using our standard SOM-based retrieval algorithm [12], concept-based retrieval with concept detectors trained as described in Section 2.1, and text search (c.f. Fig. 2).

The semantic concepts are mapped to the search topics using lexical analysis and synonym lists for the concepts obtained from WordNet. In 2006, we used a total of 430 semantic concepts from the LSCOM ontology. However, the LSCOM ontology is currently annotated only for the TRECVID 2005/2006 training data. Therefore, in 2007 and 2008, we used only the concept detectors available from the corresponding high-level feature extraction tasks, resulting in 36 and 53 concept detectors, respectively. In the 2008 experiments, 11 of the 48 search topics did not match to any of the available concepts. The visual examples were used instead for these topics.

For text search, we employed our own implementation of an inverted file index in 2006. For the 2007–2008 experiments, we replaced our indexing algorithm with the freely-available Apache Lucene[4] text search engine.

4.3 Results

The retrieval results for the three studied TRECVID test setups are shown in Figures 3–5. The three leftmost (lighter gray) bars show the retrieval performance of each of the single modalities: text search ('t'), content-based retrieval based on the visual examples ('v'), and retrieval based on the semantic concepts ('c'). The darker gray bars on the right show the retrieval performances of the combinations of the modalities. The median values for all submitted comparable runs from all participants are also shown as horizontal lines for comparison.

For 2006 and 2007, the shown performance measure is mean average precision (MAP), whereas in 2008 the TRECVID results are measured using mean inferred average precision (MIAP). Direct numerical comparison between different years of participation is not very informative, since the difficulty of the search tasks may vary greatly from year to year. Furthermore, the source of video data used was changed between years 2006 and 2007. Relative changes, however, and changes between different types of modalities can be very instructive.

[4] http://lucene.apache.org

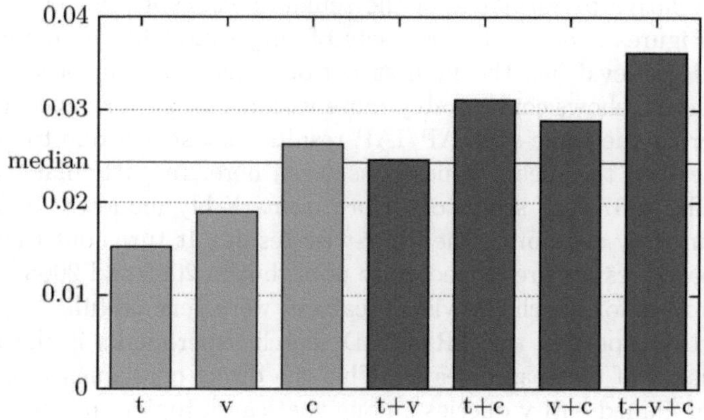

Fig. 3. MAP values for TRECVID 2006 experiments

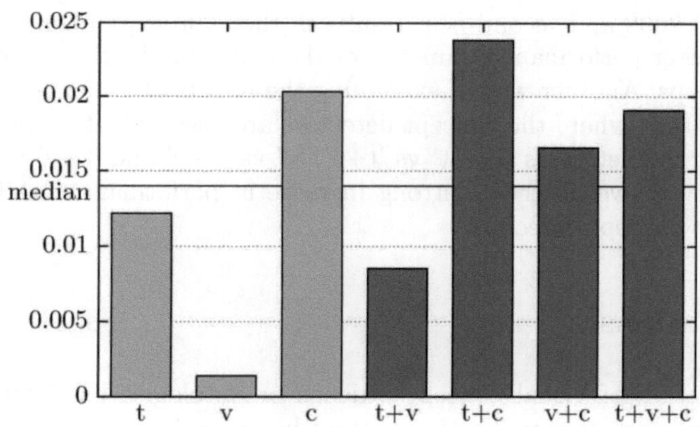

Fig. 4. MAP values for TRECVID 2007 experiments

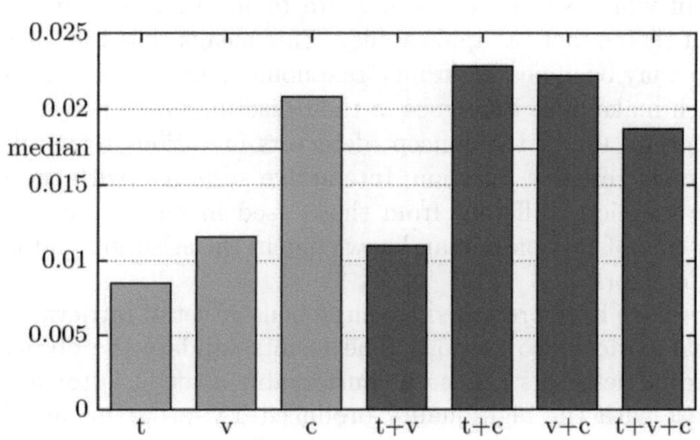

Fig. 5. MIAP values for TRECVID 2008 experiments

The good relative performance of the semantic concepts can be readily observed from Figures 3–5. In all three sets of single modality experiments, the concept-based retrieval has the highest performance. Content-based retrieval, on the other hand, shows considerably more variance in performance, especially when considering the topic-wise AP/IAP results (not shown due to space limitations) instead of the mean values considered here. In particular, the visual examples in the 2007 runs seem to perform remarkably modestly. This can be readily explained by examining the topic-wise results: It turns out that most of the content-based results are indeed quite poor, but in 2006 and 2008 there were a few visual topics for which the visual features were very useful.

A noteworthy aspect in the TRECVID search experiments is the relatively poor performance of text-based search. This is a direct consequence of both the low number of named entity queries among the search topics and the noisy text transcript resulting from automatic speech recognition and machine translation.

Of the combined runs, the combination of text search and concept-based retrieval performs reasonably well, resulting in the best overall performance in the 2007 and 2008 and second-best results in the 2006 experiments. Moreover, it reaches better performance than any of the single modalities in all three experiment setups. Another way of examining the results of the experiments is to compare the runs where the concept detectors are used with the corresponding ones without the detectors (i.e. 't' vs 't+c', 'v' vs 'v+c' and 't+v' vs 't+v+c'). Viewed this way, we observe a strong increase in performance in all cases by including the concept detectors.

5 Conclusions

The construction of visual concept lexicons or ontologies has been found to be an integral part of any effective content-based multimedia retrieval system in a multitude of recent research studies. Yet the design and construction of multimedia ontologies still remains an open research question. Currently the specification of which semantic features are to be modeled tends to be fixed irrespective of their practical applicability. This means that the set of concepts in an ontology may be appealing from a taxonomic perspective, but may contain concepts which make little difference in their discriminative power.

The appropriate use of the concept detectors in various retrieval settings is still another open research question. Interactive systems—with the user in the loop—require solutions different from those used in automatic retrieval algorithms which cannot rely on human knowledge in the selection and weighting of the concept detectors.

In this paper, we have presented a comprehensive set of retrieval experiments with large real-world video corpora. The results validate the observation that semantic concept detectors can be a considerable asset in automatic video retrieval, at least with the high-quality produced TV programs and TRECVID style search topics used in these experiments. This holds even though the performance of the individual detectors is inconsistent and still quite modest in many cases, and though the mapping of concepts to search queries was performed using a relatively naïve lexical matching approach. Similar results have been obtained in the other participants' submissions to the TRECVID search tasks as well. These findings strengthen the notion that mid-level semantic concepts provide a true stepping stone from low-level features to high-level human concepts in multimedia retrieval.

References

1. Hauptmann, A.G., Christel, M.G., Yan, R.: Video retrieval based on semantic concepts. Proceedings of the IEEE 96(4), 602–622 (2008)
2. Smeaton, A.F., Over, P., Kraaij, W.: Evaluation campaigns and TRECVid. In: MIR 2006: Proceedings of the 8th ACM International Workshop on Multimedia Information Retrieval, pp. 321–330. ACM Press, New York (2006)
3. Naphade, M., Smith, J.R., Tešić, J., Chang, S.F., Hsu, W., Kennedy, L., Hauptmann, A., Curtis, J.: Large-scale concept ontology for multimedia. IEEE MultiMedia 13(3), 86–91 (2006)
4. Lowe, D.G.: Distinctive image features from scale-invariant keypoints. International Journal of Computer Vision 60(2), 91–110 (2004)
5. Koskela, M., Laaksonen, J.: Semantic concept detection from news videos with self-organizing maps. In: Proceedings of 3rd IFIP Conference on Artificial Intelligence Applications and Innovations, Athens, Greece, June 2006, pp. 591–599 (2006)
6. Snoek, C.G.M., Worring, M.: Are concept detector lexicons effective for video search? In: Proceedings of the IEEE International Conference on Multimedia & Expo. (ICME 2007), Beijing, China, July 2007, pp. 1966–1969 (2007)
7. Natsev, A.P., Haubold, A., Tešić, J., Xie, L., Yan, R.: Semantic concept-based query expansion and re-ranking for multimedia retrieval. In: Proceedings of ACM Multimedia (ACM MM 2007), Augsburg, Germany, September 2007, pp. 991–1000 (2007)
8. Fellbaum, C. (ed.): WordNet: An Electronic Lexical Database
9. Kennedy, L.S., Natsev, A.P., Chang, S.F.: Automatic discovery of query-class-dependent models for multimodal search. In: Proceedings of ACM Multimedia (ACM MM 2005), Singapore, November 2005, pp. 882–891 (2005)
10. de Rooij, O., Snoek, C.G.M., Worring, M.: Balancing thread based navigation for targeted video search. In: Proceedings of the International Conference on Image and Video Retrieval (CIVR 2008), Niagara Falls, Canada, pp. 485–494 (2008)
11. Yilmaz, E., Aslam, J.A.: Estimating average precision with incomplete and imperfect judgments. In: Proceedings of 15th International Conference on Information and Knowledge Management (CIKM 2006), Arlington, VA, USA (November 2006)
12. Laaksonen, J., Koskela, M., Oja, E.: PicSOM—Self-organizing image retrieval with MPEG-7 content descriptions. IEEE Transactions on Neural Networks, Special Issue on Intelligent Multimedia Processing 13(4), 841–853 (2002)
13. Sjöberg, M., Muurinen, H., Laaksonen, J., Koskela, M.: PicSOM experiments in TRECVID 2006. In: Proceedings of the TRECVID 2006 Workshop, Gaithersburg, MD, USA (November 2006)
14. Koskela, M., Sjöberg, M., Viitaniemi, V., Laaksonen, J., Prentis, P.: PicSOM experiments in TRECVID 2007. In: Proceedings of the TRECVID 2007 Workshop, Gaithersburg, MD, USA (November 2007)
15. Koskela, M., Sjöberg, M., Viitaniemi, V., Laaksonen, J.: PicSOM experiments in TRECVID 2008. In: Proceedings of the TRECVID 2008 Workshop, Gaithersburg, MD, USA (November 2008)

Content-Aware Video Editing in the Temporal Domain

Kristine Slot, René Truelsen, and Jon Sporring

Dept. of Computer Science, Copenhagen University,
Universitetsparken 1, DK-2100 Copenhagen, Denmark
kristine@diku.dk, rtr@rtr.dk, sporring@diku.dk

Abstract. An extension of 2D Seam Carving [Avidan and Shamir, 2007] is presented, which allows for automatic resizing the duration of video from stationary cameras without interfering with the velocities of the objects in the scenes. We are not interested in cutting out entire frames, but instead in removing spatial information across different frames. Thus we identify a set of pixels across different video frames to be either removed or duplicated in a seamless manner by analyzing 3D space-time sheets in the videos. Results are presented on several challenging video sequences.

Keywords: Seam carving, video editing, temporal reduction.

1 Seam Carving

Video recording is increasingly becoming a part of our every day use. Such videos are often recorded with an abundance of sparse video data, which allows for temporal reduction, i.e. reducing the duration of the video, while still keeping the important information. This article will focus on a video editing algorithm, which permits unsupervised or partly unsupervised editing in the time dimension. The algorithm shall be able to reduce, without altering object velocities and motion consistency (no temporal distortion). To do this we are not interested in cutting out entire frames, but instead in removing spatial information across different frames. An example of our results is shown in Figure 1.

Seam Carving was introduced in [Avidan and Shamir, 2007], where an algorithm for resizing images without scaling the objects in the scene is introduced. The basic idea is to constantly remove the least important pixels in the scene, while leaving the important areas untouched. In this article we give a novel extension to the temporal domain, discuss related problems and perform evaluation of the method on several challenging sequences. Part of the work presented in this article has earlier appeared as a masters thesis [Slot and Truelsen, 2008].

Content aware editing of video sequences has been treated by several authors in the literature typically by using steps involving: Extract information from the video, and determine which parts of the video can be edited. We will now discuss related work from the literature. An simple approach is frame-by-frame removal: An algorithm for temporal editing by making an automated object-based extraction of key frames was developed in [Kim and Hwang, 2000], where a key frame

A.-B. Salberg, J.Y. Hardeberg, and R. Jenssen (Eds.): SCIA 2009, LNCS 5575, pp. 490–499, 2009.

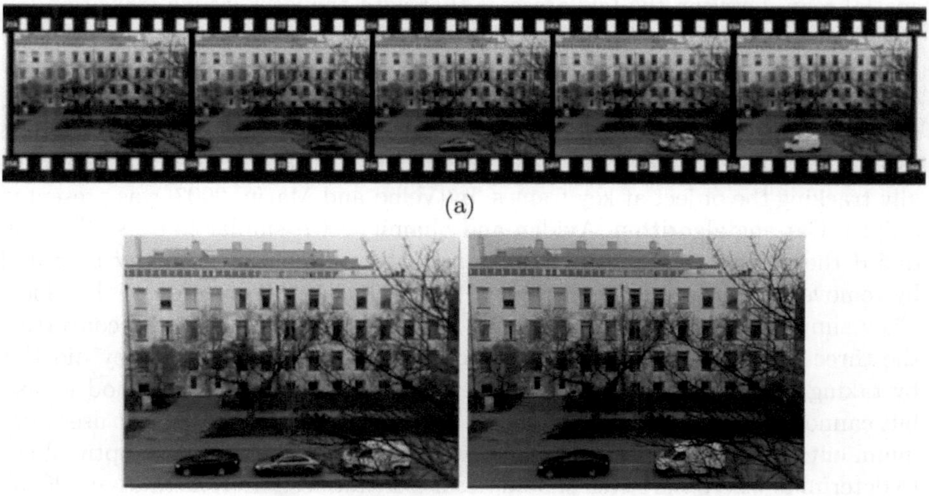

(a)

(b) (c)

Fig. 1. A sequence of driving cars where 59% of the frames may be removed seamlessly. Frames from the original (`http://rtr.dk/thesis/videos/diku_biler_orig.avi`) is shown in (a), a frame from the shortened movie in (b) (`http://rtr.dk/thesis/videos/diku_biler_mpi_91removed.avi`), and a frame where the middle car is removed in (c) (`http://rtr.dk/thesis/videos/xvid_diku_biler_remove_center_car.avi`).

is a subset of still images which best represent the content of the video. The key frames were determined by analyzing the motion of edges across frames. In [Uchihashi and Foote, 1999] was presented a method for video synopsis by extracting key frames from a video sequence. The key frames were extracted by clustering the video frames according to similarity of features such as color-histograms and transform-coefficients. Analyzing a sequence as a spatio-temporal volume was first introduced in [Adelson and Bergen, 1985]. The advantage of viewing the motion using this new perspective is clear: Instead of approaching it as a sequence of singular problems, which includes complex problems such as finding feature correspondence, object motion can instead be considered as an edge in the temporal dimension. A method for achieving automatic video synopsis from a long video sequence, was published by [Rav-Acha et al., 2007], where a short video synopsis of a video is produced by calculating the activity of each pixel in the sequence as the difference between the pixel value at some time frame, t, and the average pixel value over the entire video sequence. If the activity varies more than a given threshold it is labeled as an active, otherwise as an inactive pixel at that time. Their algorithm may change the order of events, or even break long events into smaller parts showed at the same time. In [Wang et al., 2005] was an article presented on video editing in the 3D-gradient domain. In their method, a user specifies a spatial area from the source video together with an area in the target video, and their algorithm seeks optimal spatial seam between the two areas as that with the least visible transition between them. In [Bennett and McMillan, 2003] an approach with potential for different editing options was presented. Their approach includes video stabilization, video mosaicking or object removal. Their idea differs from previous models, as they adjust the image layers in the spatio-temporal box according to some fixed points. The strength of this concept is to ease the object tracking, by manually tracking the object at key frames. In [Velho and Marín, 2007] was presented a Seam Carving algorithm [Avidan and Shamir, 2007] similar to ours. They reduced the videos by finding a surface in a three-dimensional energy map and by remove this surface from the video, thus reducing the duration of the video. They simplified the problem of finding the shortest-path surface by converting the three dimensional problem to a problem in two dimensions. They did this by taking the mean values along the reduced dimension. Their method is fast, but cannot handle crossing objects well. Several algorithms exists that uses minimum cut: An algorithm for stitching two images together using an optimal cut to determine where the stitch should occur is introduced in [Kvatra et al., 2003]. Their algorithm is only based on colors. An algorithm for resizing the spatial information is presented in [Rubenstein et al., 2008]. where a graph-cut algorithm is used to find an optimal solution, which is slow, since a large amount of data has to be maintained. In [Chen and Sen, 2008] an presented is algorithm for editing the temporal domain using graph-cut, but they do not discuss letting the cut uphold the basic rules determined in [Avidan and Shamir, 2007], which means that their results seems to have stretched the objects in the video.

2 Carving the Temporal Dimension

We present a method for reducing video sequences by iteratively removing spatio-temporal sheets of one voxel depth in time. This process is called carving, the sheets are called seams, and our method is an extension of the 2D Seam Carving method [Avidan and Shamir, 2007]. Our method may be extended to simultaneously carving both spatial and temporal information, however we will only consider temporal carving.

We detect seams whose integral minimizes an energy function, and the energy function is based on the change of the sequence in the time direction:

$$E_1(r,c,t) = \left\| \frac{I(r,c,t+1) - I(r,c,t)}{1} \right\|, \tag{1}$$

$$E_2(r,c,t) = \left\| \frac{I(r,c,t+1) - I(r,c,t-1)}{2} \right\|, \tag{2}$$

$$E_{g(\sigma)}(r,c,t) = \left\| \left(I \star \frac{dg_\sigma}{dt} \right)(r,c,t) \right\|. \tag{3}$$

The three energy functions differ by their noise sensitivity, where E_1 is the most and $E_{g(\sigma)}$ is the least for moderate values of σ. A consequence of this is also that the information about motion is spread spatially proportionally to the objects speeds, where E_1 spreads the least and $E_{g(\sigma)}$ the most for moderate values of σ. This is shown in Figure 2.

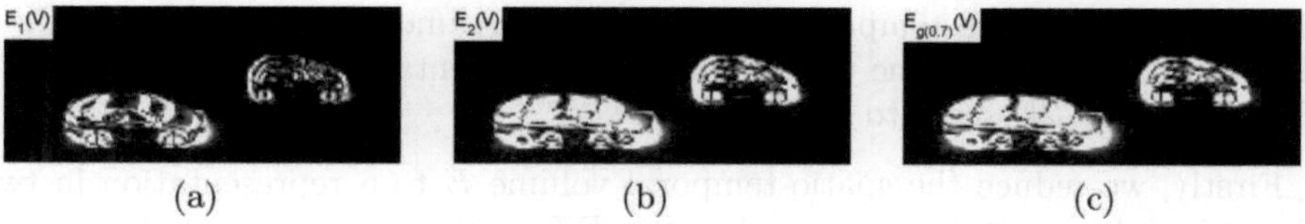

Fig. 2. Examples of output from (a) E_1, (b) E_2, and (c) $E_{g(0.7)}$. The response is noted to increase spatially from left to right.

To reduce the video's length we wish to identify a seam which is equivalent to selecting one and only one pixel from each spatial position. Hence, given an energy map $E \in \mathbb{R}^3 \rightarrow \mathbb{R}$ we wish to find a seam $S \in \mathbb{R}^2 \rightarrow \mathbb{R}$, whose value is the time of each pixel to be removed. We assume that the sequence has (R, C, T) voxels. An example of a seam is given in Figure 3.

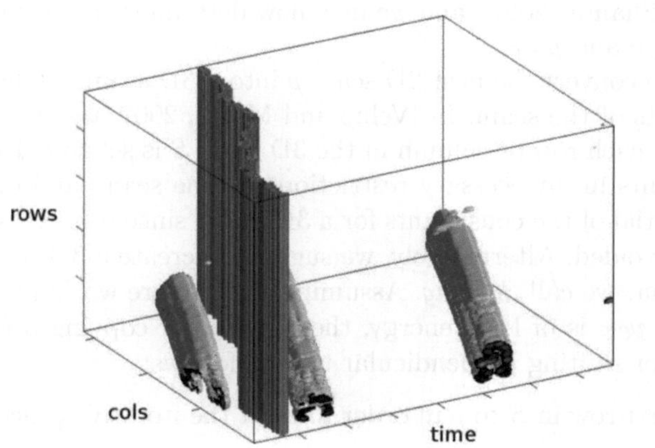

Fig. 3. An example of a seam found by choosing one and only one pixel along time for each spatial position

To ensure temporal connectivity in the resulting sequence, we enforce regularity of the seam by applying the following constraints:

$$|S(r,c)-S(r-1,c)| \leq 1 \wedge |S(r,c)-S(r,c-1)| \leq 1 \wedge |S(r,c)-S(r-1,c-1)| \leq 1. \tag{4}$$

We consider an 8-connected neighborhood in the spatial domain, and to optimize the seam position we consider the total energy,

$$\mathbb{E}_p = \min_S \left(\sum_{r=1}^{R} \sum_{c=1}^{C} E(r,c,S(r,c))^p \right)^{\frac{1}{p}}. \tag{5}$$

A seam intersecting an event can give visible artifacts in the resulting video, wherefore we use $p \rightarrow \infty$, and terminate the minimization, when $\mathbb{E}_\infty$ exceeds a break limit b. Using these constraints, we find the optimal seam as:

1. Reduce the spatio-temporal volume E to two dimensions.
2. Find a 2D seam on the two dimensional representation of E.
3. Extend the 2D seam to a 3D seam.

Firstly, we reduce the spatio-temporal volume E to a representation in two dimensions by projection onto either the $\mathcal{RT}$ or the $\mathcal{CT}$ plane. To distinguish between rows with high values and rows containing noise when choosing a seam, we make an improvement to [Velho and Marín, 2007], by using the variance

$$M_{\mathcal{CT}}(c,t) = \frac{1}{R-1}\sum_{r=1}^{R}(E(r,c,t) - \mu(c,t))^2. \tag{6}$$

and likewise for $M_{\mathcal{RT}}(r,t)$. We have found that the variance is a useful balance between the noise properties of our camera and detection of outliers in the time derivative.

Secondly, we find a 2D seam $p_{\cdot\mathcal{T}}$ on $M_{\cdot\mathcal{T}}$ using the method described by [Avidan and Shamir, 2007], and we may now determine the seam of least energy of the two, $p_{\mathcal{CT}}$ and $p_{\mathcal{RT}}$.

Thirdly, we convert the best 2D seam p into a 3D seam, while still upholding the constraints of the seam. In [Velho and Marín, 2007] the 2D seam is copied, implying that each row or column in the 3D seam S is set to p. However, we find that this results in unnecessary restrictions on the seam, and does not achieve the full potential of the constraints for a 3D seam, since it is areas of high energy may not be avoided. Alternatively, we suggest to create a 3D seam S from a 2D seam p by what we call *Shifting*. Assuming that we are working with the case of having found $p_{\mathcal{CT}}$ is of least energy, then instead of copying p for every row in S, we allow for shifting perpendicular to r as follows:

1. Set the first row in S to p in order to start the iterative process. We call this row $r = 1$.
2. For each row r from $r = 2$ to $r = R$ we determine which values are legal for the row r while still upholding the constraints to row $r - 1$ and to the neighbor elements in the row r.
3. We choose the legal possibility which gives the minimum energy in E and insert in the 3D seam S in the r'th row.

The method of Shifting is somewhat inspired from the *sum-of-pairs Multiple Sequence Alignment (MSA)* [Gupta et al., 1995], but our problem is more complicated, since the constraints must be upheld to achieve a legal seam.

3 Carving Real Sequences

By locating seams in a video, it is possible to both reduce and extend the duration of the video by either removing or copying the seams. The consequence

Fig. 4. Seams have been removed between two cars, making them appear to have driven with shorter distance. (a) Part of the an original frame, and (b) The same frame after having removed 30 seams.

Fig. 5. Two people working at a blackboard (http://rtr.dk/thesis/videos/events_overlap_orig_456f.avi), which our algorithm can reduce by 33% without visual artifacts (http://rtr.dk/thesis/videos/events_overlap_306f.avi)

of removing one or more seams from a video is that the events are moved close together in time as illustrated in Figure 4.

In Figure 1 we see a simple example of a video containing three moving cars, reduced until the cars appeared to be driving in convoy. Manual frame removal may produce a reduction too, but this will be restricted to the outer scale of the image, since once a car appears in the scene, then frames cannot be removed without making part of or the complete cars increase in speed. For more complex videos such as illustrated in Figure 5, there does not appear to be any good seam to the untrained eye, since there are always movements. Nevertheless it is still possible to remove 33% of the video without visible artifacts, since the algorithm can find a seam even if only a small part of the characters are standing still.

Many consumer cameras automatically sets brightness during filming, which for the method described so far introduces global energy boosts, luckily, this may be detected and corrected by preprocessing: If the brightness alters through the video, an editing will create some undesired edges as illustrated in Figure 6(a)(a), because the pixels in the current frame are created from different frames in the original video. By assuming that the brightness change appears somewhat evenly throughout the entire video, we can observe a small spatial neighborhood φ of the video, where no motion is occurring, and find an adjustment factor $\Delta(t)$ for

(a) The brightness edge is visible between the two cars to the right.

(b) The brightness edge is corrected by our brightness correction algorithm.

Fig. 6. An illustration of how the brightness edge can inflict a temporal reduction, and how it can be reduced or maybe even eliminated by our brightness correction algorithm

(a)

(b)

(c)

Fig. 7. Four selected frames from the original video (a) (`http://rtr.dk/thesis/videos/diku_crossing_243f.avi`), a seam carved video with a stretched car (b), and a seam carved video with spatial split applied (c) (`http://rtr.dk/thesis/videos/diku_crossing_142f.avi`)

each frame t in the video. If $\varphi(t)$ is the color in the neighborhood in the frame t, then we can adjust the brightness to be as in the first frame by finding

$$\Delta(t) = \varphi(1) - \varphi(t),$$

and then subtract $\Delta(t)$ from the entire frame t. This corrects brightness problem as seen in Figure 6(b).

For sequences with many co-occurring events, it becomes seemingly more difficult to find good cuts through the video. E.g. when objects appear that move in opposing directions, then no seams may exist that does no violate our constraints. E.g. in Figure 7(a), we observe an example of a road with cars moving in opposite directions, whose energy map consists of perpendicular moving objects as seen in Figure 8(a). In this energy map it is impossible to locate a connected 3D seam without cutting into any of the moving objects, and the consequence can be seen in Figure 7(b), where the car moving left has been stretched. For this particular traffic scene, we may perform *Spatial Splitting*, where the sequence is split into two spatio temporal volumes, which is possible if no event crosses between the two volume boxes. A natural split in the video from Figure 7(a) will be between the two lanes. We now have two energy maps as seen in Figure 8, where we notice that the events are disjunctive, and thus we are able to easily find legal seams. By stitching the video parts together after editing an equal number of seams, we get a video as seen in Figure 7(c), where we both notice that the top car is no longer stretched, and at the same time that to move the cars moving right drive closer.

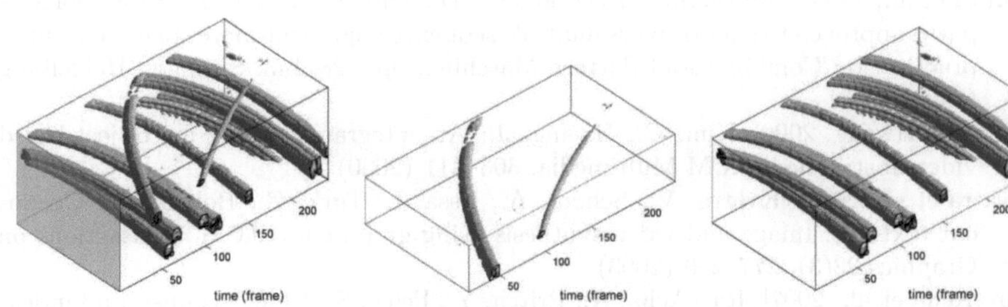

(a) The energy map of the video in Figure 7(a). (b) The top part of the split box. (c) The bottom part of the split box.

Fig. 8. When performing a split of a video we can create energy maps with no perpendicular events, thus allowing much better seams to be detected

4 Conclusion

By locating seams in a video, it is possible to both reduce and extend the duration of the video by either removing or copying the seams. The visual outcome, when removing seams, is that objects seems to have been moved closer together. Likewise, if we copy the seams, then we will experience that the events are moved further apart in time.

We have developed a fast seam detection heuristics called *Shifting*, which presents a novel solution for minimizing energy in three dimensions. The method does not guarantee a local nor global minimum, but the tests have shown that the method is still able to deliver a stable and strongly reduced solution.

Our algorithm has worked on gray scale videos, but may easily be extended to color by (1)–(3). Our implementation is available in Matlab, and as such only a proof of concept not useful for handling larger videos, and even with a translation into a more memory efficient language, a method using a sliding time window is most likely needed for analysing large video sequences, or the introduction of some degree of user control for artistic editing.

References

[Adelson and Bergen, 1985] Adelson, E.H., Bergen, J.R.: Spatiotemporal energy models for the perception of motion. J. of the Optical Society of America A 2(2), 284–299 (1985)

[Avidan and Shamir, 2007] Avidan, S., Shamir, A.: Seam carving for content-aware image resizing. ACM Trans. Graph. 26(3) (2007)

[Bennett and McMillan, 2003] Bennett, E.P., McMillan, L.: Proscenium: a framework for spatio-temporal video editing. In: MULTIMEDIA 2003: Proceedings of the eleventh ACM international conference on Multimedia, pp. 177–184. ACM, New York (2003)

[Chen and Sen, 2008] Chen, B., Sen, P.: Video carving. In: Short Papers Proceedings of Eurographics (2008)

[Gupta et al., 1995] Gupta, S.K., Kececioglu, J.D., Schffer, A.A.: Making the shortest-paths approach to sum-of-pairs multiple sequence alignment more space efficient in practice. In: Combinatorial Pattern Matching, pp. 128–143. Springer, Heidelberg (1995)

[Kim and Hwang, 2000] Kim, C., Hwang, J.: An integrated sceme for object-based video abstraction. ACM Multimedia, 303–311 (2000)

[Kvatra et al., 2003] Kvatra, V., Schödl, A., Essa, I., Turk, G., Bobick, A.: Graph-cut textures: Image and video synthesis using graph cuts. ACM Transactions on Graphics 22(3), 277–286 (2003)

[Rav-Acha et al., 2007] Rav-Acha, A., Pritch, Y., Peleg, S.: Video synopsis and indexing. Proceedings of the IEEE (2007)

[Rubenstein et al., 2008] Rubenstein, M., Shamir, A., Avidan, S.: Improved seam carving for video editing. ACM Transactions on Graphics (SIGGRAPH) 27(3) (2008) (to appear)

[Slot and Truelsen, 2008] Slot, K., Truelsen, R.: Content-aware video editing in the temporal domain. Master's thesis, Dept. of Computer Science, Copenhagen University (2008), `www.rtr.dk/thesis`

[Uchihashi and Foote, 1999] Uchihashi, S., Foote, J.: Summarizing video using a shot importance measure and a frame-packing algorithm. In: the International Conference on Acoustics, Speech, and Signal Processing (Phoenix, AZ), vol. 6, pp. 3041–3044. FX Palo Alto Laboratory, Palo Alto (1999)

[Velho and Marín, 2007] Velho, L., Marín, R.D.C.: Seam carving implementation: Part 2, carving in the timeline (2007), http://w3.impa.br/~rdcastan/SeamWeb/Seam%20Carving%20Part%202.pdf

[Wang et al., 2005] Wang, H., Xu, N., Raskar, R., Ahuja, N.: Videoshop: A new framework for spatio-temporal video editing in gradient domain. In: CVPR 2005: Proceedings of the 2005 IEEE Computer Society Conference on Computer Vision and Pattern Recognition (CVPR 2005), Washington, DC, USA, vol. 2, p. 1201. IEEE Computer Society, Los Alamitos (2005)

High Definition Wearable Video Communication*

Ulrik Söderström and Haibo Li

Digital Media Lab, Dept. Applied Physics and Electronics,
Umeå University,
SE-90187, Umeå, Sweden
{ulrik.soderstrom,haibo.li}@tfe.umu.se

Abstract. High definition (HD) video can provide video communication which is as crisp and sharp as face-to-face communication. Wearable video equipment also provide the user with mobility; the freedom to move. HD video requires high bandwidth and yields high encoding and decoding complexity when encoding based on DCT and motion estimation is used. We propose a solution that can drastically lower the bandwidth and complexity for video transmission. Asymmetrical principal component analysis can initially encode HD video into bitrates which are low considering the type of video (< 300 kbps) and after a startup phase the bitrate can be reduced to less than 5 kbps. The complexity for encoding and decoding of this video is very low; something that will save battery power for mobile devices. All of this is done only at the cost of lower quality in frame areas which aren't considered semantically important.

1 Introduction

As much as 65% of communication between people is determined by non-verbal cues such as facial expressions and body language. Therefore, face-to-face meetings are indeed essential. It is found that face-to-face meetings were more personal and easier to understand than phone or email. It is easy to see that face-to-face meetings are clearer than email since you can get direct feedback; email is not real-time communication. Face-to-face meetings were also seen as more productive and the content easier to remember. But, face-to-face does not need to be in person. Distance communication through video conference equipment is a human-friendly technology that provides the face-to-face communications that people need in order to work together productively, without having to travel. The technology also allows people who work at home or teleworkers to collaborate as if they actually were in the office. Even if there are several benefits with video conferencing it is not very popular. In most cases, video phones have not been a commercial success, but there is a market on the corporate side. Video conferencing with HD resolution can give the impression of face-to-face communication even over networks.

* The wearable video equipment used in this work is constructed by Easyrig AB.

A.-B. Salberg, J.Y. Hardeberg, and R. Jenssen (Eds.): SCIA 2009, LNCS 5575, pp. 500–512, 2009.

HD video conference essentially can eliminate the distance and make the world connected. On a communication link with HD resolution you can look people in the eye and see whether they follow your argument or not.

Two key expressions for video communication are anywhere and anytime. Anywhere means that communication can occur at any location, regardless of the available network, and anytime means that the communication can occur regardless of the surrounding network traffic or battery power. To achieve this there are several technical challenges:

1. The usual video format for video conference is CIF (352x288 pixels) with a framerate of 15 fps. 1080i video (1920x1080 pixels) has a framerate of 25 fps. Every second there is $\approx$ 26 times more data for a HD resolution video than a CIF video.
2. The bitrate for HD video grows so large that it is impossible to achieve communication over several networks. Even with a high-speed wired connection the bitrate may be too low since communication data is very sensitive to delays.
3. Most of the users want to have high mobility; having the freedom to move while communicating.

A solution for HD video conferencing is to use the H.264 [1, 2] video compression standard. This standard can compress the video to high quality video. There are however two major problems with H.264:

1. The complexity for H.264 coding is quite high. High complexity means high battery consumption; something that is becoming a problem with mobile battery-driven devices. The power consumption is directly related to the complexity so high complexity will increase the power usage.
2. The bitrate for H.264 encoding is very high. The vision of providing video communication anywhere cannot be fulfilled with the bitrates required for H.264. The transmission power is related t the bitrate so low bitrate will save battery power.

H.264 encoding cannot provide video neither anywhere or anytime. The question we try to answer in this article is if principal component analysis (PCA) [3] video coding [4, 5] can fulfill the requirements for providing video anywhere and anytime.

The bitrate for PCA video coding can be really low; below 5 kbps. The complexity for PCA encoding is linearly dependent on the number of pixels in the frames; when HD resolution is used the complexity will increase and consume power. PCA is extended into asymmetrical PCA (aPCA) which can reduced the complexity for both encoding and decoding [6, 7]. aPCA can encode the video by using only a subset of the pixels while still decoding the entire frame. By combining the pixel subset and full frames it is possible to relieve the decoder of some complexity as well. For PCA and aPCA it is essential that the facial features are positioned on approximately the same pixel positions in all frames so a wearable video equipment is very important for coding based on PCA.

aPCA enables protection of certain areas within the frame; areas which are important. This area is chosen as the face of the person in the video. We will show how aPCA outperforms encoding with discrete cosine transform (DCT) of the video when it comes to quality for the selected region. The rest of the frame will have poorer reconstruction quality with aPCA compared to DCT encoding. For H.264 video coding it is also possible to protect a specific area by selecting a region of interest (ROI); similarly to aPCA. For encoding of this video the bitrate used for the background is very low and the quality of this area is reduced. So the bitrate for H.264 can be lowered without sacrificing quality for the important area but not to the same low bitrate as aPCA. Video coding based on PCA has the benefit of a much lower complexity for encoding and decoding compared to H.264 and this is a very important factor. The reduced complexity can be achieved at the same time as the bitrate for transmission is reduced. This lowers the power consumption for encoding, transmission and decoding.

1.1 Intracoded and Intercoded Frames

H.264 encoding uses transform coding with discrete cosine transform (DCT) and motion estimation through block matching. There are, at least, two different coding types associated with H.264; intracoded and intercoded frames. An intracoded frame is compressed as an image, which it is. Intercoded frames encode the differences from the previous frame. Since frames which are adjacent in time usually share large similarities in appearance it is very efficient to only store one frame and the differences between this frame and the others. Only the first frame in a sequence is encoded through DCT. For the following frames only the changes between the current and first frame is encoded. The number of frames between intracoded frames are called the group of pictures (GOP). A large GOP size means fewer intracoded frames and lower bitrate.

2 Wearable Video Equipment

Recording yourself with video usually requires that another person carries the camera or that you use a tripod to place the camera on. When the camera is placed on a tripod the movements that you can make are restricted since the camera cannot move; except for the movements that can be controlled remotely. A wearable video equipment allows the user to move freely and have both hands free for use while the camera follows the movements of the user. The equipment is attached to the back of the person wearing it so the camera films the user from the front. The equipment that we have used is built by the company Easyrig AB and resembles a backpack; it is worn on the back (Figure 1). It consists of a backpack, an aluminium arm and a mounting for a camera at the tip of the arm.

3 High Definition (HD) Video

High-definition (HD) video refers to a video system with a resolution higher than regular standard-definition video used in TV broadcasts and DVD-movies. The

Fig. 1. Wearable video equipment

display resolutions for HD video are called 720p (1280x720), 1080i and 1080p (both 1929x1080). i stands for interlaced and p for progressive. Each interlaced frame is divided into two parts where each part only contains half the lines of the frame. The two parts contain either odd or even lines and when they are displayed the human eye perceives that the entire frame is updated. TV-transmissions that have HD resolution use either 720p or 1080i; in Sweden it is mostly 1080i. The video that we use as HD video has a resolution of 1440x1080 (HD anamorphic). It is originally recorded as interlaced video with 50 interlace fields per second but it is transformed into progressive video with 25 frames per second.

4 Wearable Video Communication

Wearable video communication enables the user to move freely; the users mobility is largely increased compared to regular video communication.

The wearable equipment is described in section 2 and video recorded with this equipment is efficiently encoded with principal component analysis (PCA). PCA [3] is a common tool for extracting compact model of faces [8]. A model of a persons facial mimic is called personal face space, facial mimic space or personal mimic space [9, 10]. This space contain the face of the same person but with several different facial expressions. This model can be used to encode video and images of human faces [11, 12] or the head and shoulders of a person [4, 13] to extremely low bitrates.

A space that contains the facial mimic is called Eigenspace Φ and it is constructed as

$$\phi_i = \sum_j b_{ij}(\mathbf{I} - \mathbf{I}_{\underline{0}}) \tag{1}$$

where $\mathbf{I}$ are the original frames and $\mathbf{I}_{\underline{0}}$ is the mean of all video frames. b_{ij} are the Eigenvectors from the the covariance matrix $(\mathbf{I} - \mathbf{I}_{\underline{0}})^T(\mathbf{I} - \mathbf{I}_{\underline{0}})$. The Eigenspace Φ consists of the principal components ϕ_j (Φ=$\{\phi_j \; \phi_{j+1} \; ... \; \phi_N\}$). Encoding of a video frame is done through projection of the video frame onto the Eigenspace Φ.

$$\alpha_j = \phi_j(\mathbf{I} - \mathbf{I}_{\underline{0}})^T \tag{2}$$

where $\{\alpha_j\}$ are projection coefficients for the encoded video frame. The video frame is decoded by multiplying the projection coefficients $\{\alpha_j\}$ with the Eigenspace Φ.

$$\hat{\mathbf{I}} = \mathbf{I}_{\underline{0}} + \sum_{j=1}^{M} \alpha_j \phi_j \tag{3}$$

where M is a selected number of principal components used for reconstruction ($M < N$). The extent of the error incurred by using fewer components (M) than possible (N) is examined in [5]. With asymmetrical PCA (aPCA) one part of the image can be to encode the video and a different part can be decoded [6, 7]. Asymmetrical PCA uses pseudo principal components; information where not the entire frame is a principal component. Parts of the video frames are considered to be important; they are regarded as foreground $\mathbf{I}^f$. The Eigenspace for the foreground Φ^f is constructed according to the following formula:

$$\phi_j^f = \sum_j b_{ij}^f(\mathbf{I}^f - \mathbf{I}_{\underline{0}}^f) \tag{4}$$

where b_{ij}^f are the Eigenvectors from the the covariance matrix $(\mathbf{I}^f - \mathbf{I}_{\underline{0}}^f)^T(\mathbf{I}^f - \mathbf{I}_{\underline{0}}^f)$ and $\mathbf{I}_{\underline{0}}^f$ is the mean of the foreground.

A space which is spanned by components where only the foreground is orthogonal can be created. The components spanning this space are called pseudo principal components and this space has the same size as a full frame:

$$\phi_j^p = \sum_j b_{ij}^f(\mathbf{I} - \mathbf{I}_{\underline{0}}) \tag{5}$$

Encoding is performed using only the foreground:

$$\alpha_j^f = (\mathbf{I}^f - \mathbf{I}_{\underline{0}}^f)^T \phi_j^f \tag{6}$$

where $\{\alpha_j^f\}$ are coefficients extracted using information from the foreground $\mathbf{I}^f$. By combining the pseudo principal components Φ^p and the coefficients $\{\alpha_j^f\}$ full frame video can be reconstructed.

$$\hat{\mathbf{I}}^p = \mathbf{I}_{\underline{0}} + \sum_{j=1}^{M} \alpha_j^f \phi_j^p \tag{7}$$

where M is the selected number of pseudo components used for reconstruction.

By combining the two Eigenspaces Φ^p and Φ^f we can reconstruct frames with full frame size and reduce the complexity for reconstruction. Only a few principal components of Φ^p are used to reconstruct the entire frame. More principal components from Φ^f is used to add details to the foreground.

$$\hat{\mathbf{I}}' = \mathbf{I}_{\underline{0}} + \sum_{j=1}^{P} \alpha_j \phi_j^p + \sum_{j=P+1}^{M} \alpha_j \phi_j^f \tag{8}$$

The result is reconstructed frames with slightly lower quality for the background but with the same quality for the foreground $\mathbf{I}^f$ as if only Φ^p was used for reconstruction. By adjusting the parameter P it is possible to control the bitrate needed for transmission of Eigenimages. Since P decides how many Eigenimages of Φ^p that are used for decoding it also decides how many Eigenimages of Φ^p that needs to be transmitted to the decoder. Φ^f has a much smaller spatial size than Φ^p and transmission of an Eigenimage from Φ^f requires fewer bits than transmission of an Eigenimage from Φ^p.

A third space $\Phi^{p^{bg}}$ which contain only the background and not the entire frame is easily created. This is a space with pseudo principal components; this space is exactly the same as Φ^p without information from the foreground $\mathbf{I}^f$.

$$\phi_j^{p^{bg}} = \sum_j b_{ij}^f (\mathbf{I}^{bg} - \mathbf{I}_{\underline{0}}^{bg}) \tag{9}$$

where $\mathbf{I}^{bg}$ is frame $\mathbf{I}$ minus the pixels from the foreground $\mathbf{I}^f$. This space is combined with the space from the foreground to create reconstructed frames.

$$\hat{\mathbf{I}}' = \mathbf{I}_{\underline{0}} + \sum_{j=1}^{M} \alpha_j \phi_j^f + \sum_{j=1}^{P} \alpha_j \phi_j^{bg} \tag{10}$$

The result is exactly the same as for Eq. (8); high foreground quality, lower background quality, reduced decoding complexity and reduced bitrate for Eigenspace transmission.

When both the encoder and decoder have access to the model of facial mimic the bitrate needed for this video is extremely low (<5 kbps). If the model needs to be transmitted between the encoder and decoder almost the entire bitrate need consists of bits for model transmission.

The complexity for encoding through PCA is linearly dependent on the spatial resolution. The complexity for PCA encoding is dependent on the number of pixels K in the frame. This complexity can be reduced for aPCA since K is a much smaller value for aPCA compared to PCA.

5 HD Video with H.264

As a comparison of HD video encoded with aPCA we encode the video sequence with H.264 as well. We use the same software for encoding of the entire video as for encoding of the Eigenimages; but we also enable motion estimation. The entire video is encoded with H.264 with a target bitrate of 300 kbps. To get this bitrate we encode the video with a quantization step of 29. We compare the quality of the foreground and background separately since they have different qualities when aPCA is used. With standard H.264 encoding the quality for the background and foreground are approximately equal.

The complexity for H.264 encoding is linearly dependent on the frame size. Most of the complexity for H.264 encoding comes from motion estimation through block matching. The blocks has to be matched for several positions and the blocks can move both in horizontal and vertical direction. The complexity for H.264 encoding is dependent on K and the displacement D in square (D^2). When the resolution is increased the number of displacements are increased. Imagine a line in a video with CIF resolution. This line will consist of a number of pixels, e.g., 5. If the same line is described in HD resolution the number of pixels in the line will increase to almost 19. If the same movement between frames is used in CIF and HD the displacement in pixels is much higher for HD video. When motion estimation is used for H.264 video the complexity grows high because of D^2. So even if the complexity is only linearly dependent on the number of pixels K the complexity grows more faster than linearly for high resolution video.

6 HD Video at Low Bitrates

aPCA can be utilized by the decoder to decode parts of the same frame with different spatial resolution. Since the same part of the frame $\mathbf{I}^f$ is used for encoding in both cases, the decoder can choose to decode either $\mathbf{I}^f$ or the entire frame $\mathbf{I}$. The decoded video can also be a combination of $\mathbf{I}^f$ and $\mathbf{I}$. This is described in detail in [7]. How Φ^f and Φ^p are combined can be selected by a number of parameters, such as quality, complexity or bitrate. In this article we will focus on bitrate and complexity.

6.1 Bitrate Calculations

The bitrate that we select as a target for video transmission is 300 kbps. The video needs to be transmitted below this bitrate at all times. The frame size for the video is 1440x1080 ($\mathbf{I}$). The foreground in this video is 432x704 ($\mathbf{I}^f$) (Figure 2). After YUV 4:1:1 compression the number of pixels in the foreground is 456192. The entire frame $\mathbf{I}$ consists of 2332800 pixels and the frame area which is not foreground is 1876608 pixels. The video has a framerate of 25 fps but this has only slight impact on the bitrate for aPCA since each frame is encoded to a few coefficients. The bitrate for these coefficients is easily kept below 5 kbps. Audio is an important part of communication but we will not discuss this in our work. There are several codecs that can provide audio with good quality at a bitrate which can be used. We use 300 kbps for transmission of the Eigenimages (Φ^p and Φ^f) and the coefficients $\{\alpha_j^f\}$ between sender and receiver.

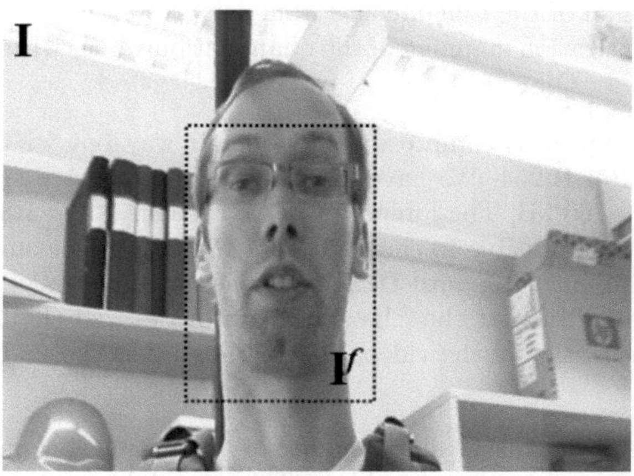

Fig. 2. Frame with the foreground shown

Transmission of the Eigenimages ϕ_j means transmission of images. The Eigenimages have too large size $\approx$ 7,5 MB (1440x1080 resolution minus the foreground) to be transmitted without compression. Since they are images we could use image compression but the images share large similarities in appearance; the facial mimic is independent between the images but it is the same face with similar background. Globally the images are not only uncorrelated but also independent and doesn't share any similarities. Image or video compression based on DCT divides the frames into blocks and encodes each block individually. Even though the frames are independent globally it is possible to find local similarities so to consider the images as a sequence will provide higher compression. We want to remove the complexity associated with motion estimation and only encode the images through DCT.

We use the H.264 video compression without any motion estimation; this encoding uses both intracoding and intercoding. The first image is intracoded and the subsequent images are intercoded but without motion estimation. The mean image is only one image so we will use the JPEG [14] standard for compression of it. The mean image is in fact compressed in the same manner as in [5].

To make the compression more efficient we first use quantization of the images. In our previous article we discussed the usage of pdf-optimized or uniform quantization extensively and came to the conclusion that it is sufficient to use uniform quantization [5]. So, in this work we will use uniform quantization. In our previous work we also examined the effect of high compression and loss of orthogonality between the Eigenimages. To retain high visual quality on the reconstructed frames we will not use so high compression that the loss of orthogonality becomes an important factor. The compression is achieved through the following steps:

- Quantization of the Eigenimages. Φ^Q=Q(Φ)
- The Eigenimages are compressed. Φ^{Comp}=C(Φ^Q)
- Reconstruction of the Eigenimages from compressed video. $\hat{\Phi}^Q$=C′(Φ^{Comp})
- Inverse quantization mapping of the quantization values with the reconstruction values. $\hat{\Phi}$=Q′($\hat{\Phi}^Q$)

The mean image $\mathbf{I}_0$ is compressed in a similar way but we use JPEG compression instead of H.264. We have 295 kbps for Eigenimage transmission and this is equal to ≈ 60 kB. The foreground $\mathbf{I}^f$ have a size of ≈ 1,8 MB when it is uncompressed. It is possible to choose from a wide range of compression grades when it comes to encoding with DCT. We select a compression ratio based on reconstruction quality that the Eigenimages provides and the bitrate which is needed for transmission of the video; the compression is chosen by the following criteria.

- A compression ratio that allow the use of a bitrate below our needs.
- A compression ratio that provide sufficiently high reconstruction of video when compressed Eigenimages are used for encoding and decoding of video.

The first criteria decides how fast the Eigenimages can be transmitted; e.g., how fast high quality video can be decoded. The second criteria decides the quality of reconstructed video.

7 aPCA Decoding of HD Video

The face is the most important information in the video so Eigenimages ϕ_j^f for the foreground $\mathbf{I}^f$ is transmitted first. The bitrate for the compressed Eigenimages $\phi^{f^{Comp}}$ is 13 kbps but the bitrate for the first Eigenimage is higher since it is intracoded. The background is larger in spatial size so the bitrate for this is 42 kbps. Transmission of 10 Eigenimages for the foreground ϕ^{Comp^f}, 1 pseudo Eigenimage for the background ϕ^{Comp^p} plus the mean for both areas can be done within 1 second. After ≈ 220 ms the first Eigenimage and the mean for the foreground is available and decoding of the video can start. All the other Eigenimages are intercoded and a new image arrives every 34th ms. After ≈ 520 ms the decoder has 10 Eigenimages for the foreground. The mean and the first Eigenimage for the background needs ≈ 460 ms for transmission and a new

Fig. 3. Frame reconstructed with aPCA (25 ϕ_j^f and 5 ϕ_j^p are used)

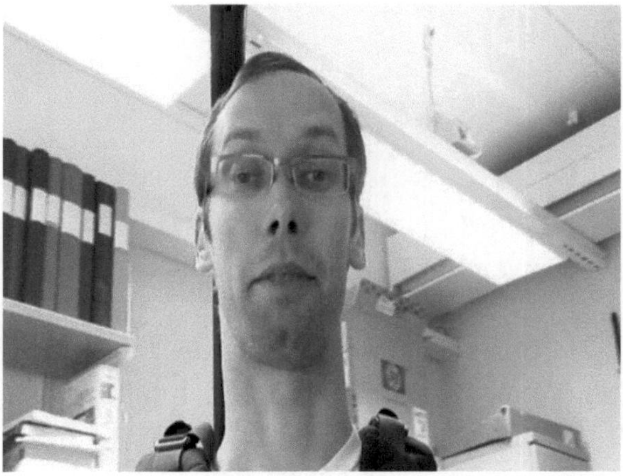

Fig. 4. Frame encoded with H.264

Eigenimage for the background can be transmitted every 87th ms. The quality of the reconstructed video is increased as more Eigenimages arrive. There doesn't have to be a stop to the quality improvement; more and more Eigenimages can be transmitted. But when all Eigenimages that the decoder wants to use for decoding has arrived only the coefficients needs to be transmitted so the bitrate is then below 5 kbps. The Eigenimages can also be updated; something we examined in [5]. The Eigenspace may need to be updated because of loss of alignment between the model and the new video frames.

The average results measured in psnr for the video sequences are shown in table 1 and table 2. Table 1 show the results for the foreground and 2 show the results for the background. The results in the tables are for full decoding

Table 1. Reconstruction quality for the foreground

	Rec. qual. PSNR [dB]		
	Y	U	V
H.264	36.4	36.5	36.5
aPCA	44.2	44.3	44.3

Table 2. Reconstruction quality for the background

	Rec. qual. PSNR [dB]		
	Y	U	V
H.264	36.3	36.5	36.6
aPCA	29.6	29.7	29.7

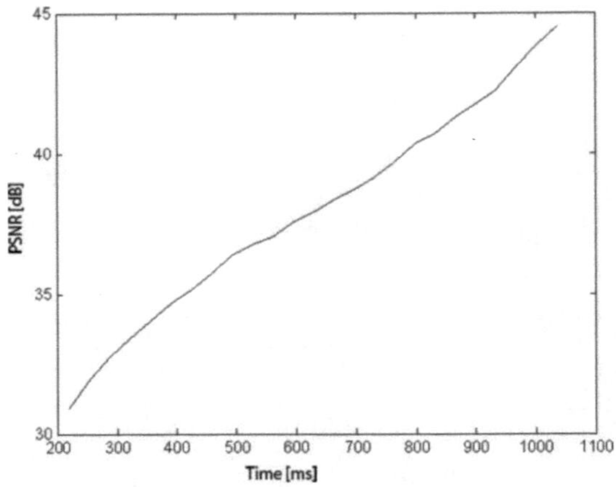

Fig. 5. Foreground quality (Y-channel) over time

quality (25 ϕ_j^f and 5 ϕ_j^p). Figure 5 show how the foreground quality of the Y-channel is increased over time for aPCA. Figure 6 show the same progress for the background. An example of a frame reconstructed with aPCA is shown in figure 3. A reconstructed frame from H.264 encoding is shown in figure 4.

As it can be seen from the tables and the figures the background quality is always lower for aPCA compared with H.264. This will not change even if all Eigenimages are used for reconstruction; the background is always blurred. The exception is when the background is homogenous but the quality of this background with H.264 encoding is also very good.

The foreground quality for aPCA is better than H.264 already when 10 Eigenimages (after $\approx$ 1 second) are used for reconstruction and just improves after that.

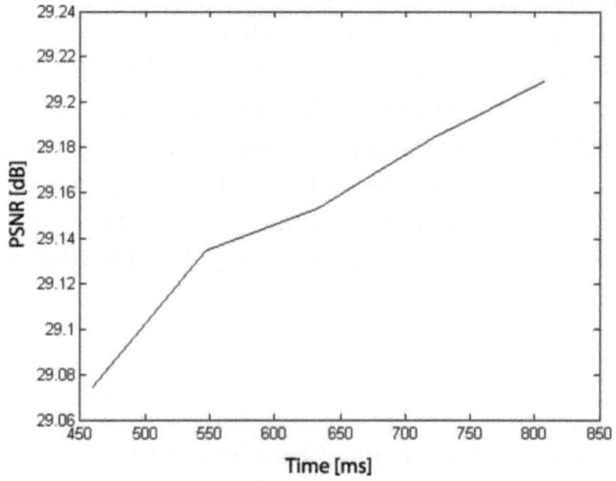

Fig. 6. Background quality (Y-channel) over time

That the quality doesn't increase linearly depends on the fact that the Eigenimages that are added to reconstruction have different mimics. The most important mimic is the first so it should improve the quality the most and the subsequent ones should improve the quality less and less. But the 5th expression may improve some frames with really bad reconstruction quality and thus increase the quality more than the 1st Eigenimage. It may also improve the mimic for several frames; the most important mimic can be visible in fewer frames than another mimic which is not as important based on the variance.

8 Discussion

The use of aPCA for compression for video with HD resolution can reduce the bitrate for transmission vastly after an initial transmission of Eigenimages. The available bitrate can also be used to improve the reconstruction quality further. A drawback with any implementation based on PCA is that it is not possible to reconstruct a changing background with high quality; it will always be blurred due to motion.

The complexity for both encoding and decoding is reduced vastly when aPCA is used compared to DCT encoding with motion estimation. This can be an extremely important factor since the power consumption is reduced and any device that is driven by batteries will have longer operating time. Since the bitrate also can be reduced the devices can save power on lower transmission costs as well.

Initially there are no Eigenimages available at the decoder side and no video can be displayed. This initial delay in video communication cannot be dealt with by buffering if the video is used in online communication such as a video telephone conversation. This shouldn't have to be a problem for video conference applications since you usually don't start communicating immediately. And a second is enough time to wait for good quality video.

There are possibilities of combining PCA or aPCA with DCT encoding such as H.264 and this will be a hybrid codec. For an initial period the frames can be encoded with H.264 and transmitted between the encoder and decoder. The fames are available at both the encoder and decoder so they can both perform PCA for the images and produce the same Eigenimages. All other frames can then be encoded with the Eigenimages to very low bitrates with low encoding and decoding complexity.

References

[1] Schäfer, R., et al.: The emerging h.264 avc standard. EBU Technical Review 293 (2003)
[2] Wiegand, T., Sullivan, G.J., Bjontegaard, G., Luthra, A.: Overview of the h.264/avc video coding standard. IEEE Trans. Circuits Syst. Video Technol. 13(7), 560–576 (2003)
[3] Jolliffe, I.: Principal Component Analysis. Springer, New York (1986)
[4] Söderström, U., Li, H.: Full-frame video coding for facial video sequences based on principal component analysis. In: Proceedings of Irish Machine Vision and Image Processing Conference 2005 (IMVIP 2005), August 30-31, 2005, pp. 25–32 (2005), www.medialab.tfe.umu.se
[5] Söderström, U., Li, H.: Representation bound for human facial mimic with the aid of principal component analysis. EURASIP Journal of Image and Video Processing, special issue on Facial Image Processing (2007)
[6] Söderström, U., Li, H.: Asymmetrical principal component analysis for video coding. Electronics letters 44(4), 276–277 (2008)
[7] Söderström, U., Li, H.: Asymmetrical principal component analysis for efficient coding of facial video sequences (2008)
[8] Turk, M., Pentland, A.: Eigenfaces for recognition. Journal of Cognitive Neuroscience 3, 71–86 (1991)
[9] Ohba, K., Clary, G., Tsukada, T., Kotoku, T., Tanie, K.: Facial expression communication with fes. In: International conference on Pattern Recognition, pp. 1378–1378 (1998)
[10] Ohba, K., Tsukada, T., Kotoku, T., Tanie, K.: Facial expression space for smooth tele-communications. In: FG 1998: Proceedings of the 3rd International Conference on Face & Gesture Recognition, p. 378 (1998)
[11] Torres, L., Prado, D.: A proposal for high compression of faces in video sequences using adaptive eigenspaces. In: 2002 International Conference on Image Processing, 2002. Proceedings, vol. 1, pp. I–189– I–192 (2002)
[12] Torres, L., Delp, E.: New trends in image and video compression. In: Proceedings of the European Signal Processing Conference (EUSIPCO), Tampere, Finland, September 5-8 (2000)
[13] Söderström, U., Li, H.: Eigenspace compression for very low bitrate transmission of facial video. In: IASTED International conference on Signal Processing, Pattern Recognition and Application (SPPRA) (2007)
[14] Wallace, G.K.: The jpeg still picture compression standard. Communications of the ACM 34(4), 30–44 (1991)

Regularisation of 3D Signed Distance Fields

Rasmus R. Paulsen, Jakob Andreas Bærentzen, and Rasmus Larsen

Informatics and Mathematical Modelling, Technical University of Denmark
Richard Petersens Plads, Building 321, DK-2800 Kgs. Lyngby, Denmark
{rrp,jab,rl}@imm.dtu.dk

Abstract. Signed 3D distance fields are used a in a variety of domains. From shape modelling to surface registration. They are typically computed based on sampled point sets. If the input point set contains holes, the behaviour of the zero-level surface of the distance field is not well defined. In this paper, a novel regularisation approach is described. It is based on an energy formulation, where both local smoothness and data fidelity are included. The minimisation of the global energy is shown to be the solution of a large set of linear equations. The solution to the linear system is found by sparse Cholesky factorisation. It is demonstrated that the zero-level surface will act as a membrane after the proposed regularisation. This effectively closes holes in a predictable way. Finally, the performance of the method is tested with a set of synthetic point clouds of increasing complexity.

1 Introduction

A signed 3D distance field is a powerful and versatile implicit representation of 2D surfaces embedded in 3D space. It can be used for a variety of purposes as for example shape analysis [15], shape modelling [2], registration [9], and surface reconstruction [13]. A signed distance field consists of distances to a surface that is therefore implicitly defined as the zero-level of the distance field. The distance is defined to be negative inside the surface and positive outside. The inside-outside definition is normally only valid for closed and non-intersecting surfaces. However, as will be shown, the applied regularisation can to a certain degree remove the problems with non-closed surfaces. Commonly, the distance field is computed from a sampled point set with normals using one of several methods [14,1]. However, a distance field computed from a point set is often not well regularised and contains discontinuities. Especially, the behaviour of the distance field can be unpredictable in areas with sparse sampling or no points at all. It is desirable to regularise the distance field so the behaviour of the field is well defined even in areas with no underlying data. In this paper, regularisation is done by applying a constrained smoothing operator to the distance field. In the following, it is described how that can be achieved.

2 Data

The data used is a set of synthetic shapes represented as point sets, where each point also has a normal. It is assumed that there are consistent normal directions

A.-B. Salberg, J.Y. Hardeberg, and R. Jenssen (Eds.): SCIA 2009, LNCS 5575, pp. 513–519, 2009.

over the point set. There exist several methods for computing consistent normals over unorganised point sets [12].

3 Methods

The signed distance field is represented as a uniform voxel volume. In theory, it is possible to use a multilevel tree-like structure, as for example octrees. However, this complicates matters and is beyond the scope of this work. Initially, the signed distance to the point set is computed using a method similar to the method described in [13]. For each voxel, the five closest (to the voxel centre) input points are found using the standard Euclidean metric. Secondly, the distance to the five points are computed as the projected distance from the voxel centre to the line spanned by the point and its associated normal as seen in Fig. 1. Finally, the distance is chosen as the average of the five distances.

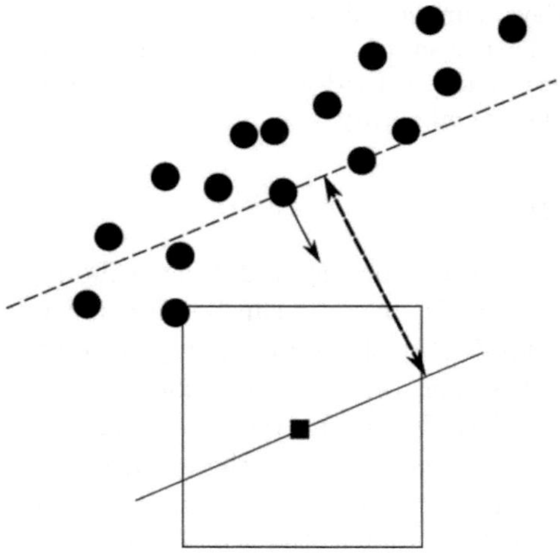

Fig. 1. Projected distance. The distance from the voxel centre (little solid square) to the point with the normal is shown as the dashed double ended arrow.

The zero-level iso-surface can now be extracted using a standard iso-surface extractor as marching cubes [16] or Bloomenthals algorithm [4]. However, this surface will neither be smooth nor behave predictable in areas with no input points. This is mostly critical if the input points do not represent shapes that are topologically equivalent to spheres. In the following, marching cubes is used when more than two distinct iso-surfaces exist and the Bloomenthal polygoniser is used if only one surface needs to be extracted.

In order to define the behaviour of the surface, we define an energy function. In this work, we choose a simple energy function based on the difference of neighbouring voxels. This classical energy has been widely used in for example Markov Random Fields [3]:

$$E(d_i) = \frac{1}{n} \sum_{i \sim j} (d_i - d_j)^2, \tag{1}$$

here d_i is the voxel value at position i and $i \sim j$ is the neighbours of the voxel at position i. For simplicity a one dimensional indexing system is used instead of the cumbersome (x, y, z) system. In this paper, a 6-neighbourhood system is used, so the number of neighbours are $n = 6$, except at the edge of the volume. From statistical physics and using the Gibbs measure it is known that this energy terms induces a Gaussian prior on the voxel values. A global energy for the entire field can now be defined as:

$$E_G = \sum_i E(d_i) \tag{2}$$

Minimising this energy is trivial. It is equal to diffusion and it can therefore be done by convolving the volume using Gaussian kernels. However, the result would be a voxel volume with the same value (the average) in all voxels. This is obviously not very interesting. In order to preserve the information stored in the point set, the energy term in Eq. (1) is changed to:

$$E_C(d_i) = \alpha_i \beta (d_i - d_i^o)^2 + (1 - \alpha_i \beta) \frac{1}{n} \sum_{i \sim j} (d_i - d_j)^2. \tag{3}$$

Here d_i^o is the original distance estimate in voxel i, α_i is a local confidence measure, and β is a global parameter that controls overall smoothing. Obviously, α should be one where there is complete confidence in the estimated distance and zero where maximum regularisation is needed. In this work, we use a simple distance based estimation of α. It is based on the Euclidean distance from the voxel centre to the nearest input point d_i^E. Here $\alpha_i = 1 - \min(d_i^E / d_{\max}^E, 1)$, where $d_{\max}^E$ is a user defined maximum Euclidean distance. A sampling density estimate is computed to estimate $d_{\max}^E$. The average μ_l and standard deviation σ_l of the closest point neighbours distances are estimated from the input point set. The distance is calculated by for each point locating the closest point and computing the Euclidean distance between the two. In this work a value of $d_{\max}^E = 3\mu_l$ was found to be suitable. A discussion of other confidence measures used for data captured using range scanners can be found in [8]. The global regularisation parameter β is set to 0.5. It is mostly useful in case of Gaussian-like noise in the input data.

A global energy can now be defined using the local energy from Eq. (3):

$$E_{G,C} = \sum_i E_C(d_i). \tag{4}$$

The minimisation of this energy is not as trivial as the minimisation of Eq. (2). Initially, it can be observed that the local energy in Eq. (3) is minimised by:

$$d_i = \alpha_i\beta\, d_i^o + (1 - \alpha_i\beta)\, \frac{1}{n}\sum_{i\sim j} d_j. \tag{5}$$

This can be rearranged into:

$$\frac{n_i d_i}{1 - \alpha_i\beta} - \sum_{i\sim j} d_j = \frac{n_i \alpha_i \beta}{1 - \alpha_i\beta} d_i^o, \tag{6}$$

If N is the number of voxel in the volume, we now have N linear equations, each with up to six unknowns (six except for the border voxels). It can therefore be cast into to the linear system $\mathbf{A}x = \mathbf{b}$:

$$\begin{bmatrix} \frac{n_1}{1-\alpha_1\beta} & -1 & \dots -1 \dots \\ -1 & \frac{n_2}{1-\alpha_2\beta} & -1 \dots & \\ \vdots & & & \\ -1 & & & \\ \vdots & & \ddots & \end{bmatrix} x = \begin{bmatrix} \frac{n_1\alpha_1\beta}{1-\alpha_1\beta} d_1^o \\ \frac{n_2\alpha_2\beta}{1-\alpha_2\beta} d_2^o \\ \vdots \\ \\ \frac{n_N\alpha_N\beta}{1-\alpha_N\beta} d_N^o \end{bmatrix},$$

where $x_i = d_i$ and $\mathbf{A}$ is a sparse tridiagonal matrix with fringes [17] having dimensions NxN. The number of neighbours of a voxel determines the number of -1 in each row in $\mathbf{A}$ (normally six). The column indexes of the -1 depend on the ordering of the voxel volume. In our case the index is computed as $i = x_t + y_t \cdot N_x + z_t \cdot N_x \cdot N_y$, where (x_t, y_t, z_t) are the voxel displacement compared to the current voxel and (N_x, N_y, N_z) is the volume dimensions. Some special care is needed for edge and corner voxels that do not have six neighbours. Furthermore, $\mathbf{A}$ is symmetric and positive definite. Several approaches to the solution of these types of problems exist. An option is to use the iterative conjugate gradients [11], but due to its $O(n^2)$ complexity, it is not suitable for large volumes [6]. Multigrid solvers are normally very fast, but require problem-dependent design decisions [7]. An alternative is to use sparse direct Cholesky solvers [5]. A sparse Cholesky solver initially factors the matrix $\mathbf{A}$, such that the solution is found by back-substitution. This is especially efficient in dynamic system where the right hand side changes and the solution can be found extremely efficient by using the pre-factored matrix to do the back substitution. However, this is not the case in our problem, but the sparse Cholesky approach still proved efficient. A standard sparse Cholesky solver (CHOLMOD) is used to solve the system [10]. With this approach, the estimation and regularisation of the distance field is done in less than two minutes for a final voxel volume of approximately (100, 100, 100) on a standard dual core, 2.4 GHz, 2GB RAM PC.

4 Results

The described approach has been applied to different synthetically defined shapes. In Figure 2, a sphere that has been cut by one, two and three planes

Fig. 2. The zero level iso-surface extracted when the input cloud is a sphere that has one, two, or three cuts

Fig. 3. The zero level iso-surface extracted when the input cloud is two cylinders that are moved away from each other

can be seen. The input points are seen together with the extracted zero-level iso-surface of the regularised distance field.

It can be seen that the zero-level exhibits a membrane-like behaviour. This is not surprising since it can be proved that Eq. (1) is a discretisation of the membrane energy. Furthermore, it can be seen that the zero-level follow the input points. This is due to the local confidence estimates α.

In Figure 3, the input consists of the sampled points on two cylinders. It is visualised how the zero-level of the regularised distance field behaves when the two cylinders are moved away from each other. When they are close, the iso-surface connects the two cylinders and when they are far away from each other, the iso-surface encapsulates each cylinder separately. Interestingly, there is a topology change in the iso-surface when comparing the situation with the close cylinders and the far cylinders. This adds an extra flexibility to the method, when seen as a surface fairing approach. Other surface fairing techniques uses an already computed mesh [18] and topology changes are therefore difficult to handle.

Finally, the method has been applied to some more complex shapes as seen in Figure 4.

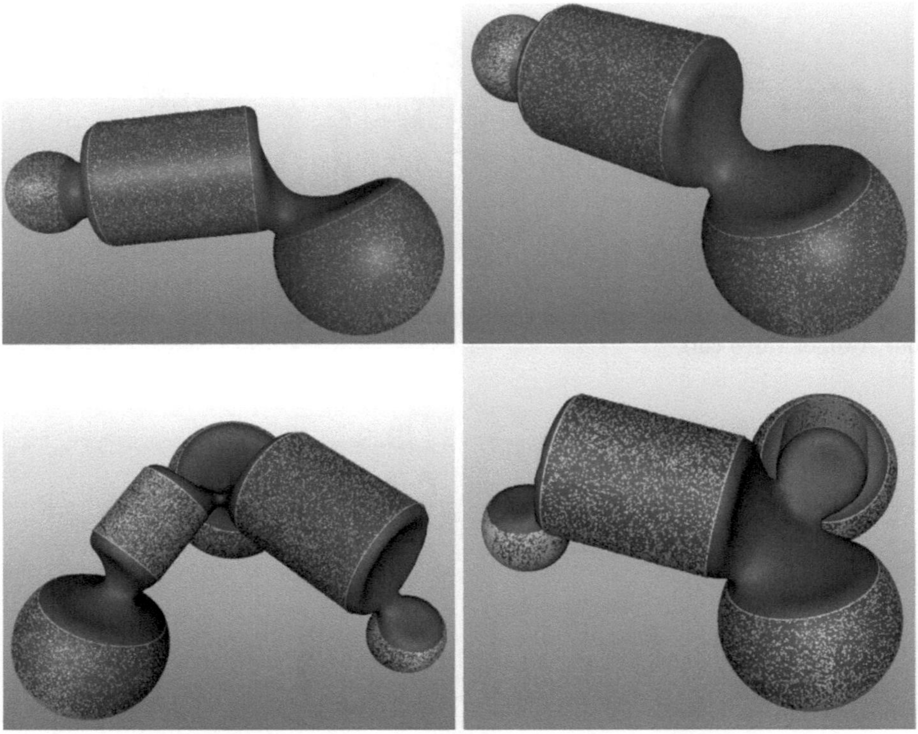

Fig. 4. The zero level iso-surface extracted when the input cloud is complex

5 Conclusion

In this paper, a regularisation scheme is presented together with a mathematical framework for fast and efficient estimation of a solution. The approach described can be used for pre-processing distance field before further processing. An obvious use for the approach is surface reconstruction of unorganised point clouds. It should noted, however, that the result of the regularisation in a strict sense is not a distance field, since it will not have global unit gradient length. If a distance field with unit gradient is needed, it can be computed based on the regularised zero-level using one of several update strategies as described in [14].

Acknowledgement

This work was in part financed by a grant from the Oticon Foundation.

References

1. Bærentzen, J.A., Aanæs, H.: Computing discrete signed distance fields from triangle meshes. Technical report, Informatics and Mathematical Modelling, Technical University of Denmark, DTU, Richard Petersens Plads, Building 321, DK-2800 Kgs, Lyngby (2002)

2. Bærentzen, J.A., Christensen, N.J.: Volume sculpting using the level-set method. In: International Conference on Shape Modeling and Applications, pp. 175–182 (2002)
3. Besag, J.: On the statistical analysis of dirty pictures. Journal of the Royal Statistical Society, Series B 48(3), 259–302 (1986)
4. Bloomenthal, J.: An implicit surface polygonizer. In: Graphics Gems IV, pp. 324–349 (1994)
5. Botsch, M., Bommes, D., Kobbelt, L.: Efficient Linear System Solvers for Mesh Processing. In: Martin, R., Bez, H.E., Sabin, M.A. (eds.) IMA 2005. LNCS, vol. 3604, pp. 62–83. Springer, Heidelberg (2005)
6. Botsch, M., Sorkine, O.: On Linear Variational Surface Deformation Methods. IEEE Transactions on Visualization and Computer Graphics, 213–230 (2008)
7. Burke, E.K., Cowling, P.I., Keuthen, R.: New models and heuristics for component placement in printed circuit board assembly. In: Proc. Information Intelligence and Systems, pp. 133–140 (1999)
8. Curless, B., Levoy, M.: A volumetric method for building complex models from range images. In: Proceedings of ACM SIGGRAPH, pp. 303–312 (1996)
9. Darkner, S., Vester-Christensen, M., Larsen, R., Nielsen, C., Paulsen, R.R.: Automated 3D Rigid Registration of Open 2D Manifolds. In: MICCAI 2006 Workshop From Statistical Atlases to Personalized Models (2006)
10. Davis, T.A., Hager, W.W.: Row modifications of a sparse cholesky factorization. SIAM Journal on Matrix Analysis and Applications 26(3), 621–639 (2005)
11. Golub, G.H., Van Loan, C.F.: Matrix Computations. Johns Hopkins University Press (1996)
12. Hoppe, H., DeRose, T., Duchamp, T., McDonald, J., Stuetzle, W.: Surface reconstruction from unorganized points. In: ACM SIGGRAPH, pp. 71–78 (1992)
13. Jakobsen, B., Bærentzen, J.A., Christensen, N.J.: Variational volumetric surface reconstruction from unorganized points. In: IEEE/EG International Symposium on Volume Graphics (September 2007)
14. Jones, M.W., Bærentzen, J.A., Sramek, M.: 3D Distance Fields: A Survey of Techniques and Applications. IEEE Transactions On Visualization and Computer Graphics 12(4), 518–599 (2006)
15. Leventon, M.E., Grimson, W.E.L., Faugeras, O.: Statistical shape influence in geodesic active contours. In: IEEE Conference on Computer Vision and Pattern Recognition, 2000, vol. 1 (2000)
16. Lorensen, W.E., Cline, H.E.: Marching cubes: A high resolution 3D surface construction algorithm. Computer Graphics (SIGGRAPH 1987 Proceedings) 21(4), 163–169 (1987)
17. Press, W.H., Teukolsky, S.A., Vetterling, W.T., Flannery, B.P.: Numerical recipes in C: the art of scientific computing. Cambridge University Press, Cambridge (2002)
18. Schneider, R., Kobbelt, L.: Geometric fairing of irregular meshes for free-form surface design. Computer Aided Geometric Design 18(4), 359–379 (2001)

An Evolutionary Approach for Object-Based Image Reconstruction Using Learnt Priors*

Péter Balázs and Mihály Gara

Department of Image Processing and Computer Graphics,
University of Szeged,
Árpád tér 2., H-6720, Szeged, Hungary
{pbalazs,gara}@inf.u-szeged.hu

Abstract. In this paper we present a novel algorithm for reconstructing binary images containing objects which can be described by some parameters. In particular, we investigate the problem of reconstructing binary images representing disks from four projections. We develop a genetic algorithm for this and similar problems. We also discuss how prior information on the number of disks can be incorporated into the reconstruction in order to obtain more accurate images. In addition, we present a method to exploit such kind of knowledge from the projections themselves. Experiments on artificial data are also conducted.

1 Introduction

The aim of *Computerized Tomography* (CT) is to obtain information about the interior of objects without damaging or destroying them. Methods of CT (like filtered backprojection or algebraic reconstruction techniques) often require several hundreds of projections to obtain an accurate reconstruction of the studied object [8]. Since the projections are usually produced by X-ray, gamma-ray, or neutron imaging, the acquisition of them can be expensive, time-consuming or can (partially or fully) damage the examined object. Thus, in many applications it is impossible to apply reconstruction methods of CT with good accuracy. In those cases there is still a hope to get a satisfactory reconstruction by using *Discrete Tomography* (DT) [6,7]. In DT one assumes that the image to be reconstructed contains just a few grey-intensity values that are known beforehand. This extra information allows one to develop algorithms which reconstruct the image from just few (usually not more than four) projections.

When the image to be reconstructed is binary we speak of *Binary Tomography* (BT) which has its main applications in angiography, electron microscopy, and non-destructive testing. BT is a relatively new field of research, and for a large variety of images the reconstruction problem is still not satisfactorily solved. In this paper we present a new approach for reconstructing binary images representing disks from four projections. The method is more general in the sense that it can be adopted to similar reconstruction tasks as well. The

* This work was supported by OTKA grant T048476.

A.-B. Salberg, J.Y. Hardeberg, and R. Jenssen (Eds.): SCIA 2009, LNCS 5575, pp. 520–529, 2009.

paper is structured as follows. In Sect. 2 we give the preliminaries. In Sect. 3 we outline an object-based genetic reconstruction algorithm. The algorithm can use prior knowledge to grade up the reconstruction. Section 4 describes a method to collect such information when it is not explicitly given. In Sect. 5 we present experimental results. Finally, Sect. 6 is for the conclusion.

2 Preliminaries

The reconstruction of 3D binary objects is usually done slice-by-slice, i.e, by integrating together the reconstructions of 2D slices of the object. Such a 2D binary slice can be represented by a 2D binary function $f(x, y)$. The *Radon transformation* $\mathcal{R}f$ of f is then defined by

$$[\mathcal{R}f](s, \vartheta) = \int_{-\infty}^{\infty} f(x, y)du \ , \quad (1)$$

where s and u denote the variables of the coordinate system obtained by a rotation of angle ϑ. For a fixed angle ϑ we call $\mathcal{R}f$ as the *projection* of f defined by the angle ϑ (see Fig. 1). The reconstruction problem can be stated mathematically as follows. Given the functions $g(s, \vartheta_1), \ldots, g(s, \vartheta_n)$ (where n is a positive integer) find a function f such that

$$[\mathcal{R}f](s, \vartheta_i) = g(s, \vartheta_i) \quad (i = 1, \ldots, n) \ . \quad (2)$$

3 An Object-Based Genetic Reconstruction Algorithm

3.1 Reconstruction with Optimization

In this work we concentrate on the reconstruction of binary images representing disjoint disks inside a ring (see again Fig. 1). Such images were introduced for testing the effectiveness of reconstruction algorithms developed for neutron tomography [9,10,11]. For the reconstruction we will use just four projections. Our aim is to find a function f that satisfies (2) with the given angles $\vartheta_1 = 0°$, $\vartheta_2 = 45°$, $\vartheta_3 = 90°$, and $\vartheta_4 = 135°$. In practice, instead of finding the exact function f, we are usually satisfied with a good approximation of it. On the other

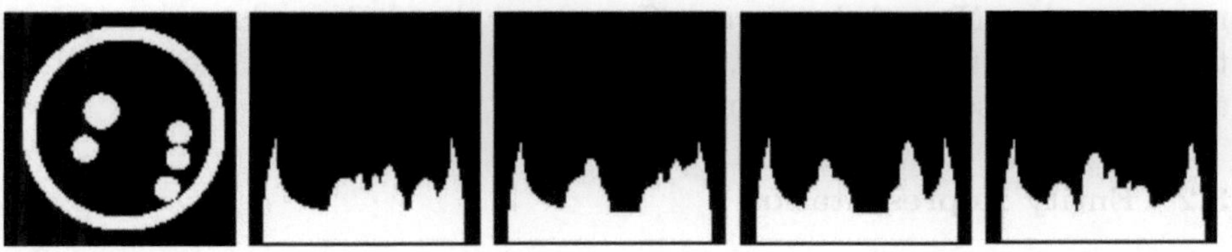

Fig. 1. A binary image and its projections defined by the angle $\vartheta = 0°$, $\vartheta = 45°$, $\vartheta = 90°$, and $\vartheta = 135°$ (from left to right, respectively)

hand – especially if the number of projections is small – there can be several different functions which (approximately) satisfy (2). Fortunately, with additional knowledge of the image to be reconstructed some of them can be eliminated, which might yield that the reconstructed image will be close to the original one. For this purpose we rewrite the reconstruction task as an optimization problem where the aim is to find the minimum of the objective functional

$$\Phi(f) = \lambda_1 \cdot \sum_{i=1}^{4} ||\mathcal{R}f(s, \vartheta_i) - g(s, \vartheta_i)|| + \lambda_2 \cdot \varphi(c_f, c) \; . \tag{3}$$

The first term in the right hand side of (3) guarantees that the projections of the reconstructed image will be close to the prescribed ones. In the second term we can keep control over the number of disks in the image to be reconstructed. We will use this prior information to obtain more accurate reconstructions. Here, c_f is the number of disks in the image f. Finally, λ_1 and λ_2 are suitably chosen scaling constants. With the aid of them we can also express whether the projections or the prior information is more reliable.

In DT (3) is usually solved by simulated annealing (SA) [12]. In [9] two different approaches were presented to reconstruct binary images representing disks inside a ring with SA. The first one is a pixel-based method where in each iteration a single pixel value is inverted to obtain a new proposed solution. Although this method can be applied in more general (i.e., also in the case when the image does not represent disks), it has some serious drawbacks: it is quite sensitive to noise, it can not exploit geometrical information of the image to be reconstructed, and it needs 10-16 projections for an accurate reconstruction. The other method of [9] is a parameter-based one in which the image is represented by the centers and radii of the disks, and the aim is to find the proper setting of these parameters. This algorithm is less sensitive to noise, easy to extend to direct 3D reconstruction, but its accuracy decreases drastically as the complexity of the image (i.e. the number of disks in it) increases. Furthermore, the number of disks should be given before the reconstruction. In this paper we design an algorithm that can benefit the advantages of both reconstruction methods. However, instead of using SA to find an approximately good solution, we will describe an evolutionary approach. Evolutionary computation [2] proved to be successful in many large-scale optimization tasks. Unfortunately, the pixel-based representation of the image makes evolutionary algorithms difficult to use in binary image reconstruction. Nevertheless, some efforts have already been done to overcome this problem in tricky ways [3,5,14]. Our idea is a more natural one, we will use a parameter-based representation of the image.

3.2 Entity Representation

We assume that there exists a ring which center coincides the center of the image, and there are some disjoint disks inside this ring (the ring and each of the disks are disjoint as well) (see, e.g., Fig. 1). The outer ring can be represented as the difference of two disks, and therefore the whole image can be described by a list of triplets $\langle(x_1, y_1, r_1), \ldots, (x_n, y_n, r_n)\rangle$ where $n \geq 3$. Here, (x_i, y_i) and r_i $(i = 1, \ldots, n)$ denote the center and the radius of the ith disk, respectively (the bottom-left corner of the image is $(0, 0)$). Since the first two elements of the list stand for the outer ring, $x_1 = x_2$, $y_1 = y_2$, and $r_1 > r_2$ do always hold. Moreover, the point (x_1, y_1) is the center of the image.

The evolutionary algorithm seeks for the optimum by a population of entities. Each entity is a suggestion for the optimum, and its fitness is simply measured by the formula of (3) (smaller values belong to better solutions). The entities of the actual population are modified with the mutation and crossover operators. These are described in the followings in more detail.

3.3 Crossover

Crossover is controlled by a global probability parameter p_c. During the crossover each entity e is assigned a uniform randomly chosen number $p_e \in [0, 1]$. If $p_e < p_c$ then the entity is subject to crossover. In this case we randomly choose an other entity e' of the population and try to cross it with e. Suppose that e and e' are described by the lists $\langle(x_1, y_1, r_1), \ldots, (x_n, y_n, r_n)\rangle$ and $\langle(x'_1, y'_1, r'_1), \ldots, (x'_k, y'_k, r'_k)\rangle$, respectively ($e$ and e' can have different number of disks, i.e., k is not necessarily equal to n). Then the two offsprings are presented by $\langle(x_1, y_1, r_1), \ldots, (x_t, y_t, r_t), (x'_{s+1}, y'_{s+1}, r'_{s+1}), \ldots, (x'_k, y'_k, r'_k)\rangle$ and $\langle(x'_1, y'_1, r'_1), \ldots, (x'_s, y'_s, r'_s), (x_{t+1}, y_{t+1}, r_{t+1}), \ldots, (x_n, y_n, r_n)\rangle$ where $3 \leq t \leq n$ and $3 \leq s \leq k$ are chosen from uniform random distributions. As special cases an offspring can inherit all or none of the innner disks of one of its parents (the method guarantees that the outer rings in both parent images are kept). A crossover is valid if the ring and all of the disks are pairwisely disjoint in the image. Though, in some cases it can happen that both offsprings are invalid. In this case we repeat to choose s and t randomly until at least one of the offsprings is valid or we reach the maximal number of allowed attempts a_c. Figure 2 shows an example for the crossover. The list of the two parents are ⟨(50, 50, 40.01), (50, 50, 36.16), (41.29, 27.46, 8.27), (65.12, 47.3, 5.65), (54.69, 55.8, 5), (56.56, 73.38, 5.04), (46.49, 67.41, 5)⟩ and ⟨(50, 50, 45.6), (50, 50, 36.14), (40.33, 24.74, 7.51), (24.17, 54.79, 7.59), (74.35, 46.37, 10.08)⟩. The offsprings are ⟨(50, 50, 45.6), (50, 50, 36.14), (40.33, 24.74, 7.51), (24.17, 54.79, 7.59), (54.69, 55.8, 5), (56.56, 73.38, 5.04), (46.49, 67.41, 5)⟩ and ⟨(50, 50, 40.01), (50, 50, 36.16), (41.29 27.46, 8.27), (65.12, 47.3, 5.65), (74.35, 46.37, 10.08)⟩.

3.4 Mutation

During the mutation an entity can change in three different ways:

(1) the number of disks increases/decreases by 1,
(2) the radius of a disk changes by at most 5 units, or
(3) the center of a disk moves inside a circle having a radius of 5 units.

For each type of the above mutations we set global probability thresholds, p_{m_1}, p_{m_2}, and p_{m_3}, respectively, which have the same roles as p_c has for crossover. For

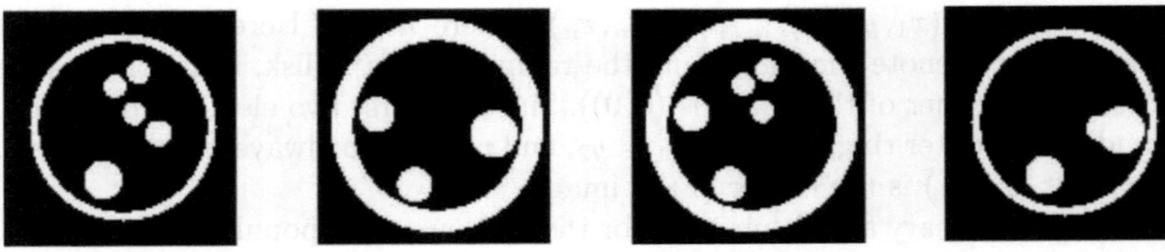

Fig. 2. An example for crossover. The images are the two parents, a valid, and an invalid offspring (from left to right).

Fig. 3. Examples for mutation. From left to right: original image, decreasing and increasing the number of disks, moving the center of a disk, and resizing a disk.

the first type of mutation the number of disks is increased and decreased with equal $0.5 - 0.5$ probability. If the number of disks is increased then we add a new element to the end of the list. If this newly added element intersects any element of the list (except itself) then we do a new attempt. We repeat this method until we succeed or the maximal number of attempts a_m is reached. When the number of disks should be decreased then we simply delete one element of the list (which cannot be among the first two elements since the ring should be unchanged).

In the case when the radius of a disks had to be changed then this disk is randomly chosen from the list and we modify its radius by a randomly chosen value from the interval $[-5, 5]$. The disk to modify can be one of the disks describing the ring, as well. Finally, if we move the center of a disk then it is done again with uniform random distribution in a given interval. In this case the ring can not be subject to change. In the last two types of mutation we do not take another attempts if the mutated entity is not valid. Figure 3 shows examples of the several mutation types.

3.5 Selection

During the genetic process the population consists of a fixed number (say γ) of entities, and only entities with the best fitness values will survive to the next generation. In each iteration we first apply the crossover operator with which we obtain μ_1 (valid) offsprings. In this stage all the parents and offsprings are present in the population. With the aid of the mutation operators we obtain μ_2 new entities from the $\gamma + \mu_1$ entities and we also add them to the population. Finally, from the $\gamma + \mu_1 + \mu_2$ number of entities we only keep γ having the best fitness values and they will form the next generation.

4 Guessing the Number of Disks

Our final aim is to design a reconstruction algorithm that can cleverly use the knowledge of the number of disks present in the image. The method developed in [9] assumes that this information is available beforehand. In contrary, we try to exploit it from the projections themselves, thus, making or method more flexible, and more widely applicable. Our preliminary investigations showed that decision trees can help to gain structural information from the projections of a binary image [1]. Therefore we again used C4.5 tree classifiers for this task [13].

With the aid of the generator algorithm of DIRECT [4] we generated 1100-1100 images having 1, 2, ..., 10 disks inside the outer ring. All of them were of size of 100×100 and the number of projections were 100 from each directions. We used 1000 images from each set to train the tree, and the remaining 100 to test the accuracy of the classification. Our decision variables were the number of local optima and their values in all four projection vectors. In this way we had $4(1+6) = 28$ variables for each training example and we classified those examples into 10 classes (depending on the number of disks in the image from which the projections arose). If the number of local maxima was less than 6 then we simply set the corresponding decision variable to be 0, if this number was greater than six, then the remaining values were omitted. Table 1 shows the results of classification of the decision tree on the test data set. Although the tree built during the learning was not able to predict the exact number of disks with good accuracy (except if the image contained just a very few disks) its classification can be regarded quite accurate if we allow an error of 1 or 2 disks. This observation turns out to be useful to add information on the number of disks into the fitness function of our genetic algorithm. We set the term $\varphi(c_f, c)$ in the fitness function in the following way

$$\varphi(c_f, c) = 1 - \frac{t_{c_f,c}}{\sum_{i=1}^{10} t_{i,c}} \tag{4}$$

where c is the class given by the decision tree by using the projections, and t_{ij} denotes the element of Table 1 in the i-th row and the j-th column. For example,

Table 1. Predicting the number of disks by a decision tree from the projection data

(a)	(b)	(c)	(d)	(e)	(f)	(g)	(h)	(i)	(j)	<-classified as
100										(a): class 1
	92	8								(b): class 2
	8	75	16	1						(c): class 3
		23	49	23	3	2				(d): class 4
			2	21	45	22	5	5		(e): class 5
			6	22	35	24	7	5	1	(f): class 6
				8	25	26	22	14	5	(g): class 7
				3	12	16	30	23	16	(h): class 8
					5	15	18	25	37	(i): class 9
						7	20	29	44	(j): class 10

if on the basis of the projection vectors the decision tree predicts that the image to be reconstructed has five inner disks (class (e)) then for an arbitrary image f $\varphi(c_f, 5)$ is equal to 1.0, 1.0, 0.9871, 0.7051, 0.7307, 0.7179, 0.8974, 0.9615, 1.0, and 1.0 for $c_f = 1, \ldots, 10$, respectively.

5 Experimental Results

In order to test the efficacy of our method we conducted the following experiment. We designed 10 test images with increasing structural complexity having 1, 2, ..., 10 disks inside the ring. We tried to reconstruct each image 10 times by our approach with no information about the number of disks, 10 times with the information defined by (4), and finally 10 times when we assumed that the number of disks is known in advance (by setting φ to be 0.0 if the reconstructed image had the same number of disks as it was expected and 1.0 otherwise).

The initial population consisted of 200-200 entities from the classes 3 to 9 (i.e. we used $\gamma = 1400$). For the random generation of the entities we again used the algorithm of DIRECT [4]. The threshold parameters for the operators were set to $p_c = 0.05$, $p_{m_1} = 0.05$, and $p_{m_2} = p_{m_3} = 0.25$. The maximal number of attempts were $a_c = 50$ for the crossover and $a_m = 1000$ for the mutation of the first type. We found the best results with setting $\lambda_1 = 0.000025$ and $\lambda_2 = 0.015$. We set the reconstruction process to terminate after 3000 generations. Figure 4 represents the best reconstruction results achieved by the three methods.

To the numerical evaluation of the accuracy of our method we used the relative mean error (RME) that was defined in [9] as

$$RME = \frac{\sum_i |f_i^o - f_i^r|}{\sum_i f_i^o} \cdot 100\% \tag{5}$$

where f_i^o and f_i^r denote the ith pixel of the original and the reconstructed image, respectively. Thus, the smaller the RME value is, the better the reconstruction is. The numerical results are given in Table 2 and - for the sake of transparency – they are also shown on a graph (see Fig. 5). On the basis of this experiment we can deduce that all three variants of our method perform quite well for simple images (say, for images having less than 5-6 disks), and give results that can be suitable for practical applications, as well. Just for a comparison, the best reconstruction obtained by our method using four projections for the test image having 4 inner disks gives an RME of 1.95%, while the pixel-based method of [9] on an image having the same complexity yields an RME of 12.57% by using eight (!) projections (cf. [9] for more sophisticated comparisons). For more complex images the reconstruction becomes more inaccurate. However, the best results are usually achieved by the decision tree approach, and it still gives images of relatively good quality. Regarding the reconstruction time we found that it is about 10 minutes for images having few (say, 1-3 inner disks), 30 minutes if there are more than 3 disks, and 1 hour for images having 8-10 disks (on an Intel Celeron 2.8GHz processor with 1.5GB of memory).

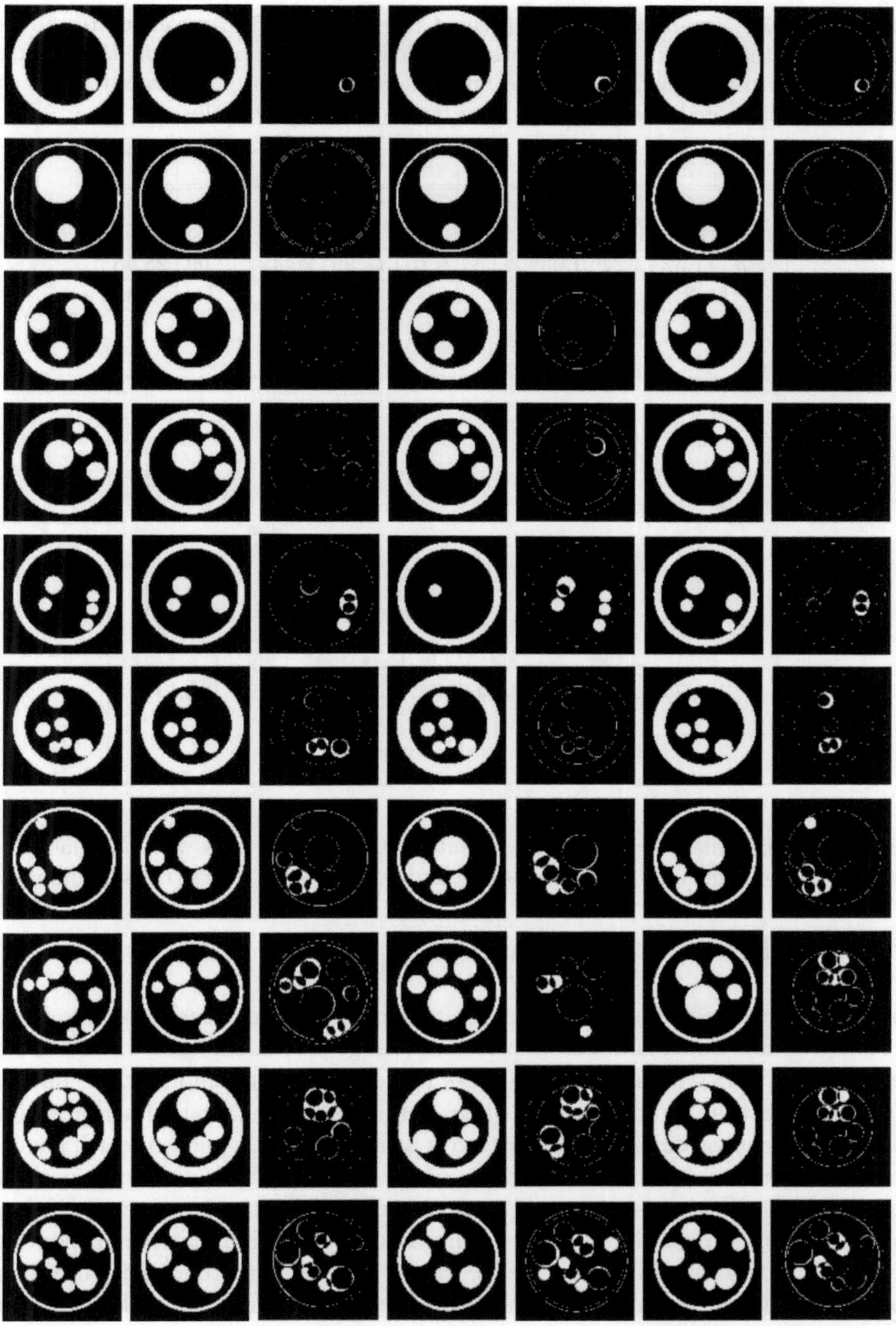

Fig. 4. Reconstruction with the genetic algorithm. From left to right: Original image, reconstruction with no prior information, the difference image, reconstruction with fix prior information, the difference image, and reconstruction with the decision tree approach and the difference image.

Table 2. RME (rounded to two digits) of the best out of 10 reconstructions as it depends on the number of inner disks (first row) with no prior information (second row), fix (third row), and learnt prior information (fourth row). In the latter case the number of disks predicted by the decision tree is given in the fifth row.

1	2	3	4	5	6	7	8	9	10
1.92	8.66	0.78	2.29	13.86	7.72	19.63	29.00	12.06	33.51
3.60	4.50	3.01	7.16	4.27	5.51	22.31	11.20	17.05	39.52
4.75	11.32	1.22	1.95	8.08	6.15	17.98	26.42	12.09	28.48
1	2	3	5	5	8	5	10	7	10

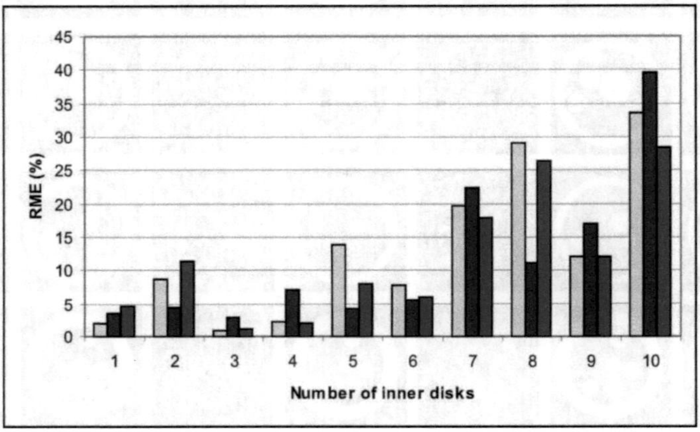

Fig. 5. Relative mean error of the best out of 10 reconstructions with no prior information (left column), fix priors (middle column), and learnt priors (right column)

6 Conclusion and Further Work

We have developed an evolutionary algorithm for object-based binary image reconstruction which can handle prior knowledge also in the case when it is not explicitly given. We used decision trees for learning prior information, but the framework is easy to adapt to use other classifiers, as well. Experimental results show that each variant of our algorithm is promising, but some work still have to be done. We found that the repetition of the mutation and crossover operators until a valid offspring is generated can take quite a long time - especially if there are many disks in the image. Our future aim is to develop faster mutation and crossover operators. In our further work we also want to tune the parameters of our algorithm to achieve more accurate reconstructions. This includes finding more sophisticated attributes that can be used in the decision tree for describing the number of disks present in the image. The study of noise-sensitivity and possible 3D extensions of our method form also parts of our further research.

References

1. Balázs, P., Gara, M.: Decision trees in binary tomography for supporting the reconstruction of *hv*-convex connected images. In: Blanc-Talon, J., Bourennane, S., Philips, W., Popescu, D., Scheunders, P. (eds.) ACIVS 2008. LNCS, vol. 5259, pp. 433–443. Springer, Heidelberg (2008)
2. Bäck, T., Fogel, D.B., Michalewicz, T. (eds.): Evolutionary Computation 1. Institute of Physics Publishing, Bristol (2000)
3. Batenburg, K.J.: An evolutionary algorithm for discrete tomography. Disc. Appl. Math. 151, 36–54 (2005)
4. DIRECT - DIscrete REConstruction Techniques. A toolkit for testing and comparing 2D/3D reconstruction methods of discrete tomography, `http://www.inf.u-szeged.hu/~direct`
5. Di Gesù, V., Lo Bosco, G., Millonzi, F., Valenti, C.: A memetic algorithm for binary image reconstruction. In: Brimkov, V.E., Barneva, R.P., Hauptman, H.A. (eds.) IWCIA 2008. LNCS, vol. 4958, pp. 384–395. Springer, Heidelberg (2008)
6. Herman, G.T., Kuba, A. (eds.): Discrete Tomography: Foundations, Algorithms and Applications. Birkhäuser, Boston (1999)
7. Herman, G.T., Kuba, A. (eds.): Advances in Discrete Tomography and its Applications. Birkhäuser, Boston (2007)
8. Kak, A.C., Slaney, M.: Principles of Computerized Tomographic Imaging. IEEE Press, New York (1988)
9. Kiss, Z., Rodek, L., Kuba, A.: Image reconstruction and correction methods in neutron and x-ray tomography. Acta Cybernetica 17, 557–587 (2006)
10. Kiss, Z., Rodek, L., Nagy, A., Kuba, A., Balaskó, M.: Reconstruction of pixel-based and geometric objects by discrete tomography. Simulation and physical experiments. Elec. Notes in Discrete Math. 20, 475–491 (2005)
11. Kuba, A., Rodek, L., Kiss, Z., Ruskó, L., Nagy, A., Balaskó, M.: Discrete tomography in neutron radiography. Nuclear Instr. Methods in Phys. Research A 542, 376–382 (2005)
12. Metropolis, N., Rosenbluth, A., Rosenbluth, M., Teller, E.: Equation of state calculation by fast computing machines. J. Chem. Phys. 21, 1087–1092 (1953)
13. Quinlan, J.R.: C4.5: Programs for Machine Learning. Morgan Kaufmann, San Francisco (1993)
14. Valenti, C.: A genetic algorithm for discrete tomography reconstruction. Genet. Program Evolvable Mach. 9, 85–96 (2008)

Disambiguation of Fingerprint Ridge Flow Direction—Two Approaches

Robert O. Hastings

School of Computer Science and Software Engineering,
The University of Western Australia, Australia
http://www.csse.uwa.edu.au/~bobh

Abstract. One of the challenges to be overcome in automated fingerprint matching is the construction of a ridge pattern representation that encodes all the relevant information while discarding unwanted detail. Research published recently has shown how this might be achieved by representing the ridges and valleys as a periodic wave. However, deriving such a representation requires assigning a consistent unambiguous direction field to the ridge flow, a task complicated by the presence of singular points in the flow pattern. This disambiguation problem appears to have received very little attention.

We discuss two approaches to this problem — one involving construction of branch cuts, the other using a divide-and-conquer approach, and show how either of these techniques can be used to obtain a consistent flow direction map, which then enables the construction of a phase based representation of the ridge pattern.

1 Introduction

A goal that has until recently eluded researchers is the representation of a fingerprint in a form that encodes only the information relevant to the task of fingerprint matching, i.e. the details of the ridge pattern, while omitting extraneous detail.

Level 1 detail, which refers to the ridge flow pattern and forms the basis of the Galton-Henry classification of fingerprints into arch patterns, loops, whorls etc., (Maltoni et al., 2003, p 174) is encapsulated in the ridge orientation field. Level 2 detail, which refers to details of the ridges themselves, especially instances where ridges bifurcate or terminate, is the primary tool of fingerprint based identification, and it is not so obvious how best to represent this. A popular approach has been to define ridges as continuous lines defining the ridge axes. For example, Ratha et al. (1995) convert the grey-scale image into a binary image, then thin the ridges to construct a "skeleton ridge map" which they then represent by a set of chain codes. Shi and Govindaraju (2006) employ chain codes to represent the ridge edges rather than the axes of the ridges — that is, the ridges are allowed to have a finite width. This avoids the need for a thinning step, but still requires that the image be binarised.

A.-B. Salberg, J.Y. Hardeberg, and R. Jenssen (Eds.): SCIA 2009, LNCS 5575, pp. 530–539, 2009.

Some problems with the skeleton image representation are:

1. The output of thinning is critically dependent on the chosen value of the binarisation threshold, and is also highly sensitive to noise.
2. It is not immediately clear how one might quickly determine the degree of similarity of two given chain codes.

An alternative is to represent the ridges via a scalar field, the value at each point specifying where that point is relative to the local ridge and valley axes. The phase representation, discussed in the next section, is a way to achieve this.

2 Representation of the Ridges as a Periodic Wave

Except near core and delta points, the ridges resemble the peaks or troughs of a periodic wave train. This suggests that the pattern might be modeled using, say, the cosine of some smoothly varying phase quantity. Two fingerprint segments could then be compared directly with one another by taking the point-wise correlation of the cosine values. There are two major difficulties with this approach:

1. Any wave model must somehow incorporate the Level 2 detail (minutiae), meaning that the wave phase must be discontinuous at these points. Recently published research describes a phase representation in which the minutiae appear as **spiral points** in the phase field (Sect. 2.1).
2. Deriving a phase field implies the assignment of a direction to the wave. Whilst it is relatively easy to obtain the ridge orientation, disambiguating this into a consistent flow field is a non-trivial task. The challenge of disambiguation is the main theme of this paper, and is discussed in Sect. 3.

2.1 The Larkin and Fletcher Phase Representation

Larkin and Fletcher (2007) propose a finger ridge pattern representation based on phase, with the grey-scale intensity taking the form:

$$I(x,y) - a(x,y) = b(x,y)\cos[\psi(x,y)] + n(x,y), \tag{1}$$

where I is image intensity at each point, a is the offset, or "DC component", b is the wave amplitude, ψ is the phase term and n is a noise term. The task is to determine the parameters a, b and ψ; this is termed **demodulation.**

After first removing the offset term $a(x,y)$ by estimating this as the mid-value of a localised histogram, they define a **demodulation operator** $\$$ and apply this to the remainder. They show that, neglecting the noise term:

$$\$\{b(x,y)\cos[\psi(x,y)]\} \approx -i\exp[i\beta(x,y)]b(x,y)\sin[\psi(x,y)], \tag{2}$$

where β is the direction of the wave normal. Comparison of (1) and (2) shows that the right hand sides are in the ratio:

$$-i\exp[i\beta(x,y)]\tan[\psi(x,y)], \tag{3}$$

so that provided we know β we can use (3) to determine the phase term ψ.

By the Helmholtz Decomposition Theorem (Joseph, 2006), ψ can be decomposed into a continuous component ψ_c and a spiral component ψ_s, where ψ_s is generated by summing a countable number of spiral phase functions centred on n points and defined as [1]:

$$\psi_s(x,y) = \sum_i p_i \arctan\left(\frac{y-y_i}{x-x_i}\right). \tag{4}$$

The points $\{(x_i, y_i)\}$ are the locations of spirals in the phase field; each has an associated "polarity" $p_i = \pm 1$. These points can be located using the Poincaré Index, defined as the total rotation of the phase vector when traversing a closed curve surrounding any point (Maltoni, Maio et al., 2003, p 97). This quantity is equal to $+2\pi$ at a positive phase vortex, -2π at a negative vortex and zero everywhere else. The residual phase component $\psi_c = \psi - \psi_s$ contains no singular points, and can therefore be unwrapped to a continuous phase field.

Referring to (3), note that replacement of β by $\beta + \pi$ implies a negation of ψ, so that, in order to derive a continuous ψ field, we must disambiguate the ridge flow direction to obtain a continuous wave normal across the image.

3 Disambiguating the Ridge Orientation

Deriving the orientation field is comparatively straightforward. We use the methodology of Bazen and Gerez (2002), who compute orientation via Principal Component Analysis applied to the squared directional image gradients. The output from this analysis is an expression for the **doubled angle** $\phi = 2\theta$, with θ being the orientation. This reflects the fact that θ has an inherent ambiguity of π.

Inspection of the orientation field around a core or delta point (Fig. 1) reveals that, in tracing a closed curve around the point, the orientation vector rotates through an angle of π in the case of a core, and through $-\pi$ for a delta. The Poincaré Index of ϕ is therefore 2π at a core, -2π at a delta, and zero elsewhere.

Larkin and Fletcher note in passing that their technique requires that the orientation field be unwrapped to a direction field, but Fig. 2 illustrates the difficulty inherent in determining a consistent direction field. The difficulty arises from the presence of a singular point (in this case a core). This unwrapping task appears to received scant attention to date, perhaps because there has been no clear incentive for doing so prior to the publication of the ridge phase model. Sherlock and Monro (1993) discuss the unwrapping of the **orientation** field (which they term the "direction field"), but this is a different and much simpler task, because the orientation, expressed as the doubled angle ϕ, contains only singular points rather than lines of discontinuity.

[1] In this paper the arctan function is understood to be implemented via $\arctan(y/x) = atan2(y, x)$, where the *atan2* function returns a single angle in the correct quadrant determined by the signs of the two input arguments.

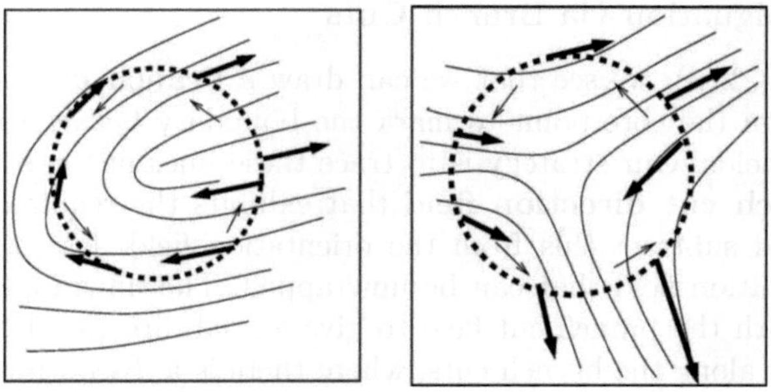

(a) Core (b) Delta

Fig. 1. Closed loop surrounding a singular point, showing the orientation vector (dark arrows) at various points around the curve

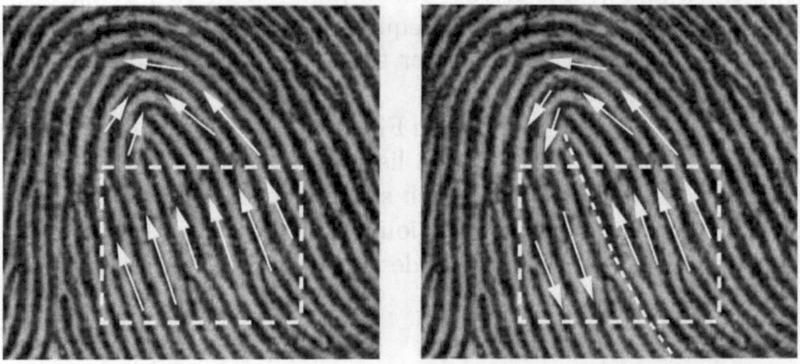

(a) An unsuccessful attempt to assign a consistent flow direction (b) Flow direction correctly assigned with consistency across the image

Fig. 2. A flow direction that is consistent within a region (dashed rectangle) cannot always be extended to the rest of the image without causing an inconsistency (a). Reversing the direction over part of the image resolves this inconsistency (b).

We discuss here two possible approaches to circumventing this difficulty:

1. Construct a flow pattern in which regions of oppositely directed flow are separated by branch cuts, as illustrated in Fig. 2(b).
2. Bypass the problem of singular points by subdividing the image into a number of sub-images. It is always possible to do this in such a way that no cores or deltas occur within any of the sub-images.

Both of these approaches were tried, and the methods and results are now described in detail. While the first approach is perhaps the more appealing, it does possess some shortcomings, as will be seen.

3.1 Disambiguation via Branch Cuts

Examining Fig. 2(b), we see that we can draw a **branch cut** (sloping dashed line) down from the core point to mark the boundary between two oppositely directed flow fields. Our strategy is to trace these lines in the orientation field, define a **branch cut direction field** that exhibits the required discontinuity properties, and subtract this from the orientation field, leaving a continuous residual orientation field that can be unwrapped. The unwrapped field is then recombined with the branch cut field to give a final direction field that is continuous except along the branch cuts, where there is a discontinuity of π.

We define a **dipole field** ϕ_d on a line segment (illustrated in Fig. 3) as follows:

$$\phi_d(x, y, x_1, y_1, x_2, y_2) = \arctan\left(\frac{y - y_1}{x - x_1}\right) - \arctan\left(\frac{y - y_2}{x - x_2}\right), \tag{5}$$

where (x_1, y_1) and (x_2, y_2) are the start and end points of the line. If ϕ_d is defined to lie between $-\pi$ and π, this gives a discontinuity of 2π only along the line itself (Fig. 3(a)). This is precisely what is required, except that we must later divide by 2 to give a discontinuity of π rather than 2π. There is also a **phase spiral** at each end of the dipole (Fig. 3(b)).

Branch cuts such as the one shown in Fig. 2(b) are constructed by commencing at a singular point and constructing a list of nodes $\{(x_i, y_i)\}$. The first node is the location of the singular point; each subsequent node is located by drawing a straight line segment from the previous node following the ridge orientation (which is already known). Further nodes are added to the list until the image

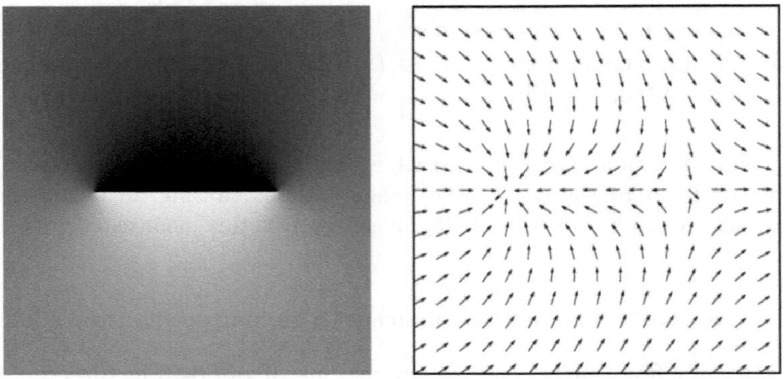

(a) Phase dipole field (grey scale). (b) Phase dipole field, shown in vector form.

Fig. 3. Phase field around a phase dipole. The positive end of the dipole is on the left, the negative on the right. Grey-scale values in (a) range from $-\pi$ (black) to $+\pi$(white); direction values in (b) increase anticlockwise with zero towards the right. Note from (a) that the field is continuous everywhere except at the two poles and along the line between them. The linear discontinuity is not apparent in (b), because the directions of π and $-\pi$ are equivalent.

border is reached. Each core point is the source of one branch cut, while three branch cuts emanate from each delta point (see for example Fig. 4(b)). A **branch cut phase field** ϕ_b is then defined for each individual branch cut:

$$\phi_b(x,y) = \sum_{i=1}^{n-1} \phi_d(x, y, x_i, y_i, x_{i+1}, y_{i+1}). \tag{6}$$

Positive and negative dipole phase spirals cancel at each node except for the first and last nodes, leaving only a linear discontinuity of 2π along each segment of the cut, plus a positive phase spiral at the start of the branch cut and a negative spiral at the end. In most cases the end node of a branch cut is outside the image so that it can be ignored (see however Sect. 4, where this is presented as one of the shortcomings of the branch cut based method of disambiguation).

Although Φ_N contains phase spirals at the same locations as ϕ, the Poincaré Index does not have the correct value at the delta points, because the three branch cuts emanating from the point contribute a total of $3 \times 2\pi = 6\pi$ to the Index, whereas for ϕ the value of the Index at a delta is -2π. To correct this, we define an additional **spiral field** ϕ_s:

$$\phi_s(x,y) = \sum_i \arctan\left(\frac{y-y_i}{x-x_i}\right), \tag{7}$$

where (x_i, y_i) is the location of the ith flow singularity and the summation is taken for all the core and delta points. The **nett branch cut phase field** Φ_N is now defined by:

$$\Phi_N = 2\phi_s - \sum_j \phi_{bj}, \tag{8}$$

where the index j refers to the jth branch cut. Inspection of (8) shows that:

- At a core, the Poincaré Index of Φ_N is $2 \times 2\pi - 2\pi = 2\pi$.
- At a delta, the Poincaré Index of Φ_N is $2 \times 2\pi - 3 \times 2\pi = -2\pi$.

This matches the behaviour of ϕ, meaning that Φ_N may now be subtracted from ϕ giving a residual field ϕ_c that can be unwrapped (giving $\phi_c^{'}$). A field $\phi^{'}$ is then generated by adding $\phi_c{'}$ back to ϕ_s. Finally the result is halved. The resultant direction field $\theta^{'}$ now possesses the desired discontinuity properties, viz. a discontinuity of π exists along each branch cut, and the Poincaré Index is $\pm\pi$ at a core or delta respectively.

3.2 Disambiguation via "Divide and Conquer"

The second method of obtaining a consistent flow direction does not require the construction of branch cuts but instead proceeds by progressively subdividing the image into a number of rectangular sub-images.

If the orientation field contains just one singular point P, we divide the image into four slightly overlapping rectangles [2] surrounding P. If there are two or

[2] The sub-images must be allowed to overlap slightly, because the Poincaré Index is obtained by taking differences, so that there is the risk of overlooking a minutia that lies close to the border of a sub-image.

more singular points, partitioning is applied recursively by further subdividing any sub-image that contains a singular point. Each of the final sub-images is free of singular points, allowing a consistent flow direction field (and hence a wave normal direction field) to be assigned by directly unwrapping the orientation (Fig. 5(a)), though the directions may not match where the sub-images adjoin.

To avoid counting a minutia twice that occurs in a region of overlap, we set the minimum distance between minutiae to be $\overline{\lambda}$, the standard fingertip ridge spacing. Two minutiae closer than this distance are counted as one. To provide for generous overlap, the overlap distance is set at $3\overline{\lambda}$. [3]

4 Results

Ten-print images from the NIST14 and NIST27 Special Fingerprint Databases, supplied by the U.S. National Institute of Standards, formed the raw inputs for our work. In the results presented here, image regions identified as background are shown in dark grey or black. Segmentation of the image into foreground (discernible ridge pattern) and background (the remainder) is an important task, but is outside the scope of this paper.

4.1 Flow Disambiguation Using Branch Cuts

Figure 4 shows a portion of a typical input image and the results of various stages of deriving a ridge phase representation using the branch cut approach. For simplicity only a central subset of the image, most of which was segmented as foreground, was used for illustration.

Figure 4(f) illustrates that the output cosine of total phase is an acceptable representation over most of the image, but this method suffers from some drawbacks:

- Small inaccuracies in the placement of the branch cuts result in the generation of some spurious features on and near the branch cuts.
- Uncertainties in the orientation estimate in any region traversed by the cut may result in misplacement of later segments of the cut. This problem is not apparent in the example shown here, where the print was of sufficiently high quality to obtain an accurate orientation field over most of the image.
- Branch cuts were easily traced for the simple loop pattern shown here — other patterns are not so straightforward, eg. a tented arch pattern contains a core and a delta connected by a single branch cut; twin loop patterns contain spiraling branch cuts which may be very difficult to trace accurately. The model would need modification in order to handle these more difficult cases.

[3] The standard fingertip ridge spacing is about 0.5mm (Maltoni, Maio et al. p 83). In our images this corresponds to about 10 pixels.

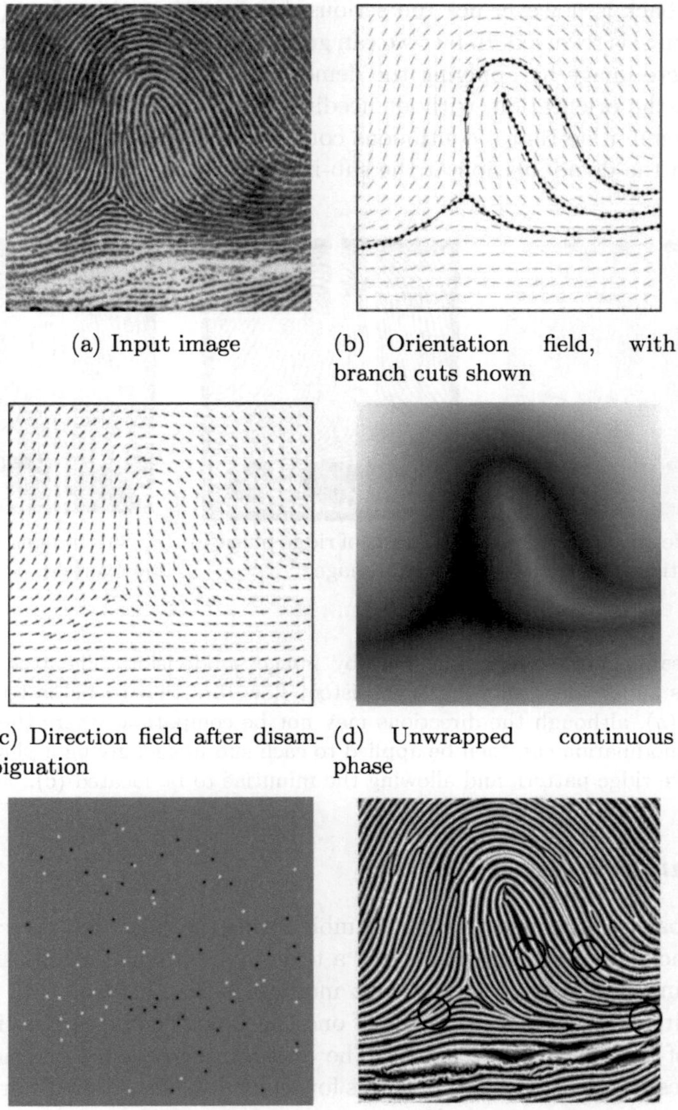

(a) Input image (b) Orientation field, with branch cuts shown

(c) Direction field after disambiguation (d) Unwrapped continuous phase

(e) Spiral phase points (f) Cosine of total phase.

Fig. 4. Results from disambiguation via branch cuts. White and black dots in (e) represent positive and negative spiral points respectively. Circled regions in (f) indicate where some artifacts appear on and around the branch cuts.

4.2 Flow Disambiguation via Image Subdivision

Figure 5 shows the results of flow disambiguation using image subdivision. Because flow direction is not necessarily consistent between neighbouring sub-images, the resultant phase sub-images cannot in general be combined into one.

This drawback is however not too serious, because the value of $\cos(\psi)$ is unaffected when ψ is reversed. In fact we can generate a suitable image of $\cos(\psi)$ from the complete image by applying the demodulation formula using $\beta = \theta + \pi/2$, where θ is the orientation, without needing to disambiguate θ. It is only in locating the minutiae that a continuous consistent ψ field is needed, requiring us to perform the demodulation at the sub-image level.

(a) Sample fingerprint image partitioned into sub-images. (b) Cosine of ridge phase in each sub-image. (c) Sub-images with minutiae overlaid.

Fig. 5. Disambiguating the ridge flow by image subdivision. The test image from Fig. 4(a) is subdivided, allowing a consistent flow direction to be assigned for each sub-image (a), although the directions may not be compatible where the sub-images adjoin. Demodulation can then be applied to each sub-image, giving a phase representation of the ridge pattern and allowing the minutiae to be located (c).

5 Summary

Two approaches are presented for disambiguating the ridge flow direction — one using branch cuts, and one employing a technique of image subdivision.

The primary advantage of the first method is that it leads to a description of the entire ridge pattern in terms of one continuous carrier phase image, plus a listing of the spiral phase points. The disadvantage is that certain classes of print possess ridge orientation patterns for which it is very difficult or impossible to construct branch cuts, and, even where these can be constructed, certain unwanted artifacts may appear on and near the branch cuts.

The second method does not suffer from these deficiencies. It cannot be used to generate a continuous carrier phase image for the entire pattern — nevertheless we can still obtain a continuous map of the **cosine** of the phase, and demodulation can be employed on the sub-images to locate the minutiae.

This phase based representation appears to be a more useful way of describing the ridge pattern than a means such as a skeleton ridge map described by chain codes, because the value of the cosine of the phase offers a natural means by which one portion of a fingerprint pattern can be directly compared with another via direct correlation, facilitating fingerprint matching.

Acknowledgments. The assistance of my supervisors, Dr Peter Kovesi and Dr Du Huynh, in proof-reading the manuscript and contributing many constructive suggestions is gratefully acknowledged. This research was supported by a University Postgraduate award.

References

Bazen, A.M., Gerez, S.H.: Systematic Methods for the Computation of the Directional Fields and Singular Points of Fingerprints. IEEE Trans. Pattern Analysis and Machine Intelligence 24(7), 905–919 (2002)

Joseph, D.: Helmholtz Decomposition Coupling Rotational to Irrotational Flow of a Viscous Fluid, `www.pnas.org/cgi/reprint/103/39/14272.pdf?ck=nck` (retrieved May 6, 2008)

Larkin, K.G., Fletcher, P.A.: A Coherent Framework for Fingerprint Analysis: Are Fingerprints Holograms? Optics Express 15(14), 8667–8677 (2007)

Maltoni, M., Maio, D., Jain, A.K., Prabhakar, S.: Handbook of Fingerprint Recognition. Springer, Heidelberg (2003)

Ratha, N.K., Chen, S., Jain, A.K.: Adaptive Flow Orientation-Based Feature Extraction in Fingerprint Images. Pattern Recognition 28(11), 1657–1672 (1995)

Sherlock, B.G., Monro, D.M.: A Model for Interpreting Fingerprint Topology. Pattern Recognition 26(7), 1047–1054 (1993)

Shi, Z., Govindaraju, V.: A Chaincode Based Scheme for Fingerprint Feature Extraction. Pattern Recognition Letters 27, 462–468 (2006)

Similarity Matches of Gene Expression Data Based on Wavelet Transform

Mong-Shu Lee, Mu-Yen Chen, and Li-Yu Liu

Department of Computer Science & Engineering,
National Taiwan Ocean University,
Keelung, Taiwan R.O.C.
{mslee,chenmy,M93570030}@mail.ntou.edu.tw

Abstract. This study presents a similarity-determining method for measuring regulatory relationships between pairs of genes from microarray time series data. The proposed similarity metrics are based on a new method to measure structural similarity to compare the quality of images. We make use of the Dual-Tree Wavelet Transform (DTWT) since it provides approximate shift invariance and maintain the structures between pairs of regulation related time series expression data. Despite the simplicity of the presented method, experimental results demonstrate that it enhances the similarity index when tested on known transcriptional regulatory genes.

Keywords: Wavelet transform, Time series gene expression.

1 Introduction

Time series data, such as microarray data, have become increasingly important in numerous applications. Microarray series data provides us with a possible means for identifying transcriptional regulatory relationships among various genes. To identify such regulation among genes is challenging because these gene time series data result from complex activation or repressed exertion of proteins. Several methods are available for extracting regulatory information from the time series microarray data including simple correlation analysis [5], edge detection [7], the event method [13], and the spectral component correlation method [15]. Among these approaches, correlation-based clustering is perhaps the most popular one for this purpose in this occasion. This method utilizes the common Pearson correlation coefficient to measure the similarity between two expression series profiles and to determine whether or not two genes exhibit a regulatory relationship. Four cases are considered in the evaluation of a pair of similar time series expression data.

(1) **Amplitude scaling:** two time series gene expressions have similar waveform but with different expression strengths.
(2) **Vertical shift:** two time series gene expressions have the same waveform but the difference between their expression data is constant.
(3) **Time delay** (horizontal shift)**:** A time delay exists between two time series gene expressions.
(4) **Missing value** (noisy)**:** Some points are missing from the time series data because of the noisy nature of microarray data.

A.-B. Salberg, J.Y. Hardeberg, and R. Jenssen (Eds.): SCIA 2009, LNCS 5575, pp. 540–549, 2009.

Generally, the similarity in cases (1) and (2) can typically be solved by using the Pearson correlation coefficient (and the necessary normalization of each sequence according to its mean). However, the time delay problem caused by the regulatory gene on the target gene significantly degrades the performance of the Pearson correlation-based approach.

Over the last decade or so, the discrete wavelet transform (DWT) has been successfully adopted to various problems of signal and image processing, including data compression [20], image segmentation [17], and ECG signal classification [9]. The wavelet transform is fast, local in the time and the frequency domain, and provides multi-resolution analysis of real-world signals and images. However, the DWT also has some disadvantages that limit its range of applications. A major problem of the common DWT is its lack of shift invariance, which is such that, on small shifts, the input signal can abruptly vary in the distribution of energy between wavelet coefficients on various scales. Some other wavelet transforms have been studied recently to solve these problems, such as the over-complete wavelet transform which discards all down-sampling in DWT to ensure shift invariance. Unfortunately, this method has a very large computational cost that is often not desirable in applications. Several authors [6, 19] have proposed that in a formulation in which two dyadic wavelet bases form a Hilbert transform pair, the DWT can provide the answer to some of the aforementioned limitations. As an alternative, The Kingsburg's dual-tree wavelet transform (DTWT) [11, 12] achieves approximate shift invariance and has been applied to motion estimation [18], texture synthesis [10] and image denoising [24].

Wavelets have recently been used in the similarity analysis of time series because they can extract compact feature vectors and support similarity searches on different scales [3]. Chan and Fu [2] proposed an efficient time series matching strategy based on wavelets. The Haar wavelet transform is first applied and the first few coefficients of the transform sequences are indexed in an R-tree for similarity searching. Wu et al. [23] comprehensively compared DFT (discrete Fourier transform) with DWT transformations, but only in the context of time series databases. Aghili et al. [1] examined the effectiveness of the integration of DFT/DWT for sequence similarity of biological sequence databases.

Recently, Wang et al. [22] have developed a measure of structure similarity (SSIM) for evaluating image quality. The SSIM metrics models perception implicitly by taking into accounts high-level HVS (human visual system) characteristics. The simple SSIM algorithm provides excellently predicting the quality of various distorted images. The proposed approach to comparing similar time series data is motivated by the fact that the DTWT provides shift invariance, enabling the extracting the global shape of the data waveform, and therefore, such measures are to catch the structural similarity between time series expression data. The goal of this study is to extend the current SSIM approach to the dual-tree wavelet transform domain, and base it on a similarity metrics, creating the dual-tree wavelet transform SSIM. This work reveals that the DTWT-SSIM metric can be used for matching gene expression time series data. The regulation-related gene data are modelled by the familiar scaling and shifting transformations, indicating that the introduced DTWT-SSIM index is stable under these transformations. Our experimental results show that the proposed similarity measure outperforms the traditional Pearson correlation coefficient on Spellman's yeast data set.

In Section 2, we briefly give some background information about DWT and DTWT. In section 3, we present our proposed method for the DTWT based similarity measure. We then describe the sensitivity of the DTWT-SSIM metric under the general linear transformation. Experimental tests on a database of gene expression data, and comparison with the Pearson correlation are reported in Section 4. This demonstrates that our results are similar to the spectral component correlation method [15]. Finally, we draw the conclusions of our work in Section 5.

2 Dual-Tree Wavelet Transform

As shown in Fig. 1, in the one-dimensional DTWT, two real wavelet trees are used, each capable of perfect reconstruction. One tree generates the real part of the transform and the other one is used to generate the complex part. In Fig. 1, $h_0(n)$ and $h_1(n)$ are the low-pass and high-pass filters, respectively, of a Quadrature Mirror Filter (QMF) pair in the analysis branch. In the complex part, $\{g_0(n), g_1(n)\}$ is another QMF pair in the analysis branch. All filter pairs considered here are orthogonal and real-valued. Each tree yields a valid set of real DWT detail coefficients u_i and v_i , and altogether form the complex coefficients $d_i = u_i + jv_i$. Similarly, Sa_i and Sb_i is the pair of scaling coefficients of the DWT, as shown in Fig. 1.

A three-level decomposition of DTWT and DWT is applied to the test signal $T(n)$ and its shifted version $T(n-3)$, shown in Fig. 2(a) and (b), respectively, to demonstrate the shift invariance property of DTWT. Fig. 2(c) and (e) show the reconstruction signals $T(n)$ from the wavelet coefficients on the third level of the DWT and DTWT, respectively. Fig. 2(d) and (f) show the counterparts of the shifted signal $T(n-3)$. Comparing Figs. 2(a), (c), and (e) with Figs. 2(b), (d), and (f), they indicate that the shape of the DTWT-reconstructed signal remains mostly unchanged. However, the shape of the DWT-reconstructed signal varies significantly. These results clearly illustrate the characteristics of the shift invariance of the dual-tree wavelet transform. This property helps to simplify some applications.

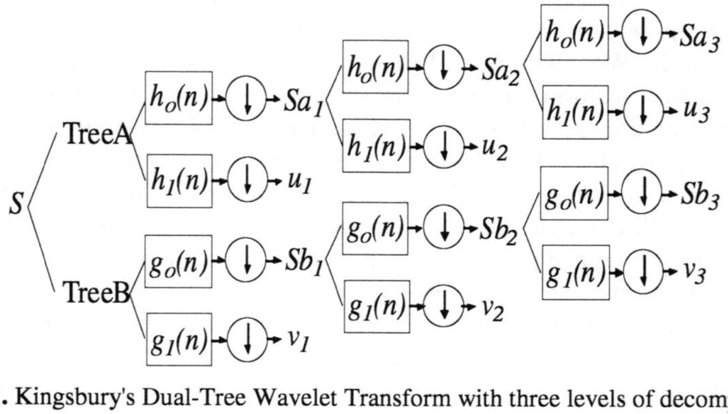

Fig. 1. Kingsbury's Dual-Tree Wavelet Transform with three levels of decomposition

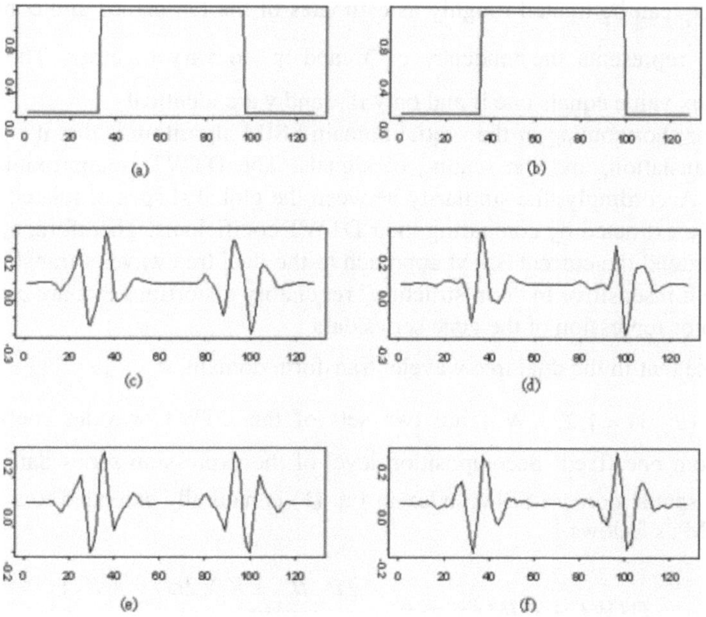

Fig. 2. (a) Signal T(n). (b) Shifted version of (a), T(n-3). (c), (d) are the reconstructed signals using the level 3 DWT coefficients of (a) and (b), respectively. (e), (f) are the reconstructed signals using the level 3 DTWT coefficients of (a) and (b), respectively.

3 DTWT-SSIM Measure

3.1 DTWT-SSIM Index

The proposed application of the DTWT to evaluate the similarity among time series data is inspired by the success of the spatial domain structural similarity (SSIM) index algorithm in image processing [22]. The use of the SSIM index to quantify image quality has been studied. The principle of the structural approach is that the human visual system is highly adapted and can extract structural information (about the objects) from a visual scene. Hence, a metric of structure similarity is a good approximation of a similar shape in time series data. In the spatial domain, the SSIM index that quantizes the luminance, contrast and structure changes between two image patches $x = \{x_i \mid i = 1, \ldots, M\}$ and $y = \{y_i \mid i = 1, \ldots, M\}$, and is defined as [22]

$$S(x,y) = \frac{(2\mu_x\mu_y + C_1)(2\sigma_{xy} + C_2)}{(\mu_x^2 + \mu_y^2 + C_1)(\sigma_x^2 + \sigma_y^2 + C_2)} \tag{1}$$

where C_1 and C_2 are two small positive constants; $\mu_x = \frac{1}{M}\sum_{i=1}^{M} x_i$, $\sigma_x^2 = \frac{1}{M}\sum_{i=1}^{M}(x_i - \mu_x)^2$, and $\sigma_{xy} = \frac{1}{M}\sum_{i=1}^{M}(x_i - \mu_x)(y_i - \mu_y)$.

μ_x and σ_x can be treated roughly as estimates of the luminance and contrast of x, while σ_{xy} represents the tendency of x and y to vary together. The maximum SSIM index value equals one if and only if x and y are identical.

A major shortcoming of the spatial domain SSIM algorithm is that it is very sensitive to translation, and the scaling of signals. The DTWT is approximately shift-invariant. Accordingly, the similarity between the global shapes of related time series data can be extracted by comparing their DTWT coefficients. Therefore, an attempt is made to extend the current SSIM approach to the dual tree wavelet transform domain and make it insensitive to "non-structure" regulatory distortions that are caused by the activation or repression of the gene series data.

Suppose that in the dual tree wavelet transform domain, $d_x = \{d_{x,i} \mid i = 1, 2, ..., N\}$ and $d_y = \{d_{y,i} \mid i = 1, 2, ..., N\}$ are two sets of the DTWT wavelet coefficients extracted from one fixed decomposition level of the expression series data x and y. Now, the spatial domain SSIM index in Eq. (1) is naturally extended to a DTWT domain SSIM as follows.

$$DTWT - SSIM(x, y) = \frac{(2\mu_{d_x}\mu_{d_y} + K_1)(2\sigma_{d_x d_y} + K_2)}{(\mu_{d_x}^2 + \mu_{d_y}^2 + K_1)(\sigma_{d_x}^2 + \sigma_{d_y}^2 + K_2)}$$

$$= \frac{\left(2\mu_{|d_x|}\mu_{|d_y|} + K_1\right)\left(\left(2\sum_{i=1}^{N}(|d_{x,i}| - \mu_{|d_x|})(|d_{y,i}| - \mu_{|d_y|})\right) + K_2\right)}{\left((\mu_{|d_x|})^2 + (\mu_{|d_y|})^2 + K_1\right)\left(\left(\sum_{i=1}^{N}(|d_{x,i}| - \mu_{|d_x|})^2 + \sum_{i=1}^{N}(|d_{y,i}| - \mu_{|d_y|})^2\right) + K_2\right)}$$

$$= \frac{\left(2\sum_{i=1}^{N}(|d_{x,i}|)(|d_{y,i}|)\right) + K_2}{\left(\sum_{i=1}^{N}(|d_{x,i}|)^2 + \sum_{i=1}^{N}(|d_{y,i}|)^2\right) + K_2}. \quad (2)$$

The third equality in Eq. (2) derives from the fact that the dual-tree wavelet coefficients of x and y are zero mean ($\mu_{|d_x|} = \mu_{|d_y|} = 0$), because the DTWT coefficients are normalized after the time series gene data taking DTWT. Herein $|d_x| = |d_{x,i}|$ denotes the magnitude (absolute value) of the complex numbers $d_x = d_{x,i}$, and K_1, K_2 are two small positive constants to avoid instability when the denominator is very close to zero. (We have $K_1 = K_2 = 0.3$ in the experiment).

3.2 Sensitivity Measure

The linear transformation is a convenient way to model the regulation-related gene expression that was described in the Introduction section. The general linear transformation is commonly written in the vector notations with coordinates in the $\mathbb{R}^n$. Now, the scaling and shifting (including vertical and horizontal) relationships that follow from regulation is described in terms of matrices and the following coordinate system as follow.

Let $x=[x_1,x_2,...,x_n]$ and $y=[y_1,y_2,...,y_n]$ be two gene expression data, we define $y=Ax+B$ by

$$[y_1,y_2,...,y_n]^T = A[x_1,x_2,...,x_n]^T + B^T$$

where matrix $A = [a_{ij}]_{i,j=1}^n$ and vector B specify the desired relation. For example, by defining $A=\begin{bmatrix}1 & 0 & \cdots & 0\\ 0 & 1 & \cdots & 0\\ \vdots & & \ddots & \vdots\\ 0 & 0 & \cdots & 1\end{bmatrix}$ and $B = [b_1, b_2, \quad \cdots, \quad b_n]$, this transformation can carry out vertical shifting.

Similarly, the scaling operation is $A=\begin{bmatrix}r & 0 & \cdots & 0\\ 0 & r & \cdots & 0\\ \vdots & & \ddots & \vdots\\ 0 & 0 & \cdots & r\end{bmatrix}$, $B = [0,0, \quad \cdots, \quad 0]$.

The condition number $\kappa(A)$ denotes the sensitivity of a specified linear transformation problem. Define the condition number $\kappa(A)$ as $\kappa(A) = \| A \|_\infty \| A^{-1} \|_\infty$, where A is a $n \times n$ matrix and $\| A \|_\infty = \max_{1\le i\le n} \sum_{j=1}^{n} | a_{ij} |$.

For a non-singular matrix, $\kappa(A) = \| A \|_\infty \| A^{-1} \|_\infty \ge \| A \cdot A^{-1} \|_\infty = \| I \|_\infty = 1$. Generally, matrices with a small condition number, $\kappa(A) \cong 1$, are said to be well- conditioned. Clearly, the scaling and shifting transformation matrices are well-conditioned. Furthermore, the composition matrix of these well-conditioned transformations still satisfies $\kappa(A) \cong 1$. Let A_1 and A_2 be two such transformations. Applying $\kappa(A_1A_2) \le \kappa(A_1)\kappa(A_2)$, we establish that the composition of two such transformations also satisfies $\kappa(A_1A_2) \cong 1$. Fig. 3 and Table 1 present an example comparison of the stability of DTWT-SSIM index and Pearson coefficient under shifting and scaling transformations. Figure 3 shows the original waveform SIN and some distorted SIN waveforms with various scaling and shifting factors. The similarity index between the original SIN and the distorted SIN waveforms is then evaluated using the proposed DTWT-SSIM and Pearson-correlated metrics. The results presented in Table 1 reveal that except in the scaling case, the DTWT-SSIM unlike the Pearson metric, which decreases sharply, is more stable than the Pearson metric, due to its steady decrease as distortion increases.

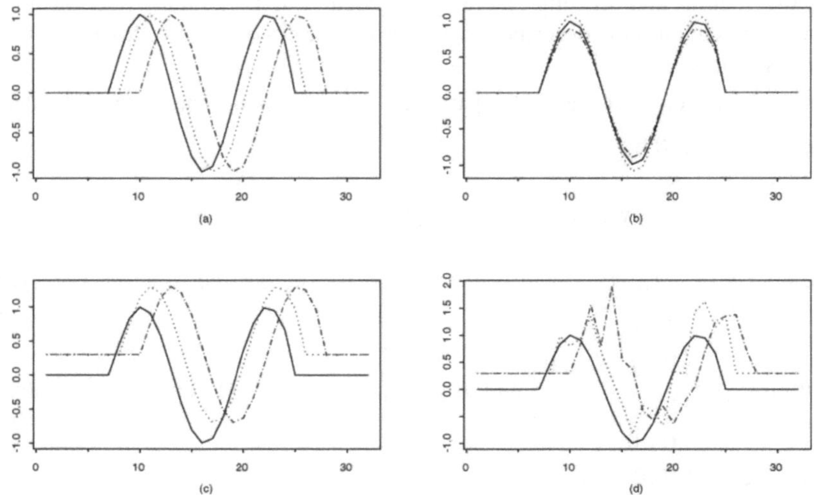

Fig. 3. Original signal SIN (the solid line) and distorted SIN signals with various scaling and shifting factors (the dashed lines). (a) The horizontal shift factors are 1 and 3 units, respectively. (b) The scaling factors are 0.9 and 1.1 respectively. (c) H. shift factor 1 unit + V. shift 0.3 units and H. shift factor 3 units + V. shift 0.3 units. (d) H. shift factor 1 unit + V. shift 0.3 units + noise and H. shift factor 3 units + V. shift0.3 units + noise. (H: Horizontal, V: Vertical)

4 Test Results

A time series expression data similarity comparison experiment was performed using the regulatory gene pairs from [4] and [21], to demonstrate the efficiency of SSIM in the DTWT domain. The gene pairs are extracted by a biologist from the Cho and Spellman alpha and cdc28 datasets. Filkov et al. [8] formed a subset of 888 known transcriptional regulation pairs, comprising 647 activations and 241 inhibitions. The data set is available from the web site at http://www.cs.sunysb.edu/~skiena/gene/jizu/. The alpha data set used in this experiment, contained 343 activations and 96 inhibitions. After all the missing data (noise) were replaced by zeros, the known regulation subsets were analyzed using the proposed algorithm.

The Q-shift version of the DTWT, with three levels of decomposition, was applied to the gene pair to be compared, to evaluate the DTWT-SSIM measure and thus determine gene similarity. The amount of energy is well-known to increase toward the low frequency sub-bands after decomposing the original data into several sub-bands with general wavelet transforms. Therefore, the DTWT-SSIM index was calculated in Eq. (2) using only the lowest sub-band and sequence of normalized wavelet coefficients.

The traditional Pearson correlation and DTWT-SSIM analysis were performed on each pair of 343 known regulations. The proposed DTWT-SSIM method was able to detect many regulatory pairs that were missed by the traditional correlation method due to small correlation value. Numerous visually dissimilar gene pairs have a high DTWT-SSIM index. Table 2 plots the distribution of the two similarity index among the 343 known regulatory pairs. The result demonstrates that less than 11% (36/343) had a Pearson coefficient greater than 0.5 between the activator and activated. However, the DTWT-SSIM index increases the similarity between the known activating relationships by up to 57% (198/343), and the ratio is very close to the result of the spectral component correlation method [15].

Table 1. Similarity comparisons between the original SIN and the distorted SIN waveforms in Fig. 3 using DTWT-SSIM and Pearson metrics

Various scaling and shifting factors in Fig. 3		Pearson coefficient	DTWT-SSIM index
Fig. 3(a)	H. shift 1 unit	0.8743	0.974
	H. shift 3 units	0.1302	0.7262
Fig. 3(b)	Scaling factor: 0.9	1	0.9945
	Scaling factor: 1.1	1	0.9955
Fig. 3(c)	H. shift 1 unit +V. shift 0.3 units	0.8743	0.974
	H. shift 3 units +V. shift 0.3 units	0.1302	0.7263
Fig. 3(d)	H. shift 1 unit +V. shift 0.3 units+ noise	0.8897	0.952
	H. shift 3 units +V. shift 0.3 units+ noise	0.2086	0.5755

Table 2. The cumulative distribution of Pearson and DTWT-SSIM similarity measures among the 343 pairs

Similarity index range	Pearson	DTWT-SSIM
>0	173/343	336/343
>0.25	97/343	291/343
>0.33	72/343	265/343
>0.5	36/343	198/343
>0.75	9/343	107/343

The number of false dismissals that occurred in the experiment is considered to determine the effectiveness of these two similarity metrics. If the margin of DTWT-SSIM and the Pearson metrics of the pair expression data exceed 0.5, then the Pearson coefficient is regarded as a false dismissal. For instance, the DTWT-SSIM index of the gene pair is highly correlated with each other but the Pearson metric is negative or lowly correlated. Similarly, if the margin of the Pearson and DTWT-SSIM metrics of the pair expression data exceed 0.5, then the DTWT-SSIM index is regarded as a false dismissal. 177 out of 343 pairs are false dismissals, based on the Pearson coefficient, while only two out of 343 pairs are false dismissals, based on the DTWT-SSIM.

5 Conclusion

This study presented a new similarity metric, called the DTWT-SSIM index, which not only can be easily implemented but also can enhance the similarity between activation pairs of gene expression data. The traditional Pearson correlation coefficient does not perform well with gene expression time series because of time shift and noise problems. In our dual-tree wavelet transform-based approach, the shortcoming of the space domain SSIM method was avoided by exploiting the almost shift-invariant property of DTWT. This effectively solves the time shift problem. The proposed DTWT-SSIM index was demonstrated to be more stable than the Pearson correlation coefficient when the signal waveform underwent scaling and shifting. Therefore, the DTWT-SSIM measure captures the shape similarity between the time series regulatory pairs. The concept is also useful for other important image processing tasks, including image matching and recognition [16].

References

[1] Aghili, S.A., Agrawal, D., Abbadi, A.: Sequence similarity search using discrete Fourier and wavelet transformation techniques. International Journal on Artificial Intelligence Tools 14(5), 733–754 (2005)

[2] Chan, K.P., Fu, A.: Efficient time series matching by wavelets. In: ICDE, pp. 126–133 (1999)

[3] Chiann, C., Morettin, P.: A wavelet analysis for time series. Journal of Nonparametric Statistics 10(1), 1–46 (1999)

[4] Cho, R.J., Campbell, M.J., Winzeler, E.A., Steinmetz, L., Conway, A., Wodicka, L., Wolfsberg, T.G., Gabrielian, A.E., Landsman, D., Lockhart, D.J., Davis, R.W.: A genome-wide transcriptional analysis of the mitotic cell cycle. Molecular Cell 2, 65–73 (1998)

[5] Eisen, M.B., Spellman, P.T., Brown, P.O.: Cluster analysis and display of genome-wide expression patterns. Proceedings of the National Academy of Sciences of the United States of America 96(19), 10943–10943 (1999)

[6] Fernandes, F., Selesnick, I.W., Spaendonck, V., Burrus, C.S.: Complex wavelet transforms with allpass filters. Signal Processing 83, 1689–1706 (2003)

[7] Filkov, V., Skiena, S., Zhi, J.: Identifying gene regulatory networks from experiomental data. In: Proceeding of RECOMB, pp. 124–131 (2001)

[8] Filkov, V., Skiena, S., Zhi, J.: Analysis techniques for microarray time-series data. Journal of Computational Biology 9(2), 317–330 (2002)

[9] Froese, T., Hadjiloucas, S., Galvao, R.K.H.: Comparison of extrasystolic ECG signal classifiers using discrete wavelet transforms. Pattern Recognition Letters 27(5), 393–407 (2006)

[10] Hatipoglu, S., Mitra, S., Kingsbury, N.: Image texture description using complex wavelet transform. In: Proc. IEEE Int. Conf. Image Processing, pp. 530–533 (2000)

[11] Kingsbury, N.: Image Processing with Complex Wavelets. Phil. Trans. R. Soc. London. A 357, 2543–2560 (1999)
[12] Kingsbury, N.: Complex wavelets for shift invariant analysis and filtering of signals. Appl. Comput. Harmon. Anal. 10(3), 234–253 (2001)
[13] Kwon, A.T., Hoos, H.H., Ng, R.: Inference of transcriptional regulation relationships from gene expression data. Bioinformatics 19(8), 905–912 (2003)
[14] Kwon, O., Chellappa, R.: Region adaptive subband image coding. IEEE Transactions on Image Processing 7(5), 632–648 (1998)
[15] Liew, A.W.C., Hong, Y., Mengsu, Y.: Pattern recognition techniques for the emerging field of bioinformatics: A review. Pattern Recognition 38, 2055–2073 (2005)
[16] Lee, M.-S., Liu, L.-Y., Lin, F.-S.: Image Similarity Comparison Using Dual-Tree Wavelet Transform. In: Chang, L.-W., Lie, W.-N. (eds.) PSIVT 2006. LNCS, vol. 4319, pp. 189–197. Springer, Heidelberg (2006)
[17] Liang, K.H., Tjahjadi, T.: Adaptive scale fixing for multiscale texture segmentation. IEEE Transactions on Image Processing 15(1), 249–256 (2006)
[18] Magarey, J., Kingsbury, N.G.: Motion estimation using a complex-valued wavelet transform. IEEE Transactions on Image Processing 46, 1069 (1998)
[19] Selesnick, I.: The design of approximate Hilbert transform pairs of wavelet bases. IEEE Trans. on Signal Processing 50, 1144–1152 (2002)
[20] Shapiro, J.M.: Embedded image coding using zerotrees of wavelet coefficients. IEEE Trans. Signal Proc. 41(12), 3445–3462 (1993)
[21] Spellman, P., Sherlock, G., Zhang, M.Q., Iyer, V.R., Anders, K., Eisen, M.B., Brown, P.O., Botstein, D., Futcher, B.: Comprehensive identification of cell cycle-regulated genes of the yeast Saccharomyces cerevisiae by microarray hybridization. Molecular Biology of the Cell 9, 3273–3297 (1998)
[22] Wang, Z., Bovik, A., Sheikh, H., Simoncelli, E.: Image quality assessment: from error visibility to structural similarity. IEEE Trans. Image Processing 13, 600–612 (2004)
[23] Wu, Y., Agrawal, D., Abbadi, A.: A comparison of DFT and DWT based similarity search in time series database. CIKM, 488–495 (2000)
[24] Ye, Z., Lu, C.: A complex wavelet domain Markov model for image denoising. In: Proc. IEEE Int. Conf. Image Processing, pp. 365–368 (2003)

Simple Comparison of Spectral Color Reproduction Workflows

Jérémie Gerhardt* and Jon Yngve Hardeberg

Gjøvik University College,
2802 Gjøvik, Norway
jeremie.gerhardt@first.fraunhofer.de

Abstract. In this article we compare two workflows for spectral color reproduction : colorant separation (CS) followed by halftoning by scalar error diffusion (SED) of its resulting multi-colorant channel image and a second workflow by spectral vector error diffusion (sVED). Identical filters are used in both SED and sVED to diffuse the error. Gamut mapping is performed as pre-processing and the reproductions are compared to the gamut mapped spectral data. The inverse spectral Yule-Nielsen modified Neugebauer (YNSN) model is used for the colorant separation. To bring the improvement of the YNSN model upon the regular Neugebauer model into the sVED halftoning the n factor is introduced in the sVED halftoning. The performances of both workflows are evaluated in term of spectral and color differences but also visually with the dot distributions obtained by the two halftoning techniques. Experimental results have shown close performances for the compared workflows in term of color and spectral differences but visually cleaner and more stable dot distributions obtained by sVED.

Keywords: Spectral color reproduction, spectral gamut mapping, colorant separation, halftoning, spectral vector error diffusion.

1 Introduction

With a color reproduction system it is possible to acquire the color of a scene or object under a given illuminant and reproduce it. With proper calibration and characterization of the devices involved, and not considering the problems related to color gamut limitations, it is theoretically possible to reproduce a color which will be perceived identically to the original color of the scene or object. For example, a painting and its color reproduction viewed side by side will appear identical under the illuminant used for its color acquisition even if the spectral properties of the painting pigments are different from those of the print inks. This phenomenon is called metamerism. On the other hand, if we change the illumination, then most probably the reproduction will no longer be

* Jérémie Gerhardt is working since the 1st August 2008 in Fraunhofer Institute FIRST-ISY, in Berlin, Germany (http://www.first.fraunhofer.de). This work was part of his PhD thesis done in the Norwegian Color Research Laboratory at HiG.

A.-B. Salberg, J.Y. Hardeberg, and R. Jenssen (Eds.): SCIA 2009, LNCS 5575, pp. 550–559, 2009.

perceived similar to the original. This problem can be solved in a spectral color reproduction system.

Multispectral color imaging offers the great advantage of providing the full spectral color information of the scene or object surface. Color acquisition system records the color of a scene or object surface under a given illuminant, but a multispectral color acquisition system can record the spectral reflectance and allows us to simulate the color of the scene under any illuminant. In an ideal case, after acquiring a spectral image we would like to display it or print it. For that we basically have two options: either to calculate the color rendering of our spectral image for a given illuminant and to display/print it, or to reproduce the image spectrally. This is a challenging task when for example we have made the spectral acquisition of a 2 century old painting and the colorants used at that time are not available anymore or we have lost the technical knowledge to produce them.

Multi-colorant printers offer the possibility to print the same color by various colorant combinations, i.e. metameric print is possible (note that this was already possible with a *cmyk* printer when the grey component of a *cmy* colorant combination was replaced by black ink k). This is an advantage for colorant separation [1],[2],[3] and it allows for example to select colorant combinations minimizing colorant coverage or to optimize the separation for a given illuminant. In spectral colorant separation we aim to reduce the spectral difference between a spectral target and its reproduction, i.e. we want to reduce the metamerism. This task is performed by inverting the spectral Yule-Nielsen modified Neugebauer printer model [4],[5],[6].

Once the colorant separation has been performed the resulting multi-colorant image still has to be halftoned, and this channel independently. An alternative solution for the reproduction of spectral image is to combine the colorant separation and the halftoning in a single step: halftoning by spectral vector error diffusion [7],[8] (sVED). In our experiment we introduce the Yule-Nielsen n factor in the sVED halftoning technique. Identical n factor value is used at the different stages of the workflows (see diagram in Figure 1).

In the following section we will compare the reproduction of spectral data by two possible workflows for a simulated six colorants printer. The first workflow (WF1) is divided in two steps: colorant separation (CS) and halftoning by colorant channels using scalar error diffusion (SED). The second workflow (WF2) will hafltone directly the spectral image by sVED. The first step involved in the reproduction process, which is common to the two compared approaches, is a gamut mapping operation: spectral gamut mapping (sGM) is performed as pre-processing. It is the reproduction of the gamut mapped spectral data which is compared.

2 Experiment

The spectral images we reproduce are spectral patches. They consist of spectral image of size 512 × 512 pixels, each patch having a single spectral reflectance

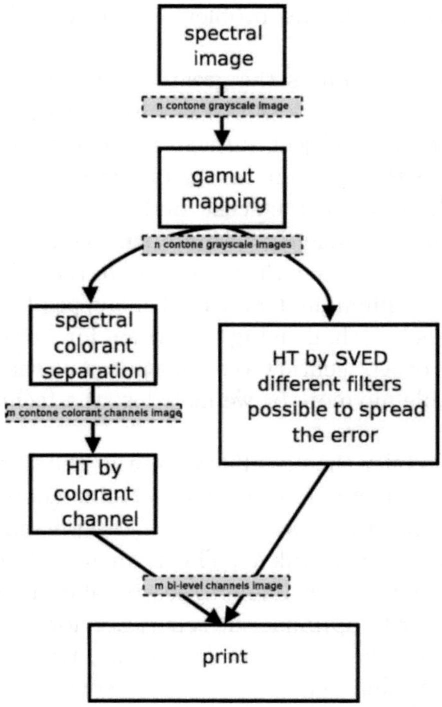

Fig. 1. Illustration of two possible workflows for the reproduction of spectral data with a m colorant printer. The diagram illustrates how is transformed a spectral image into a multi bi-level colorants image.

value. The spectral reflectance targets correspond to spectral reflectance measurements extracted from a painting called La Madeleine [9].

The spectral reflectance targets have been obtained by measuring the spectral reflectances of a painting at different locations, see in Figure 2 to the left an image of the painting. Twelve samples have been selected and their spectral reproduction simulated. See in Figure 2 to the right an illustration of where the measurements were taken. The spectral reflectance corresponding to these locations are shown in Figure 4 (a). According to the workflows in Figure 1 the first step is the spectral gamut mapping. Comparison of the two workflows is based on the reproduction of the gamut mapped data.

2.1 Spectral Gamut Mapping

The reproduction of the spectral patches are simulated for our 6 colorants printers, see in Figure 3 the spectral reflectance the colorants. After the gamut mapping operation an original spectral reflectance $\mathbf{r}$ is replaced by its gamut mapped version $\mathbf{r}'$ such that:

$$\mathbf{r}' = \mathbf{P}\mathbf{w}' \tag{1}$$

Fig. 2. Painting of *La Madeleine*, the 12 black spots correspond of the location where the spectral reflectances were taken

where **P** is the matrix of Neugebauer primaries (the NPs are all possible the binary combination between the available colorant of a printing system, here we have $2^6 = 64$ NPs) and the vector of weights $\mathbf{w}'$ is obtained while solving a convex optimization problem:

$$\min_{\mathbf{w}'} ||\mathbf{r} - \mathbf{P}\mathbf{w}'|| \tag{2}$$

with the constraints on the weights $\mathbf{w}'$:

$$\sum_{i=0}^{2^m-1} w_i = 1 \text{ and } 0 \leq w_i \leq 1 \tag{3}$$

and m being the number of colorant. The n factor is taking into account in the sGM operation by rising **r** and **P** to the power $1/n$ before the optimization. In this article the n factor has been set to $n = 2$. As opposed to the inversion of the YNSN model by optimization we do not use the Demichel [10] equations in our gamut mapping operation [4].

The gamut mapped spectral reflectances are displayed in Figure 4 (b). Color and spectral differences between measured spectral reflectances and gamut mapped spectral reflectances are displayed in Table 1.

2.2 WF1: Colorant Separation and Scalar Error Diffusion by Colorant Channel

For the WF1 the colorant separation (CS) is performed for the 12 gamut mapped spectral reflectances using the linear regression iteration method (LRI) presented by [5]. From the 12 colorant combinations obtained we create 12 patches of 6 channels each and size 512 × 512 pixels. The final step is the halftoning operation which is performed channel independently. We use scalar error diffusion

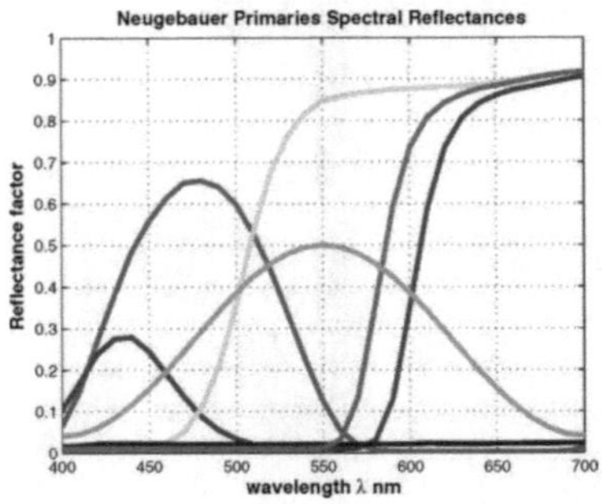

Fig. 3. Spectral reflectances of the six colorants of our simulated printing system

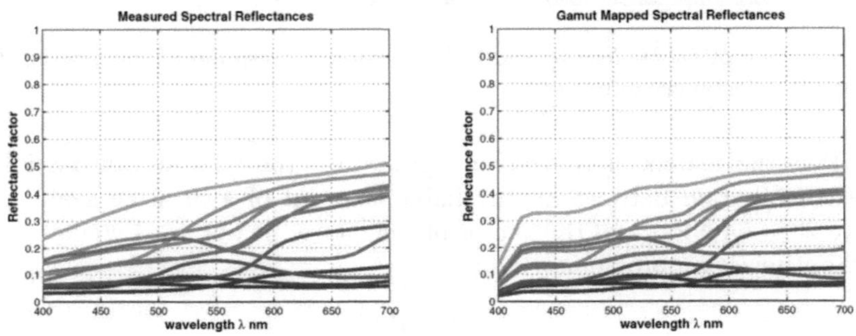

(a) Spectral reflectance measurements (b) Gamut mapped spectral reflectances

Fig. 4. Spectral reflectance measurements of the 12 samples in (a) and their gamut mapped version for our 6 colorants printer in (b). For each spectral reflectance displayed above the RGB color corresponds to its color rendering for illuminant D50 and CIE 1931 2° standard observer.

(SED) halftoning technique [11] with Jarvis [12] filter to diffuse the error in the halftoning algorithm.

Each pixel of an halftoned image can be described by a multi-binary colorant combination, combination corresponding to a NP. The spectral reflectance of each patch is estimated by counting the NPs pixel's occurrences and then considering a unitary area for each patch, see the following equation:

$$R(\lambda) = \left(\sum_{i=0}^{2^m-1} s_i P_i^{1/n}(\lambda) \right)^n \tag{4}$$

where s_i is the area occupied by the ith Neugebauer primary P_i and n the so-called n factor. Differences between the gamut mapped spectral reflectances and their simulated reproduction by CS and SED are presented in all left columns of each pair of column in Table 2.

Table 1. Differences between the spectral reflectance measurements and their gamut mapped version to our 6 colorants printer

	ΔE^*_{ab}			ΔE^*_{94}	
Samples	A	D50	FL11	D50	sRMS
1	3.0	4.2	6.1	3.1	0.014
2	3.5	4.9	6.9	3.5	0.014
3	2.4	3.1	4.8	2.5	0.013
4	2.9	4.1	5.7	2.9	0.009
5	1.2	1.3	2.8	0.7	0.009
6	2.1	2.9	3.8	2.0	0.006
7	1.3	1.4	0.8	1.2	0.016
8	1.8	1.3	1.7	1.1	0.005
9	3.5	2.6	3.3	1.8	0.023
10	2.5	2.7	2.4	1.7	0.007
11	4.6	5.7	5.3	2.8	0.011
12	1.1	1.7	3.2	1.2	0.013
Av.	2.5	3.0	3.9	2.0	0.012
Std	1.1	1.5	1.9	0.9	0.005
Max	4.6	5.7	6.9	3.5	0.023

2.3 WF2: Spectral Vector Error Diffusion

For this workflow we have created 12 spectral patches of size 512 × 512 pixels. Each spectral image have 31 channels since each spectral reflectance in our experiment is described by 31 discrete values equally spaced from 400nm to 700nm.

The spectral patches are halftoned by sVED using [12] filter as in WF1 for the SED halftoning. For each pixel of a spectral image the distance to each NP is calculated, the smallest distance given colorant combination at the processed pixel. This operation is performed in a raster scan path mode, see in Figure 5 the diagram of the sVED algorithm.

Here the colorant combination selected is directly a binary combination of the 6 colorants available in our printing system, this corresponding to a command for the printer to lay down (if 1) or not (if 0) a drop of ink at the pixel position. Once the output is selected the difference between the processed pixel (i.e. a spectral reflectance) and the spectral reflectance of the closest NP is weighted and spread to the neighbors pixels according to the filter size.

As for the sGM operation, the CS in WF1 and while estimating the spectral reflectance of an halftoned patch we take into account the n factor in the sVED algorithm: all spectral reflectances of each patch and the NPs are raised to the power $1/n$ before to perform to halftoning.

The spectral reflectance of each patch is estimated by counting the NP pixel's occurrences and then considering a unitary area for each patch, see Equation 4. Differences between the gamut mapped spectral reflectances and their simulated reproduction by sVED are presented in all right columns of each pair of column in Table 2.

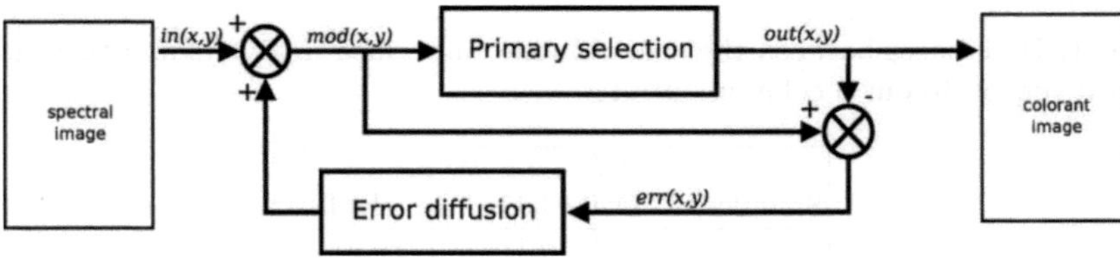

Fig. 5. The process of spectral vector error diffusion halftoning. *in*(x, y), *mod*(x, y), *out*(x, y) and *err*(x, y) are vector data representing at the position (x, y) in the image the spectral reflectance of the image, the modified spectral reflectance, the spectral reflectance of the chosen primary and the spectral reflectance error.

Table 2. Differences between the gamut mapped spectral reflectances and their reproduction by CS and SED (left columns of each double column) and by sVED (right columns of each double column). The differences in bold tell us which workflow gives the smallest difference for a given sample at a given illumination condition.

	ΔE^*_{ab}						ΔE^*_{94}		sRMS	
Samples	A		D50		FL11		D50			
1	0.57	**0.46**	0.51	**0.38**	**0.43**	0.57	0.26	**0.25**	**0.0021**	0.0036
2	**0.38**	0.43	0.35	**0.33**	**0.31**	0.52	**0.21**	0.23	**0.0018**	0.0039
3	**0.15**	0.41	**0.15**	0.29	**0.15**	0.50	**0.11**	0.23	**0.001**	0.004
4	0.68	**0.52**	0.58	**0.46**	**0.57**	0.63	0.35	**0.24**	**0.0021**	0.0028
5	**0.24**	0.40	**0.25**	0.30	**0.21**	0.52	**0.16**	0.19	**0.0019**	0.0041
6	0.64	**0.51**	0.59	**0.44**	**0.59**	0.61	0.37	**0.30**	**0.0011**	0.0021
7	0.43	**0.37**	0.41	**0.26**	**0.37**	0.48	0.31	**0.22**	**0.0031**	0.0043
8	**0.15**	0.44	**0.19**	0.28	**0.16**	0.61	**0.18**	0.27	**0.0004**	0.0013
9	0.74	**0.60**	0.96	**0.58**	0.82	**0.80**	0.81	**0.45**	0.0037	**0.0027**
10	1.01	**0.73**	1.21	**0.72**	1.08	**0.93**	0.85	**0.51**	0.0019	**0.0015**
11	1.65	**0.67**	1.81	**0.65**	1.81	**0.88**	1.06	**0.43**	0.0038	**0.0018**
12	**0.31**	0.71	**0.34**	0.60	**0.36**	0.99	**0.29**	0.35	**0.0029**	0.0057
Av.	0.58	0.52	0.61	0.44	0.57	0.67	0.41	0.31	0.0021	0.0032
Std	0.43	0.13	0.49	0.16	0.48	0.18	0.31	0.11	0.0011	0.0013
Max	1.65	0.73	1.81	0.72	1.81	0.99	1.06	0.51	0.0038	0.0058

3 Results and Discussion

The first analysis of the results, by looking at the color and spectral differences between the gamut mapped data and their simulated reproductions (see Table 2) does not help to decide between WF1 or WF2. We can only observe that the average performance of the WF2 is slightly better than for the WF1 with a smaller standard deviation and a minimum maximum for all chosen illuminant.

To evaluate visually the quality of the reproduction we have created color images of the halftoned patches. Each pixel of an halftone patch (i.e. a spectral reflectance of a NP) is replaced by its RGB color rendering value for the illuminant D50 and the CIE 1931 2° standard observer. As illustration two of the 12 patches are displayed in Figure 6 for samples 1 and 2. For all tested sample we

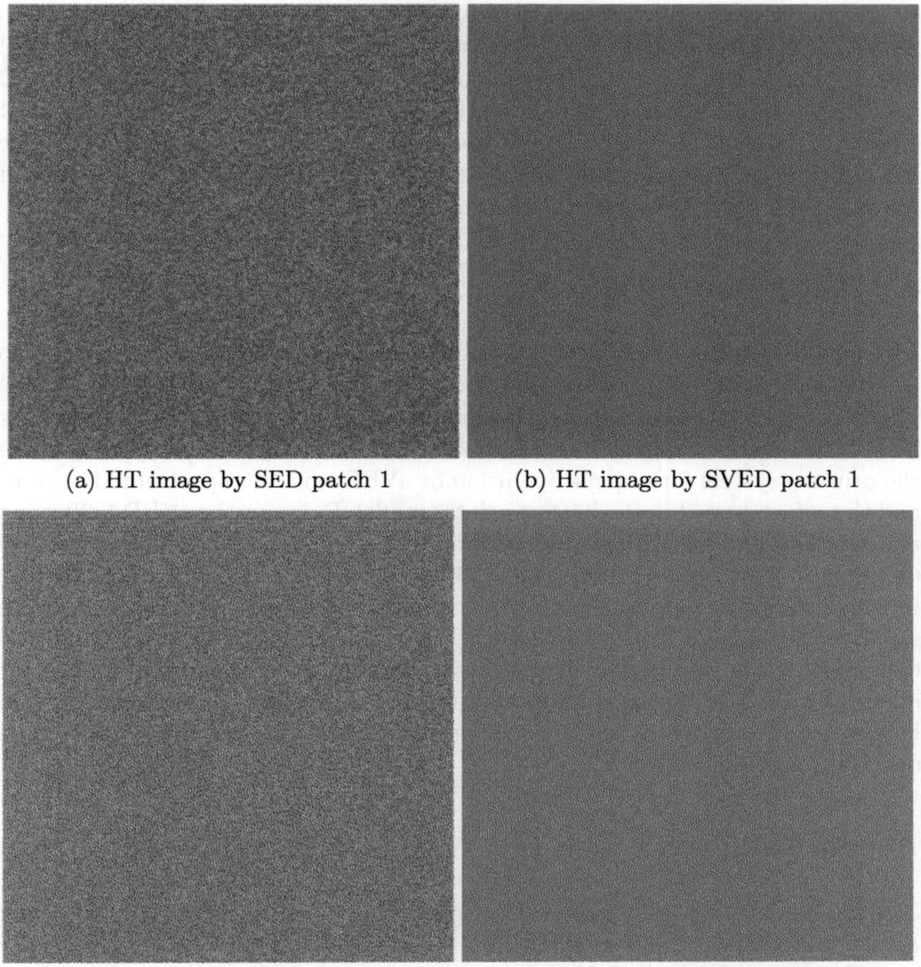

(a) HT image by SED patch 1 (b) HT image by SVED patch 1

(c) HT image by SED patch 2 (d) HT image by SVED patch 2

Fig. 6. Color renderings of the HT images for WF1 to the left and WF2 to the right, patches 1 Figure (a) and Figure (b), patches 2 Figure (c) and Figure (d)

can observe much more pleasant spatial distributions of the NPs when halftoning by sVED has been used. The spatial NPs distribution being extremely noisy when SED halftoning was performed.

A known problem with sVED or simply VED halftoning is the slowness of error diffusion. In case of color/spectral reflectance reproduction of single patch with a single value a border effect is visible because of the path the filter is following. This border effect is also visible with SED but less stronger. The introduction of the n factor before the sVED have shown a real improvement of the sVED algorithm by reaching faster a stable spatial dots distribution and a reduced border effect.

To complete the comparison of the two proposed WFs it will be necessary to reproduce spectral images. The confrontation of sVED (including the n factor) with an image will allow to compare completely the two WFs. First the sVED itself and how it behaves when the path followed the filter is crossing region of different contents (i.e. very different spectral reflectance) and how fast a stable dot distribution is reached. Secondly the computational cost and complexity of the two WFs can be evaluated and compared [13],[14].

4 Conclusion

The experimentation carried out in this article has allowed to compare two workflows for the reproduction of spectral images. One involving the inverse YNSN model for the colorant separation, process followed by the halftoning by SED. The second workflow has seen the use of the same parameters describing the printing system for a single operation by sVED: the NPs spectral reflectances and the n factor used in the inverse printer model. Doing so the sVED halftoning and the colorant separation were performed both in $1/n$ space. The possibility of spectral color reproduction by sVED has been already shown, but with the introduction of the n factor we have have observed a clear improvement of the sVED performances in term of error visiblity by reaching faster a stable dot distribution. The slowness of error diffusion being a major drawback when vector error diffusion is the chosen halftoning technique. Further experiments have to be conducted in order to evaluate the performance on spectral images other than spectral patches.

Acknowledgment

Jérémie Gerhardt is now flying of his own wings, but he would like to thank his two directors to have selected him for his research work on spectral color reproduction and for all the helpful discussions and feedback on his work, Jon Yngve Hardeberg at HIG (Norway) and especially Francis Schmitt at ENST (France) who left us too early.

References

1. Ostromoukhov, V.: Chromaticity Gamut Enhancement by Heptatone Multi-Color Printing. In: IS&T SPIE, pp. 139–151 (1993)
2. Agar, A.U.: Model Based Color Separation for CMYKcm Printing. In: The 9th Color Imaging Conference: Color Science and Engineering: Systems, Technologies, Applications (2001)
3. Jang, I., Son, C., Park, T., Ha, Y.: Improved Inverse Characterization of Multi-colorant Printer Using Colorant Correlation. J. of Imaging Science and Technology 51, 175–184 (2006)
4. Gerhardt, J., Hardeberg, J.Y.: Spectral Color Reproduction Minimizing Spectral and Perceptual Color Differences. Color Research & Application 33, 494–504 (2008)

5. Urban, P., Grigat, R.: Spectral-Based Color Separation Using Linear Regression Iteration. Color Research & Application 31, 229–238 (2006)
6. Taplin, L., Berns, R.S.: Spectral Color Reproduction Based on a Six-Color Inkjet Output System. In: The Ninth Color Imaging Conference, pp. 209–212 (2001)
7. Gerhardt, J., Hardeberg, J.Y.: Spectral Colour Reproduction by Vector Error Diffusion. In: Proceedings CGIV 2006, pp. 469–473 (2006)
8. Gerhardt, J.: Reproduction spectrale de la couleur: approches par modélisation d'imprimante et par halftoning avec diffusion d'erreur vectorielle, Ecole Nationale Supérieur des Télécommunications, Paris, France (2007)
9. Dupraz, D., Ben Chouikha, M., Alquié, G.: Historic period of fine art painting detection with multispectral data and color coordinates library. In: Proceedings of Ninth International Symposium on Multispectral Colour Science and Application (2007)
10. Demichel, M.E.: Le procédé 26, 17–21 (1924)
11. Ulichney, R.: Digital Halftoning. MIT Press, Cambridge (1987)
12. Jarvis, J.F., Judice, C.N., Ninke, W.H.: A Survey of Techniques for the Display of Continuous-Tone Pictures on Bilevel Displays. Computer Graphics and Image Processing 5, 13–40 (1976)
13. Urban, P., Rosen, M.R., Berns, R.S.: Fast Spectral-Based Separation of Multispectral Images. In: IS&T SID Fifteenth Color Imaging Conference, pp. 178–183 (2007)
14. Li, C., Luo, M.R.: Further Accelerating the Inversion of the Cellular Yule-Nielsen Modified Neugebauer Model. In: IS&T SID Sixteenth Color Imaging Conference, pp. 277–281 (2008)

Kernel Based Subspace Projection of Near Infrared Hyperspectral Images of Maize Kernels

Rasmus Larsen[1], Morten Arngren[1,2], Per Waaben Hansen[2], and Allan Aasbjerg Nielsen[3]

[1] DTU Informatics, Technical University of Denmark
Richard Petersens Plads, Building 321, DK-2800 Kgs. Lyngby, Denmark
{rl,ma}@imm.dtu.dk
[2] FOSS Analytical AS, Slangerupgade 69, DK-3400 Hillerød, Denmark
pwh@foss.dk
[3] DTU Space, Technical University of Denmark
Richard Petersens Plads, Building 321, DK-2800 Kgs. Lyngby, Denmark
aa@space.dtu.dk

Abstract. In this paper we present an exploratory analysis of hyperspectral 900-1700 nm images of maize kernels. The imaging device is a line scanning hyper spectral camera using a broadband NIR illumination. In order to explore the hyperspectral data we compare a series of subspace projection methods including principal component analysis and maximum autocorrelation factor analysis. The latter utilizes the fact that interesting phenomena in images exhibit spatial autocorrelation. However, linear projections often fail to grasp the underlying variability on the data. Therefore we propose to use so-called kernel version of the two afore-mentioned methods. The kernel methods implicitly transform the data to a higher dimensional space using non-linear transformations while retaining the computational complexity. Analysis on our data example illustrates that the proposed kernel maximum autocorrelation factor transform outperform the linear methods as well as kernel principal components in producing interesting projections of the data.

1 Introduction

Based on work by Pearson [1] in 1901, Hotelling [2] in 1933 introduced principal component analysis (PCA). PCA is often used for linear orthogonalization or compression by dimensionality reduction of correlated multivariate data, see Jolliffe [3] for a comprehensive description of PCA and related techniques.

An interesting dilemma in reduction of dimensionality of data is the desire to obtain simplicity for better understanding, visualization and interpretation of the data on the one hand, and the desire to retain sufficient detail for adequate representation on the other hand.

Schölkopf et al. [4] introduce kernel PCA. Shawe-Taylor and Cristianini [5] is an excellent reference for kernel methods in general. Bishop [6] and Press et al. [7] describe kernel methods among many other subjects.

A.-B. Salberg, J.Y. Hardeberg, and R. Jenssen (Eds.): SCIA 2009, LNCS 5575, pp. 560–569, 2009.

The kernel version of PCA handles nonlinearities by implicitly transforming data into high (even infinite) dimensional feature space via the kernel function and then performing a linear analysis in that space.

The maximum autocorrelation factor (MAF) transform proposed by Switzer [11] defines maximum spatial autocorrelation as the optimality criterion for extracting linear combinations of multispectral images. Contrary to this PCA seeks linear combinations that exhibit maximum variance. Because the interesting phenomena in image data often exhibit some sort of spatial coherence spatial autocorrelation is often a better optimality criterion than variance. A kernel version of the MAF transform has been proposed by Nielsen [10].

In this paper we shall apply kernel MAF as well as kernel PCA and ordinary PCA and MAF to find interesting projections of hyperspectral images of maize kernels.

2 Data Acquisition

A hyperspectral line-scan NIR camera from Headwall Photonics sensitive from 900-1700nm was used to capture the hyperspectral image data. A dedicated NIR light source illuminates the sample uniformly along the scan line and an advanced optic system developed by Headwall Photonics disperses the NIR light onto the camera sensor for acquisition. A sledge from MICOS GmbH moves the sample past the view slot of the camera allowing it to acquire a hyperspectral image. In order to separate the different wavelengths an optical system based on the Offner principle is used. It consists of a set of mirrors and gratings to guide and spread the incoming light into a range of wavelengths, which are projected onto the InGaAs sensor.

The sensor has a resolution of 320 spatial pixels and 256 spectral pixels, i.e. a physical resolution of 320×256 pixels. Due to the Offner dispersion principle (the convex grating) not all the light is in focus over the entire dispersed range. This means that if the light were dispersed over the whole 256 pixel wide sensor the wavelengths at the periphery would be out of focus. In order to avoid this the light is only projected onto 165 pixels instead and the top 91 pixels are disregarded. This choice is a trade-off between spatial sampling resolution and focus quality of the image.

The camera acquires 320 pixels and 165 bands for each frame. The pixels are represented in 14 bit resolution with 10 effective bits In Fig. 1 average spectra for a white reference and dark background current images are shown. Note the limited response in the 900-950 nm range.

Before the image cube is subjected to the actual processing a few preprocessing step are conducted. Initially the image is corrected for the reference light and dark background current. A reference and dark current image are acquired and the mean frame is applied for the correction. In our case the hyperspectral data are kept as reflectance spectra throughout the analysis.

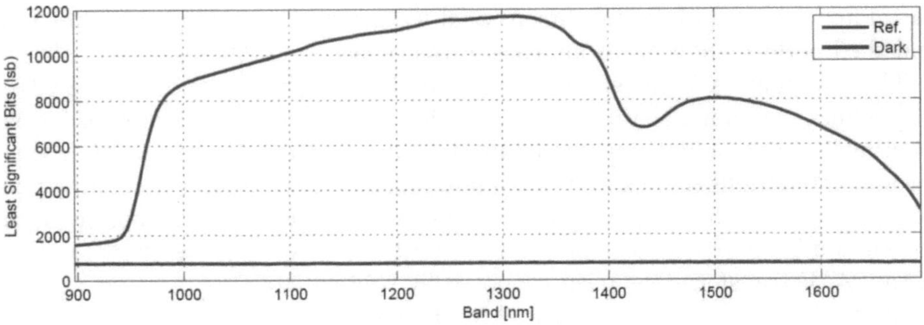

Fig. 1. Average spectra for white reference and dark background current images

2.1 Grain Samples Dataset

For the quantitative evaluation of the kernel MAF method a hyperspectral image of eight maize kernels is used as the dataset. The hyperspectral image of the maize samples are comprised of the front and back-side of the kernels on a black background (NCS-9000) appended as two separate cropped images as depicted in Fig. 2(a). In Fig. 2(b) an example spectrum is shown. The kernels are not

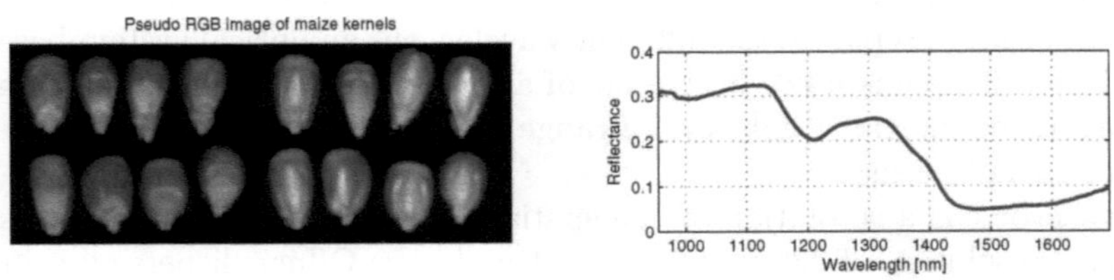

(a) (b)

Fig. 2. (a) Front (left) and back (right) images of eight maize kernels on a dark background. The color image is constructed as an RGB combination of NIR bands 150, 75, and 1; (b) reflectance spectrum of the pixel marked with red circle in (a).

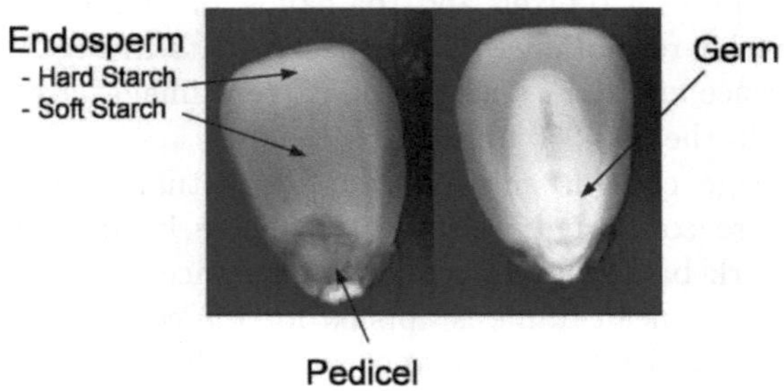

Fig. 3. Maize kernel constituents front- and backside (pseudo RGB)

fresh from harvest and hence have a very low water content and are in addition free from any infections. Many cereals in general share the same compounds and basic structure. In our case of maize a single kernel can be divided into many different constituents on the macroscopic level as illustrated in Fig. 3.

In general, the structural components of cereals can be divided into three classes denoted *Endosperm*, *Germ* and *Pedicel*. These components have different functions and compounds leading to different spectral profiles as described below.

Endosperm. The endosperm is the main storage for starch (~66%), protein (~11%) and water (~14%) in cereals. Starch being the main constituent is a carbohydrate and consists of two different glucans named *Amylose* and *Amylopectin*. The main part of the protein in the endosperm consists of *zein* and *glutenin*. The starch in maize grains can be further divided into a soft and a hard section depending on the binding with the protein matrix. These two types of starch are typically mutually exclusive, but in maize grain they both appear as a special case as also illustrated in figure 3.

Germ. The germ of a cereal is the reproductive part that germinates to grow into a plant. It is the embryo of the seed, where the scutellum serves to absorb nutrients from the endosperm during germination. It is a section holding proteins, sugars, lipids, vitamins and minerals [13].

Pedicel. The pedicel is the flower stalk and has negligible interest in terms of production use. For a more detailed description of the general structure of cereals [12].

3 Principal Component Analysis

Let us consider an image with n observations or pixels and p spectral bands organized as a matrix X with n rows and p columns; each column contains measurements over all pixels from one spectral band and each row consists of a vector of measurements x_i^T from p spectral bands for a particular observation $X = [x_1^T x_2^T \dots x_n^T]^T$. Without loss of generality we assume that the spectral bands in the columns of X have mean value zero.

3.1 Primal Formulation

In ordinary (primal also known as R-mode) PCA we analyze the sample variance-covariance matrix $S = X^T X/(n-1) = 1/(n-1)\sum_{i=1}^{n} x_i x_i^T$ which is p by p. If $X^T X$ is full rank $r = \min(n,p)$ this will lead to r non-zero eigenvalues λ_i and r orthogonal or mutually conjugate unit length eigenvectors u_i ($u_i^T u_i = 1$) from the eigenvalue problem

$$\frac{1}{n-1} X^T X u_i = \lambda_i u_i. \tag{1}$$

We see that the sign of u_i is arbitrary. To find the principal component scores for an observation x we project x onto the eigenvectors, $x^T u_i$. The variance of these scores is $u_i^T S u_i = \lambda_i u_i^T u_i = \lambda_i$ which is maximized by solving the eigenvalue problem.

3.2 Dual Formulation

In the dual formulation (also known as Q-mode analysis) we analyze $XX^T/(n-1)$ which is n by n and which in image applications can be very large. Multiply both sides of Equation 1 from the left with X

$$\frac{1}{n-1} XX^T(Xu_i) = \lambda_i(Xu_i) \quad \text{or} \quad \frac{1}{n-1} XX^T v_i = \lambda_i v_i \tag{2}$$

with v_i proportional to Xu_i, $v_i \propto Xu_i$, which is normally not normed to unit length if u_i is. Now multiply both sides of Equation 2 from the left with X^T

$$\frac{1}{n-1} X^T X(X^T v_i) = \lambda_i(X^T v_i) \tag{3}$$

to show that $u_i \propto X^T v_i$ is an eigenvector of S with eigenvalue λ_i. We scale these eigenvectors to unit length assuming that v_i are unit vectors $u_i = X^T v_i/\sqrt{(n-1)\lambda_i}$.

We see that if $X^T X$ is full rank $r = \min(n,p)$, $X^T X/(n-1)$ and $XX^T/(n-1)$ have the same r non-zero eigenvalues λ_i and that their eigenvectors are related by $u_i = X^T v_i/\sqrt{(n-1)\lambda_i}$ and $v_i = Xu_i/\sqrt{(n-1)\lambda_i}$. This result is closely related to the Eckart-Young [8,9] theorem.

An obvious advantage of the dual formulation is the case where $n < p$. Another advantage even for $n \gg p$ is due to the fact that the elements of the matrix $G = XX^T$, which is known as the Gram[1] matrix, consist of inner products of the multivariate observations in the rows of X, $x_i^T x_j$.

3.3 Kernel Formulation

We now replace x by $\phi(x)$ which maps x nonlinearly into a typically higher dimensional feature space. The mapping by ϕ takes X into Φ which is an n by q ($q \geq p$) matrix, i.e. $\Phi = [\phi(x_1)^T \phi(x_2)^T \ldots \phi(x_n)^T]^T$ we assume that the mappings in the columns of Φ have zero mean. In this higher dimensional feature space $C = \Phi^T\Phi/(n-1) = 1/(n-1)\sum_{i=1}^n \phi(x_i)\phi(x_i)^T$ is the variance-covariance matrix and for PCA we get the primal formulation $1/(n-1)\Phi^T\Phi u_i = \lambda_i u_i$ where we have re-used the symbols λ_i and u_i from above. For the corresponding dual formulation we get re-using the symbol v_i from above

$$\frac{1}{n-1} \Phi\Phi^T v_i = \lambda_i v_i. \tag{4}$$

As above the non-zero eigenvalues for the primal and the dual formulations are the same and the eigenvectors are related by $u_i = 1/(\sqrt{(n-1)\lambda_i})\, \Phi^T v_i$, and $v_i = \Phi\, u_i/\sqrt{(n-1)\lambda_i}$. Here $\Phi\Phi^T$ plays the same role as the Gram matrix above and has the same size, namely n by n (so introducing the nonlinear mappings in ϕ does not make the eigenvalue problem in Equation 4 bigger).

[1] Named after Danish mathematician Jørgen Pedersen Gram (1850-1916).

Kernel Substitution. Applying kernel substitution also known as the kernel trick we replace the inner products $\phi(x_i)^T\phi(x_j)$ in $\Phi\Phi^T$ with a kernel function $\kappa(x_i, x_j) = \kappa_{ij}$ which could have come from some unspecified mapping ϕ. In this way we avoid the explicit mapping ϕ of the original variables. We obtain

$$Kv_i = (n-1)\lambda_i v_i \tag{5}$$

where $K = \Phi\Phi^T$ is an n by n matrix with elements $\kappa(x_i, x_j)$. To be a valid kernel K must be symmetric and positive semi-definite, i.e., its eigenvalues are non-negative. Normally the eigenvalue problem is formulated without the factor $n-1$

$$Kv_i = \lambda_i v_i. \tag{6}$$

This gives the same eigenvectors v_i and eigenvalues $n-1$ times greater. In this case $u_i = \Phi^T v_i/\sqrt{\lambda_i}$ and $v_i = \Phi u_i/\sqrt{\lambda_i}$.

Basic Properties. Several basic properties including the norm in feature space, the distance between observations in feature space, the norm of the mean in feature space, centering to zero mean in feature space, and standardization to unit variance in feature space, may all be expressed in terms of the kernel function without using the mapping by ϕ explicitly [5,6,10].

Projections onto Eigenvectors. To find the kernel principal component scores from the eigenvalue problem in Equation 6 we project a mapped x onto the primal eigenvector u_i

$$\begin{aligned}\phi(x)^T u_i &= \phi(x)^T \Phi^T v_i/\sqrt{\lambda_i} = \phi(x)^T \left[\phi(x_1)\ \phi(x_2)\ \cdots\ \phi(x_n)\right] v_i/\sqrt{\lambda_i} \\ &= \left[\kappa(x, x_1)\ \kappa(x, x_2)\ \cdots\ \kappa(x, x_n)\right] v_i/\sqrt{\lambda_i},\end{aligned} \tag{7}$$

or in matrix notation $\Phi U = KV\Lambda^{-1/2}$ (U is a matrix with u_i in the columns, V is a matrix with v_i in the columns and $\Lambda^{-1/2}$ is a diagonal matrix with elements $1/\sqrt{\lambda_i}$), i.e., also the projections may be expressed in terms of the kernel function without using ϕ explicitly. If the mapping by ϕ is not column centered the variance of the projection must be adjusted, cf. [5,6].

Kernel PCA is a so-called memory-based method: from Equation 7 we see that if x is a new data point that did not go into building the model, i.e., finding the eigenvectors and -values, we need the original data $x_1, x_2, \ldots, x_n$ as well as the eigenvectors and -values to find scores for the new observations. This is not the case for ordinary PCA where we do not need the training data to project new observations.

Some Popular Kernels. Popular choices for the kernel function are stationary kernels that depend on the vector difference $x_i - x_j$ only (they are therefore invariant under translation in feature space), $\kappa(x_i, x_j) = \kappa(x_i - x_j)$, and homogeneous kernels also known as radial basis functions (RBFs) that depend on the Euclidean distance between x_i and x_j only, $\kappa(x_i, x_j) = \kappa(\|x_i - x_j\|)$. Some of the most often used RBFs are ($h = \|x_i - x_j\|$)

- multiquadric: $\kappa(h) = (h^2 + h_0^2)^{1/2}$,
- inverse multiquadric: $\kappa(h) = (h^2 + h_0^2)^{-1/2}$,
- thin-plate spline: $\kappa(h) = h^2 \log(h/h_0)$, or
- Gaussian: $\kappa(h) = \exp(-\frac{1}{2}(h/h_0)^2)$,

where h_0 is a scale parameter to be chosen. Generally, h_0 should be chosen larger than a typical distance between samples and smaller than the size of the study area.

4 Maximum Autocorrelation Factor Analysis

In maximum autocorrelation factor (MAF) analysis we maximize the autocorrelation of linear combinations, $a^T x(r)$, of zero-mean original (spatial) variables, $x(r)$. $x(r)$ is a multivariate observation at location r and $x(r + \Delta)$ is an observation of the same variables at location $r + \Delta$; Δ is a spatial displacement vector.

4.1 Primal Formulation

The autocovariance R of a linear combination $a^T x(r)$ of zero-mean $x(r)$ is

$$\begin{aligned} R &= \mathrm{Cov}\{a^T x(r), a^T x(r + \Delta)\} && (8) \\ &= a^T \mathrm{Cov}\{x(r), x(r + \Delta)\} a && (9) \\ &= a^T C_\Delta a && (10) \end{aligned}$$

where C_Δ is the covariance between $x(r)$ and $x(r + \Delta)$. Assuming or imposing second order stationarity of $x(r)$, C_Δ is independent of location, r. Introduce the multivariate difference $x_\Delta(r) = x(r) - x(r + \Delta)$ with variance-covariance matrix $S_\Delta = 2\,S - (C_\Delta + C_\Delta^T)$ where S is the variance-covariance matrix of x defined in Section 3. Since

$$\begin{aligned} a^T C_\Delta a &= (a^T C_\Delta a)^T && (11) \\ &= a^T C_\Delta^T a && (12) \\ &= a^T (C_\Delta + C_\Delta^T) a/2 && (13) \end{aligned}$$

we obtain

$$R = a^T (S - S_\Delta/2) a. \tag{14}$$

To get the autocorrelation ρ of the linear combination we divide the covariance by its variance $a^T S a$

$$\begin{aligned} \rho &= 1 - \frac{1}{2} \frac{a^T S_\Delta a}{a^T S a} && (15) \\ &= 1 - \frac{1}{2} \frac{a^T X_\Delta^T X_\Delta a}{a^T X^T X a} && (16) \end{aligned}$$

where the n by p data matrix X is defined in Section 3 and X_Δ is a similarly defined matrix for x_Δ with zero-mean columns. C_Δ above equals $X^T X_\Delta/(n-1)$. To maximize ρ we must minimize the Rayleigh coefficient $a^T X_\Delta^T X_\Delta a/(a^T X^T X a)$ or maximize its inverse.

Unlike linear PCA, the result from linear MAF analysis is scale invariant: if x_i is replaced by some matrix transformation $T x_i$ corresponding to replacing X by XT, the result is the same.

4.2 Kernel MAF

As with the principal component analysis we use the kernel trick to obtain an implicit non-linear mapping for the MAF transform. A detailed account of this is given in [10].

5 Results and Discussion

To be able to carry out kernel MAF and PCA on the large amounts of pixels present in the image data, we sub-sample the image and use a small portion termed the training data only. We typically use in the order 10^3 training pixels (here ~3,000) to find the eigenvectors onto which we then project the entire image termed the test data kernelized with the training data. A Gaussian kernel $\kappa(x_i, x_j) = \exp(-\|x_i - x_j\|^2/2\sigma^2)$ with σ equal to the mean distance between the training observations in feature space is used.

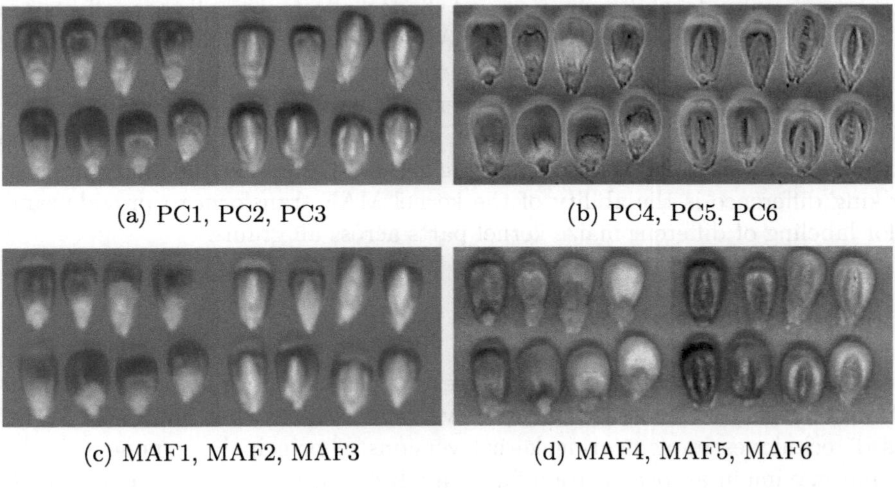

(a) PC1, PC2, PC3 (b) PC4, PC5, PC6

(c) MAF1, MAF2, MAF3 (d) MAF4, MAF5, MAF6

Fig. 4. Linear principal component projections of front and back sides of 8 maize kernels shown as RGB combination of factors (1,2,3) and (4,5,6) (two top panels), and corresponding linear maximum autocorrelation factor projections (bottom two panels)

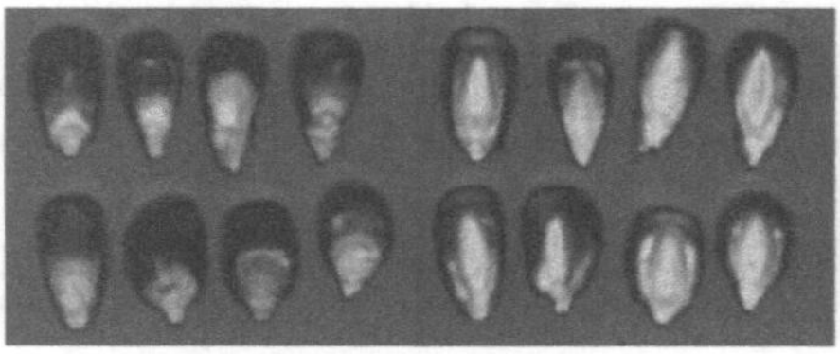

(a) kPC1, kPC2, kPC3

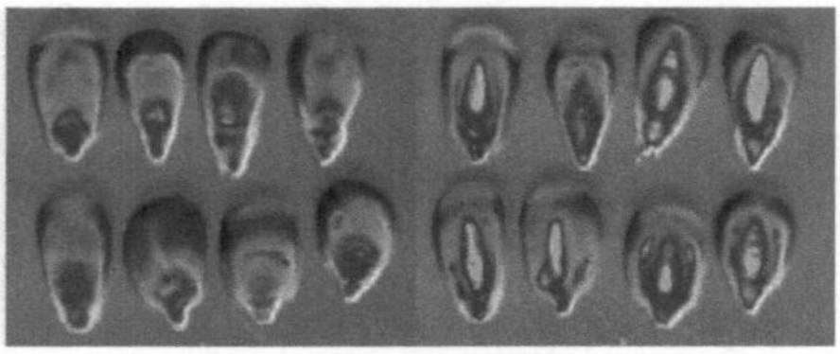

(b) kPC4, kPC5, kPC6

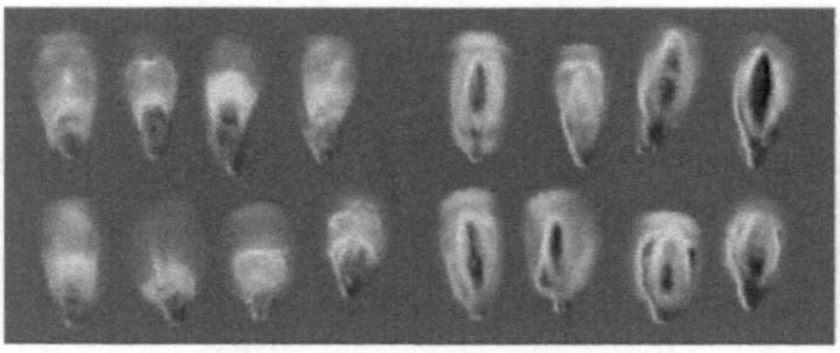

(c) kMAF1, kMAF2, kMAF3

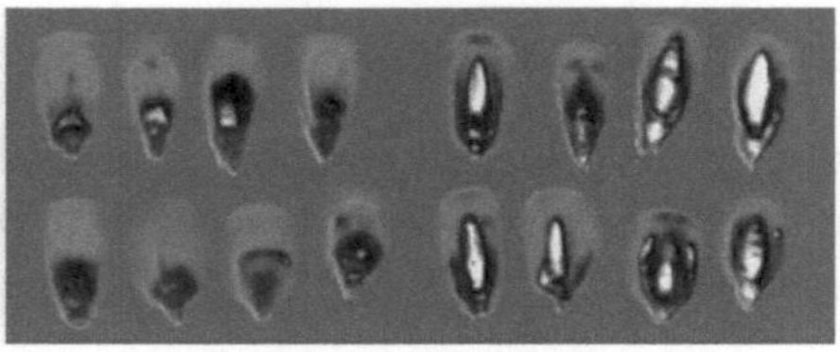

(d) kMAF4, kMAF5, kMAF6

Fig. 5. Non-linear kernel principal component projections of front and back sides of 8 maize kernel shown as RGB combination of factors (1,2,3) and (4,5,6) (two top panels), and corresponding non-linear kernel maximum autocorrelation factor projections (bottom two panels)

In Fig. 4 linear PCA and MAF components are shown as RGB combination of factors (1,2,3) and (4,5,6) are shown. The presented images are scaled linearly between ± 3 standard deviations. The linear transforms both struggle with the background noise, local illumination and shadow effects, i.e., all these effects are enhanced in some of the first 6 factors. Also the linear methods fail in labeling the same kernel parts in same colors. On the other hand the kernel based factors shown in Fig. 5 have a significantly better ability to suppress background noise, illumination variation and shadow effect. In fact this is most pronounced in the kernel MAF projections. When comparing kernel PCA and kernel MAF the most striking difference is the ability of the kernel MAF transform to provide same color labeling of different maize kernel parts across all grains.

6 Conclusion

In this preliminary work on finding interesting projections of hyperspectral near infrared imagery of maize kernels we have demonstrated that non-linear kernel based techniques implementing kernel versions of principal component analysis and maximum autocorrelation factor analysis outperform the linear variants by their ability to suppress background noise, illumination and shadow effects. Moreover, the kernel maximum autocorrelation factors transform provides a superior projection in terms of labeling different maize kernels parts with same color.

References

1. Pearson, K.: On lines and planes of closest fit to systems of points in space. Philosofical Magazine 2(3), 559–572 (1901)
2. Hotelling, H.: Analysis of a complex of statistical variables into principal components. Journal of Educational Psychology 24, 417–441, 498–520 (1933)
3. Jolliffe, I.T.: Principal Component Analysis, 2nd edn. Springer, Heidelberg (2002)
4. Schölkopf, B., Smola, A., Müller, K.-R.: Nonlinear component analysis as a kernel eigenvalue problem. Neural Computation 10(5), 1299–1319 (1998)
5. Shawe-Taylor, J., Cristianini, N.: Kernel Methods for Pattern Analysis. Cambridge University Press, Cambridge (2004)
6. Bishop, C.M.: Pattern Recognition and Machine Learning. Springer, Heidelberg (2006)
7. Press, W.H., Teukolsky, S.A., Vetterling, W.T., Flannery, B.P.: Numerical Recipes: The Art of Scientific Computing, 3rd edn. Cambridge University Press, Cambridge (2007)
8. Eckart, C., Young, G.: The approximation of one matrix by another of lower rank. Psykometrika 1, 211–218 (1936)
9. Johnson, R.M.: On a theorem stated by Eckart and Young. Psykometrika 28(3), 259–263 (1963)
10. Nielsen, A.A.: Kernel minimum noise fraction transformation (2008) (submitted)
11. Switzer, P.: Min/Max Autocorrelation factors for Multivariate Spatial Imagery. In: Billard, L. (ed.) Computer Science and Statistics, pp. 13–16 (1985)
12. Hoseney, R.C.: Principles of Cereal Science and Technology. American Association of Cereal Chemists (1994)
13. Belitz, H.-D., Grosch, W., Schieberle, P.: Food Chemistry, 3rd edn. Springer, Heidelberg (2004)

The Number of Linearly Independent Vectors in Spectral Databases

Carlos Sáenz, Begoña Hernández, Coro Alberdi, Santiago Alfonso, and José Manuel Diñeiro

Departamento de Física, Universidad Pública de Navarra, Campus Arrosadia, 31006 Pamplona, Spain

Abstract. Linear dependence among spectra in spectral databases affects the eigenvectors obtained from principal component analysis. This affects the values of usual spectral and colorimetric metrics. The effective dependence is proposed as a tool to quantify the maximum number of linearly independent vectors in the database. The results of the proposed algorithm do not depend on the selection of the first seed vector and are consistent with the results based on reduction of the bivariate coefficient of determination.

Keywords: Spectral databases, effective dependence, linear correlation, collinearity.

1 Introduction

Spectral databases are used in many applications within the context of spectral colour science. Dimensionality reduction techniques like principal component analysis (PCA), incomplete component analysis (ICA) and others are used to describe spectral information with a reduced number of basis functions. Applications of these techniques are found in many fields and require a detailed evaluation of their performance. Testing the performance of these methods usually involve spectral databases from two complementary but different points of view. The set of basis functions or vectors are obtained from a particular spectral database, called the Training set, using some specific spectral or colorimetric metrics. Then the performance of the basis functions in order to reconstruct spectral or colorimetric information is checked with the help of a second spectral database, the Test set. Numerical results depend on the used databases [1] and metrics, in this scenario some authors recommend the simultaneous use of several metrics to evaluate the quality of the data reconstruction [2,3].

Spectral databases may differ because of the measurement technique, wavelength limits, wavelength interval or number of data points in their spectra. Even more important differences are found because of the origin of the samples used to construct the database. Some databases have been obtained from color atlases or color collections, others correspond to measurements of natural objects or to samples specifically created with some purpose. Recently the principal characteristics of some frequently used spectral databases have been reviewed [4].

A.-B. Salberg, J.Y. Hardeberg, and R. Jenssen (Eds.): SCIA 2009, LNCS 5575, pp. 570–579, 2009.

Some of the most frequently used spectral databases, like Munsell or NCS, have been measured in collections of color samples. These color collections have been constructed according to some specific colorimetric or perceptual criteria, say uniformly distributed samples in the color space. No spectral criteria were used in their construction. In fact, we do not actually posses a criterion that allows us to talk for instance about uniformly distributed spectra.

In this work we will analyze the possibility of using the linear dependence between spectra as a measure of the amount of spectral information contained in the database. A parameter of this kind, independent of particular choices of spectral or colorimetric measures, could be a valuable indicator of the 'spectral diversity' within the database.

2 Spectral Databases and Linear Dependence

2.1 Effect of Linear Dependence in *RMSE* and ΔE^*

Let us suppose that we have a spectral database formed by q spectra $\mathbf{r}_i$, $i = 1, 2 \ldots q$, representing the reflectance factor $\mathbf{r}$ of q samples measured at n wavelengths. In general any spectrum $\mathbf{r}_j$ can be obtained from the other spectra $\mathbf{r}_i$, $i \neq j$ in the database as:

$$\mathbf{r}_j = \sum_{\substack{i=1 \\ i \neq j}}^{q} w_i \mathbf{r}_i + \mathbf{e}_j = \hat{\mathbf{r}}_j + \mathbf{e}_j \tag{1}$$

Where w_i are the appropriate weights. In (1) the vector $\hat{\mathbf{r}}_j$ is the estimated value of $\mathbf{r}_j$ that can be obtained from the remaining vectors in the database and $\mathbf{e}_j = \mathbf{r}_j - \hat{\mathbf{r}}_j$ is an error term. Respect to the spectral information in $\mathbf{r}_j$ the error term $\mathbf{e}_j$ represents the intrinsic information contained in $\mathbf{r}_j$ that can not be reproduced by the rest of spectra. In general, an accepted measure of the spectral similarity/difference is the $RMSE_j$ value between the original and estimated vectors defined as

$$RMSE_j = \sqrt{\frac{1}{n}\sum_{k=1}^{n} (\mathbf{r}_{kj} - \hat{\mathbf{r}}_{kj})^2} = \sqrt{\frac{1}{n}\sum_{k=1}^{n} \mathbf{e}_{kj}^2} \tag{2}$$

Where the index k identifies each of the n measured wavelengths. If se are interested in colorimetric information, the tristimulus values must be also computed. For a given illuminant $\mathbf{S}$, the X_i tristimulus value of $\mathbf{r}_j$ is:

$$X_j = K \sum_{k=1}^{n} r_{kj} S_k \bar{x}_k \tag{3}$$

Where K is a normalization factor, r_{kj} the reflectance factor at wavelength k and $\bar{x}$ the color matching function. The tristimulus Y_j and Z_j are obtained using the color matching functions $\bar{y}$ and $\bar{z}$ respectively. Substituting (1) in (3) the tristimulus value X_j can be obtained as a function of the tristimulus values of the other spectra in the database as:

$$\begin{aligned} X_j &= K \sum_{k=1}^{n} \hat{r}_{kj} S_k \bar{x}_k + K \sum_{k=1}^{n} e_{kj} S_k \bar{x}_k = \\ &= \sum_{\substack{i=1 \\ i \neq j}}^{q} w_i X_i + K \sum_{k=1}^{n} e_{kj} S_k \bar{x}_k = \\ &= \hat{X}_j + X e_j \end{aligned} \tag{4}$$

Which is an obvious consequence of the linearity of (3). In this expression $\hat{X}_j$ is the estimation of X_j that we can obtain solely with the vectors in the database and Xe_j is the tristimulus value associated to the error term $\mathbf{e}_j$. Therefore the tristimulus values of $\mathbf{r}_j$ are a linear combination of the tristimulus values or the other spectra in the database plus an extra term that depends on $\mathbf{e}_j$.

If the error term $\mathbf{e}_j$ is sufficiently small, in the sense that all e_{kj} are small, then $RMSE_j$ will be also small. Furthermore, Xe_j will be also small and $X_j \approx \hat{X}_j$. The same argument can be extended to the other tristimulus values. If true and stimated tristimulus values are very similar, then ΔE^* color differences between true and estimated spectra are expected to be small too.

All these arguments are well known and linear models are extensively used in spectral and color reproduction and estimation; where in general spectra are reconstructed using a limited number of basis vectors. An interesting and ever present problem is that there is no evident relationship between the spectral reconstruction accuracy, measured with $RMSE$ or other spectral metric, and the color reproduction accuracy determined with a particular color difference measure. This means that we do not have a clear criterion to quantify what does mean a sufficiently small error term $\mathbf{e}_j$. Furthermore colorimetric results are sensible to the illuminant $\mathbf{S}$ used in the calculations.

When the error term $\mathbf{e}_j$ vanishes in (1) then $\mathbf{r}_j$ is an exact linear combination of other spectra in the database. In this case we have $RMSE_j = 0$ and identical tristimulus values in (4) and therefore color differences between the original and reconstructed spectra vanish. It could be said that in this situation $\mathbf{r}_j$ does not provide additional spectral or color information respect to the remaining vectors in the database.

In general the number of spectra q in the database is higher than the number of sampled wavelengths n. If $\mathbf{X}$ is the n x q matrix where each column is a spectrum of the database, then upper limit to the number of linearly independent vectors in $\mathbf{X}$ is $rank(\mathbf{X}) = min(n, q) = n$ assuming that $q > n$.

PCA is affected by collinearity [5] and the effect on the basis vectors can be noticeable. Since only few basis vectors are usually retained, the spectral and colorimetric reconstruction accuracy will be also affected. In order to show this effect we have performed the following experiment that resembles the standard Training and Test databases approach. We have used the Munsell colors measured by the Joensuu Color Group [6]. The Munsell dataset consists in 1269 reflectance factor spectra measured with a Perkin-Elmer lambda 9 UV/VIS/NIR spectrophotometer at 1 nm intervals between 380 and 800 nm. We have randomly split the Munsell database in two, a Training database **A** with q_A spectra and a Test database **B** with q_B spectra. Then we have randomly selected a single vector from **A** to serve as seed vector in order to generate linearly dependent vectors. We have iteratively added to **A** vectors proportional to the seed vector, thus increasing q_A in each iteration. The proportionality constant has been uniformly sampled in the range [0,1]. After every iteration we have used PCA to obtain the first n_b eigenvectors. Using these eigenvectors we have obtained $\hat{\mathbf{A}}$ and $\hat{\mathbf{B}}$, the estimations of **A** and **B**. The process has been repeated for different random partitions of the Munsell database.

The effect that the addition of such vectors has on the first two principal components can be seen in Fig. 1 for an example with $q_A = 10$. The seed spectrum (reduced by a factor two) has been also included for comparison. Due to the randomness of the multiplicative constant, eigenvectors do not always evolve in the same direction, being these changes rather unpredictable even though we are modifying the database in the simplest way. A similar situation is found for the other eigenvectors. We can also see the effect on the $RMSE$ and ΔE^* between original and reconstructed data sets in Fig. 2. As the number of linearly dependent vectors in the Training set **A** increases, eigenvectors evolve to explain the resulting changes in the correlation matrix. This produces the reduction in the mean $RMSE$ and ΔE^* values between **A** and $\hat{\mathbf{A}}$. On the contrary, maximum $RMSE$ and ΔE^* differences increase slightly because the reconstruction accuracy of the original vectors in the database deteriorates accordingly. Respect to the Test database, the new vectors added to **A** do not improve either the mean nor the maximum $RMSE$ and ΔE^* values between **B** and $\hat{\mathbf{B}}$ and these parameters are roughly constant during all the process. In the presence of linear correlation, the minimization of $RMSE$ or ΔE^* in the Training database does not guarantee optimal results in the Test database. Similar conclusions are obtained if we repeat the process for different initial **A** and **B** sets although details may differ, sometimes substantially.

2.2 Effective Dependence

In the previous examples the collinearity within the data is *a priori* known, by construction. In a real situation collinearity will be distributed over the entire sample set in an unknown manner. Therefore it is interesting to posses a measure of the amount of collinearity or linear dependence between variables for the entire spectral set. Although bivariate correlation is accurately defined through the Pearson correlation coefficient we do not have a single, widely accepted measure of linear dependence in the case of multivariate data.

In a recent paper Peña and Rodriguez [7] have proposed two new descriptive measures for multivariate data: the effective variance and the effective dependence. Their main objective was to define a dependence measure that could be used to compare data sets with different number of variables. In particular, if **X** is the n x p matrix having p variables and n observations of each variable, then the effective dependence $D_e(\mathbf{X})$ is defined as:

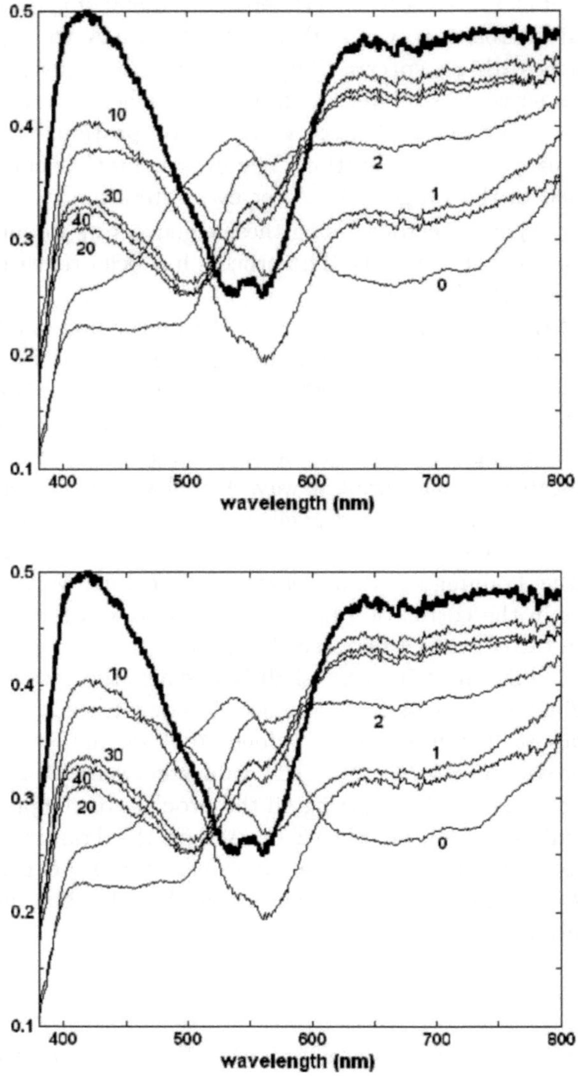

Fig. 1. Changes in the first (top) and second (bottom) eigenvectors after de addition of 1,2,10,20,30 and 40 vectors proportional to a single seed vector belonging to the original set. The seed vector (dark line) has been reduced by a factor 2.

$$D_e(\mathbf{X}) = 1 - |\mathbf{R_X}|^{1/p} \tag{5}$$

Where $|\mathbf{R_X}|$ is the determinant of the correlation matrix $\mathbf{R_X}$ of $\mathbf{X}$. Authors demonstrate that $D_e(\mathbf{X})$ satisfies the main properties of a dependence measure and of particular interest in our discussion:

a) $0 \leq D_e(\mathbf{X}) \leq 1$, and $D_e(\mathbf{X}) = 1$ if and only if we can find a vector $\mathbf{a} \neq 0$ and $\mathbf{b}$ such $\mathbf{a}'\mathbf{X} + \mathbf{b} = \mathbf{0}$. This means that $D_e(\mathbf{X}) = 1$ implies that there exists

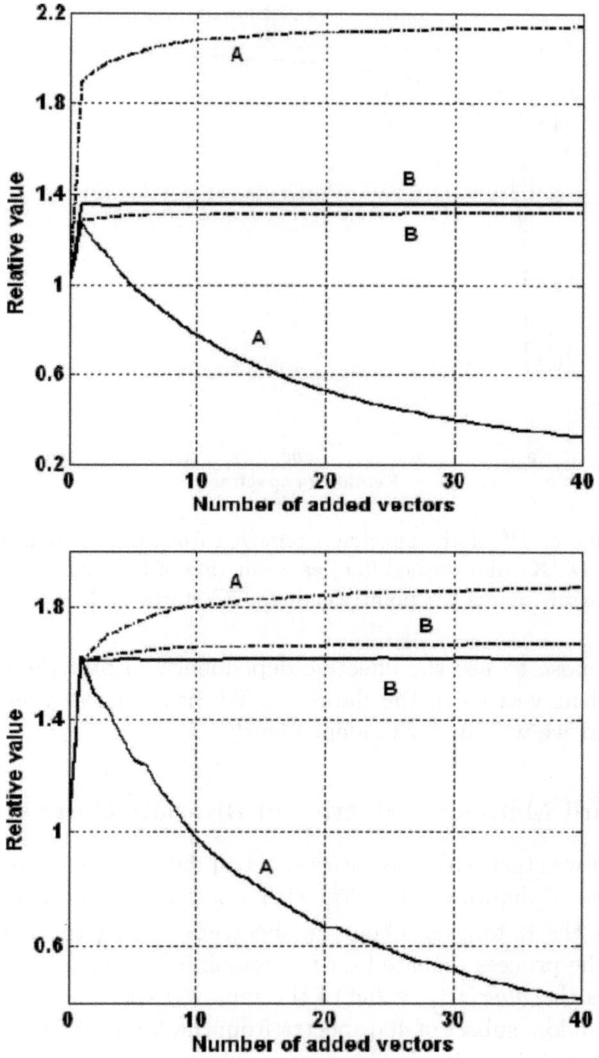

Fig. 2. $RMSE$ (top) and ΔE^* (bottom) as a function of the number linearly dependent vectors added to the training database. Solid lines are mean values and dot dashed lines maximum values. Letters A and B refer to the training and test databases respectively. All parameters have been normalized to the first value.

collinearity within the data. Also $D_e(\mathbf{X}) = 0$ if and only if the covariance matrix of $\mathbf{X}$ is diagonal.

b) Let $\mathbf{Z}' = [\mathbf{X}'\mathbf{Y}']$ be a random vector of dimension $p + q$ where $\mathbf{X}$ and $\mathbf{Y}$ are random variables of dimension p and q respectively, then $D_e(\mathbf{Z}) \geq D_e(\mathbf{X})$ if and only if $D_e(\mathbf{Y} : \mathbf{X}) > D_e(\mathbf{X})$ where $D_e(\mathbf{Y} : \mathbf{X})$ is the additional correlation introduced by $\mathbf{Y}$. Analogously $D_e(\mathbf{Z}) \leq D_e(\mathbf{X})$ if and only if $D_e(\mathbf{Y} : \mathbf{X}) < D_e(\mathbf{X})$.

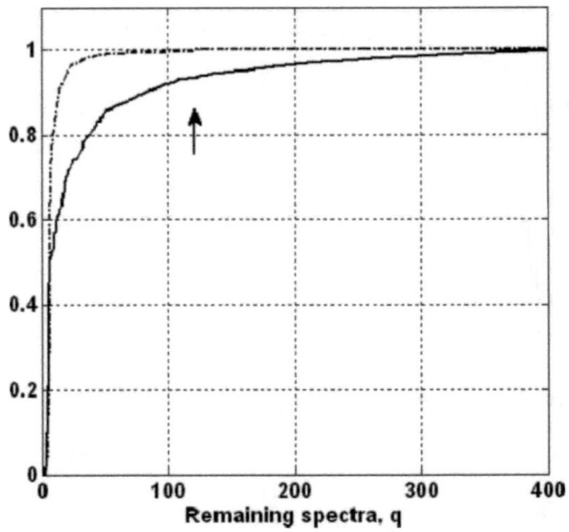

Fig. 3. The value of R^2 of the spectrum removed from the training database (solid line) and of the $D_e(\mathbf{X})$ (dot dashed line) as a function of the number of the remaining spectra q. The arrow marks the point where $D_e(\mathbf{X})$ starts to decrease.

We now propose to use the effective dependence to find the number of linearly independent vectors in the database. We have investigated two different approaches that we will analyze independently.

2.3 Backward Method: Reduction of Bivariate Correlation

In this method we start with the entire spectral database and calculate the pair wise coefficients of determination R^2_{ij} with $i \neq j$ between all possible pairs of spectra within the database. Then the spectrum having the maximum R^2_{ij} is removed and the process repeated for the remaining spectra.

Fig. 3 shows the $max(R^2_{ij})$ value of the removed spectrum during the entire process for a random subset of 400 spectra from the Munsell database. The value of $D_e(\mathbf{X})$ after each iteration is also shown. The reduction process starts at the rightmost value in the figure ($n = 400$) and continues to the left. We can observe that in this example the first occurrence of $D_e(\mathbf{X}) < 0$ happens where the number of remaining spectra in the database is $q_1 = 120$ and $max(R^2_{ij}) = 0.9349$.

Further reduction in q also implies a reduction in the value of the effective dependence. Notice that the effective dependence decreases monotonically with the number of remaining spectra in the database q.

Since R^2_{ij} is a bivariate statistic we can not assume that this procedure is the most effective to reduce global collinearity within the database. Therefore the value q_1 must be regarded as a lower limit to the number of linearly independent vectors in the original spectral database.

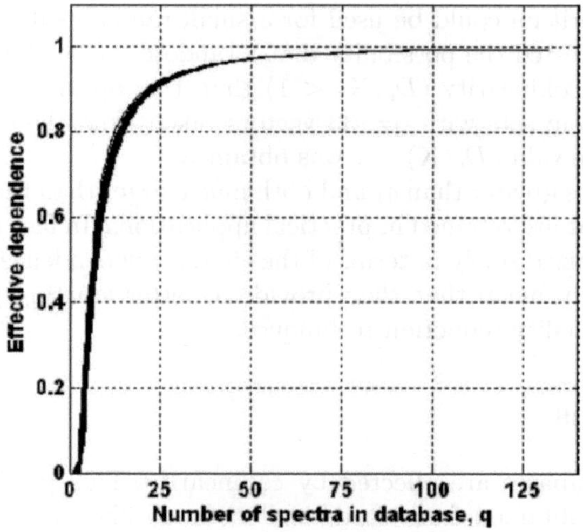

Fig. 4. The effective dependence as a function of the spectra in the data base

2.4 Forward Method: $D_e(\mathbf{X})$ Minimization

The second approach is based in the properties of the effective dependence and consists in finding the subset of spectra of the original database that minimizes $D_e(\mathbf{X})$ and maximizes the number of spectra. The algorithm begins with a single spectrum, the seed spectrum. Then the value of $D_e(\mathbf{X})$ resulting after the addition of a second spectrum is computed for all remainging spectra in the database. The spectrum providing the minimum increment to $D_e(\mathbf{X})$ is retained increasing the number of spectra in one. The the process is repeated, adding new vectors, until $D_e(\mathbf{X}) = 1$ is obtained. Let it be q_2 the number of spectra in the optimized set inmediatly before $D_e(\mathbf{X}) = 1$.

In order to apply this method, we must select an initial spectrum, the seed spectrum, from the data set. Lacking of a good reason to choose a particular one we have repeated the process using all vectors as seed vectors. In principle this would led to different solutions, having different number of spectra q_2. The solution or solutions having maximum q_2 inform us about the maximum number of independent vectors in the original dataset.

We have performed the experiment over the same subset of the preceding section, with 400 vectors. In Fig. 4 we show the evolution of the effective dependence during the construction of the 'optimized' sets. The 400 curves corresponding to the 400 possible seed vectors have been plotted. It can be seen that the rate of change in the effective dependence depends only slightly on the seed vector and $D_e(\mathbf{X})$ values rapidly converge in all cases, giving very similar number of vectors q_2 in the optimized set. In particular, for this dataset, we have obtained $q_2 = 133$ vectors in 338 cases and $q_2 = 134$ vectors in 62 cases. This suggests that the choice of the initial seed vector is of little relevance. This fact is of practical importance since the forward algorithm is time consuming. Therefore, for large databases the algorithm could be used for a small random subset of seed spectra. We have also tested the possibility that a random set having $q = q_2$ spectra could exhibit less collinearity ($D_e(\mathbf{X}) < 1$) than the 'optimized' set. We have created 5000 random sets with q=133 vectors taken from the original dataset and in all cases the value $D_e(\mathbf{X}) = 1$ was obtained.

As expected, q_2 is greater than q_1 and both much larger than the usual number of basis vectors that are retained in practical applications. In fact the 'optimized' data sets are optimized solely in terms of the effective dependence measure. This does not necessarily mean that they provide a better starting point to apply standard dimensionality reduction techniques.

3 Conclusions

Most spectral databases are affected by collinearity. This produces a bias in the basis vectors obtained from statistical methods like principal component analysis. This bias need not to be a drawback, since it accounts for the distributional properties of the original data, which may be necessary for the particular application. However collinearity may affect the results when different spectral databases, with different origin, are compared.

The effective dependence provides a measure of the degree of collinearity within a spectral database. The maximum number of spectra that can be retained before the effective dependence becomes unity inform us about the quantity of independent information contained. The properties of the effective dependence allow a forward construction algorithm that gives solution having a number of vectors that are almost independent on the seed vector used to start the process. The results obtained are in agreement with the simpler and more intuitive backward algorithm based in the removal of those spectra having high bivariate correlations.

Several practical aspects need further investigation: the properties of the optimized sets with regard to the spectral and colorimetric reconstruction, the relationship between the effective dependence and the number of sampled wavelengths or how to use the 'effective number of spectra' to compare different spectral data sets.

References

1. Sáenz, C., Hernández, B., Alberdi, C., Alfonso, S., Diñeiro, J.M.: The effect of selecting different training sets in the spectral and colorimetric reconstruction accuracy. In: Ninth International Symposium on Multispectral Colour Science and Application, MCS 2007, Taipei, Taiwan (2007)
2. Imai, F.H., Rosen, M.R., Berns, R.S.: Comparative study of metrics for spectral match quality. In: Cgiv 2002: First European Conference on Colour in Graphics, Imaging, and Vision, Conference Proceedings, pp. 492–496 (2002)
3. Viggiano, J.S.: Metrics for evaluating spectral matches: A quantitative comparison. In: Cgiv 2004: Second European Conference on Color in Graphics, Imaging, and Vision - Conference Proceedings, pp. 286–291 (2004)

4. Kohonen, O., Parkkinen, J., Jaaskelainen, T.: Databases for spectral color science. Color Research and Application 31(5), 381–390 (2006)
5. Jolliffe, I.T.: Principal component analysis, 2nd edn. Springer series in statistics. Springer, New York (2002)
6. Spectral Database, University of Joensuu Color Group, `http://spectral.joensuu.fi`
7. Peña, D., Rodriguez, J.: Descriptive measures of multivariate scatter and linear dependence. Journal of Multivariate Analysis 85(2), 361–374 (2003)

A Clustering Based Method for Edge Detection in Hyperspectral Images

V.C. Dinh[1,2], Raimund Leitner[2], Pavel Paclik[3], and Robert P.W. Duin[1]

[1] ICT Group, Delft University of Technology, Delft, The Netherlands
[2] Carinthian Tech Research AG, Villach, Austria
[3] PR Sys Design, Delft, The Netherlands

Abstract. Edge detection in hyperspectral images is an intrinsically difficult problem as the gray value intensity images related to single spectral bands may show different edges. The few existing approaches are either based on a straight forward combining of these individual edge images, or on finding the outliers in a region segmentation. As an alternative, we propose a clustering of all image pixels in a feature space constructed by the spatial gradients in the spectral bands. An initial comparative study shows the differences and properties of these approaches and makes clear that the proposal has interesting properties that should be studied further.

1 Introduction

Edge detection plays an important role in image processing and analyzing systems. Success in detecting edges may have a great impact on the result of subsequent image processing, e.g. region segmentation, object detection, and may be used in a wide range of applications, from image and video processing to multi/hyper-spectral image analysis. For hyperspectral images, in which channels may provide different or even conflicting information, edge detection becomes more important and essential.

Edge detection in gray-scale images has been thoroughly studied and is well established. But for color images, especially multi-channel images like hyperspectral images, this topic is much less developed since even defining edges for those images is already a challenge [1]. Two main approaches to detect edges in multi-channel images based on monochromatic [2,3] and vector techniques [4,5,6] have been published. The first detects edges in each individual band, and then combines the results over all bands. The latter, which has been proposed recently, treats each pixel in a hyperspectral image as a vector in the spectral domain, then performs edge detection in this domain. This approach is more efficient than the first one since it does not suffer from the localization variability of edge detection result in the individual channel. Therefore, in the scope of this paper, we mainly focus on the vector based approach.

Zenzo [4] proposed a method to extend the edge detection for gray-scale images to multi-channel images. The main idea is to find the direction for a point x for which its vector in the spectral domain has the maximum rate of change.

A.-B. Salberg, J.Y. Hardeberg, and R. Jenssen (Eds.): SCIA 2009, LNCS 5575, pp. 580–587, 2009.

Therefore, the largest eigenvalue of the covariance matrix of the set of partial derivatives at a pixel is selected as its edge magnitude. A thresholding method can be applied to reveal the edges. However, this method is sensitive to small texture variations as gradient-based operators are sensitive even to small changes. Moreover, determining the scale for each channel is another problem since the derivatives taken for different channels are often scaled differently.

Inspired by the work of using morphological edge detectors for the edge detection in gray-scale images [7], Trahanias et al. [5] suggested vector-valued ranking operators to detect edges in color images. First, they divided the image into small windows. For each window, they ordered the vector-valued data of pixels belonging to this window in increasing order based on the R-ordering algorithm [8]. Then, the vector range (VR), which can be considered as the edge strength, of every pixel is calculated as the deviation of the vector outlier in the highest rank to the vector median in the window. Different from Trahanias et al.'s method, Evans et al. [6] defined the edge strength of a pixel as the maximum distance between any two pixels within the window. Therefore, it helps to localize edge locations more precisely. However, the disadvantage of this method is neighborhood pixels often have same edge strength values since the window's space to find the edge strength of the two pixels are highly overlapping. As a result, it may create multiple responds for a single edge and the method is sensitive to noise.

These three methods could also be classified as model based or non-statistical approaches as they are designed by assuming a model of edges. Typical model based method can be mentioned as Canny's method [9], in which edges are assumed to be step functions corrupted by additive Gaussian noise. This assumption is often wrong for natural images which have highly structured statistical properties [10,11,12]. For a hyperspectral dataset, the number of channels can be up to hundreds, while the number of pixels in each channel can be easily up to millions. Therefore, how to exploit statistical information in both spatial and spectral domains of hyperspectral images is a challenging issue. However, there have been not much works on hyperspectral edge detection cornering this issue until now. Initial work on statistical based approach for edge detection in color image can be mentioned as Huntsberger et al. [13]. They considered each pixel as a point in the feature space. A clustering algorithm is applied for a fuzzy segmentation of the image and then outliers of the clusters are considered as edges. However, this method performs image segmentation rather than edge detection and often produces discontinuous edges.

This paper proposes as an alternative a clustering based method for edge detection in hyperspectral images that could overcome the problem of Huntsberger et al.'s method. It is well-known that the pixel intensity is good for measuring the similarity among pixels, and therefore it is good for the purpose of image segmentation. But it is not good for measuring the abrupt changes to find the edges. The pixel gradient value is much more appropriate for that. Therefore, in our approach, we first consider each pixel as a point in the spectral space composed of gradient values in all image bands, instead of intensity values. Then, a clustering algorithm is applied in the spectral space to classify edge and non-edge pixels in the image. Finally, a thresholding strategy similar to Canny's method is used to refine the results.

The rest of this paper is organized as follows: Section 2 presents the proposed method for edge detection in hyperspectral images. To demonstrate its effectiveness, experimental results and comparisons with other typical methods are given in Section 3. In Section 4, some concluding remarks are drawn.

2 Clustering Based Edge Detection in Hyperspectral Images

First, the spatial derivatives of each channel in a hyperspectral images are determined. From [14,1], it is well-known that the use of fixed convolution masks of 3x3 size pixels is not suitable for the complex problem of determining discontinuities in image functions. Therefore, we use the 2-D Gaussian blur convolution to determine the partial derivatives. The advantage of using the Gaussian function is that we could reduce the effect of noise, which commonly occurs in hyperspectral images.

After the spatial derivatives of each channel are determined, gradient magnitudes of the pixels are calculated using the hypotenuse functions. Then each pixel can be considered as a point in the spectral space, which includes gradient magnitudes over all channels of the hyperspectral images. The problem of finding edges in the hyperspectral images could be considered as the same problem as classifying points in a spectral space into two classes: edge and non-edge points. We then use a clustering method based on the k-means algorithm for this classification purpose.

One important factor in designing the k-means algorithm is determining the number of clusters N. Formally, N should be two as we distinguish edges and non-edges. However, in fact, the number of non-edge pixels often dominates the pixel population (from 75% to 95%). Therefore, setting the number of clusters to two often results in losing edges since points in spectral space tend to belong to non-edge clusters rather than edge clusters. In practise, N should be set to be larger than two. In this case, the cluster with the highest population is considered as the non-edge cluster. The remaining $N-1$ clusters are merged together and considered as the edge cluster. In our experiments, the number of clusters N is set in the range of [4.0,8.0]. Experiments show that the edge detection results do not change much when N is in this range.

After applying the k-means algorithm to classify each point in spectral space into one of N clusters, a combined classifier method proposed by Paclik et al. [15] is applied to remove noise as well as isolated edges. The main idea of this method is to combine the results of two separate classifiers in spectral domain and spatial domain. This combining process is repeated until achieving stable results. In the proposed method, the results of two classifiers are combined using the maximum combination rule.

A thresholding algorithm as in the Canny edge detection method [9] is then applied to refine results from the clustering step, e.g. to make the edges thinner.

There are two different threshold values in the thresholding algorithm: a lower threshold and a higher threshold. Different from Canny's method, in which the threshold values are based on gradient intensity, the proposed threshold values are determined based on the confidence of a pixel belonging to the non-edge cluster. A pixel in the edge cluster is considered as a "true" edge pixel if its confidence to the non-edge cluster is smaller than the lower threshold. A pixel is also considered as an edge pixel if it satisfies two criteria: its confidence to the non-edge cluster is in a range between the two thresholds and it has a spatial connection with an already established edge pixel. The remaining pixels are considered as non-edge pixels. Confidence of a pixel belonging to a cluster used in this step is obtained from the clustering step.

The proposed algorithm is briefly described as followings:

Algorithm 1. Edge detection for hyperspectral images

Input: A hyperspectral image I, number of clusters N.
Output: Detected edges of the image as an image map.
Step 1:

- Smoothing the hyperspectral image using Gaussian blur convolution.
- Calculating pixel gradient values in each image channel.
- Forming pixel as a point composed of gradient values over all bands in a feature space. The number of dimensions in the feature space is equal to the number of bands in the hyperspectral images.

Step 2: Applying the k-means algorithm to classify points into N clusters.
Step 3: Refining the clustering result using the combined classifier method.
Step 4: Selecting the highest population cluster as non-edge cluster, merge other clusters as an edge cluster.
Step 5: Mapping the thresholding algorithm to refine results from Step 4.

3 Experimental Results

3.1 Datasets

Two typical hyperspectral datasets from [16] have been used for evaluating the performance of the proposed method. The first is a hyperspectral image of Washington DC Mall. The second is the "Flightline C1 (FLC1)" dataset taken from the southern part of Tippecanoe County, Indiana by an airborne scanner [16]. The properties of the two datasets are shown in the Table 1.

Since the spatial resolution of the two datasets is too large for handling it directly, we split the first dataset into 20 small parts of size 128*153 and carry out experiments with each of the small ones. Similarly, we split the second dataset into 3 small parts of size 316*220.

These two datasets are significantly diverse to evaluate the edge detector's performance. The first contains various types of regions, i.e. roofs, roads, paths,

Table 1. Properties of datasets used in experiments

Dataset	No. channels	Spatial Resolution	Response(μm)
DC Mall	191	1280*307	0.4-2.4
FLC1	12	949*220	0.4-1.0

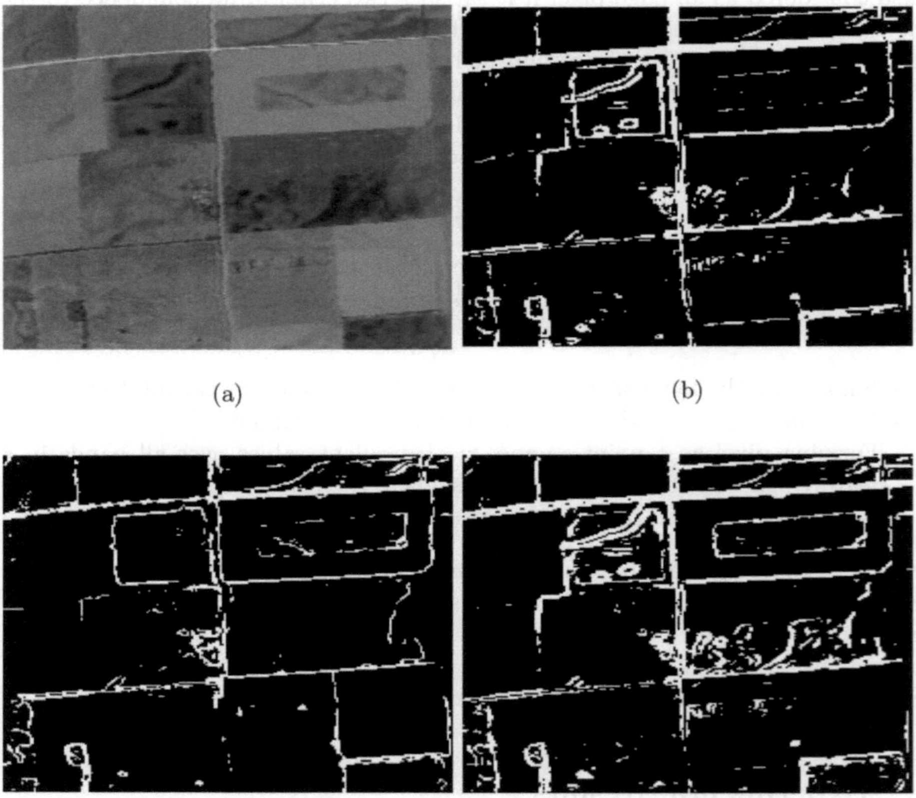

(a) (b) (c) (d)

Fig. 1. Edge detection results on FLC1 dataset: dataset represented using PCA (a); edge detection results from Zenzo's method (b), Huntsberger's method (c), and the proposed method (d)

trees, grass, water, and shadows and has a large number of channels, while the second contains much simpler scene and has a moderate number of channels.

To provide the intuitive representations of these datasets, PCA is used. For each dataset, the first three principle components extracted by PCA are used to compose a RGB image. The first, second, and the third most important component corresponds to the Red, Green, Blue channels, respectively. Color representation of the two dataset are shown in Fig. 1(a) and Fig. 2(a).

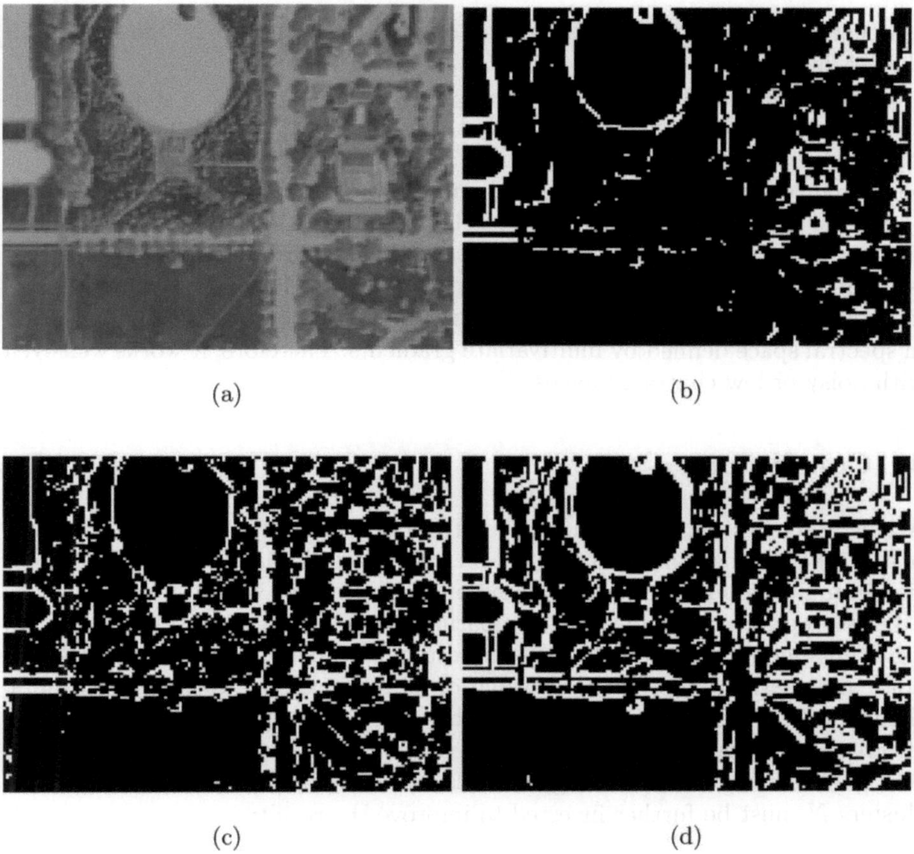

Fig. 2. Edge detection results on DC Mall dataset: dataset represented using PCA (a); edge detection results from Zenzo's method (b), Huntsberger's method (c), and the proposed method (d)

3.2 Results

In order to evaluate the effectiveness of the proposed method, we have compared it with two typical edge detection methods: Zenzo's method [4], a gradient based method, and a method presented by Huntsberger [13], an intensity clustering based method.

To provide a fair comparison, we carry out experiments with different parameter values for each edge detection method in both datasets and select the most suitable one. Moreover, we fix the parameter values for each method and use them for all datasets. For Zenzo's method, the threshold t was set to the value that satisfies the number of pixels of which gradient strengths larger that t is equal to 25% of the total number of pixels in spatial domain of the hyperspectral image. For Huntsberger 's method, the number of clusters is set to 5, and the confident value of pixels with respect to the background cluster is set to 0.55. For the proposed method, we apply Gaussian blur convolution for every channel of hyperspectral images with the standard deviation equal to 1. The number of clusters is set to 6.

Experimental results on the two datasets are shown in Fig. 1 and Fig. 2(b)-2(d). It can be seen from the figures that Huntsberger's method performs worst: losing edges and creating discontinuous edges. Therefore, we will focus on the performance between Zenzo's method and the proposed method.

For the first dataset, which contains simple images, the two methods produce similar results. But for the second dataset, which contains a complex image, it is clear that the proposed method can preserve more local edges than Zenzo's method. It is because the proposed method makes use of statistical information in spectral space defined by multivariate gradients. Therefore, it works well even with noisy or low contrast images.

4 Conclusions

A clustering based method for edge detection in hyperspectral images is proposed. The proposed method enables the use of multivariate statistical information in multi-dimensional space. Based on pixel gradient values, it also provides a better representation of edges comparing to those based on intensity values, e.g. Huntsberger's method [13]. As the results, the method reduces the effect of noise and preserves more edge information in the images. Experimental results, though still at preliminary work, show that the proposed method could be used effectively for edge detection in hyperspectral images. More thorough investigation in stabilizing the clustering methods and how to determine the number of clusters N must be further invested to improve the results.

Acknowledgements

The authors would like to thank Sergey Verzakov, Yan Li, and Marco Loog for their useful discussions. This research is supported by the CTR, Carinthian Tech Research AG, Austria, within the COMET funding programme.

References

1. Koschan, A., Abidi, M.: Detection and classification of edges in color images. Signal Processing Magazine, Special Issue on Color Image Processing 22, 67–73 (2005)
2. Robinson, G.: Color edge detection. Optical Engineering, 479–484 (1977)
3. Hedley, M., Yan, H.: Segmentation of color images using spatial and color space information. Journal of Electronic Imaging 1, 374–380 (1992)
4. Di Zenzo, S.: A note on the gradient of a multi-image. Computer Vision, Graphics, and Image Processing, 116–125 (1986)
5. Trahanias, P., Venetsanopoulos, A.: Color edge detection using vector statistics. IEEE Transactions on Image Processing 2, 259–264 (1993)
6. Evans, A., Liu, X.: A morphological gradient approach to color edge detection. IEEE Transactions on Image Processing 15(6), 1454–1463 (2006)

7. Haralick, R., Sternberg, S., Zhuang, X.: Image analysis using mathematical morphology. IEEE Transactions on Pattern Analysis and Machine Intelligence 9(4), 532–550 (1987)
8. Barnett, V.: The ordering of multivariate data. J. Royal Statist., 318–343 (1976)
9. Canny, J.: A computational approach to edge detection. IEEE Transactions on Pattern Analysis and Machine Intelligence, 679–698 (1986)
10. Field, D.: Relations between the statistics and natural images and the responses properties of cortical cells. Journal of Optical Society of America A(4), 2379–2394 (1987)
11. Zhu, S.C., Mumford, D.: Prior learning and gibbs reaction-diffusion. IEEE Transactions on Pattern Analysis and Machine Intelligence 19(11), 1236–1250 (1997)
12. Konishi, S., Yuille, A.L., Coughlan, J.M., Zhu, S.C.: Statistical edge detection: Learning and evaluating edge cues. IEEE Transactions on Pattern Analysis and Machine Intelligence 25(1), 57–74 (2003)
13. Huntsberger, T., Descalzi, M.: Color edge detection. Pattern Recognition Letter, 205–209 (1985)
14. Marr, D., Hildreth, E.: Theory of edge detection. Proceedings of Royal Society of London, 187–217 (1980)
15. Paclik, P., Duin, R.P.W., van Kempen, G.M.P., Kohlus, R.: Segmentation of multispectral images using the combined classifier approach. Journal of Image and Vision Computing 21, 473–482 (2005)
16. Landgrebe, D.: Signal theory methods in multispectral remote sensing. John Wiley and Sons, Chichester (2003)

Contrast Enhancing Colour to Grey

Ali Alsam

Sør-Trøndelag University College, Trondheim,
Norway

Abstract. A spatial algorithm to convert colour images to greyscale is presented. The method is very fast and results in increased local and global contrast. At each image pixel, three weights are calculated. These are defined as the difference between the blurred luminance image and the colour channels: red, green and blue. The higher the difference the more weight is given to that channel in the conversion. The method is multi-resolution and allows the user to enhance contrast at different scales. Results based on three colour images show that the method results in higher contrast than luminance and two spatial methods: Socolinsky and Wolff [1,2] and Alsam and Drew [3].

1 Introduction

Colour images contain information about the intensity, hue and saturation of the physical scenes that they represent. From this perspective, the conversion of colour images to black and white has long been defined as: The operation that maps RGB colour triplets to a space which represents the luminance in a colour-independent spatial direction. As a second step, the hue and saturation information are discarded, resulting in a single channel which contains the luminance information.

In the colour science literature, there are, however, many standard colour spaces that serve to separate luminance information from hue and saturation. Standard examples include: CIELab, HSV, LHS, YIQ etc. But the luminance obtained from each of these colour spaces is different.

Assuming the existence of a colour space that separates luminance information perfectly, we obtain a greyscale image that preserves the luminance information of the scene. Since this information has real physical meaning related to the intensity of the light signals reflected from the various surfaces, we can redefine the task of converting from colour to black and white as: An operation that aims at preserving the luminance of the scene.

In recent years, research in image processing has moved away from the idea of preserving the luminance of a single image pixel to methods that include spatial context, thus including simultaneous contrast effects. Including the spatial context means that we need to generate the intensity of an image pixel based on its neighbourhood. Further, for certain applications, preserving the luminance information per se might not result in the desired output. As an example, an equi-luminous image may easily have pixels with very different hue and saturation. However, equating grey with luminance results in a flat uniform grey. So we wish to retain colour regions while best preserving achromatic information.

A.-B. Salberg, J.Y. Hardeberg, and R. Jenssen (Eds.): SCIA 2009, LNCS 5575, pp. 588–596, 2009.

To proceed, we state that a more encompassing definition of colour to greyscale conversion is: An operation that reduces the number of channels from three to one while preserving certain, user defined, image attributes. As an example, Bala and Eschbach [4], introduced an algorithm to convert colour images to greyscale while preserving colour edges. Socolinsky and Wolff [1,2], developed a technique for multichannel image fusion with the aim of preserving *contrast*. More recently, Alsam and Drew [3] introduced the idea of defining contrast as the maximum change in any colour channel along the x and y directions. In general, we can state that the literature of spatial colour to grey is based on the idea of preserving the differences between colour and grey regions in the original image.

In this paper, a new approach to the problem of converting colour images to grey is taken. The approach is based on the photographic definition of what is an optimal, or beautiful, black and white image. During the preparation work for this article, I surveyed the views of many professional photographers. Their response was exclusively that a black and white image is aesthetically more beautiful than the colour original because it has higher global and local contrast. This view is supported in the vision science literature [5] where: It is well known that contrast between black and white is greater than that between red-green or blue-yellow. Based on this, in this paper, an optimal conversion from colour to black and white image is defined as an algorithm that converts colour values to grey while maximizing the local contrast. A new definition of contrast is presented and the conversion is performed to optimize it.

2 Background

As stated in the introduction, the best transformation from a multi-channel image to greyscale depends on the given definition. It is possible, however, to divide the solution domain into two groups. In the first, we have global projection based methods. In the second, we have spatial methods.

Global methods can further be divided into image independent and image dependent algorithms. Image independent algorithms, such as the calculation of luminance, assume that the transformation from colour to grey is related to the cone sensitives of the human eye. Based on that, the luminance approach is defined as a weighted sum of the red, green and blue values of the image without any measure of the image content. Further, the weights assigned to the red, green and blue channels are derived from vision studies where it is known that the eye is more sensitive to green than to red and blue.

To improve upon the performance of the image-independent averaging methods, we can incorporate statistical information about the image's colour, or multi-spectral, information. Principal component analysis (PCA) achieves this by considering the colour information as vectors in an n-dimensional space. The covariance matrix of all the colour values in the image, is analyzed using PCA and the principal vector with the largest principal value is used to project the image data onto the vector's one dimensional space [6]. Generally speaking, using PCA, more weight is given to channels with more intensity. It has, however, been shown that PCA shares a common problem with the global averaging techniques [2]: The contrast between adjacent pixels in the grey reproduction is always less than the original. This problem becomes more noticeable when the number of channels increases [2].

Spatial methods are based on the assumption that the transformation from colour to greyscale needs to be defined such that differences between pixels are preserved. Bala and Eschbach [4], introduced a two step algorithm. In the first step the luminance image is calculated based on a global projection. In the second, the chrominance edges that are not present in the luminance are added to the luminance. Similarly, Grundland and Dodgson [7], introduced an algorithm that starts by transforming the image to *YIQ* colour space. The Y-channel is assumed to be the luminance of the image and treated separately from the the chrominance *IQ* plane. Based on the chrominance information in the *IQ* plane, they calculate a single vector: The predominant chromatic change vector. The final greyscale image is defined as a weighted sum of the luminance Y and the projection of the 2-dimensional *IQ* onto the predominant vector.

Socolinsky and Wolff [1,2], developed a technique for multichannel image fusion with the aim of preserving *contrast*. In their work, these authors use the Di Zenzo structure-tensor matrix [8] to represent contrast in a multiband image. The interesting idea added to [8] was to suggest re-integrating the gradient produced in Di Zenzo's approach into a single, representative, grey channel encapsulating the notion of contrast. Connah et al. [9] compared six algorithms for converting colour images to greyscale. Their findings indicate that the algorithm presented by Socolinsky and Wolff [1,2] results in visually preferred rendering.

The Di Zenzo matrix allows us to represent contrast at each image pixel by utilising a 2×2 symmetric matrix whose elements are calculated based on the derivatives of the colour channels in the horizontal and vertical directions. Socolinsky and Wolff defined the maximum absolute colour contrast to be the square root of the maximum eigenvalue of the Di Zenzo matrix along the direction of the associated eigenvector. In [1], Socolinsky and Wolff noted that the key difference between contrast in the greyscale case and that in a multiband image is that, in the latter, there is no preferred *orientation* along the maximum contrast direction. In other words, contrast is defined along a line, not a vector.

To resolve the resulting sign ambiguity, Alsam and Drew [3] introduced the idea of defining contrast as the maximum change in any colour channel along the x and y directions. Using the maximum change resolves the sign ambiguity and results in a very fast algorithm that was shown to produce better results than those achieved by Socolinsky and Wolff [1,2].

3 Contrast Enhancing

RGB colour images are commonly converted to greyscale using a weighted sum of the form:

$$Gr(x,y) = \alpha R(x,y) + \beta G(x,y) + \gamma B(x,y) \tag{1}$$

where α, β and γ are positive scalars that sum to one.

At the very heart of the algorithm presented in this article is the question: Which local weights $\alpha(x,y)$, $\beta(x,y)$ and $\gamma(x,y)$ would result in maximizing the contrast of the greyscale image pixel Gr(x,y)? To answer this question we need to first define contrast.

In the image processing literature, contrast, for a single channel, is define as the deviation from the mean of an $n \times n$ neighborhood. As an example the contrast at the red pixel $R(x, y)$ is:

$$C_r(x, y) = R(x, y) - \sum_{i=1}^{n} \sum_{j=1}^{n} \lambda(i, j) R(i, j) \tag{2}$$

where $\lambda(i, j)$ are the weights assigned to each image pixel. We note that contrast as defined in (2) represents the high frequency elements of the red channel.

The main contribution of this paper is to define contrast enhancing weights based on the original colour image and a greyscale version calculated as a weighted sum. The author's argument is as follows: The greyscale scale image defined in Equation (1), is a weighted average of all three colour values, red, green and blue at pixel (x, y). To arrive at a similar formulation as in Equation (2), we calculate the difference between red, green and blue at pixel (x, y) and the average of an $n \times n$ neighborhood calculated based on the greyscale image Gr, i.e.:

$$C_{rg}(x, y) = |R(x, y) - \sum_{i=1}^{n} \sum_{j=1}^{n} \lambda(i, j) Gr(i, j)| + \kappa \tag{3}$$

$$C_{gg}(x, y) = |G(x, y) - \sum_{i=1}^{n} \sum_{j=1}^{n} \lambda(i, j) Gr(i, j)| + \kappa \tag{4}$$

$$C_{bg}(x, y) = |B(x, y) - \sum_{i=1}^{n} \sum_{j=1}^{n} \lambda(i, j) Gr(i, j)| + \kappa \tag{5}$$

where κ is a small positive scalar used to avoid division with zero. The scalar κ can also be used as a regularization factor where to larger the value the more the closer the resultant weights $C_{rg}(x, y)$, $C_{gg}(x, y)$ and $C_{bg}(x, y)$ are to each other. The weights, $C_{rg}(x, y)$, $C_{gg}(x, y)$ and $C_{bg}(x, y)$ represent the level of high frequency, based on the individual channels, lost when converting an RGB colour image to grey. Thus, if we use those weights to convert the colour image to black and white we get a greyscale representation that gives more weight to the channel that loses most information in the conversion. In other words: The greyscale value $Gr(x, y)$ is the average of the three channels and the weights $C_{rg}(x, y)$, $C_{gg}(x, y)$ and $C_{bg}(x, y)$ are the spatial difference from the average. Using those would, thus, increase the contrast of $Gr(x, y)$. The formulation given in Equations (3), (4), (5), however, suffers from a main drawback: For a flat region, one with a single colour, the weights , $C_{rg}(x, y)$, $C_{gg}(x, y)$ and $C_{bg}(x, y)$ will not have a spatial meaning. Said differently, contrast at a single pixel or a region with no colour change is not defined. To resolve this problem we modify the weights $C_{rg}(x, y)$, $C_{gg}(x, y)$ and $C_{bg}(x, y)$:

$$C_{Rg}(x, y) = |D(x, y) \times (R(x, y) - \sum_{i=1}^{n} \sum_{j=1}^{n} \lambda(i, j) Gr(i, j))| + \kappa \tag{6}$$

$$C_{Gg}(x,y) = |D(x,y) \times (G(x,y) - \sum_{i=1}^{n}\sum_{j=1}^{n}\lambda(i,j)Gr(i,j))| + \kappa \quad (7)$$

$$C_{Bg}(x,y) = |D(x,y) \times (B(x,y) - \sum_{i=1}^{n}\sum_{j=1}^{n}\lambda(i,j)Gr(i,j))| + \kappa \quad (8)$$

where the spatial weights $D(x,y)$ are defined as:

$$\begin{aligned} D(x,y) = \; & \|R(x,y) - \textstyle\sum_{i=1}^{n}\sum_{j=1}^{n}\lambda(i,j)R(i,j)\| \\ & + \|G(x,y) - \textstyle\sum_{i=1}^{n}\sum_{j=1}^{n}\lambda(i,j)G(i,j)\| \\ & + \|B(x,y) - \textstyle\sum_{i=1}^{n}\sum_{j=1}^{n}\lambda(i,j)B(i,j)\| \end{aligned} \quad (9)$$

Introducing the difference $D(x,y)$ into the calculation of the weights $C_{Rg}(x,y)$, $C_{Gg}(x,y)$ and $C_{Bg}(x,y)$ means that contrast is only enhanced at regions with colour transition.

Finally, based on $C_{Rg}(x,y)$, $C_{Gg}(x,y)$ and $C_{Bg}(x,y)$ we define the weights: $\alpha(x,y)$, $\beta(x,y)$ and $\gamma(x,y)$ as:

$$\alpha(x,y) = \frac{C_{Rg}(x,y)}{C_{Rg}(x,y) + C_{Gg}(x,y) + C_{Bg}(x,y)} \quad (10)$$

$$\gamma(x,y) = \frac{C_{Gg}(x,y)}{C_{Rg}(x,y) + C_{Gg}(x,y) + C_{Bg}(x,y)} \quad (11)$$

$$\beta(x,y) = \frac{C_{Bg}(x,y)}{C_{Rg}(x,y) + C_{Gg}(x,y) + C_{Bg}(x,y)} \quad (12)$$

For completeness, we modify the conversion given in Equation (1) from colour to grey:

$$Gr(x,y) = \alpha(x,y)R(x,y) + \beta(x,y)G(x,y) + \gamma(x,y)B(x,y) \quad (13)$$

4 Experiments

Figure 1, London photo, shows a colour image with the luminance rendering to its right. In the second, third, fourth and fifth rows the difference maps defined in Equation (9) are shown in the first column and the results achieved with the present method in the second. These results are achieved by blurring the luminance image by: 5×5, 10×10, 15×15 and 25×25 Gaussian kernels respectively. As seen, the contrast increases with the increasing size of the kernel.

In Figure 2, two women, the same layout as in Figure 1 is used. Again, we notice that the contrast increases with the increasing size of the kernel. We note, however, that finer details are better preserved at lower scales. This suggests that the method can be used to combine results at different scales. The best way to combine different scales is, however, left as future work.

In Figure 3, daughter and father, the colour original is shown at the top left corner and the luminance rendition is shown at the top right corner. In the

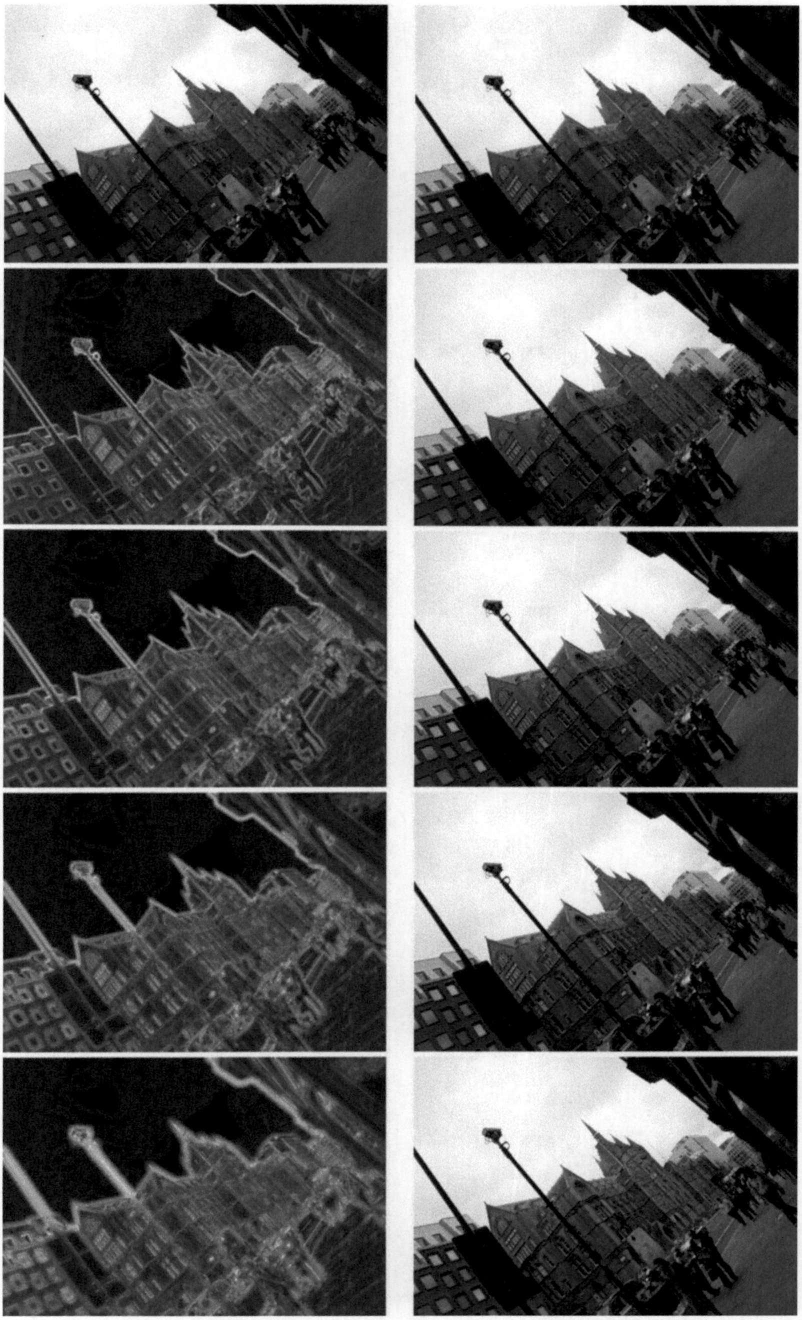

Fig. 1. London photo: top row a colour image with the luminance rendering to its right. In the second, third, fourth and fifth rows the difference maps defined in Equation (9) are shown in the first column and the results achieved with the present method in the second. These results are achieved by blurring the luminance image by: 5 × 5, 10 × 10, 15 × 15 and 25 × 25 Gaussian kernels respectively.

Fig. 2. Two women: top row a colour image with the luminance rendering to its right. In the second, third, fourth and fifth rows the difference maps defined in Equation (9) are shown in the first column and the results achieved with the present method in the second. These results are achieved by blurring the luminance image by: 5×5, 10×10, 15×15 and 25×25 Gaussian kernels respectively.

Fig. 3. Daughter and father: top row a colour image with the luminance rendering to its right. In the second row, the results obtained by Socolinsky and Wolff are shown in the first column and those achieved by Alsam and Drew are shown in the second column. The results obtained with the present method based on a 5 × 5 and 15 × 15 Gaussian kernels are shown in the first and second columns, the third row, respectively.

second row, the results obtained by Socolinsky and Wolff [1,2] are shown to the left and those achieved by Alsam and Drew [3] to the right. In the third row the present method is shown with a blurring of 5 × 5 to the left and 15 × 15 to the right. We note that the present method achieves the highest contrast out of all other methods.

5 Conclusions

Starting with the idea that a black and white image can be optimized to have higher contrast than the colour original, a spatial contrast-enhancing algorithm to convert colour images to greyscale was presented. At each image pixel, three spatial weights are calculated. These are derived to increase the difference between the resulting greyscale value and the mean of the luminance at the given image pixel. Results based on general photographs show that the method results in visually preferred rendering. Given that contrast is defined at different spatial scales, the method can be used to combine contrast in a pyramidal fashion.

References

1. Socolinsky, D.A., Wolff, L.B.: A new visualization paradigm for multispectral imagery and data fusion. In: CVPR, pp. I:319–324 (1999)
2. Socolinsky, D.A., Wolff, L.B.: Multispectral image visualization through first-order fusion. IEEE Trans. Im. Proc. 11, 923–931 (2002)
3. Alsam, A., Drew, M.S.: Fastcolour2grey. In: 16th Color Imaging Conference: Color, Science, Systems and Applications, Society for Imaging Science & Technology (IS&T)/Society for Information Display (SID) joint conference, Portland, Oregon, pp. 342–346 (2008)
4. Bala, R., Eschbach, R.: Spatial color-to-grayscale transform preserving chrominance edge information. In: 14th Color Imaging Conference: Color, Science, Systems and Applications, pp. 82–86 (2004)
5. Hunt, R.W.G.: The Reproduction of Colour, 5th edn. Fountain Press, England (1995)
6. Lillesand, T.M., Kiefer, R.W.: Remote Sensing and Image Interpretation, 2nd edn. Wiley, New York (1994)
7. Grundland, M., Dodgson, N.A.: Decolorize: Fast, contrast enhancing, color to grayscale conversion. Pattern Recognition 40(11), 2891–2896 (2007)
8. Di Zenzo, S.: A note on the gradient of a multi-image. Comp. Vision, Graphics, and Image Proc. 33, 116–125 (1986)
9. Connah, D., Finlayson, G.D., Bloj, M.: Seeing beyond luminance: A psychophysical comparison of techniques for converting colour images to greyscale. In: 15th Color Imaging Conference: Color, Science, Systems and Applications, pp. 336–341 (2007)

On the Use of Gaze Information and Saliency Maps for Measuring Perceptual Contrast

Gabriele Simone, Marius Pedersen, Jon Yngve Hardeberg, and Ivar Farup

Gjøvik University College, Gjøvik, Norway

Abstract. In this paper, we propose and discuss a novel approach for measuring perceived contrast. The proposed method comes from the modification of previous algorithms with a different local measure of contrast and with a parameterized way to recombine local contrast maps and color channels. We propose the idea of recombining the local contrast maps using gaze information, saliency maps and a gaze-attentive fixation finding engine as weighting parameters giving attention to regions that observers stare at, finding them important. Our experimental results show that contrast measures cannot be improved using different weighting maps as contrast is an intrinsic factor and it's judged by the global impression of the image.

1 Introduction

Contrast is a difficult and not very well defined concept. A possible definition of contrast is the difference between the light and dark parts of a photograph, where less contrast gives a flatter picture, and more a deeper picture. Many other definitions of contrast are also given, it could be the difference in visual properties that makes an object distinguishable or just the difference in color from point to point. As various definitions of contrast are given, measuring contrast is very difficult. Measuring the difference between the darkest and lightest point in an image does not predict perceived contrast since perceived contrast is influenced by the surround and the spatial arrangement of the image. Parameters such as resolution, viewing distance, lighting conditions, image content, memory color etc. will effect how observers perceive contrast.

First, we briefly introduce some of the contrast measures present in literature. However none of these take the visual content into account. Therefore we propose the use of gaze information and saliency maps to improve the contrast measure. A psychophysical experiment and statistical analysis are reported.

2 Background

The very first measure of global contrast, in the case of sinusoids or other periodic patterns of symmetrical deviations ranging from the maximum luminance (L_{max}) to minimum luminance (L_{min}), is the Michelson [1] formula proposed in 1927: $C^M = \frac{L_{max}-L_{min}}{L_{max}+L_{min}}$ King-Smith and Kulikowski [2] (1975), Burkhardt [3] (1984) and Whittle [4] (1986) follow a similar concept replacing L_{max} or L_{min} with L_{avg}, which is the mean luminance in the image.

A.-B. Salberg, J.Y. Hardeberg, and R. Jenssen (Eds.): SCIA 2009, LNCS 5575, pp. 597–606, 2009.

These definitions are not suitable for natural images since one or two points of extreme brightness or darkness can determine the contrast of the whole image, resulting in high contrast while perceived contrast is low. To overcome to this problem, local measures which take account of neighboring pixels, have been developed later.

Tadmor and Tolhurst [5] proposed in 1998 a measure based on the Difference Of Gaussian (D.O.G.) model. They propose the following criteria to measure the contrast in a *pixel (x,y)*, where *x* indicates the row and *y* the column:

$$c^{DOG}(x,y) = \frac{R_c(x,y) - R_s(x,y)}{R_c(x,y) + R_s(x,y)},$$

where R_c is the output of the so called central component and R_s is the output of the so called surround component. The central and surround components are calculated as:

$$R_c(x,y) = \sum_i \sum_j Centre(i-x, j-y) I(i,j),$$

$$R_s(x,y) = \sum_i \sum_j Surround(i-x, j-y) I(i,j),$$

where *I(i,j)* is image pixel at position *(i,j)*, while *Centre(x,y)* and *Surround(x,y)* are described by bi-dimensional Gaussian functions:

$$Centre(x,y) = \exp\left[-\left(\frac{x}{r_c}\right)^2 - \left(\frac{y}{r_c}\right)^2\right],$$

$$Surround(x,y) = 0.85\left(\frac{r_c}{r_s}\right)^2 \exp\left[-\left(\frac{x}{r_s}\right)^2 - \left(\frac{y}{r_s}\right)^2\right],$$

where r_c and r_s are their respective radiuses, parameters of this measure. In their experiments, using 256x256 images, the overall image contrast is calculated as the average local contrast of 1000 pixel locations taken randomly.

In 2004 Rizzi et al. [6] proposed a contrast measure, referred here as RAMMG, working with the following steps:

- It performs a pyramid subsampling of the image to various levels in the CIELAB color space.
- For each level, it calculates the local contrast in each pixel by taking the average of absolute value difference between the lightness channel value of the pixel and the surrounding eight pixels, thus obtaining a contrast map of each level.
- The final overall measure is a recombination of the average contrast for each level: $C^{RAMMG} = \frac{1}{N_l}\sum_l^{N_l}\overline{c}_l$, where N_l is the number of levels and $\overline{c}_l$ is the mean contrast in the level *l*.

In 2008 Rizzi et al. [7] proposed a new contrast measure, referred here as RSC, based on the previous one from 2004 [6] . It works with the same pyramid subsampling as Rizzi et al. but:

- It computes in each pixel of each level the DOG contrast instead of the simple 8-neighborhood local contrast.
- It computes the DOG contrast separately for the lightness and the chromatic channels, instead of only for the lightness; the three measures are then combined with different weights.

The final overall measure can be expressed by the formula:

$$C^{RSC} = \alpha \cdot C_{L*}^{RSC} + \beta \cdot C_{a*}^{RSC} + \gamma \cdot C_{b*}^{RSC},$$

where α, β and γ represent the weighting of each channel.

Pedersen et al. [8] evaluated five different contrast measures in relation to observers perceived contrast. The results indicate room for improvement for all contrast measures, and the authors proposed using region-of-interest as one possible way for improving contrast measures, as we will do in this paper.

In 2009 Simone et al. [9] analyzed in details the previous measures proposed by Rizzi et al. [6,7] and they developed a framework for measuring perceptual contrast that takes account lightness, chroma information and weighted pyramid levels. The overall final measure of contrast is given by equation: $C^{MLF} = \alpha \cdot C_1 + \beta \cdot C_2 + \gamma \cdot C_3$, where α, β and γ are the weights of each color channel.

The overall contrast in each channel is defined as follows: $C_i = \frac{1}{N_l} \sum_l^{N_l} \lambda_l \cdot \overline{c}_l$, where N_l is the number of levels, $\overline{c}_l$ is the mean contrast in the level l, λ_l is the weight assigned to each level l, and i indicates the applied channel.

In this framework α, β, γ, and λ can assume values from particular measures taken from the image itself as for example the variance of pixel values in each channel separately. In this framework RAMMG and RSC previously developed can be considered just special cases with uniform weighting of levels and uniform weighting of channels.

Eye tracking has been used in a number of different color imaging research projects with great success, allowing researchers to obtain information on where observers gaze. Babcock et al. [10] examined differences between rank order, paired comparison, and graphical rating tasks by using an eye tracker. The results showed a high correlation of the spatial distributions of fixations across the three tasks. Peak areas of attention gravitated toward semantic features and faces. Bai et al. [11] evaluated S-CIELAB, an image difference metric, on images produced by the Retinex method by using gaze information. The authors concluded that the frequency distribution of gazing area in the image gives important information on the evaluation of image quality. Pedersen et al. [12] used a similar approach to improve image difference metrics.

Endo et al. [13] showed that individual distribution of gazing points were very similar among observers for the same scenes. The results also indicate that each image has a particular gazing area, particularly images containing human faces.

While Mackworth and Morandi [14] found that a few regions in the image dominated the data. Informative areas had a tendency to receive clusters of fixations. Half to two-thirds of the image receive few or no fixations, these areas (for example texture) were predictable, containing common objects and not very informative. While more recent research by Underwood and Foulsham [15] found that highly salient objects attracted fixations earlier than less conspicuous objects. Walther and Koch [16] introduced a model for computing salient objects, which Sharma et al. [17] modified to account for a high level feature, human faces .While Rajashekar et al. [18] proposed a gaze-attentive fixation finding engine (GAFFE) that uses a bottom-up model for fixation selection in natural scenes. Testing showed that GAFFE correlated well with observers, and could be used to replace eye tracking experiments.

Assuming that the whole image is not weighted equally when we rate contrast, some areas will be more important than other. Because of this we propose to use region-of-interest to improve contrast measures.

3 Experiment Setup

In order to investigate perceived contrast a psychophysical experiment with 15 different images (Figure 1) was set up asking observers to judge perceptual contrast in images while recording their eye movements.

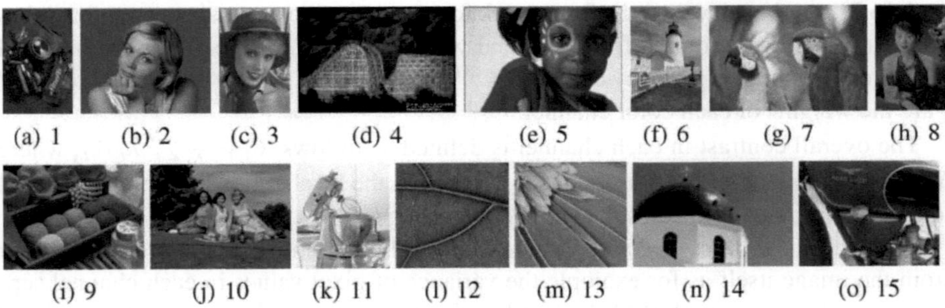

Fig. 1. Images 1 to 15 were used in the experiment, each representing different characteristics. The dataset is similar to the one used by Pedersen et al. [8]. Images 1 and 2 provided by Ole Jakob Bøe Skattum, image 10 is provided by CIE, images 8 and 9 from ISO 12640-2 standard, images 3, 5, 6 and 7 from Kodak PhotoCD, images 4, 11, 12, 13, 14 and 15 from ECI Visual Print Reference.

17 observers were asked to rate the contrast in the 15 images. Nine of the observers were considered experts, i.e. had experience in color science, image processing, photography or similar, and eight were considered non-experts with none or little experience in these fields. Observers rated contrast on a scale from 1 to 100, where 1 was the lowest contrast and 100 maximum contrast. Each image was shown for 40 seconds with the rest of the screen black, and the observers stated the perceived contrast within this time limit. The experiment was carried out on a calibrated CRT monitor, LaCIE electron 22 blue II, in a gray room with the observers seated approximately 80 cm from the screen. The lights were dimmed and measured to approximately 17 lux. During the experiment the observer's gaze position was recorded using a SMI iView X RED, a contact free gaze measurement device. The eye tracker was calibrated in nine points for each observer before commencing the experiment.

4 Weighting Maps

Previous studies have shown that there is still room for improvement for contrast measures [8,7]. We propose to use gaze information, saliency maps and a gaze-attentive fixation finding engine to improve contrast measure. Regions that draw attention should be weighted higher than regions that observers do not look at or pay attention to.

4.1 Gaze Information Retrieval

Gaze information have been used by researches to improve image quality metrics, the region-of-interest have been used as a weighting map for the metrics. We use a similar approach, and apply gaze information as a weighting map for the contrast measures. From the eye tracking data a number of different maps have been calculated, among them time used at one pixel multiplied with the number of times the observer fixated on this pixel, the number of fixations at the same pixel, mean time at each pixel and time. All of these have been normalized by the maximum value in the map, and a Gaussian filter corresponding to the 2-degree visual field of the human eye was applied to the map to even out differences [11] and to simulate that we look at an area rather than one particular pixel [19].

4.2 Saliency Map

Gathering gaze information is time consuming, and because of this we have investigated other ways to obtain similar information. One possibility is saliency maps, which is a map that represents visual saliency of a corresponding visual scene. One proposed model was introduced by Walther and Koch [16] for bottom-up attention to salient objects, and this has been adopted for the saliency maps used in this study. The saliency map has been computed at level one (i.e. the size of the saliency map is equal to original images) and seven fixations (i.e. giving the seven most salient regions in the image), for the other parameters standard values in the SaliencyToolbox [16] have been used.

4.3 A Gaze-Attentive Fixation Finding Engine

Rajashekar et al. [18] proposed "gaze-attentive fixation finding engine" (GAFFE) based on statistical analysis of image features for fixation selection in natural scenes. The GAFFE uses four foveated low-level image features: luminance, contrast, luminance-bandpass and contrast-bandpass to compute the simulated fixations of a human observer. The GAFFE maps have been computed for 10, 15 and 20 fixations, where the first fixation has been removed since this always is placed in the center resulting in a total of 9, 14 and 19 fixations. A Gaussian filter corresponding to the 2-degree visual field of the human eye was applied to simulate that we look at an area rather than at one single point and a larger filter (approximately 7-degree visual field) was also tested.

5 Results

This section analyzes the results of the gaze maps, saliency maps and GAFFE maps when applied to contrast measures.

5.1 Perceived Contrast

The perceived contrast for the 15 images (Figure 1) from 17 observers were gathered. After investigation of the results we found that the data cannot be assumed to be normally distributed, and therefore a special care must be given to the statistical analysis. One common method for statistical analysis is the Z-score [20], this require the data to be normally distributed, and in this case this analysis will not give valid results. Just using the mean opinion score will also result in problems, since the dataset cannot be assumed to be normally distributed. Because of this we use the rank from each observer to carry out a Wilcoxon signed rank test, a non-parametric statistical hypothesis test. This test does not make any assumption on the distribution, and it's therefore an appropriate statistical tool for analyzing this data set.

The 15 images have been grouped into three groups based on the Wilcoxon signed rank test: high, medium and low contrast. From the signed rank test observers can differentiate between the images with high and low contrast, but not between high/low and medium contrast. Images 5, 9 and 15 have high contrast while images 4, 6, 8 and 13 have low contrast. This is further used to analyze the performance of the different contrast measures and weighting maps.

5.2 Contrast Algorithm

The contrast measures used are the ones proposed by Rizzi et al [6,7]. RAMMG and RSC. Both measures were used in their extended form in the framework, explained above, developed by Simone et al. [9] with particular measures taken from the image itself as weighting parameters. The most important issues are:

- The overall measure of each channel is a weighted recombination of the average contrast for each level.
- The final measure of contrast is defined by a weighted sum of the overall contrast of the three channels.

In this new approach each contrast map of each level is weighted pixelwise with its relative gaze information or saliency map or gaze-attentive fixation finding engine (Figure 2).

We have tested many different weighting maps, and due to page limitations we cannot show all results. We will show results for fixations only, fixations multiplied with time, saliency, 10 fixation GAFFE map (GAFFE10), 20 fixations big Gaussian GAFFE

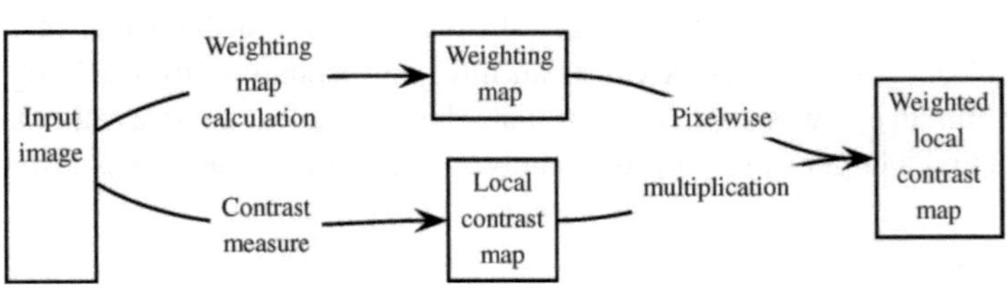

Fig. 2. Framework for using weighting maps with contrast measures. As weighting maps we have used gaze maps, saliency maps and GAFFE maps.

Table 1. Resulting p values for RAMMG maps. We can see that the different weighting maps have the same performance as no map at a 5% significance level, indicating that weighting RAMMG with maps does not improve predicted contrast.

Map	fixation only	fixation × time	saliency	GAFFE10	GAFFEBG20	no map
fixation only	1.000	1.000	0.625	0.250	0.125	0.500
fixation × time	1.000	1.000	1.000	0.250	0.375	0.500
saliency	0.625	1.000	1.000	0.250	1.000	0.625
GAFFE10	0.250	0.250	0.250	1.000	0.063	1.000
GAFFEBG20	0.125	0.375	1.000	0.063	1.000	0.063
no map	0.500	0.500	0.625	1.000	0.063	1.000

Table 2. Resulting p values for RSC maps. None of the weighting maps are significantly different from no map, indicating that they have the same performance at a 5% significance level. There is a difference between salicency maps and gaze maps (fixation only and fixation × time), but since these are not significantly different from no map they do not increase the contrast measure's ability to predict perceived contrast. Gray cells indicate significant difference at a 5% significance level.

Map	fixation only	fixation × time	saliency	GAFFE10	GAFFEBG20	no map
fixation only	1.000	1.000	0.016	0.289	0.227	0.500
fixation × time	1.000	1.000	0.031	0.508	0.227	1.000
saliency	0.016	0.031	1.000	1.000	0.727	0.125
GAFFE10	0.289	0.508	1.000	1.000	0.688	0.727
GAFFEBG20	0.227	0.227	0.727	0.688	1.000	0.344
no map	0.500	1.000	0.125	0.727	0.344	1.000

map (GAFFEBG20) and no map. The maps that were excluded are time only, mean time, 15 fixation GAFFE map, 20 fixations GAFFE map, 10 fixations big Gaussian GAFFE map, 15 fixations big Gaussian GAFFE map, and 6 combinations of gaze maps and GAFFE maps. All of these maps that have been excluded show no significant difference from no map, or have a lower performance than no map.

In order to test the performance of the contrast measures with different weighting maps and parameters, an extensive statistical analysis has been carried out. First, the images have been divided into two groups: "high contrast" and "low contrast" based on the user rating. Only the images having a statistically significant difference in user rated contrast were taken into account. The two groups have gone through the Wilcoxon rank sum test for each set of parameters of the algorithms. The obtained p values from this test rejected the null hypothesis that the two groups are the same, therefore indicating that the contrast measures are able to differentiate between the two groups of images with perceived low and high contrast. Thereafter these p values have been used for a sign test to compare each map against each other for all parameters and each set of parameters against each other for all maps. The results from this analysis indicate whether using a weighting map is significantly different from using no map, or if a parameter is significantly different from other parameters. In case of a significant difference further analysis is carried out to indicate whether the performance is better or worse for the tested weighting map or parameter.

5.3 Discussion

As we can see from Table 1 and Table 2, using maps is not significantly different from not using them as they have the same performance at a 5% significance level. We can

Table 3. Resulting *p* values for RAMMG parameters. Gray cells indicate significant difference at a 5% significance level. RAMMG parameters are the following: color space (CIELAB or RGB), pyramid weight, and the three last parameters are channel weights. "var" indicates the variance.

Parameters	LAB-1-1-0-0	LAB-1-0.33-0.33-0.33	RGB-4-var1-var2-var3	LAB-4-0.33-0.33-0.33	LAB-4-0.5-0.25-0.25	LAB-4-var1-var2-var3
LAB-1-1-0-0	1.000	0.092	0.000	0.002	0.000	0.000
LAB-1-0.33-0.33-0.33	0.092	1.000	0.012	0.012	0.001	0.001
RGB-4-var1-var2-var3	0.000	0.012	1.000	1.000	0.500	0.500
LAB-4-0.33-0.33-0.33	0.002	0.012	1.000	1.000	1.000	1.000
LAB-4-0.5-0.25-0.25	0.000	0.001	0.500	1.000	1.000	1.000
LAB-4-var1-var2-var3	0.000	0.001	0.500	1.000	1.000	1.000

Table 4. Resulting *p* values for RSC parameters. Gray cells indicate significant difference at a 5% significance level. RSC parameters are the following: color space (CIELAB or RGB), radius of the centre Gaussian, radius of the surround Gaussian, pyramid weight, and the three last parameters are channel weights. "m" indicates the mean.

Parameters	LAB-1-2-1-0.33-0.33-0.33	LAB-1-2-1-0.5-0.25-0.25	LAB-1-2-1-1-0-0	RGB-1-2-4-0.33-0.33-0.33	RGB-2-4-4-m1-m2-m3	RGB-2-3-4-m1-m2-m3	LAB-2-3-4-0.5-0.25-0.25
LAB-1-2-1-0.33-0.33-0.33	1.000	1.000	0.000	0.454	0.000	0.000	0.289
LAB-1-2-1-0.5-0.25-0.25	1.000	1.000	0.000	0.454	0.000	0.000	0.289
LAB-1-2-1-1-0-0	0.000	0.000	1.000	0.000	0.581	0.774	0.000
RGB-1-2-4-0.33-0.33-0.33	0.454	0.454	0.000	1.000	0.000	0.000	0.004
RGB-2-4-4-m1-m2-m3	0.000	0.000	0.581	0.000	1.000	0.219	0.000
RGB-2-3-4-m1-m2-m3	0.000	0.000	0.774	0.000	0.219	1.000	0.000
LAB-2-3-4-0.5-0.25-0.25	0.289	0.289	0.000	0.004	0.000	0.000	1.000

see only a difference between salicency maps and gaze maps (fixation only and fixation × time), but since these are not significantly different from no map they do not increase the ability of the contrast measures to predict perceived contrast. The contrast measures with the use of maps have been tested in the framework developed by Simone et al. [9] with different settings shown in Table 3 and Table 4. For RAMMG the standard parameters (LAB-1-1-0-0-0 and LAB-1-0.33-0.33-0.33) perform significantly worse than the other parameters in the table. For RSC we noticed that three parameters are significantly different from the standard parameters (LAB-1-2-1-0.33-0.33-0.33 and LAB-1-2-1-0.5-0.25-0.25) but after further analysis of the underlying data these ones perform worse than the standard parameters.

(a) Original (b) Relative local contrast map (c) Saliency weighted local contrast map

Fig. 3. The original, the relative local contrast map and saliency weighted local contrast map

We can see from Figure 3 that using a saliency map for weighting discards relevant information used by the observer to judge perceived contrast since contrast is a complex feature and it is judged by the global impression of the image.

5.4 Validation

In order to validate the results with other dataset we have carried out the same analysis for 25 images, each with four contrast levels, from the TID2008 database [21]. The score from the two contrast measure have been computed for all 100 images and a similar statistical analysis is carried out as above but for four groups (very low contrast, low, high and very high contrast). The results from this analysis supports the findings from the first dataset, where using weighting maps did not improve the performance of the contrast measures.

6 Conclusion

The results in this paper shows that weighting maps, from gaze information, saliency maps or GAFFE maps does not improve contrast measures to predict perceived contrast in digital images. This suggests that region-of-interest cannot be used to improve contrast measures as contrast is an intrinsic factor and it's judged by global impression of the image. This indicates that further work on contrast measures should be carried out accounting for the global impression of the image while preserving the local information.

References

1. Michelson, A.: Studies in Optics. University of Chicago Press (1927)
2. King-Smith, P.E., Kulikowski, J.J.: Pattern and flicker detection analysed by subthreshold summation. J. Physiol. 249(3), 519–548 (1975)
3. Burkhardt, D.A., Gottesman, J., Kersten, D., Legge, G.E.: Symmetry and constancy in the perception of negative and positive luminance contrast. J. Opt. Soc. Am. A 1(3), 309 (1984)
4. Whittle, P.: Increments and decrements: luminance discrimination. Vision Research (26), 1677–1691 (1986)
5. Tadmor, Y., Tolhurst, D.: Calculating the contrasts that retinal ganglion cells and lgn neurones encounter in natural scenes. Vision Research 40, 3145–3157 (2000)
6. Rizzi, A., Algeri, T., Medeghini, G., Marini, D.: A proposal for contrast measure in digital images. In: CGIV 2004 – Second European Conference on Color in Graphics, Imaging and Vision (2004)
7. Rizzi, A., Simone, G., Cordone, R.: A modified algorithm for perceived contrast in digital images. In: CGIV 2008 - Fourth European Conference on Color in Graphics, Imaging and Vision, Terrassa, Spain, IS&T, June 2008, pp. 249–252 (2008)
8. Pedersen, M., Rizzi, A., Hardeberg, J.Y., Simone, G.: Evaluation of contrast measures in relation to observers perceived contrast. In: CGIV 2008 - Fourth European Conference on Color in Graphics, Imaging and Vision, Terrassa, Spain, IS&T, June 2008, pp. 253–256 (2008)
9. Simone, G., Pedersen, M., Hardeberg, J.Y., Rizzi, A.: Measuring perceptual contrast in a multilevel framework. In: Rogowitz, B.E., Pappas, T.N. (eds.) Human Vision and Electronic Imaging XIV, vol. 7240. SPIE (January 2009)

10. Babcock, J.S., Pelz, J.B., Fairchild, M.D.: Eye tracking observers during rank order, paired comparison, and graphical rating tasks. In: Image Processing, Image Quality, Image Capture Systems Conference (2003)
11. Bai, J., Nakaguchi, T., Tsumura, N., Miyake, Y.: Evaluation of image corrected by retinex method based on S-CIELAB and gazing information. IEICE trans. on Fundamentals of Electronics, Communications and Computer Sciences E89-A(11), 2955–2961 (2006)
12. Pedersen, M., Hardeberg, J.Y., Nussbaum, P.: Using gaze information to improve image difference metrics. In: Rogowitz, B., Pappas, T. (eds.) Human Vision and Electronic Imaging VIII (HVEI 2008), San Jose, USA. SPIE proceedings, vol. 6806. SPIE (January 2008)
13. Endo, C., Asada, T., Haneishi, H., Miyake, Y.: Analysis of the eye movements and its applications to image evaluation. In: IS&T and SID's 2nd Color Imaging Conference: Color Science, Systems and Applications, pp. 153–155 (1994)
14. Mackworth, N.H., Morandi, A.J.: The gaze selects informative details with pictures. Perception & psychophyscics 2, 547–552 (1967)
15. Underwood, G., Foulsham, T.: Visual saliency and semantic incongruency influence eye movements when inspecting pictures. The Quarterly Journal of Experimental Psychology 59, 1931–1949 (2006)
16. Walther, D., Koch, C.: Modeling attention to salient proto-objects. Neural Networks 19, 1395–1407 (2006)
17. Sharma, P., Cheikh, F.A., Hardeberg, J.Y.: Saliency map for human gaze prediction in images. In: Sixteenth Color Imaging Conference, Portland, Oregon (November 2008)
18. Rajashekar, U., van der Linde, I., Bovik, A.C., Cormack, L.K.: Gaffe: A gaze-attentive fixation finding engine. IEEE Transactions on Image Processing 17, 564–573 (2008)
19. Henderson, J.M., Williams, C.C., Castelhano, M.S., Falk, R.J.: Eye movements and picture processing during recognition. Perception & Psychophysics 65, 725–734 (2003)
20. Engeldrum, P.G.: Psychometric Scaling, a toolkit for imaging systems development. Imcotek Press, Winchester (2000)
21. Ponomarenko, N., Lukin, V., Egiazarian, K., Astola, J., Carli, M., Battisti, F.: Color image database for evaluation of image quality metrics. In: International Workshop on Multimedia Signal Processing, Cairns, Queensland, Australia, October 2008, pp. 403–408 (2008)

A Method to Analyze Preferred MTF for Printing Medium Including Paper

Masayuki Ukishima[1,3], Martti Mäkinen[2], Toshiya Nakaguchi[1], Norimichi Tsumura[1], Jussi Parkkinen[3], and Yoichi Miyake[4]

[1] Graduate School of Advanced Integration Science, Chiba University, Japan
ukishima@graduate.chiba-u.jp
[2] Department of Physics and Mathematics, University of Joensuu, Finland
[3] Department of Computer Science and Statistics, University of Joensuu, Finland
[4] Research Center for Frontier Medical Engineering, Chiba University, Japan

Abstract. A method is proposed to analyze the preferred Modulation Transfer Function (MTF) of printing medium like paper for the image quality of printing. First, the spectral intensity distribution of printed image is simulated by changing the MTF of medium. Next, the simulated image is displayed on a high-precision LCD to reproduce the appearance of printed image. An observer rating evaluation experiment is carried out to the displayed image to discuss what the preferred MTF is. The appearance simulation of printed image was conducted on particular printing conditions: several contents, ink colors, a halftoning method and a print resolution (dpi). The experiments on different printing conditions can be conducted since our simulation method is flexible about changing conditions.

Keywords: MTF, printing, LCD, sharpness, granularity.

1 Introduction

Image quality of the printed image is mainly related to its tone reproduction, color reproduction, sharpness and granularity. These characteristics are significantly affected by a phenomenon called dot gain which makes the tone appear to be darker. There are two types of dot gain: mechanical dot gain and optical dot gain. Mechanical dot gain is the physical change in dot size as the results of ink amount, strength and tack. Emmel et al. have tried to model mechanical dot gain effect using a combinatorial approach based on Pólya's counting theory [1]. Optical dot gain (or the Yule-Nielsen effect) is a phenomenon in printing whereby printed dots are perceived bigger than intended, which is caused by the light scattering phenomenon in the medium layer, where the portion of light transmitted from ink outputs from medium and vice versa as shown in Fig. 1. Optical dot gain causes difficulty to predict the spectral reflectance of print and it produces the reduction in the sharpness of image. It also contributes the reduction in the granularity of image caused by the microscopic distribution of ink dots. The light scattering phenomenon can be quantified by the Modulation

A.-B. Salberg, J.Y. Hardeberg, and R. Jenssen (Eds.): SCIA 2009, LNCS 5575, pp. 607–616, 2009.

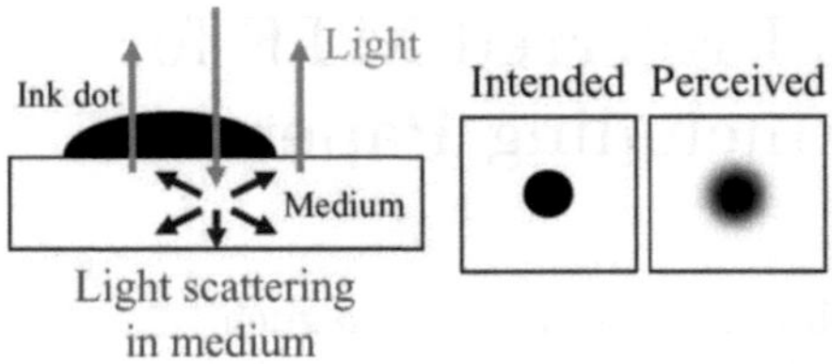

Fig. 1. Optical dot gain

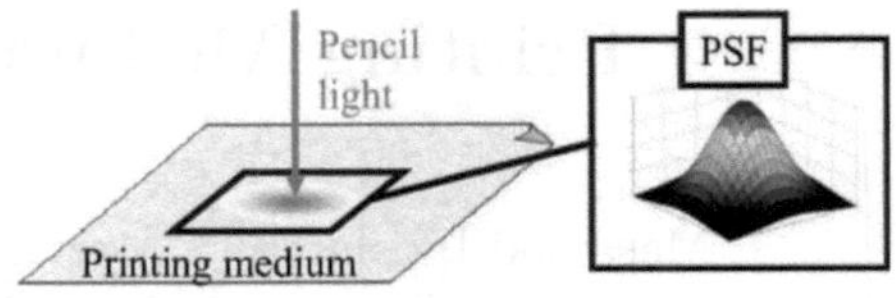

Fig. 2. PSF

Transfer Function (MTF) of medium. The MTF is defined as the absolute value of Fourier transformed Point Spread Function (PSF). The PSF is the impulse response of the system. In this case, the impulse signal is the pencil light like laser and the system is the printing medium as shown in Fig. 2. Because of importance for the image quality control, several researchers have studied the methods to measure and analyze the MTF or PSF of the printing medium [2,3,4]. However, discussions have not been done enough about the relationship between the preferred MTF and the printing conditions such as contents, spectral characteristics of inks, halftoning methods, the mechanical dot gain and the printing resolution (dpi). A main objective of this research is constructing a framework of method to simply evaluate the effects of MTF to the printed image. First, we propose a method to simulate the spectral intensity distribution of printed image by changing the MTF of printing medium. Next, we discuss the preferred MTF on particular conditions of printing through the observer rating evaluation experiment which carried out to the simulated print image displayed on a high-precision LCD.

2 Modulation Transfer Function

2.1 MTF of Linear System

Let a lens system is considered as shown in Fig. 3. For simplicity, we assume that the transmittance of lens is one and the phase transfer of system can be ignored. The output intensity distribution $o(x, y)$ through the lens is given by

$$\begin{aligned} o(x, y) &= i(x, y) * \mathrm{PSF}(x, y) \\ &= \mathfrak{F}^{-1}\{I(u, v)\mathrm{MTF}(u, v)\}, \end{aligned} \tag{1}$$

where (x, y) indicates space coordinates, (u, v) indicates spatial frequency coordinates, $i(x, y)$ is the input intensity distribution whose Fourier transformation is $I(u, v)$, $\mathrm{PSF}(x, y)$ and $\mathrm{MTF}(u, v)$ are the PSF and MTF of the lens system, respectively, $*$ indicates convolution integral operation and $\mathfrak{F}^{-1}$ indicates inverse Fourier transform operation. If the $\mathrm{MTF}(u, v) = 1$, the input signal is perfectly transfered through the system: $o(x, y) = i(x, y)$. However, if the value of $\mathrm{MTF}(u, v)$ is decreased as the increase of (u, v), the function $o(x, y)$ becomes to be blurred because of the loss of information at the high spatial frequency area. Therefore, the higher MTF is generally preferred in the linear system, and it is the best case that $\mathrm{MTF}(u, v) = 1$.

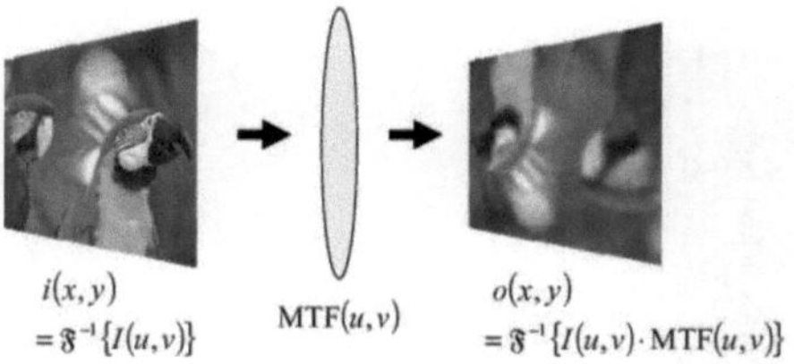

Fig. 3. Lens system

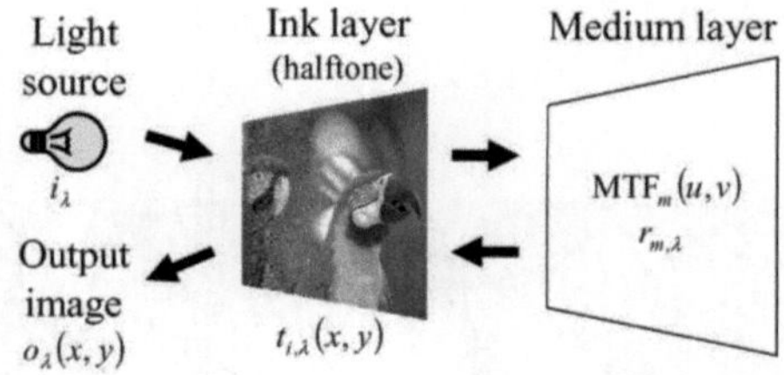

Fig. 4. Printing system

2.2 MTF of Nonliner System Like Printing Medium

Let a printing system is considered as shown in Fig. 4 given by

$$o_\lambda(x,y) = i_\lambda \mathfrak{F}^{-1}\{\mathfrak{F}\{t_{i,\lambda}(x,y)\}\mathrm{MTF}_m(u,v)\} r_{m,\lambda} t_{i,\lambda}(x,y), \quad (2)$$

where the suffix λ indicates wavelength, $o_\lambda(x,y)$ is the spectral intensity distribution of output light, i_λ is the spectral intensity of input incidence assumed spatial uniformity, $t_{i,\lambda}(x,y)$ is the spectral transmittance distribution of ink, $\mathrm{MTF}_m(u,v)$ is the MTF of printing medium like paper assumed wavelength independency, $r_{m,\lambda}$ is the spectral reflectance of medium assumed spatial uniformity, and $\mathfrak{F}$ indicates Fourier transform operation. Equation (2) is called the reflection image model [7], where the incident light transmits the ink layer, the light is scattered and reflected by the medium layer and transmits the ink layer again. Equation (2) assumes the two layers (ink and medium) are perfectly separable optically, the scattering and reflection phenomena in ink can be ignored, therefore multi reflections between two layers can also be ignored. What is preferred MTF of the medium for image quality in this system? In the case of lens system in previous subsection, the information of image is comprised in the incident distribution $i(x,y)$ and, generally, the information should perfectly be reproduced through the system. On the other hand, in the case of printing system, the information of image is comprised in the ink layer as a halftone image. The half tone image should not be always to reproduce perfectly since it is the microscopic distribution of ink dots causing unpleasant graininess. However, too low MTF may cause the reduction of sharpness of image. Therefore the optimal MTF may exist for the best image quality depending on the printing conditions such as contents, ink colors, halftoning methods and values of the print resolution (dpi). Note that the MTF of medium is different from the MTF of printer. The MTF of printer is the modulation transfer between the input data to the printer and the output response corresponding to $o_\lambda(x,y)$. Several methods to measure the MTF of printer has been proposed [5,6].

3 Apperance Simulation of Printed Image on LCD

A method is considered in this section to simulate the apperance of the printed image using the 8-bit [0-255] digital color (RGB) image whose resolution is 256 × 256.

(a) $g_j(x, y)$ (b) $h_j(x, y)$

Fig. 5. Digital halftoning

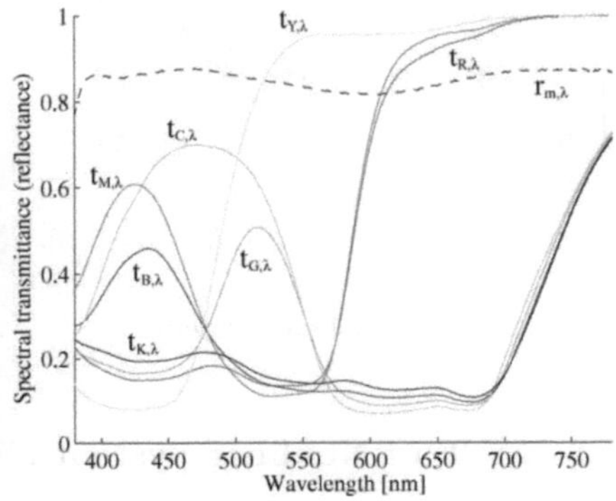

Fig. 6. Spectral transmittance of ink

3.1 Producing Color Halftone Digital Image

Assuming that one pixel of the image is printed by four ink dots of 2 × 2, the digital image is upsampled from 256 × 256 to 512 × 512 by the nearest neighbor interpolation [8]. The upsampled image $f_j(x, y)$ where $j = R, G$ and B is transformed to the CMY image $g_k(x, y)$ where $k = C, M$ and Y:

$$\begin{aligned} g_C(x, y) &= 255 - f_R(x, y) \\ g_M(x, y) &= 255 - f_G(x, y) \\ g_Y(x, y) &= 255 - f_B(x, y). \end{aligned} \tag{3}$$

The color digital halftone image $h_k(x, y)$ is produced applying the error diffusion method of Floyd and Steinberg [9] to g_C, g_M and g_Y, respectively. Figure 5 shows the examples of $g_j(x, y)$ and $h_j(x, y)$. We used the error diffusion method in this subsection, however, the use of any other halftoning methods do not affect the simulation method described in following subsections. In the real scene of printing, the color change process form RGB to CMY is more complex since it needs the dot gain correction and the gamut mapping from the RGB profile (e.g. sRGB profile) to the print profile. Therefore, the process in this sub-section should be modified as the future work.

3.2 Estimating Spectral Transmittance of Inks

Assuming spatial uniformity of ink transmittance for solid prints, the light scattering effect in the printing medium can be ignored mathematically in Eq. (2):

$$\mathfrak{F}^{-1}\{\mathfrak{F}\{t_{i,\lambda}\}\mathrm{MTF}_m(u, v)\} = t_{i,\lambda},$$

and it is derived that

$$\begin{aligned} t_{i,\lambda} &= \sqrt{r_\lambda / r_{m,\lambda}} \\ r_\lambda &= o_\lambda / i_\lambda, \end{aligned} \tag{4}$$

where r_λ is the reflectance of solid print. Therefore, $t_{i,\lambda}$ can be estimated from the measured values of r_λ and $r_{m,\lambda}$. In this research, seven solid patches were printed on a glossy paper (XP-101, CANON) such as cyan, magenta, yellow, red, green, blue and black using a inkjet printer (W2200, CANON) which is set cyan, magenta and yellow inks (BCI-1302 C, M and Y, CANON). The patches of red, green and blue were printed using two of the three inks, respectively. The patch of black was printed using the three inks simultaneously. The spectral reflectance r_λ of each solid patch and the spectral reflectance $r_{m,\lambda}$ of the unprinted paper were measured using a spectrophotometer (Lambda 18, Perkin Elmer). Figure 6 shows the estimated $t_{i,\lambda}$ using Eq. (4).

3.3 Optical Propagation Simulation in Print

The digital halftone image $h_j(x, y)$ produced in Subsection 3.1 can be rewritten to the form $h_{x,y}(C, M, Y)$ having one of the following eight values at each position $[x, y]$: $(1, 0, 0)$, $(0, 1, 0)$, $(0, 0, 1)$, $(0, 1, 1)$, $(1, 0, 1)$, $(1, 1, 0)$, $(1, 1, 1)$ and $(0, 0, 0)$ corresponding to the colors of cyan, magenta, yellow, red, green, blue, black and white (no inks), respectively. By allocating $t_{i,\lambda}$ of each ink estimated in the previous subsection to $h_{x,y}(C, M, Y)$, the spectral transmittance distribution of ink $t_{i,[x,y]}(\lambda)$ can be produced, where $t_{i,[x,y]}(\lambda)$ can be rewritten to the same form in Eq. (2) that is $t_{i,\lambda}(x, y)$. Note that there is no inks at the locations $[x_w, y_w]$ where $h_{x_w,y_w}(C, M, Y) = (0, 0, 0)$, therefore, $t_{i,\lambda}(x_w, y_w) = 1$.

Now we have the components $r_{m,\lambda}$ and $t_{i,\lambda}(x, y)$ of Eq. (2). If we define the other components i_λ and $\mathrm{MTF}_m(u, v)$, the output spectral intensity distribution of the print $o_\lambda(x, y)$ can be calculated. The incidence i_λ was assumed to be CIE D65 standard illuminant since we used the LCD whose color temperature is $6500K$ described in detail in next subsection. We defined the one dimensional MTF of medium given by

$$\mathrm{MTF}_m(u) = \frac{d}{\sqrt{d^2 + u^2}} \tag{5}$$

where d is a parameter to define the shape of MTF curve. Equation (5) well approximates the MTF of paper as shown in Fig. 7 where this is a example of glossy paper's MTF measured in our previous research [4]. Using Eq. (5), we produced seven types of MTF curve as shown in Fig. 8. Each parameter d is decided in condition that the following formula is equal to 10, 25, 40, 55, 70, 85, 100[%], where such parameters d are 0.212, 0.756, 1.57, 2.74, 4.62, 8.47 and ∞.

$$\frac{\int_0^{10} \mathrm{MTF}_m(u)du}{10} \times 100 \tag{6}$$

Assuming spatial isotropy, two dimensional $\mathrm{MTF}_m(u, v)$ was produced using each one dimensional $\mathrm{MTF}_m(u)$. Finally, the function $o_\lambda(x, y)$ was calculated by Eq. (2) for each λ.

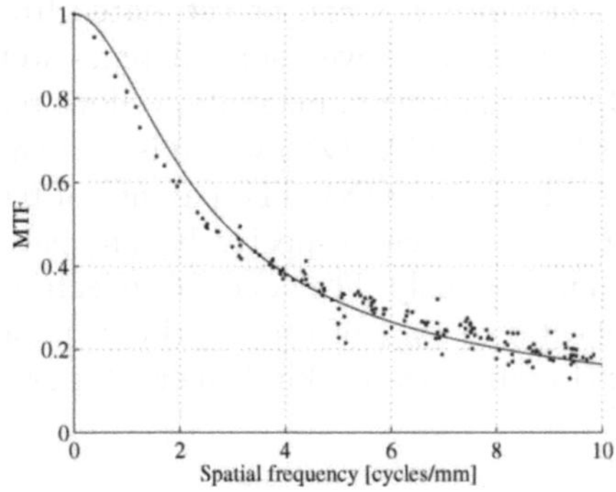

Fig. 7. MTF of a glossy paper

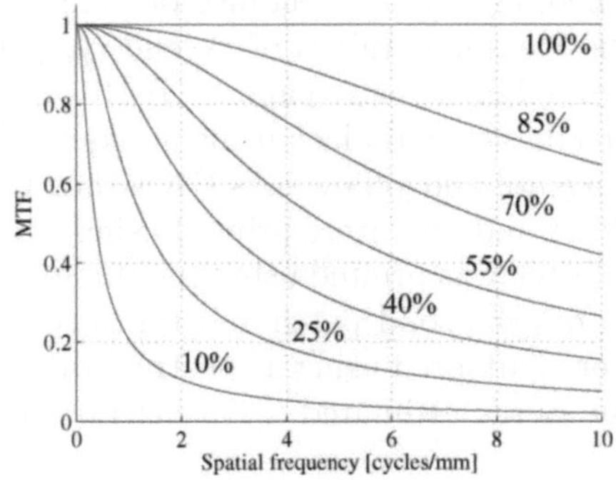

Fig. 8. Generated MTFs

3.4 Display on LCD and Viewing Distance

The output intensity distribution of the print $o_\lambda(x, y)$ can be rewritten to the form $o_{x,y}(\lambda)$. The spectral function $o_{x,y}(\lambda)$ is converted to CIE RGB tristimulus values given by

$$\begin{aligned} R_{x,y} &= \int_{380}^{780} o_{x,y}(\lambda)\bar{r}(\lambda)d\lambda \\ G_{x,y} &= \int_{380}^{780} o_{x,y}(\lambda)\bar{g}(\lambda)d\lambda \\ B_{x,y} &= \int_{380}^{780} o_{x,y}(\lambda)\bar{b}(\lambda)d\lambda, \end{aligned} \tag{7}$$

where $\bar{r}(\lambda)$, $\bar{g}(\lambda)$ and $\bar{b}(\lambda)$ are color matching functions [10]. The tristimulus values are displayed on the LCD after the gamma correction given by

$$V'_{x,y} = 255 \times \{V_{x,y}\}^{\frac{1}{\gamma}}, \tag{8}$$

where V is R, G or B and γ is the gamma value of LCD. An high-precision LCD (CG-221, EIZO) was used, where the color mode was set to sRGB mode whose gamma value $\gamma = 2.2$ and the color temperature is $6500K$. The examples of simulated images are shown in Fig. 9, where the subcaptions (a)-(c) correspond to the applied MTF percentages.

In this simulation, one ink dot is expressed by one pixel of LCD. However, the ink dot size is practically quite smaller than the pixel size. If the printer whose resolution is 600dpi is assumed, the ink dot size is 4.08×10^{-2} [mm/dot]. On the other hand, the pixel size of the LCD is 2.49×10^{-1} [mm/pixel]. In order to approximate the appearance of the simulated image to that of the real print, the viewing angles between these were conformed as shown in Fig. 10 by adjusting the viewing distance from the LCD given by

$$d_d = s_d d_p / s_p, \tag{9}$$

Fig. 9. Simulated print images

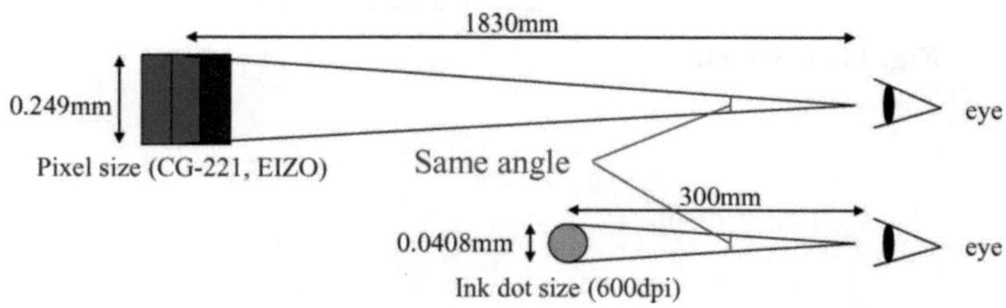

Fig. 10. Viewing distance

where d_d and d_p are the viewing distance from the LCD and real print, respectively, s_d is one pixel size of the LCD and s_p is one ink dot size of the real print. Assuming the distance d_p is equal to 300 [mm], the distance d_d becomes to be equal to about 1830 [mm].

We used not the real print but the LCD for simulation because of several reasons. The objective of this research is to analyze the effects caused by the MTF of medium. However, if we use real medium, other characteristics except the MTF are also changed such as the mechanical dot gain and the color, opacity and granularity of medium. The simulation-based evaluation on display using Eq. (2) can change only the MTF characteristic. The simplicity of observer rating experiment is another advantage using the display. The reason to use the LCD as a display is that the MTF of LCD itself hardly decreases until its Nyquist frequency [11]. Therefore, the MTF of device can be ignored.

4 Observer Rating Evaluation

To analyze the preferred MTF of printing medium, an observer rating evaluation test is carried out. Two images simulated in Section 3 are displayed on the LCD simultaneously. We defined seven types of MTF in Subsection 3.3, therefore ${}_7C_2 = 21$ combinations exist. Subjects evaluate the total image quality of the two images and select the better one. Thurstone's paired comparison method [12] is carried out to the obtained data and the psychological scale are obtained. Three contents were used such as Lenna, Parrots and Pepper [13] as shown in

(a) Lenna (b) Parrots

(c) Pepper

Fig. 11. Contents

Table 1. Paired comparison result (Lenna)

	10%	25%	40%	55%	70%	85%	100%
10%	0.50	0.85	0.80	0.65	0.70	0.50	0.40
25%	0.15	0.50	0.45	0.40	0.35	0.25	0.10
40%	0.20	0.55	0.50	0.45	0.25	0.20	0.20
55%	0.35	0.60	0.55	0.50	0.20	0.15	0.00
70%	0.30	0.65	0.75	0.80	0.50	0.15	0.00
85%	0.50	0.75	0.80	0.85	0.85	0.50	0.10
100%	0.60	0.90	0.80	1.00	1.00	0.90	0.50

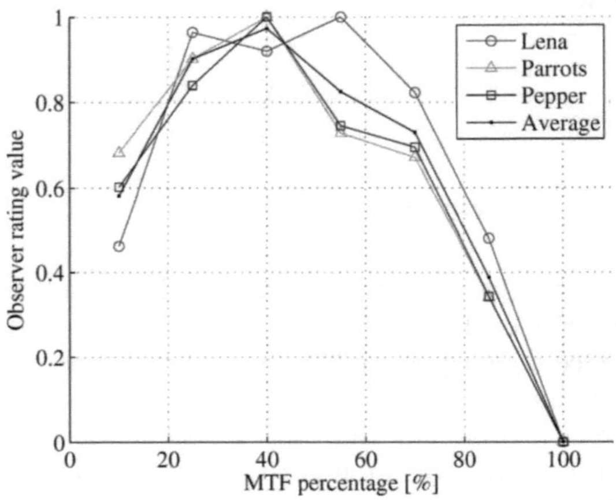

Fig. 12. Observer rating values

Fig. 11. The number of subjects were twenty. The viewing distance was set to 1830 [mm]. The evaluation was conducted in a dark room.

Table 1 shows an example of measured result whose content is Lenna, where these percentages are the MTF coverages. For example, the probability, (row, column) = (2,4) = 0.40, indicates that the 40 % of observers evaluated that the MTF coverage of 55% is better than that of 25% for the image quality. If the probability is 0.00 or 1.00, it was converted to 0.01 or 0.99 since Thurstone's method cannot calculate the psychological scale in that case [12]. Figure 12 shows the observer rating value of each MTF percentage. The result shows that too low MTF is not preferred and too high MTF is also not preferred. We consider too low MTF causes too low sharpness and too high MTF causes too high granularity caused by the microscopic distribution of ink dots. As the dependence on the contents, the rating results of Parrots and Pepper were similar, however, the rating result of Lenna was different from others. Parrots and Pepper have a commonality about the color histogram compared to Lenna. Therefore, it is a possibility that the color histogram affects the preffered MTF of printing medium. The case of using grayscale image should be tested to separate the MTF effects to color and other characteristics such as tone, sharpness and granularity. As the average observer rating value to all contents, it was the best case that the MTF percentage is 40%. However it may significantly depend on the resolution of the print which is 600 dpi in this case. In the case of higher resolution, the granularity of the image is smaller therefore the preferred MTF may become to be higher.

5 Conclusion

A method was proposed to simulate the spectral intensity distribution of printed image by changing the MTF of printing medium like paper. The simulated image was displayed on a high-precision LCD to simulate the appearance of image printed on particular conditions: using three contents, dye-based inks, the error diffusion method as the halftoning and a print resolution (600dpi). An observer rating evaluation experiment was carried out to the displayed image to discuss what the preferred MTF is for the image quality of printed image. Thurstone's paired comparison method was adopted as the observer rating evaluation method because of the simplicity of evaluation and high reliability. The main achievement of this research is that a framework was constructed to simply evaluate the effects of MTF to the printed image. Our simulation method is flexible about changing the printing conditions such as contents, ink colors, halftoning methods and the printing resolution (dpi). As future works, we intend to carry out the same kind of experiments on different printing conditions. The case of using grayscale image should be tested to separate the MTF effects to color and other characteristics such as tone, sharpness and granularity. The cases of using other halftoning methods should be tested such as on-demand dither methods and density pattern methods. The simulated printing resolution (dpi) can be changed by changing the viewing distance from the LCD or by using other LCDs having different pixel size (pixel pitch). In this paper, one ink dot of printed image was expressed by one pixel on the LCD. If one ink dot is expressed by multiple pixels on the LCD, the shape of ink dots can be simulated, which can express the mechanical dot gain. We also intend to carry out the physical evaluation using the simulated microscopic spectral intensity distribution $o_\lambda(x, y)$.

References

1. Emmel, P., Herch, R.D.: Modeling Ink Spreading for Color Prediction. J. Imaging Sci. Technol. 46(3), 237–246 (2002)
2. Inoue, S., Tsumura, N., Miyake, Y.: Measuring MTF of Paper by Sinusoidal Test Pattern Projection. J. Imaging Sci. Technol. 41(6), 657–661 (1997)
3. Atanassova, M., Jung, J.: Measurement and Analysis of MTF and its Contribution to Optical Dot Gain in Diffusely Reflective Materials. In: Proc. IS&T's NIP23: 23rd International Conference on Digital Printing Technologies, Anchorage, pp. 428–433 (2007)

4. Ukishima, M., Kaneko, H., Nakaguchi, T., Tsumura, N., Kasari, M.H., Parkkinen, J., Miyake, Y.: Optical dot gain simulation of inkjet image using MTF of paper. In: Proc. Pan-Pacific Imaging Conference 2008 (PPIC 2008), Tokyo, pp. 282–285 (2008)
5. Jang, W., Allebach, J.P.: Characterization of printer MTF. In: Cui, L.C., Miyake, Y. (eds.) Image Quality and System Performance III. SPIE Proc., vol. 6059, pp. 1–12 (2006)
6. Lindner, A., Bonnier, N., Leynadier, C., Schmitt, F.: Measurement of Printer MTFs. In: Proc. SPIE, San Jose, California. Image Quality and System Performance VI, vol. 7242 (2009)
7. Inoue, S., Tsumura, N., Miyake, Y.: Analyzing CTF of Print by MTF of Paper. J. Imaging Sci. Technol. 42(6), 572–576 (1998)
8. Gonzalez, R.C., Woods, R.E.: Digital Image Processing, 2nd edn., pp. 64–66. Prentice-Hall, Inc., New Jersey (2002)
9. Ulichhey, R.: Digital Halftoning. MIT Press, Cambridge (1987)
10. Ohta, N., Robertson, A.A.: Colorimetry: Fundamentals And Applications. Wiley–Is&t Series in Imaging Science and Technology (2006)
11. Ukishima, M., Nakaguchi, T., Kato, K., Fukuchi, Y., Tsumura, N., Matsumoto, K., Yanagawa, N., Ogura, T., Kikawa, T., Miyake, Y.: Sharpness Comparison Method for Various Medical Imaging Systems. Electronics and Communications in Japan, Part 2 90(11), 65–73 (2007); Translated from Denshi Joho Tsushin Gakkai Ronbunshi J89-A(11), 914–921 (2006)
12. Thurstone, L.L.: The Measurement of Values. Psychol. Rev. 61(1), 47–58 (1954)
13. `http://www.ess.ic.kanagawa-it.ac.jp/app_images_j.html`

Efficient Denoising of Images with Smooth Geometry

Agnieszka Lisowska

University of Silesia, Institute of Informatics,
ul. Bedzinska 39, 41-200 Sosnowiec, Poland
alisow@ux2.math.us.edu.pl
http://www.math.us.edu.pl/al/eng_index.html

Abstract. In the paper the method of smooth geometry image denoising has been presented. It is based on smooth second order wedgelets proposed in this paper. Smooth wedgelets (and second order wedgelets) are defined as wedgelets with smooth edges. Additionally, smooth borders of quadtree partition have been introduced. The first kind of smoothness is defined adaptively whereas the second one is fixed once for the whole estimation process. The proposed kind of wedgelets has been applied to image denoising. As follows from experiments performed on benchmark images this method gives far better results of denoising of images with smooth geometry than the other state-of-the-art methods.

Keywords: Image denoising, wedgelets, second order wedgelets, smooth edges, multiresolution.

1 Introduction

Image denoising plays very important role in image processing. It follows from the fact that images are obtained mainly from different electronic devices. It causes that many kinds of noise generated by these devices are present on such images. It is well known fact that medical images are characterized by Gaussian noise and astronomical images are corrupted by Poisson noise, to mention a few kinds of noise. Determination of the noise characteristic is not difficult and may be done automatically. The main problem is related to defining of efficient methods of image denoising.

In the case of the most commonly generated Gaussian noise there is a wide spectrum of denoising methods. These methods are based on wavelets due to the fact that noise is characterized by high frequency what can be suppressed just by wavelets. Image denoising by wavelets is very similar to compression — in order to perform denoising a forward transform is applied, some coefficients are replaced by zero and then the inverse transform is applied [1]. The standard method was improved in many ways, to mention an introduction of different kinds of thresholds or different kinds of thresholding [2], [3].

Recently, also geometrical wavelets have been introduced to image denoising. Since they give better results in image coding than classical wavelets they are also

A.-B. Salberg, J.Y. Hardeberg, and R. Jenssen (Eds.): SCIA 2009, LNCS 5575, pp. 617–625, 2009.

applied in image estimation. There is a wide spectrum of geometrical wavelets, for example the ones based on frames like ridgelets [4], curvelets [5], bandelets [6], or the ones based on dictionaries like wedgelets [7], beamlets [8], platelets [9]. As presented in the literature [10], [11], [12] some of them give better results of image denoising than classical wavelets. Especially, in [12] a comparative study is presented from which follows that adaptive methods based on wedgelets are competitive in image denoising to the other wavelets-based methods.

In the paper the improvement of wedgelets-based technique of image estimation has been proposed. It was motivated by the observation that edges present in images are of different smoothness kind. Because second order wedgelets always estimate smooth edges by sharp step functions it may introduce additional errors. In order to avoid it second order wedgelets with smooth edges and smooth borders are used in image denoising. From the experiments performed on the set of benchmark images follows that the proposed method assures better results of image denoising than the leading state-of-the-art methods.

2 Geometrical Image Denoising

The problem of image denoising is related to image estimation. Instead of approximating of original image F one needs to estimate it basing only on a version contaminated by noise

$$I(x_1, x_2) = F(x_1, x_2) + \sigma Z(x_1, x_2), \quad x_1, x_2 \in [0, 1], \tag{1}$$

where Z is an additive zero-mean Gaussian noise. Image F can be quite effectively estimated by multiresolution techniques thanks to the fact that a frequency of added noise is usually higher than that of the original image.

2.1 Multiresolution Denoising

The great majority of denoising methods is based on wavelets. It follows from the fact that wavelets are efficient in removing high frequency signal (especially a noise) from an image. However, these methods tend to slightly smoothen edges present on images. So, similarly as in the case of image coding, geometrical wavelets have been introduced to image estimation. Thanks to the possibility of catching changes of signal in different directions they are more efficient in image denoising than classical wavelets. For example, the denoising method based on curvelets [10] is characterized by very efficient estimation nearby edges giving very accurate denoising results. However, as shown in [11], [12], adaptive geometrical wavelets can assure even better estimation results than curvelets. The methods based on wedgelets [11] or second order wedgelets [12] are very efficient in proper reconstruction of image geometry. Below we describe the wedgelets-based methods in more details.

2.2 Geometrical Denoising

Consider an image F defined on dyadic discrete support of size $N \times N$ pixels (dyadic means that $N = 2^n$ for $n \in \mathbb{N}$). To such an image a quadtree partition may be assigned. Consider then any square S from that partition and any line segment b (called beamlet [8]) connecting any two points (not lying on the same border side) from the border of the square. The wedgelet is defined as [7]

$$W(x,y) = \mathbf{1}\{y \leq b(x)\}, \quad (x,y) \in S. \tag{2}$$

Similarly, consider any segment of second degree curve (as ellipse, parabola or hyperbola) $\hat{b}$ (called second order beamlet [13], [14]) connecting any two points from the border of the square S. The second order wedgelet is defined as [13], [14]

$$\hat{W}(x,y) = \mathbf{1}\{y \leq \hat{b}(x)\}, \quad (x,y) \in S. \tag{3}$$

Taking into account all possible squares from the quadtree partition (of different locations and scales) and all possible beamlet connections one obtains the wedgelets' dictionary. Taking additionally all possible curvatures of second order beamlets one obtains the second order wedgelets' dictionary. It is assumed that the wedgelets' dictionary is included in the second order wedgelets' dictionary (with the parameter reflecting curvature equals to zero). Because wedgelet is a special case of second order wedgelet in the rest of the paper the dictionary of second order wedgelets is considered. Additionally, second order wedgelets are called for simplicity as s-o wedgelets.

Such defined set of functions can be used adaptively in image approximation or estimation. It is performed in the way that s-o wedgelets are adapted to a geometry from an image. Depending on image content appropriate s-o wedgelets are used in approximation. The process is performed in two steps. In the first step a so-called full quadtree is built. Each node of the quadtree represents the best s-o wedgelet within the appropriate square in the mean of Mean Square Error (MSE) sense. In the second step the tree is pruned in order to solve the following minimization problem

$$P = min\{||F - F_{\hat{W}}||_2^2 + \lambda^2 K\}\,, \tag{4}$$

where minimum is taken within all possible quad-splits of an image, F denotes the original image, $F_{\hat{W}}$ its s-o wedgelet approximation, K the number of s-o wedgelets used in approximation and λ is the penalization factor. Indeed, we are interested in obtaining the sparsest image approximation assuring the best quality in the mean of MSE metric sense of the approximation. The minimization problem can be solved by the use of the bottom-up tree pruning algorithm [7].

The algorithm of s-o wedgelets-based image denoising is very similar to image approximation. The only difference is that instead of original image approximation a noised image approximation is performed. However, the additional problem has to be solved. Because the approximation algorithm is dependent on parameter λ, the optimal value of it should be obtained. It is done in the way that the second step of the approximation is repeated for different values of λ. As the optimal one is chosen the one for which the dependency between λ and number of s-o wedgelets has a saddle point [12].

3 Image Denoising with Smooth Wedgelets

Many images are characterized by presence of edges with different kinds of smoothness. In the case of artificial images very often edges are rather sharp and well defined. However, in the case of still images some edges are sharp and the others are more or less smooth. The approximation of smooth edges by s-o wedgelets causes that MSE increases. It leads to false edges detection what degenerates denoising results. However, the problem may be solved by introducing smooth s-o wedgelets.

3.1 Smooth Wedgelets Denoising

Consider any s-o wedgelet like the ones presented in Fig. 1 (a), (c). Smooth s-o wedgelet is defined by introducing smooth connection between two s-o wedgelets defined within the square support (see Fig. 1 (b), (d)). In other words, instead of step discontinuity we introduce linear continuity between two constant areas represented by s-o wedgelets. In this way we introduce additional parameter to the s-o wedgelets' dictionary. The parameter, denoted as R, reflects the half of the length of smoothness of the edge. For $R = 0$ we obtain just s-o wedgelet, and the larger the value of R the longer the smoothness. This approach is symmetrical. It means that the smoothness is equally elongated on both sides of the original edge (marked in Fig. 1 (b), (d)).

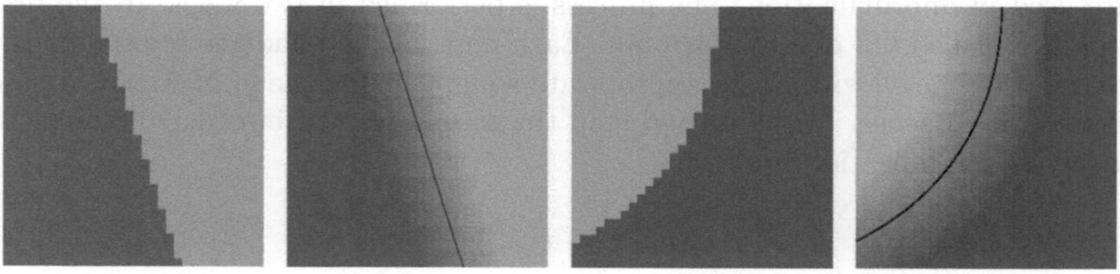

(a) (b) (c) (d)

Fig. 1. (a) Wedgelet, (b) smooth wedgelet, (c) s-o wedgelet, (d) smooth s-o wedgelet

Because wedgelets-based algorithms are known to have large time complexity the additional parameter causes that the computation time is not acceptable. To overcome that problem the following method of finding optimal smooth s-o wedgelets is proposed. Consider any square S from the quadtree partition. Firstly, find an optimal wedgelet within it. Secondly, basing on it find the best s-o wedgelet in the neighborhood, like proposed in [14]. And finally, basing on the best s-o wedgelet find optimal smooth s-o wedgelet basing on it by incrementing the value of R and computing new values of constant areas. While you find better approximation do incrementation, otherwise stop. This method not necessary assures that the best smooth s-o wedgelet is found but great improvement in the approximation is done anyway. After processing of all nodes of the quadtree the bottom-up tree pruning may be applied.

Smooth s-o wedgelets are used in image denoising in the same way as s-o wedgelets are. The algorithm works according to the following steps:

1. Find the best smooth s-o wedgelet matching for every node of the quadtree.
2. Apply the bottom-up tree pruning algorithm to find the optimal approximation.
3. Repeat step 2 for different values of λ and choose as the final result the one which gives the best result of denoising.

The most problematic step of the algorithm is to find the optimal value of λ. In the case of original image approximation the value of λ may be set as the one for which RD dependency (in other words the plot of number of wedgelets versus MSE) has the saddle point. Since we do not know the original image we have to use the plot of λ versus number of wedgelets and the saddle point of that dependency [11], [12].

3.2 Double Smooth Wedgelets

When we deal with images with smooth geometry we can additionally apply the postprocessing step in order to improve the results of denoising performed by smooth s-o wedgelets. Because all quadtree-based techniques lead to blocking artifacts, especially in smooth images, in the postprocessing step we perform smoothing between neighboring blocks. The length of smoothing is represented by parameter R_S. It is defined in the same way as parameter R. However, the differences between them are meaningful. Indeed, parameter R works in adaptive way, it means that depending on an estimated image its value changes and different values of R lead to different values of wedgelet parameters (represented by constant areas). Typically, different segments of approximated image are characterized by different values of R. On the other hand parameter R_S is constant and does not depend on the image content. Once fixed for a given image, it never changes.

Taking into account above considerations we can define double smooth s-o wedgelet as a smooth s-o wedgelet with smooth borders. An example of image approximation by such wedgelets is presented in Fig. 2. As one can see the more smoothness is used the better approximation we obtain.

4 Experimental Results

The experiments presented in this section were performed on the set of benchmark images presented in Fig. 3. All the described methods were implemented in Borland C++ Builder 6 environment. The images were artificially noised by Gaussian noise with zero mean and eight different values of variances (presented in the paper after normalization). This set of images was submitted to denoising process with the use of three different methods, namely based on wedgelets, s-o

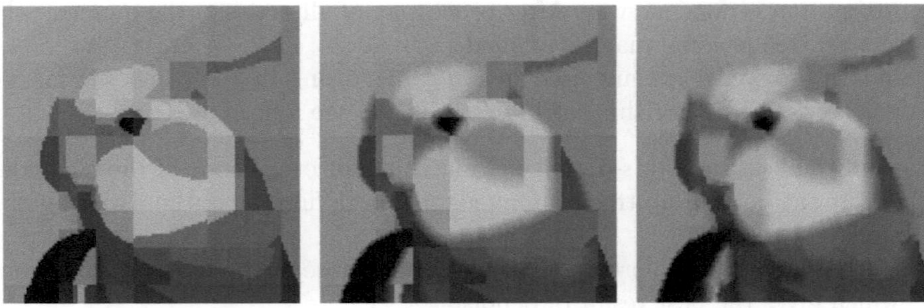

(a) (b) (c)

Fig. 2. The segment of "bird" approximated by (a) s-o wedgelets, (b) smooth s-o wedgelets, (c) double smooth s-o wedgelets

wedgelets and smooth s-o wedgelets (with and without the postprocessing). Additionally, we assumed that $R_S = 1$. As follows from experiments larger values of R_S give better results of denoising only for very smooth images (like "chromosome"). Setting the parameter to one causes that in nearly all tested images an improvement is visible. It should be mentioned also that we applied smooth borders only for square sizes larger than 4×4 pixels.

Fig. 3. The benchmark images: "balloons", "monarch", "peppers", "bird", "objects", "chromosome"

In Table 1 the numerical results of image denoising are presented. From that table follows that the proposed method (denoted as wedgelets2S) assures better denoising results than the state-of-the-art reference methods (for further comparisons, like between wavelets and wedgelets see [12]). More precisely, in the case of images without smooth geometry (like "balloons") the improvement of denoising method based on smooth s-o wedgelets is rather small. However, in the

Table 1. Numerical results of image denoising with the help of the following methods: wedgelets, s-o wedgelets (wedgelets2), smooth s-o wedgelets (wedgelets2S) and double smooth s-o wedgelets (wedgelets2SS)

Image	Method	Noise variance							
		0.001	0.005	0.010	0.015	0.022	0.030	0.050	0.070
balloons	wedgelets	30.50	26.10	24.03	23.17	22.29	21.72	20.60	19.94
	wedgelets2	30.40	25.92	24.00	23.12	22.26	21.71	20.67	19.97
	wedgelets2S	29.99	26.36	24.45	23.35	22.49	21.93	20.75	20.05
	wedgelets2SS	29.89	26.57	24.84	23.80	22.98	22.44	21.24	20.42
monarch	wedgelets	30.47	26.20	24.34	23.27	22.33	21.63	20.50	19.70
	wedgelets2	30.38	26.21	24.39	23.40	22.37	21.71	20.56	19.71
	wedgelets2S	29.15	25.97	24.37	23.45	22.50	21.80	20.59	19.81
	wedgelets2SS	28.69	25.88	24.50	23.71	22.91	22.23	21.02	20.29
peppers	wedgelets	31.71	27.44	25.82	24.89	24.10	23.41	22.43	21.75
	wedgelets2	31.56	27.31	25.81	24.79	24.04	23.37	22.36	21.68
	wedgelets2S	31.82	27.77	26.21	25.28	24.47	23.72	22.63	21.95
	wedgelets2SS	31.82	28.39	26.92	26.03	25.11	24.36	23.11	22.35
bird	wedgelets	34.24	30.24	28.76	28.05	27.35	26.82	25.71	25.21
	wedgelets2	34.07	30.24	28.76	28.02	27.29	26.79	25.66	25.09
	wedgelets2S	34.61	30.70	29.25	28.54	27.74	27.24	26.01	25.38
	wedgelets2SS	34.90	31.41	30.00	29.08	28.28	27.70	26.47	25.72
objects	wedgelets	33.02	28.36	26.90	25.89	25.16	24.43	23.51	22.73
	wedgelets2	32.84	28.27	26.72	25.82	25.15	24.34	23.47	22.66
	wedgelets2S	33.36	29.46	27.85	26.84	25.96	25.26	24.13	23.24
	wedgelets2SS	33.46	29.98	28.36	27.41	26.43	25.69	24.51	23.51
chromosome	wedgelets	36.45	32.78	31.48	30.40	29.56	29.07	28.31	27.15
	wedgelets2	36.29	32.69	31.31	30.31	29.56	29.29	28.32	27.12
	wedgelets2S	38.00	34.67	33.24	32.43	31.30	30.71	29.52	28.17
	wedgelets2SS	38.78	35.34	33.91	33.03	31.73	31.17	29.94	28.56

case of images with typical smooth geometry (like "chromosome" and "objects") the improvement is substantial and can oscillate round 1.6 dB. For images with smooth and non-smooth geometry the improvement depends on the amount of smooth geometry within an image.

However, applying the method of denoising based on smooth s-o wedgelets (and wedgelets in general) causes that the so-called blocking artifacts are visible. Even if the denoising results are competitive in the mean of PSNR values the visible false edges lead to uncomfortable perceiving such images by human observer. To overcome that inconvenience also the double smooth s-o wedgelets were applied to image denoising (denoted as wedgelets2SS). As follows from Table 1 that method additionally improves the results of denoising quite substantially.

Additionally, in Fig. 4 the sample result of denoising is presented. As one can see the method based on s-o wedgelets introduces false edges in the very smooth image. Applying smooth s-o wedgelets causes that the edges are better represented. However, some blocking artifacts are visible. The double smooth s-o wedgelets reduce slightly that problem.

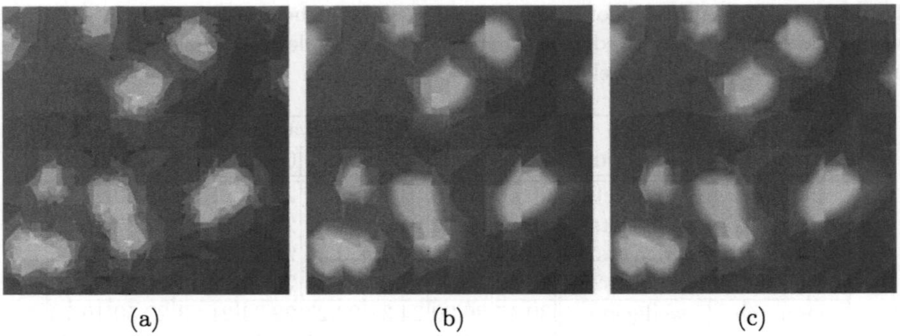

Fig. 4. Sample image (contaminated by Gaussian noise with variance equals to 0.015) denoised by: (a) s-o wedgelets, (b) smooth s-o wedgelets, (c) double smooth s-o wedgelets ($R_S = 1$)

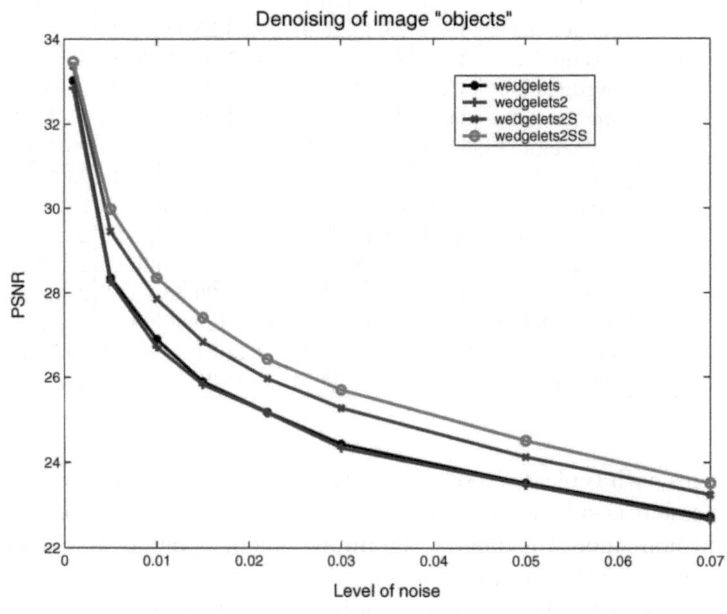

Fig. 5. Typical level-of-noise-PSNR dependency for the presented methods

Finally, in Fig. 5 there is presented typical dependency between four described methods of denoising. The plot was generated for image "objects", but for the remaining images the dependency is very similar.

5 Summary

In the paper smooth s-o wedgelets and their additional postprocessing have been introduced. Though the postprocessing step is well known and used in different approximation (estimation) methods based on quadtrees or similar image partitions it was never used in wedgelets-based image approximations (estimations), especially in denoising. In the case of images with smooth geometry it substantially improves the results of denoising, both — visually and computationally.

By comparing of denoising methods based on classical wavelets and wedgelets one can conclude that the former ones give better visual quality of reconstruction than the latter ones. But it is quite opposite if we consider computational quality. In fact, both of them have disadvantages — wavelets-based methods tend to smooth sharp edges and wedgelets-based methods produce false edges. The proposed method seems to overcome both inconveniences thanks to the adaptivity and postprocessing, respectively.

References

1. Donoho, D.L., Johnstone, I.M.: Ideal Spatial Adaptation via Wavelet Shrinkage. Biometrica 81, 425–455 (1994)
2. Donoho, D.L.: Denoising by Soft Thresholding. IEEE Transactions on Information Theory 41(3), 613–627 (1995)
3. Donoho, D.L., Vetterli, M., de Vore, R.A., Daubechies, I.: Data Compression and Harmonic Analysis. IEEE Transactions on Information Theory 44(6), 2435–2476 (1998)
4. Candès, E.: Ridgelets: Theory and Applications, PhD Thesis, Departament of Statistics, Stanford University, Stanford, USA (1998)
5. Candès, E., Donoho, D.: Curvelets — A Surprisingly Effective Nonadaptive Representation for Objects with Edges Curves and Surfaces Fitting. In: Cohen, A., Rabut, C., Schumaker, L.L. (eds.). Vanderbilt University Press, Saint-Malo (1999)
6. Mallat, S., Pennec, E.: Sparse Geometric Image Representation with Bandelets. IEEE Transactions on Image Processing 14(4), 423–438 (2005)
7. Donoho, D.L.: Wedgelets: Nearly–Minimax Estimation of Edges. Annals of Statistics 27, 859–897 (1999)
8. Donoho, D.L., Huo, X.: Beamlet Pyramids: A New Form of Multiresolution Analysis, Suited for Extracting Lines, Curves and Objects from Very Noisy Image Data. In: Proceedings of SPIE, vol. 4119 (2000)
9. Willet, R.M., Nowak, R.D.: Platelets: A Multiscale Approach for Recovering Edges and Surfaces in Photon Limited Medical Imaging, Technical Report TREE0105, Rice University (2001)
10. Starck, J.-L., Candès, E., Donoho, D.L.: The Curvelet Transform for Image Denoising. IEEE Transactions on Image Processing 11(6), 670–684 (2002)
11. Demaret, L., Friedrich, F., Führ, H., Szygowski, T.: Multiscale Wedgelet Denoising Algorithm. In: Proceedings of SPIE, Wavelets XI, San Diego, vol. 5914, pp. 1–12 (2005)
12. Lisowska, A.: Image Denoising with Second-Order Wedgelets. International Journal of Signal and Imaging Systems Engineering 1(2), 90–98 (2008)
13. Lisowska, A.: Effective Coding of Images with the Use of Geometrical Wavelets. In: Proceedings of Decision Support Systems Conference, Zakopane, Poland (2003) (in Polish)
14. Lisowska, A.: Geometrical Wavelets and their Generalizations in Digital Image Coding and Processing, PhD Thesis, University of Silesia, Poland (2005)

Kernel Entropy Component Analysis Pre-images for Pattern Denoising

Robert Jenssen and Ola Storås

Department of Physics and Technology, University of Tromsø, 9037 Tromsø, Norway
Tel.: (+47) 776-46493; Fax: (+47) 776-45580
robert.jenssen@uit.no

Abstract. The recently proposed kernel entropy component analysis (kernel ECA) technique may produce strikingly different spectral data sets than kernel PCA for a wide range of kernel sizes. In this paper, we investigate the use of kernel ECA as a component in a denoising technique previously developed for kernel PCA. The method is based on mapping noisy data to a kernel feature space, for then to denoise by projecting onto a kernel ECA subspace. The denoised data in the input space is obtained by computing pre-images of kernel ECA denoised patterns. The denoising results are in several cases improved.

1 Introduction

Kernel entropy component analysis was proposed in [1][1]. The idea is to represent the input space data set by a projection onto a kernel feature subspace spanned by the k kernel principal axes which corresponds to the largest contributions of Renyi entropy with regard to the input space data set. This mapping may produce a radically different kernel feature space data set than kernel PCA, depending on the kernel size used.

Recently, kernel PCA [2] has been used for denoising by mapping a noisy input space data point into a Mercer kernel feature space, for then to project the data point onto the leading kernel principal axes obtained using kernel PCA based on clean training data. This is the actual denoising. In order to represent the input space denoised pattern, i.e. the pre-image of the kernel feature space denoised pattern, a method for finding the pre-image is needed. Mika et al. [3] proposed such a method using an iterative scheme. More recently, Kwok and Tsang [4] proposed an algebraic method for finding the pre-image, and reported positive results compared to [3]. This method has also been used in [5].

In this paper, we introduce kernel ECA for pattern denoising. Clean training data is used to obtain the "entropy subspace" in the kernel feature space. A noisy input pattern is mapped to kernel space and then projected onto this subspace. This removes the noise in a different manner as opposed to using kernel PCA

[1] In [1], this method was referred to as kernel maximum entropy data transformation. However, kernel entropy component analysis (kernel ECA) is a more proper name, and will be used subsequently.

A.-B. Salberg, J.Y. Hardeberg, and R. Jenssen (Eds.): SCIA 2009, LNCS 5575, pp. 626–635, 2009.

for this purpose. Subsequently, Kwok and Tsang's [4] method for finding the pre-image, i.e. the denoised input space pattern, is employed. Positive results are obtained.

This paper is organized as follows. In Section 2, we review the kernel ECA method, and in Section 3, we explain how to use kernel ECA for denoising in combination with Kwok and Tsang's [4] pre-image method. We report experimental results in Section 4 and conclude the paper with Section 5.

2 Kernel Entropy Component Analysis

We first discuss how to perform kernel ECA based on a sample of data points. This is referred to as in-sample kernel ECA. Thereafter, we discuss how to project an out-of-sample data point onto the kernel ECA principal axes.

2.1 In-Sample Kernel ECA

The Renyi (quadratic) entropy is given by $H(p) = -\log V(p)$, where $V(p) = \int p^2(\mathbf{x})d\mathbf{x}$ and $p(\mathbf{x})$ is the probability density function generating the input space data set, or sample, $\mathcal{D} = \mathbf{x}_1, \ldots, \mathbf{x}_N$. By incorporating a Parzen window density estimator $\hat{p}(\mathbf{x}) = \frac{1}{N}\sum_{\mathbf{x}_t \in \mathcal{D}} k_\sigma(\mathbf{x}, \mathbf{x}_t)$, [1] showed that an estimator for the Renyi entropy is given by

$$\hat{V}(p) = \frac{1}{N^2}\mathbf{1}^T\mathbf{K}\mathbf{1}, \tag{1}$$

where element (t, t') of the *kernel matrix* $\mathbf{K}$ equals $k_\sigma(\mathbf{x}_t, \mathbf{x}_{t'})$. The parameter σ governs the width of the window function. If the Parzen window is positive semi-definite, such as for example the Gaussian function, then a direct link to Mercer kernel methods is made (see for example [6]). In that case $\hat{V}(p) = \|\mathbf{m}\|^2$, where $\mathbf{m} = \frac{1}{N}\sum_{\mathbf{x}_t \in \mathcal{D}} \phi(\mathbf{x}_t)$ and $\phi(\mathbf{x}_1), \ldots, \phi(\mathbf{x}_N)$ represents the input data mapped to a Mercer kernel feature space. Note that centering of the kernel matrix does not make sense when estimating Renyi entropy. Centering means that $\mathbf{m} = \mathbf{0}$, which again results in $\hat{V}(p) = 0$. Therefore, the kernel matrix is not centered in kernel ECA.

The kernel matrix may be eigendecomposed as $\mathbf{K} = \mathbf{E}\mathbf{D}\mathbf{E}^T$, where $\mathbf{D}$ is a diagonal matrix storing the eigenvalues $\lambda_1, \ldots, \lambda_N$ and $\mathbf{E}$ is a matrix with the corresponding eigenvectors $\mathbf{e}_1, \ldots, \mathbf{e}_N$ as columns. Re-writing Eq. (1), we then have

$$\hat{V}(p) = \frac{1}{N^2}\sum_{i=1}^{N}\left(\sqrt{\lambda_i}\mathbf{e}_i^T\mathbf{1}\right)^2, \tag{2}$$

where $\mathbf{1}$ is a $(N \times 1)$ ones-vector and $\sqrt{\lambda_1}\mathbf{e}_1^T\mathbf{1} \geq, \ldots, \geq \sqrt{\lambda_N}\mathbf{e}_N^T\mathbf{1}$.

Let the kernel feature space data set be represented by $\mathbf{\Phi} = \phi(\mathbf{x}_1), \ldots, \phi(\mathbf{x}_N)$. As shown for example in [7], the projection of $\mathbf{\Phi}$ onto the ith principal axis $\mathbf{u}_i$ in kernel feature space defined by $\mathbf{K}$ is given by $P_{\mathbf{u}_i}\mathbf{\Phi} = \sqrt{\lambda_i}\mathbf{e}_i^T$. This reveals an interesting property of the Renyi entropy estimator. The ith term in Eq. (2) in fact corresponds to the squared sum of the projection onto the ith principal axis in kernel feature space. The first terms of the sum, i.e. the largest values, will contribute most to the entropy of the input space data set. Note that each term depends both on an eigenvalue and on the corresponding eigenvector.

Kernel entropy component analysis represents the input space data set by a projection onto a kernel feature subspace U_k spanned by the k principal axes corresponding to the largest "entropy components", that is, the eigenvalues and eigenvectors comprising the first k terms in Eq. (2). If we collect the chosen k eigenvalues in a $(k \times k)$ diagonal matrix $\mathbf{D}_k$ and the corresponding eigenvectors in the $(N \times k)$ matrix $\mathbf{E}_k$, then the kernel ECA data set is given by

$$\mathbf{\Phi}_{eca} = P_{U_k}\mathbf{\Phi} = \mathbf{D}_k^{\frac{1}{2}}\mathbf{E}_k^T. \tag{3}$$

The ith column of $\mathbf{\Phi}_{eca}$ now represents $\mathbf{\Phi}(\mathbf{x}_i)$ projected onto the subspace. We refer to this as in-sample kernel ECA, since $\mathbf{\Phi}_{eca}$ represents each data point in the original input space sample data set $\mathcal{D}$. We may refer to $\mathbf{\Phi}_{eca}$ as a *spectral* data set, since it is composed of the eigenvalues (spectrum) and eigenvectors of the kernel matrix. The value of k is a user-specified parameter. For an input data set which is composed of subgroups (as revealed by training data), [1] discusses how kernel ECA approximates the "ideal" situation by selecting the value of k equal to the number of subgroups.

In contrast, kernel principal component analysis projects onto the leading principal axes, as determined solely by the largest eigenvalues of the kernel matrix. The kernel matrix may be centered or non-centered[2]. We denote the kernel matrix used in kernel PCA $\mathbf{K}' = \mathbf{V\Delta V}^T$. The kernel PCA mapping is given by $\mathbf{\Phi}_{pca} = \mathbf{\Delta}_k^{\frac{1}{2}}\mathbf{V}_k^T$, using the k *largest eigenvalues* of $\mathbf{K}'$ and the corresponding eigenvectors.

2.2 Out-of-Sample Kernel ECA

In a similar manner as in kernel PCA, out-of-sample data points may also be projected into the kernel ECA subspace obtained based on the sample $\mathcal{D}$. Let the out-of-sample data point be denoted by $\mathbf{x} \to \phi(\mathbf{x})$. The principal axis $\mathbf{u}_i$ in the kernel feature space defined by $\mathbf{K}$ is given by $\mathbf{u}_i = \frac{1}{\sqrt{\lambda_i}}\boldsymbol{\phi}\mathbf{e}_i$, where $\boldsymbol{\phi} = [\phi(\mathbf{x}_1), \ldots, \phi(\mathbf{x}_n)]$ [7] . Moreover, the projection of $\phi(\mathbf{x})$ onto the direction $\mathbf{u}_i$ is given by

$$P_{\mathbf{u}_i}\phi(\mathbf{x}) = \mathbf{u}_i^T\phi(\mathbf{x}) = \frac{1}{\sqrt{\lambda_i}}\mathbf{e}_i^T\mathbf{k}_\mathbf{x}, \tag{4}$$

where $\mathbf{k}_\mathbf{x} = [k_\sigma(\mathbf{x}, \mathbf{x}_1), \ldots, k_\sigma(\mathbf{x}, \mathbf{x}_N)]^T$. The projection $P_{U_k}\phi(\mathbf{x})$ of $\phi(\mathbf{x})$ onto the subspace U_k spanned by the k principal axes as determined by kernel ECA is then

$$P_{U_k}\phi(\mathbf{x}) = \sum_{i=1}^{k} P_{\mathbf{u}_i}\phi(\mathbf{x})\mathbf{u}_i = \sum_{i=1}^{k} \frac{1}{\sqrt{\lambda_i}}\mathbf{e}_i^T\mathbf{k}_\mathbf{x}\frac{1}{\sqrt{\lambda_i}}\boldsymbol{\phi}\mathbf{e}_i = \boldsymbol{\phi}\mathbf{M}\mathbf{k}_\mathbf{x}, \tag{5}$$

[2] Most researchers center the kernel matrix in kernel PCA. But [7] shows that centering is not really necessary. In this paper we consider both.

where $\mathbf{M} = \sum_{i=1}^{k} \frac{1}{\lambda_i} \mathbf{e}_i \mathbf{e}_i^T$ is symmetric. If using kernel PCA, then $\mathbf{D}_k^{\frac{1}{2}}$ and $\mathbf{E}_k$ is replaced by $\mathbf{\Delta}_k^{\frac{1}{2}}$ and $\mathbf{V}_k^T$ and $\mathbf{M}$ is adjusted accordingly. See [4] a detailed analysis of centered kernel PCA.

3 Denoising and Pre-image Mapping

Kernel ECA may produce a strikingly different spectral data set than kernel PCA, as will be illustrated in next section. We want to take advantage of this property for denoising. Given clean training data, the kernel ECA subspace U_k may be found. When utilizing kernel ECA for denoising, a noisy out-of-sample data point $\mathbf{x}$ is projected onto U_k, resulting in $P_{U_k}\phi(\mathbf{x})$. If the subspace U_k represents the clean data appropriately, this operation will remove the noise. The final step is the computation of the pre-image $\hat{\mathbf{x}}$ of $P_{U_k}\phi(\mathbf{x})$, yielding the input space denoised pattern. Here, we will adopt Kwok and Tsang's [4] method for finding the pre-image. The method presented in [4] assumes that the pre-image lies in the span of its n nearest neighbors. The nearest neighbors of $\hat{\mathbf{x}}$ will be equal to the kernel feature space nearest neighbors of $P_{U_k}\phi(\mathbf{x})$, which we denote $\phi(\mathbf{x}_n) \in \mathcal{D}_n$. The algebraic method for finding the pre-image needs Euclidean distance constraints between $\hat{\mathbf{x}}$ and the neighbors $\mathbf{x}_n \in \mathcal{D}_n$. Kwok and Tsang [4] show how to obtain these constraints in kernel PCA via Euclidean distances in the kernel feature space, using an invertible kernel such as the Gaussian. In the following, we show how to obtain the relevant kernel ECA Euclidean distances. We use a Gaussian kernel function. The pseudo-code for kernel ECA pattern denoising is summarized as

Pseudo-Code of Kernel ECA Pattern Denoising

- Based on noise free training data $\mathbf{x}_1, \ldots, \mathbf{x}_N$ determine $\mathbf{K}$ and the kernel ECA projection $\mathbf{\Phi}_{eca} = \mathbf{D}_k^{\frac{1}{2}} \mathbf{E}_k^T$ onto the subspace U_k.
- For a noisy pattern $\mathbf{x}$ do
  1. Project $\phi(\mathbf{x})$ onto U_k by $P_{U_k}\phi(\mathbf{x})$
  2. Determine the feature space Euclidean distances from $P_{U_k}\phi(\mathbf{x})$ to its n nearest neighbors $\phi(\mathbf{x}_n) \in \mathcal{D}_n$
  3. Translate the feature space Euclidean distances into input space Euclidean distances
  4. Find the pre-image $\hat{\mathbf{x}}$ using the input space Euclidean distances (Kwok and Tsang [4])

3.1 Euclidean Distances Based on Kernel ECA

We need the Euclidean distances between $P_{U_k}\phi(\mathbf{x})$ and $\phi(\mathbf{x}_n) \in \mathcal{D}_n$. These are obtained by

$$\tilde{d}^2\left[P_{U_k}\phi(\mathbf{x}), \phi(\mathbf{x}_n)\right] = \|P_{U_k}\phi(\mathbf{x})\|^2 + \|\phi(\mathbf{x}_n)\|^2 - 2P_{U_k}^T\phi(\mathbf{x})\phi(\mathbf{x}_n), \quad (6)$$

where $\|\phi(\mathbf{x}_n)\|^2 = K_{nn} = k_\sigma(\mathbf{x}_n, \mathbf{x}_n)$. Based on the discussion in 2.2, we have

$$\|P_{U_k}\phi(\mathbf{x})\|^2 = (\boldsymbol{\phi}\mathbf{M}\mathbf{k}_\mathbf{x})^T (\boldsymbol{\phi}\mathbf{M}\mathbf{k}_\mathbf{x}) = \mathbf{k}_\mathbf{x}^T\mathbf{M}\mathbf{K}\mathbf{M}\mathbf{k}_\mathbf{x}, \tag{7}$$

since $\mathbf{M}^T = \mathbf{M}$ and $\boldsymbol{\phi}^T\boldsymbol{\phi} = \mathbf{K}$. Moreover

$$P_{U_k}^T\phi(\mathbf{x})\phi(\mathbf{x}_n) = (\boldsymbol{\phi}\mathbf{M}\mathbf{k}_\mathbf{x})^T \phi(\mathbf{x}_n) = \mathbf{k}_\mathbf{x}^T\mathbf{M}\boldsymbol{\phi}^T\phi(\mathbf{x}_n) = \mathbf{k}_\mathbf{x}^T\mathbf{M}\mathbf{k}_{\mathbf{x}_n}, \tag{8}$$

where $\boldsymbol{\phi}^T\phi(\mathbf{x}_n) = \mathbf{k}_{\mathbf{x}_n} = [k_\sigma(\mathbf{x}_n, \mathbf{x}_1), \ldots, k_\sigma(\mathbf{x}_n, \mathbf{x}_N)]^T$. Hence, we obtain a formula for finding the Euclidean distance as $\tilde{d}^2 [P_{U_k}\phi(\mathbf{x}), \phi(\mathbf{x}_n)] = \mathbf{k}_\mathbf{x}^T\mathbf{M}\mathbf{K}\mathbf{M}\mathbf{k}_\mathbf{x} + K_{nn} - 2\mathbf{k}_\mathbf{x}^T\mathbf{M}\mathbf{k}_{\mathbf{x}_n}$.

We may translate the feature space Euclidean distance $\tilde{d}^2 [P_{U_k}\phi(\mathbf{x}), \phi(\mathbf{x}_n)]$ into an equivalent input space Euclidean distance which we may denote d_n^2. Since we use a Gaussian kernel function, we have

$$\exp\left(-\frac{1}{2\sigma^2}d_n^2\right) = \frac{1}{2}\left[\|\phi(\mathbf{x})\|^2 - \tilde{d}^2 [P_{U_k}\phi(\mathbf{x}), \phi(\mathbf{x}_n)] + \|\phi(\mathbf{x}_n)\|^2\right], \tag{9}$$

where $\|\phi(\mathbf{x})\|^2 = \|\phi(\mathbf{x}_n)\|^2 = K_{nn} = 1$. Hence, d_n^2 is given by

$$d_n^2 = -2\sigma^2 \log\frac{1}{2}\left[2 - \tilde{d}^2 [P_{U_k}\phi(\mathbf{x}), \phi(\mathbf{x}_n)]\right]. \tag{10}$$

3.2 The Kwok and Tsang Pre-image Method

Ideally, the pre-image $\hat{\mathbf{x}}$ should obey the distance constraints d_i^2, $i = 1, \ldots, n$, which may be represented by a column vector $\mathbf{d}^2$. However, as pointed out by [4] and others, in general there is no exact pre-image in the input space, so a solution obeying these distance constraints may not exist. Hence, we must settle with an approximation. Using the method in [4], the neighbors are collected in the $(d \times n)$ matrix $\mathbf{X} = [\mathbf{x}_1, \ldots, \mathbf{x}_n]$. These are centered at their centroid $\overline{\mathbf{x}}$ by a centering matrix $\mathbf{H}$. Assuming that the training patterns span a q-dimensional space, a singular value decomposition is performed $\mathbf{XH} = \mathbf{U}\boldsymbol{\Lambda}\mathbf{V}^T = \mathbf{UZ}$, where $\mathbf{Z} = [\mathbf{z}_1, \ldots, \mathbf{z}_n]$ is $(q \times n)$ and $\mathbf{d}_0^2 = \left[\|\mathbf{z}_1\|^2, \ldots, \|\mathbf{z}_n\|^2\right]^T$ represents the squared Euclidean distance of each $\mathbf{x}_n \in \mathcal{D}_n$ to the origin. The Euclidean distance between $\hat{\mathbf{x}}$ and $\mathbf{x}_n$ is required to resemble d_n^2 in a least-square sense. The pre-image is then obtained as $\hat{\mathbf{x}} = -\frac{1}{2}\mathbf{U}\left(\mathbf{Z}\mathbf{Z}^T\right)\mathbf{Z}(\mathbf{d}^2 - \mathbf{d}_0^2) + \overline{\mathbf{x}}$.

4 Experiments

We always use $n = 7$ neighbors in the experiments. When using centered kernel PCA, we denoise as described in [4].

Landsat Image. We consider the Landsat multi-spectral satellite image, obtained from [8]. Each pixel is represented by 36 spectral values. Firstly, we extract the classes *red soil* and *cotton* yielding a two-class data set. The data is normalized to unit variance in each dimension, since we use a spherical kernel function.

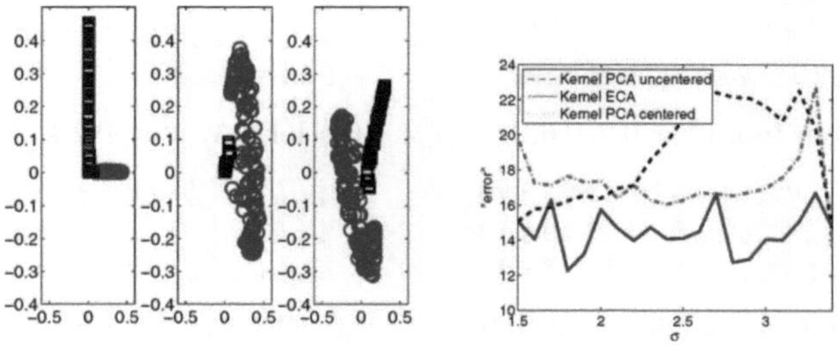

(a) $\sigma = 2.8$ (b)

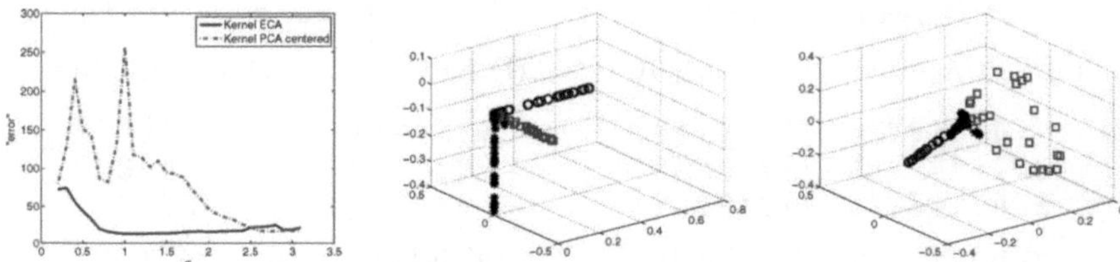

(c) (d) $\sigma = 1.5$ (e) $\sigma = 1.5$

Fig. 1. Denoising results for the Landsat image, using two and three classes

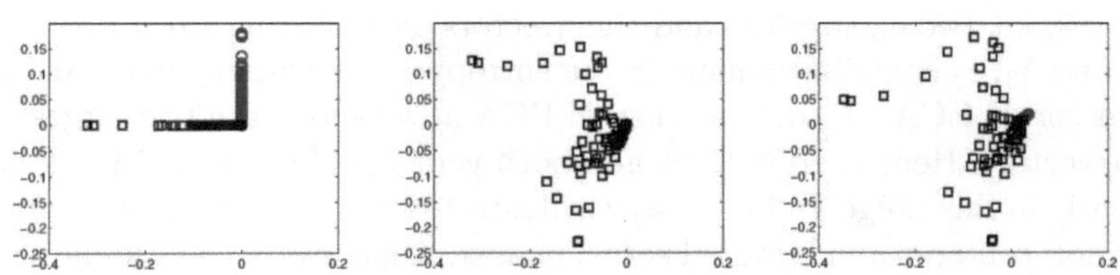

(a) USPS 69 Kernel ECA (b) USPS 69 Kernel PCA non-centered (c) USPS 69 Kernel PCA centered

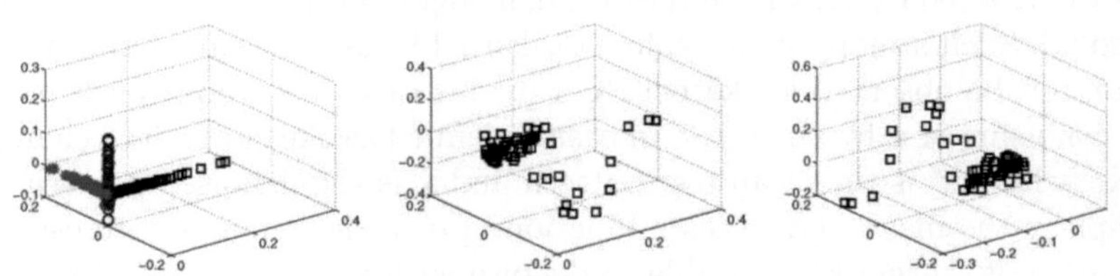

(d) USPS 369 Kernel ECA (e) USPS 369 Kernel PCA non-centered (f) USPS 369 Kernel PCA centered

Fig. 2. Examples of Kernel ECA and kernel PCA mapping for USPS handwritten digits

The clean training data is represented by 100 data points drawn randomly from each class. We add Gaussian noise with variance $v^2 = 0.2$ to 50 random test data points (25 from each class, not in the training set). Since there are two classes, we

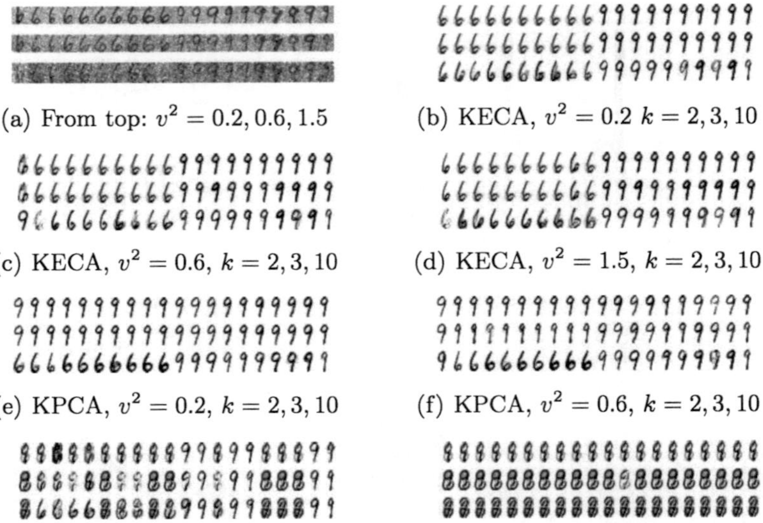

(a) From top: $v^2 = 0.2, 0.6, 1.5$
(b) KECA, $v^2 = 0.2$ $k = 2, 3, 10$
(c) KECA, $v^2 = 0.6$, $k = 2, 3, 10$
(d) KECA, $v^2 = 1.5$, $k = 2, 3, 10$
(e) KPCA, $v^2 = 0.2$, $k = 2, 3, 10$
(f) KPCA, $v^2 = 0.6$, $k = 2, 3, 10$
(g) KT, $v^2 = 0.2$, $k = 2, 3, 10$
(h) KT, $v^2 = 0.6$, $k = 2, 3, 10$

Fig. 3. Denoising of USPS digits 6 and 9

use $k = 2$, i.e. two eigenvalues and eigenvectors. For a kernel size $1.6 < \sigma < 3.3$, $\lambda_1, \mathbf{e}_1$ and $\lambda_3, \mathbf{e}_3$ contributes most to the entropy of the training data, and is thus used in kernel ECA. In contrast, kernel PCA always uses the two largest eigenvalues/vectors. Hence kernel ECA and both versions of kernel PCA will denoise differently in this range. In Fig. 1 (a) we illustrate from left to right $\mathbf{\Phi}_{eca}$ and $\mathbf{\Phi}_{pca}$ (using non-centered and centered kernel matrix, respectively.) The kernel size $\sigma = 2.8$ is used and the classes are marked by different symbols for clarity. Observe how kernel ECA produces a data set with an angular structure, in the sense that each class is distributed radially from the origin, in angular directions which are almost orthogonal. Such a mapping is typical for kernel ECA. The same kind of separation can not be observed for kernel PCA in this case. We quantify the goodness of the denoising of $\mathbf{x}$ by an "error" measure defined as the sum of the elements in $|\mathbf{x}' - \hat{\mathbf{x}}|$, where $\mathbf{x}'$ is the clean test pattern and $\hat{\mathbf{x}}$ is the denoised pattern. Fig. 1 (b) displays the mean "error" as a function of σ in the range of interest. Of the three methods, kernel ECA is able to denoise with the least error. Secondly, we create a three-class data set by extracting the classes *cotton*, *damp grey soil* and *vegetation stubble*. Fig. 1 (c) shows the denoising error in this case (300 training data, 100 test data) for kernel ECA and centered kernel PCA. Fig. 1 (d) and (e) show $\mathbf{\Phi}_{eca}$ and (centered) $\mathbf{\Phi}_{pca}$ for $\sigma = 1.5$ (omitting non-centered kernel PCA in this case to save space). Kernel ECA uses $\lambda_1, \mathbf{e}_1$, $\lambda_3, \mathbf{e}_3$ and $\lambda_4, \mathbf{e}_4$. Also in this case, kernel ECA separates the classes in angular directions. This seems to have a postitive effect on the denoising.

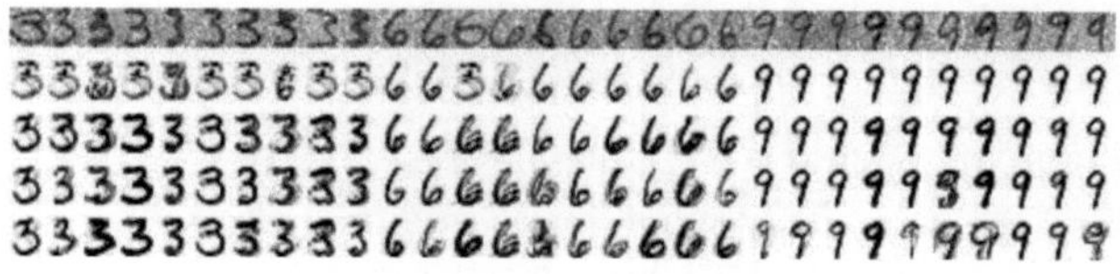

(a) KECA, $\sigma = 3.0$, k=3,8,10,15.

(b) KPCA, $\sigma = 3.0$, k=3,8,10,15.

Fig. 4. Denoising of USPS digits 3, 6 and 9

USPS Handwritten Digits. Denoising experiments are conducted on the (16×16) USPS handwritten digits, obtained from [8], and represented by (256×1) vectors. We concentrate on two and three class problems. In the former case, the data set is composed of the digits six and nine, denoted USPS 69. In the latter case, we use the digits three, six and nine, denoted USPS 369. For USPS 69, we use $k = 2$, since there are two classes. For $\sigma > 3.7$ the the two top kernel ECA eigenvalues are λ_1 and λ_2, which are the same as the two top kernel PCA eigenvalues. Hence, the denoising results will be equal for kernel ECA and non-centered kernel PCA in this case. However, for $\sigma \leq 3.7$, the two top kernel ECA eigenvalues are always different from λ_1 and λ_2, so that kernel ECA and both versions of kernel PCA will be different. As an example, Fig. 2 (a) shows $\mathbf{\Phi}_{eca}$ for $\sigma = 2.8$. Here, $\lambda_1, \mathbf{e}_1$ and $\lambda_7, \mathbf{e}_7$ is used. Notice also for this data set the typical angular separation provided by kernel ECA. In contrast, Fig. 2 (b) and (c) show non-centered and centered $\mathbf{\Phi}_{pca}$, respectively. Notice how one class (the "nine"s, marked by squares) dominate the other class, especially in (b). Fig. 3 (a) shows ten test digits from each class, with noise added. From top to bottom panel, we have noise variances $v^2 = 0.2, 0.6, 1.5$. Since there are two classes, we initially perform the denoising using $k = 2$. However, we also show results with more dimensions added to the subspace U_k, for $k = 3$ and $k = 10$. Fig. 3 (b), (c) and (d) show the kernel ECA results (denoted KECA) for $\sigma = 2.8$ for $v^2 = 0.2, 0.6, 1.5$, respectively. The top panel in each subfigure corresponds to $k = 2$, the middle panel corresponds to $k = 3$, and in the bottom panel $k = 10$ is used. For all noise variances kernel ECA performs very robustly. The results are very good for $k = 2$, so the inclusion of more dimensions in the subspace U_k does not seem to be necessary. Notice that the shape of the denoised patterns are quite similar. It seems as if the method produces a kind of prototype for each class. This behavior will be further studied below. Fig. 3 (e) and (f) show the non-centered kernel PCA results (denoted KPCA) for $\sigma = 2.8$ for $v^2 = 0.2, 0.6$, respectively. In both cases, for $k = 2$ and $k = 3$, the "nine"

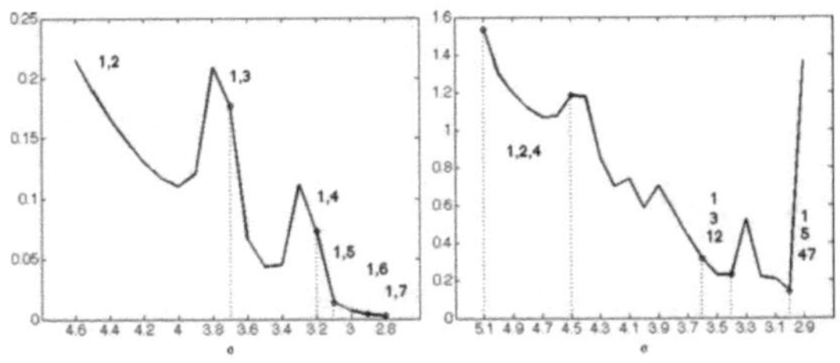

(a) (b)

Fig. 5. Computing the Cauchy-Schwarz class separability criterion as a function of σ

class totally dominates the "six" class. For each noisy pattern, this means that the nearest neighbors of $P_{U_k}\phi(\mathbf{x})$ will always belong to the "nine" class. If we project onto more principal axes, the method improves, as shown in the bottom panel of each figure for $k = 10$. Clearly, however, for small subspaces U_k kernel ECA performs significantly better than non-centered kernel PCA. Fig. 3 (g) and (h) show the centered kernel PCA results (denoted KT after Kwok and Tsang). In this case, the KT results are much inferior to kernel ECA. Including more principal axes improves the results somewhat, but more dimensions are clearly needed.

When it comes to USPS 369, for $\sigma \leq 5.1$, the three top kernel ECA eigenvalues are always different than λ_1, λ_2 λ_3, such that kernel ECA and both versions of kernel PCA will be different. For example, for $\sigma = 3.0$ kernel ECA uses $\lambda_1, \mathbf{e}_1$, $\lambda_5, \mathbf{e}_5$ and $\lambda_{47}, \mathbf{e}_{47}$, producing a data set with a clear angular structure as shown in Fig. 2 (d). In contrast, Fig. 2 (e) and (f) show non-centered and centered $\mathbf{\Phi}_{pca}$, respectively. The data is not separated as clearly as in kernel ECA. This has an effect on the denoising. Fig. 4 (a) shows the kernel ECA results for $\sigma = 3.0$ for $v^2 = 0.2$ and $k = 3, 8, 10, 15$ (from top to bottom.) Using only $k = 3$, kernel ECA for the most part provides reasonable denoising results, but has some problems distinguishing between the "six" class and the "three" class. In this case, it helps to expand the subspace U_k by including a few more dimensions. At $k = 8$, for instance, the results are very good by visual inspection. Fig. 4 (b) shows the corresponding non-centered kernel PCA results (centered kernel PCA omitted due to space limitations.) Also in this case, the "nine" class dominates the other two classes. When using $k = 15$ principal axes, the results starts to improve, in the sense that all classes are represented. As a final experiment on the USPS 69 and USPS 369 data, we measure the sum of the cosine of the angle between all pairs of class mean vectors of the kernel ECA data set $\mathbf{\Phi}_{eca}$ as a function of σ. This is equivalent to computing the Cauchy-Schwarz divergence between the class densities as estimated by Parzen windowing [1], and may hence be considered a class separability criterion. We require that the top k entropy components must account for at least 50% of the total sum of the entropy components, see Eq. (2). Fig. 5 (a) shows the result for USPS 69 using $k = 2$. The eigenvalues/vectors used in a region of σ are indicated by the numbers above the graph. In this case, the stopping criterion is met for $\sigma = 2.8$, which yields the smallest value, i,e, the best separation using $\lambda_1, \mathbf{e}_1$ and $\lambda_7, \mathbf{e}_7$. Fig. 5 (b) shows the result for USPS 369 using $k = 3$. In this case, the best result is obtained for $\sigma = 3.0$ using $\lambda_1, \mathbf{e}_1$, $\lambda_5, \mathbf{e}_5$ and $\lambda_{47}, \mathbf{e}_{47}$. These experiments indicate that such a class separability criterion makes sense in kernel ECA, providing the angular structure observed on $\mathbf{\Phi}_{eca}$, and may be developed into a method for selecting an appropriate σ. This is however an issue which needs further attention in future work.

Finally, we remark that in preliminary experiments not shown here, it appear as if kernel ECA may be a more beneficial alternative to kernel PCA if the number of classes in the data set is relatively low. If there are may classes, more eigenvalues and eigenvectors, or principal components, will be needed by both methods, and as the number of classes grows, the two methods will likely share more and more components.

5 Conclusions

Kernel ECA may produce strikingly different spectral data sets than kernel PCA, separating the classes angularly, in terms of the kernel feature space. In this paper, we have exploited this property, by introducing kernel ECA for pattern denoising using the pre-image method proposed in [4]. This requires kernel ECA pre-images to be computed, as derived in this paper. The different behavior of kernel ECA vs. kernel PCA have in our experiments a positive effect on the denoising results, as demonstrated on real data and on toy data.

References

1. Jenssen, R., Eltoft, T., Girolami, M., Erdogmus, D.: Kernel Maximum Entropy Data Transformation and an Enhanced Spectral Clustering Algorithm. In: Advances in Neural Information Processing Systems 19, pp. 633–640. MIT Press, Cambridge (2007)
2. Schölkopf, B., Smola, A.J., Müller, K.-R.: Nonlinear Component Analysis as a Kernel Eigenvalue Problem. Neural Computation 10, 1299–1319 (1998)
3. Mika, S., Schölkopf, B., Smola, A., Müller, K.R., Scholz, M., Rätsch, G.: Kernel PCA and Denoising in Feature Space. In: Advances in Neural Information Processing Systems, 11, pp. 536–542. MIT Press, Cambridge (1999)
4. Kwok, J.T., Tsang, I.W.: The Pre-Image Problem in Kernel Methods. IEEE Transactions on Neural Networks 15(6), 1517–1525 (2004)
5. Park, J., Kim, J., Kwok, J.T., Tsang, I.W.: SVDD-Based Pattern Denoising. Neural Computation 19, 1919–1938 (2007)
6. Jenssen, R., Erdogmus, D., Principe, J.C., Eltoft, T.: The Laplacian PDF Distance: A Cost Function for Clustering in a Kernel Feature Space. In: Advances in Neural Information Processing Systems 17, pp. 625–632. MIT Press, Cambridge (2005)
7. Shawe-Taylor, J., Cristianini, N.: Kernel Methods for Pattern Analysis. Cambridge University Press, Cambridge (2004)
8. Murphy, R., Ada, D.: UCI Repository of Machine Learning databases. Tech. Rep., Dept. Comput. Sci. Univ. California, Irvine (1994)

Combining Local Feature Histograms of Different Granularities*

Ville Viitaniemi and Jorma Laaksonen

Department of Information and Computer Science, Helsinki University of Technology,
P.O. Box 5400, FI-02015 TKK, Finland
{ville.viitaniemi,jorma.laaksonen}@tkk.fi

Abstract. Histograms of local features have proven to be powerful representations in image category detection. Histograms with different numbers of bins encode the visual information with different granularities. In this paper we experimentally compare techniques for combining different granularities in a way that the resulting descriptors can be used as feature vectors in conventional vector space learning algorithms. In particular, we consider two main approaches: fusing the granularities on SVM kernel level and moving away from binary or hard to soft histograms. We find soft histograms to be a more effective approach, resulting in substantial performance improvement over single-granularity histograms.

1 Introduction

In supervised image category detection the goal is to predict whether a novel test image belongs to a category defined by a training set of positive and negative example images. The categories can correspond, for example, to the presence or absence of a certain object, such as a dog. In order to automatically perform such a task based on the visual properties of the images, one must use a representation for the properties that can be extracted automatically from the images.

Histograms of local features have proven to be powerful image representations for image classification and object detection. Consequently their use has become commonplace in image content analysis tasks. This paradigm is also known by the name Bag of Visual Words (BoV) in analogy with the successful Bag of Words paradigm in text retrieval. In this analogue, images correspond to documents and different local feature values to words.

Use of local image feature histograms for supervised image classification and characterisation can be divided into several steps:

1. Selecting image locations of interest.
2. Describing each location with suitable visual descriptors (e.g. SIFT).
3. Characterising the distribution of the descriptors within each image with a histogram.

* Supported by the Academy of Finland in the *Finnish Centre of Excellence in Adaptive Informatics Research* project.

A.-B. Salberg, J.Y. Hardeberg, and R. Jenssen (Eds.): SCIA 2009, LNCS 5575, pp. 636–645, 2009.

4. Using the histograms as feature vectors representing the images in a supervised vector space algorithm, such as SVM.

All parts of the BoV pipeline are subject of continuous study. However, for this paper we regard the beginning of the chain (steps 1, 2 and partially also 3) as given. The alternative techniques we describe and evaluate in the subsequent sections place themselves and extend the step 3 of the pipeline. They can be regarded as different histogram creation and post-processing techniques that build on top of the readily-existing histogram codebooks used in our baseline implementation. Step 4 is once again regarded as given for the current studies.

The process of forming histograms loses much information about the details of the descriptor distribution. Information reduction step is, however, necessary in order to be able to perform the fourth step using conventional methods. In the histogram representation the continuous distance between two visual descriptors is reduced to a single binary decision: whether the descriptors are deemed similar (i.e. fall into the same histogram bin) or not.

Selecting the number of bins used in the histograms—i.e. the histogram size—directly determines how coarsely the visual descriptors are quantised and subsequently compared. In this selection, there is a trade-off involved. A small number of bins leads to visually rather different descriptors being regarded as similar. On the other other hand, too numerous bins result in visually rather similar descriptors ending up in different histogram bins and regarded as dissimilar. The latter problem is not caused by the histogram representation itself, but the desire to use the histograms as structureless feature vectors in the step 4 above so that conventional learning algorithms can be used.

Earlier [8] we have performed category detection experiments where we have compared ways to select a codebook for a single histogram representation, with varying histogram sizes. For the experiments we used the images and category detection tasks of the publicly available VOC2007 benchmark. In this paper we extend these experiments by proposing and evaluating methods for simultaneously taking information from several levels of descriptor-space granularity into account while still retaining the possibility to use the produced image representations as feature vectors in conventional vector space learning methods. In the first of the considered methods, histograms of different granularities are concatenated with weights, corresponding to a multi-granularity kernel function in the SVM. This approach is closely related to the pyramid matching kernel method of [4]. We also propose two ways of modifying the histograms so that the descriptor-space similarity of the histogram bins and descriptors of the interest points are better taken into account: the post smoothing and soft histogram techniques.

The rest of this paper is organised as follows. Our baseline BoV implementation and its proposed improvements are described in Sections 2 through 5. Section 6 details the experiments that compare the algorithmic variants. In Section 7 we summarise the experiments and draw our conclusions.

2 Baseline System

In this section we describe our baseline implementation of the Bag of Visual Words pipeline of Sect. 1. In the first stage, a number of interest points are identified in each image. For these experiments, the interest points are detected with a combined Harris-Laplace detector [6] that outputs around 1200 interest points on average per image for the images used in this study. In step 2 the image area around each interest point is individually described with a 128-dimensional SIFT descriptor [5], a widely-used and rather well-performing descriptor that is based on local edge statistics.

In step 3 each image is described by forming a histogram of the SIFT descriptors. We determine the histogram bins by clustering a sample of the interest point SIFT descriptors (20 per image) with the Linde-Buzo-Gray (LBG) algorithm. In our earlier experiments [8] we have found such codebooks to perform reasonably well while the computational cost associated with the clustering still remains manageable. The LBG algorithm produces codebooks with sizes in powers of two. In our baseline system we use histograms with sizes ranging from 128 to 8192. In some subsequently reported experiments we also employ codebook sizes 16384 and 32768.

In the final fourth step the histogram descriptors of both training and test images are fed into supervised probabilistic classifiers, separately for each of the 20 object classes. As classifiers we use weighted C-SVC variants of the SVM algorithm, implemented in the version 2.84 of the software package LIBSVM [2]. As the kernel function g we employ the exponential χ^2-kernel

$$g_{\chi^2}(\mathbf{x}, \mathbf{x}') = \exp\left(-\gamma \sum_{i=1}^{d} \frac{(x_i - x_i')^2}{x_i + x_i'}\right) . \tag{1}$$

The free parameters of the C-SVC cost function and the kernel function are chosen on basis of a search procedure that aims at maximising the six-fold cross validated area under the receiver operating characteristic curve (AUC) measure in the training set. To limit the computational cost of the classifiers, we perform random sampling of the training set. Some more details of the SVM classification stage can be found in [7].

In the following we investigate techniques for fusing together information from several histograms. To provide comparison reference for these techniques, we consider the performance of post-classifier fusion of the detection results based on the histograms in question. For classifier fusion we employ Bayesian Logistic Regression (BBR) [1] that we have found usually to perform at least as well as other methods we have evaluated (SVM, sum and product fusion mechanism) for small sets of similar features.

3 Speed-Up Technique

For the largest codebooks, the creation of histograms becomes impractically time-consuming if implemented in a straightforward fashion. Therefore, a speed-up structure is employed to facilitate fast approximate nearest neighbour search.

The structure is formed by creating a succession of codebooks diminishing in size with the k-means clustering algorithm. The structure is employed in the nearest-neighbour search of vector $\mathbf{v}$ by first determining the closest match of $\mathbf{v}$ in the smallest of the codebooks, then in the next larger codebook. This way a match is found in successively larger codebooks, and eventually among the original codebook vectors. The time cost of this search algorithm is proportional to the logarithm of the codebook size. In our evaluations, the approximative algorithm comes rather close to the full search in terms of both MSE quantisation error and category detection MAP. Despite some degradation of performance, the speed-up structure is necessary as it facilitates the practical use of larger codebooks than would otherwise be feasible. The technique of soft histogram forming (Section 5) is able to make use of such large codebooks.

4 Multi-granularity Kernel

In this section we describe the first one of the considered techniques for combining descriptor similarity on various level of granularity. In this technique we extend the kernel of the SVM to take into account not only a single SIFT histogram H, but a whole set of histograms $\{H_i\}$. To form the kernel, we evaluate the multi-granularity distance d_m between two images as a weighted sum of distances d_i in different granularities i, i.e. evaluated by a means of the distance

$$d_{\mathrm{m}} = \sum_i w_i d_i, \quad w_i = N_i^{1/K}. \tag{2}$$

The distance d_i is evaluated as the χ^2 distance between two histograms of granularity i. In the formula for weight w_i, N_i is the number of bins in histogram i and K is a free parameter of the method that can be thought to correspond to the dimensionality of the space the histograms quantise. Value $K = \infty$ corresponds to unweighted concatenation of the histograms. The distance values d_{m} are used to form a kernel for SVM through exponential function, just as in the baseline technique:

$$g_{\mathrm{m}} = \exp(-\gamma d_{\mathrm{m}}). \tag{3}$$

The described technique is related to the pyramid match kernel introduced in [4]. Also there the image similarity is a weighted sum of similarities of histograms of different granularities. However, the authors of [4] use histogram intersection as the similarity measure. They use similarities directly as kernel values, leading to also the kernel being a linear combination of similarity values. In our method this is not the case. Another difference is that in [4] the descriptor space is partitioned to histogram bins with a fixed grid whereas we employ data-adaptive clustering. Furthermore, the bins in our histograms are not hierarchically related, i.e. bins in larger histograms are not mere subdivisions of the bins in smaller histograms.

The functional form of our weighting scheme is borrowed from [4]. Despite the seemingly similar form of the weighting function, their weighting scheme results in different relative weights being assigned to distances in different resolutions. This is because their histogram intersection measure is invariant to the number of histogram bins whereas our distance measure is not.

5 Histogram Smoothing Methods

In this section we describe and evaluate methods that try to leverage from the knowledge that we possess of the descriptor-space similarity of the histogram bins. In the baseline method for creating histograms, two descriptors falling into different histogram bins are considered equally different, regardless of whether the codebook vectors of the histogram bins are neighbours or far from each other in the descriptor space.

5.1 Post-smoothing

Our first remedy is a post-processing method of the binary histograms that is subsequently denoted *post-smoothing*. In this method a fraction λ of the hit count c_i of histogram bin i is spread to its n_{nbr} closest neighbours. Among the neighbours, the hit count is distributed in proportion to inverse squared distance from the originating histogram bin. This histogram smoothing scheme has the convenient property that it can be applied to readily created histograms without the need to redo the hit counting which is relatively time-consuming. Alternatively, this smoothing scheme could be implemented as a modification to the SVM kernel function.

5.2 Soft Histograms

The latter of the described methods (denoted the *soft histogram* method from here on) specifically redefines the way the histograms are created. Hard assignments of descriptors to histogram bins are replaced with soft ones. Thus each descriptor increments not only the hit count of the bin whose codebook vector is closest to the descriptor, but the counts of all the n_{nbr} closest bins. The increments are no longer binary, but are determined as a function to the closeness of the codebook vectors of the histogram bins to the descriptor.

We evaluated several proportionality functions for distributing bin increments Δ_i among the k histogram bins nearest to the descriptor $\mathbf{v}$:

1. inverse Euclidean distance : $\Delta_i \propto \|\mathbf{v}_i - \mathbf{v}\|^{-1}$
2. squared inverse Euclidean distance : $\Delta_i \propto \|\mathbf{v}_i - \mathbf{v}\|^{-2}$
3. (negative) exponential of Euclidean distance : $\Delta_i \propto \exp(-\alpha_{\mathrm{exp}} \frac{\|\mathbf{v}_i - \mathbf{v}\|}{d_0})$
4. Gaussian : $\Delta_i \propto \exp(-\alpha_{\mathrm{g}} \frac{\|\mathbf{v}_i - \mathbf{v}\|^2}{d_0^2})$

Here the normalisation term d_0 is the average distance between two neighbouring codebook vectors.

6 Experiments

6.1 Category Detection Task and Experimental Procedures

In this paper we consider the supervised image category detection problem. Specifically, we measure the performance of several algorithmic variants for the task using images and categories defined in the PASCAL NoE Visual Object Classes (VOC) Challenge 2007 collection [3]. In the collection there are altogether 9963 photographic images of natural scenes. In the experiments we use the half of them (5011 images) denoted "trainval" by the challenge organisers.

Each of the images contains at least one occurrence of the defined 20 object classes, including e.g. several types of vehicles (bus,car,bicycle etc.), animals and furniture. The presences of these objects in the images were manually annotated by the organisers. In many images there are objects of several classes present. In the experiments (and in the "classification task" of VOC Challenge) each object class is taken to define an image category.

In the experiments the 5011 images are partitioned approximately equally into training and test sets. Every experiment was performed separately for each of the 20 object classes. The category detection accuracy is measured in terms of non-interpolated average precision (AP). The AP values were averaged over the 20 object classes and six different train/test partitionings. The resulting average MAP values tabulated in the result tables had 95% confidence intervals of the order 0.01 in all the experiments. This means that, for some pairs of techniques with nearly the same MAP values, the order of superiority can not be stated very confidently on basis of a single experiment. However, in the experiments the discussed techniques are usually evaluated with several different histogram codebook sizes and other algorithmic variations. Such experiment series usually lead to rather definitive conclusions. Moreover, because of systematic differences between the six trials, the confidence intervals arguably underestimate the reliability of the results for the purpose of comparing various techniques. The variability being similar for all the results, we do not clutter the tables of results with confidence intervals.

The row χ^2 of Table 1 summarises the category detection performance of the baseline system for different codebook sizes. A fact worth noting is that for the baseline histograms, the performance seemingly saturates at codebook size around 4096 and starts to degrade for larger codebooks.

Our multi-granularity kernel employs the χ^2 distance measure whereas histogram intersection is used in [4]. It is therefore of interest to know if there is essential difference in the performances of the distance measures. Our experiments with histograms of single granularity (Table 1) point to the direction that for category detection, the exponential χ^2-kernel might be more suitable measure of histogram similarity than histogram intersection, although we did not explicitly evaluate this in the case of multiple granularities. It seems safe to say that at least the use of the χ^2 distance does not make the multi-granularity kernel any weaker. This belief is supported by the anecdotal evidence of the χ^2-distance and exponential kernels often working well in image processing applications.

Table 1. Comparison of the MAP performance of χ^2 and histogram intersection distance measures for single-granularity histograms

	size						
	128	256	512	1024	2048	4096	8192
χ^2	0.357	0.376	0.387	0.397	0.400	0.404	0.398
HI	0.333	0.353	0.359	0.367	0.387	0.380	0.381

6.2 Multi-granularity Kernel

In Table 2 we show the classification performance of the multi-granularity kernel technique. The different columns correspond to combinations of increasing sets of histograms. In the experiments we use LBG codebooks with sizes from 128 to 8192. The upper rows of the table correspond to different values of the weighting parameter K. The MAP values can be compared against the individual-granularity baseline values (row "indiv.") for the largest of the involved histograms, and also against the performance of post-classifier fusion of the histograms in question (row "fusion"). From the table one can observe that better performance is obtained by combining distances of multiple granularities already in the kernel calculation —just as the proposed technique does—rather than fusing the classifier outcomes later. Both methods for combining several granularities perform clearly better than the best one of the individual granularities. No weighting parameter value K was found that would significantly outperform the unweighted sum of distances ($K = \infty$).

In the tabulated experiments the speedup structure of Section 3 was not used. We repeated some of the experiments using the speedup structure with essentially no difference in MAP performance. The additional experiments also reveal that the inclusion of histograms larger than 8192 bins no longer improves the MAP.

6.3 Histogram Smoothing

For post-smoothing of histograms, we evaluated the category detection MAP for several values of λ and n_{nbr}. In the experiments the 2048 unit LBG histogram was used as a starting point. The best parameter value combination we

Table 2. MAP performance of the multi-granularity kernel technique

K	128–256	128–512	128–1024	128–2048	128–4096	128–8192
1	0.376	0.385	0.391	0.395	0.398	0.399
2	**0.382**	0.394	0.402	0.409	0.413	0.414
4	**0.382**	**0.399**	0.407	0.413	0.418	0.421
∞	0.379	0.396	0.409	**0.418**	**0.423**	**0.425**
-4	0.377	**0.399**	**0.411**	0.417	0.422	0.422
indiv.	0.376	0.387	0.397	0.400	0.404	0.398
fusion (BBR)	0.380	0.396	0.404	0.409	0.414	0.415

Table 3. MAP performance of different smoothing functions of the soft histogram technique for LBG codebook with 2048 codebook vectors

	n_{nbr} 3	5	8	10	15
inverse Euclidean	0.426	0.427	-	0.421	-
inverse squared Euclidean	0.426	0.429	0.427	-	
negexp ($\alpha_{\text{exp}} = 3$)	0.428	0.433	0.435	0.435	0.433
Gaussian ($\alpha_{\text{g}} = 0.3$)	0.428	0.432	0.435	0.435	0.432

tried resulted in MAP 0.407 that is a slight improvement over the baseline MAP 0.400. The soft histogram technique, discussed next, provided clearly better performance which made more thorough testing of post-smoothing unappealing.

For the soft histogram technique, Table 3 compares the four different functional forms of smoothing functions for LBG codebook of size 2048. Among these, the exponential and Gaussian seem to provide somewhat better performance than the others. We evaluated the effect of the parameters α_{exp} and α_{g} to the detection performance and found the peak in performance to be broad in the parameter values. In these experiments, as well as in all subsequent ones, we use the value $n_{\text{nbr}} = 10$. The Gaussian functional form was chosen for the subsequent experiments of the two almost equally well performing functional forms of the exponential family.

In Table 4, a selection of MAP accuracies of the Gaussian soft histogram technique is shown for several different histogram sizes. The results for larger codebook sizes (512 and beyond) are obtained using the speed-up technique of Section 3. The results can be compared with the MAP of hard assignment baseline histograms on column "hard". It can be seen that the improvement brought by the soft histogram technique is substantial, except for the smallest histograms. This is intuitive since in small histograms the centers of the different histogram bins are far apart in the descriptor space and should therefore not be considered similar. For hard assignment histograms, the performance peaks with

Table 4. MAP performance of the soft histogram technique for different codebook sizes (rows) and different values of parameter α_{g} (columns)

	hard	α_{g} 0.05	0.1	0.2	0.3	0.5	1
256	0.376	-	-	0.376	0.381	0.385	0.384
512	0.388	-	-	-	-	0.406	-
1024	0.393	-	-	-	0.419	-	-
2048	0.400	0.423	0.429	0.433	0.435	0.433	0.423
4096	0.403	-	-	0.438	-	-	-
8192	0.395	0.443	0.445	0.448	0.445	0.434	0.419
16384	0.392	0.450	0.451	0.451	-	-	-
32768	0.387	-	-	0.448	-	-	-

Table 5. The percentage of non-zero bin counts in various-sized histograms collected using either hard (conventional) or soft assignment to histogram bins

	512	1024	2048	4096	8192	16384
Hard histograms	53.47	35.33	21.45	12.11	6.72	3.63
Soft histograms	86.74	72.55	55.96	37.58	26.15	17.07

Table 6. MAP performance of combining soft histograms with the multi-granularity kernel technique

K	128–256	128–512	128–1024	128–4096	128–8192	128–16384	128–32768
4	0.383	0.398	0.407	0.419	0.422	0.426	0.428
∞	0.377	0.395	0.408	0.427	0.432	0.437	0.442
indiv.	0.385	0.406	0.419	0.438	0.448	0.451	0.448
fusion (BBR)	0.385	0.405	0.416	0.433	0.442	0.447	0.447

histograms of size 4096. The soft histogram technique makes larger histograms than this beneficial, the observed peak being at size 16384.

The improved accuracy brought by the histogram smoothing techniques comes with the price of sacrificing some sparsity of the histograms. Table 5 quantifies this loss of sparsity. This could be of importance from the point of view of computational costs if the classification framework represents the histograms in a way that benefits from sparsity (which is not the case in our implementation).

Table 6 presents the results of combining soft histograms with the multi-granularity kernel technique. From the results, it is evident that combining these two techniques does not bring further performance gain over the soft histograms. On the contrary, the MAP values of the combination are clearly lower than those of the largest soft histograms included in the combination (row "indiv.").

7 Conclusions

In this paper we have investigated methods of combining information in local feature histograms of several granularities in the descriptor space. The presented methods are such that the resulting histogram-like descriptors can be used as feature vectors in conventional vector space learning methods (here SVM), just as the histograms would be.

The methods have been evaluated in a set of image category detection tasks. By using the best one of the methods, a significant improvement of MAP from 0.404 to 0.451 was obtained in comparison with the best-performing histogram of a single granularity. Of the techniques, the soft assignment of descriptors to histogram bins resulted in clearly the best performance. Histogram smoothing as post-processing improved the performance only slightly over the single-granularity histograms. The multi-granularity kernel technique was better than the baseline of single-granularity histograms with maximum MAP 0.425, but clearly inferior to soft histograms. Combining soft histograms with the multi-granularity kernel technique did not result in performance gain, supporting the conclusion that the both techniques leverage on the same information and are thus redundant. The soft histogram technique adds some computational cost in comparison with individual hard histograms as it becomes beneficial to use larger histograms, and the generated histograms are less sparse.

The issue of the generalisability of the described techniques is not addressed by the experiments of this paper. It seems plausible that this kind of smoothing methods would be usable also in other kinds of image analysis tasks and also with other local descriptors than just SIFT.

The selection of the parameters of the methods is another open issue. Currently we have demonstrated that there exists parameter values (such as α_g in the soft histogram technique) that result in good performance. Finding such values has not been addressed here. Reasonably good parameter values could in practice be picked e.g. by cross-validation.

Of the discussed methods, the best performance was obtained by the soft histogram technique. However, the LBG codebooks for the histograms were generated with a conventional hard clustering algorithm. Using also here an algorithm specifically targeted at soft clustering instead—such as fuzzy c-means—could be beneficial. Yet, this is not so self-evident as the category detection performance is not the immediate target function optimised by the clustering algorithms.

References

1. Madigan, D., Genkin, A., Lewis, D.D.: BBR: Bayesian logistic regression software (2005), `http://www.stat.rutgers.edu/~madigan/BBR/`
2. Chang, C., Lin, C.: LIBSVM: a library for support vector machines (2001), `http://www.csie.ntu.edu.tw/~cjlin/libsvm`
3. Everingham, M., Van Gool, L., Williams, C.K.I., Winn, J., Zisserman, A.: The PASCAL Visual Object Classes Challenge 2007 (VOC 2007) (2007), `http://www.pascal-network.org/challenges/VOC/voc2007/workshop/index.html`
4. Grauman, K., Darrell, T.: The pyramid match kernel: Efficient learning with sets of features. Journal of Machine Learning Research (2007)
5. Lowe, D.G.: Distinctive image features from scale-invariant keypoints. International Journal of Computer Vision 60(2), 91–110 (2004)
6. Mikolajcyk, K., Schmid, C.: Scale and affine point invariant interest point detectors. International Journal of Computer Vision 60(1), 68–86 (2004)
7. Viitaniemi, V., Laaksonen, J.: Improving the accuracy of global feature fusion based image categorisation. In: Falcidieno, B., Spagnuolo, M., Avrithis, Y., Kompatsiaris, I., Buitelaar, P. (eds.) SAMT 2007. LNCS, vol. 4816, pp. 1–14. Springer, Heidelberg (2007)
8. Viitaniemi, V., Laaksonen, J.: Experiments on selection of codebooks for local image feature histograms. In: Sebillo, M., Vitiello, G., Schaefer, G. (eds.) VISUAL 2008. LNCS, vol. 5188, pp. 126–137. Springer, Heidelberg (2008)

Extraction of Windows in Facade Using Kernel on Graph of Contours

Jean-Emmanuel Haugeard*, Sylvie Philipp-Foliguet, Frédéric Precioso, and Justine Lebrun

ETIS, CNRS, ENSEA, Univ Cergy-Pontoise,
6 avenue du Ponceau, BP 44,F 95014 Cergy Pontoise, France
{jean-emmanuel.haugeard,sylvie.philipp,
frederic.precioso,justine.lebrun}@ensea.fr

Abstract. In the past few years, street-level geoviewers has become a very popular web-application. In this paper, we focus on a first urban concept which has been identified as useful for indexing then retrieving a building or a location in a city: the windows. The work can be divided into three successive processes: first, object detection, then object characterization, finally similarity function design (kernel design). Contours seem intuitively relevant to hold architecture information from building facades. We first provide a robust window detector for our unconstrained data, present some results and compare our method with the reference one. Then, we represent objects by fragments of contours and a relational graph on these contour fragments. We design a kernel similarity function for structured sets of contours which will take into account the variations of contour orientation inside the structure set as well as spatial proximity. One difficulty to evaluate the relevance of our approach is that there is no reference database available. We made, thus, our own dataset. The results are quite encouraging regarding what was expected and what provide methods the literature.

Keywords: Relational graph of segments, kernel on graphs, window extraction, inexact graph matching.

1 Introduction

Several companies, like Blue Dasher Technologies Inc., EveryScape Inc., Earthmine Inc., or Google^{TM} provide their street-level pictures either to specific clients or as a new world wide web-service. However, none of these companies exploits the visual content, from the huge amount of data they are acquiring, to characterize semantic information and thus to enrich their system.

Among many approaches proposed to address object retrieval task, local features are commonly considered as the most relevant data description. Powerful object retrieval methods are based on local features such as Point of Interest (PoI) [1] or region-based descriptions [2]. Recent works consider not anymore a

* The images are acquired by the STEREOPOLIS mobile mapping system of IGN. Copyright images: ©IGN for iTOWNS project.

A.-B. Salberg, J.Y. Hardeberg, and R. Jenssen (Eds.): SCIA 2009, LNCS 5575, pp. 646–656, 2009.

single signature vector as object description but a set of local features. Several strategies are then possible, either consider these sets as unorganized (bags of features) or put some explicit structure on these sets of features. Efficient kernel functions have been designed to represent similarity between bags [3]. In [4], Gosselin et al. investigate the kernel framework on sets of features using sets of PoI. In [4], the same authors address multi-object retrieval with color-based regions as local descriptions. Based on the same region-based local features, Lebrun et al. [5] presented a method introducing a rigid structure in the data representation since they consider objects as graphs of regions. Then, they design dedicated kernel functions to efficiently compare graphs of regions.

Edge fragments appear to be relevant key-support for architecture information on building facades. However, a pixel set from a contour is not as informative as a pixel set from a region. Regarding previous works [6], [7] which consider exclusively or mainly, contour fragments as the information supports, this lack of intrinsic information requires to emphasize underlying structure of the objects in the description. Independently, Shotton et al. and Opelt et al. proposed several approaches to build contour fragment descriptors dedicated to a specific class of object. Basically, they learn a model of distribution of the contour fragments for a specific class of objects. Although, they can be more discriminative for the learned class, they are not robust to noisy contours found in real images. Indeed, to learn a class, they must select clean contours from segmentation masks. Ferrari et al. [11] use the properties of perceptual grouping of contours.

Following this same last idea, we propose to design a kernel similarity function for structured sets of contours. First, objects are represented by fragments of contours and a relational graph on these contour segments. The graph vertices are contour segments extracted from the image and characterized by their orientation to the horizontal axis. The graph edges represent the spatial relationships between contour segments.

This paper is organized as follows. First, we extract window candidate using the accumulation of gradients. We describe inital method and present our improvement on the automatic setting of the scale of extraction. Then, we focus on similarity functions between objects characterized by an attributed relational graph of segments of contours. To compare these graphs, we adapt kernels on graphs [8], [9] in order to define a kernel on paths more powerful than previous ones.

2 Extraction of Window Candidates

In this section, we explain the extraction of window candidates. We are inspired by the work of Lee et al. [10] that uses the properties of windows and facades and we propose a new algorithm.

2.1 Accumulation of Gradient

In 2004, Lee et al. [10] proposed a profile projection method to extract windows. They exploited both the fact that windows are horizontally and vertically aligned in the facade and that they have usually rectangular shape. Results are good and accurate on a simple database, where walls are not textured, windows are regularly aligned and there is no occlusion nor shadows. In the context of old historical cities like Paris, images are much more complex: windows are not always aligned (figure 1a), textures are not uniform, there are illumination variations, there may be occlusions due to trees, cars, etc. Since they are organized in floors, windows are usually horizontally aligned. We propose thus to firstly find the floors and then to work on them separately to extract the windows, or at least rectangles which are candidates to be windows. Moreover we improve this method by completely automatizing the extraction of window candidates by determining the correct scale of analysis.

Floor and Window Candidate Detection
In order to find the floors, the vertical gradients are computed (figure 1b), and their norms are horizontally accumulated to form an horizontal histogram (figure 1c). High values of this histogram correspond more or less to window positions whereas low values correspond to wall (or roof). The histogram is thus threshold to its average value, the facade is so split into floors (figure 1d). The process is repeated in the other direction (horizontal gradients,vertical projection) separately for each floor, giving the window candidates.

Automatic Window Candidate Extraction
As we need an accurate set of edges to perform the recognition, we used the optimal operators of smoothing and derivation of Shen-Castan [12] (optimal in the Canny sense). The operators of Canny family depend on a parameter linked to the size of the filter (size of the Gaussian for Canny filter for example) or, which is equivalent, to the level of detail of the edge detection. We will denote β this parameter.

If the smoothing is too strong, some edges disappear (figure 2) whereas if the smoothing is too weak, there is too much noise (texture between windows). Thus, the number of extracted floors p_β depends on β, but it does not regularly evolve with β (cf. figure 2d). It passes by a plateau which usually constitutes a good compromise.

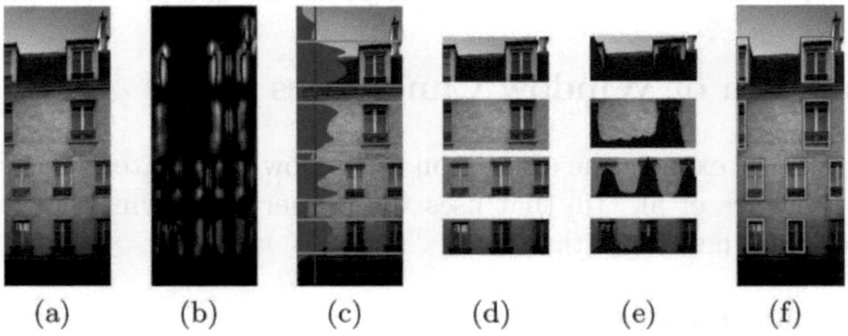

(a) (b) (c) (d) (e) (f)

Fig. 1. Window candidate extraction. (a) Example of facade where the windows are not vertically aligned. **(b)** Vertical gradient norms. **(c)** Horizontal projection. **(d)** Split into 4 floors. **(e)** Vertical projection. **(f)** Window candidates.

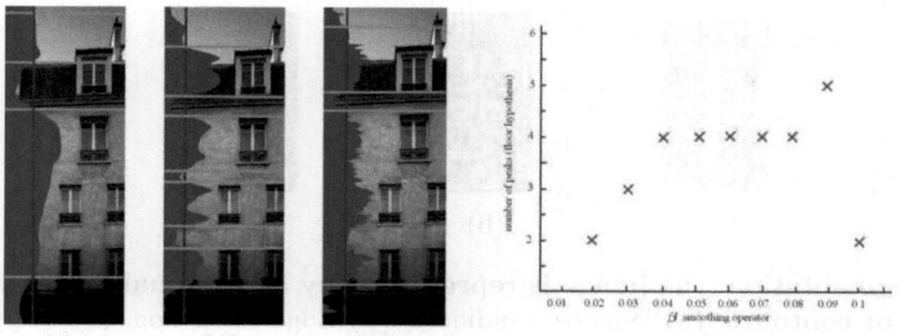

(a) (b) (c) (d)

Fig. 2. The number of floors depends on the smoothing and derivation parameter β. (a) Strong smoothing. **(b)** Good compromise. **(c)** Weak smoothing. **(d)** Evolution of number of floors according to β.

In order to determine the value of β corresponding to this plateau, we compute a score S_{β_i} for each value β_i (β_i grows between 0 and 1). The idea is to maximize this score depending on the stability of the histogram and the amplitude of peaks H_{pj}.

$$S_{\beta_i} = \begin{cases} \overbrace{\frac{p_{\beta_{i-1}}}{p_{\beta_i}}}^{Stability} \cdot \overbrace{\frac{\sum_{j=1}^{p_{\beta_i}} maxH_{pj}}{p_{\beta_i}}}^{average\ peak\ amplitude} & \text{if } p_{\beta_{i-1}} < p_{\beta_i} \\ \frac{\sum_{j=1}^{p_{\beta_i}} maxH_{pj}}{p_{\beta_{i-1}}} & \text{else} \end{cases}$$

with p_{β_i} the number of peaks for β_i.

For each image, a value of β is evaluated to extract window candidates in each floor.

To summarize, the algorithm of window candidate extraction is:

Algorithm 1. Automatic Windows Extraction

Require: rectified facade image I_0

Initialization: $\beta_0 \leftarrow 0.02$

repeat

1) Compute vertical gradient norms
2) Project and accumulate horizontally these vertical gradient norms
3) Calculate evaluation score S_{β_i}
4) $\beta_i \leftarrow \beta_i + 0.01$

until $\beta_i = 0.3$

Choose $\beta_t = argmax_{\beta_i} \quad S_{\beta_i}$

Cut into floors with β_t according to the peaks. Compute the histogram of horizontal gradient norms on each floor with β_t and search the peaks of this vertical projection

Rectangles are window candidates

Fig. 3. Segmentation: the image is represented by a relational graph of line segment of contours. (a) Window candidate. **(b)** Edge extraction. **(c)** Polygonalization.

2.2 Representation of Window Candidates by Attributed Relational Graphs

After this first step, we have extracted rectangles which are candidates for defining windows. Of course, because of the complexity of the images, there are many mismatches and a step of classification is necessary to remove outliers. In each rectangle, edges are extracted, extended and polygonalized (figure 3). In order to consider the set of edges as a whole, we represent it by an Attributed Relational Graph (ARG). Each line segment is a vertex of this graph and the relative positions of line segments are represented by the edges of the graph. The topological information (such as parallelism, proximity) can be considered only for the nearest neighbors of each line segment. We use a Voronoi diagram to find the segments that are the closest to a given segment. An edge in the ARG represents the adjacency of two Voronoi regions that is to say the proximity of two line segments.

In order to be robust to scale changes, a vertex is only characterized by its direction (horizontal or vertical). If Θ is the angle between line segment C_i and the horizontal axis ($\Theta \in [0, 180[$), C_i is represented by $v_i = \begin{pmatrix} cos(2\Theta) \\ sin(2\Theta) \end{pmatrix}$.

Edge (v_i, v_j) represents the adjacency between line segments C_i and C_j. It is characterized by the relative positions of the centres of gravity of C_i and C_j, denoted $g_{C_i}(Xg_{C_i}, Yg_{C_i})$ and $g_{C_j}(Xg_{C_j}, Yg_{C_j})$. Edge (v_i, v_j) is then characterized by $e_{ij} = \begin{pmatrix} Xg_{C_j} - Xg_{C_i} \\ Yg_{C_j} - Yg_{C_i} \end{pmatrix}$.

3 Classification and Graph Matching with Kernel

In order to classify the window candidates into true windows and false positives, we chose to use machine learning techniques. Support Vector Machines (SVM) are state-of-the-art large margin classifiers which have demonstrated remarkable performances in image retrieval, when associated with adequate kernel functions.

The problem of classifying our candidates can be considered as a problem of inexact graph matching. The problem is twofold : first, find a similarity measure between graphs of different sizes and second, find the best match between graphs in an "acceptable" time. For the second problem, we opted for the "branch and bound" algorithm, which is more suitable with kernels involving "max" [5]. For the first problem, recent approaches propose to consider graphs as sets of paths [8], [9].

3.1 Graph Matching

Recent approaches of graph comparison propose to consider graphs as sets of paths. A path h in a graph $G = (V, E)$ is a sequence of vertices of V linked by edges of $E : h = (v_0, v_1, ..., v_n)$, $v_i \in V$.

Kashima et al. proposed [9] to compare two graphs G and G' by a kernel comparing all possible paths of same length of both graphs.

The problem of this kernel is its high computational complexity. If this is acceptable with graphs of chemical molecules, which have symbolic values, it is unaffordable with our attributed graphs.

Other kernels on graphs were proposed by Lebrun et al. [5], which are faster than Kashima kernel:

$$K_{Lebrun}(G,G') = \frac{1}{|V|}\sum_{i=1}^{|V|} max\ K_C(h_{v_i}, h'_{s(v_i)}) + \frac{1}{|V'|}\sum_{i=1}^{|V'|} max\ K_C(h_{s(v'_i)}, h'_{v'_i}).$$

$$\text{with} \begin{cases} h_{v_i} \text{ is a path of } G \text{ whose first vertex is } v_i \\ h'_{s(v_i)} \text{ is a path of } G' \text{ whose first vertex is the most similar to } v_i \end{cases}$$

Each vertex v_i is the starting point of one path and this path is matched with a path starting with the vertex $s(v_i)$ of G' the most similar to v_i. This property is interesting for graphs of regions because regions carry a lot of information, but in our case of graphs of line segments, the information is more included in the structure of the graph (the edges) than in the vertices.

We propose a new kernel that removes this constraint of start (h_{v_i} path starting from v_i):

$$K_{struct}(G,G') = \frac{1}{|V|}\sum_{i=1}^{|V|} max\ K_C(h_{v_i}, h') + \frac{1}{|V'|}\sum_{i=1}^{|V'|} max\ K_C(h, h'_{v'_i}). \quad (1)$$

Concerning the kernels on paths, several K_C were proposed [5] (sum, product...). We tested all these kernels and the best results were obtained with this one, where e_j denotes edge (v_{j-1}, v_j) :

$$K_C(h_{v_i}, h') = K_v(v_i, v'_0) + \sum_{j=1}^{|h|} K_e(e_j, e'_j)\ K_v(v_j, v'_j).$$

K_v and K_e are the minor kernels which define the vertex similarity and the edge similarity. We propose these minor kernels:

Fig. 4. Example: structures and scale edge problem. Is the segment of contour on the right in graph G' a contour of the object or not?

$$K_e(e_j, e'_j) = \frac{\langle e_j, e'_j \rangle}{||e_j||.||e'_j||} + 1. \text{ and } K_v(v_j, v'_j) = \frac{\langle v_j, v'_j \rangle}{||v_j||.||v'_j||} + 1.$$

Our kernel aims at comparing sets of contours, from the point of view of their orientation and their relative positions. However, some paths may have a strong similarity but provide no structural information; for example, paths whose all vertices represent segment almost parallel. To deal with this problem, we can increase the length of paths, but the complexity of calculation becomes quickly unaffordable. To overcome this problem, we add in K_C a weight $O_{i,j}$ that penalizes the paths whose segment orientations do not vary.

$$O_{i,j} = sin(\phi_{ij}) = \sqrt{\tfrac{1}{2}(1 - \langle v_i, v_j \rangle)}.$$

with ϕ_{ij} the angle between vertices i and j.

Moreover the perceptual grouping of sets of contours is crucial for the recognition. For example in figure 4, graphs G' and G'' have almost the same structure as graph G, but the rightmost contour is further away in graph G' than in the two other graphs. The question is: has this contour to be clustered with the other contours to form an object or not? To model this information, we add a scale factor Se_i.

$$Se_i = min(\frac{||e_i||}{||e_{i-1}||} \cdot \frac{||e'_{i-1}||}{||e'_i||}, \frac{||e_{i-1}||}{||e_i||} \cdot \frac{||e'_i||}{||e'_{i-1}||}).$$

Our final kernel K_C becomes ($Se_i \in [0,1]$ et $O_{i,j} \in [0,1]$):

$$K_C(h_{v_i}, h') = K_v(v_i, v'_0) + \sum_{j=1}^{|h|} Se_j O_{j,j-1} K_e(e_j, e'_j)\, K_v(v_j, v'_j). \qquad (2)$$

4 Experiments and Discussions

In this section, we first compare our algorithm of window extraction to Lee algorithm [10]. Then we evaluate our kernel and the interest of the weights proposed in this paper.

(a) Lee (b) Our method (c) Lee (d) Our method

Fig. 5. Comparison of Lee and our method on complex cases. (a) (b) windows are not vertically aligned. (c) (d) chimneys induce false detection with Lee.

4.1 Window Candidate Extraction

Institut Geographique National (IGN) is currently initiating a data acquisition of Paris facades. The aim of our work is to extract and recognize objects present in the images (cars, windows, doors, pedestrians ...) of this large database. We have tested our algorithm on Paris facade database and compared it with Lee and Nevatia algorithm [10] (we denote it Lee in the figures). Images are rectified before processing. On simple cases we get results similar to Lee, but in more complex cases, when windows are not exactly aligned or when there is noise due to chimneys, drainpipes, etc (figures 5 and 6), we obtain better results. Moreover, our algorithm is automatic, it chooses by itself the correct scale of analysis to properly extract the contours.

4.2 Classification of Window Candidates

We tested our method to remove the false detections on a database of 300 images, for which we had the ground-truth : 70 windows and 230 false detections

(a) Method of Lee (b) Our method

Fig. 6. Comparison of Lee and our method on a complex case: windows are not exactly horizontally aligned and there is a lot of distractors

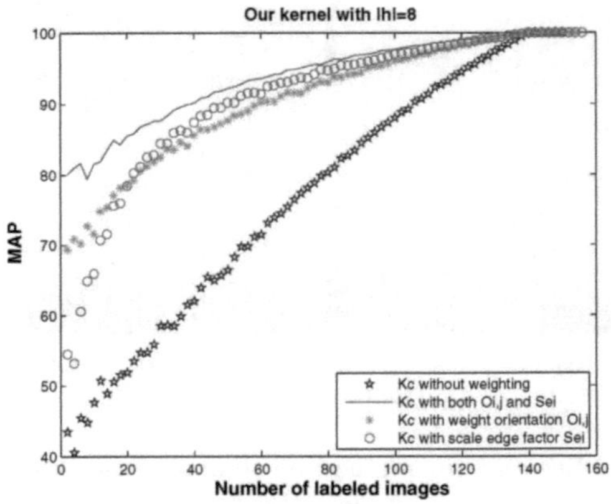

Fig. 7. Comparaison of versions kernels on paths with weighting by scale and orientation of the contours

(negative examples). Each image contains between 10 and 30 line segments. We tried paths of lengths between 3 and 10. With a 3-length edges, we do not fully take advantage of the structure of the graph, and with a 10-length edges, the time complexity becomes problematic. We opted for a compromise : $|h| = 8$.

Each retrieval session is initialized with one image containing an example of window. We simulated an active learning scheme, where the user annotates a few images at each iteration of relevance feedback, thanks to the interface (cf. Fig. 8). The system labels at each iteration one image as window or false detection, and the system updates the ranking of the database according to these new labels. The

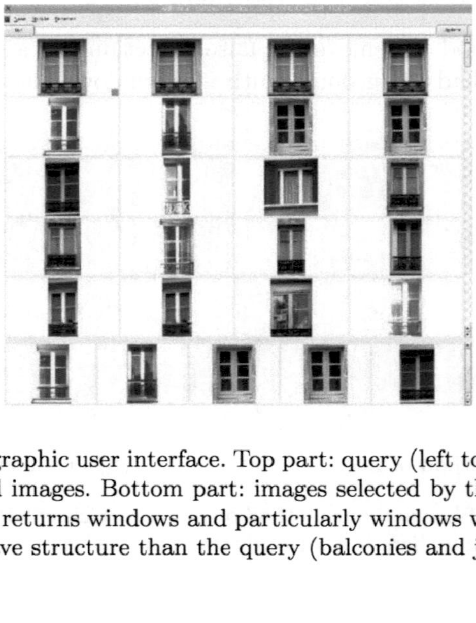

Fig. 8. The RETIN graphic user interface. Top part: query (left top image with a green square) and retrieved images. Bottom part: images selected by the active learner. We note that the system returns windows and particularly windows which are in the same facade or have the save structure than the query (balconies and jambs).

whole process is iterated 100 times with different initial images and the Mean Average Precision (MAP) is computed from all these sessions (figure 7).

We compared our kernels with and without the various weights proposed in section 3. With only one example of window and one negative example, we obtain 42 % of correct classification with the kernel without weighting. This percentage goes up to 54% with the scale weighting, to 69% with the orientation weighting, and to 80 % with both weightings. Results with weightings are much more improved after a few steps of relevance feedback than without weighting, to reach 90 % with 40 examples (20 positive and 20 negative), instead of 100 examples without weighting. Figure 8 shows that we are also able to discriminate between various types of window, the most similar being the windows of the same facade or of the same number of jambs.

5 Conclusions

We have proposed an accurate detection of contours from images of facades. Its main interest, apart the accuracy of detection is that it is automatic, since it adapts its parameter to the correct scale smoothing of analysis. We have also shown that objects extracted from images can be represented by a structured set of contours. The new kernel we have proposed is able to take into account orientations and proximity of contours in the structure. With this kernel, the system retrieves the most similar windows from facade database. The next step is to free oneself from the step of window candidate extraction, and to be able to recognize a window as a sub-graph of the graph of all contours of the image. This process involving perceptual grouping will then be extended to another type of objects like cars for example.

Acknowledgments. This work is supported by ANR (the french National Research Agency) within the scope of the iTOWNS research project (ANR 07-MDCO-007-03). Copyright images: ©IGN for iTOWNS project.

References

1. Mikolajczyk, K., Tuytelaars, T., Schmid, C., Zisserman, A., Matas, J., Schaffalitzky, F., Kadir, T., Gool, L.V.: A comparison of affine region detectors. International Journal of Computer Vision (2005)
2. Carson, C., Belongie, S., Greenspan, H., Malik, J.: Blobworld: Image segmentation using expectation-maximization and its application to image querying. IEEE Transactions on Pattern Analysis and Machine Intelligence (2004)
3. Shawe-Taylor, J., Cristianini, N.: Kernel methods for Pattern Analysis. Cambridge University Press, Cambridge (2004)
4. Gosselin, P.-H., Cord, M., Philipp-Foliguet, S.: Kernel on Bags for multi-object database retrieval. In: ACM International Conference on Image and Video Retrieval, pp. 226–231 (2007)
5. Lebrun, J., Philipp-Foliguet, S., Gosselin, P.-H.: Image retrieval with graph kernel on regions. In: IEEE International Conference on Pattern Recognition (2008)

6. Shotton, J., Blake, A., Cipolla, R.: Contour-Based Learning for Object Detection. In: 10th IEEE International Conference on Computer Vision (2005)
7. Opelt, A., Pinz, A., Zisserman, A.: A Boundary-Fragment-Model for Object Detection. In: Leonardis, A., Bischof, H., Pinz, A. (eds.) ECCV 2006. LNCS, vol. 3952, pp. 575–588. Springer, Heidelberg (2006)
8. Suard, F., Rakotomamonjy, A., Bensrhair, A.: Détection de piétons par stéréovision et noyaux de graphes. In: 20th Groupe de Recherche et d'Etudes du Traitement du Signal et des Images (2005)
9. Kashima, H., Tsuboi, Y.: Kernel-based discriminative learning algorithms for labeling sequences, trees and graphs. In: International Conference on Machine Learning (2004)
10. Lee, S.C., Nevatia, R.: Extraction and Integration of Window in a 3D Building Model from Ground View Image. In: IEEE Computer Society Conference on Computer Vision and Pattern Recognition (2004)
11. Ferrari, V., Fevrier, L., Jurie, F., Schmid, C.: Groups of Adjacent Contour Segments for Object Detection. IEEE Transactions on Pattern Analysis and Machine Intelligence (2008)
12. Shen, J., Castan, S.: An Optimal Linear Operator for Step Edge Detection. Graphical Models and Image Processing (1992)

Multi-view and Multi-scale Recognition of Symmetric Patterns

Dereje Teferi and Josef Bigun

Halmstad University, SE 30118 Halmstad, Sweden
{Dereje.Teferi,Josef.Bigun}@hh.se

Abstract. This paper suggests the use of symmetric patterns and their corresponding symmetry filters for pattern recognition in computer vision tasks involving multiple views and scales. Symmetry filters enable efficient computation of certain structure features as represented by the generalized structure tensor (GST). The properties of the complex moments to changes in scale and multiple views including in-depth rotation of the patterns and the presence of noise is investigated. Images of symmetric patterns captured using a low resolution low-cost CMOS camera, such as a phone camera or a web-cam, from as far as three meters are precisely localized and their spatial orientation is determined from the argument of the second order complex moment I_{20} without further computation.

1 Introduction

Feature extraction is a crucial research topic in computer vision and pattern recognition having numerous applications. Several feature extraction methods have been developed and published in the last few decades for general and/or specific purposes. Early methods such as Harris detector [3] use stereo matching and corner detection to find corner like singularities in local images whereas more recent algorithms use extraction of other features from gradient of images [4,7] or orientation radiograms [5] with the intention of achieving invariance or resilience to certain adverse effects in vision, e.g. rotation, scale, view and noise level changes, to match against a database of image features.

In this paper, the strength of symmetric filters in localizing and detecting the orientation of known symmetric patterns such parabola, hyperbola, circle and spiral etc in varying scales and spatial and in-depth rotation is investigated. The design of the pattern via coordinate transformation by analytic functions and their detection by symmetry filters is discussed. These patterns are non-trivial and often do not occur in natural environments. Because they are non-trivial, they can be used as artificial markers to recognize certain points of interest in an image. Symmetry derivatives of Gaussians are used as filters to extract features from their second order moments that are able to localize as well as detect the local orientation of these special patterns simultaneously. Because of the ease of detection, these patterns are used for example in vehicle crash tests by using the known patterns as markers on artificial test driver for automatic tracking [2]

A.-B. Salberg, J.Y. Hardeberg, and R. Jenssen (Eds.): SCIA 2009, LNCS 5575, pp. 657–666, 2009.

and in ffingerprint recognition by using the symmetry filters to detect core and delta points (minutia points) in fingerprints[6].

2 Symmetry Features

Symmetry features are discriminative features that are capable of detecting local orientations in an image. The most notorious patterns that contain such features are lines (linear symmetry), that can be detected by eigen analysis of the ordinary 2D structure tensor. However, with some care even other patterns such as parabolic, circular or spiral (logarithmic), or hyperbolic shapes can be detected but by eigen analysis of the generalized structure tensor [1,2] which is summarized below.

First, we revise the structure tensor S which enables to determine the dominant direction of ordinary line patterns (if any) and the fitting error through the analysis of its eigenvalues and their corresponding eigenvectors. S is computed as:

$$S = \begin{pmatrix} \iint (\omega_x)^2 |F|^2 & \iint (\omega_x \omega_y)|F|^2 \\ \iint (\omega_x \omega_y)|F|^2 & \iint (\omega_y)^2 |F|^2 \end{pmatrix} \tag{1}$$

$$= \begin{pmatrix} \iint (D_x f)^2 & \iint (D_x f)(D_y f) \\ \iint (D_x f)(D_y f) & \iint (D_y f)^2 \end{pmatrix} \tag{2}$$

Where $F = F(\omega_x, \omega_y)$ is the fourier transform of f and the eigenvectors k_{max}, k_{min} corresponding to the eigenvalues λ_{min}, λ_{max} represent the inertia extremes and the corresponding axes of inertia of the power spectrum $|F|^2$ respectively.

The second order complex moment I_{mn} of a function h, where m, n are non negative integers and $m + n = 2$ is calculated as,

$$I_{mn} = \iint (x + iy)^m (x - iy)^n h(x, y) dx dy \tag{3}$$

It turns out that I_{20} and I_{11} are related to the eigenvalues and eigenvectors of the structure tensor S as follows:

$$I_{20}|F|^2 = (\lambda_{max} - \lambda_{min}) e^{i2\varphi_{min}} \tag{4}$$

$$I_{11}|F|^2 = \lambda_{max} + \lambda_{min} \tag{5}$$

$$|I_{20}| = \lambda_{max} - \lambda_{min} \leq \lambda_{max} + \lambda_{min} = I_{11} \tag{6}$$

Here $\lambda_{max} \geq \lambda_{min} \geq 0$. If $\lambda_{min} = 0$ then $|I_{20}| = I_{11}$ which signifies the existence of a perfect linear symmetry which is also the unique occasion where the inequality in Eq. (6) is fulfilled with equality, i.e. $|I_{20}| = I_{11}$. Thus a measure of linear symmetry (LS) can be written as:

$$LS = \frac{|I_{20}|}{I_{11}} = \frac{\lambda_{max} - \lambda_{min}}{\lambda_{max} + \lambda_{min}} e^{i2\varphi_{min}} \tag{7}$$

In practice this is a normalization of I_{20} with I_{11}. The value of LS falls within [0, 1] where $LS = 1$ for perfect linear symmetry and 0 for complete lack of linear symmetry (balanced directions or lack of direction).

The Generalized structure tensor (GST) is similar in its essence with the ordinary structure tensor but its target patterns are "lines" in curvilinear coordinates, ξ and η. For example, using $\xi(x,y) = log\sqrt{x^2+y^2}$ and $\eta(x,y) = tan^{-1}(x,y)$ as coordinates, "oriented lines" in the log-polar coordinate system ($a\xi(x,y) + b\eta(x,y) = constant$), GST will simultaneously estimate evidence for presence of circles, spirals and parabolas etc. In GST, the I_{20} and I_{11} interpretations remain unchanged except that they are now with respect to lines in curvilinear coordinates, with the important restriction that the allowed curves for coordinate definitions must be drawn from harmonic curve family. [2] has shown that as the consequence of local orthogonality of ξ and η the complex moments I_{20} and I_{11} of the harmonic patterns can be computed in the cartesian coordinates system without the need for coordinate transformation as:

$$I_{20} = \iint e^{iarg((D_\xi - iD_\eta)\xi)^2}[D_x + iD_y f]^2 dxdy \tag{8}$$

$$I_{11} = \iint |(D_x + iD_y)f|^2 dxdy \tag{9}$$

where $\eta = \eta(x,y)$ and $\xi = \xi(x,y)$ represent a pair of harmonic coordinate transformations. Such pairs of harmonic transformations satisfy the following constraint: $\xi(x,y) = constant_1$ and $\eta(x,y) = constant_2$ are orthogonal to each other i.e. $D_x\xi = D_y\eta$ and $D_y\xi = -D_x\eta$.

Thus, the measure of linear symmetry in the harmonic coordinate system by the generalized structure tensor is in fact the analogue of the measure of linear symmetry by the ordinary structure tensor but in a cartesian coordinate system. The advantage is that we can use the same theoretical and practical machinery to detect the presence and quantify the orientation of for example parabolic symmetry (PS), circular symmetry (CS), hyperbolic symmetry (HS) drawn in cartesian coordinates depending on the analytic function $q(z)$ used to define the harmonic transformation. Some of these patterns are shown in Figure 1 where the iso-curves represent a line as $a\xi + b\eta = constant$ for predetermined ξ and η.

Harmonic transformation pairs can be readily obtained as the real and imaginary parts of (complex) analytic functions by restricting us further to $q(z)$ such that $\frac{dq}{dz} = z^{\frac{n}{2}}$. Thus we have,

$$q(z) = \begin{cases} \frac{1}{\frac{n}{2}+1} z^{\frac{n}{2}+1} \text{ if } n \neq -2 \\ log(z), \quad \text{if } n = -2 \end{cases} \tag{10}$$

Each of the curves generated by the real and imaginary parts of $q(z)$ can then be detected by symmetry filters Γ shown in the fourth row of Figure 1. The gray values and the superimposed arrows respectively show the magnitude and orientation of the filter that can be used for detection.

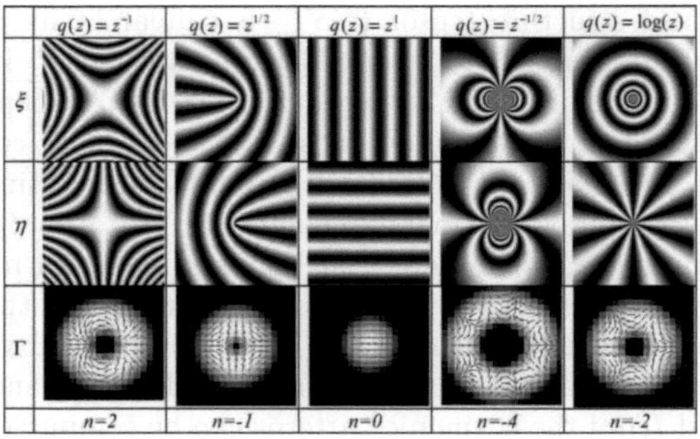

Fig. 1. First row: Example harmonic function $q(z)$, second and third rows show the real and the imaginary parts ξ and η of the $q(z)$ where $z = x + iy$. The fourth row shows the filters that can be used to detect the patterns in row 2 and 3. The last row shows the order of symmetry.

$$\Gamma^{\{n,\sigma^2\}} = \begin{cases} (D_x + iD_y)^n g & \text{if } n \geq 0 \\ (D_x - iD_y)^{|n|} g & \text{if } n < 0 \end{cases} \tag{11}$$

Here $g(x, y) = \frac{1}{2\pi\sigma^2} e^{-\frac{x^2+y^2}{2\sigma^2}}$ is the Gaussian and n is the order of symmetry. For $n = 0$, Γ is an ordinary Gaussian. Moreover, $(D_x + iD_y)$ and $\left(\frac{-1}{\sigma.^2}\right)^p (x + iy)^p$ behave identically when acting on and multiplied to, a Gaussian respectively [2,1]. Due to this elegant property of Gaussians functions, the symmetry filters in the above equation can be rewritten as:

$$\Gamma^{\{n,\sigma^2\}} = \begin{cases} \left(-\frac{1}{\sigma^2}\right)^n (x + iy)^n g & \text{if } n \geq 0 \\ \left(-\frac{1}{\sigma^2}\right)^{|n|} (x - iy)^{|n|} g & \text{if } n < 0 \end{cases} \tag{12}$$

3 In-Depth (Non-planar) Rotation of Symmetric Patterns

Recognition of a pattern when rotated spatially in 3D is a challenging issue and requires resilient features. To test the strength of the symmetry filters in recognizing patterns viewed from different angles, we rotated the patterns geometrically using ray tracing as follows.

Suppose we are looking at the world plane W from point O through an image plane I in a pin-hole camera model as in Figure 2. Note that, if the image plane I is parallel to the world plane W, we would see a zoomed version of the world image depending on how far the image plane is from the world plane. When W is not parallel to I, then the image plane is a skewed and zoomed version of the world plane.

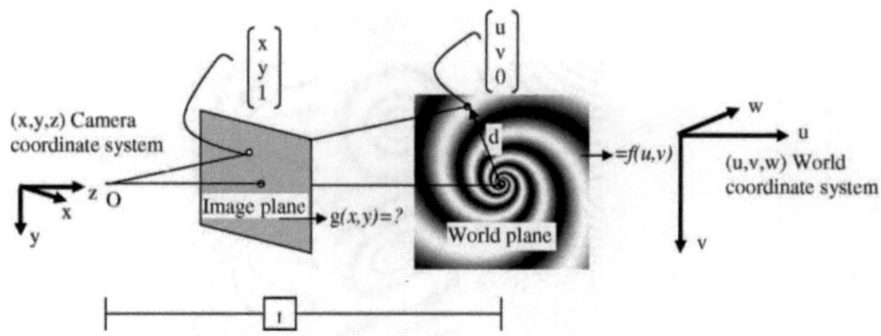

Fig. 2. Ray tracing for non-planar rotation

A point P represented in the world coordinates as d, transfers to the camera coordinates as $R(t+d)$ if both t and d are in world coordinates. Here R is a rotation matrix aligning the world coordinate axes with the camera coordinate axes and t is the translation vector aligning the origin of the world coordinate system to the origin of the camera coordinate system. The rotation matrix R of the world plane is the product of the rotation matrices around each axis R_x, R_y, and R_z relative to the world coordinates. As an example R_x is given as:

$$R_{x(\alpha)} = \begin{pmatrix} 1 & 0 & 0 \\ 0 & cos(\alpha) & -sin(\alpha) \\ 0 & sin(\alpha) & cos(\alpha) \end{pmatrix} \tag{13}$$

similarly R_y and R_z are defined and the overall rotation matrix R is given as:

$$R = R_{x(\alpha)} * R_{y(\beta)} * R_{z(\gamma)} \tag{14}$$

The normal n to the world plane is the 3rd row of the rotation matrix R expressed in the camera coordinates.

To find the distance vector from O to the world plane W, we can proceed in two ways as $L^T n$ and $t^T n$. Because both measure the same distance, they are equal, i.e. $L^T n = t^T n$

$$L = \tau \begin{pmatrix} x \\ y \\ 1 \end{pmatrix} = \tau L_s \Rightarrow \tau L_s^T n = t^T n \tag{15}$$

where $L_s = (x, y, 1)^T$. Thus

$$\tau = \frac{t^T n}{L_s^T n} \tag{16}$$

$$\Rightarrow L = \left(\frac{t^T n}{L_s^T n} L_s \right) \tag{17}$$

$$d = R(L - t) \tag{18}$$

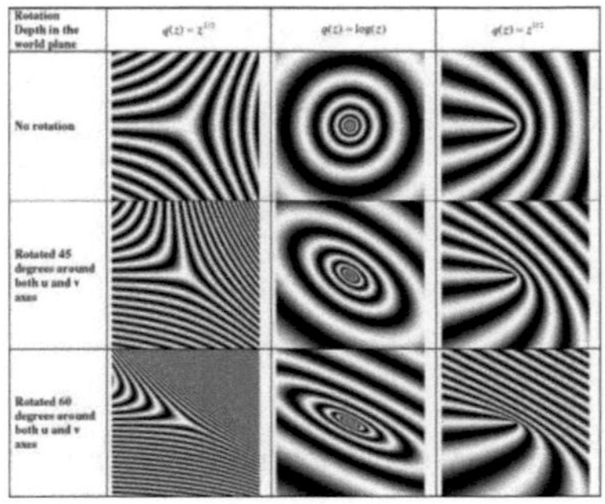

Fig. 3. Illustration of in-depth rotation of symmetric patterns

Accordingly, $g(x, y) = f(u, v)$, where $d = (u, v, 0)$. The last two rows of Figure 3 show the results of some of the symmetric patterns painted on the world plane but observed by the camera in the image plane.

4 Experiment

4.1 Recognition of Symmetric Patterns Using Symmetry Filters

We used the filter designed as in Eq. 12 to detect the family of patterns f generated by the analytic function $q(z)$, where

$$f = cos(k_1\Re(q(z)) + k_2\Im(q(z))) + 1 \tag{19}$$

Here $q(z)$ is given by Eq. 10 and $n \in -4, -3, -2, -1, 0, 1, 2$

The following steps are applied on the image to detect the pattern and its local orientation:

1. Compute the square of the derivative image h_k by convolving the image f with a symmetry filter of order 1, $\Gamma^{\{1,\sigma_1^2\}}$ and pixelwise squaring of the complex valued result as:
$h_k =< \Gamma_k^{\{1,\sigma_1^2\}}, f_k >^2$. Here σ_1 controls the extension of the interpolation function, i.e. the size of the derivative filter Γ^{1,σ_1} that is modeling the expected noise in the image;
2. Compute I_{20} by convolving the complex image h_k of step 1 with the appropriate complex filters from Eq 12 according to their pattern family defined by n and by their expected spatial extension controlled by σ_2. That is:
$I_{20} =< \Gamma_k^{\{m,\sigma_2^2\}}, h_k >$.
3. Compute the magnitude image I_{11} by convolving the magnitude of the complex image h_k with the magnitude of the symmetry filters from Eq 12 as:
$I_{11} =< |\Gamma_k^{\{m,\sigma_2^2\}}|, |h_k| >$;

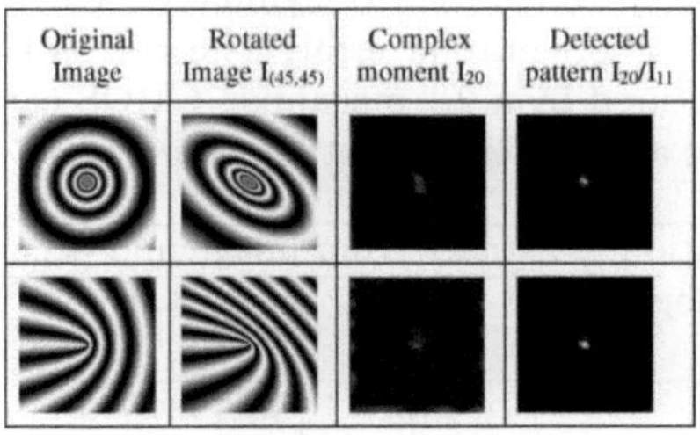

Fig. 4. Detection of symmetric patterns using symmetry derivatives of Gaussians on simulated rotated patterns

4. Compute the certainty image and detect the position and orientation of the symmetry pattern from its local maxima. The argument of I_{20} at locations characterized by high response of the certainty image, I_{11} yields the group orientation of the pattern;

The strength of the filters in detecting patterns and their rotated version is tested by applying the in-depth rotation of the symmetric patterns as discussed in the previous section. Figure 4 illustrates the detection results of circular and parabolic patterns rotated 45° around the x and y axes.

The color of the I_{20} image corresponding to the high response on the detected pattern (last column) indicates the spatial orientation of the symmetric pattern. The filters are also tested on real images captured with low-cost off the shelf CMOS camera. The result shows that symmetry filters detect these patterns from distance of up to 3 meters and in-depth rotation of up to 45 degrees, see Table 1. Similar result is achieved with web cameras and phone cameras as well. The color of the I_{20} once again indicates the spatial orientation of the symmetric pattern detected.

4.2 Recognition of Symmetric Patterns Using Scale Invariant Feature Transform-SIFT

Lowe [4] proposed features known as SIFT to match images representing different views of the same scene by using histograms of gradient directions. The features extracted are often used for matching between different views severed by scale and in-depth local rotation as well as illumination changes. SIFT feature matching is one of the most popular object detection methods.

The SIFT approach uses the following four steps to extract the location of a singularity and its corresponding feature vector from an image and store them for subsequent matching.

1. Scale-space extrema detection: this is the first step where all candidate points that are presumably scale invariant are extracted using arguments from scale-space theory. The implementation is done using Difference of Gaussian (*DoG*) function by successively subtracting images from its Gaussian smoothed version within an octave;

Table 1. Average results of recognition of symmetric patterns from multiple views. d=localization error and α=orientation error. The test is performed on 12 different images, e.g. Figure 5 captured by a 2.1 megapixel CMOS camera. Each of the images are subjected to zooming and in-depth rotation as in Figure 4 but naturally.

Rotation (in-depth)	Distance from image and accuracy			
	2 meters		3 meters	
	d	α	d	α
0°	±1 pixel	±2°	±1 pixel	±5°
30°	±1 pixel	±3°	±1 pixel	±8°
45°	±2 pixel	±6°	±2 pixel	±12°
60°	±3 pixel	±15°	±4 pixel	±20°

2. Keypoint localization: the candidate points from step 1 that are poorly localized and sensitive to noise, especially those around edges, are removed;
3. Orientation assignment: in this step, orientation is assigned to all key points that have passed the first two steps. The orientation of the local image around the key point in the neighborhood is computed using image gradients;
4. Extracting keypoint descriptors: the histograms of image gradient directions are created for non-overlapping subsets of the local image around the key point. The histograms are concentrated to a feature vector representing the structure in the neighborhood of the key points to which the global orientation computed in step 3 is attached.

The SIFT demo software[1] can be used to extract the necessary features to automatically recognize patterns in an image such as those shown in Figure 5. To this end, we used real images (containing symmetric patterns), e.g. the 2nd and 3rd rows of Figure 4, so that a set of SIFT features could be collected for each image. However, keypoint extraction failed often presumably. The method returned a few key points or in some cases failed to return any key point at all in the extraction of the SIFT features.

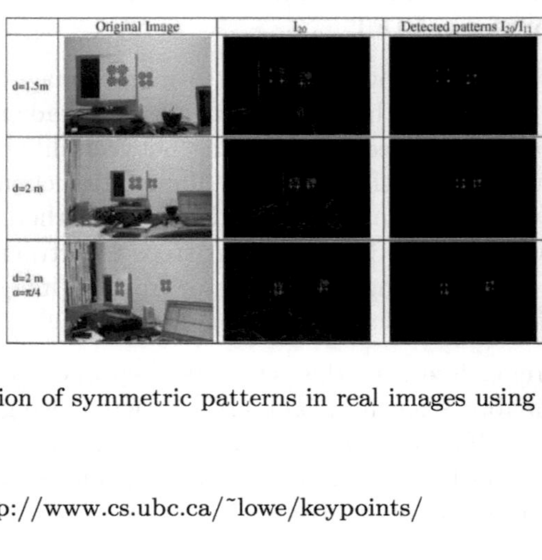

Fig. 5. Detection of symmetric patterns in real images using symmetry filters

[1] SIFT Demo http://www.cs.ubc.ca/~lowe/keypoints/

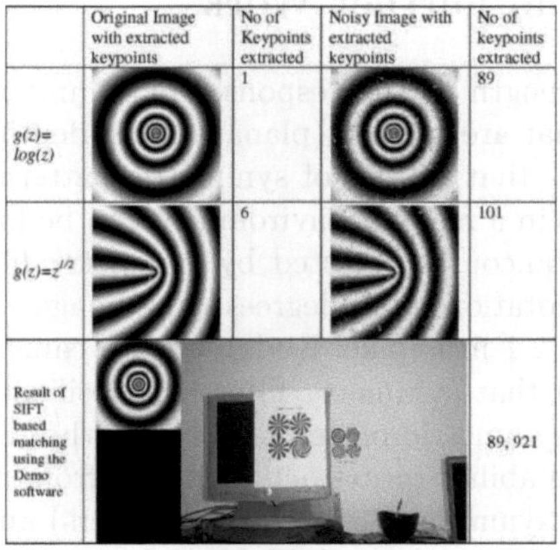

Fig. 6. Extraction and matching of keypoints on Symmetric patterns and their noisy counterparts using SIFT

SIFT features are often successful in extraction of discriminative features in images and are widely used in computer vision. The key points at which these features are extracted are essentially based on their lack of linear symmetry (orientation of lines) in the respective neighborhood, e.g. to detect corner like structures. These keypoints as well as the corresponding features are organized in such a way that they could be matched against keypoints in other images with similar local structure. However, the lack of linear symmetry does not describe the presence of a specific model of curves in the neighborhood such as parabolic, circular, spiral, hyperbolic etc. In our case, lack of linear symmetry in addition to existence of known types of curve families as well as their orientation can be precisely determined, as demonstrated in Figure 4. Although these patterns are structurally different, SIFT keypoints consider them as the same often with only one key point - the center of the pattern leaving the description of the neighborhood type to histograms of gradient angles(SIFT features). The center of the pattern is chosen as a key point by SIFT since that is where there is a lack of linear symmetry. However, SIFT features apparently cannot be used to identify what patterns are represented around the key point because all orientations equally exist in the local neighborhood for all curve families despite their obvious differences in shape.

Two of the images from Figure 1 are used to test the capability of SIFT features in detecting the patterns in real images. Additive noise is applied to the images to study the change in extraction of keypoints as well as the corresponding SIFT features. The clean images returned 1 and 6 key points and the noisy images returned 89 and 101 key points, see Figure 6. Although, 89 and 101 key points are extracted from the two noisy images, none of these points actually match to the patterns in the real scene which contains these patterns, last row of Figure 6.

5 Conclusion and Further Work

In conclusion, the strength of the responses of symmetry filters in detecting symmetric patterns that are rotated (planar and in-depth) is investigated. It is shown via experiments that images of symmetric patterns (see Figure 5) used as artificial landmarks in a realistic environment can be localized and their spatial orientation simultaneously detected by symmetry filters from as far as 3 meters and in-depth rotation of 45 degrees. The images are captures by a low resolution commercial 2.1 megapixel Kodak CMOS camera. The results of this experiment illustrated that symmetry filters are resilient to in-depth rotation and scale changes in symmetric patterns. On the other hand, it is shown that SIFT features lack the ability to extract keypoints from these patterns as they look for lack of linear symmetry (existence of corners) and not the presence of certain types of known symmetries. SIFT feature extraction fails because all orientations equally exist around the center of the image which makes it difficult for SIFT feature to find differences in the gradients in the local neighborhood.

The findings of this work can be applied for automatic camera calibration where symmetric patterns are used as artificial markers in a non-planar arrangement in a world coordinate system to automatically determine the intrinsic and extrinsic parameter matrices of a camera by point correspondence. Other possible applications include generic object detection and encoding and decoding of numbers using local orientation and shape of symmetric images.

References

1. Bigun, J.: Vision with direction. Springer, Heidelberg (2006)
2. Bigun, J., Bigun, T., Nilsson, K.: Recognition of symmetry derivatives by the generalized structure tensor. IEEE Transactions on Pattern Analysis and Machine Intelligence 26(12), 1590–1605 (2004)
3. Harris, C., Stephens, M.: A combined corner and edge detector. In: Fourth Alvey Vision Conference, Manchester, UK, pp. 147–151 (1988)
4. Lowe, D.G.: Distinctive image features from scale-invariant key points. International Journal of Computer Vision 60(2), 91–110 (2004)
5. Michel, S., Karoubi, B., Bigun, J., Corsini, S.: Orientation radiograms for indexing and identification in image databases. In: European Conference on Signal Processing (Eupsico), Trieste, September 1996, pp. 693–696 (1996)
6. Nilsson, K., Bigun, J.: Localization of corresponding points in fingerprints by complex filtering. Pattern Recognition Letters 24, 2135–2144 (2003)
7. Schmid, C., Mohr, R.: Local gray value invariants for image retrieval. IEEE Transactions on Pattern Analysis and Machine Intelligence 19(5), 530–534 (1997)

Automatic Quantification of Fluorescence from Clustered Targets in Microscope Images

Harri Pölönen, Jussi Tohka, and Ulla Ruotsalainen

Tampere University of Technology, Tampere, Finland

Abstract. A cluster of fluorescent targets appears as overlapping spots in microscope images. By quantifying the spot intensities and locations, the properties of the fluorescent targets can be determined. Commonly this is done by reducing noise with a low-pass filter and separating the spots by fitting a Gaussian mixture model with a local optimization algorithm. However, filtering smears the overlapping spots together and lowers quantification accuracy, and the local optimization algorithms are uncapable to find the model parameters reliably. In this sudy we developed a method to quantify the overlapping spots accurately directly from the raw images with a stochastic global optimization algorithm. To evaluate the method, we created simulated noisy images with overlapping spots. The simulation results showed the developed method produced more accurate spot intensity and location estimates than the compared methods. Microscopy data of cell membrane with caveolae spots was also succesfully quantified with the developed method.

1 Introduction

Fluorescence microscopy is used to examine various biological structures such as cell membrane. Due to the diffraction limit, targets smaller than the optical resolution of the microscope system appear as a spot-shaped intensity distibutions in the image. A group of closely located targets with a mutual distance near the Rayleigh limit appear as a cluster of overlapping spots. The locations (with sub-pixel accuracy) and intensities of these small targets or spots are the point of interests in many applications[1],[2],[3]. A common approach to this quantification is first to reduce the noise by filtering and then fitting Gaussian mixture model[4],[5],[6]. A low-pass filter is used in order to eliminate the high frequency noise and a Gaussian kernel is also commonly used to simplify the fitting of the mixture model to the filtered image. Another common point of interest is to estimate the number of individual spots in the image, which we will not discuss here.

When imaging small targets such as cell membrane, subtle properties and variations are to be detected, and therefore the best possible accuracy must be achieved in the image processing and analysis. Although the widely applied low-pass filter makes the image visually more appealing to the human eye due to noise reduction (see Fig. 2), valuable information is lost during filtering and the accuracy of the quantification of the spots is weakened. Also, fitting the mixture

A.-B. Salberg, J.Y. Hardeberg, and R. Jenssen (Eds.): SCIA 2009, LNCS 5575, pp. 667–675, 2009.

model to the image is not as straightforward as it is often assumed, and the fitting may introduce errors to the results if not performed properly.

In this study, we developed a procedure to to quantify the overlapping spots from the raw microscope images reliably and accurately using Gaussian mixture models and a differential evolution algorithm. We show with simulated data that this new method produces significant improvements in both the spot intensity and the location estimates. We do not filter the image, which makes the mixture model parameter estimation more challenging due to the several local optima, and we present a variant of the differential evolution algorithm that is able to determine the optimal parameters for the model.

2 Methods

2.1 Model Description

We model a raw microscope image of mutually overlapping spots with a mixture model of Gaussian components. We create an image C_θ according to mixture model parameters θ and determine the fitness of the parameters by mean squared error between the raw image D and the created image. The value of a pixel (i,j) in the image C_θ is defined by the probability density function of the mixture model with k components multiplied by the spot intensity ρ_p as

$$C_\theta(i,j) = \sum_{p=1}^{k} \frac{\rho_p}{2\pi\sqrt{|\Sigma|}} \exp\left(-\frac{1}{2}((i,j)-\mu_p)^T \Sigma^{-1}((i,j)-\mu_p)\right) , \tag{1}$$

where μ_p is the centroid location of the component p and Σ is the covariance matrix.

The covariance Σ in the Equation (1) is kept fixed as in [5] and is determined according to the microscope settings as

$$\Sigma = \begin{pmatrix} 0.21\frac{\lambda}{A} & 0 \\ 0 & 0.21\frac{\lambda}{A} \end{pmatrix} , \tag{2}$$

where λ is the emission wavelength of the used fluorophore and A denotes the numerical aperture of the used solvent (water, oil). It is shown in [5] that this fixed shape of the Gaussian component corresponds well to the true spot shape, i.e. the point spread function, produced by a small fluorescent target.

The location and intensity of each spot, i.e. Gaussian component in the model, are estimated together with the level of background fluorescence. The parameter set to be optimised is thereby

$$\theta = (\mu_1, \rho_1, \ldots, \mu_k, \rho_k, \beta) , \tag{3}$$

where β is the background fluorescence level. The number of components k equals to the number of mutually overlapping spots and the total number of estimated parameters is $3k+1$.

If we denote the observed image pixel (i, j) value as $D(i, j)$, the mean squared fitness function $f(\theta|D)$ can then be defined as

$$f(\theta|D) = \frac{1}{nm} \sum_{i=1}^{n} \sum_{j=1}^{m} (D(i,j) - C_\theta(i,j) - \beta)^2 , \quad (4)$$

where n and m are the image dimensions. The best parameter set $\hat{\theta}$ is then found by solving the optimization problem

$$\hat{\theta} = \min_\theta f(\theta|D) . \quad (5)$$

2.2 Modified Differential Evolution Algorithm (DE)

Although the number of parameters in Equation (4) is not huge, it is challenging to find the parameters that minimise the squared error with high accuracy. Due to the noise in image, the parameter space is not smooth as with the filtered image but severely multimodal instead. This causes deterministic optimization algorithms to stuck easily to local optima near initial guess producing erroneous parameter estimates.

To find the optimal parameters $\hat{\theta}$, we apply a modification of differential evolution algorithm[7], which is a population-based search algorithm. Here a population member is a parameter set θ defined in Equation (3) . Unlike e.g. in genetic algorithms, in differential evolution the population is improved one member at a time and not in generation cycles. A new population candidate member θ_c is constructed from randomly chosen current population members θ_1, θ_2 and θ_3 by linear combination

$$\theta_c = \theta_1 + K \cdot (\theta_2 - \theta_3) , \quad (6)$$

where $K \in \mathbb{R}$ is a convergence control parameter. If θ_c has a better fit, i.e. smaller mean squared error in terms of observed image, than θ_4, the candidate θ_c replaces θ_4 in the population immediately. This procedure is repeated until all the population members are equal, and thereby the algorithm has converged.

The K in Equation (6) controls the convergence rate. With high values ($K \approx$ 1.0 or above), the algorithm is very exploratory and searches the parameter space thoroughly having a good capability to finally end up near global optimum but the search may be very slow. With low values ($K \approx 0.5$ or lower) the algorithm converges faster but has a risk of converging prematurely to a local optimum. With a constant K differential evolution has also a risk of stagnation[8], where the population neither evolves nor converges but rather repeats the same set of parameter values all over again.

In this study, we developed a modification of the above described algorithm to avoid the stagnation problem (see Fig. 1 for pseudo-code) and improve the performance. In our modification the convergence rate parameter K from univariate distribution in the interval $[0.5, 1.5]$ in each candidate θ_c creation by Equation (6), a new value for K is randomly chosen. This guarantees that the algorithm will not stagnate because a different K in each candidate calculation makes the candidates θ_c different even with the same components $\theta_1, \theta_2, \theta_3$ in Equation (6).

Our modification of differential evolution algorithm includes also an additional randomization step to improve the robustness of the algorithm. When the algorithm has converged and all the population members are equal, all but two population members are renewed by applying random mutations to the parameters. In practise, we multiplied each parameter of every population member with a unique random number drawn from a normal distribution with mean 1 and standard deviation 0.5. The motivation is to make the algorithm jump out of a local optimum. The algorithm is then rerun and if there is no improvement, it is assumed that the global optimum is achieved. Otherwise, if the best fit of the population was improved after the randomization, the randomization is repeated until no improvement is found. Thereby the algorithm is always run at least twice.

In our modification the population size was dependent on the number of mixture components. We used population size $30k$, where k is the number of components in the model. This is justified by the fact, that a model with more components is more complicated to estimate, and the increased population size provides more diversity to the population. We didn't include any mutation operator to the algorithm.

```
Initialize population
REPEAT
    Choose random population members θ1, θ2, θ3, θ4
    Set random K, construct a candidate θC := θ1 + K · (θ2 − θ3)
    IFf(θC) < f(θ4)
       Replace θ4 by θC in population
    ENDIF
UNTIL All population members are equal
Randomize population and rerun the algorithm until the achieved fit is
equal in two consequtive runs
```

Fig. 1. Pseudo-code for modified differential evolution algorithm

2.3 Other Methods

As a widely used reference method to quantify the overlapping spots we use low-pass filtering and Gaussian mixture model fitting with a local non-linear deterministic algorithm. Similarly as in e.g. [5] to find the mixture model parameters we use the Levenberg-Marquardt algorithm implemented in Matlab as *lsqnonlin* function. We also tested the performance of the differential evolution algorithm on the filtered data, and the performance of the *lsqnonlin* function on the raw data. Thereby the following three methods were evaluated:

- Ref A: Filtered image and local optimization
- Ref B: Filtered image and differential evolution algorithm
- Ref C: Noisy image and local optimization

The method A reprensents the common approach. The method B is used to test the inaccuracies produced by the local optimization algorithm in comparison to the differential evolution optimization. The method C is used to evaluate the effect of image filtering in Ref A in comparison to using the raw image. In this paper, we wanted to compare the accuracy of the methods, and therefore the correct number of components i.e. spots in each image was given to each algorithm. In practice, a spot detection method should also be implemented to determine the correct number of components.

The filter kernel in Ref A and Ref B was set as Gaussian with identity matrix as covariance matrix i.e. diagonal elements equal to one and off-diagonal elements equal to zero. In the methods A and B the fixed covariance parameter in the mixture model 1 was thereby modified to

$$\Sigma = \begin{pmatrix} 1 + 0.21\frac{\lambda}{A} & 0 \\ 0 & 1 + 0.21\frac{\lambda}{A} \end{pmatrix} \tag{7}$$

to better fit to the filtered image.

The accuracy of the deterministic optimization algorithm *lsqnonlin* is highly dependent on the quality of the initial guess. Here, we chose the k highest local maxima in the image as an initial guess for spot centroid locations, and the sum of their surrounding eight pixels as an initial guess for spot intensity.

3 Experimental Results

3.1 Simulated Data

Simulated data was created by placing spots to overlap each other partially. The shape of a spot was determined by the theoretical point spread function defined by the Bessel function of first kind, J_1, as

$$P(r) = \left(\frac{2J_1(ra)}{r}\right)^2 \qquad \text{with} \qquad a = \frac{2\pi A}{\lambda}\,. \tag{8}$$

Thereby, value of pixel (i, j) of a spot is defined by $P(r)$, where r is the distance between pixel centre to the spot centroid.

Artificial spots were located to overlap each other partially, more spesifically with a distance equal to the Rayleigh limit [9]. In cases with more than two overlapping spots, each spot had a neighbor with a distance equal to Rayleight limit, and other spots were farther away. This way, two spots never had smaller mutual distance than the Rayleigh limit and the spots were resolvable. Finally a constant background level value was added to every pixel (including pixels with spot intensity).

After creating the simulated image with point spread function spots, Poisson noise was implemented to simulate shot noise. For each pixel, we drew a random value from a Poisson distribution with parameter λ equal to the pixel value (multiplied by a factor α, and used this random value as the "noisy" pixel value.

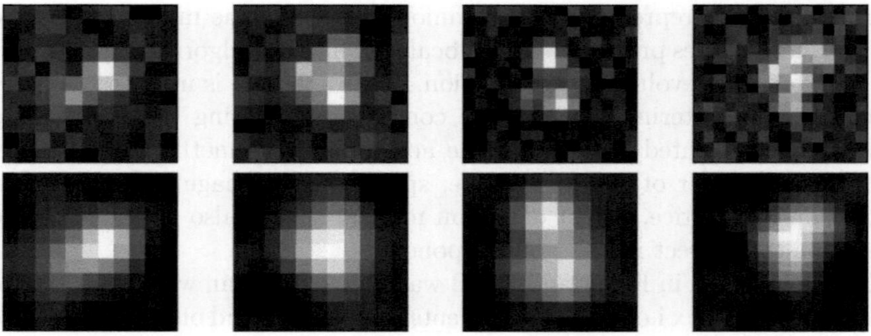

Fig. 2. Simulated data with 2 to 5 overlapping spots (left to right). Top row shows raw images with noise, bottom row shows the same images low-pass filtered.

This simulates the number of emitted photons collected by the ccd camera. With the noise multiplier α the signal-to-noise ratio of the images could be controlled.

In our simulated images we chose the following parameters: numerical aperture $A = 1.45$, emission wavelength $\lambda = 507nm$ and image pixel size $87nm$. These follow the setting that our collaborators have used in their biological studies. These values produced the Rayleigh limit of

$$d = 0.61\frac{\lambda}{A} = 213nm \approx 2.45 \text{ pixels} , \tag{9}$$

which was used as a distance between centroids of overlapping spots. Three different values were used as spot intensities: 1000, 2000 and 3000 and the background level was set to 2000 in every image. The signal-to-noise ratio was set to be 2.0 in every image by controlling the parameter α.

Four simulated images each with a unique number of overlapping spots were created and quantified with all the methods. The easiest image had clusters of two mutually overlapping spots while the other images had three, four and in the most difficult case, five overlapping spots per cluster. There were 1000 clusters in each image. Examples of simulated overlapping spots can be seen in Figure 2.

3.2 Results with Simulated Data

The quantification errors with simulated images can be seen in Tables 1 and 2. The spot intensity error in Table 1 is calculated as the error between the

Table 1. Median errors in spot intensities (percent)

	METHOD			
Spots	Ref A	Ref B	Ref C	New
2	34.4	34.4	6.8	6.5
3	32.2	32.2	7.7	7.0
4	31.4	30.7	9.2	7.4
5	29.5	28.3	13.5	8.2

Table 2. Median errors in spot locations (pixels)

	METHOD			
Spots	Ref A	Ref B	Ref C	New
2	0.199	0.199	0.130	0.124
3	0.255	0.250	0.145	0.134
4	0.304	0.274	0.176	0.147
5	0.436	0.313	0.246	0.165

estimated spot intensities and the true spot intensities in comparison to the true intensities. Perfect estimation results would produce zero percent error. The error in location in figures in Table 2 is calculated as the distance (norm) between the true spot location and the estimated spot location. Both the tables present median values within each image. Median values were used instead of mean values, because in some rare cases (less than one percent of the quantifications) deterministic optimization failed severely producing completely unrealistic results like spot intensity larger than 10^{11}. These extreme values would affect the calculated mean error and therefore median error is more representational in this case.

As can be seen in Table 1, the proposed method was the most accurate in comparison to the other methods in quantifying the spot intensities. Note that the largest error source based on these simulation results was the filtering, because the estimates obtained from filtered images (Ref A and Ref B) were significantly worse than those obtained without filtering (Ref C and New). This was expected because the filtering causes loss of information together with noise reduction. The improvement achieved by the stochastic optimization algorithm was especially notable with the raw data and with more complicated overlapping.

The results in estimation of spot locations in Table 2 are rather consistent with the intensity estimation results. However, it seems that the filtering increased less the error in location estimates than in intensity estimates. Nevertheless,

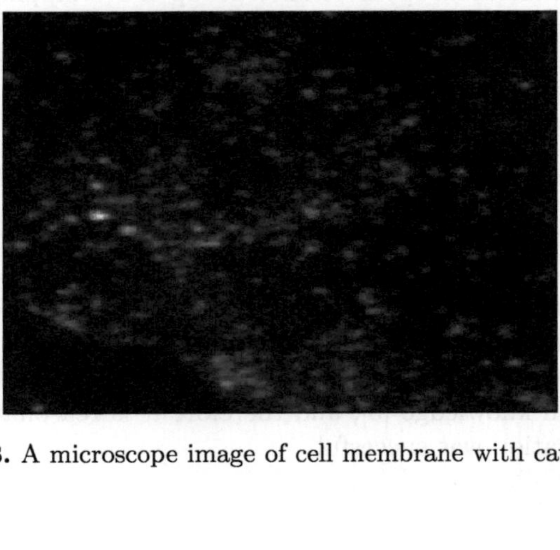

Fig. 3. A microscope image of cell membrane with caveolae

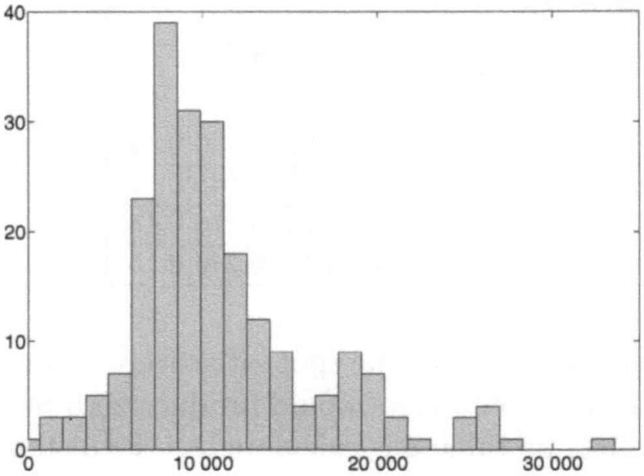

Fig. 4. Histogram of estimated intensities from a real microscope image

also in this case the new method improved the results significantly and in more complicated cases the choice of optimization algorithm seems to be crucial. The values in the Table 2 are stated in pixel units and can be converted to nanometers by multiplying with the chosen pixel size 87nm to give some reference to the possible accuracy improvement with real microscopy data.

3.3 Results with Microscopy Data

To show that the developed method is applicable to a real microscope data, we quantify an image of a cell membrane with fluorescent caveolin-1 protein spots. The image was acquired by Institute of Biomedicine at University of Helsinki and the data has been described in detail in [10]. An example of such an image can be seen in Figure 3. The intensity of spots is quantified to estimate the amount of fluorescently tagged proteins within a corresponding cell membrane invagination.

The number of individual spots within a group of overlapping spots was determined by increasing the number of components in the model iteratively until the addition didn't cause a significant improvement in the fitness of the model. Due to the fixed covariance matrix, the risk of overfitting was not so severe and the difference between significant and insignificant improvements was usually quite evident. Here, five percent improvement (or greater) was judged significant.

The results of intensity quantification with the developed method from the raw microscope image can be seen in Figure 4. There were 219 spots in total, of which 84 were overlapping with another spot. Thereby a significant portion of information would have been lost if overlapping spots have been left out of the study or quantified with poor accuracy. It can be seen in Figure 4 that the estimated intensities from clusters (at about 9000, 18000 and 27000) as expected based on biological knowledge [3], and therefore it is reasonable to assume the intensity quantification was succesful.

4 Conclusion

The widely applied method to quantify fluorescence microscopy images with filtering and local optimization was found to be unoptimal for spot intensity and sub-pixel location estimation. Filtering causes significant errors especially in spot intensity estimation and reduces accuracy in the location estimation as well. Thereby the quantification should be done from the raw images, and in this study we introduced a procedure to perform such a task. The raw image quantification requires a more robust optimization algorithm and we applied a stochastic global optimization algorithm. The results with simulated data show that significant improvements were achieved in both intensity and location esimates with the developed method. Also the quantification of the microscope data of cell membrane with caveolae was succesful.

Acknowledgements

The work was financially supported by the Academy of Finland under the grant 213462 (Finnish Centre of Excellence Program (2006 - 2011)). JT received additional support from University Alliance Finland Research Cluster of Excellence STATCORE. HP received additional support from Jenny and Antti Wihuri Foundation.

References

[1] Schmidt, T., Schütz, G.J., Baumgartner, W., Gruber, H.J., Schindler, H.: Imaging of single molecule diffusion. Proceedings of the National Academy of Sciences of the United States of America 93(7), 2926–2929 (2006)

[2] Schutz, G.J., Schindler, H., Schmidt, T.: Single-molecule microscopy on model membranes reveals anomalous diffusion. Biophys. J. 73(2), 1073–1080 (1997)

[3] Pelkmans, L., Zerial, M.: Kinase-regulated quantal assemblies and kiss-and-run recycling of caveolae. Nature 436(7047), 128–133 (2005)

[4] Anderson, C., Georgiou, G., Morrison, I., Stevenson, G., Cherry, R.: Tracking of cell surface receptors by fluorescence digital imaging microscopy using a charge-coupled device camera. Low-density lipoprotein and influenza virus receptor mobility at 4 degrees c. J. Cell Sci. 101(2), 415–425 (1992)

[5] Thomann, D., Rines, D.R., Sorger, P.K., Danuser, G.: Automatic fluorescent tag detection in 3D with super-resolution: application to the analysis of chromosome movement. J. Microsc. 208(Pt 1), 49–64 (2002)

[6] Mashanov, G.I.I., Molloy, J.E.E.: Automatic detection of single fluorophores in live cells. Biophys. J. 92, 2199–2211 (2007)

[7] Price, K.V., Storn, R.M., Lampinen, J.A.: Differential evolution - A practical approach to global optimization. Natural computing series. Springer, Heidelberg (2007)

[8] Lampinen, J., Zelinka, I.: On stagnation of the differential evolution algorithm. In: 6th international Mendel Conference on Soft Computing, pp. 76–83 (2000)

[9] Inoue, S.: Handbook of optics. McGraw-Hill Inc., New York (1995)

[10] Jansen, M., Pietiäinen, V.M., Pölönen, H., Rasilainen, L., Koivusalo, M., Ruotsalainen, U., Jokitalo, E., Ikonen, E.: Cholesterol Substitution Increases the Structural Heterogeneity of Caveolae. J. Biol. Chem. 283, 14610–14618 (2008)

Bayesian Classification of Image Structures

D. Goswami[1], S. Kalkan[2], and N. Krüger[3]

[1] Dept. of Computer Science, Indian School of Mines University, India
dibyendusekharg@gmail.com
[2] BCCN, University of Göttingen, Germany
sinan@bccn-goettingen.de
[3] Cognitive Vision Lab, Univ. of Southern Denmark, Denmark
norbert@mip.sdu.dk

Abstract. In this paper, we describe work on Bayesian classifiers for distinguishing between homogeneous structures, textures, edges and junctions. We build semi–local classifiers from hand-labeled images to distinguish between these four different kinds of structures based on the concept of intrinsic dimensionality. The built classifier is tested on standard and non-standard images.

1 Introduction

Different kinds of image structures coexist in natural images: homogeneous image patches, edges, junctions, and textures. A large body of work has been devoted to their extraction and parametrization (see, e.g., [1,2,3]). In an artificial vision system, such image structures can have rather different roles due to their implicit properties. For example, processing of local motion at edge-like structures faces the aperture problem [4] while junctions and most texture-like structures give a stronger motion constraint. This has consequences also for the estimation of the global motion. It has turned out (see, e.g., [5]) to be advantageous to use different kinds of constraints (i.e., line constraints for edges and point constraints for junctions and textures) for these different image structures. As another example, in stereo processing, it is known that it is impossible to find correspondences at homogeneous image patches by direct methods (i.e., triangulation based methods based on pixel correspondences) while textures, edges and junctions give good indications for feature correspondences. Also, it has been shown that there is a strong relation between the different 2D image structures and their underlying depth structure [6,7]. Therefore, it is important to classify image patches according to their junction–ness, textured-ness, edge–ness or homogeneous–ness.

In many hierarchical artificial vision systems, later stages of visual processing are discrete and sparse, which requires a transition from signal-level, continuous, pixel-wise image information to sparse information to which often a higher semantic can be associated. During this transition, the continuous signal becomes discretisized; *i.e.*, it is given discrete labels. For example, an image pixel whose contrast is above a given threshold is labeled as edge. Similarly, a pixel is classified as junction if, for example, the orientation variance in the neighborhood is high enough.

A.-B. Salberg, J.Y. Hardeberg, and R. Jenssen (Eds.): SCIA 2009, LNCS 5575, pp. 676–685, 2009.

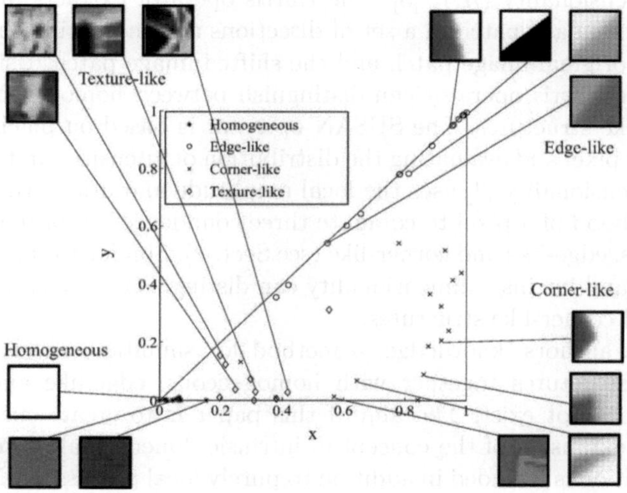

Fig. 1. How a set of 54 patches map to the different areas of the intrinsic dimensionality triangle. Some examples from these patches are also shown. The horizontal and vertical axes of the triangle denote the contrast and the orientation variances of the image patches, respectively.

The parameters of this discretization process are mostly set by its designer to perform best on a set of standard test images. However, it is neither trivial nor ideal to manually assign discrete labels to image structures since the domain is continuous. Hence, one benefits from *build*ing classifiers to give discrete labels to continuous signals. In this paper, we use hand-labeled image regions to *learn* the probability distributions of the features for different image structures and use this distribution to determine the type of image structure at a pixel. The local 2D structures that we aim to classify are listed below (examples of each structure is given in Fig. 1):

- Homogeneous image structures, which are signals of uniform intensities.
- Edge–like image structures, which are low-level structures that constitute the boundaries between homogeneous or texture-like signals.
- Junction-like structures, which are image patches where two or more edge-like structures with significantly different orientations intersect.
- Texture-like structures, which are often defined as signals which consist of repetitive, random or directional structures. In this paper, we define texture as 2D structures which have low spectral energy and high variance in local orientation (see Fig. 1 and Sect. 2).

Classification of image structures has been extensively studied in the literature, leading to several well-known feature detectors such as Harris [1], SUSAN [2] and intrinsic dimensionality (*iD*)[1] [8]. The Harris operator extracts image features by shifting the image patch in a set of directions and measuring the correlation between the original image patch and the shifted image patch. Using this measurement, the Harris operator can distinguish between homogeneous, edge-like and corner-like structures. The SUSAN operator is based on placing a circular mask at each pixel and evaluating the distribution of intensities in the mask. The intrinsic dimensionality [8] uses the local amplitude and orientation variance in the neighborhood of a pixel to compute three confidences according to its being homogeneous, edge-like and corner-like (see Sect. 2). Similar to the Harris operator, SUSAN and intrinsic dimensionality can distinguish between homogeneous, edge-like and corner-like structures.

Up to the authors' knowledge, a method for simultaneous classification of texture-like structures together with homogeneous, edge-like and corner-like structures does not exist. The aim of this paper is to create such a classifier based on an extansion of the concept of intrinsic dimensionality in which semi-local information is included in addition to purely local processing. Namely, from a set of hand-labeled images[2], we learn local as well as semi–local classifiers to distinguish between homogeneous, edge-like, corner-like as well as texture-like structures. We present results of the built classifier on standard as well as non-standard images.

The paper is structured as following: In Sect. 2, we describe the concept of intrinsic dimensionality. In Sect. 3, we introduce our method for classifying image structures. Results are given in Sect. 4 with a conclusion in Sect. 5.

2 Intrinsic Dimensionality

When looking at the spectral representation of a local image patch (see Fig. 2(a,b)), we see that the energy of an i0D signal is concentrated in the origin (Fig. 2(b)-top), the energy of an i1D signal is concentrated along a line (Fig. 2(b)-middle) while the energy of an i2D signal varies in more than one dimension (Fig. 2(b)-bottom).

Recently, it has been shown [8] that the structure of the *iD* can be understood as a triangle that is spanned by two measures: origin variance and line variance. Origin variance describes the deviation of the energy from a concentration at the origin while line variance describes the deviation from a line structure (see Fig. 2(b) and 2(c)); in other words, origin variance measures non-homogeneity of the signal whereas the line variance measures the junctionness. The corners of the triangle then correspond to the 'ideal' cases of *iD*. The surface of the triangle corresponds to signals that carry aspects of the three 'ideal' cases, and the distance from the corners of the triangle indicates the similarity (or dissimilarity) to *ideal* i0D, i1D and i2D signals.

---

1 *iD* assigns the names *intrinsically zero dimensional (i0D)*, *intrinsically one dimensional (i1D)* and *intrinsically two dimensional (i2D)* respectively to homogeneous, edge-like and junction-like structures.

2 The software to label images is freely available for public use at `http://www.mip.sdu.dk/covig/software/label_on_web.html`

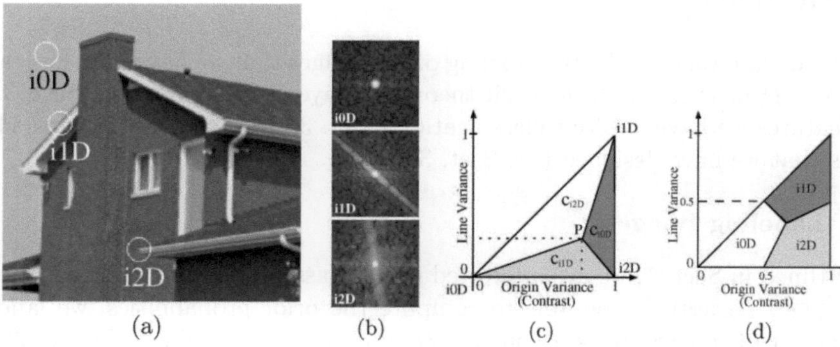

Fig. 2. Illustration of the intrinsic dimensionality (Sub-figures (a,b,c) taken from [8]). (a) Three image patches for three different intrinsic dimensions. (b) The 2D spatial frequency spectra of the local patches in (a), from top to bottom: i0D, i1D, i2D. (c) The topology of *iD*. Origin variance is variance from a point, i.e., the origin. Line variance is variance from a line, measuring the junctionness of the signal. c_{iND} for $N = 0, 1, 2$ stands for confidence for being i0D, i1D and i2D, respectively. Confidences for an arbitrary point P is shown in the figure which reflect the areas of the sub-triangles defined by P and the corners of the triangle. (d) The decision areas for local image structures.

As shown in [8], this triangular interpretation allows for a *continuous formulation* of *iD* in terms of 3 confidences assigned to each discrete case. This is achieved by first computing two measurements of origin and line variance which define a point in the triangle (see Fig. 2(c)). The bary-centric coordinates (see, e.g., [9]) of this point in the triangle directly lead to a definition of three confidences that add up to one. These three confidences reflect the volume of the areas of the three sub-triangles which are defined by the point in the triangle and the corners of the triangle (see Fig. 2(c)). For example, for an arbitrary point P in the triangle, the area of the sub-triangle i0D-P-i1D denotes the confidence for i2D as shown in Fig. 2(c). That leads to the decision areas for i0D, i1D and i2D as seen in Fig. 2(d). For the example image in Fig. 2, computed *iD* is shown in Fig. 3.

Fig. 3. Computed *iD* for the image in Fig. 2, black means zero and white means one. From left to right: $c_{i0D}, c_{i1D}, c_{i2D}$ and highest confidence marked in gray, white and black for i0D, i1D and i2D, respectively.

3 Methods

In this section, we describe the labeling of the images that we have used for learning and testing (Sect. 3.1), the basic theory for Bayesian classification (Sect. 3.2), the features we have used for classification (Sect. 3.3), as well as the three classifiers that we have designed (see Sect. 3.4).

3.1 Labeling Images

As outlined in Sect. 1, we are interested in the classification of four image structures (*i.e.*, classes). To be able to compute the prior probabilities, we labeled a large set of images using a software that we developed. The software allows for the labeling arbitrary regions in an image, which are saved and then used for computing the prior probabilities (as well as evaluating the performance of learned classifiers that will be introduced in 3.4) for classifying image structures. Fig. 4 shows a few examples of labeled images patches.

We labeled only image patches that were close to be the 'ideal' cases of their class because we did not want to make decisions about the class of an image patch which might be carrying aspects of different kinds of image structures. We would like a Bayesian classifier to make manifestations about the type of 'non-ideal' image patches based on what it has learned about the 'ideal' image structures.

3.2 Bayesian Classification

If C_i, for $i = 1, \ldots, 4$, represents on the the four classes, and X is the feature vector extracted for the pixel whose class has to be found, then the probability that the pixel belongs to a particular class C_i is given by the posterior probability $P(C_i|X)$ of that class C_i given the feature vector X (using Bayes' Theorem):

$$P(C_i|X) = \frac{P(X|C_i)P(C_i)}{P(X)}, \tag{1}$$

where $P(C_i)$ is the prior probability of the class C_i; $P(X|C_i)$ is the probability of feature vector X, given the pixel belongs to the class C_i; and, $P(X)$ is the total probability of the feature vector X (*i.e.*, $\sum_i P(X|C_i)P(C_i)$).

Fig. 4. Images with various classes labeled. The colors blue, red, yellow and green correspond to homogeneous, edge-like, junction-like and texture-like structures, respectively.

A Bayesian classifier first computes $P(C_i|X)$ using equation 1. Then, the classifier gives the label C_m to a given feature vector X_0 if $P(C_m|X_0)$ is maximal, *i.e.*, $C_m = \arg\max_i \{ P(C_i|X)\}$. The prior probabilities $P(C_i)$, $P(X)$ and the conditional probability $P(X|C_i)$ are computed from the labeled images. The prior probabilities $P(C_i)$ are 0.5, 0.3, 0.02 and 0.18 respectively for homogeneous, texture-like, corner-like and edge-like structures. An immediate conclusion from these probabilities is that corners are the least frequent image structures whereas homogeneous structures are abundant.

3.3 Features for Classification

As can be seen from Fig. 1, image structures have different neighborhood patterns. The type of an image structure at a pixel can be *estimated* from the signal information in the neighborhood. For this reason, we utilize the neighborhood of a given pixel for computing features that will be used for estimating the class of the pixel.

Now we define three features for each pixel P in the image. For two of these we define a neighborhood which is a ring of radius r[3]:

- **Central feature** $(x_{central}, y_{central})$**:** The co-ordinates of pixel $p = (p_x, p_y)$ in the *iD* triangle (see Sect. 2): $x_{central} = 1 - i0D_p$, $y_{central} = i1D_p$. The central feature has been used in [8] to distinguish between edges, corners and homogeneous image patches based on the barycentric co-ordinates. As we show in this work, it can also be used in a Bayesian classifier to characterize also texture, however not surprisingly with a large degree of misclassification in particular between texture and junctions.
- **Neighborhood mean feature** (x_{nmean}, y_{nmean})**:** The mean value of the co-ordinates (x, y) in the *iD* triangle of all the pixels in the circular neighborhood of the pixel P. More formally, $x_{nmean} = \frac{1}{N}\sum_{i=1}^{N} 1 - i0D_i$, $y_{nmean} = \frac{1}{N}\sum_{i=1}^{N} i1D_i$.
- **Neighborhood variance feature** (x_{nvar}, y_{nvar})**:** The variance value of the co-ordinates (x, y) in the *iD* triangle of all the pixels in the neighborhood of pixel P. So, $x_{nvar} = i0D_{nvar}, y_{nvar} = i1D_{nvar}$, where $i0D_{nvar}$ and $i1D_{nvar}$ are respectively the variance in the values of i0D and i1D in the neighborhood of pixel P.

The motivation behind using these three features is the following. The central feature represents the classical *iD*concept as outlined in [8] and has already been used for classification (however, not in a Bayesian sense). The neighborhood mean represent the mean *iD*value in the ring neighborhood. For edge-like structures it can be assumed that there will be *iD*values representing edges (at the

[3] The radius r has to be chosen depending on the frequency the signal is investigated at. In our case, we chose a radius of 3 pixels which reflects that the spatial features at that distance, although still sufficiently local, give new information in comparison to the iD values at the center pixel.

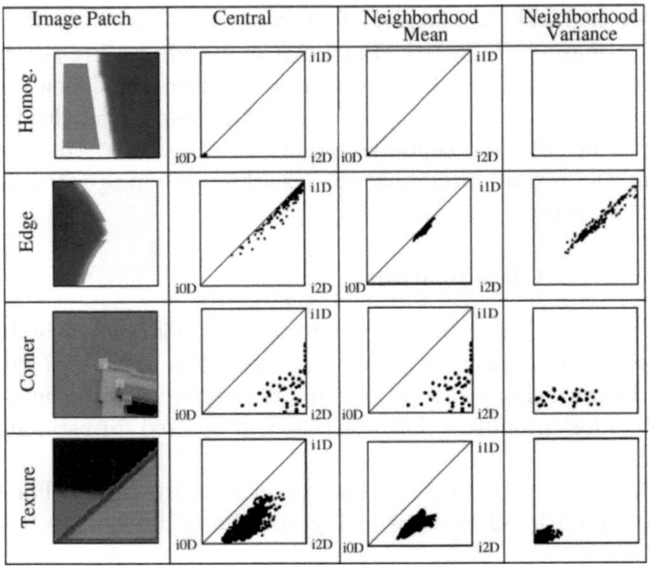

Fig. 5. The distributions of the features for each of the individual classes

prolongation of the edge at the center) as well as homogeneous image patches orthogonal to the edge. For junctions, there will be a more distributed pattern at the i2D corner while for textures, we will expect rather similar iD values on the ring due to the repetitive nature of texture. These thoughts will also be reflected in the neighborhood variance feature. Hence the two last features should give complementary information to the central feature. This is becoming clear looking at the distribution of these features over example structures as outlined in the next paragraph.

Fig. 5 shows the distribution of the features for selected regions in different images, and the total distribution of the features for each type of image structure is given in Fig. 6 (computed from a set of 65 images). The labeling process led to 91.500 labeled pixels which included 45.000 homogeneous, 20.000 edge-like, 1.500 corner-like and 25.000 texture-like pixels.

By observing the central feature distributions in Fig. 6, we see that many points labeled as corners have overlapping regions with textures and edges. However, we see from Fig. 6 that the neighborhood mean as well as the neighborhood variance can further help to distinguish between the four classes. Another important observation from Fig. 6 is that the neighborhood variance divides the points into two distinct divisions: the high variance classes (edges and corners) and the low variance classes (homogeneous and texture). This is due to the fact that edges and corners have, by definition, more variance in their neighborhood.

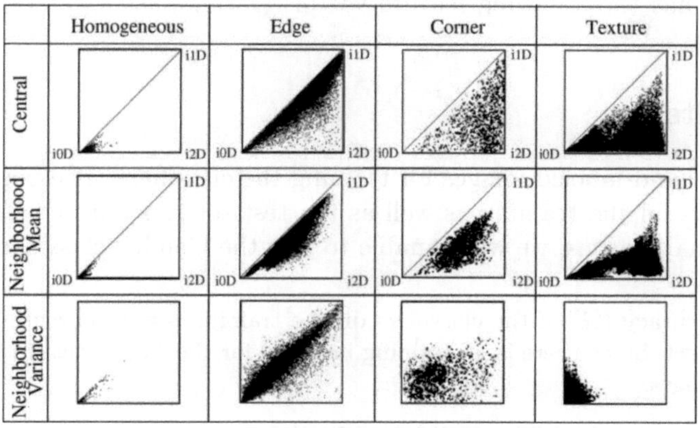

Fig. 6. The cumulative distribution of the features collected from a set of 65 images. There are 91,500 labeled pixels in total, which includes 45,000 homogeneous, 20,000 edge-like, 1,500 corner-like and 25,000 texture-like pixels.

3.4 The Classifiers

We design five classifiers:

- **Naive classifier (NaivC):** Classifier just using the iD based on barycentric co-ordinates, which is only able to distinguish junctions, homogeneous image patches and edges.
- **Central Bayesian Classifier (CentC):** The first and elementary Bayesian Classifier that we built is based on (x, y) co-ordinates of the pixel in the iD triangle, where $x = 1 - i0D_P$ and $y = i1DP$. Our experiments with this classifier showed that though it is good at detecting edges and the other classes, its detection of corners is poor: It could only detect only about 35% of the corners in the training set of images and only 20% in the test set. With the intention of building a better classifier, therefore, we decided to enhance the performance of the classifier by taking into account the features of the neighborhood of a pixel.
- **Classifier using neighborhood mean (NmeanC):** Our next classifier (NmeanC) is based on the central and neighborhood mean features of a pixel; *i.e.*, classifier NmeanC has the following feature vector: $(x_{central}, y_{central}, x_{nmean}, y_{nmean})$.
- **Classifier using neighborhood variance (NvarC):** Though classifier NmeanC is much better than the CentC, it made many errors in the detection of corners. We can observe from figure 6 that there is some overlap between the neighborhood mean distributions of corners and edges, and also corners and textures. With this observation, we build a classifier taking into account the central and neighborhood variance features of a pixel; *i.e.*, classifier NvarC has the following feature vector: $(x_{central}, y_{central}, x_{nvar}, y_{nvar})$.
- **Classifier using all features (CombC):** CombC consists of all three features: central, neighborhood mean and neighborhood variance; *i.e.*, classifier

CombC has the following feature vector: $(x_{central}, y_{central}, x_{nmean}, y_{nmean}, x_{nvar}, y_{nvar})$.

4 Results

We used 85 hand-labeled images for training the classifiers. The performance of the classifiers on the training as well as the test set is given in table 1. Due to computational reasons, we were unable to test the CombC classifier.

Table 1. Accuracy (%) of the classifiers on the training set (in parentheses) and the non-training set. Since there is no training involved for the NaivC classifier, it is tested on all the images.

Class	NaivC	CentC	NmeanC	NvarC
Homogeneous	95	85 (88)	98 (99)	95 (99)
Edge	70	80 (85)	90 (95)	89 (97)
Corner	70	20 (35)	70 (97)	86 (98)
Texture	–	75 (83)	77 (96)	73 (90)

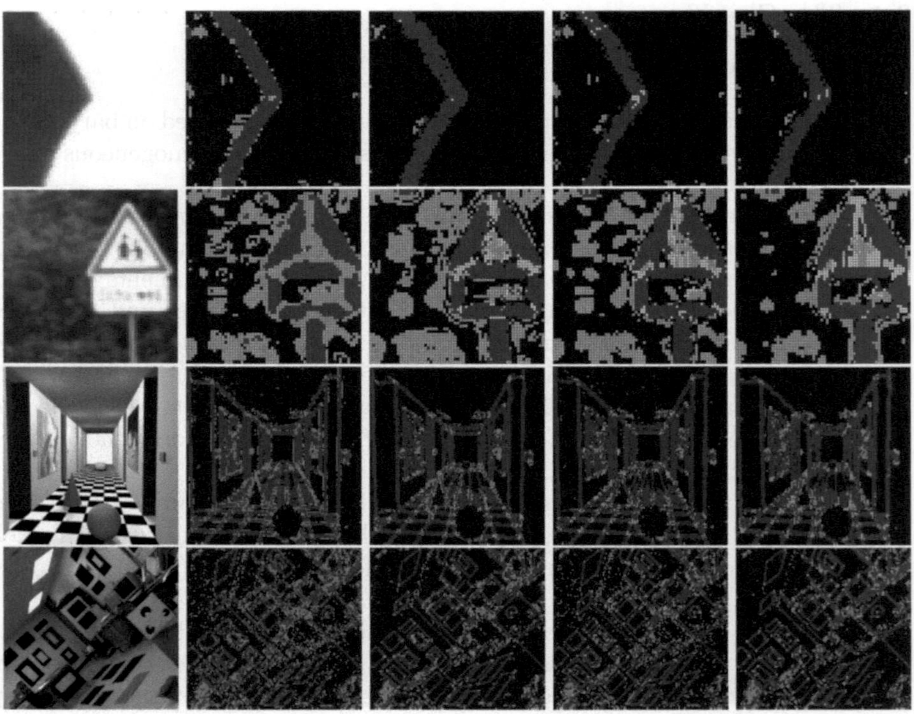

(a) (b) (c) (d) (e)

Fig. 7. Responses of the classifiers on a subset of the non-training set. Colors blue, red, light blue and yellow respectively encode homogeneous, edge-like, texture-like and corner-like structures.

We observe that the classifiers NmeanC, NvarC and CombC are good edge as well as corner detectors. Comparing NmeanC, NvarC and CombC against CentC, we can see that inclusion of neighborhood in the features improves the detection of corners drastically, and other image structures quite significantly (both on the training and non-training sets). Fig. 7 provide the responses of the classifiers on the non-training set. A surprising results is that combination of neighborhood variance and neighborhood mean features (CombC) performs worse than neighborhood variance feature (NvarC).

5 Conclusion

In this paper, we have introduced simultaneous classification of homogeneous, edge-like, corner-like and texture-like structures. This approach goes beyond current feature detectors (like Harris [1], SUSAN [2] or intrinsic dimensionality [8]) that distinguish only between up to three different kinds of image structures. The current paper has proposed and demonstrated a probabilistic extension to one of such approaches, namely the intrinsic dimensionality.

Acknowledgements. This work is supported by the EU Drivsco project (FP6-IST-FET-016276-2).

References

1. Harris, C.G., Stephens, M.J.: A combined corner and edge detector. In: Proc. Fourth Alvey Vision Conference, Manchester, pp. 147–151 (1988)
2. Smith, S., Brady, J.: SUSAN - a new approach to low level image processing. Int. Journal of Computer Vision 23(1), 45–78 (1997)
3. Mikolajczyk, K., Schmid, C.: A performance evaluation of local descriptors. IEEE Trans. Pattern Anal. Mach. Intell. 27(10), 1615–1630 (2005)
4. Kalkan, S., Calow, D., Wörgötter, F., Lappe, M., Krüger, N.: Local image structures and optic flow estimation. Network: Computation in Neural Systems 16(4), 341–356 (2005)
5. Rosenhahn, B., Sommer, G.: Adaptive pose estimation for different corresponding entities. In: Van Gool, L. (ed.) DAGM 2002. LNCS, vol. 2449, pp. 265–273. Springer, Heidelberg (2002)
6. Grimson, W.: Surface consistency constraints in vision. CVGIP 24(1), 28–51 (1983)
7. Kalkan, S., Wörgötter, F., Krüger, N.: Statistical analysis of local 3D structure in 2D images. In: IEEE Int. Conference on Compter Vision and Pattern Recognition (CVPR), vol. 1, pp. 1114–1121 (2006)
8. Felsberg, M., Kalkan, S., Krüger, N.: Continuous dimensionality characterization of image structures. Image and Vision Computing (2008) (in press)
9. Coxeter, H.: Introduction to Geometry, 2nd edn. Wiley & Sons, Chichester (1969)

Globally Optimal Least Squares Solutions for Quasiconvex 1D Vision Problems

Carl Olsson, Martin Byröd, and Fredrik Kahl

Centre for Mathematical Sciences
Lund University, Lund, Sweden
{calle,martin,fredrik}@maths.lth.se

Abstract. Solutions to non-linear least squares problems play an essential role in structure and motion problems in computer vision. The predominant approach for solving these problems is a Newton like scheme which uses the hessian of the function to iteratively find a local solution. Although fast, this strategy inevitably leeds to issues with poor local minima and missed global minima.

In this paper rather than trying to develop an algorithm that is guaranteed to always work, we show that it is often possible to verify that a local solution is in fact also global. We present a simple test that verifies optimality of a solution using only a few linear programs. We show on both synthetic and real data that for the vast majority of cases we are able to verify optimality. Further more we show even if the above test fails it is still often possible to verify that the local solution is global with high probability.

1 Introduction

The most studied problem in computer vision is perhaps the (2D) least squares triangulation problem. Even so no efficient globally optimal algorithm has been presented. In fact studies indicate (e.g. [1]) that it might not be possible to find an algorithm that is guaranteed to always work. On the other hand, under the assumption of Gaussian noise the $\mathbb{L}_2$-norm is known to give the statistically optimal solution. Although this is a desirable property it is difficult to develop efficient algorithms that are guaranteed to find the globally optimal solution when projections are involved. Lately researchers have turned to methods from global optimization, and a number of algorithms with guaranteed optimality bounds have been proposed (see [2] for a survey). However these algorithms often exhibit (worst case) exponential running time and they can not compare with the speed of local, iterative methods such as bundle adjustment [3,4,5].

Therefore a common heuristic is to use a minimal solver to generate a starting guess for a local method such as bundle adjustment [3]. These methods are often very fast, however since they are local the success depends on the starting point. Another approach is to minimize some algebraic criteria. Since these typically don't have any geometric meaning this approach usually results in poor reconstructions.

A different approach is to use the maximum residual error rather than the sum of squared residuals. This yields a class of quasiconvex problems where it

A.-B. Salberg, J.Y. Hardeberg, and R. Jenssen (Eds.): SCIA 2009, LNCS 5575, pp. 686–695, 2009.

is possible to devise efficient global optimization algorithms [6]. This was done in the context of 1D cameras in [7].

Still, it would be desirable to find the statistically optimal solution. In [8] it was shown that for the 2D-triangulation problem (with spherical 2D-cameras) it is often possible to verify that a local solution is also global using a simple test. It was shown on real datasets that for the vast majority of all cases the test was successful. From a practical point of view this is of great value since it opens up the possibility of designing systems where bundle adjustment is the method of choice and only turning to more expensive global methods when optimality can not be verified.

In [9] a stronger condition was derived and the method was extended to general quasiconvex muliview problems (with 2D pinhole cameras).

In this paper we extend this approach 1D multiview geometry problems with spherical cameras. We show that for most real problems we are able to verify that a local solution is global. Further more in case the test fails we show that it is possible to relax the test to show that the solution is global with large probability.

2 1D-Camera Systems

Before turning to the least squares problem we will give a short review of 1D-vision (see [7]). Throughout the paper we will use spherical 1D-cameras.

We start by considering a camera that is located at the origin with zero angle to the Y axis (see figure 1). For each 2D-point (X, Y) our camera gives a direction in which the point has been observed. The direction is given in the form of an angle θ with respect to a reference axis (see figure 1). Let $\Pi : \mathbb{R}^2 \mapsto [0, \pi^2/4]$ be defined by

$$\Pi(X, Y) = \text{atan}^2 \frac{X}{Y}. \tag{1}$$

if $Y > 0$, (otherwise we let $\Pi(X, Y) = \infty$). The function $\Pi(X, Y)$ measures the squared angle between the Y-axis and the vector $U = (X\ Y)^T$. Here we have explicitly written $\Pi(X, Y)$ to indicate that Π always takes values in $\mathbb{R}^2$, however throughout the paper we will use both $\Pi(X, Y)$ and $\Pi(U)$. Now, suppose that we have a measurement of a point with angle $\theta = 0$. Then Π can be interpreted as the squared angular distance between the point (X,Y) and the measurement. If the measurement θ is not zero we let $R_{-\theta}$ be a rotation $-\theta$ then $\Pi(R_{-\theta}U)$ can be seen as the squared angular distance $(\phi - \theta)^2$.

Next we introduce the camera parameters. The camera may be located anywhere in $\mathbb{R}^2$ with any orientation with respect to a reference coordinate system. In practice we have two coordinate systems, the camera- and the reference- coordinate system. To relate these two we introduce a similarity transformation P that takes points coordinates in the reference system and transforms then into coordinates in the camera system. We let

$$P = \begin{pmatrix} a & -b & c \\ b & a & d \end{pmatrix} \tag{2}$$

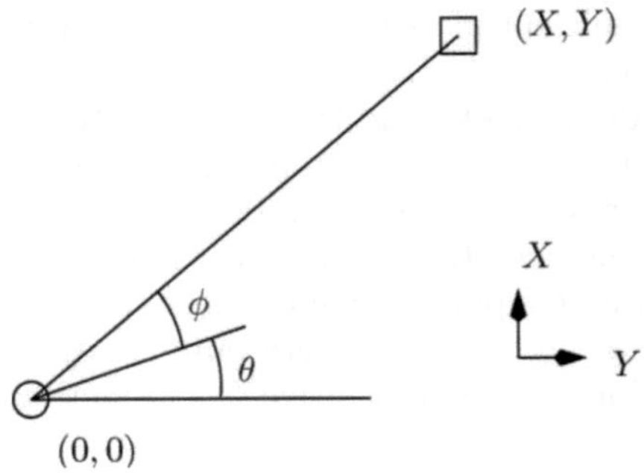

Fig. 1. 1D camera geometry for a calibrated camera

The parameters (a, b, c, d) are what we call the inner camera parameters and they determine the orientation and position of the camera. The squared angular error can now be written

$$\Pi\left(R_{-\theta} P \begin{bmatrix} U \\ 1 \end{bmatrix}\right) \tag{3}$$

In the remaining part of the paper the concept of quasiconvexity will be important. A function f is said to be quasiconvex if its sublevel sets $S_\phi(f) = \{x; f(x) \leq \phi\}$ are convex. In the case of traingulation (as well as resectioning) we see that the squared angular errors (3) can be written as the composition of the projection Π and two affine functions

$$X_i(x) = a_i^T x + \tilde{a_i} \tag{4}$$

$$Y_i(x) = b_i^T x + \tilde{b_i} \tag{5}$$

(here i denotes the i'th error residual). It was shown in [7] that functions of this type are quasiconvex. The advantage of quasiconvexity is that a function with this property can only have a single local minimum, when using the $\mathbb{L}_\infty$-norm. This class of problems include, among others, camera resectioning and triangulation.

In this paper, we will use the theory of quasiconvexitivity as a stepping stone to verify global optimality under the L_2 Norm. Our approach closely parallels that of [8] and [9]. However while [8] considered spherical 2D cameras only for the triangulation problem and [9] considered 2D-pinhole cameras for general multiview problems, we will consider 1D-spherical cameras.

3 Theory

In this section we will give sufficient conditions for global optimality. If x^* is a global minimum then there is an open set containing x^* where the Hessian of f is positive semidefinite. Recall that a function is convex if and only if its Hessian is positive semidefinite. The basic idea which was first introduced in [8] is the following: If we can find a convex region $\mathcal{C}$ containing x^* that is large enough to include all globally optimal solutions and we are able to show that the Hessian of f is convex on this set, then x^* must be the globally optimal solution.

3.1 The Set $\mathcal{C}$

The first step is to determine the set $\mathcal{C}$. Suppose that for our local candidate solution x^* we have $f(x^*) = \phi_{max}^2$. Then clearly any global optimum must fulfill $f_i(x) \leq \phi_{max}^2$ for all residuals, since otherwise our local solution is better. Hence we take the region $\mathcal{C}$ to be

$$\mathcal{C} = \{x \in \mathbb{R}^n, f_i(x) \leq \phi_{max}^2\}. \tag{6}$$

It is easily seen that this set is convex since this is the intersection of the sublevel sets $S_{\phi_{max}^2}(f_i)$ which are known to be convex since the residuals f_i are quasiconvex. Hence if we can show that the Hessian of f is positive definite on this set we may conclude that x^* is the global optimum.

Note that the condition $f_i(x) \leq \phi_{max}^2$ is somewhat pessimistic. Indeed it assumes that the entire error may occur in one residual which is highly unlikely under any reasonable noise model. In fact we will show that it it possible to replace ϕ_{max}^2 with a stronger bound to show that x^* is with high probability the global optimum.

3.2 Bounding the Hessian

The goal of this section is to show that the Hessian of f is positive semidefinite on the set $\mathcal{C}$. To do this we will find a constant matrix H that acts as a lower bound on $\nabla^2 f(x)$ for all $x \in \mathcal{C}$. More formally we will construct H such that $\nabla^2 f(x) \succeq H$ on $\mathcal{C}$, that is if H is positive semidefinite than so is $\nabla^2 f(x)$. We begin by studying the 1D-projection mapping Π. The Hessian of Π is

$$\nabla^2 \Pi(X,Y) = \frac{2}{(X^2+Y^2)^2} \begin{pmatrix} Y^2 - XY\text{atan}\frac{X}{Y} & (X^2-Y^2)\text{atan}\frac{X}{Y} - XY \\ (X^2-Y^2)\text{atan}\frac{X}{Y} - XY & X^2 + XY\text{atan}\frac{X}{Y} \end{pmatrix} \tag{7}$$

To simplify notation we introduce the measurement angle $\phi = \text{atan}\frac{X}{Y}$ and the radial distance to the camera center $r = \sqrt{X^2+Y^2}$. After a few simplifications one obtains

$$\nabla^2 \Pi(X,Y) = \frac{1}{r^2} \begin{pmatrix} 1 + \cos(2\phi) - 2\phi\sin(2\phi) & -\sin(2\phi) - 2\phi\cos(2\phi) \\ -\sin(2\phi) - 2\phi\cos(2\phi) & 1 - \cos(2\phi) + 2\phi\sin(2\phi) \end{pmatrix} \tag{8}$$

In the case of 3D to 2D projections Hartley et.al. [8] obtained a similar 3×3 matrix. Using the same arguments it may be seen our matrix can be bounded by the diagonal matrix

$$H(X,Y) = \nabla^2 \Pi(X,Y) = \frac{2}{r^2} \begin{pmatrix} \frac{1}{4} & 0 \\ 0 & -4\phi^2 \end{pmatrix} \tag{9}$$

To see this we need to show that the eigenvalues of $\nabla^2 \Pi(X,Y) - H(X,Y)$ are all positive. Taking the trace of this matrix we see that the sum of the eigenvalues are $\frac{1}{r^2}(3/2 + 8\phi^2)$ which is always positive. We also have the determinant

$$\det(\nabla^2 \Pi(X,Y) - H(X,Y)) = -1 + (1+16\phi^2)(\cos(2\phi) - 2\phi\sin(2\phi)) \tag{10}$$

It can be shown (see [8]) that this expression is positive if $\phi \leq 0.3$. Hence for $\phi \leq 0.3$, $H(X,Y)$ is a lower bound on $\nabla^2 \Pi(X,Y)$.

Now, the error residuals $f_i(x)$ of our class of problems are related to the projection mapping via an affine change of coordinates

$$f_i(x) = \Pi(a_i^T x + \tilde{a}_i, b_i^T x + \tilde{b}_i). \tag{11}$$

It was noted in [9] that since the coordinate change is affine the Hessian of f_i is can be bounded by H. To see this we let W_i be the matrix containing a_i and b_i as columns Using the chain rule we obtain the Hessian

$$\nabla^2 f_i(x) = W_i \nabla^2 \Pi(a_i^T x + \tilde{a}_i, b_i^T x + \tilde{b}_i) W_i^T. \tag{12}$$

And since $\nabla^2 \Pi$ is bounded by H we obtain

$$\nabla^2 f(x) \succeq \sum_i W_i H(a_i^T x + \tilde{a}_i, b_i^T x + \tilde{b}_i) W_i^T = \sum_i \frac{2}{r_i^2}\left(\frac{a_i a_i^T}{4} - 4\phi_i^2 b_i b_i^T\right). \tag{13}$$

The matrix appearing on the right hand side of (13) seems easier to handle however it still depends on x through r and ϕ. This dependence may be removed by using bound of the type

$$\phi \leq \phi_{\max} \tag{14}$$

$$r_{i,\min} \leq r_i \leq r_{i,\max} \tag{15}$$

The first bound is readily obtained since $x \in \mathcal{C}$. In the second one we need to find an upper and lower bound on the radial distance in every camera. We shall see later that this can be cast as a convex problem which can be solved efficiently. As is [9] we now obtain the bound

$$\nabla^2 f(x) \succeq \sum_i \left(\frac{1}{2r_{i,\max}^2} a_i a_i^T - 8\frac{\phi_{\max}^2}{r_{i,\min}^2} b_i b_i^T\right). \tag{16}$$

Hence if the minimum eigenvalue of the right hand side is non negative the function f will be convex on the set $\mathcal{C}$.

3.3 Bounding the Radial Distances r_i

In order to be able to use the criterion (13) we need to be able to compute bounds on the radial distances. The k'th radial distance may be written

$$r_k(x) = \sqrt{(a_k^T x + \tilde{a}_k)^2 + (b_k^T x + \tilde{b}_k)^2} \tag{17}$$

Since $x \in \mathcal{C}$ we know that (see [7])

$$(a_k^T x + \tilde{a}_k)^2 + (b_k^T x + \tilde{b}_k)^2 \leq (1 + \tan^2(\phi_{\max}))(b_k^T x + \tilde{b}_k)^2 \tag{18}$$

and obviously

$$(a_k^T x + \tilde{a}_k)^2 + (b_k^T x + \tilde{b}_k)^2 \geq (b_k^T x + \tilde{b}_k)^2 \tag{19}$$

The bound (15) can be obtained by solving the linear programs

$$r_{k,\max} = \max \sqrt{(1+\tan^2(\phi_{\max}))}(b_k^T x + \tilde{b}_k) \quad (20)$$

$$\text{s.t } |a_i^T x + \tilde{a}_i| \leq \tan(\phi_{\max})(b_i^T x + \tilde{b}_i), \quad \forall i \quad (21)$$

and

$$r_{k,\min} = \min\ (b_k^T x + \tilde{b}_k) \quad (22)$$

$$\text{s.t } |a_i^T x + \tilde{a}_i| \leq \tan(\phi_{\max})(b_i^T x + \tilde{b}_i), \quad \forall i. \quad (23)$$

At first glance this may seem as a quite rough estimate, however since $\phi_{\max}$ is usually small this bound is good enough. By using SOCP-programming instead of linear programming it is possible to improve these bounds, however since linear programming is faster we prefer to use the looser bounds.

To summarize, the following steps are performed in order to verify optimality:

1. Compute a local minimizer x^* (e.g. with bundle adjustment).
2. Compute maximum/minimum radial depths over $\mathcal{C}$.
3. Test if the convexity condition in (16) holds.

4 A Probabilistic Approach

In practice, the constraints $f_i(x) \leq \phi_{max}^2$ are often overly pessimistic. In fact what is assumed here is that the entire residual error ϕ_{max}^2 could (in worst case) arise from a single error residual, which is not very likely. Assume that $\hat{x}_i$ is the point measurements that would be obtained in a noise free system and that x_i is the real measurement. Under the assumption of independent Gaussian noise we have

$$\hat{x}_i - x_i = r_i, \quad r_i \sim \mathcal{N}(0,\sigma). \quad (24)$$

Since r_i has zero mean, an unbiased estimate of σ is given by

$$\hat{\sigma} = \sqrt{\frac{1}{m-d}}\phi_{max}, \quad (25)$$

where m is the number of residuals and d denotes the number of degrees of freedom in the underlying problem (for example, $d = 2$ for 2D triangulation and $d = 3$ for 2D calibrated resectioning). As before, we are interested in finding a bound for each residual. This time, however, we are satisfied with a bound that holds *with high probability*. Specifically, given $\hat{\sigma}$, we would like to find $L(\hat{\sigma})$ so that

$$P[\forall i : -L(\hat{\sigma}) \leq r_i \leq L(\hat{\sigma})] \geq P_0 \quad (26)$$

for a given confidence level P_0. To this end, we make use of a basic theorem in statistics which states that $\frac{X}{\sqrt{Y_\gamma/\gamma}}$ is T-distributed with γ degrees of freedom, when X is normal with mean 0 and variance 1, Y is a chi squared random variable with γ degrees of freedom and X and Y are independent. A further basic fact from statistics states that, $\hat{\sigma}^2(m-d)/\sigma^2$ is chi squared distributed with $\gamma = m - d$ degrees of freedom. Thus,

$$\frac{r_i}{\hat{\sigma}} = \frac{r_i/\sigma}{\sqrt{\hat{\sigma}^2/\sigma^2}} \tag{27}$$

fulfills the requirements to be T distributed apart from a small dependence between r_i and $\hat{\sigma}$. This dependence, however, vanishes with enough residuals and in any case leads to a slightly more conservative bound.

Given a confidence level β we can now *e.g* do a table lookup for the T distribution to get t_γ^β so that

$$P[-t_\gamma^\beta \leq \frac{r_i}{\hat{\sigma}} \leq t_\gamma^\beta] \geq \beta. \tag{28}$$

Multiplying through with $\hat{\sigma}$ we obtain $L(\hat{\sigma}) = \hat{\sigma} t_\gamma^\beta$. Given a confidence level β_0 for all r_i, we assume that the $r_i/\hat{\sigma}$ are independent and thus set $\beta = \beta_0^{1/m}$ to get

$$P[\forall i : -t_\gamma^\beta \leq \frac{r_i}{\hat{\sigma}} \leq t_\gamma^\beta] \geq \beta_0. \tag{29}$$

The independence assumption is again only approximately correct, but similarly yields a slightly more conservative bound than necessary.

5 Experiments

In this section we demonstrate our theory on a few experiments. We used two real datasets to verify the theory. The first one is measurements of measurements performed at a ice hockey rink. The set contains 70 1D-images (with 360 degree field of view) and 14 reflectors. Figure 2 shows the setup, the motion of the cameras and the position of the reflectors.

The structure and motion was obtained using the L_∞ optimal methods from [7]. We first picked 5 cameras and solved structure and motion for these cameras and the viewed reflectors. We then added the remaining cameras and reflectors using alternating resection and triangulation. Finally we did bundle adjustment to obtain locally optimal L_2 solutions. We then ran our test on all (14) triangulation and (70) resectioning subproblems in this and in every case we were able to verify that these subproblems were in fact globally optimal. Figure refhockey2 shows one instance of the triangulation problem and one instance of the resectioning problem. The L_2 angular errors where roughly the same ($\approx$ 0.1-0.2 for both triangulation and resectioning) throughout the sequence.

In the hockey rink dataset the the cameras are placed so that the angle measurements can take roughly any value in $[-\pi, \pi]$. In our next dataset we wanted to test what happens if the measurements are restricted to a smaller interval. It is well known that for example resectioning is easier if one has measurements in vide spread directions. Therefore we used a data set where the the cameras do not have a 360 field of view and where there are not reflectors in every direction. Figure 4 shows the setup. We refer to this data set as the the coffee room

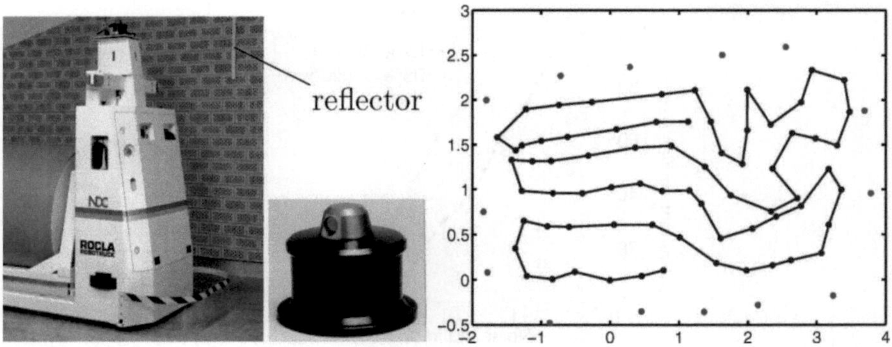

Fig. 2. Left: A laser guided vehicle. Middle: A laser scanner or angle meter. Right: positions of the reflectors and motion for the vehicle.

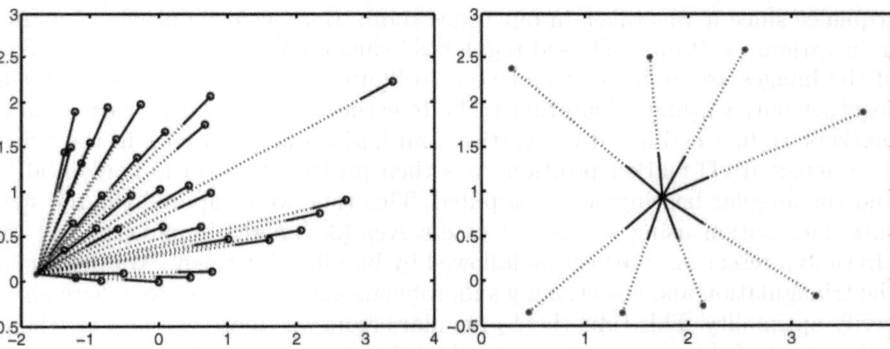

Fig. 3. Left: An instance of the triangulation problem. The reflector is visible from 36 positions with the total angular L_2-error of 0.15 degrees. Right: An instance of the resectioning problem. The camera detected 8 reflectors with the total angular L_2-error of 0.12 degrees.

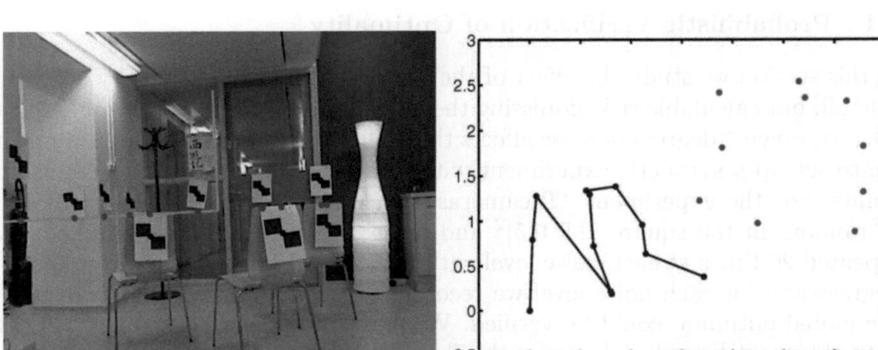

Fig. 4. Left: An images from the coffee room sequence. The green lines are estimated horizontal and vertical directions in the image, the blue dots are detected markers and red dots are the estimated bearings to the markers. Right: Positions of the markers and motion for the camera.

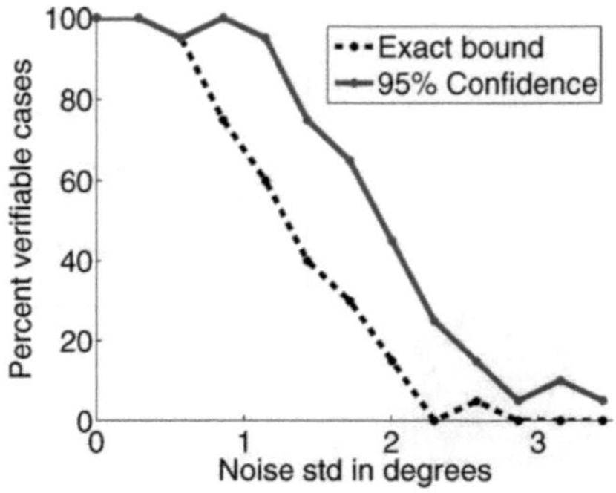

Fig. 5. Proportion of instances where global optimality could be verified versus image noise

sequence since it was taken in our coffee room. Here we have places 10 markers in in various positions and used regular 2D-cameras to obtain 13 images. (Some of the images are difficult to make out in figure 4 since they where taken close together only varying orientation.) The to estimate the angular bearings to the markers we first estimated the vertical and horizontal green lines in the figures. The detected 2D-marker positions was then projected onto the horizontal line and the angular bearings was computed. This time we computed the the structure and motion using a minimal case solver (3-cameras 5-markers) and then alternated resection-intersection followed by bundle adjustment. We then ran all the triangulation and resectioning subproblems and in all cases we where able to verify optimality. This time the L_2 angular errors was more varied. For triangulation most of the errors where around 0.5-1 degree whereas for resectioning the most of error where smaller ($\approx$ 0.1-0.2). Although in one camera L_2-error was as large as 3.2 degrees, however we were still able to verify that the resection was optimal.

5.1 Probabilistic Verification of Optimality

In this section we study the effect of the tighter bound one obtains by accepting a small, but calculable risk of missing the global optimum. Here, we would like to see how varying degrees of noise affects the ability to verify a global optimum and hence set up a synthetic experiment with randomly generated 1D cameras and points. For the experiment, 20 cameras and 1 point were generated uniformly at random in the square $[0.5, 0.5]^2$ and noise was added. The experiment was repeated 20 times at each noise level with noise standard deviation from 0 to 3.5 degrees and for each noise level we recorded the proportion of instances where the global optimum could be verified. We performed the whole procedure once with the exact bound and once with a bound set at a 95% confidence level. The result is shown in Figure 5. As expected, the tighter 95% bound allows one to verify a substantially larger proportion of cases.

6 Conclusions

Global optimization of the reprojection errors in L_2 norm is desirable, but difficult and no really practical general purpose algorithm exists. In this paper we have shown in the case of 1D vision how local optima can be checked for global optimality and found that in practice, local optimization paired with clever initialization is a powerful approach which often finds the global optimum. In particular our approach might be used in a system to filter out only the truly difficult local minima and pass these on to a more sophisticated but expensive global optimizer.

Acknowledgments

This work has been funded by the European Research Council (GlobalVision grant no. 209480), the Swedish Research Council (grant no. 2007-6476) and the Swedish Foundation for Strategic Research (SSF) through the programme Future Research Leaders. Travel funding has been recieved from The Royal Swedich Academy of Sciences and the Foundation Stiftelsen J.A. Letterstedts resesitpendiefond.

References

1. Stewénius, H., Schaffalitzky, F., Nistér, D.: How hard is three-view triangulation really? In: Int. Conf. Computer Vision, Beijing, China, pp. 686–693 (2005)
2. Hartley, R., Kahl, F.: Optimal algorithms in multiview geometry. In: Yagi, Y., Kang, S.B., Kweon, I.S., Zha, H. (eds.) ACCV 2007, Part I. LNCS, vol. 4843, pp. 13–34. Springer, Heidelberg (2007)
3. Triggs, B., McLauchlan, P.F., Hartley, R.I., Fitzgibbon, A.W.: Bundle adjustment – A modern synthesis. In: Triggs, B., Zisserman, A., Szeliski, R. (eds.) ICCV-WS 1999. LNCS, vol. 1883, pp. 298–372. Springer, Heidelberg (2000); in conjunction with ICCV 1999
4. Engels, C., Stewénius, H., Nistér, D.: Bundle adjustment rules. In: Photogrammetric Computer Vision (PCV) (2006)
5. Kai, N., Steedly, D., Dellaert, F.: Out-of-core bundle adjustment for large-scale 3D reconstruction. In: Conf. Computer Vision and Pattern Recognition, Minneapolis, USA (2007)
6. Hartley, R., Kahl, F.: Critical configurations for projective reconstruction from multiple views. Int. Journal Computer Vision 71, 5–47 (2007)
7. Åström, K., Enqvist, O., Olsson, C., Kahl, F., Hartley, R.: An L_∞ approach to structure and motion problems in 1d-vision. In: Int. Conf. Computer Vision, Rio de Janeiro, Brazil (2007)
8. Hartley, R., Seo, Y.: Verifying global minima for L_2 minimization problems. In: Conf. Computer Vision and Pattern Recognition, Anchorage, USA (2008)
9. Olsson, C., Kahl, F., Hartley, R.: Projective Least Squares: Global Solutions with Local Optimization. In: Proc. Int. Conf. Computer Vision and Pattern Recognition (2009)

Spatio-temporal Super-Resolution Using Depth Map

Yusaku Awatsu, Norihiko Kawai, Tomokazu Sato, and Naokazu Yokoya

Graduate School of Information Science, Nara Institute of Science and Technology,
8916-5 Takayama, Ikoma, Nara 630-0192, Japan
`http://yokoya.naist.jp/`

Abstract. This paper describes a spatio-temporal super-resolution method using depth maps for static scenes. In the proposed method, the depth maps are used as the parameters to determine the corresponding pixels in multiple input images by assuming that intrinsic and extrinsic camera parameters are known. Because the proposed method can determine the corresponding pixels in multiple images by a one-dimensional search for the depth values without the planar assumption that is often used in the literature, spatial resolution can be increased even for complex scenes. In addition, since we can use multiple frames, temporal resolution can be increased even when large parts of the image are occluded in the adjacent frame. In experiments, the validity of the proposed method is demonstrated by generating spatio-temporal super-resolution images for both synthetic and real movies.

Keywords: Super-resolution, Depth map, View interpolation.

1 Introduction

A technology that enables users to virtually experience a remote site is called telepresence [1]. In a telepresence system, it is important to provide users with high spatial and high temporal resolution images in order to make users feel like they are existing at the remote site. Therefore, many methods that increase spatial and temporal resolution have been proposed.

The methods that increase spatial resolution can be generally classified into methods that use one image as input [2,3] and methods that require multiple images as input [4,5,6,7]. The methods using one image are further classified into two types: ones that need a database [2] and ones that do not [3]. The former method increases the spatial resolution of the low resolution image based on previous learning of the correlation between various pairs of low and high resolution images. The latter method increases the spatial resolution by using a local statistic. These methods are effective for limited scenes but largely depend on the database and the scene. The methods using multiple images increase the spatial resolution by corresponding pixels in the multiple images that are taken from different positions. These methods determine pixel values in the super-resolved image by blending the corresponding pixel values [4,5,6] or minimizing

A.-B. Salberg, J.Y. Hardeberg, and R. Jenssen (Eds.): SCIA 2009, LNCS 5575, pp. 696–705, 2009.

the difference between the pixel values in an input image and the low resolution image generated from the estimated super-resolved image [7]. Both methods require the correspondence of pixels with sub-pixel accuracy. However, in these methods, the target scene is quite limited because the constraints of objects in the target scene such as planar constraint are often used in order to correspond the points with sub-pixel accuracy.

The temporal super-resolution method increases the temporal resolution by generating interpolated frames between the adjacent frames. Methods have been proposed that generate an interpolated frame by morphing that uses the movement of the points between adjacent frames [8,9]. Generally, the quality of the generated image by morphing largely depends on the number of corresponding points between the adjacent frames. Therefore, especially when many corresponding points do not exist due to occlusions, the methods rarely obtain good results.

The methods that simultaneously increase the spatial and temporal resolution by integrating the images from multiple cameras have been proposed [10,11]. These methods are effective for dynamic scenes but require a high-speed camera that can capture the scene faster than ordinary cameras. Therefore, the methods cannot be applied to a movie taken by an ordinary camera.

In this paper, by paying attention to the fact that determination of dense corresponding points is essential for spatio-temporal super-resolution, we propose the method that determines corresponding points of multiple images with sub-pixel accuracy by one-dimensionally searching for the corresponding points using the depth value of each pixel as a parameter. In this research, each pixel in multiple images is corresponded with high accuracy without the strong constraints for a target scene such as the planar assumption by a one-dimensional search of depth under the condition that intrinsic and extrinsic camera parameters are known. In work similar to our method, the spatial super-resolution method that uses a depth map has already been proposed [12]. However, this method needs stereo-pair images and does not increase the temporal resolution. Our advantages are that: (1) a stereo camera is not needed but only a general camera is needed, (2) the temporal resolution is increased by applying the proposed spatial super-resolution method to a virtual viewpoint located between temporally adjacent viewpoints of input images, and (3) corresponding points are densely determined by considering occlusions based on the estimated depth map.

2 Generation of Spatio-temporal Super-Resolved Images Using Depth Maps

This section describes the proposed method which generates spatio-temporal super-resolved images by corresponding pixels in each frame using depth maps. Here, in this research, a target scene is assumed to be static and camera position and posture of each frame and initial depth maps are given by some other methods like structure from motion and multi-baseline stereo. In the proposed method, the spatial resolution is increased by minimizing the energy function, which is based on the image consistency and the depth smoothness. The temporal resolution is also increased by the same framework with the spatial super-resolution method.

2.1 Energy Function Based on Image Consistency and Depth Smoothness

Energy function E_f for the target f-th frame is defined by the sum of two different kinds of energy terms:

$$E_f = E_{If} + wE_{Df}, \tag{1}$$

where E_{If} is the energy for the consistency between the pixel values in the super-resolved image of the target f-th frame and those in the input images of each frame, E_{Df} is the energy for the smoothness of the depth map, and w is the weight. In the following, the energies E_{If} and E_{Df} are described in detail.

(1) Energy E_{If} for Consistency
The energy E_{If} is defined based on the plausibility of the super-resolved image of the f-th frame using multiple input images from the a-th frame to the b-th frame ($a \leq f \leq b$) as follows:

$$E_{If} = \frac{\sum_{n=a}^{b} |\mathbf{N}(\mathbf{O}_n)(\mathbf{g}_n - \mathbf{m}_{nf})|^2}{\sum_{n=a}^{b} |\mathbf{O}_n|^2}. \tag{2}$$

Here, $\mathbf{g}_n = (g_{n1}, \cdots, g_{np})^T$ is a vector notation of pixel values in an input image of the n-th frame and $\mathbf{m}_{nf} = (m_{nf1}, \cdots, m_{nfp})^T$ is a vector notation of pixel values in the image of the n-th frame simulated by the estimated super-resolved image and the depth map of the f-th frame (Fig. 1). $\mathbf{N}(\mathbf{O}_n)$ is a $p \times p$ diagonal matrix whose on-diagonal element is the same as the element of vector $\mathbf{O}_n$. Although E_{If} is basically calculated based on the difference between the input image $\mathbf{g}_n$ and the simulated image $\mathbf{m}_{nf}$, some pixels in the simulated image $\mathbf{m}_{nf}$ do not correspond to pixels in the f-th frame due to occlusions and projection to the outside of the image. Therefore, by using the mask image $\mathbf{O}_n = (O_{n1}, \cdots, O_{np})$ whose element is 0 or 1, the energies of the non-corresponding pixels are not calculated in Eq. (2). Here, the simulated low-resolution image $\mathbf{m}_{nf}$ is generated as follows:

$$\mathbf{m}_{nf} = \mathbf{H}_{fn}(\mathbf{z}_f)\mathbf{s}_f, \tag{3}$$

where $\mathbf{s}_f = (s_{f1}, \cdots, s_{fq})^T$ is a vector notation of pixel values in the super-resolved image and $\mathbf{z}_f = (z_{f1}, \cdots, z_{fq})^T$ is a vector notation of depth values corresponding to the pixels in the super-resolved image $\mathbf{s}_f$. $\mathbf{H}_{fn}(\mathbf{z}_f)$ is the transformation matrix that generates the simulated low-resolution image of n-th frame from the super-resolved image of the f-th frame by using the depth map $\mathbf{z}_f$. $\mathbf{H}_{fn}(\mathbf{z}_f)$ is represented as follows:

$$\mathbf{H}_{fn}(\mathbf{z}_f) = \left[\alpha_1\mathbf{h}_1, \cdots, \alpha_i\mathbf{h}_i, \cdots, \alpha_p\mathbf{h}_p\right]^T, \tag{4}$$

where α_i is a normalization factor and $\mathbf{h}_i$ is a q-dimensional vector.

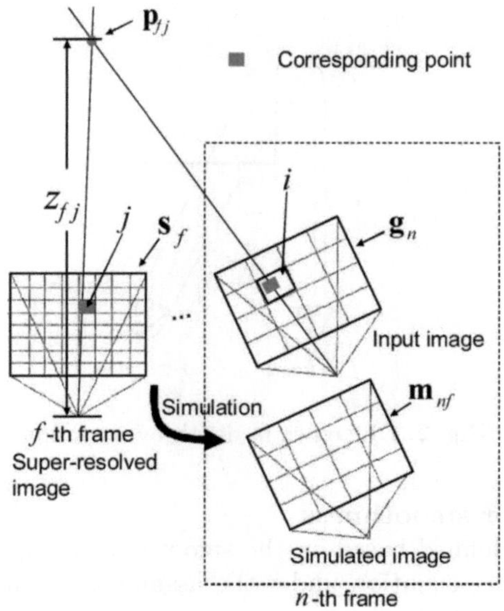

Fig. 1. Relationship between an input image and a super-resolved image

$$\mathbf{h}_i = \left(h_{i1}, \cdots, h_{ij}, \cdots, h_{iq}\right)^T. \tag{5}$$

Here, h_{ij} is a scalar value (1 or 0) that indicates the existence of correspondence between the j-th pixel in the super-resolved image and the i-th pixel in the input image. h_{ij} is calculated based on the estimated depth map as follows:

$$h_{ij} = \begin{cases} 0; d_n(\mathbf{p}_{fj}) \neq i \;\; or \;\; z'_{fj} > z_{ni} + C \\ 1; otherwise, \end{cases} \tag{6}$$

where $\mathbf{p}_{fj}$ indicates the three-dimensional coordinate in the scene corresponding to the j-th pixel in the super-resolved image as shown in Fig. 1 and $d_n(\mathbf{p})$ indicates the index of pixels in the n-th frame onto which $\mathbf{p}$ is projected. As shown in Fig. 2, z'_{fj} is the depth value in the n-th frame converted from the depth value z_{fj} in the f-th frame and z_{ni} is the corresponding depth value in the n-th frame. C is a threshold for determining occlusion.

The normalization factor α_i in Eq. (4) is the number of pixels in the super-resolved image that are projected onto the i-th pixel in the simulated image $\mathbf{m}_{nf}$. α_i is defined as follows using $\mathbf{h}_i$:

$$\alpha_i = \begin{cases} 0 & ; |\mathbf{h}_i| = 0 \\ \frac{1}{|\mathbf{h}_i|^2} & ; |\mathbf{h}_i| > 0. \end{cases} \tag{7}$$

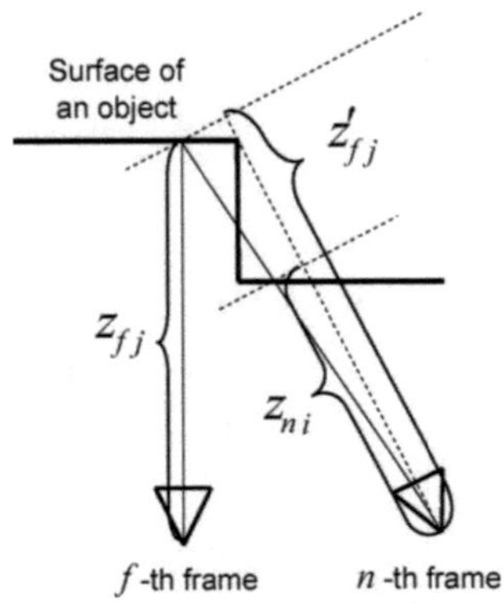

Fig. 2. Difference in depth by occlusion

(2) Energy E_{Df} for smoothness

The energy E_{Df} is defined based on the smoothness of the depth in the target frame as the following equation under the assumption that the depth along x and y direction is smooth in the target scene.

$$E_{Df} = \sum_j ((\frac{\partial^2 z_{fj}}{\partial x^2})^2 + 2(\frac{\partial^2 z_{fj}}{\partial x \partial y})^2 + (\frac{\partial^2 z_{fj}}{\partial y^2})^2), \tag{8}$$

2.2 Spatial Super-Resolution by Depth Optimization

In this research, a super-resolved image is generated by minimizing the energy E_f whose parameters are pixel and depth values in the super-resolved image. As shown in Eq. (2), E_{If} is calculated based on the difference between the input image $\mathbf{g}_n$ and the simulated image $\mathbf{m}_{nf}$. Here, whereas $\mathbf{g}_n$ is invariant, $\mathbf{m}_{nf}$ depends on the pixel values $\mathbf{s}_f$ and the depth values $\mathbf{z}_f$. Because it is difficult to minimize the energy by simultaneously updating the pixel and depth values in this research, the energy E_f is minimized by repeating the following two processes until the energy converges: (i) update of the pixel values $\mathbf{s}_f$ in the super-resolved image keeping the depth values $\mathbf{z}_f$ in the target frame fixed, (ii) update of the depth values $\mathbf{z}_f$ in the target frame keeping the pixel values $\mathbf{s}_f$ in the super-resolved image fixed.

In process (i), the transformation matrix $\mathbf{H}_{fn}(\mathbf{z}_f)$ for the pixel correspondence between the super-resolved image and the input image is invariant because the depth values $\mathbf{z}_f$ in the target frame are fixed. The energy E_{Df} for depth smoothness is also constant. Therefore, in order to minimize the total energy E_f, the pixel values $\mathbf{s}_f$ in the super-resolved image are updated so as to minimize the energy E_{If} for the image consistency. Here, each pixel value s_{fj} in the super-resolved image is updated in a way similar to method [7] as follows:

$$s_{fj} \leftarrow s_{fj} + \frac{\sum_{n=a}^{b}((g_{ni} - m_{nfi})O_{ni})}{\sum_{n=a}^{b} O_{ni}} \tag{9}$$

In process (ii), the depth values $\mathbf{z}_f$ are updated by fixing the pixel values $\mathbf{s}_f$ in the super-resolved image. In this research, because each pixel value in the simulated image $\mathbf{m}_{nf}$ discontinuously changes by the change in the depth $\mathbf{z}_f$, it is difficult to differentiate the energy E_f with respect to depth. Therefore, each depth value is updated by discretely moving the depth within a small range so as to minimize the energy E_f.

2.3 Temporal Super-Resolution by Setting a Virtual Viewpoint

In this research, a temporal interpolated image is generated by applying completely the same framework with the spatial super-resolution to a virtual viewpoint located between temporally adjacent viewpoints of input images. Here, because camera position and posture and a depth map, which are used for spatial super-resolution, are not given for an interpolated frame, it is necessary to set these values.

The position of the interpolated frame is determined by averaging the positions of the adjacent frames. If we want to generate multiple interpolated frames, the positions of adjacent frames are divided internally according to the number of interpolated frames. The posture of the interpolated frame is also determined by interpolating roll, pitch and yaw parameters of adjacent frames. The depth map of the interpolated frame is generated by averaging the depth maps of the adjacent frames.

3 Experiments

In order to demonstrate the effectiveness of the proposed method, spatio-temporal super-resolution images are generated for both synthetic and real movies.

3.1 Spatio-temporal Super-Resolution for a Synthetic Movie

In this experiment, a movie taken in a virtual environment as shown in Fig. 3 was used as input. Here, true camera position and posture of each frame were used as input. As for the initial depth values, Gaussian noise equivalent to an average of one pixel projection error on an image was added to the true depth values and the depth values were used as input. Table 1 shows parameters, and all 31 input frames are used for spatio-temporal super-resolution. In this experiment, a PC (CPU: Xeon 3.4GHz, Memory: 3GB) was used and it took about five minutes to generate one super-resolved image.

Table 1. Parameters in experiment

Input movie	320 240[pixels]	31[frames]
Output movie	640 480[pixels]	61[frames]
Weight w	100	
Threshold C	1[m]	

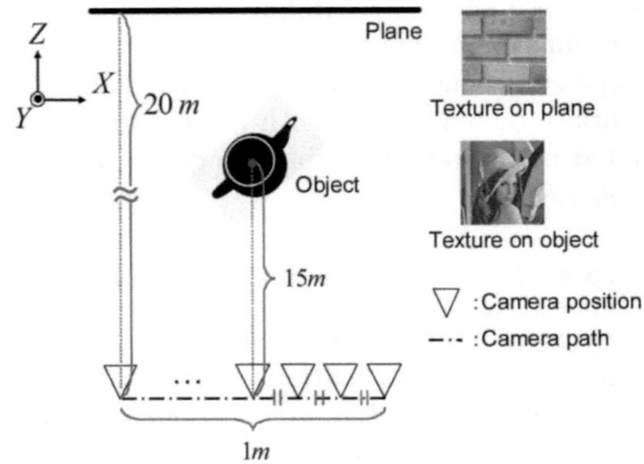

Fig. 3. Experimental environment

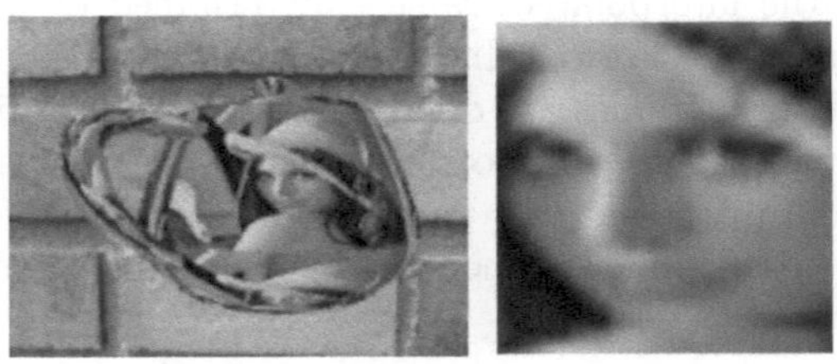

(a) Input image (Bilinear interpolation)

(b) Super-resolved image

(c) Ground truth image

Fig. 4. Comparison of images

Figure 4 shows the enlarged input image by bilinear interpolation (a), the super-resolved image generated by the proposed method (b) and a ground truth image (29-th frame) (c). The right part of each figure is a close-up of the same

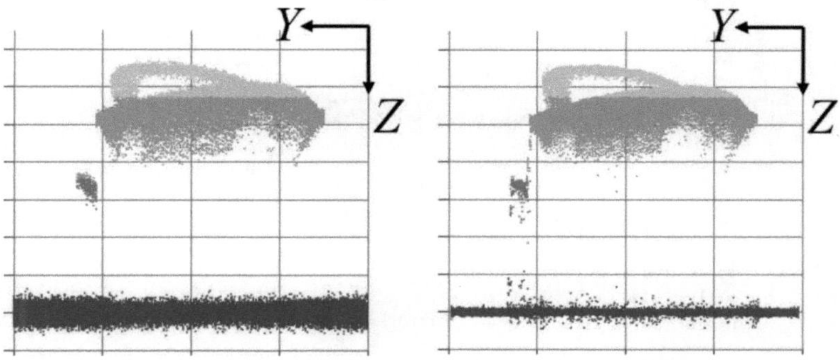

(a) Initial depth (YZ plane) (b) Optimized depth (YZ plane)

Fig. 5. Change in depth

region. From Fig. 4, the quality of the image is improved by super-resolution of the proposed method. Figure 5 shows the initial depth values and the depth values after energy minimization. From this figure, the depth values become smooth from the noisy ones.

Next, the spatio-temporal super-resolved images generated by the proposed method were evaluated quantitatively by calculating PSNR (Peak Signal to Noise Ratio) using the ground truth images. Here, as comparison movies, the following two movies were used.

(a) A movie in which the spatial resolution is enlarged by bilinear interpolation and the temporal resolution is the ground truth
(b) A movie in which the interpolation frame is generated by using the adjacent previous frame and the spatial resolution is the ground truth

Figure 6 shows PSNR between the ground truth images and the images by each method. Here, as for movie (b), PSNR only for the interpolated frames is shown because the interpolated frame in movie (b) is the same as the ground truth image. From this figure, the super-resolved images by the proposed method obtained higher PSNR than movie (a). In the interpolated frames, the super-resolved images by the proposed method also obtained higher PSNR than movie (b). However, in the proposed method, the improvement effectiveness of the image quality is small around the first and last frames. This is because there are only a few frames that are taken at spatially close positions from the observed position of the target frame.

3.2 Super-Resolution for a Real Image Sequence

In this experiment, a video movie was taken by Sony HDR-FX1 (1920 × 1080 pixels) from the air and we used a movie that was scaled to 320 × 240 pixels by averaging pixel values as input. As camera position and posture, we used the parameters estimated by structure from motion based on feature point tracking

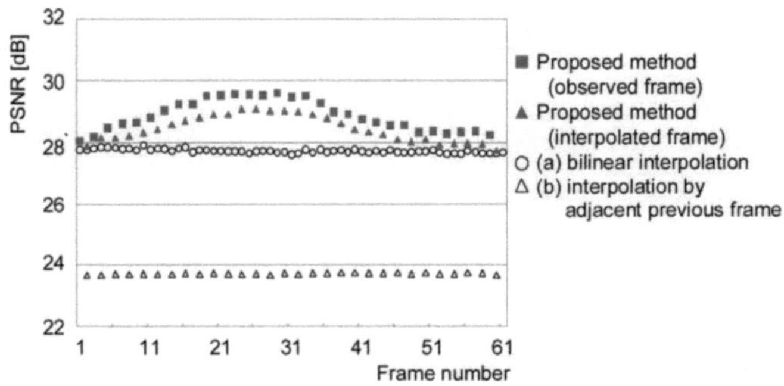

Fig. 6. Comparison of PSNR between the ground truth images and the images by each method

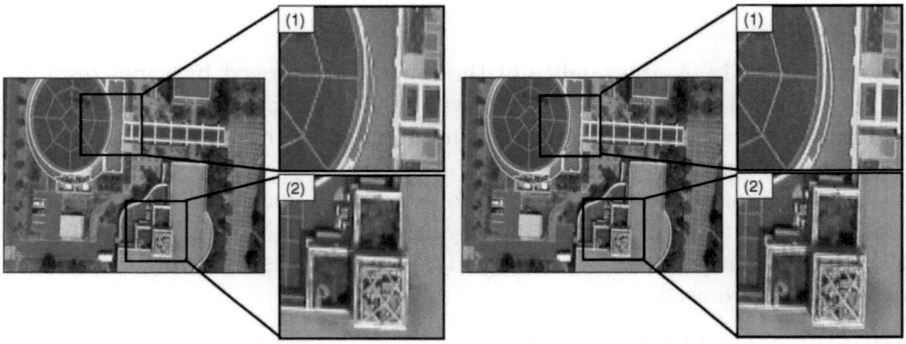

(a) Input image (b) Super-resolved image

Fig. 7. Comparison of input and super-resolved images

[13]. As initial depth maps, we used the interpolated depth map estimated by multi-baseline stereo for interest points [14]. Figure 7 shows the input image of the target frame and the super-resolved image (640 × 480 pixels) generated by using eleven frames around the target frame. From this figure, both the improved part ((1) in this figure) and the degraded part ((2) in this figure) can be observed. We consider that this is because the energy converges to a local minimum because the initial depth values are largely different from the ground truth due to the depth interpolation.

4 Conclusion

In this paper, we have proposed a spatio-temporal super-resolution method by simultaneously determining the corresponding points among many images by using the depth map as a parameter under the condition that camera parameters are given. In an experiment using a simulated video sequence, super-resolved images were quantitatively evaluated by RMSE using the ground truth image and the effectiveness of the proposed method was demonstrated by comparison with other methods. In addition, a real movie was also super-resolved by the proposed method. In future work, the quality of the super-resolved image should be improved by increasing the accuracy of correspondence of points by optimizing the camera parameters.

Acknowledgments. This research was partially supported by the Ministry of Education, Culture, Sports, Science and Technology, Grant-in-Aid for Scientific Research (A), 19200016.

References

1. Ikeda, S., Sato, T., Yokoya, N.: Panoramic Movie Generation Using an Omnidirectional Multi-camera System for Telepresence. In: Proc. Scandinavian Conf. on Image Analysis, pp. 1074–1081 (2003)
2. Freeman, W.T., Jones, T.R., Pasztor, E.C.: Example-based Super-Resolution. IEEE Computer Graphics and Applications 22, 56–65 (2002)
3. Hong, M.C., Stathaki, T., Katsaggelos, A.K.: Iterative Regularized Image Restoration Using Local Constraints. In: Proc. IEEE Workshop on Nonlinear Signal and Image Processing, pp. 145–148 (1997)
4. Zhao, W.Y.: Super-Resolving Compressed Video with Large Artifacts. In: Proc. Int. Conf. on Pattern Recognition, vol. 1, pp. 516–519 (2004)
5. Chiang, M.C., Boult, T.E.: Efficient Super-Resolution via Image Warping. Image and Vision Computing, 761–771 (2000)
6. Ben-Ezra, M., Zomet, A., Nayar, S.K.: Jitter Camera: High Resolution Video from a Low Resolution Detector. In: Proc. IEEE Conf. on Computer Vision and Pattern Recognition, pp. 135–142 (2004)
7. Irani, M., Peleg, S.: Improving Resolution by Image Registration. Graphical Models and Image Processing 53(3), 231–239 (1991)
8. Yamazaki, S., Ikeuchi, K., Shingawa, Y.: Determining Plausible Mapping Between Images Without a Priori Knowledge. In: Proc. Asian Conf. on Computer Vision, pp. 408–413 (2004)
9. Chen, S.E., William, L.: View Interpolation for Image Synthesis. In: Proc. Int. Conf. on Computer Graphics and Interactive Techniques, vol. 1, pp. 279–288 (1993)
10. Shechtman, E., Caspi, Y., Irani, M.: Space-Time Super-Resolution. IEEE Trans. on Pattern Analysis and Machine Intelligence 27(4), 531–545 (2005)
11. Imagawa, T., Azuma, T., Sato, T., Yokoya, N.: High-spatio-temporal-resolution image-sequence reconstruction from two image sequences with different resolutions and exposure times. In: ACCV 2007 Satellite Workshop on Multi-dimensional and Multi-view Image Processing, pp. 32–38 (2007)
12. Kimura, K., Nagai, T., Nagayoshi, H., Sako, H.: Simultaneous Estimation of Super-Resolved Image and 3D Information Using Multiple Stereo-Pair Images. In: IEEE Int. Conf. on Image Processing, vol. 5, pp. 417–420 (2007)
13. Sato, T., Kanbara, M., Yokoya, N., Takemura, H.: Camera parameter estimation from a long image sequence by tracking markers and natural features. Systems and Computers in Japan 35, 12–20 (2004)
14. Sato, T., Yokoya, N.: New multi-baseline stereo by counting interest points. In: Proc. Canadian Conf. on Computer and Robot Vision, pp. 96–103 (2005)

A Comparison of Iterative 2D-3D Pose Estimation Methods for Real-Time Applications

Daniel Grest, Thomas Petersen, and Volker Krüger

Aalborg University Copenhagen, Denmark
Computer Vision Intelligence Lab
{dag,vok}@cvmi.aau.dk

Abstract. This work compares iterative 2D-3D Pose Estimation methods for use in real-time applications. The compared methods are available for public as C++ code. One method is part of the openCV library, namely POSIT. Because POSIT is not applicable for planar 3D-point configurations, we include the planar POSIT version. The second method optimizes the pose parameters directly by solving a Non-linear Least Squares problem which minimizes the reprojection error. For reference the Direct Linear Transform (DLT) for estimation of the projection matrix is inlcuded as well.

1 Introduction

This work deals with the 2D-3D pose estimation problem. Pose Estimation has the aim to find the rotation and translation between an object coordinate system and a camera coordinate system. Given are correspondences between 3D points of the object and their corresponding 2D projections in the image. Additionally the internal parameters focal length and principal point have to be known.

Pose Estimation is an important part of many applications as for example structure-from-motion [11], marker-based Augmented Reality and other applications that involve 3D object or camera tracking [7]. Often these applications require short processing time per image frame or even real-time constraints[11]. In that case pose estimation algorithms are of interest, which are accurate and fast. Often, lower accuracy is acceptable, if less processing time is used by the algorithm. Iterative methods provide this feature.

Therefore we compare three popular methods with respect to their accuracy under strict time constraints. The first is POSIT, which is part of openCV [6]. Because POSIT is not suited for planar point configurations, we take the planar version of POSIT also into the comparison (taken from [2]. The second method we call CamPoseCalib (CPC) from the class name of the BIAS library [8]. The third method is the Direct Linear Transform for estimation of the projection matrix (see section 2.3.2 of [7]), because it is well known, used often as a reference [9] and easy to implement.

A.-B. Salberg, J.Y. Hardeberg, and R. Jenssen (Eds.): SCIA 2009, LNCS 5575, pp. 706–715, 2009.

Even though pose estimation is studied long since, new methods have been developed recently. In [9] a new linear method is developed and a comparison is given, which focuses on linear methods. We compare here iterative algorithms, which are available in C++, under the constraint of fixed computation time as required in real-time applications.

2 2D-3D Pose Estimation

Given are correspondences between 3D-points $\boldsymbol{p}_i$, which project into a camera image at position $\boldsymbol{p'}_i$ (see Fig. 1). Pose estimation from these 2D-3D correspondences is about finding the rotation and translation between camera and object coordinate systems.

2.1 CamPoseCalib (CPC)

The approach of *CamPoseCalib* is to estimate the relative rotation and translation of an object from an initial position and orientation (pose) to a new pose. The correspondences $(\boldsymbol{p}_i, \tilde{\boldsymbol{p}}'_i)$ are given for the new pose. Figure 1 illustrates this. The method was originally published in [1]. Details about the implementation used can be found in [5].

The algorithm can be formulated as a non-linear least squares problem, which minimizes the reprojection error d:

$$\hat{\boldsymbol{\theta}} = \arg\min_{\boldsymbol{\theta}} \sum_{i=1}^{m} (\boldsymbol{r}_i(\boldsymbol{\theta}))^2 \qquad (1)$$

for m correspondences. The residui functions $\boldsymbol{r}_i(\theta)$ represent the reprojection error $d = \boldsymbol{r}_i(\theta)^2 = r_x^2 + r_y^2$ and $\boldsymbol{\theta} = (\theta_x, \theta_y, \theta_z, \theta_\alpha, \theta_\beta, \theta_\gamma)^T$ are the 6 pose parameters, three for translation and three angles of rotation around the world axes.

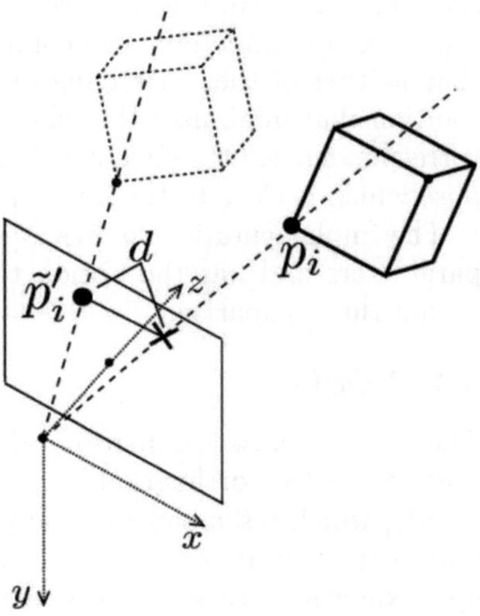

Fig. 1. *CamPoseCalib* estimates the pose by minimizing the reprojection error d between initial projected points from given correspondences $(\boldsymbol{p}_i, \tilde{\boldsymbol{p}}'_i)$

More specifically, the residui functions give the difference between moved, projected 3D point $\boldsymbol{m'}(\boldsymbol{p}_i, \boldsymbol{\theta})$ and the target point:

$$\boldsymbol{r}_i(\boldsymbol{\theta}) = \boldsymbol{m'}(\boldsymbol{p}_i, \boldsymbol{\theta}) - \tilde{\boldsymbol{p}}' \qquad (2)$$

The projection with pixel scales s_x, s_y and principal point $(c_x, c_y)^T$ is:

$$\boldsymbol{m'}(\boldsymbol{p}, \boldsymbol{\theta}) = \begin{pmatrix} s_x \frac{m_x(\boldsymbol{p},\boldsymbol{\theta})}{m_z(\boldsymbol{p},\boldsymbol{\theta})} + c_x \\ s_y \frac{m_y(\boldsymbol{p},\boldsymbol{\theta})}{m_z(\boldsymbol{p},\boldsymbol{\theta})} + c_y \end{pmatrix} \qquad (3)$$

where $\boldsymbol{m}(\boldsymbol{\theta}, \boldsymbol{p}) = (m_x, m_y, m_z)^T$ is the rigid motion in 3D:

$$\boldsymbol{m}(\boldsymbol{\theta}, \boldsymbol{p}) = (\theta_x, \theta_y, \theta_z)^T + R_x(\theta_\alpha)R_y(\theta_\beta)R_z(\theta_\gamma)\boldsymbol{p} \tag{4}$$

In order to avoid Euler angle problems, a compositional approach is used, that accumulates a rotation matrix during the overall optimization, rather than the rotation angles around camera axes x, y, z, which are estimated each iteration. More details in page 38-43 of [5].

The solution to the optimization problem is found by the *Levenberg-Marquardt (LM)* algorithm, which estimates the change in parameters in each iteration by:

$$\boldsymbol{\Delta\theta} = -(J^T J + \lambda I)^{-1} J^T \boldsymbol{r}(\boldsymbol{\theta_t}) \tag{5}$$

where I is the identity matrix and J is the Jacobian with the partial derivatives of the residui functions (see page 21 of [5]).

The inversion of $J^T J$ requires $det(J^T J) > 0$, which is achieved by 3 correspondences, because each correspondence gives two rows in the Jacobian and there are 6 parameters. The configuration requirement of 3D and 2D points is, that neither of them are lying on a line. However, due to the LM extension a solution that minimizes the reprojection error is always found, even for a single correspondence. Of course it will not give the correct new pose, but it returns a pose which is close to the initial pose.

The implementation in BIAS [8] also allows to optimize the internal camera parameters and has the option to estimate an initial guess, both is not used within this comparison.

2.2 POSIT

The second pose estimation algorithm uses a scaled orthographic projection (SOP), which resembles the real perspective projection at convergence. The SOP approximation leads to a linear equation system, which gives the rotation and translation directly , without the need of a starting pose. A scale value is introduced for each correspondence, which is iteratively updated. We give a brief overview of the method here. More details about POSIT can be found in [4,3].

Figure 2 illustrates this. The correspondences are $\boldsymbol{p}_i, \boldsymbol{p'}_i$. The SOP of $\boldsymbol{p}_i$ is here shown as $\hat{\boldsymbol{p}}'_i$ with a scale value of 0.5. The POSIT algorithm estimates the rotation by finding the values for $\boldsymbol{i}, \boldsymbol{j}, \boldsymbol{k}$ in the object coordinate system, whose origin is $\boldsymbol{p}_0$. The translation between object and camera system is $O\boldsymbol{p}_0$.

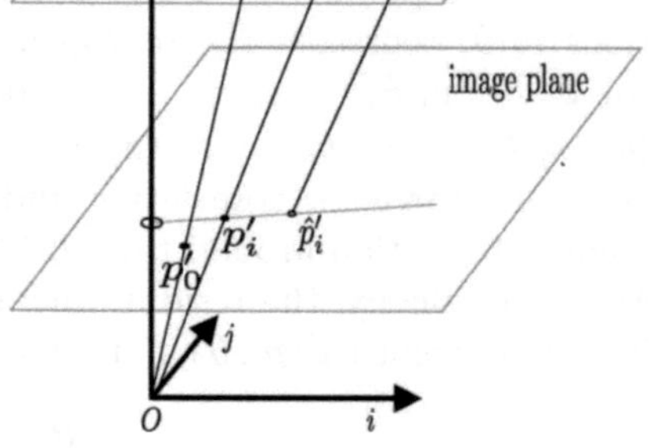

Fig. 2. POSIT estimates the pose by using a scaled orthographic projection (SOP) from given correspondences $\boldsymbol{p}_i, \boldsymbol{p'}_i$. The SOP of $\boldsymbol{p}_i$ is here shown as $\hat{\boldsymbol{p}}'_i$ with a scale value of 0.5.

For each SOP 2D-point a scale value can be found such that the SOP $\hat{\boldsymbol{p}}'_i$ equals the correct perspective projection $\boldsymbol{p}'_i$. The POSIT algorithm refines iteratively these scale values. Initially the scale value (w in the following) is set to one.

The POSIT algorithm works as follows:

1. Initially set the unknown values $w_i = 1$ for each correspondence.
2. Estimate pose parameters from the linear equation system
3. Estimate new values w_i by $w_i = \frac{\boldsymbol{p}_i^T \boldsymbol{k}}{t_z} + 1$
4. Repeat from step 2 until the change in w_i is below a threshold or maximum iterations are reached

The initially chosen $w_i = 1$ approximates the real configuration of camera position and scene points well, if the fraction of object elongation to camera distance is small.

If the 3D points lie in one plane the POSIT algorithm needs to be altered. A description of the co-planar version of POSIT can be found in [10].

3 Experiments

There are several experiments on synthetic data conducted, whose purpose is to reveal the advantages and disadvantages of the different methods. We use implementations as available for the public for download of *CamPoseCalib* [8] and the two *POSIT* methods from Daniel DeMenthons homepage [2]. The C++ sources are compiled with *Microsoft's Visual Studio 2005 C++ compiler* in standard release mode settings. The POSIT method is also part of openCV [6]. Experiments showed, that the openCV version is about two times faster than our compilation. However we chose to use our self compiled version, because we want to compare the algorithms rather than binary realeases or compilers.

In order to resemble a realistic setup, we chose the following values for all experiments. Some values are changed as stated in the specific tests.

- 3D points are randomly distributed in a 10x10x10 box
- camera is positioned 25 units away, facing the box
- internal camera parameters are $s_x = s_y = 882$, $c_x = 600$ and $c_y = 400$, which corresponds to a real camera with 49 degree opening angle in y-direction and an image resolution of 1200x800 pixels
- the number of correspondences is 10.
- Gaussian noise is added to the 2D positions with a variance of 0.2 pixels
- each test is run 100 times with varying 3D points

The accuracy is measured in the following tests by comparing the estimated translation and rotation of the camera to the known groundtruth.

The translation error is measured as the Euclidean distance between estimated camera position and real camera position divided by the distance of the camera to the center of the 3D points. For example in the first test, an translation error of 100% means 25 units difference.

The rotational error is measured as the Euclidean distance between the rotation quaternions representing the real and the estimated orientation.

3.1 Test 1: Increasing Noise

In many applications the time for pose estimation is bound by an upper limit. Therefore, we compare here the accuracy of different methods, which are given the same calculation time. The time chosen for each iterative algorithm is the same time as for the non-iterative DLT.

Normal distributed noise is added to the 2D positions with changing variance. The following settings are used:

- 2D-noise is increased from 0 to 3.3 pixels standard deviation (variance 10)
- The initial pose guess for CPC: rotation is two degrees off and position is 3.4% away from the real position
- Initial scale value of POSIT is 1 for all points
- Number of iterations for CPC is 9 and for POSIT 400

The initial guess for CPC is 2 degrees and 0.034 units off. This resembles a tracking scenario as in augmented reality applications.

In Figure 10 the accuracy of all methods is shown with boxplots. A boxplot shows the median (red horizontal line within boxes) instead of the mean, as well as the outliers (red crosses). The blue boxes denote the first and third quartile (the median is the second quartile).

The left column shows the difference in estimated camera position, the right column the difference in orientation as the Euclidian length of the difference rotation quaternion. The top row shows CPC, which accuracy is better than POSIT (middle row) and DLT (bottom row).

3.2 Test 2: Point Cloud Morphed to Planarity

In many applications the spatial configuration of the 3D points is unknown as in structure-from-motion. Especially interesting is the case, where the points lie in a plane or are close to a plane. In order to test the performance of the different algorithms, the point cloud is transformed into a plane by reducing its thickness each time by 30%.

Figure 3 illustrates the test. The plane is chosen not to face the camera directly (the plane normal is not aligned with the optical axis), because a correct pose is in that case also found, if the camera is on the opposite side of the plane. Because the POSIT algorithm can't handle coplanar points, the planar POSIT version is tested in addition to CPC and DLT.

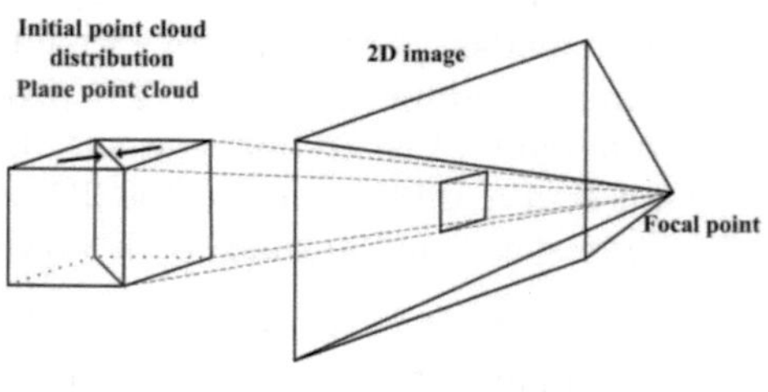

Fig. 3. Test 2: Initial box shaped point cloud distribution is changed into planarity

Figure 4 shows the translation error versus the thickness of the box (rotational errors are similar). As visible, the DLT error increases greatly when the box gets thinner than 0.2 and fails to give correct results for a thickness smaller than

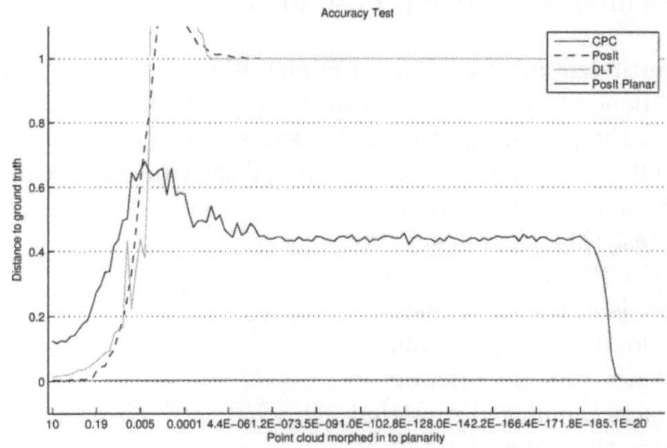

Fig. 4. Test 2: Point cloud is morphed into planarity. Shown is the mean of 100 runs.

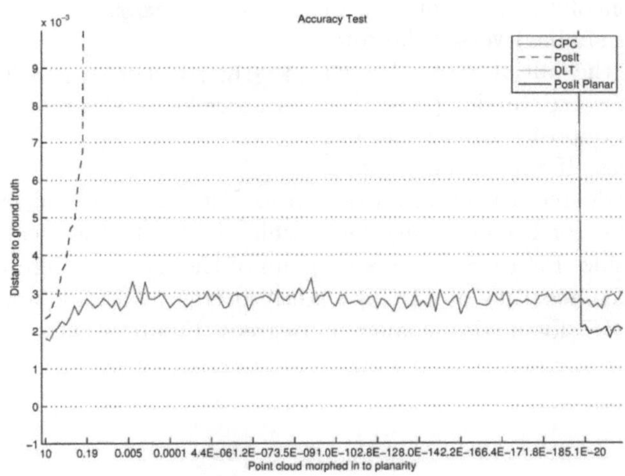

Fig. 5. Test 2: Point cloud is morphed into planarity. Shown is a closeup of the same values as in Fig. 4.

1E-05 (the algorithm returns $(0,0,0)^T$ as position in that case). The normal POSIT algorithm performs similar to the DLT. Interesting to note is, that the planar POSIT algorithm works only correctly, if the 3D points are very close to coplanar (a thickness of 1E-20). Important is the observation, that there is a thickness range, where non of the POSIT algorithms estimates a correct result.

The CPC algorithm is unaffected by a change in the thickness, while the accuracy of the planar POSIT is slightly better for nearly coplanar points as visible on in Figure 5.

3.3 Test 3: Different Starting Pose for CPC

The iterative optimization of CPC requires an initial guess of the pose. The performance of CPC depends on how close these initial parameters are to the real ones. Further there is the possibility, that CPC gets stuck in a local minimum during optimization. Often a local minimum is found, if the camera is positioned exactly on the opposite side of the 3D points.

In order to test this dependency, the initial guess of CPC is changed, such that the camera is at the same distance to the point cloud circling around it. Figure 6 illustrates this, the orientation of the initial guess is changed such that the camera faces the point cloud at all times.

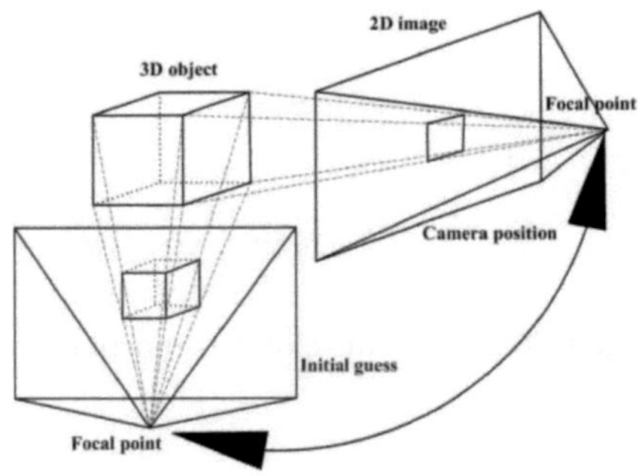

Fig. 6. Test 3 illustrated. The initial camera pose for CPC is rotated on a circle.

Figure 7 shows the mean and standard deviation of the rotational error (translation is similar) versus the rotation angle of the initial guess. Higher angles mean a worse starting point. The initial pose is opposite to the real one for 180 degrees. If the initial guess is worse than 90 degrees the accuracy decreases. For angles around 180 degrees the deviation and error becomes very high, which is due to the local minimum on the opposite side. Figure 8 shows a close-up of the mean of Figure 7. Here it is visible, that the accuracy of CPC is slightly better than CPC and significantly better than DLT for angles smaller 90 degrees. Figure 9 shows the mean and

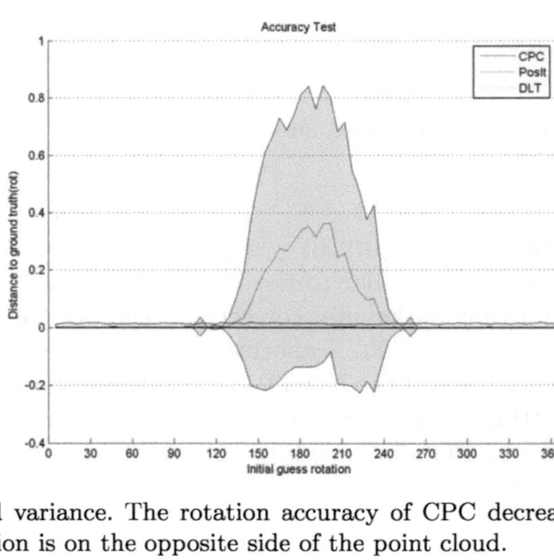

Fig. 7. Mean and variance. The rotation accuracy of CPC decreases significantly, if the starting position is on the opposite side of the point cloud.

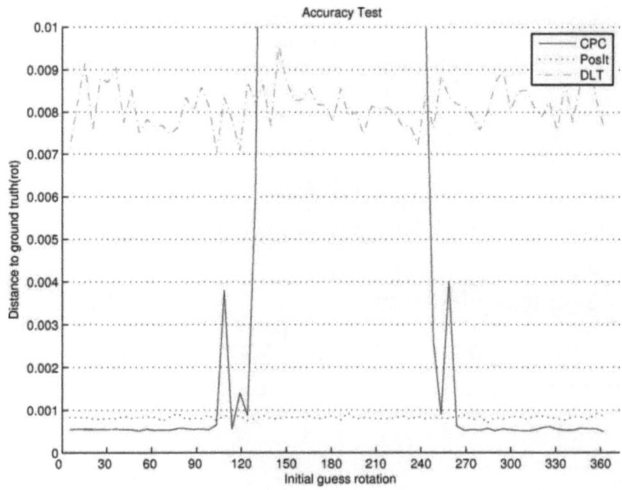

Fig. 8. A closeup of the values of figure 7. The accuracy of CPC is better than the other methods for an initial angle that is within 90 degrees of the actual rotation.

standard deviation of the computation time for CPC, POSIT and DLT. If the initial guess is worse than 30 degrees, CPC uses more time because of the LM iterations. However, even in worse cases it is only 2 times slower.

From the accuracy and timing results for this test it can be concluded, that CPC is the more accurate method compared to POSIT, if given the same time and an initial guess which is within 30 degrees of the real one.

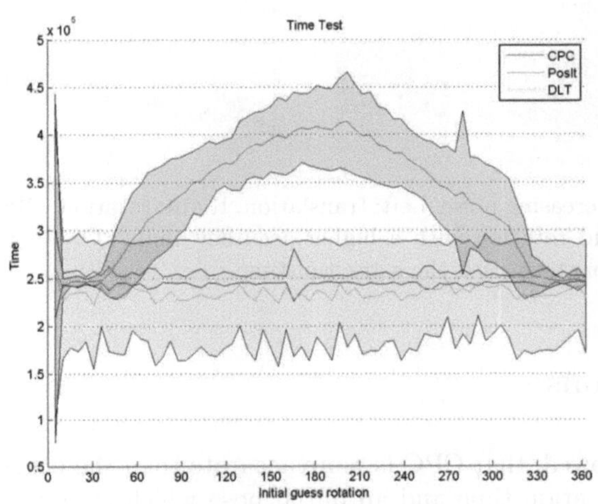

Fig. 9. Timings. Mean and variance.

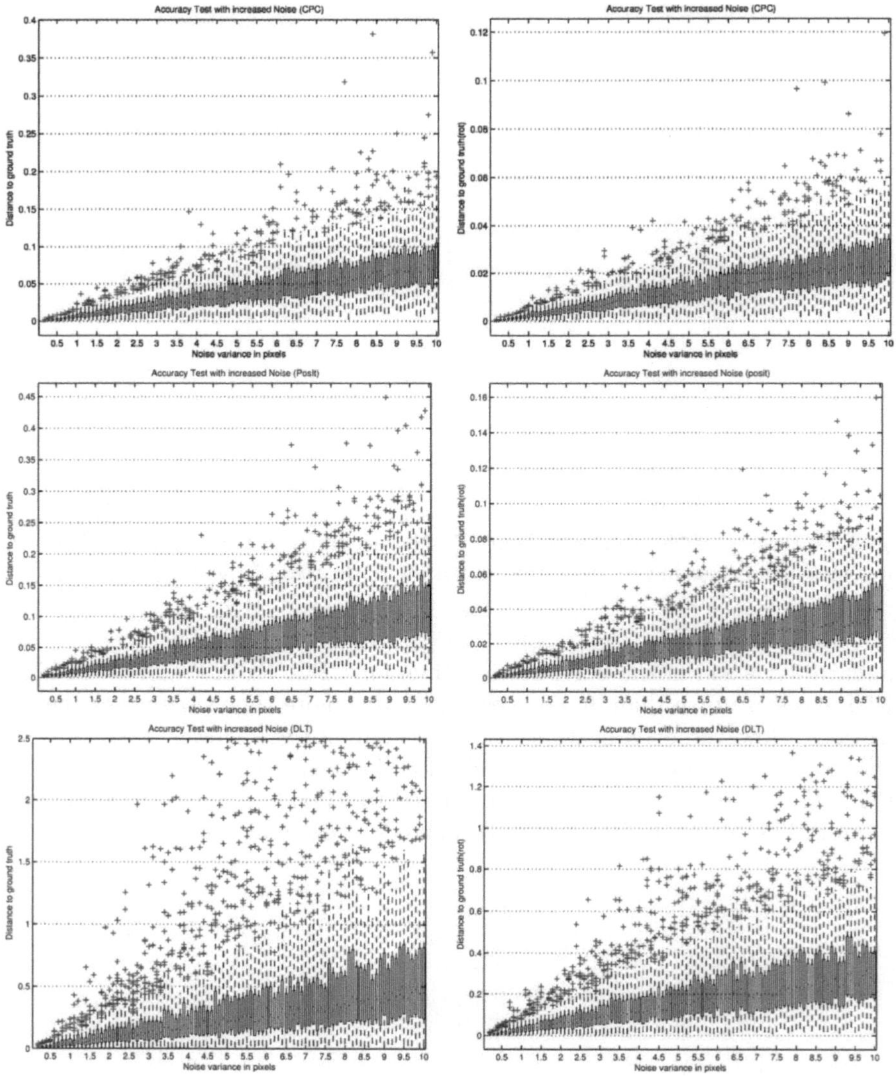

Fig. 10. Test 1: Increasing noise. Left: translation. Right: rotation. CPC (top) estimates the translation and rotation with a higher accuracy than POSIT (middle) and DLT (bottom). All algorithms used the same run-time.

4 Conclusions

The first test showed, that CPC is more accurate than the other methods given the same computation time and an initial pose which is only 2 degrees off the real one, which is similar to the changes in real time tracking scenarios. CPC is also more accurate if the starting angle is within 30 degrees as test 3 showed.

POSIT has the advantage, that it is not in the need of a starting pose and is available as an highly optimized version in openCV.

In test 2 the point cloud was changed into a planar surface. Here the POSIT algorithms gave inaccurate results for a box thickness from 0.2 to 1E-19 making the POSIT methods not applicable for applications where the 3D configuration of points is close to co-planar as in structure-from-motion applications.

The planar version of POSIT was most accurate, if the 3D points are arranged exactly in a plane. Additionally it can return 2 solutions: camera positions on both sides of the plane. This is advantageous because in applications where a planar marker is observed, the pose with smaller reprojection error is not necessarily the correct one, because of noisy measurements.

References

1. Araujo, H., Carceroni, R., Brown, C.: A Fully Projective Formulation to Improve the Accuracy of Lowe's Pose Estimation Algorithm. Journal of Computer Vision and Image Understanding 70(2) (1998)
2. De Menthon, D.: (2008), `http://www.cfar.umd.edu/~daniel`
3. David, P., Dementhon, D., Duraiswami, R., Samet, H.: SoftPOSIT: Simultaneous Pose and Correspondence Determination. Int. J. Comput. Vision 59(3), 259–284 (2004)
4. DeMenthon, D.F., Davis, L.S.: Model-Based Object Pose in 25 Lines of Code. International Journal of Computer Vision 15, 335–343 (1995)
5. Grest, D.: Marker-Free Human Motion Capture in Dynamic Cluttered Environments from a Single View-Point. PhD thesis, MIP, Uni. Kiel, Kiel, Germany (2007)
6. Intel. openCV: Open Source Computer Vision Library (2008), `opencvlibrary.sourceforge.net`
7. Lepetit, V., Fua, P.: Monocular Model-Based 3D Tracking of Rigid Objects: A Survey. Foundations and Trends in Computer Graphics and Vision 1(1), 1–104 (2005)
8. MIP Group Kiel. Basic Image AlgorithmS (BIAS) open-source-library, C++ (2008), `www.mip.informatik.uni-kiel.de`
9. Moreno-Noguer, F., Lepitit, V., Fua, P.: Accurate Non-Iterative O(n) Solution to the PnP Problem. In: ICCV, Brazil (2007)
10. Oberkampf, D., DeMenthon, D.F., Davis, L.S.: Iterative pose estimation using coplanar feature points. CVIU 63(3), 495–511 (1996)
11. Williams, B., Klein, G., Reid, I.: Real-time SLAM Relocalisation. In: Proc. of Internatinal Conference on Computer Vision (ICCV), Brazil (2007)

A Comparison of Feature Detectors with Passive and Task-Based Visual Saliency

Patrick Harding[1,2] and Neil M. Robertson[1]

[1] School of Engineering and Physical Sciences, Heriot-Watt Univ., UK
[2] Thales Optronics Ltd., UK
{pjh3,nmr3}@hw.ac.uk

Abstract. This paper investigates the coincidence between six interest point detection methods (SIFT, MSER, Harris-Laplace, SURF, FAST & Kadir-Brady Saliency) with two robust "bottom-up" models of visual saliency (Itti and Harel) as well as "task" salient surfaces derived from observer eye-tracking data. Comprehensive statistics for all detectors vs. saliency models are presented in the presence and absence of a visual search task. It is found that SURF interest-points generate the highest coincidence with saliency and the overlap is superior by 15% for the SURF detector compared to other features. The overlap of image features with task saliency is found to be also distributed towards the salient regions. However the introduction of a specific search task creates high ambiguity in knowing how attention is shifted. It is found that the Kadir-Brady interest point is more resilient to this shift but is the least coincident overall.

1 Introduction and Prior Work

In Computer Vision there are many methods of obtaining distinctive "features" or "interest points" that stand out in some mathematical way relative to their surroundings. These techniques are very attractive because they are designed to be resistant to image transformations such as affine viewpoint shift, orientation change, scale shift and illumination. However despite their robustness they do not *necessarily* relate in a meaningful way to the human interpretation of what in an image is distinctive. Let us consider a practical example of why this might be important. An image processing operation should only be applied if it aides the observer to perform an interpretation task (enhancement algorithms) or does not destroy the key details within the image (compression algorithms). We may wish to predict the effect of an image processing algorithm on a human's ability to interpret the image. Interest points would be a natural choice to construct a metric given their robustness to transforms. But in order to use these feature points we must first determine (a) how well the interest-point detectors coincide with the human visual system's impression of images i.e. what is visually salient, and (b) how the visual salience changes in the presence of a task such as "find all cars in this image". This paper seeks to address these problems. First let us consider the interest points and then explain in more detail what we mean by feature points and visual salience.

A.-B. Salberg, J.Y. Hardeberg, and R. Jenssen (Eds.): SCIA 2009, LNCS 5575, pp. 716–725, 2009.

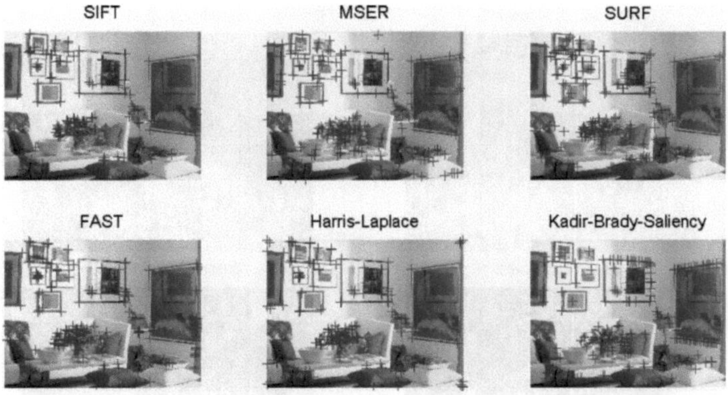

Fig. 1. An illustration of distribution of the interest-point detectors used in this paper

Interest Point Detection: The interest points chosen for analysis are: SIFT [1], MSER [2], Harris-Laplace [3], SURF [4], FAST [5,6] and Kadir-Brady Saliency [7].

These are shown superimposed on one of our test images in Figure 1[1]. These schemes are well-known detectors of regions that are suitable for transformation into robust regional descriptors that allow for good levels of scene-matching via orientation, affine and scale shifts. This set represents a spread of different working mechanisms for the purposes of this investigation. These algorithms have been assessed in terms of mathematical resilience [8,9]. But what we are interested in is how well they correspond to visually salient features in the image. Therefore we are not investigating descriptor robustness or repeatability (which has been done extensively – see e.g. [8]), nor trying to *select* keypoints based on modelled saliency (such as the efforts in [10]) but rather we want to ascertain how well interest-point locations naturally correspond to *saliency* maps generated under passive and task conditions. This is important because if the interest-points coincide with salient regions at a higher-than coincidence level, they are attractive for two reasons. First, they may be interpreted as primitive saliency detectors *and* secondly can be stored robustly for matching purposes.

Visual Salience: There exist tested models of "bottom-up" saliency, which accurately predict human eye-fixations under passive observation conditions. In this paper, two models were used, those of saliency by Itti Koch and Neibur [11] and the model by Harel, Koch, and Perona [12]. These models are claimed to be based on observed psycho-visual processes in assessing the saliency of the images. They each create a "Saliency Map" highlighting the pixels in order of ranked saliency using intensity shading values. An example of this for Itti and Harel saliency is shown in Figure 2. The Itti model assesses center-surround differences in *Colour*, *Intensity* and *Orientation* across scale and assigns values to feature maps based on outstanding attributes. Cross scale differences are also examined to give a multi-scale representation of the local saliency. The maps for each channel (*Colour*, *Intensity* and

[1] Note: these algorithms all act on *greyscale* images. In this paper, colour images are converted to grey values by forming a weighted sum of the RGB components (0.2989 R + 0.5870 G + 0.1140 B).

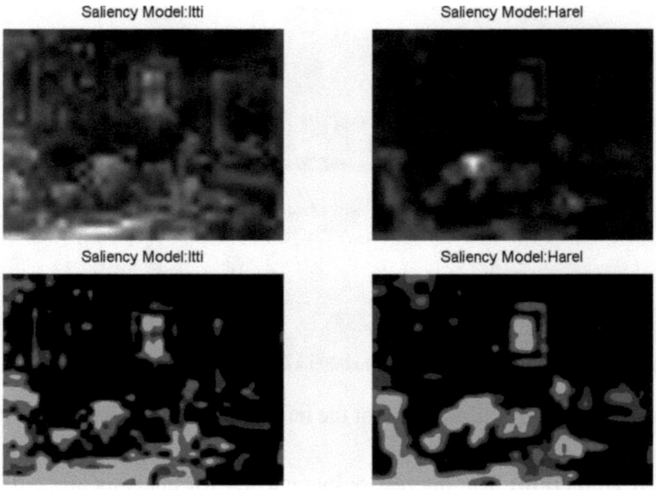

Fig. 2. An illustration of the passive saliency maps on one of the images in the test set. (Top left) Itti Saliency Map, (Top right) Harel Saliency map (Bottom left) thresholded Itti, (Bottom right) thrsholded Harel. Threshold levels are 10, 20, 30, 40 & 50% of image pixels ranked in saliency, represented at descending levels of brightness.

Orientation) are then combined by normalizing and weighting each map according to the local values. Homogenous areas are ignored and "interesting" areas are highlighted. The maps from each channel are then combined into "conspicuity maps" via cross-scale addition. These are combined into a final saliency map by normalization and summed with an equal weighting of 1/3 importance. The model is widely known and is therefore included in this study. However, the combination weightings of the map are arbitrary at 1/3 and it is not the most accurate model at predicting passive eye-scan patterns [12]. The Harel *et al.* method uses a similar basic feature extraction method but then forms *activation* maps in which "unusual" locations in a feature map are assigned high values of activation. Harel uses a Markovian graph-based approach based on a ratio-based definition of dissimilarity. The output of this method is an activation measure derived from pairwise contrast. Finally, the activation maps are normalized using another Markovian algorithm which acts as a mass concentration algorithm, prior to additive combination of the activation maps. Testing of these models in [12] found that the Itti and Harel models achieved, respectively, 84% and 96-98% of the ROC area of a human-based control experiment based on eye-fixation data under passive observation conditions. Harel *et al.* explain that their model is apparently more robust at predicting human performance than Itti because it (a) acts in a center-bias manner, which corresponds to a natural human tendency, and (b) it has superior robustness to differences in the size of salient regions in their model compared to the scale differences in Itti's.

Both models offer high coincidence with eye-fixation from passive viewing observed under strict conditions. The use of both models therefore provides a pessimistic (Itti) and optimistic (Harel) estimation of saliency for passive attentional guidance for each image.

The impact of tasking on visual salience: There is at present no corresponding model of task performance on the saliency map of an image but there has been much work performed in this field, often using eye-tracker data and object learning [13,14,15,16]. It is known that an observer acting under the influence of a specific task will perceive the bottom-up effects mentioned earlier *but will impose constraints* on his observation in an effort to priority-filter information. These impositions will result from experience and therefore are partially composed of memory of likely target positions under similar scenarios. (In Figure 4 the green regions show those areas which *became* salient due to a task constraint being imposed.)

2 Experimental Setup

Given that modeling the effect of tasking on visual salience is not readily quantifiable, in this paper eye-tracker data is used to construct a "task probability surface". This is shown (along with eye-tracker points) in Figure 3, where the higher values represent the more salient locations, as shown in Figure 2. The eye-tracker data generated by Henderson and Torralba [16] is used to generate the "saliency under task" of each test image. This can then be used to gauge the resilience of the interest-points to top down factors based on real task data. The eye tracker data gives the coordinates of the fixation points attended to by the participants. This data, collected under a search-task condition, is the "total task saliency", which is composed of both the bottom-up factors as well as the top down factors.

Task Probability Surface Construction: The three tasks used to generate the eye-tracker data were: (a) "count people", (b) "count cups" and (c) "count paintings". There are 36 street scene images, used for the people search, and 36 indoor scene images, used for both the cup and painting search. The search target was not always present in the images. A group of eight observers was used to gather the eye-tracker data for each image with an accuracy of fixation of +/- 4 pixels. (Full details in [17].)

To construct the task surfaces for all 108 search scenarios over the 72 images, the eye tracker data from all eight participants was combined into a single data vector. Then for each pixel in a mask of the same size as the image, the Euclidean distance to each eye-point was calculated and placed into ranked order. This ordered distance vector was then transformed into a value to be assigned to the pixel in the mask using the formula $P = \sum_{i=1}^{N} d_i / i^2$ in which, d is the distance to eye point, i and N is the number of fixations from all participants. The closer the pixel to an eye-point cluster, the lower the P value is assigned. When the pixel of the mask coincides with an eye-point there is a notable dip compared to all other neighbours because d_1 in the above P-formula is 0. To avoid this problem, pixels at coordinates coinciding with the eye-tracker data are exchanged for the mean value of the eight nearest neighbours, or the mean of valid neighbours at image boundary regions. The mask is then inverted and normalised to give a probabilistic task saliency map in which high intensity represents high task saliency, as shown in Figure 3. This task map is based on the ground truth of the eye-tracker data collected from the whole observer set focusing their priority on a

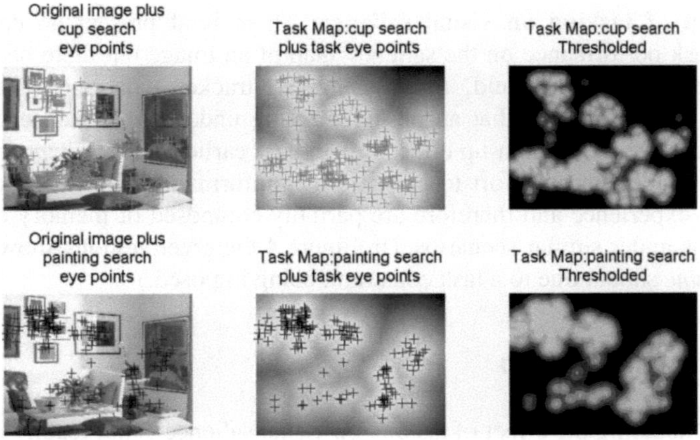

Fig. 3. Original image with two sets of eye tracking data superimposed representing two different search tasks. *Green points = cup search, Blue points =painting search.* (Centre top) Task Map derived from cup search eye-tracker data, (Centre bottom) Task Map generated from painting search eye-tracker data. (Top right) Thresholded cup search. (Bottom right) Thresholded painting search.

particular search task. It should be noted that the constructed maps are derived from a mathematically plausible probability construction (the closer the eye-point to a cluster, the higher the likelihood of attention). However, the formula does not explicitly model biological attentional tail off away from eye-point concentrations, which is a potential source of error in subsequent counts.

Interest-points vs. Saliency: The test image data set for this paper comprises 72 images and 108 search scenarios (3x36 tasks) performed by 8 observers. In doing so, the bottom-up and task maps can be directly compared. The Itti and Harel saliency models were used to generate bottom-up saliency maps for all 72 images. These are interpreted as the likely *passive* viewing eye-fixation locations. Using the method described previously, the corresponding task saliency maps were then generated for all 108 search scenarios. Finally, the interest-point detectors were applied to the 72 images (an example in Figure 1). The investigation was to determine how well the interest-points match up with each viewing scenario surface – passive viewing and search task in order to assess interest-point coincidence with visual salience. We perform a count of the inlying and out lying points of the different interest-points in both the bottom-up and task saliency maps. Each of these saliency maps are thresholded at different levels i.e. the X% *most salient* pixels of each map for each image is counted as being above threshold X and the interest-points lying within threshold are counted. This method of thresholding allows for comparison between the bottom-up and the task probability maps even though they have different underlying construction mechanisms. X = 10, 20, 30, 40 and 50% were chosen since these levels clearly represent the "more salient" half of the image to different degrees. This quantising of the saliency maps into contour-layers of equal-weighted saliency is another possible source of error in our experimental setup, although it is plausible. An example of thresholding is shown in Figure 2. In summary, the following steps were performed:

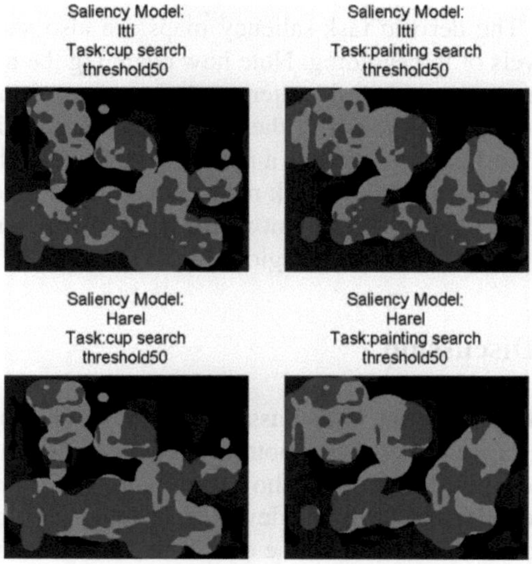

Fig. 4. An illustration of the overlap of the thresholded passive and task-directed saliency maps. Regions in neither map are in *Black*. Regions in the passive saliency map exclusively are in *Blue*. Regions exclusively in the task map *Green*. Regions in both passive and task-derived maps are in *Red*. The first row shows Itti saliency for cup search (left) and painting search (right) task data. The second row shows the same for the Harel saliency model. For Harel vs. "All Tasks" at 50% threshold the average % coverages are: Black – 30%, Blue – 20%, Green – 20%, Red – 30%, (+/- 5%). For Harel (at 50%), there is a 20% attention shift away from the bottom-up-only case due to the influence of a visual search task.

1. The interest-points were collected for the whole image set of 72 images.
2. The Itti and Harel saliency maps were collected for the entire image set.
3. The task saliency map surfaces were calculated across the image set (36 x people search and 2 x 36 for cup and painting task on the same image set).
4. The saliency maps were thresholded to 10, 20, 30, 40 and 50% of the map areas.
5. The number of each of the interest-points lying within the thresholded saliency maps was counted.

It can be seen in Figure 1 that the interest points are generally clustered around visually "interesting" objects i.e. those which stand out from their immediate surroundings. *This paper analyses whether they coincide with measurable visual saliency.* For each image, the number of points generated by each interest point detector was limited to be equal or slightly above the total number of eye-tracker data points from all observers attending the image under task. For the 36 images with two tasks applied, the number of "cup search" task eye-points was used for this purpose.

The bottom-up models of visual saliency are illustrated in Figure 2, both in their raw map form and at the different chosen levels of thresholding. In Figure 3 the eye-tracker patterns from all eight observers are shown superimposed upon the image for two different tasks. The derived task-saliency maps are also shown, as are the task maps at different levels of thresholding. Note how changing the top down information (in this case varying the search task) alters the visual search pattern considerably. Figure 4 shows the different overlaps of the search task maps and the bottom-up saliency maps at 50% thresholding. There is a noticeable difference between the bottom-up models of passive viewing and the task maps. Note that the green-shaded pixels in these maps show where the task constraint is diverting overt attention away from the naturally/contextually/passively salient regions of the image.

3 Results and Discussion

Coincidence of Interest Points with Passive Saliency: The full count of interest-point overlap with the two models of bottom-up saliency at different surface area thresholds across the entire image set is shown in Figure 5. In comparing the interest-point overlap at the different threshold levels it is important to consider what the numbers mean in context. In this case, the chance level would correspond to a set of randomly distributed data points across the image, which would tend to the threshold level over an infinite number of images. Therefore at the thresholds in this investigation the chance levels are 10, 20, 30, 40, and 50% overlap corresponding to the threshold levels. If the distribution of interest-points is notably above (or below) chance levels, the interest-point detectors are concentrated in regions of saliency (or anti-saliency/background) and they can be considered statistical saliency detectors. Considering first the Itti model, it is clear that in general the mean percentages of data points are distributed in favour of lying within salient regions. For example, the SURF model (best performer) has 29% of SURF interest-points lying within the top 10% of

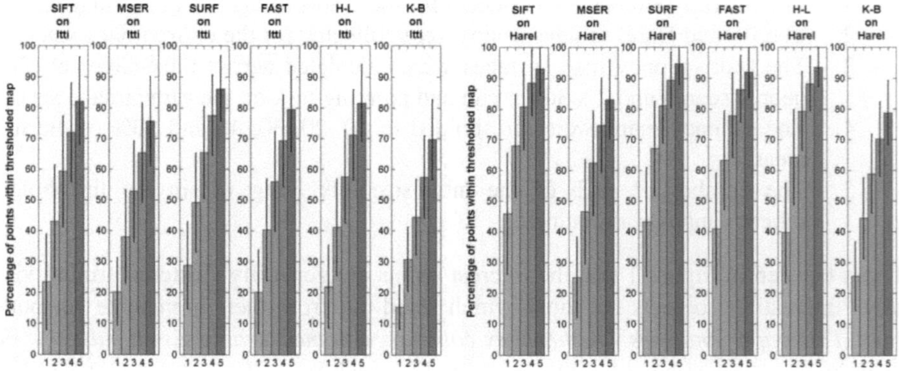

Fig. 5. The results of the *bottom up saliency map by Itti* (left) *and Harel* (right) models computed using the entire data set in comparison to the interest-point detectors. The bar indices 1 to 5 correspond to the 10 to 50 surface percentage coverage of the masks. The main axis is the percentage of interest points over the whole image set that lie within the saliency maps at the different threshold levels. The bars indicate average overlap at each threshold. Errors are gathered across the 72 image set: *standard deviation* is plotted in *black*.

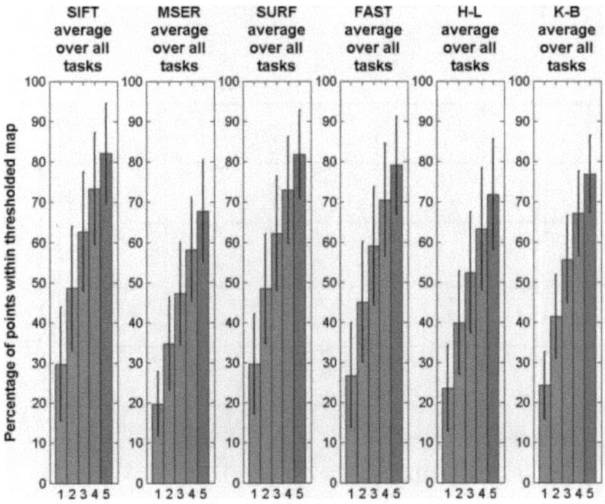

Fig. 6. The overlap of the interest-points with the *task probability surfaces* across the all 108 search scenarios. The bar indices 1 to 5 correspond to the 10 to 50 surface percentage coverage of the masks. The main axis is the percentage of interest points over the whole image set that lie within the task maps at the different threshold levels. The bars indicate average overlap at each threshold. Errors are gathered across all 108 tasks: *standard deviation* is plotted in *blue*.

ranked saliency points, 49% of SURF points distributed towards the top 20% of saliency points and 86% of the SURF points lie within the top 50% of saliency points. Overlap with the Harel model is better than for the Itti map. This is interesting because the Harel model was found to be more robust than Itti's model in predicting eye-fixation points under passive viewing conditions. The overlap levels of the SIFT and SURF are almost identical for Harel, with 46%, 68% and 93% of SIFT points overlapping the 10%, 20% and 50% saliency thresholds, respectively. All of the values are well above mere coincidence with very strong distribution towards the salient parts of the image. They are therefore a statistical indicator of saliency. For each saliency surface class, the overlaps of SIFT, SURF, FAST and Harris-Laplace are similar while the MSER and Kadir-Brady detectors have lower overlap.

Coincidence of Interest Points with Task-Based Saliency: The interest-point overlap with levels of the thresholded task maps is illustrated in Figure 6: bottom up *and* task data is combined in Figure 7. As illustrated in Figure 4, the imposition of a task can promote some regions that are "medium" or even "marginally" salient under passive conditions to being "highly" salient under task. The interest-points remain fixed for all of the images. This section therefore needs to consider the chance overlap levels as before, but also how the attention-shift due to task-imposition impacts upon the count relative to the passive condition. The detectors are again well above chance level in all cases, with both SIFT and SURF the strongest performers, with 30%, 48% and 83% of SIFT points overlapping the 10%, 20% and 50% thresholds respectively. In the task overlap test, the Kadir-Brady detector performs at a similar level of overlap to the others - in contrast to the passive case, where it has the poorest overlap. The Kadir-Brady "information saliency" detector clearly does highlight regions that might be of interest under task, while not picking out points that are the best overlap with bottom-up models. K-B saliency is not the best performer under task and there is not

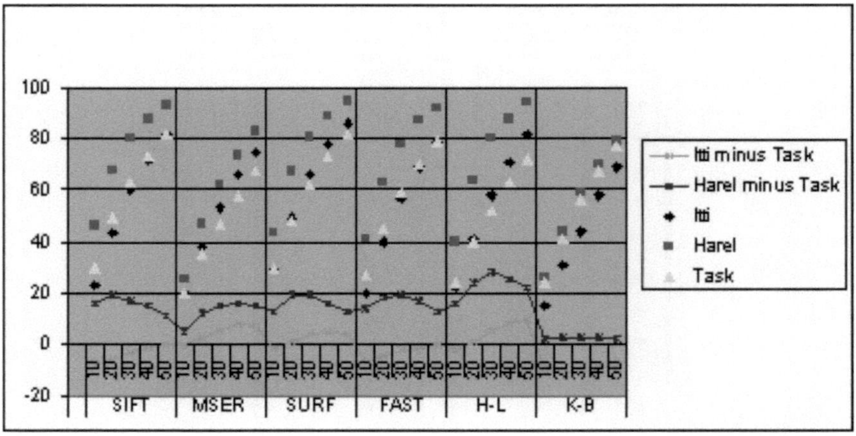

Fig. 7. The *average* percentage overlaps of the interest-points at different threshold levels of the two bottom-up and the task saliency surfaces. The difference between the passive and task cases is plotted to emphasise the *overlap difference* resulting from the application of "task".

enough information in this test to draw strong inference as to why this favourable shift should take place.

Looking at Figure 4 this should not be surprising since there exist conditions where the bottom-up and task surface overlap changes significantly: between 8% and 20% shift (*Green*, "only task" case in Figure 4) for coverage of 10% and 50% of surface area. Figure 7 reveals that the average Itti vs. interest-points overlap is overall very similar to the aggregate average task vs. interest-points overlap (between approx. *+/- 7% at most* for SIFT and SURF) implying that any attention shift due to task is directed towards other interest-points that do not overlap with the thresholded bottom-up saliency. Considering the Harel vs. task data, the task factors do reduce the surface overlap compared to the Harel surfaces by around 12% to 20% for the best performers, but very low for the Kadir-Brady. The initial high coincidence with the Harel surfaces (Figure 5) may cause this drop-off, especially since there is a task-induced shift of around 20% in some cases by the addition of a task (Figure 4).

4 Conclusion

In this paper the overlap between six well-known interest point detection schemes, two parametric models of bottom up saliency and task information derived from observer eye-tracking search experiments under task were compared. It was found that for both saliency models the SURF interest-point detector generated the highest coincidence with saliency. The SURF algorithm is based on similar techniques to the SIFT algorithm, but seeks to optimize the detection and descriptor parts using the best of available techniques. SIFT's Gaussian filters for scale representation are approximated using box filters and a fast Hessian detector is used in the case of SURF. Interestingly, the overlap performance was *superior* for the supposedly more robust saliency model for passive viewing, Graph Based Visual Saliency by Harel *et al.* Interest-points coinciding with bottom-up visually-salient information are valuable because of the robust description that can be applied to them for scene matching.

However, under task the attentional guidance surface is shifted in an unpredictable way. Even though statistical coincidence between Interest-points and the task surface remain well above chance levels, there is still no way of knowing what is being shifted where. The comparison of Kadir-Brady information-theoretic saliency with verified passive visual saliency models shows that Kadir-Brady is not in fact imitating the mechnisms of the human visual system, although it does pick out task-relevant pieces of information at the same level as other detectors.

References

1. Lowe, D.G.: Distinctive Image Features from Scale-Invariant Interest points. International Journal of Computer Vision 60, 91–110 (2004)
2. Matas, J., Chum, O., Urban, M., Pajdla, T.: Robust wide baseline sterio from maximally stable extremal regions. In: Proc. of British Machine Vision Conference, pp. 384-393 (2002)
3. Mikolajczyk, K., Schmid, C.: An Affine Invariant Interest Point Detector. In: Heyden, A., Sparr, G., Nielsen, M., Johansen, P. (eds.) ECCV 2002. LNCS, vol. 2350, pp. 128–142. Springer, Heidelberg (2002)
4. Bay, H., Tuytelaars, T., Van Gool, L.: SURF: Speeded Up Robust Features. In: Leonardis, A., Bischof, H., Pinz, A. (eds.) ECCV 2006. LNCS, vol. 3951, pp. 404–417. Springer, Heidelberg (2006)
5. Rosten, E., Drummond, T.: Fusing points and lines for high performance tracking. In: 10th IEEE International Conference on Computer Vision, vol. 2, pp. 1508–1511 (2005)
6. Rosten, E., Drummond, T.: Machine learning for high-speed corner detection. In: Leonardis, A., Bischof, H., Pinz, A. (eds.) ECCV 2006. LNCS, vol. 3951, pp. 430–443. Springer, Heidelberg (2006)
7. Kadir, T., Brady, M.: Saliency, Scale and Image Description. Int. Journ. Comp. Vision 45(2), 83–105 (2001)
8. Mikolajczyk, K., Tuytelaars, T., Schmid, C., Zisserman, A., Matas, J., Schaffalitzky, F., Kadir, T., Van Gool, L.: A comparison of affine region detectors. Int. Journ. Comp. Vision 65(1/2), 43–72 (2005)
9. Mikolajczyk, K., Schmid, C.: A performance evaluation of local descriptors. IEEE Transactions on Pattern Analysis & Machine Intelligence 27(10), 1615–1630 (2005)
10. Gao, K., Lin, S., Zhang, Y., Tang, S., Ren, H.: Attention Model Based SIFT Keypoints Filtration for Image Retrieval. In: Proc. ICIS 2008, pp. 191–196 (2008)
11. Itti, L., Koch, C., Niebur, E.: A Model of Saliency-Based Visual Attention for Rapid Scene Analysis. IEEE Transactions on Pattern Analysis and Machine Intelligence 20(11), 1254–1259 (1998)
12. Harel, J., Koch, C., Perona, P.: Graph-Based Visual Saliency. In: Advances in Neural Information Processing Systems, vol. 19, pp. 545–552 (2006)
13. Navalpakkam, V., Itti, L.: Search goal tunes visual features optimally. Neuron 53(4), 605–617 (2007)
14. Navalpakkam, V., Itti, L.: Modeling the influence of task on attention. Vision Research 45(2), 205–231 (2005)
15. Peters, R.J., Itti, L.: Beyond bottom-up: Incorporating task-dependent influences into a computational model of spatial attention. In: Proc. IEEE Conference on Computer Vision and Pattern Recognition, pp. 1–8 (2007)
16. Torralba, A., Oliva, A., Castelhano, M., Henderson, J.M.: Contextual Guidance of Attention in Natural scenes: The role of Global features on object search. Psychological Review 113(4), 766–786 (2006)

Grouping of Semantically Similar Image Positions

Lutz Priese, Frank Schmitt, and Nils Hering

Institute for Computational Visualistics,
University Koblenz-Landau, Koblenz
{priese,fschmitt,nilshering}@uni-koblenz.de

Abstract. Features from the Scale Invariant Feature Transformation (SIFT) are widely used for matching between spatially or temporally displaced images. Recently a topology on the SIFT features of a single image has been introduced where features of a similar semantics are close in this topology. We continue this work and present a technique to automatically detect groups of SIFT positions in a single image where all points of one group possess a similar semantics. The proposed method borrows ideas and techniques from the Color-Structure-Code segmentation method and does not require any user intervention.

Keywords: Image analysis, segmentation, semantics, SIFT.

1 Introduction

Let I be a 2-dimensional image. We regard I as a mapping $I : Loc \rightarrow Val$ that maps coordinates (x, y) from Loc (usually $Loc = [0, N-1] \times [0, M-1]$) to values $I(x, y)$ in Val (usually $Val = [0, 2^n[$ or $Val = [0, 2^n[^3$). We present a new technique to automatically detect groups $G_1, ..., G_l$ of coordinates, i.e., $G_i \subseteq Loc$, where all coordinates in a single group represent positions of a similar semantics in I. Take, e.g., an image of a building with trees. We are searching for sets $G_1, ..., G_l$ of coordinates with different semantics. E.g., there shall be coordinates for crossbars in windows in some set G_i, for window panes in another set G_j, inside the trees in a third set G_k, etc.. G_i, G_j, G_k form three different semantic classes (for crossbars, panes, trees in this example) for some i, j, k with $1 \leq i, j, k \leq l$. Obviously, such an automatic grouping of semantics can be an important step in many image analysis applications and is a rather ambitious programme. In this paper we propose a solution for SIFT features. Our technique is based on ideas from the CSC segmentation method.

2 SIFT

SIFT (Scale Invariant Feature Transformation) is an algorithm for an extraction of "interesting" image points, the so called SIFT features. SIFT was developed by David Lowe, see [2] and [3]. The SIFT algorithm follows the scale space approach

A.-B. Salberg, J.Y. Hardeberg, and R. Jenssen (Eds.): SCIA 2009, LNCS 5575, pp. 726–734, 2009.

and computes scale- and orientation-invariant points of interest in images. SIFT features consist of a coordinate in the image, a scale, a main orientation, and a 128-dimensional description vector. SIFT is commonly used for matching objects between spatially (e.g. in stereo vision) or temporally displaced images. It may also be used for object recognition where in a data base characteristic classes of features of known objects are stored and features from an image are matched with this data base to detect objects.

Slot and Kim use class keynotes of SIFT features in [5] for object class detection. Those class keynotes have been found by a clustering of similar features. They use spatial locations, orientations and scales as similarity criteria to cluster the features. The regions in which the clustering takes place (the spatial locations) are selected manually. In those regions clusters are built by a grouping via a low variance criteria in scale orientation space.

Mathematically speaking, a SIFT feature f is a tuple $f = (l_f, s_f, o_f, v_f)$ of four attributes: l_f for the location of the feature in x,y-coordinates in the image, s_f for the scale, o_f for the main orientation, v_f for the 128-dimensional vector. The range of o_f is $[0, 2\pi[$. The range of s_f depends on the size of the image and is about $0 \leq i \leq 100$ in our examples. The Euclidean distance $d_E(f, f')$ of two SIFT features f, f' is simply the Euclidean distance between the two 128-dimensional vectors v_f and $v_{f'}$.

3 CSC

Let $I : Loc \rightarrow Val$ be some image. A *region* R in I is a connected set of pixels of I. Connected means that any two pixels in R may be connected by a path of neighbored pixels that will not leave R. A region R is called a *segment* if in addition all pixels in R possess similar values in Val. A *segmentation* $\mathcal{S}$ is a partition $\mathcal{S} = \{S_1, ..., S_k\}$ with

1. $I = S_1 \cup ... \cup S_k$,
2. $S_i \cap S_j = \emptyset$ for $1 \leq i \neq j \leq k$,
3. each $S_i \in \mathcal{S}$ is a segment of I.

$\mathcal{S}$ is a *semi segmentation* if only 1 and 3 hold.

The CSC (Color Structure Code) is a rather elaborated region growing segmentation technique with a merge phase first and a split phase after that. It was developed by Priese and Rehrmann [4]. The algorithm is logically steered by an overlapping hexagonal topology. In the merge phase two already constructed overlapping segments S_1, S_2 of some level n may be merged into one new segment if S_1 and S_2 are similar enough. Otherwise, the overlap $S_1 \cap S_2$ is split between S_1 and S_2. In region growing algorithms without overlapping structures two similar segments with a common border may be merged. However, possessing a common substructure $S_1 \cap S_2$ leads to much robuster results than merging in case of a common border. Although the CSC gives a segmentation it operates with semi segmentations on different scales.

We will exploit the idea of merging overlapping sets for a segmentation in the following for a grouping of semantics.

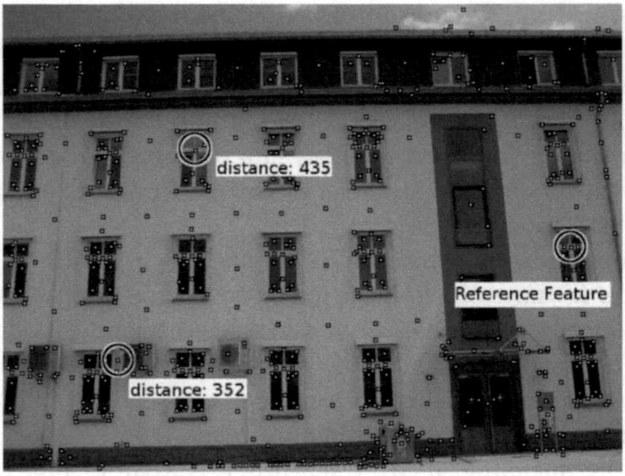

Fig. 1. Euclidean distances between appropriate and external features

4 A Topology on SIFT Features

To group semantically similar SIFT features we are looking for a topology where those semantically similar features become neighbors. Unfortunately, the Euclidean distance gives no such topology. Two SIFT features f_1, f_2 of the same image I with a very similar semantics may possess a rather large Euclidean distance $d_E(f_1, f_2)$ while for a third SIFT feature f_3 with a very different semantics $d_E(f_1, f_3) < d_E(f_1, f_2)$ may hold, compare Fig. 1. Thus, the Euclidean distance is not the optimal measure for the semantic distance of SIFT features. To overcome this problem we have introduced a new topology $\mathcal{T}$ on SIFT features in [1]. A *7-distance* $d_7(f, f')$ between f and f' has been introduced as the sum of the seven largest values of the 128 differences in $|v_f - v_{f'}|$. Let $f = (l, s, o, v)$ be some SIFT feature and let $f_i = (l_i, s_i, o_i, v_i)$ denote the i-th closest SIFT feature to f in the image with respect to d_E. For some set N of SIFT features we denote by μ_N^s (μ_N^o) the mean value of N in the coordinate for scale (orientation). The following algorithm computes a neighborhood $N(f)$ for f with three thresholds t_s, t_o, t_v by:

$N :=$ *empty list*; *insert* f *into* N; $i := 0; fault := 0;$
repeat
  $i := i + 1;$
  if $|(s, o, v) - (s_i, o_i, v_i)| \leq (t_s, t_o, t_v)$ and $(\mu_N^s \leq 0.75$ or $|s - s_i| \leq 2 \cdot \mu_N^s)$
  and $(\mu_N^o \leq 0.01$ or $|o - o_i| \leq 5 \cdot \mu_N^o)$
    then *insert* f_i *into* N*; update* μ_N^s *and* μ_N^o
    else $fault := fault + 1$
until $fault = 3$.

Thus, the Euclidean distance gives candidates f_i for $N(f)$ and the 7-distance excludes some of them. This semantic neighborhood defines a topology $\mathcal{T}$ on

SIFT features where the location of the SIFT features in the image plays no role.

5 Grouping of Semantics

5.1 The Problem

We want a grouping of the locations of SIFT features with the "same" semantics. The obvious approach is to group the SIFT features themselves and not their locations. Thus, the first task is:

Let $\mathcal{F}_I$ be the set of all SIFT features in a single image I detected by the SIFT algorithm. Find a partition $\mathcal{G} = \{G_1, ..., G_l\}$ of $\mathcal{F}_I$ s.t.

1. $\mathcal{F}_I = G_1 \cup ... \cup G_l$,
2. l is rather small, and
3. G_i consists of SIFT features of a similar semantics, for $1 \leq i \leq l$.

Each $G \in \mathcal{G}$ represents one semantic class. We do not claim that $G_i \cap G_j = \emptyset$ holds for $G_i \neq G_j$.

$loc(\mathcal{G}) := \{loc(G) | G \in \mathcal{G}\}$ becomes the wanted grouping of locations of a similar semantics in I where $loc(G)$ is the set of all positions of the features in G.

The topology $\mathcal{T}$ was designed to approach this task. All features inside a neighborhood $N(f)$ are usually of the same semantics as f. Let T_C be a known set of all SIFT features with a common semantics C as a ground truth and suppose f, f' are two features in T_C. Unfortunately, in general $N(f) \neq N(f')$ and $N(f) \neq T_C$ holds. $N(f)$ is usually smaller than T_C and may sometimes contain features not in T_C at all. Thus, computing $N(f)$ does not solve our task but will be the initial step towards a solution.

5.2 The Solution

One may imagine $\mathcal{F}_I$ as some sparse image $\mathcal{F}_I : Loc \to \mathbb{R}^{130}$ into a high dimensional value space with

$$\mathcal{F}_I(p) = \begin{cases} (s_f, o_f, v_f) & : \text{ for some } f \in \mathcal{F}_I \text{ with } l_f = p, \\ \text{undefined} & : \text{ if } \nexists f \in \mathcal{F}_I \text{ with } l_f = p. \end{cases}$$

Thus, the task of grouping semantics is similar to the task of computing a semi segmentation. The main difference is that $\mathcal{F}_I$ is rather sparse and connectivity of a segment plays no role. As a consequence, a region in $\mathcal{F}_I$ is simply any subset of $\mathcal{F}_I$ and a segment in $\mathcal{F}_I$ is a subset of features of $\mathcal{F}_I$ with a pairwise similar semantics. We will devise the segmentation technique CSC into a grouping algorithm for sparse images.

In a first step $N(f)$ is computed for any SIFT feature f in the image. $\mathcal{N} := \{N(f) | f \in \mathcal{F}_I\}$ is a semi segmentation of $\mathcal{F}_I$. However, there are too many overlapping segments in $\mathcal{N}$. $\mathcal{N}$ serves just as an initial grouping.

In the main step overlapping groups G, G' will be merged if they are similar enough. Here similarity is measured by the overlap rate $\frac{|G \cap G'|}{min(|G|,|G'|)}$. In contrast to the CSC we do not apply a split phase where $G \cap G'$ becomes distributed between G and G' in case that G and G' are not similar enough to be merged. The reason is that the rare cases where a SIFT feature is put into several semantic classes may be of interest for the following image analysis. In short, our algorithm AGS (**A**utomatic **G**rouping of **S**emantics) may be described as:

$\mathcal{H} := \mathcal{N}$;
(1) $\mathcal{G}$:= *empty list*;
for $0 \leq i < |\mathcal{H}|$ do $G := \mathcal{H}[i]$;
  for $0 \leq j < |\mathcal{H}|, i \neq j$ do
    if $G = \mathcal{H}[j]$
      then *remove* $\mathcal{H}[j]$ *from* $\mathcal{H}$
      else if G *and* $\mathcal{H}[j]$ *are similar* then $G := G \cup \mathcal{H}[j]$
  end for;
  insert G *into* $\mathcal{G}$
end for;
if $\mathcal{H} \neq \mathcal{G}$ then $\mathcal{H} := \mathcal{G}$; goto line (1) else end.

6 Some Examples

We present some pairs of images (Fig. 3 to 7) in the Appendix where the AGS algorithm has been applied. The left images show the coordinates of all features as detected by the SIFT algorithm. In a few cases two features with different scale or main orientation may be present at the same coordinate. The right ones show locations of some groups as computed by AGS. All features of one group are marked by the same symbol. Only groups consisting of at least five features are regarded in those examples. The number of such groups found by the AGS are given in #group and the semantics of the presented groups is named. Obviously, the results of this version of the AGS depend highly on the results of SIFT (as AGS regards solely detected SIFT features). The following qualitative observations are typical: The AGS algorithm works well on images with many symmetric edges (as in images of buildings). However, the quality is not good on very heterogeneous images with only very few symmetric edges (as in Fig. 5 where only one group with more than four elements is detected). In images with a larger crowd of people the AGS failed, e.g., to group features inside human faces.

7 Quantitative Evaluation

7.1 SIFT

Let $\mathcal{G} = \{G_1, ..., G_n\}$ be the set of SIFT features groups as computed by the AGS. Let $L_i := loc(G_i)$. Thus, $loc(\mathcal{G}) = \{L_1, ..., L_n\}$ is the found grouping

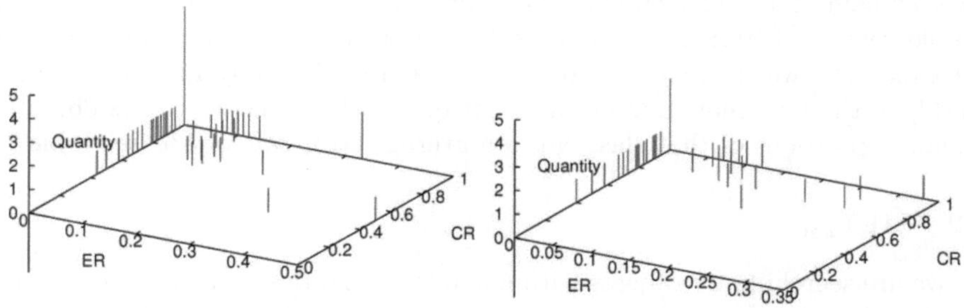

(a) SIFT (b) SIFT_{noi}

Fig. 2. Distribution of CR and ER

of locations of the same semantics. We now present a quantitative evaluation of $loc(\mathcal{G})$. We have manually annotated the SIFT locations for some semantic classes $(C_1, ..., C_n)$ in a set A of images as a ground truth. Let GT_i be the annotated ground truth for one semantic class C_i. Our evaluation tool computes the semantic grouping $\mathcal{G}$ of the AGS and compares each L in $loc(\mathcal{G})$ with GT_i by an

- coverability rate $CR(L, GT_i) := \frac{|L \cap GT_i|}{|GT_i|}$, and
- error rate $ER(L, GT_i) := \frac{|L - GT_i|}{|L|}$.

At the moment we have annotated the semantics "crossbar", "lower pane left" and "lower pane right" in windows to the corresponding feature positions in twenty-five images with buildings. This gives three sets of ground truth features, namely $GT_1 = \text{Crossbar}$, $GT_2 = \text{PaneLeft}$ and $GT_3 = \text{PaneRight}$.

For each image and each ground truth GT_i, $1 \leq i \leq 3$, we choose the group L in $loc(\mathcal{G})$ with the highest coverability rate $CR(L, GT_i)$. We show mean and standard deviation of the coverability and error rate over all three groups and all 25 images in table 1a. Figure 2a shows graphically the distribution of CR and ER over the 25×3 ground truth feature sets. The chosen parameters for $N(f)$ are $t_o = 0.5$, $t_s = 2.0$, $t_v = 500$ and the overlap rate for similarity of two groups in the AGS has been set to 0.75. Only groups with at least two members have been regarded.

In one of the 25 images there are only two windows whose crossbar features are not grouped. A single mistake in such small groups gives high errors rates.

Table 1. Evaluation of AGS algorithm on 25 manually annotated images

(a) Evaluation Lowe-SIFT

	CR	ER
mean	0.8589	0.0504
standard deviation	0.1951	0.0939

(b) Evaluation SIFT_{noi}

	CR	ER
mean	0.8939	0.0411
standard deviation	0.166	0.079

This explains the bad results in some images in figure2a. However, even this simple version of AGS gives good results in our analysis of the semantic classes "crossbar", "lower pane left" and "lower pane right". On average, the locations $loc(G)$ of the best matching group G for one of those classes covers 86% of all semantic positions of that class with an average error rate of 5%, see table 1a.

7.2 SIFT$_{noi}$

As we are searching for objects with a similar semantics in a single image those objects should possess the same orientation, at least in our application scenario of buildings. Thus, the orientation invariance of SIFT is even unwanted here. We therefore have implemented a variant SIFT$_{noi}$ - *noi* stands for **n**o **o**rientation **i**nvariance - where the orientation normalization in the SIFT algorithm is skipped. As a consequence, the main orientation o_f plays no role and the algorithm for $N(f)$ has to be adopted, ignoring o_f and the threshold t_o. We have further changed the parameter t_v to 450 for SIFT$_{noi}$. The results of our AGS with this SIFT$_{noi}$ variant are slightly better and shown in table 1b and figure 2b. The mean of the coverability rate increases to 89% while at the same time the error rate decreases to 4%.

8 Résumé

We have presented a completely automatic approach to the detection of groups of image positions with similar semantics. Obviously, such a grouping is helpful in many image analysis tasks.

This work is by no means completed. There are many variants of the AGS algorithm worth to be studied. One may modify the computation of $N(f)$ for a feature f. To decrease the error rate, a kind of splitting phase should be tested where in case of a high overlap rate between two groups G, G' the union $G \cup G'$ may be refined by starting with $G'' := G \cap G'$ and adding to G'' only those features in $(G \cup G') - G''$ that are "similar" enough to G''. The AGS method presented in this paper uses Lowe-SIFT features and a novel variant of SIFT$_{noi}$ features without orientation invariance. AGS works well in images with many symmetries – as in the examples with buildings – but less good in chaotic images. This is mainly caused by the fact that both SIFT features are designed to react on symmetries. Therefore, a next task is the extension of AGS to other feature classes and combinations of different feature classes.

References

1. Hering, N., Schmitt, F., Priese, L.: Image understanding using self-similar sift features. In: International Conference on Computer Vision Theory and Applications (VISAPP), Lisboa, Portugal (to be published, 2009)
2. Lowe, D.: Object recognition from local scale-invariant features. In: Proc. of the International Conference on Computer Vision ICCV, Corfu, pp. 1150–1157 (1999)

3. Lowe, D.: Distinctive image features from scale-invariant keypoints. International Journal of Computer Vision 20, 91–110 (2003)
4. Rehrmann, V., Priese, L.: Fast and robust segmentation of natural color scenes. In: Chin, R.T., Pong, T.-C. (eds.) ACCV 1998. LNCS, vol. 1351, pp. 598–606. Springer, Heidelberg (1997)
5. Slot, K., Kim, H.: Keypoints derivation for object class detection with sift algorithm. In: Rutkowski, L., Tadeusiewicz, R., Zadeh, L.A., Żurada, J.M. (eds.) ICAISC 2006. LNCS, vol. 4029, pp. 850–859. Springer, Heidelberg (2006)

Appendix

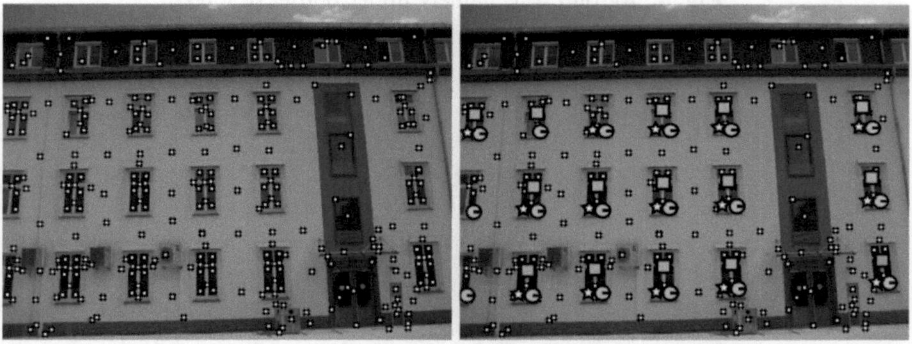

Fig. 3. #group = 10; shown are crossbars, lower right pane, lower left pane

Fig. 4. #group = 21; shown are upper border of pane, lower border of post

Fig. 5. #group = 1, namely upper border of forest

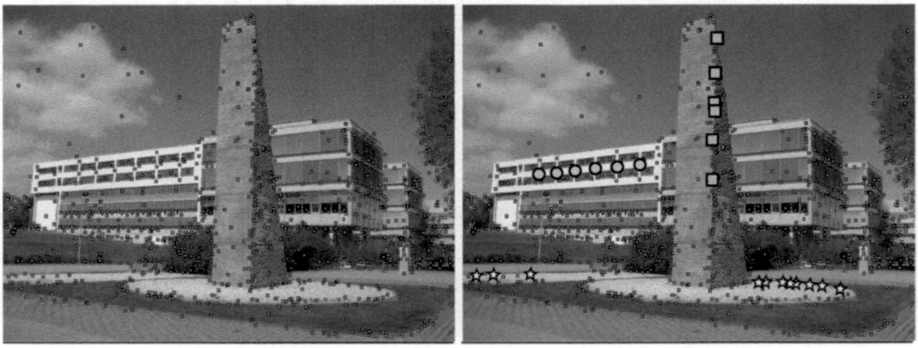

Fig. 6. #group = 24; shown are window interspace, monument edge and grass change

Fig. 7. #group = 7; shown are three different groups of repetitive vertical elements

Recovering Affine Deformations of Fuzzy Shapes

Attila Tanács[1,*], Csaba Domokos[1], Nataša Sladoje[2,**], Joakim Lindblad[3], and Zoltan Kato[1]

[1] Department of Image Processing and Computer Graphics, University of Szeged, Hungary
{tanacs,dcs,kato}@inf.u-szeged.hu

[2] Faculty of Engineering, University of Novi Sad, Serbia
sladoje@uns.ns.ac.yu

[3] Centre for Image Analysis, Swedish University of Agricultural Sciences, Uppsala, Sweden
joakim@cb.uu.se

Abstract. Fuzzy sets and fuzzy techniques are attracting increasing attention nowadays in the field of image processing and analysis. It has been shown that the information preserved by using fuzzy representation based on area coverage may be successfully utilized to improve precision and accuracy of several shape descriptors; geometric moments of a shape are among them. We propose to extend an existing binary shape matching method to take advantage of fuzzy object representation. The result of a synthetic test show that fuzzy representation yields smaller registration errors in average. A segmentation method is also presented to generate fuzzy segmentations of real images. The applicability of the proposed methods is demonstrated on real X-ray images of hip replacement implants.

1 Introduction

Image registration is one of the main tasks of image processing, its goal is to find the geometric correspondence between images. Many approaches have been proposed for a wide range of problems in the past decades [1]. Shape matching is an important task of registration. Matching in this case consists of two steps: First, an arbitrary segmentation step provides the shapes and then the shapes are registered. This solution is especially viable when the image intensities undergo strong nonlinear deformations that are hard to model, *e.g.* in case of X-ray imaging. If there are clearly defined regions in the images (*e.g.* bones or implants in X-ray images), a rather straightforward segmentation method can be used to define its shape adequately. Domokos *et al.* proposed an extension [2] to the

* Authors from University of Szeged are supported by the Hungarian Scientific Research Fund (OTKA) Grant No. K75637.

** Author is financially supported by the Ministry of Science of the Republic of Serbia through the Projects ON144029 and ON144018 of the Mathematical Institute of the Serbian Academy of Science and Arts.

A.-B. Salberg, J.Y. Hardeberg, and R. Jenssen (Eds.): SCIA 2009, LNCS 5575, pp. 735–744, 2009.

parametric estimation method of Francos *et al.* [3] to deal with affine matching of crisp shapes. These parametric estimation methods have the advantage of providing accurate and computationally simple solution, avoiding both the correspondence problem as well as the need for optimization.

In this paper we extend this approach by investigating the case when the segmentation method is capable of producing fuzzy object descriptions instead of a binary result. Nowadays, image processing and analysis methods based on fuzzy sets and fuzzy techniques are attracting increasing attention. Fuzzy sets provide a flexible and useful representation for image objects. Preserving fuzziness in the image segmentation, and thereby postponing decisions related to crisp object definitions has many benefits, such as reduced sensitivity to noise, improved robustness and increased precision of feature measures.

It has been shown that the information preserved by using fuzzy representation based on area coverage may be successfully utilized to improve precision and accuracy of several shape descriptors; geometric moments of a shape are among them. In [4] it is proved that fuzzy shape representation provides significantly higher accuracy of geometric moment estimates, compared to binary Gauss digitization at the same spatial image resolution. Precise moment estimation is essential for a successful application of the object registration method presented in [2] and the advantage of fuzzy shape representations is successfully exploited in the study presented in this paper.

In Section 2 we present the outline of the previous binary registration method [2] and extend it to accommodate fuzzy object descriptions. A segmentation method producing fuzzy object boundaries is described as well. Section 3 contains experimental results obtained during the evaluation of the method. In a study of 2000 pairs of synthetic images we observe the effect of the number of quantization levels of the fuzzy membership function to the precision of image registration and we compare the results with the binary case. Finally, we apply the registration method on real X-ray images, where we segmented objects of interest by an appropriate fuzzy segmentation method. This shows the successful adjustment of the developed method to real medical image registration tasks.

2 Parametric Estimation of Affine Deformations

In this section, we first review a previously developed binary shape registration method in the continuous space [2]. Since digital images are discrete, an approximative formula by discretization of the space is derived. The main contribution of this paper is in using a fuzzy approach when performing discretization. Instead of sampling the continuous image function at uniform grid positions, and performing binary Gauss discretization, we propose to perform area coverage discretization, providing a fuzzy object representation. We also describe a segmentation method that supports our suggested approach and produces objects with fuzzy boundaries.

2.1 Basic Solution

Herein we briefly overview the affine registration approach from [2]. Let us denote the points of the *template* and the *observation* by $\mathbf{x}, \mathbf{y} \in \mathbb{P}^2$, respectively in the projective space. The projective space allows simple notation for affine transforms and assumes using of homogeneous coordinates. Since affine transformations never alter the third (homogeneous) coordinate of a point, which is therefore always equal to 1, we, for simplicity, and without loss of generality, liberally interchange between projective and Euclidean plane, keeping the simplest notation.

Let $\mathbf{A}$ denote the unknown affine transformation that we want to recover. We can define the identity relation as follows

$$\mathbf{A}\mathbf{x} = \mathbf{y} \quad \Leftrightarrow \quad \mathbf{x} = \mathbf{A}^{-1}\mathbf{y}.$$

The above equations still hold when a properly chosen function $\omega : \mathbb{P}^2 \to \mathbb{P}^2$ is acting on both sides of the equations [2]:

$$\omega(\mathbf{A}\mathbf{x}) = \omega(\mathbf{y}) \quad \Leftrightarrow \quad \omega(\mathbf{x}) = \omega(\mathbf{A}^{-1}\mathbf{y}). \tag{1}$$

Binary images do not contain radiometric information, therefore they can be represented by their characteristic function $\mathbb{1} : \mathbb{P}^2 \to \{0, 1\}$, where 0 and 1 are assigned to the elements of the background and foreground respectively. Let $\mathbb{1}_t$ and $\mathbb{1}_o$ denote the characteristic function of the *template* and *observation*. In order to avoid the need for point correspondences, we integrate over the foreground domains $\mathcal{F}_t = \{\mathbf{x} \in \mathbb{P}^2 | \mathbb{1}_t(\mathbf{x}) = 1)$ and $\mathcal{F}_o = \{\mathbf{y} \in \mathbb{P}^2 | \mathbb{1}_o(\mathbf{y}) = 1)$ of the *template* and the *observation*, respectively, yielding [2]

$$|\mathbf{A}| \int_{\mathcal{F}_t} \omega(\mathbf{x})\, d\mathbf{x} = \int_{\mathcal{F}_o} \omega(\mathbf{A}^{-1}\mathbf{y})\, d\mathbf{y}. \tag{2}$$

The Jacobian of the transformation ($|\mathbf{A}|$) can be easily evaluated as

$$|\mathbf{A}| = \frac{\int_{\mathcal{F}_o} d\mathbf{y}}{\int_{\mathcal{F}_t} d\mathbf{x}}\,.$$

The basic idea of the proposed approach is to generate sufficiently many linearly independent equations by making use of the relations in Eq. (1)–(2). Since $\mathbf{A}$ depends on 6 unknown elements, we need at least 6 equations. We cannot have a linear system because ω is acting on the unknowns. The next best choice is a system of polynomial equations. In order to obtain a system of polynomial equations from Eq. (2), the applied ω functions should be carefully selected. It was also shown in [2] that by setting $\omega(\mathbf{x}) = (x_1^n, x_2^n, 1)$ Eq. (2) becomes

$$|\mathbf{A}| \int_{\mathcal{F}_t} x_k^n\, d\mathbf{x} = \sum_{i=0}^{n} \binom{n}{i} \sum_{j=0}^{i} \binom{i}{j} q_{k1}^{n-i} q_{k2}^{i-j} q_{k3}^{j} \int_{\mathcal{F}_o} y_1^{n-i} y_2^{i-j}\, d\mathbf{y}, \tag{3}$$

where $k = 1, 2$; $n = 1, 2, 3$ and q_{ki} denote the unknown elements of the inverse transformation $\mathbf{A}^{-1}$.

2.2 Fuzzy Object Representation

The polynomial system of equations in Eq. (3) is derived in the continuous space. However, digital image space provides only limited precision for these derivations and the integral can only be *approximated* by a discrete sum over the pixels. There are many approaches for discretization of a continuous function. The easiest way to form a discrete image is by sampling the continuous function at uniform grid positions. This approach, leading to a binary image, is also known as Gauss centre point digitization, and is used in the previous study [2]. An alternative way is to perform a fuzzy discretization of the image.

A discrete fuzzy subset F of a *reference set* $X \subset \mathbb{Z}^2$ is a set of ordered pairs $F = \{((i,j), \mu_F(i,j)) \mid (i,j) \in X\}$, where $\mu_F : X \rightarrow [0,1]$ is the *membership function* of F in X. The fuzzy membership function may be defined in various ways; its values reflect the levels of belongingness of pixels to the object. One useful way to define the membership function on a reference set in case when it is an image plane, is to assign a value to each image element (pixel) that is proportional to its coverage by the imaged object. In that way, partial memberships (values strictly between 0 and 1) are assigned to the pixels on the boundary of the discrete object.

Note that in the coefficients of the system of equations in Eq. (3) are the first, second and third order geometric moments of the *template* and *observation*. In general, *moments of order* $i + j$ of a continuous shape $\mathcal{F} = \{\mathbf{x} \in \mathbb{P}^2 | \mathbb{1}(\mathbf{x}) = 1\}$ are defined as

$$m_{i,j}(\mathcal{F}) = \int_{\mathcal{F}} x_1^i x_2^j \, d\mathbf{x}. \tag{4}$$

In the discrete formulation the geometric *moments of order* $i + j$ of a discrete fuzzy set F can be used, defined as

$$\tilde{m}_{i,j}(F) = \sum_{\boldsymbol{p} \in X} \mu_F(\boldsymbol{p}) \, p_1^i p_2^j \, . \tag{5}$$

This equation can be used to estimate geometric moments of a continuous 2D shape. Asymptotic error bounds for moments of order up to 2, derived in [4], show that moment estimates calculated from a fuzzy object representation provide a considerable increase of precision as compared to estimates computed from a crisp representation, at the same spatial resolution.

If F is fuzzy representation of $\mathcal{F}$, it follows that $m_{i,j}(\mathcal{F}) \approx \tilde{m}_{i,j}(F)$. Thus, by using Eq. (4)–(5) the integrals in Eq. (3) can be approximated as

$$\int_{\mathcal{F}_t} x_k^n \, d\mathbf{x} \approx \sum_{\boldsymbol{p} \in X_t} \mu_{F_t}(\boldsymbol{p}) \, p_k^n \quad \text{and} \tag{6}$$

$$\int_{\mathcal{F}_o} y_1^{n-i} y_2^{i-j} \, d\mathbf{y} \approx \sum_{\boldsymbol{p} \in X_o} \mu_{F_o}(\boldsymbol{p}) \, p_1^{n-i} p_2^{i-j}, \tag{7}$$

and the Jacobian can be approximated as

$$|\mathbf{A}| = \frac{m_{00}(\mathcal{F}_o)}{m_{00}(\mathcal{F}_t)} \approx \frac{\tilde{m}_{00}(F_o)}{\tilde{m}_{00}(F_t)} = \frac{\sum_{\boldsymbol{p} \in X_o} \mu_{F_o}(\boldsymbol{p})}{\sum_{\boldsymbol{p} \in X_t} \mu_{F_t}(\boldsymbol{p})}. \tag{8}$$

X_t and X_o are the reference sets (discrete domains) of the (fuzzy) template and (fuzzy) observation image, respectively.

The approximating discrete system of polynomial equations can now be produced by inserting these approximations into Eq. (3):

$$|\mathbf{A}| \sum_{\mathbf{p} \in X_t} \mu_{F_t}(\mathbf{p}) p_k^n = \sum_{i=0}^{n} \binom{n}{i} \sum_{j=0}^{i} \binom{i}{j} q_{k1}^{n-i} q_{k2}^{i-j} q_{k3}^{j} \sum_{\mathbf{p} \in X_o} \mu_{F_o}(\mathbf{p}) p_1^{n-i} p_2^{i-j} .$$

Clearly, the spatial resolution of the images affects the precision of this approximation. However, sufficient spatial resolution may be unavailable in real applications or, as it is expected in case of 3D applications, may lead to too large amounts of data to be successfully processed. On the other hand, it was shown in [4] that increasing the number of grey levels representing pixel coverage by a factor n^2 provides asymptotically the same increase in precision as an n times increase of spatial resolution. Therefore the suggested approach, utilizing increased membership resolution, is a very powerful way to compensate for insufficient spatial resolution, while still preserving desired precision of moments estimates.

2.3 Segmentation Method Providing Fuzzy Boundaries

Application of the moment estimation method presented in [4] assumes a discrete representation of a shape such that pixels are assigned their corresponding pixel coverage values. Definition of such digitization is given in [5]:

Definition 1. *For a given continuous object $\mathcal{F} \subset \mathbb{R}^2$, inscribed into an integer grid with pixels $p_{(i,j)}$, the n-level quantized pixel coverage digitization of $\mathcal{F}$ is*

$$\mathcal{D}_n(\mathcal{F}) = \left\{ \left((i,j), \frac{1}{n} \left\lfloor n \frac{A(p_{(i,j)} \cap \mathcal{F})}{A(p_{(i,j)})} + \frac{1}{2} \right\rfloor \right) \middle| (i,j) \in \mathbb{Z}^2 \right\} ,$$

where $\lfloor x \rfloor$ denotes the largest integer not greater than x, and $A(X)$ denotes the area of a set X.

Even though many fuzzy segmentation methods exist in the literature, very few of them result in pixel coverage based object representations. With an intention to show the applicability of the approach, but to not focus on designing a completely new fuzzy segmentation method, we derive pixel coverage values from an Active Contour segmentation [6]. Active Contour segmentation provides a crisp parametric representation of the object contour from which it is fairly straightforward to compute pixel coverage values. Such a straightforward derivation is not always possible, if other segmentation methods are used. The main point argued for in this paper is of a general character, and does not rely on any particular choice of segmentation method.

We have modified the SnakeD plugin for ImageJ by Thomas Boudier [7] to compute pixel coverage values. The snake segmentation is semi-automatic, and requires that an approximate starting region is drawn by the operator. Once the snake has reached a steady state solution, the snake representation is rasterized. Each pixel close to the snake boundary is given partial membership to the object proportional to how large part of that pixel is covered by the segmented object. The actual computation is facilitated by a 16×16 supersampling of the pixels close to the object edge and the pixel coverage is approximated by the fraction of sub-pixels that fall inside the object.

3 Experimental Results

When working with digital images, we are limited to a finite number of levels to represent fuzzy membership values. Using a database of synthetic binary shapes, we examine the effect of the number of quantization levels to the precision of registration and compare them to the binary case. The pairs of corresponding synthetic fuzzy shapes are obtained by applying known affine transformations. Therefore the presented registration results for synthetic images are neither dependent nor affected by a segmentation method. Finally, the proposed registration method is tested on real X-ray images, incorporating the fuzzy segmentation step.

3.1 Quantitative Evaluation on Synthetic Images

The performance of the proposed algorithm has been tested and evaluated on a database of synthetic images. The dataset consists of 39 different shapes and their transformed versions, a total of 2000 images. The width and height of the images were typically between 500 and 1000 pixels. The transformation parameters were randomly selected from uniform distributions. The rotation parameter was not restricted, any value was possible from $[0, 2\pi)$. Scale parameters varied between $[0.5, 1.5]$, shear parameters between $[-1, 1]$. The maximal translation value was set to 150 pixels. The templates were binary images, *i.e.* having either 0 or 1 fuzzy membership values. The fuzzy border representations of the observation images were generated by using 16×16 supersampling of the pixels close to the object edge and the pixel coverage was approximated by the fraction of sub-pixels that fall inside the object. The fuzzy membership values of the images were quantized and represented by integer values having k-bit ($k = 1, \ldots, 8$) representation. Some typical examples of these images and their registration accuracies are shown in Fig. 1.

In order to quantitatively evaluate the results, we have defined two error measures. The first error measure (denoted by ϵ) is the average distance in pixels between the true ($\mathbf{Ap}$), and recovered ($\widehat{\mathbf{A}}\mathbf{p}$) positions of the transformed pixels over the template. This measure is used for evaluation on synthetic images, where the true transformation is known. Another measure is the absolute difference (denoted by δ) between the *registered* template image and the *observation* image.

$$\epsilon = \frac{1}{m} \sum_{\mathbf{p} \in T} \|(\mathbf{A} - \widehat{\mathbf{A}})\mathbf{p}\|, \quad \text{and} \quad \delta = \frac{|R \triangle O|}{|R| + |O|},$$

where m is the number of *template* pixels, $\triangle$ means the symmetric difference, while R and O denote the set of pixels of the *registered* shape and the *observation*

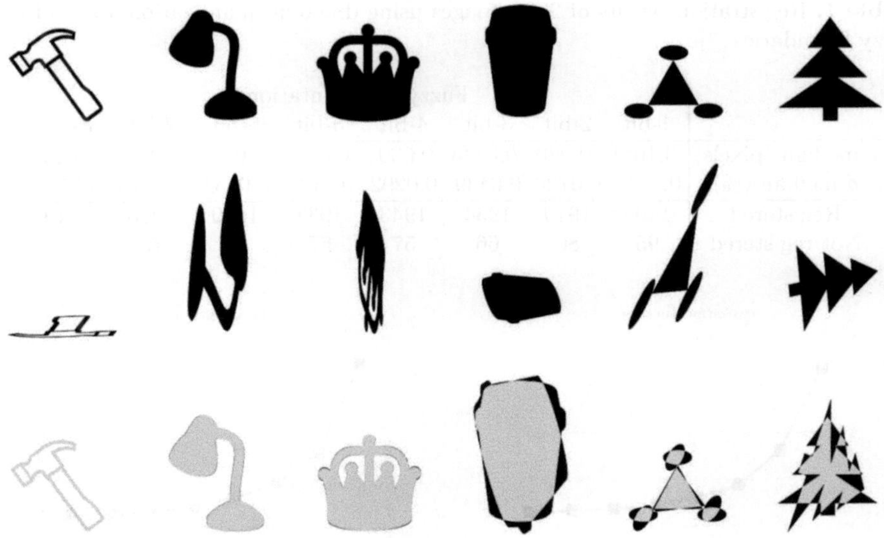

$\delta = 0.17\%$ $\delta = 0.25\%$ $\delta = 1.1\%$ $\delta = 8.87\%$ $\delta = 23.79\%$ $\delta = 25.84\%$

Fig. 1. Examples of *templates* (top row) and *observations* (middle row) images. In the third row, grey pixels show where the registered images matched each other and black pixels show the positions of registration errors.

respectively. We note that before computing the errors, the images were binarized by taking the α-cut at $\alpha = 0.5$ (in other words, by thresholding the membership function).

The medians of errors for both ϵ and δ are presented in Table 1 for different membership resolutions. For all membership resolutions, for around 5% of the images the system of equations provided no solution, *i.e.* the images were not registered. From the 56 images, there were only six whose transformed versions caused such problems. These can be seen in Fig. 2. Among the transformed versions, we found no rule to desribe when the problem occurs. Some of them caused problems for all different fuzzy membership resolutions, some of them occured for few resolutions only, randomly.

It is noticed that the experimental data confirmed the theoretical results, *i.e.* that the use of fuzzy shape representation enhances the registration, compared to the binary case. This effect can be interpreted as that the fuzzy representation "increases" the resolution of the object around its border. It also implies that registration based on fuzzy border representation may work for lower image resolutions, also where the binary approach becomes unstable.

Although based on solving a system of polynomial equations, the proposed method provides the result without any iterative optimization step or correspondence. Its time complexity is $\mathcal{O}(N)$, where N is the number of the pixels of the image. Clearly, most of the time is used for parsing the foreground pixels. All

Table 1. Registration results of 2000 images using different quantization levels of the fuzzy boundaries

	Fuzzy representation							
	1-bit	2-bit	3-bit	4-bit	5-bit	6-bit	7-bit	8-bit
ϵ median (pixels)	0.1681	0.080	0.0443	0.0305	0.0225	0.0186	0.0169	0.0147
δ median (%)	0.1571	0.0720	0.0439	0.0292	0.0196	0.0151	0.0125	0.0116
Registered	1905	1919	1934	1943	1933	1929	1925	1919
Not registered	95	80	66	57	67	71	75	81

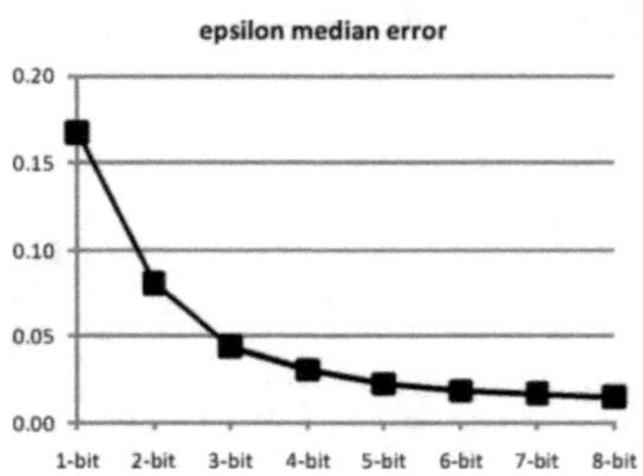

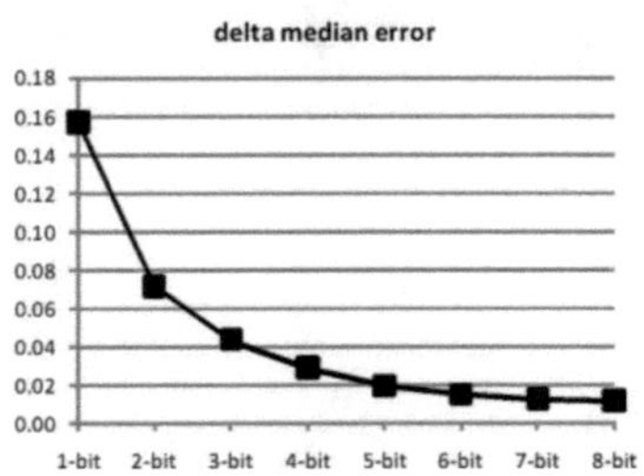

the summations can be computed in a single pass over the image. The algorithm has been implemented in Matlab 7.2 and ran on a laptop using Intel Core2 Duo processor at 2.4 GHz. The average runtime is a bit above half a second, including the computation of the discrete moments and the solution of the polynomial system. This allows real-time registration of 2D shapes.

3.2 Experiments on Real X-Ray Images

Hip replacement is a surgical procedure in which the hip joint is replaced by a prosthetic implant. In the short post-operative time, infection is a major concern. An inflammatory process may cause bone resorption and subsequent loosening or fracture, often requiring revision surgery. In current practice, clinicians assess loosening by inspecting a number of post-operative X-ray images of the patient's hip joint, taken over a period of time. Obviously, such an analysis requires the registration of X-ray images. Even visual inspection can benefit from registration as clinically significant prosthesis movement can be very small.

Fig. 2. Images where the polynomial system of equations provided no solutions in some cases. With increasing level of fuzzy discretization, the registration problem of the first three images vanished. The last three images provided problems permanently.

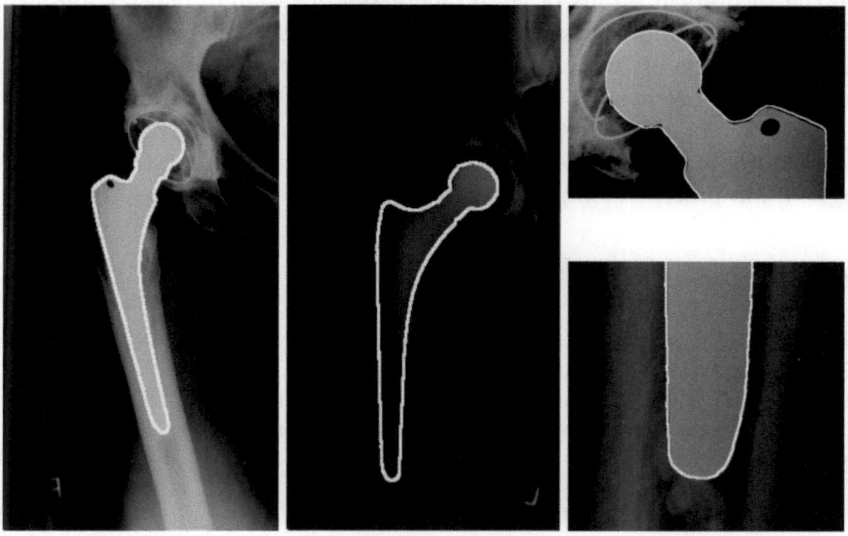

$\delta = 2.17\%$ $\delta = 4.81\%$ $\delta = 1.2\%$

Fig. 3. Real X-ray registration results. (a) and (b) show full X-ray observation images and the outlines of the registered template shapes. (c) shows a close up view of a third study around the top and bottom part of the implant.

There are two main challenges in registering hip X-ray images: One is the highly non-linear radiometric distortion [8] which makes any greylevel-based method unstable. Fortunately, the segmentation of the prosthetic implant is quite straightforward [9] so shape registration is a valid alternative here. Herein, we used the proposed fuzzy segmentation method to segment the implant. The second problem is that the true transformation is a projective one which depends also on the position of the implant in 3D space. Indeed, there is a rigid-body transformation in 3D space between the implants, which becomes a projective mapping between the X-ray images. Fortunately, the affine assumption is a good approximation here, as the X-ray images are taken in a well defined *standard position* of the patient's leg.

For for the diagnosis, the area around the implant (especially the bottom part of it) is the most important for the physician. It is where the registration must be the most precise. Fig. 3 shows some registration results. Since the best aligning transformation is not known, only the δ error measure can be evaluated. We also note, that in real applications the δ error value accumulates the registration error and the segmentation error. The preliminary results show that our approach using fuzzy segmentation and registration can be used in real applications.

4 Conclusions

In this paper we extended a binary affine shape registration method to take advantage of a discrete fuzzy representation. The tests confirmed expectations from the theoretical results of [4], on increased precision of registration if fuzzy shape representations are used. This improvement was demonstrated by a quantitative evaluation of 2000 images for different fuzzy membership discretization levels. We also presented a segmentation method based on Active Contour to generate fuzzy boundary representation of the objects. Finally, the results of a successful application of the method were shown for the registration of X-ray images of hip prosthetic implants taken during post-operative controls.

References

1. Zitová, B., Flusser, J.: Image registration methods: A survey. Image and Vision Computing 21(11), 977–1000 (2003)
2. Domokos, C., Kato, Z., Francos, J.M.: Parametric estimation of affine deformations of binary images. In: Proceedings of International Conference on Acoustics, Speech and Signal Processing, Las Vegas, Nevada, USA, pp. 889–892. IEEE, Los Alamitos (2008)
3. Hagege, R., Francos, J.M.: Linear estimation of sequences of multi-dimensional affine transformations. In: Proceedings of International Conference on Acoustics, Speech and Signal Processing, Toulouse, France, vol. 2, pp. 785–788. IEEE, Los Alamitos (2006)
4. Sladoje, N., Lindblad, J.: Estimation of moments of digitized objects with fuzzy borders. In: Roli, F., Vitulano, S. (eds.) ICIAP 2005. LNCS, vol. 3617, pp. 188–195. Springer, Heidelberg (2005)
5. Sladoje, N., Lindblad, J.: High-precision boundary length estimation by utilizing gray-level information. IEEE Transaction on Pattern Analysis and Machine Intelligence 31(2), 357–363 (2009)
6. Kass, M., Witkin, A., Terzopoulos, D.: Snakes: Active contour models. International Journal of Computer Vision 1(4), 321–331 (1988)
7. Boudier, T.: The snake plugin for ImageJ. software, `http://www.snv.jussieu.fr/~wboudier/softs/snake.html`
8. Florea, C., Vertan, C., Florea, L.: Logarithmic model-based dynamic range enhancement of hip X-ray images. In: Blanc-Talon, J., Philips, W., Popescu, D., Scheunders, P. (eds.) ACIVS 2007. LNCS, vol. 4678, pp. 587–596. Springer, Heidelberg (2007)
9. Oprea, A., Vertan, C.: A quantitative evaluation of the hip prosthesis segmentation quality in X-ray images. In: Proceedings of International Symposium on Signals, Circuits and Systems, Iasi, Romania, vol. 1, pp. 1–4. IEEE, Los Alamitos (2007)

Shape and Texture Based Classification of Fish Species

Rasmus Larsen, Hildur Olafsdottir, and Bjarne Kjær Ersbøll

DTU Informatics, Technical University of Denmark,
Richard Petersens Plads, Building 321, DK-2800 Kgs. Lyngby, Denmark
{rl,ho,be}@imm.dtu.dk

Abstract. In this paper we conduct a case study of fish species classification based on shape and texture. We consider three fish species: cod, haddock, and whiting. We derive shape and texture features from an appearance model of a set of training data. The fish in the training images were manual outlined, and a few features including the eye and backbone contour were also annotated. From these annotations an optimal MDL curve correspondence and a subsequent image registration were derived. We have analyzed a series of shape and texture and combined shape and texture modes of variation for their ability to discriminate between the fish types, as well as conducted a preliminary classification. In a linear discrimant analysis based on the two best combined modes of variation we obtain a resubstitution rate of 76 %.

1 Introduction

In connection with fishery, fishery biological research, and fishery independent stock assessment there is a need for automated methods for determination of fish species in various types of sampling systems. One technique to base such determination on is the use of automated image analysis and classification.

In conjunction with a technology project involving three departments at the Technical University of Denmark: the Departments of Informatics and Mathematical Modelling, Aquatic Systems, and Electrical Engineering, an effort is underway on researching and developing such systems.

Fish phenotype as defined by shape and color-texture both give information on fish species. Systematic description of differences in fish morphology dates back to the seminal work by d'Arcy Thompson [1]. Glasbey [2] demonstrate how a registration framework can be used to discriminate been the fish species whiting and haddock.

Modelling and automated registration of classes of biological objects with respect to shape and texture is elegantly achieved by the active appearance models [3] (AAM). The training of AAMs is based on sets of images with the objects of interests marked up by a series of corresponding landmarks. Developments of the original algorithms have aimed at alleviating the cumbersome work involved in manually annotating the training set. One such efforts is the

A.-B. Salberg, J.Y. Hardeberg, and R. Jenssen (Eds.): SCIA 2009, LNCS 5575, pp. 745–749, 2009.

minimum description length (MDL) approach to finding coordinate correspondences between surves and surfaces proposed by Davies et al [4]. A variant of this approach including curvature information was proposed by Thodberg [5].

2 Data

The study described in this article is based on a sample of 108 fish: 20 cod (torsk), 58 haddock (kuller), and 30 whiting (hviling) caugth in Kattegat. The fish were imaged using a standard color CCD camera under a standardized white light illumination. Example images are shown in Fig. 1. All fish images were mirrored to face left before further analysis.

(a) Cod, in Danish torsk (b) Whiting, in Danish hvilling (c) Haddock, in Danish kuller

Fig. 1. Example images of the three types of fish considered in the article. Note the differences in the shape of the snout as well as the absence of the thin dark line in the cod that is present in haddock and whiting.

3 Methods and Results

The fish images were contoured with the red and green curves shown in Fig. 2. Additionally, the fish eye centre was marked (the blue landmark). The two curves from the training set were input to the MDL based correspondence analysis by Thodberg [5], and the resulting landmarks recorded. Note that the landmarks are placed such that we have equi-distant sampling along the curves on the mean shape. This landmark annotated mean fish was then subjected to a Delaunay triangulation [6] and piece-wise affine warps of the corresponding triangles on each fish shape to the resulting Delaunay triangles of the mean shape constitute the training set registration. The quality of this registration is illustrated in Fig. 3. In this image each pixel is the log-transformed variance of each color

Fig. 2. The mean fish shape. The landmarks are placed according to a MDL principle.

Fig. 3. Model variance in each pixel explaining the texture variability in the training set after registration

across the training set after this registration. As can be seen the texture variation is concentrated in the fish head along the spine, and at fins.

Following this step an AAM was trained. The resulting first modes of variation are shown in Figs. 4 (shape alone), 5 (texture only), and 6 (combined shape and texture variation). The combined principal component analysis weigh the shape and texture according to the generalized variances of the two types of variation. Note, for the shape as well as for the combined model that the first factor captures a mode of variation pertaining to a bending of the fish body, i.e. a variation not related to fish specie. The second combined factor primarily captures the fish snout shape variation, and the third mode the presence/abscence of the black line along the fish body.

We next subject the principal component scores to a pairwise Fisher discriminant analysis [7] in order to evaluate the potential in discriminating between these species based on image analysis. The Fisher discriminant score explain the ability of a particular variable to discriminate between a particular pair of classes. From Table 1 we wee that it is overall most difficult to discriminate between Haddock-Whiting, texture is better for discriminating between Haddock-Cod, and combined shape and texture better for Cod-Whiting.

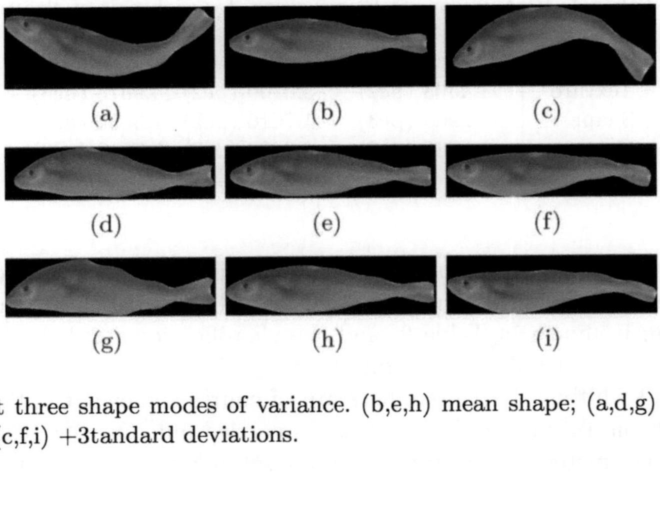

Fig. 4. First three shape modes of variance. (b,e,h) mean shape; (a,d,g) -3 standard deviations; (c,f,i) +3tandard deviations.

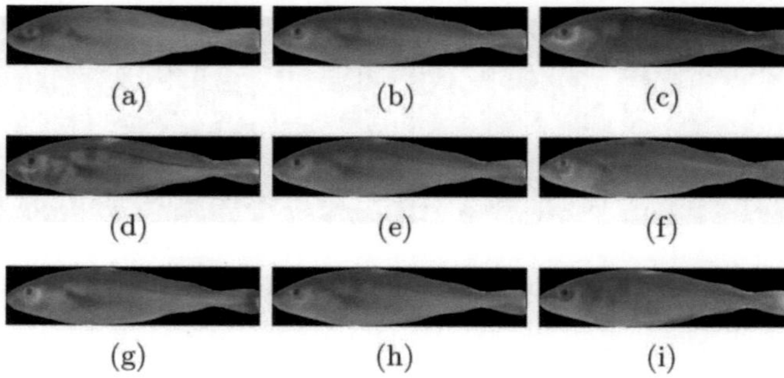

Fig. 5. First three texture modes of variance. (b,e,h) mean shape; (a,d,g) -3 standard deviations; (c,f,i) +3tandard deviations.

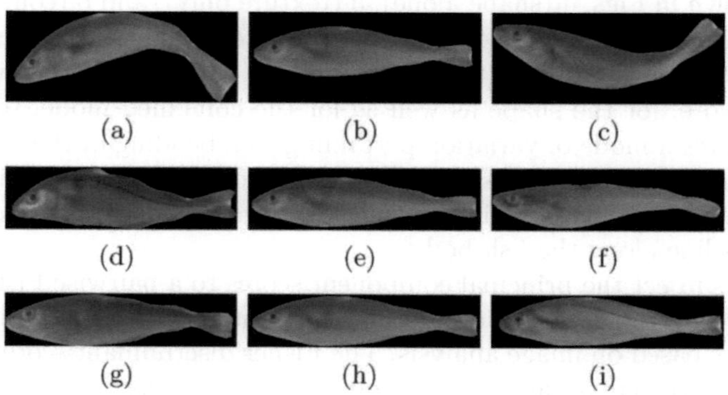

Fig. 6. First three combined shape and texture modes of variance. (b,e,h) mean shape; (a,d,g) -3 standard deviations; (c,f,i) +3tandard deviations.

Table 1. Best univariate Fisher scores for each pair of classes

	Haddock-Whiting	Haddock-Cod	Cod-Whiting
Texture	1.4303 (pc2)	5.0709 (pc2)	4.9675 (pc3)
Shape	1.2905 (pc3)	1.7616 (pc2)	1.3085 (pc4)
Combined	1.3536 (pc2)	2.6492 (pc3)	5.7519 (pc3)

Finally, the best two factors from the combined shape and texture model were applied in a linear discriminant analysis. The resubstitution matrix of the classification is shown in Table 2, and the classification result is illustrated in Fig. 7. The overall resubstitution rate is 76 %. The major confusion is between haddock and whiting. These numbers are of course somewhat optimistic given that no test on an independent test set is carried out. On the other hand the amount of parameter tuning to the training set is kept at a minimum.

Table 2. Resubstitution matrix for a linear discriminant analysis

	Cod	Haddock	Whiting
Cod	18	2	0
Haddock	2	40	16
Whiting	0	6	24

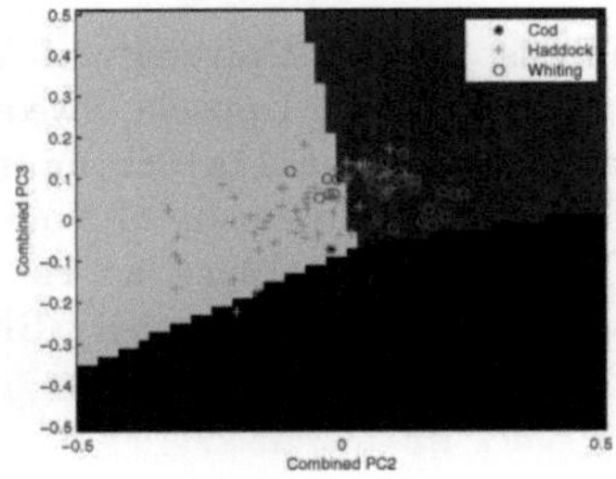

Fig. 7. Classification result for a linear discriminant analysis

4 Conclusion

In this paper we have provided an initial account of a procedure for fish species classification. We have demonstrated that to some degree shape and texture based classification can be use to discriminate between the fish species cod, haddock, and whiting.

References

1. Thompson, D.W.: On Growth and Form, 2nd edn. (1942) (1st edn. 1917)
2. Glasbey, C.A., Mardia, K.V.: A penalized likelihood approach to image warping. Journal of the Royal Statistical Society, Series B 63, 465–514 (2001)
3. Cootes, T.F., Edwards, G.J., Taylor, C.J.: Active appearance models. IEEE T. on Pattern Analysis and Machine Intelligence 23(6), 681–685 (2001)
4. Davies, R.H., Twining, C.J., Cootes, T.F., Waterton, J., Taylor, C.J.: A minimum description length approach to statistical shape modelling. IEEE Transactions on Medical Imaging (2002)
5. Thodberg, H.H.: Minimum description length shape and appearance models. In: Proc. Conf. Information Processing in Medical Imaging, pp. 51–62. SPIE (2003)
6. Delaunay, B.: Sur la sphère vide. Otdelenie Matematicheskikh i Estestvennykh Nauk, vol. 7, pp. 793–800 (1934)
7. Fisher, R.A.: The use of multiple measurements in taxonomic problems. Annals of Eugenics 7, 179–188 (1936)

Improved Quantification of Bone Remodelling by Utilizing Fuzzy Based Segmentation

Joakim Lindblad[1], Nataša Sladoje[2], Vladimir Ćurić[2], Hamid Sarve[1], Carina B. Johansson[3], and Gunilla Borgefors[1]

[1] Centre for Image Analysis, Swedish University of Agricultural Sciences, Box 337, SE-751 05 Uppsala, Sweden
{joakim,hamid,gunilla}@cb.uu.se
[2] Faculty of Engineering, University of Novi Sad, Serbia
{sladoje,vcuric}@uns.ac.rs
[3] Department of Clinical Medicine, Örebro University, SE-701 85 Örebro, Sweden
carina.johansson@oru.se

Abstract. We present a novel fuzzy theory based method for the segmentation of images required in histomorphometrical investigations of bone implant integration. The suggested method combines discriminant analysis classification controlled by an introduced uncertainty measure, and fuzzy connectedness segmentation method, so that the former is used for automatic seeding of the later. A thorough evaluation of the proposed segmentation method is performed. Comparison with previously published automatically obtained measurements, as well as with manually obtained ones, is presented. The proposed method improves the segmentation and, consequently, the accuracy of the automatic measurements, while keeping advantages with respect to the manual ones, by being fast, repeatable, and objective.

1 Introduction

The work presented in the paper is a part of a larger study aiming at improved understanding of the mechanisms of bone implants integration. The importance of this research increases together with the increased ageing of population, introducing its specific needs, which has become a characteristics of developed societies. Currently, automatic methods for quantification of bone tissue growth and modelling around the implants are in our focus. Results obtained so far are published in [9]. They address tasks of measurements of relevant quantities in 2D histological sections imaged in light microscope. While confirming the importance of the development of automatic quantification methods, in order to overcome problems of high time consumption and subjectivity of manual methods, the obtained results clearly call for further improvements and development.

In this paper we continue the study presented in [9] performed on 2D histologically stained un-decalcified cut and ground sections, with the implant in situ, imaged in light microscope. This technique, so called Exakt technique, [3], is also used for manual analysis. Observations regarding this technique are that

A.-B. Salberg, J.Y. Hardeberg, and R. Jenssen (Eds.): SCIA 2009, LNCS 5575, pp. 750–759, 2009.

it does not permit serial sectioning of bone samples with the implant in situ, but on the other hand is the state of the art when implant integration in bone tissue is to be evaluated without, e.g., extracting the device or calcifying the bone. Histological staining and subsequent colour imaging provide a lot of information, where different dyes attach to different structures of the sample, which can, if used properly, significantly contribute to the quality of the analysis results. However, variations in staining and various imaging artifacts are usually unavoidable drawbacks, that make automated quantitative analysis very difficult.

Observing that the measurements obtained by the method suggested in [9], length estimates of bone-implant contact (BIC) in particular, overestimate the manually obtained values (here considered to be the ground-truth), we found the cause of this problem in unsatisfactory segmentation results. Therefore, our main goal in this study is to improve the segmentation. For that purpose, we introduce a novel fuzzy based approach. Fuzzy segmentation methods are nowadays well accepted for handling shading, background variations, and noise/imaging artifacts. We suggest a two-step segmentation method, composed of, first, classification based on discriminant analysis (DA), as a method for automatic seeding required for the second step in the process, fuzzy connectedness (FC). We provide evaluation of the obtained results. The relevant area and length measurements derived from the images segmented by the herein proposed method show higher consistency with the manually obtained ones, compared to those reported in [9].

The paper is organized as follows: The next section contains a brief description of the previously used method, and some alternatives existing in the literature. Section 3 provides technical data on the used material. In Section 4 the proposed segmentation method is described, whereas in Section 5 we provide results of the automatic quantification and their appropriate evaluation. Section 6 concludes the paper.

2 Background

The segmentation method applied in [9] is based on supervised pixel-wise classification [4], utilizing the low intensity of the implant and the colour staining of the bone-region. RGB colour channels are supplemented with saturation (S) and value (V) channels, for improved performance. The pixel values of the three classes present in the images, *implant*, *bone*, and *soft tissue*, are assumed to be multivariate normally distributed. A number of tests carried out confirmed superiority of the approach where the classification is performed in two steps, instead of separating the three classes at the same time. For further details on the method, see [9].

The evaluation of the method exhibits overestimates of the required measurements, apparently caused by not sufficiently good segmentation. We conclude that pixel-wise classification, even though a rather natural choice and frequently used method for segmentation of colour images, relies too much on intensities/colours of individual pixels if used solely; such a method does not exploit spatial information kept in the image. We, therefore, suggest to combine spatial and intensity information existing in the image data. In addition, we want to utilize advantages of fuzzy techniques involved in segmentation. Various methods have been developed and exist in the literature; among the most frequently used ones are fuzzy c-means clustering and fuzzy connectedness. Recently, a segmentation method which combines fuzzy connectedness and fuzzy clustering is published, [5]. The method combines spatial and feature space information in the image segmentation process. The proposed algorithm is based on construction of a fuzzy connectedness relation on a membership image, obtained by some (deliberately chosen) fuzzy segmentation method; the suggested one is fuzzy c-means classification.

Motivated by the reasonably good performance of previously explored DA based classification, we suggest another combination of pixel-wise classification and fuzzy connectedness. We extend the crisp DA based classification by introducing an (un)certainty control parameter. We first use this enhanced classification to automatically generate seed regions and, in the second step, the seeded image is segmented by iterative relative FC segmentation method, as suggested in [1]. The method shows improved performance compared to the one in [9].

3 Material

Screw-shaped implants of commercially pure titanium were retrieved from rabbit bone after six weeks of integration. This study was approved by the local animal committee at Göteborg University, Sweden. The screws with surrounding bone were processed according to internal standards and guide-lines [7], resulting in 10μm un-decalcified cut and ground sections. The sections were histologically stained prior to light microscopical investigations. The histological staining method used on these sections, i.e. Toluidine blue mixed with pyronin G, results in various shades of purple stained bone tissue: old bone light purple and young bone dark purple. The soft tissue gets a light blue stain. For the suggested method, 1024x1280 24-bit RGB TIFF images were acquired by a camera

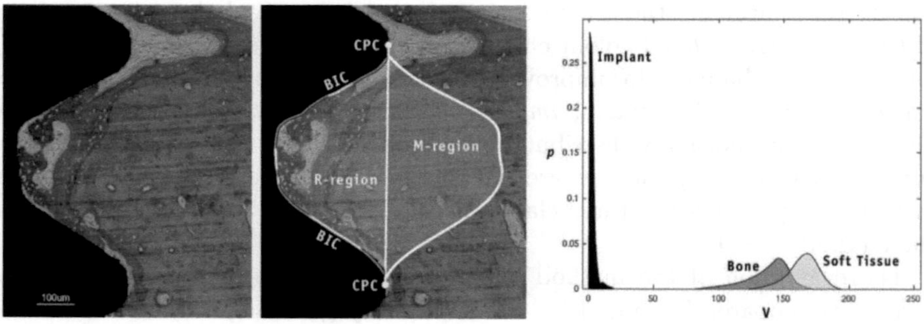

Fig. 1. Left: The screw-shaped implant (black), bone (purple), and soft tissue (light blue) are shown. Middle: Marked regions of interest. Right: Histogram of the pixel distribution in the V-channel for a sample image.

connected to a Nikon Eclipse 80i light microscope. A 10× ocular was used, giving a pixel size of 0.9μm. The regions of interest (ROIs) are marked in Fig. 1 (middle): the gulf between two centre points of the thread crests (*CPC*) denoted *R* (reference area); the area R mirrored with respect to the line connecting the two CPCs, denoted *M* (mirrored area) and regions where the bone is in contact with the screw, denoted *BIC*. Desired quantifications involve BIC length estimation and areas of different tissues in R and M; they are calculated for each thread (gulf between two CPCs) expressed as percentage of total length or area [6].

4 Method

The main result of this paper is the proposed segmentation method. Its description is given in the first part of this section. In the second part we briefly recall the types of measurements required for quantitative analysis of the bone implant integration.

4.1 Segmentation

By pure DA based classification we did not manage to overcome problems originating from artifacts resulting from preparation of specimens (visible stripes after cutting out the slices from the volume), staining of soft tissue that at some places obtained the colour of a bone, and effect of partial coverage of pixels by more than one type of tissue. All this led to unsatisfactory high misclassification rate.

There is a large overlap between the pixel values of the bone and soft tissue classes, as it is visible in the histogram in Figure 1 (right). Since all the channels exhibit similar characteristics, a perfect separation of the classes, only based on pixel intensities, is not possible. However, part of the pixels can reliably be classified using a pixel-wise DA approach. We suggest to use the DA classification when certain enough belongingness to a class can be deduced. For the remaining pixels, we suggest to utilize spatial information to address the problem of insufficient separation of the classes in the feature domain.

Automatic Seeding Based on Uncertainty in Classification. Three classes of image pixels are present in the images: implant, bone, and soft tissue. Pixel values are assumed to be multivariate normally distributed. The classification requires prior training; an expert marked different regions using a mouse based interface, after which the *RGB* values of the regions are stored as a reference.

As in [9], in addition to the three *RGB* channels, the *S* and *V* channels, obtained by a (non-linear) *HSV* transformation of *RGB*, are also considered in the feature space. For the *H* channel, it is noticed that it contains considerable amount of noise, and that the classes are not normally distributed, while the distributions of the classes are overlapping to a large extent. For these reasons, the *H* channel is not considered in the classification.

We introduce a measure of uncertainty in the classification and, with respect to that, an option for pixels to not be classified into any of the classes. A pixels may belong to the set U of non-classified (uncertain) pixels due to its low feature-based certainty u_F, or due to its spatial uncertainty. The set of seed-pixels, S, of an image I, is then defined as $S = I \backslash U$. They are assigned to appropriate classes in the early stage of the segmentation process. The decision regarding assignment of the elements of the set U is postponed. We define the uncertainty m_u of a classification to be $m_u = \frac{|U|}{|I|}$, where $|X|$ denotes the cardinality of a set X.

To determine feature-based certainty $u_F(x)$ of a pixel x, we compute posterior probabilities $p_k(x)$ for x to belong to each of the observed given classes C_k. For a multivariate normal distribution, the class-conditional density of an element x and class C_k is:

$$f_k(x) = \frac{1}{(2\pi)^{d/2} |\sum_k|^{1/2}} e^{-\frac{1}{2}(x-\mu_k)^T \sum_k^{-1}(x-\mu_k)},$$

where μ_k is the mean value of class C_k, $\sum_k$ is its covariance matrix, and d is the dimension of the space. Let $P(C_k)$ denote prior probability of a class C_k. The posterior probability of x to belong to the class C_k is then computed as

$$p_k(x) = P(C_k|x) = \frac{f_k(x)P(C_k)}{\sum_i f_i(x)P(C_i)}.$$

To avoid any class bias we assume equal prior probabilities $P(C_k)$.

To generate the sets S_k of seed points for each of the classes C_k, we first perform discriminant analysis based classification in a traditional way and obtain a crisp segmentation of the image into sets D_k. We initially set $S_k = D_k$ and then exclude, from each of the sets S_k, all the points which are considered to be uncertain, regarding belongingness to the class C_k.

We introduce a measure of feature-based certainty for x

$$u_F(x) = \frac{p_i(x)}{p_j(x)}, \quad \text{for } p_i(x) = \max_k p_k(x) \text{ and } p_j = \max_{k \neq i} p_k(x).$$

Instead of assigning pixel x to the class that provides the highest posterior probability, we define a threshold T_F, and assign the observed pixel x to the component C_i only if $u_F(x) \geq T_F$. Otherwise, $x \in U$, since its probability of belongingness is relatively similar for more than one class, and the pixel is seen as a "border case" in the feature space. Selection of T_F is discussed later in the paper. In this way, all the points x, having $p_k(x)$ as the maximal posterior probability and therefore initially assigned to $S_k = D_k$, but having $u_F(x) < T_F$ are in this step excluded form the set S_k, due to their low feature-based certainty.

Further removal of pixels from S_k is performed due to their spatial uncertainty, i.e., their position being close to a border between the classes. To detect such points, we apply erosion by a properly chosen structuring element, SE, to the sets D_k separately. The elements that do not belong to the resulting eroded set are removed from S_k and added to the set U. After this step, all seed points are detected, as $S = \bigcup_k S_k = I \setminus U$.

The amount of uncertainty affects the quality of segmentation, as confirmed by the evaluation of the method. We select the value of m_u, as given by a specific choice of T_F and SE, according to the results of empirical tests performed.

Iterative Relative Fuzzy Connectedness. We apply iterative relative fuzzy connectedness segmentation method as described in [1]. This version of the fuzzy connectedness segmentation method, originally suggested in [10], is adjusted for segmentation of multiple objects with multiple seeds.

The automatic seeding, performed as the first step of our method, provides multiple seeds for all the (three) existing objects in the image. The formulae for adjacency, affinity, and connectedness relations are, with very small adjustments, taken from [10]. For two pixels, $p, q \in I$, and their image values (intensities) $I(p)$ and $I(q)$, we compute:

$$\text{Fuzzy adjacency as} \qquad \mu_\alpha(p,q) = \frac{1}{1 + k_1 \|p - q\|_2} \quad \text{for } \|p-q\|_1 \leq n;$$

$$\text{Fuzzy affinity as} \qquad \mu_\kappa(p,q) = \mu_\alpha(p,q) \cdot \frac{1}{1 + k_2 \|I(p) - I(q)\|_2} ,$$

The value of n used in the definition of fuzzy adjacency determines the size of a neighbourhood where pixels are considered to be (to some extent) adjacent. We have tested $n \in \{1, 2, 3\}$ and concluded that they lead to similar results, and that $n = 2$ performs slightly better than the other two tested values. In addition, we use $k_1 = 0$, which leads to the following crisp adjacency relation:

$$\mu_\alpha = \begin{cases} 1, & \text{if} \quad \|p - q\|_1 \leq 2 \\ 0, & \text{otherwise} \end{cases} . \tag{1}$$

The parameter k_2, which scales the image intensities and has a very small impact on the performance of FC, is set to 2.

Algorithm 1, given in [1], is strictly followed in the implementation.

4.2 Measurements

The R- and M-regions, as well as the contact line between the implant and the tissue, are defined as described in [9]. The required measurements are: the estimate of the area of bone in R- and M- regions, relative to the area of the regions, and the estimate of the BIC length, relative to the length of the border line. Area of an object is estimated by the number of pixels assigned to the object. The length estimation is performed by using *Koplowitz and Bruckstein*'s method for perimeter estimation of digitized planar shapes (the first method of the two presented in [8]). A comparison of the results obtained by the herein proposed method with those presented in [9], as well as with manually obtained ones, is given in the following section.

5 Evaluation and Results

The automatic method is applied on three sets of images, each consisting of images of each of the 8 implant threads visible in one histological section. Training data are obtained by manual segmentation of two images from each set. In the evaluation, training images from the set being classified are not included when estimating class means and covariances in a 3-fold cross-validation fashion.

Our study includes several steps of evaluation: we evaluated the results (i) of the completed segmentation, and (ii) of the quantitative analysis of the implant integration, by comparing relevant measurements with the manually obtained ones, and with the ones obtained in [9]. Evaluation of segmentation includes separate evaluation of the automatic seeding and also of the whole two-steps process, i.e., seeding and fuzzy connectedness.

In Figure 2(a) we illustrate the performance of different discriminant analysis approaches in the seeding phase, for different levels of uncertainty m_u. As the measure of performance, *Cohen's kappa*, κ [2] is calculated for the set S and the same part (subset) of the corresponding manually segmented image. We observed two classifiers: linear (LDA), where the covariance matrices of the considered classes are assumed to be equal, and quadratic (QDA), where the

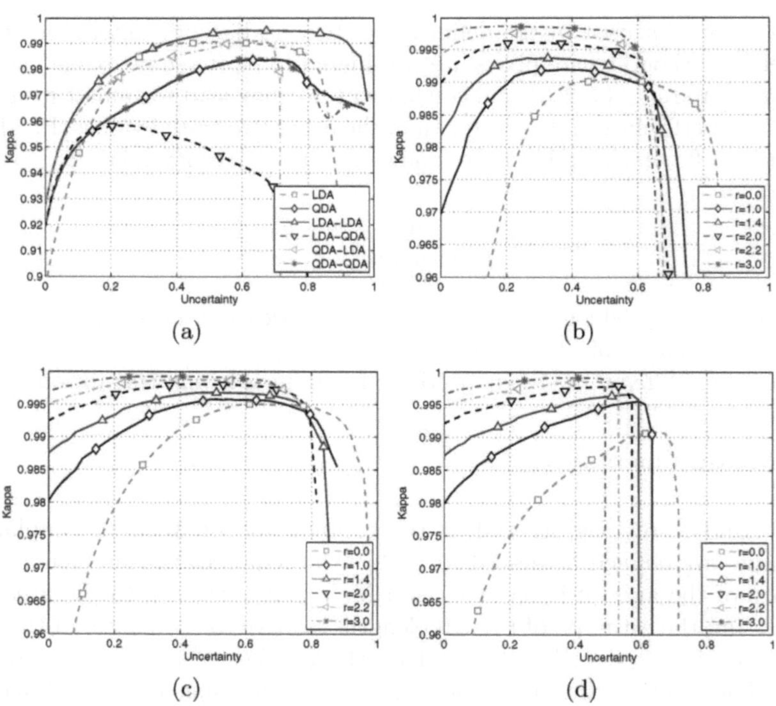

Fig. 2. Performance of DA. (a) Different DA approaches vs. different levels of m_u. (b-d) Performance for different radii r of SE, for (b) LDA, (c) LDA-LDA and (d) QDA-LDA.

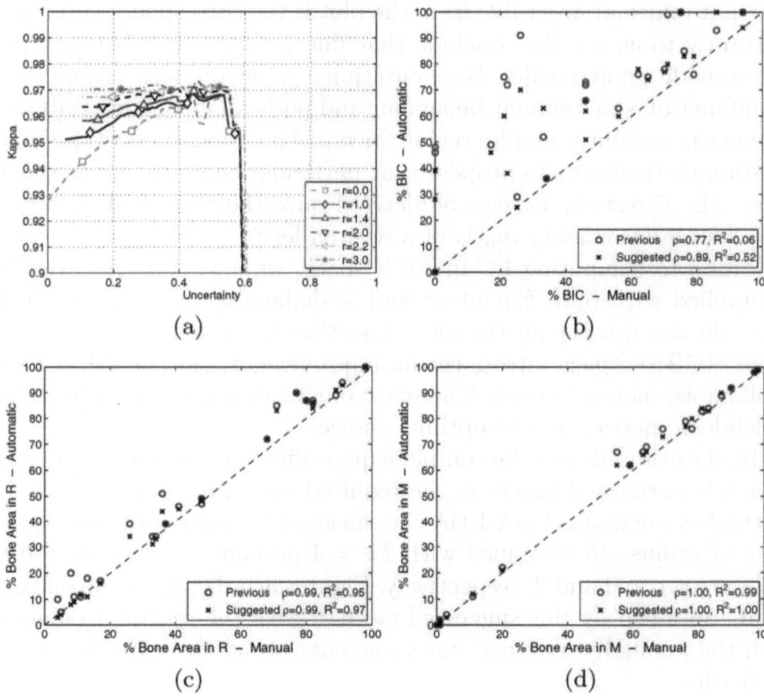

Fig. 3. Performance of the suggested method. (a) FC from LDA-LDA seeding for different m_u and radii r of SE. (b-d) Comparison of measurements from images segmented with the suggested method with those obtained by the method presented in [9].

covariance matrices of the classes are considered to be different. We observed classification into three classes in one step by both LDA, and QDA, but also by combinations of LDA and QDA used to first classify implant and non-implant regions, and then to separate bone and soft tissue. We notice that three approaches have distinctively better performance than others, for uncertainty up to 0.7 (uncertainty higher than 0.7 leaves, in our opinion, too many non-classified points): LDA-LDA provides the highest values for κ, while LDA and QDA-LDA perform slightly worse, but good enough to be considered in further evaluation.

Performance of these three DA approaches with respect to different sizes of disk-shaped structuring elements, introducing different levels of spatial uncertainty, is illustrated in Figures 2(b-d). It is clear that an increase of the size of the structuring elements leads to increased κ.

Further, we evaluate segmentation results after FC is applied, for different seed images. Figure 3(a), shows the performance for LDA-LDA. We see that κ increases with increasing size of structuring element, but beyond a radius of $\sqrt{5}$ the improvements are very small. To not loose too much of small structures in the images, we avoid larger structuring elements.

Important information visible from the plot is the corresponding optimal level of uncertainty to chose. We conclude that uncertainty levels between 25% and 50% all provide good results. Segmentations based on seeds from the QDA-LDA combination show similar behaviour and performance, but exhibiting good performance in a slightly smaller region for m_u. This robustness of the LDA-LDA combination motivates us to propose that particular combination as the method of choice. The threshold T_F can be derived once the size of SE is selected, so that the overall uncertainty m_u is at a desired level.

In addition to computing FC in RGB space, we have also observed $RGBSV$ space, supplied with both Euclidean and Mahalanobis metrics. Due to limited space, we do not present all the plots resulting from this evaluation, but only state that RGBSV space introduces no improvement, neither if Euclidean, nor if Mahalanobis, metric is used. Therefore our further tests include RGB space with Euclidean metrics, as the optimal choice.

Finally, the evaluation of the complete quantification method for bone implant integration is performed based on the required measurements, described in 4.2. The method we suggest is LDA-LDA classification for automatic seeding. Erosion by a disk of radius $\sqrt{5}$ combined with $T_F = 4$ provides $m_u \approx 0.35$. Parameters k_1 and k_2 are set to 0 and 2, respectively. Figures 3(b-d) present a comparison of the results obtained by this suggested method with the results presented in [9], and with the manually obtained measurements, which are considered to be the ground truth.

By observing the scatter plots, and additionally, considering correlation coefficients ρ between the respective method and the manual classification, as well as the coefficient of determination R^2, we conclude that the suggested method provides significant improvement of the accuracy of measurements required for quantitative evaluation of bone implant integration.

6 Conclusions

We propose a segmentation method that improves automatic quantitative evaluation of bone implant integration, compared to the previously published results. The suggested method combines discriminant analysis classification, controlled by an introduced uncertainty measure, and fuzzy connectedness segmentation. DA classification is used to define the points which are neither feature-based, nor spatially uncertain. These points are subsequently used as seed-points for the iterative relative fuzzy connectedness algorithm, which assign class belongingness to the remaining points of the images. In this way, both colour information existing in the stained histological material, and spatial information contained in the images, are efficiently utilized for segmentation. The performance of the method provides improved measurements, and overall better automatic quantification of the results obtained in the underlying histomorphometrical study.

The evaluation shows that by the described combination of DA and FC, classification performance measured by Cohen's kappa is increased from 92.7% to 97.1%, with a corresponding decrease of misclassification rate from 4.8% to 2.0%, as compared to using DA alone. Comparing feature values extracted from the segmented images with manual measurements, we observe an almost perfect match for the bone area measurements, with $R^2 \geq 0.97$. For the BIC measure, while being significantly better than previously presented method, $R^2 = 0.52$ indicates that further improvements may still be desired. Improvements may possibly be achieved by, e.g., refinement of the affinity relation used in the fuzzy connectedness segmentation, shading correction, appropriate directional filtering, or performing some fuzzy segmentation of the objects in the image, so that more precise measurements can be obtained from the resulting fuzzy representations. Our future work will certainly include some of these issues.

Acknowledgements. Research technicians Petra Hammarström-Johansson, Ann Albrektsson and Maria Hoffman are acknowledged for sample preparations. This work was supported by grants from The Swedish Research Council, 621-2005-3402 and was partly supported by the IA-SFS project RII3-CT-2004-506008 of the Framework Programme 6. Nataša Sladoje is supported by the Ministry of Science of the Republic of Serbia through the Projects ON144018 and ON144029 of the Mathematical Institute of the Serbian Academy of Science and Arts. Vladimir Ćurić is supported by the Ministry of Science of the Republic of Serbia through Project ON144016.

References

1. Ciesielski, K.C., Udupa, J.K., Saha, P.K., Zhuge, Y.: Iterative relative fuzzy connectedness for multiple objects with multiple seeds. Comput. Vis. Image Underst. 107(3), 160–182 (2007)
2. Cohen, J.: A coefficient of agreement for nominal scales. Educational and Psychological Measurement 11, 37–46 (1960)
3. Donath, K.: Die trenn-dunnschliffe-technik zur herstellung histologischer präparate von nicht schneidbaren geweben und materialien. Der Präparator 34, 197–206 (1988)
4. Duda, R.O., Hart, P.E.: Pattern Classification and Scene Analysis. Wiley, New York (1973)
5. Hasanzadeh, M., Kasaei, S., Mohseni, H.: A new fuzzy connectedness relation for image segmentation. In: Proc. of Intern. Conf. on Information and Communication Technologies: From Theory to Applications, pp. 1–6. IEEE Society, Los Alamitos (2008)
6. Johansson, C.: On tissue reactions to metal implants. PhD thesis, Department of Biomaterials / Handicap Research, Göteborg University, Sweden (1991)
7. Johansson, C., Morberg, P.: Importance of ground section thickness for reliable histomorphometrical results. Biomaterials 16, 91–95 (1995)
8. Koplowitz, J., Bruckstein, A.M.: Design of perimeter estimators for digized planar shapes. Trans. on PAMI 11, 611–622 (1989)
9. Sarve, H., Lindblad, J., Johansson, C.B., Borgefors, G., Stenport, V.F.: Quantification of bone remodeling in the proximity of implants. In: Kropatsch, W.G., Kampel, M., Hanbury, A. (eds.) CAIP 2007. LNCS, vol. 4673, pp. 253–260. Springer, Heidelberg (2007)
10. Udupa, J.K., Samarasekera, S.: Fuzzy connectedness and object definition: Theory, algorithms, and applications in image segmentation. Graphical Models and Image Processing 58(3), 246–261 (1996)

Fusion of Multiple Expert Annotations and Overall Score Selection for Medical Image Diagnosis

Tomi Kauppi[1], Joni-Kristian Kamarainen[2], Lasse Lensu[1], Valentina Kalesnykiene[3], Iiris Sorri[3], Heikki Kälviäinen[1], Hannu Uusitalo[4], and Juhani Pietilä[5]

[1] Machine Vision and Pattern Recognition Research Group (MVPR)
[2] MVPR/Computational Vision Group, Kouvola
Department of Information Technology,
Lappeenranta University of Technology (LUT), Finland
[3] Department of Ophthalmology, University of Kuopio, Finland
[4] Department of Ophthalmology, University of Tampere, Finland
[5] Perimetria Ltd., Finland

Abstract. Two problems especially important for supervised learning and classification in medical image processing are addressed in this study: i) how to fuse medical annotations collected from several medical experts and ii) how to form an image-wise overall score for accurate and reliable automatic diagnosis. Both of the problems are addressed by applying the same receiver operating characteristic (ROC) framework which is made to correspond to the medical practise. The first problem arises from the typical need to collect the medical ground truth from several experts to understand the underlying phenomenon and to increase robustness. However, it is currently unclear how these expert opinions (annotations) should be combined for classification methods. The second problem is due to the ultimate goal of any automatic diagnosis, a patient-based (image-wise) diagnosis, which consequently must be the ultimate evaluation criterion before transferring any methods into practise. Various image processing methods provide several, e.g., spatially distinct, results, which should be combined into a single image-wise score value. We discuss and investigate these two problems in detail, propose good strategies and report experimental results on a diabetic retinopathy database verifying our findings.

1 Introduction

Despite the fact that medical image processing has been an active application area of image processing and computer vision for decades, it is surprising that strict evaluation practises in other applications, e.g., in face recognition, have not been used that systematically in medical image processing. The consequence is that it is difficult to evaluate the state-of-the-art or estimate the overall maturity of methods even for a specific medical image processing problem. A step

A.-B. Salberg, J.Y. Hardeberg, and R. Jenssen (Eds.): SCIA 2009, LNCS 5575, pp. 760–769, 2009.

towards more proper operating procedures was recently introduced by the authors in the form of a public database, protocol and tools for benchmarking diabetic retinopathy detection methods [1]. During the course of work in establishing the DiaRetDB1 database and protocol, it became evident that there are certain important research problems which need to be studied further. One important problem is the optimal fusion strategy of annotations from several experts. In computer vision, ground truth information can be collected by using expert made annotations. However, in related studies such as in visual object categorisation, this problem has not been addressed at all (e.g., the recent LabelMe database [2] or the popular CalTech101 [3]). At least for medical images, this is of particular importance since the opinions of medical doctors may significantly deviate from each other or the experts may graphically describe the same finding in very different ways. This can be partly avoided by instructing the doctors while annotating, but often this is not desired since the data can be biased and grounds for understanding the phenomenon may weaken. Therefore, it is necessary to study appropriate fusion or "voting" methods.

Another very important problem arises from the fact how medical doctors actually use medical image information. They do not see it as a spatial map which is evaluated pixel by pixel or block by block, but as a whole depicting supporting information for a positive or negative diagnosis result of a specific disease. In image processing method development, on the other hand, pixel- or block-based analysis is more natural and useful, but the ultimate goal should be kept in mind, i.e., supporting the medical decision making. This issue was discussed in [1] and used in the development of the DiaRetDB1 protocol. The evaluation protocol, which simulates patient diagnosis using medical terms (specificity and sensitivity), requires a single overall diagnosis score for each test image, but it was not explicitly defined how the multiple cues should be combined into a single overall score. We address this problem throughly in this study and search for the optimal strategy to combine the cues. Also this problem is less known in medical image processing, but a well studied problem within the context of multiple classifiers or classifier ensembles (e.g., [4,5,6]).

The two problems are discussed in detail in Sections 2 and 3, and in the experimental part in Section 4 we utilise the evaluation framework (ROC graphs and equal error rate (EER) / weighted error rate (WER) error measures) to experimentally evaluate different fusion and scoring methods. Based on the discussions and the presented empirical results, we draw conclusions, define best practises and discuss the restrictions implied by our assumptions in Section 5.

2 Overall Image Score Selection for Medical Image Diagnosis

Medical diagnosis aims to diagnose the correct disease of a patient, and it is typically based on background knowledge (prior information) and laboratory tests which today include also medical imaging (e.g., ultrasound, eye fundus imaging, CT, PET, MRI, fMRI). The outcome of the tests and image or video data (observations) is typically either positive or negative evidence and the final diagnosis is based on a combination of background knowledge and test outcomes under strong Bayesian decision making for which all clinicians have been trained in the medical school [7]. Consequently, medical doctors are interested in medical image processing similar to a patient-based tool which provides a positive or negative outcome with a certain confidence. The tool confidence is typically fixed by setting the system to operate at certain *sensitivity* and *specificity* levels ([0%, 100%]), and therefore, these two terms are of special importance in medical image processing literature. The sensitivity value depends on the diseased population and specificity on the healthy population. Since these values are defined by the true positive rate (sensitivity is true positives divided by the sum of true positives and false negatives) and false positive rates (specificity is true negatives divided by the sum of true negatives and false positives), receiver operating characteristic (ROC) analysis is a natural tool to compare any methods [1]. Fixing the sensitivity and specificity values corresponds to selecting a certain operating point from the ROC. In [1], the authors introduced automatic evaluation methodology and published a tool to automatically produce the ROC graph for data where a single score value representing the test outcome (a higher score value increases the certainty of the positive outcome) is assigned to every image. The derivation of a proper image scoring method was not discussed, but is a topic in this study.

We restrict our development work to pixel- and block-based image processing schemes which are the most popular. The implication is that, for example, every pixel in an input image is classified to as a positive or negative finding, or positive finding likelihoods are directly given (see Fig. 1). To establish the final overall image score, these pixel or block values must be combined.

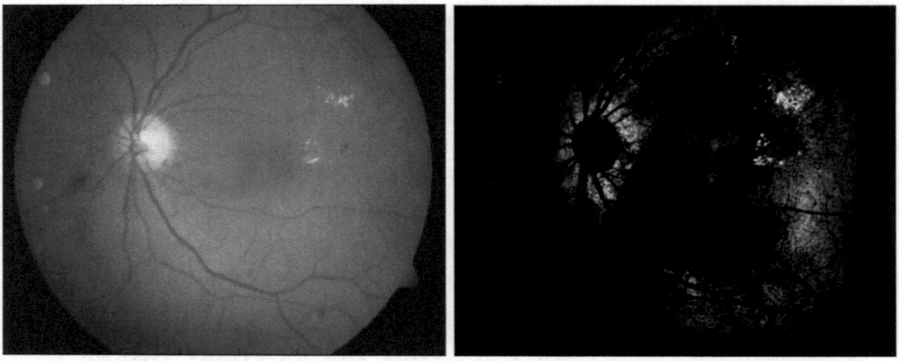

(a) (b)

Fig. 1. Example of pixel-wise likelihoods for hard exudates in eye fundus images (diabetic findings): (a) the original image (hard exudates are the small yellow spots in the upper-right part of the image); (b) probability density (likelihood) "map" for the hard exudates (estimated with a Gaussian mixture model from RGB image data)

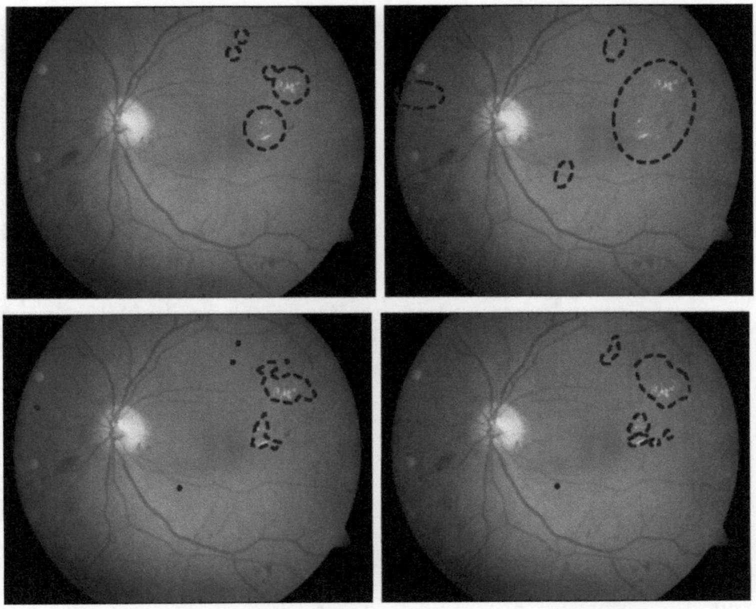

Fig. 2. Four independent expert annotations of hard exudates in one image

In the pixel- and block-based analyses, the final decision (score fusion) must be based on the fact that we have a (two-class) classification problem where the classifiers vote for positive or negative outcomes with a certain confidence. It follows that the optimal fusion strategy can easily be devised by exploring the results from a related field, combining classifiers (classifier ensembles), e.g., from the milestone study by Kittler et al. [4]. In our case, the "classifiers" act on different inputs (pixels) and therefore obey the distinct observations assumption in [4]. In addition, the classifiers have equal weights between the negative and positive outcomes. In [4], the theoretically most plausible fusion rules applicable also here were the product, sum (mean), maximum and median rules. We replaced the median rule with a more intuitive rank-order based rule for our case: "summax", i.e., the sum of some proportion of the largest values ($summax_{X\%}$). In our formulation, the maximum and sum rules can be seen as two extrema whereas summax operates between them so that X defines the operation point. Since any other straightforward strategies would be derivatives of these four, we restrict our analysis to them.

After the following discussion on fusion strategies, we experimentally evaluate all combinations of fusion and scoring strategies. Our evaluation framework and the DiaRetDB1 data is used for the purpose.

3 Fusing Multiple Medical Expert Annotations

It is recommended to collect medical ground truth (e.g., image annotations) from several experts within that specific field (e.g., ophthalmologists for eye

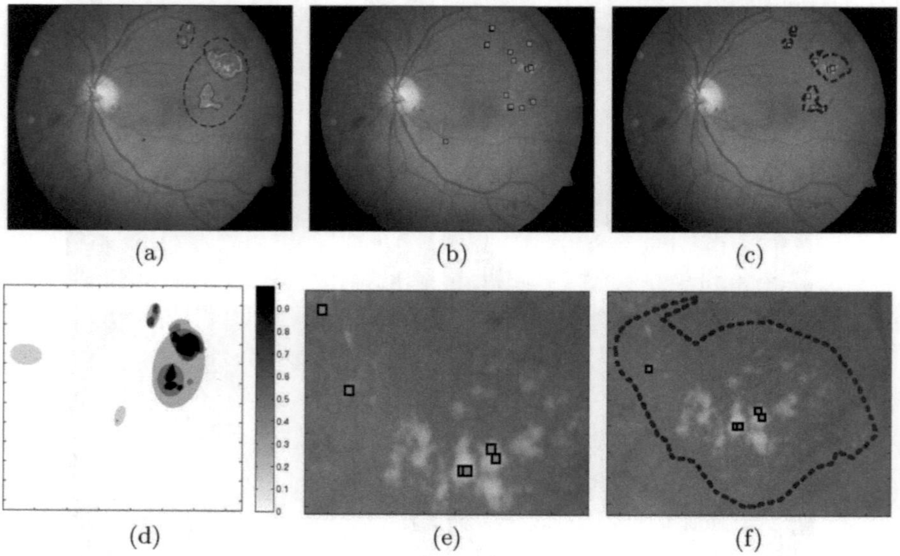

Fig. 3. Different annotation fusion approaches for the case shown in Fig. 2: (a) areas (applied confidence threshold for blue 0.25, red 0.75 and green 1.00); (b) representative points and their neighbourhoods (5×5); (c) representative point neighbourhoods masked with the areas (confidence threshold 0.75, blue colour); d) confidence map of areas in Fig. 3(a) e) close up image of representative point neighbourhoods in Fig. 3(b); f) close up image of masked representative point neighbourhoods in Fig. 3(c)

diseases). Note that this is not the practise in computer vision applications, e.g., only the eyes or bounding boxes are annotated by a single user in the face recognition databases (The FERET [8]) and rough segmentations in object category recognition (CalTech101 [3], LabelMe [2]). Multiple annotations is a necessity in medical applications where colleague consultation is the predominant working practise. Multiple annotations generate a new problem of how the annotations should be combined to a single ground truth (consultation outcome) for training a classifier. The solution certainly depends on the annotation tools provided for the experts, but it is not recommended to limit their expression power by instructions from laypersons which can harm the quality of ground truth.

For the DiaRetDB1 database, the authors introduced a set of graphical directives which are understandable for people not familiar of computer vision and graphics [1]. In the introduced directives, polygon and ellipse (circle) areas are used to annotate the spatial coverage of findings and at least one required (representative) point inside each area defining a particular spatial location that attracted expert's attention (colour, structure, etc.) With these simple but powerful directives, the independent experts produced significantly varying annotations for the same images, or even for the same finding in an image (see Fig. 2 for examples). The obvious problem is how to fuse equally trustworthy information from multiple sources to provide representative ground truth which retains

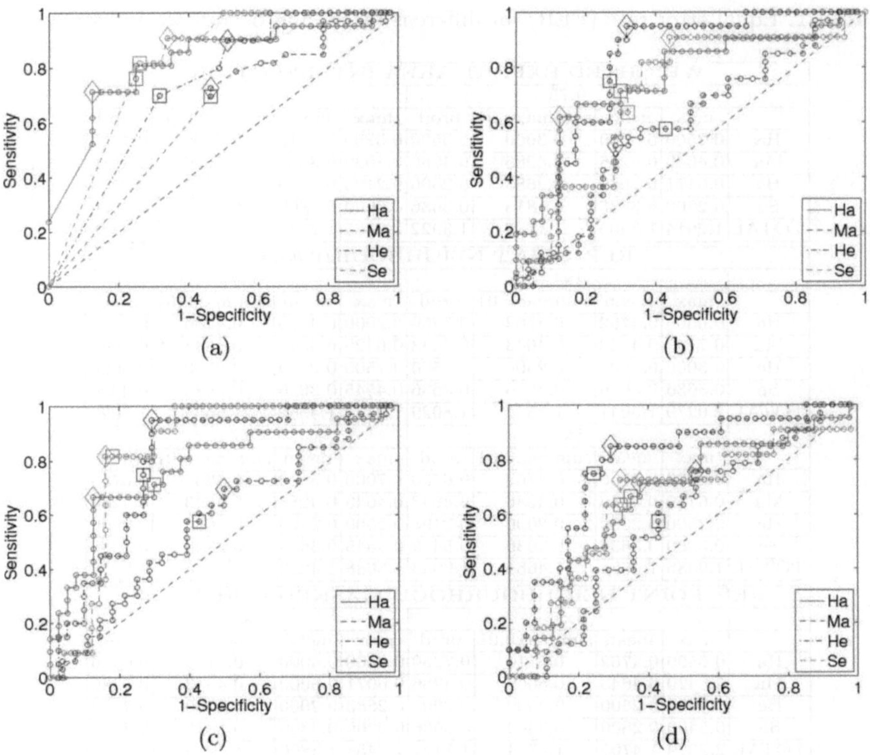

Fig. 4. Example ROC curves of "weighted expert area intersection" fusion with confidence 0.75 for two scoring rules, where EER and WER are marked with rectangle and diamond (best viewed in colours): (a) *max*; (b) *mean*; (c) $summax_{0.01}$; (d) *product*

the necessary within-class and between-class variation for supervised machine learning methods.

The available information to be fused is as follows: spatial coverage data by the polygon and ellipse areas, pixel locations (and possibly their neighbourhoods) of the representative points and the confidence levels for each marking given by each expert ("high", "moderate" or "low"). The available directives establish the available fusion strategies: intersections (sums) of the areas thresholded by a fixed average confidence (Fig. 3(a)), fixed size neighbourhoods of the representative points (Fig. 3(b)) and fixed size neighbourhoods of the representative points masked by the areas (combination of two) (Fig.3(c)).

All possible fusion strategies combined with all possible overall scoring strategies were experimentally evaluated as reported next.

4 Experiments

The experiments were conducted using the publicly available DiaRetDB1 diabetic retinopathy database [1]. The database comprises 89 colour fundus images

Table 1. Equal error rate (EER) for different fusion and overall scoring strategies

WEIGHTED EXPERT AREA INTERSECTION								
	0.75				1.00			
	max	mean	summax0.01	prod	max	mean	summax0.01	prod
Ha	0.2500	0.3000	0.3000	0.3500	0.5250	0.3810	0.4000	0.4762
Ma	0.4643	0.4286	0.4286	0.4286	0.3939	0.3636	0.3636	0.4286
He	0.3171	0.2683	0.2683	0.2500	0.2195	0.2500	0.2500	0.2500
Se	0.2600	0.3636	0.1818	0.3636	0.6600	0.2800	0.3000	0.2800
TOTAL	1.2914	1.3605	1.1787	1.3922	1.7985	1.2746	1.3136	1.4348
REP. POINT NEIGHBOURHOOD								
	1x1				3x3			
	max	mean	summax0.01	prod	max	mean	summax0.01	prod
Ha	0.6500	0.4762	0.4762	0.7250	0.7000	0.4286	0.4286	0.6750
Ma	0.7143	0.4643	0.4643	0.4643	0.6429	0.4643	0.4643	0.4643
He	0.3000	0.2000	0.2500	0.2500	0.1500	0.2000	0.2500	0.3000
Se	0.3636	0.3636	0.3636	0.3636	0.4545	0.3636	0.3636	0.3636
TOTAL	2.0279	1.5041	1.5541	1.8029	1.9474	1.4565	1.5065	1.8029
	5x5				7x7			
	max	mean	summax0.01	prod	max	mean	summax0.01	prod
Ha	0.6000	0.4762	0.4762	0.6750	0.7000	0.3810	0.5250	0.6750
Ma	0.6786	0.4286	0.4286	0.4643	0.4643	0.4286	0.4643	0.4286
He	0.2500	0.2000	0.2000	0.2195	0.2500	0.2500	0.2683	0.2000
Se	0.3800	0.3636	0.3636	0.5455	0.4545	0.3636	0.2800	0.3636
TOTAL	1.9086	1.4684	1.4684	1.9043	1.8688	1.4232	1.5376	1.6672
REP. POINT NEIGHBOURHOOD MASKED (AREA 0.75)								
	1x1				3x3			
	max	mean	summax0.01	prod	max	mean	summax0.01	prod
Ha	0.6500	0.4762	0.5714	0.7250	0.6500	0.4000	0.4762	0.6750
Ma	0.6429	0.4643	0.5000	0.4286	0.6071	0.5000	0.4643	0.4286
He	0.4000	0.2500	0.2000	0.2000	0.2683	0.2000	0.2000	0.2500
Se	0.5400	0.2800	0.3000	0.3636	0.2200	0.2800	0.2800	0.3636
TOTAL	2.2329	1.4705	1.5714	1.7172	1.7454	1.3800	1.4205	1.7172
	5x5				7x7			
	max	mean	summax0.01	prod	max	mean	summax0.01	prod
Ha	0.6500	0.5000	0.4286	0.6750	0.7250	0.4762	0.4762	0.6750
Ma	0.5152	0.4286	0.4286	0.4286	0.5455	0.4286	0.5000	0.4286
He	0.2500	0.2683	0.2500	0.2195	0.2500	0.3000	0.2195	0.2500
Se	0.2200	0.3000	0.2800	0.2800	0.4545	0.2800	0.3000	0.2727
TOTAL	1.6352	1.4969	1.3871	1.6031	1.9750	1.4848	1.4957	1.6263

of which 84 contain at least mild non-proliferative signs of diabetic retinopathy (haemorrhages (Ha), microaneurysms (Ma), hard exudates (He) and soft exudates (Se)). The images were captured with the same 50 degree field-of-view digital fundus camera[1], and therefore, the data should not contain colour distortions other than those related to the findings. The fusion and overall scoring strategies were tested using the predefined training set of 28 images and test set of 61 images.

Since this study is restricted to pixel- and block-based image processing approaches, photometric information (colour) was a natural feature for the experimental analysis. For the visual diagnosis of diabetic retinopathy, colour is also the most important single visual cue. Since the whole medical diagnosis is naturally Bayesian, we were motivated to address the classification problem with a standard statistical tool, estimating probability density functions (pdfs) of each finding given a colour observation (RGB), $p(r, g, b|finding)$. For the un-

[1] ZEISS FF 450^{plus} fundus camera with Nikon F5 digital camera.

Table 2. Weighted error rate [WER(1)] for different fusion and overall scoring strategies

WEIGHTED EXPERT AREA INTERSECTION								
	0.75				1			
	max	mean	summax0.01	prod	max	mean	summax0.01	prod
Ha	0.2054	0.2530	0.2292	0.3304	0.3577	0.3494	0.3440	0.4119
Ma	0.3685	0.3853	0.4015	0.3891	0.3685	0.3452	0.2998	0.3561
He	0.3061	0.1835	0.1713	0.2213	0.1829	0.1841	0.1963	0.1841
Se	0.2155	0.2655	0.1709	0.2964	0.4209	0.2309	0.2609	0.2718
TOTAL	1.0954	1.0872	0.9729	1.2371	1.3301	1.1097	1.1011	1.2239
REP. POINT NEIGHBOURHOOD								
	1x1				3x3			
	max	mean	summax0.01	prod	max	mean	summax0.01	prod
Ha	0.3964	0.3845	0.4417	0.5000	0.4238	0.3631	0.4018	0.5000
Ma	0.3902	0.4107	0.4015	0.3837	0.4080	0.4031	0.4042	0.3561
He	0.2476	0.1713	0.2220	0.1970	0.1482	0.1598	0.1591	0.2451
Se	0.3118	0.2755	0.3073	0.3264	0.3509	0.2809	0.2709	0.3264
TOTAL	1.3460	1.2420	1.3724	1.4070	1.3309	1.2069	1.2361	1.4275
	5x5				7x7			
	max	mean	summax0.01	prod	max	mean	summax0.01	prod
Ha	0.4190	0.3482	0.4179	0.5000	0.4113	0.3631	0.4554	0.5000
Ma	0.4302	0.4031	0.3988	0.3864	0.3231	0.3880	0.4318	0.3750
He	0.1988	0.1598	0.1854	0.1841	0.2091	0.1829	0.2207	0.1957
Se	0.2100	0.2655	0.2509	0.4127	0.3927	0.2355	0.2200	0.2709
TOTAL	1.2580	1.1766	1.2529	1.4832	1.3362	1.1695	1.3279	1.3416
REP. POINT NEIGHBOURHOOD MASKED (AREA 0.75)								
	1x1				3x3			
	max	mean	summax0.01	prod	max	mean	summax0.01	prod
Ha	0.4351	0.4369	0.4702	0.5000	0.4238	0.3631	0.4315	0.5000
Ma	0.4280	0.4291	0.4069	0.3723	0.4702	0.4383	0.4329	0.4085
He	0.2439	0.1963	0.1726	0.1976	0.1988	0.1976	0.1726	0.1963
Se	0.3609	0.2409	0.2609	0.3173	0.1555	0.2555	0.1600	0.3118
TOTAL	1.4680	1.3033	1.3106	1.3871	1.2483	1.2544	1.1970	1.4167
	5x5				7x7			
	max	mean	summax0.01	prod	max	mean	summax0.01	prod
Ha	0.4113	0.4333	0.3482	0.5000	0.3988	0.3756	0.4190	0.4524
Ma	0.4129	0.4004	0.4015	0.3544	0.4334	0.3907	0.4383	0.3701
He	0.2073	0.1963	0.2232	0.2098	0.1957	0.2713	0.1713	0.2323
Se	0.1555	0.2700	0.1855	0.2718	0.3927	0.1900	0.2609	0.2109
TOTAL	1.1870	1.3001	1.1584	1.3360	1.4207	1.2276	1.2896	1.2657

known distributions, Gaussian mixture models (GMMs) were natural models and the unsupervised Figueiredo-Jain algorithm a good estimation method [9]. We also tried the standard expectation maximisation (EM) algorithm, but since the Figueiredo-Jain always outperformed it without the need to explicitly define the number of components, it was left out from this study. For training, different fusion approaches for the expert annotations discussed in Section 3 were used to form a training set for the GMM estimates. For every test set image, our method provided a full likelihood map (see Fig. 1(b)) from which the different overall scores in Section 2 were computed.

Our interpretations of the results are based qualitatively on the produced ROC graphs and quantitatively on EER (equal error rate) and WER (weighted error rate) measures, both introduced in the evaluation framework proposed in [1]. The EER is a single point in a ROC graph and the WER takes a weighted average of the false positive and false negative rates. Here we used WER(1) which gives no preference to either failure type, i.e., a ROC point which provides the smallest average error was selected. All results are shown in Tables 1 and 2. The results indicate that better results were always achieved using the "weighted expert area intersection" fusion instead of using the "representative point neighbourhood" methods. This was at first surprising, but understandable because the areas cover the finding areas more thoroughly than the representative points which are concentrated only near the most salient points. Moreover, it is evident from the results that the product rule was generally bad for the obvious reasons discussed already in [4]. The summax rule always produced either the best results or results comparable to the best results as evident in Tables 1 and 2, and in example ROC curves in Fig. 4. Since the best performance was achieved using the "weighted expert area intersection" fusion, for which the pure sum (mean), max and product rules were clearly inferior to the summax, the summax rule should be preferred.

5 Conclusions

In this paper, the problem of fusing a united ground truth (consultation outcome) from multiple medical expert annotations (opinions) for classifier learning and the problem of forming an image-wise overall score for automatic image-based evaluation were studied. All the proposed fusion strategies and the overall scoring strategies were first discussed in the contexts of related works of different fields and then experimentally verified against a public fundus image database. As results from our more theoretical discussion and the experimental results, we conclude that the best ground truth fusion strategy is the "weighted expert area intersection" and the best overall scoring method the "summax" rule ($X = 0.01$, example proportion), both described in this study.

Acknowledgements

The authors would like to thank the Finnish Funding Agency for Technology and Innovation (TEKES) and partners of the ImageRet project[2] (No. 40039/07) for support.

References

1. Kauppi, T., Kalesnykiene, V., Kamarainen, J.K., Lensu, L., Sorri, I., Raninen, A., Voutilainen, R., Uusitalo, H., Kälviäinen, H., Pietilä, J.: The diaretdb1 diabetic retinopathy database and evaluation protocol. In: Proc. of the British Machine Vision Conference (BMVC 2007), Warwick, UK, vol. 1, pp. 252–261 (2007)
2. Russel, B., Torralba, A., Murphy, K., Freeman, W.: LabelMe: a database and web-based tool for image annotation. Int. J. of Computer Vision 77(1-3), 157–173 (2008)

[2] http://www.it.lut.fi/project/imageret/

3. Fei-Fei, L., Fergus, R., Perona, P.: One-shot learning of object categories. IEEE Trans. on PAMI 28(4) (2006)
4. Kittler, J., Hatef, M., Duin, R.P.W., Matas, J.: On combining classfiers. IEEE Trans. on Pattern Analysis and Machine Intelligence (PAMI) 20(3), 226–239 (1998)
5. Tax, D.M.J., van Breukelen, M., Duin, R.P.W., Kittler, J.: Combining multiple classifiers by averaging or by multipying. The Journal of the Pattern Recognition Society 33, 1475–1485 (2000)
6. Fumera, G., Roli, F.: A theoretical and experimental analysis of linear combiners for multiple classifier systems. IEEE Trans. on Pattern Analysis and Machine Intelligence (PAMI) 27(6), 942–956 (2005)
7. Gill, C., Sabin, L., Schmid, C.: Why clinicians are natural bayesians. British Medical Journal 330(7) (2005)
8. Phillips, P., Moon, H., Rauss, P., Rizvi, S.: The feret evaluation methodology for face recognition algorithms. IEEE Trans. on PAMI 22(10) (2000)
9. Figueiredo, M., Jain, A.: Unsupervised learning of finite mixture models. IEEE Transactions on Pattern Analysis and Machine Intelligence 24(3), 381–396 (2002)

Quantification of Bone Remodeling in SRμCT Images of Implants

Hamid Sarve[1], Joakim Lindblad[1], and Carina B. Johansson[2]

[1] Centre for Image Analysis, Swedish University of Agricultural Sciences
Box 337, 751 05 Uppsala, Sweden
{hamid,joakim}@cb.uu.se
[2] Department of Clinical Medicine, Örebro University, 701 85 Örebro, Sweden
carina.johansson@oru.se

Abstract. For quantification of bone remodeling around implants, we combine information obtained by two modalities: 2D histological sections imaged in light microscope and 3D synchrotron radiation-based computed microtomography, SRμCT. In this paper, we present a method for segmenting SRμCT volumes. The impact of shading artifact at the implant interface is reduced by modeling the artifact. The segmentation is followed by quantitative analysis. To facilitate comparison with existing results, the quantification is performed on a registered 2D slice from the volume, which corresponds to a histological section from the same sample. The quantification involves measurements of bone area and bone-implant contact percentages.

We compare the results obtained by the proposed method on the SRμCT data with manual measurements on the histological sections and discuss the advantages of including SRμCT data in the analysis.

1 Introduction

Medical devices, such as bone anchored implants, are becoming increasingly important for the aging population. We aim to improve the understanding of the mechanisms of implant integration. A necessary step for this research field is quantitative analysis of bone tissue around the implant. Traditionally, this analysis is done manually on histologically stained un-decalcified cut and ground sections (10μm) with the implant *in situ* (the so called *Exakt* technique [1]). This technique does not permit serial sectioning of bone samples with implant in situ. However, it is the state of the art when implant integration in bone tissue are to be evaluated without extracting the device or calcifying the bone. The two latter methods result in interfacial artifacts and the true interface cannot be examined.

The manual assessment is difficult and subjective: these sections are analysed both qualitatively and quantitatively with the aid of a light microscope, which consumes time and money. The desired measurements for the quantitative analysis are explained in Sect. 3.3. In our previous work [2], we present an automated method for segmentation and subsequent quantitative analysis of histological 2D sections. An experience from that work is that variations in staining and various imaging artifacts make automated quantitative analysis very difficult.

A.-B. Salberg, J.Y. Hardeberg, and R. Jenssen (Eds.): SCIA 2009, LNCS 5575, pp. 770–779, 2009.

Histological staining and subsequent color imaging provide a lot of information, where different dyes attach to different structures of the sample. X-ray imaging and computer tomography (CT) give only grey-scale images, showing the density of each part of the sample. The available information from each image element is much lower, but on the other hand the difficult staining step is avoided and the images, in general, contain significantly less variations than histological images. These last points are crucial, making automatic analysis of CT data a tractable task. In order to widen the analysis and evaluation, we combine the information obtained by the microscope with 3D SRµCT (synchrotron radiation-based computed microtomography) obtained by imaging the samples before they are cut and histologically stained. Volume data give a much better survey of the tissue surrounding the implant than one slice only. To enable a direct comparison between the two modalities, we have developed a 2D–3D multimodal registration method, presented in [3]. A slice registered according to [3] is shown in Fig. 1a and 1b.

In this work we present a segmentation method for SRµCT volumes and subsequent automatic quantitative analysis. We compare bone area and bone-implant contact measurements obtained on the 2D sections with the ones obtained on 2D slices extracted from the SRµCT volumes. In the following section we describe previous work in this field. In Sect. 3.1 the segmentation method is presented. The measurement results from the automatic method are presented in Sect. 4. Finally, in Sect. 5 we discuss the results.

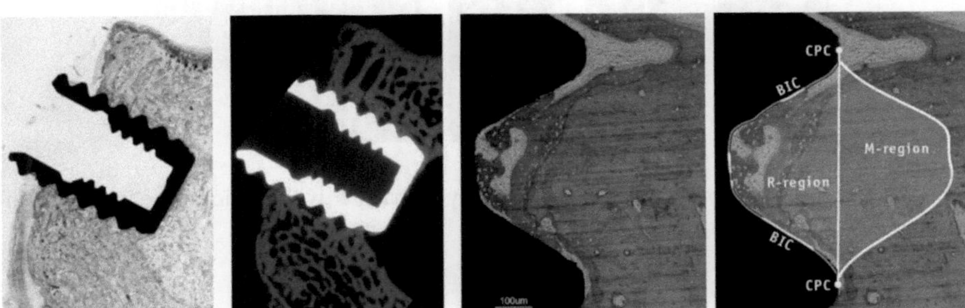

(a) (b) (c) (d)

Fig. 1. (a) A histological section (b) Corresponding registered slice extracted from the SRµCT volume (c) Histological section, single implant thread (d) Regions of interest superimposed on the thread (*CPC*=Center points of the thread crests, *R*-region=the gulf between two CPCs and *M*-region=the *R*-region mirrored with respect to the axis connecting two CPCs)

2 Background

Segmentation of CT-data is well described in the literature. Commonly used techniques for segmenting X-ray data include various thresholding or region-growing methods. Siverigh and Elliot [4] present a semi-automatic segmentation method based on connecting pixels with similar intensity. A number of works using thresholding for segmentation of X-ray data are mentioned in [5]. A method for segmentation of CT volumes of bone is proposed by Waarsing et al in [6]. They use local thresholding for segmentation and the result corresponds well to registered histological data.

CT images often suffer from various physics-based artifacts [7]. The causes of these artifacts are usually associated with the physics of the imaging technique, the imaged sample and the particular device used. A way to suppress the impact of such artifacts is to model the effect and to compensate for it [8]. When imaging very dense objects, such as the titanium implants in this study, the very high contrast between the dense object and the surrounding material leads to strong artifacts, that hide a lot of information close to the boundary of the dense object. In this study the regions of interest are close to the boundary of the dense object, which makes imaging of high density implants a very challenging task. When imaging a titanium implant in a standard μCT device, as can be seen in Fig. 2, a bright aura surrounds the implant region, making reliable discrimination between bone and soft tissue close to the implant virtually impossible.

Fig. 2. A titanium implant imaged with a SkyScan1172 μCT-device. The image to the right is an enlargement of the marked region in the image to left.

3 Material and Methods

Pure titanium screws (diam. 2.2 mm, length 3 mm), inserted in the femur condyle region of twelve-week-old rats for four weeks, are imaged using the SRμCT device of GKSS (Gesellschaft fur Kernenergieverwertung in Schiffbau und Schiffahrt mbH) at HASYLAB, DESY, in Hamburg, Germany, at beamline W2 using a photon energy of 50keV. The tomographic scans are acquired with the axis of rotation placed near the border of the detector, and with 1440 equally stepped radiograms obtained between 0°and 360°. Before reconstruction, combinations of the projection of 0°– 180° and 180°– 360°are built. A filtered back projection algorithm is used to obtain the 3D data of X-ray attenuation for the samples. The field of view of the X-ray detector is set to 6.76mm × 4.51mm (width × height) with a pixel size of 4.40μm showing a measured spatial resolution of about 11μm.

After the SRµCT-imaging, the samples are divided in the longitudinal direction of the screws. One undecalcified section with the implant *in situ* of 10µm is prepared from approximately the mid portion of each sample [9] (see Fig. 1a). The section is routinely stained in a mixture of Toluidine blue and pyronin G, resulting in various shades of purple stained bone tissue and light-blue stained soft tissue components. Finally, samples are imaged in a light-microscope, generating color images with a pixel size of about 9µm (see Fig. 1a).

3.1 Segmentation

To reduce noise, the SRµCT volume is smoothed with a bilateral filter, as described by Smith and Brady [10]. The filter smooths such that voxels are weighted by a Gaussian that extends, not only in the spatial domain, but also in the intensity domain. In this manner, the filter preserves the edges by only smoothing over intensity-homogeneous regions. The Gaussian is defined by the spatial, σ_b, and intensity standard deviation, t.

The segmentation shall classify the volume into three classes; bone tissue, soft tissue and implant. The bone implant is a low-noise high-intensity region in the volume and is easily segmented by thresholding. We use Otsu's method [11], assuming two classes with normal distribution; a tissue class (bone and soft tissue together) and an implant class.

The bone and soft tissue regions, however, are more difficult to distinguish from each other, especially in the regions close to the implant. Due to shading artifacts, the transition from implant to tissue is characterized by a low gradient from high intensity to low (see Fig. 3a). If not taken care of, this artifact leads to misclassifications.

We apply a correction by modeling the artifact and compensate for it. Representative regions with implant-to-bone tissue contact, (IB) and implant-to- soft tissue (IS) are manually extracted. A 3-4 weighted distance transform [12] is computed from the segmented implant region and intensity values are averaged for each distance d from the implant for IB and IS respectively. Based on these values, functions $b(d)$ and $s(d)$ model the intensity depending on the distance d for the two contact types for IB and IS respectively (see Fig. 3c). The corrected image, $I_c \in [0,1]$, is calculated as:

$$I_c = \frac{I - s(d)}{b(d) - s(d)} \qquad \text{for} \quad d > 1. \tag{1}$$

After artifact correction, supervised classification is used for segmenting bone and soft tissue; the respective training regions are marked and their grayscale values are saved. With an assumption of two normally distributed classes, a linear discriminant analysis, LDA, [13] is applied to identify the two classes. To reduce the effect of point noise, an $m \times m \times m$-neighborhood majority filter is applied on the whole volume after the segmentation.

For $0 < d \leq 1$ however, as seen in Fig. 3c, the intensities of the voxels are not distinguishable and they cannot be correctly classified. The classification of the voxels in this region (to either bone- or soft-tissue) is instead determined by

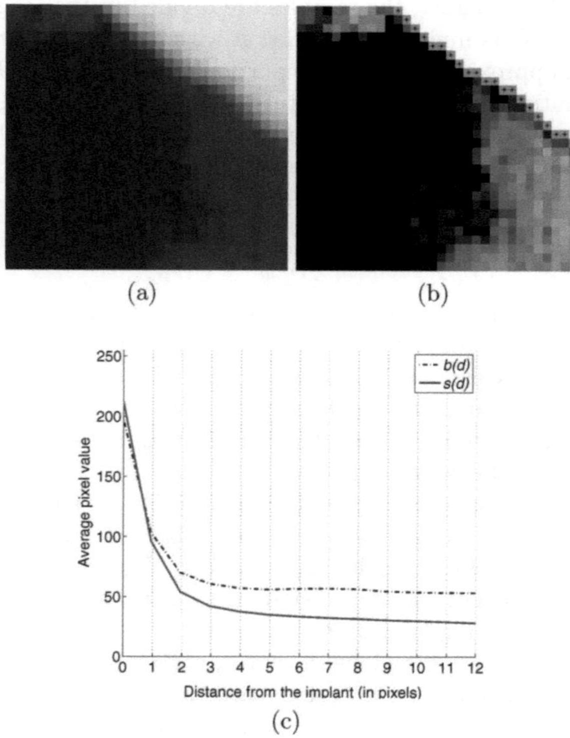

Fig. 3. (a) The implant interface region of a volume slice with implant at upper right (b) Corresponding artifact suppressed region. The marked interface region (stars) cannot be corrected (c) Plot of intensity as a function of distance from the implant for bone, $b(d)$ (dashed) and soft tissue, $s(d)$ (solid line).

the majority filter after the segmentation step. An example of shading artifact correction with the $d = 1$ region marked is shown in Fig. 3b. A segmentation example is shown in Fig. 4.

3.2 Registration

In order to find the 2D slice in the volume that corresponds to the histological section, image registration of these two data types is required. Two GPU-accelerated 2D–3D intermodal rigid-body registration methods are presented in [3]: one based on Simulated Annealing and the other on Chamfer Matching. The latter was used for registration in this work as it was shown to be more reliable. The results show good visual correspondence. In addition to the automatic registration a manual adjustment tool has been added to the method where the user can modify the registration result (six degrees of freedom, three translations and three rotations). After the pre-processing and segmentation of the volume, a slice is extracted using the coordinates found by the registration method. Note that the Chamfer matching used in [3] for registration requires a segmentation of the implant which is done by using a fixed threshold. The more difficult segmentation into bone and soft tissue is not used in the matching (the other registration approach does not include any segmentation step).

3.3 Quantitative Analysis

The current standard quantitative analysis involves measurements of bone area and bone-implant contact percentages [14]. Fig. 1c shows the regions of interest (ROIs): R=reference, inner area, is measured as the percentage of area covered by bone tissue in the R-region, i.e. the gulf between two center points of the thread crests (CPCs). In addition to R, another bone area percentage, denoted M, is measured as the bone coverage in the region in the gulf mirrored with respect to the axis connecting the two CPCs. A third important measure is *BIC*, the estimated length of the implant interface where the bone is in contact with the implant, expressed in percentage of the total length of each thread (the gulf between two CPCs).

Area is measured by summing the pixels classified as bone in the R- and M-regions. These regions are found by locating the CPCs (see [2]). BIC-length is estimated using the first of two methods for perimeter estimation of digitized planar shapes presented by Koplowitz and Bruckstein in [15]. This method requires a well defined contour, i.e. each contour-pixel shall have two neighbors only. The implant contour is extracted by dilation with a 3×3 '+'-shaped structural element on the implant region in the segmentation map. The relative overlap between the dilated implant and the bone region is defined as the bone-implant contact. Some post-processing described in [2] is applied to achieve the desired well defined contour.

4 Results

The presented method is tested on a set of five volumes. The parameters for the bilateral filter are set to $\sigma_b = 3$ and $t = 15$ and the neighborhood size of the majority filter is set to $m = 3$. This configuration is empirically assigned and gives a good trade-off between noise-suppression and edge-preservation on the analysed set of volumes. The results of the automatic and manual quantifications are shown in Fig. 5.

Classification of the histological sections is a difficult task and the inter-operator variance can be high for the manual measurements, making a direct comparison with the manual absolute measures unreliable for evaluation purposes; an important manual measurement is the judged relative order of implant integration. Hence, in addition to calculating absolute differences to measure the correspondence between the results of the automatic and manual method, we use a rank correlation technique. The three measures for each thread are ranked for both the proposed and manual method. The differences between the two ranking vectors are stored in a vector $\boldsymbol{d}$. Spearman's rank correlation [16],

$$Rs = 1 - \frac{6}{n^3 - n} \sum_{i=1}^{n} d_i \tag{2}$$

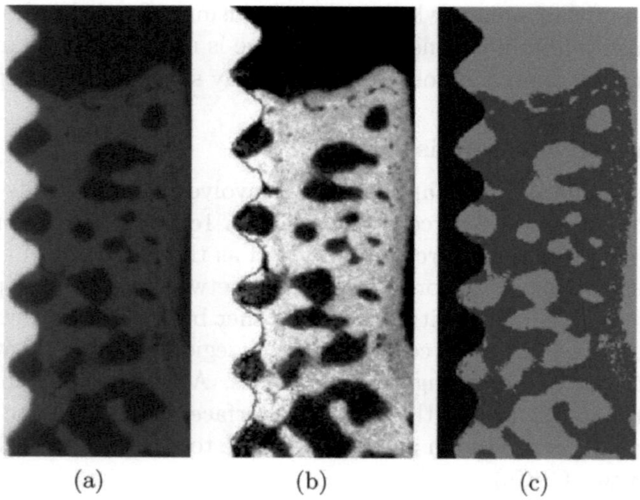

Fig. 4. (a) A slice from the SRµCT volume (b) Artifact corrected slice with the interface region marked and the implant in white to the left (c) A slice from the segmented volume, showing three classes: bone (red), soft tissue (green) and implant (blue)

where n is the number of samples, is utilized for measuring the correlation. A perfect ranking correlation implies $Rs = 1.0$.

The correlation results for all threads for all implant (five implants and ten threads each, 50 threads in total) are presented in Table 1. A two sided t-test shows that we can reject h_0 having probability $P < 0.001$ for all three measures, where h_0 is the hypothesis that the manual and automatic method do not correlate.

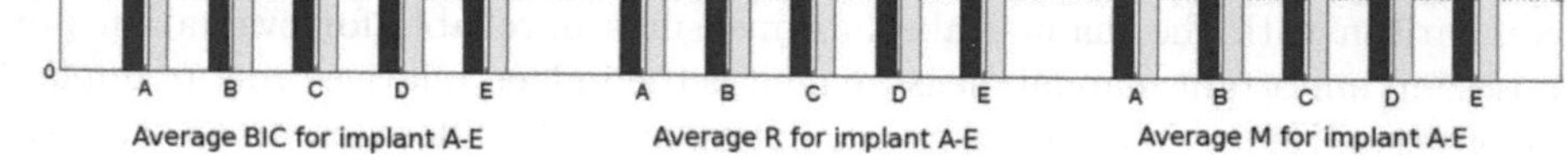

Fig. 5. Averaged absolute values for measures obtained by the automatic and manual method on five implants; the percentage of BIC, R and M averaged over all threads (10 threads per implant)

Table 1. Spearman Rank Correlation, Rs, for ranking of length and area measures (Rs_{BIC}, Rs_R and Rs_M) for all threads for all implants (50 threads in total)

Rs_{BIC}	Rs_{R}	Rs_{M}
0.5618	0.7740	0.6831

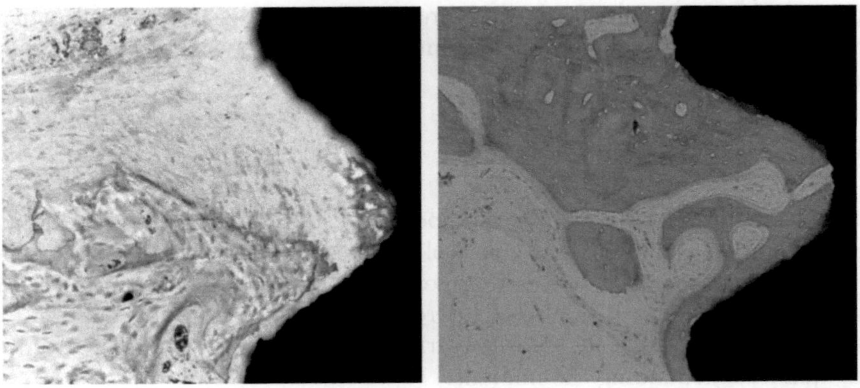

Fig. 6. Two histological sections from two different implants exemplifying variations in tissue structure. The left figure shows more immature bone and more soft tissue regions compared to the right, showing more mature bone.

5 Summary and Discussion

A method for automatic segmentation of SRμCT volumes of bone implants is presented. It involves modeling and correction of imaging artifacts. A slice is extracted from the segmented volume with the coordinates resulting from a registration of the SRμCT volume with corresponding 2D histological image. Quantitative analysis (estimation of bone areas and bone-implant-contact percentages) is performed on this slice and the obtained measurements are compared to those obtained by the manual method on the 2D histological slice. The rank correlation shows that the quantitative analysis performed by our method correlates with $Rs = 0.56$ for BIC, $Rs = 0.77$ for R and $Rs = 0.68$ for M. We note that differences between results of the two methods also include any registration errors. Spearman's rank correlation coefficient, shown in Table 1, indicates highly significant correlation ($P < 0.001$) between the automatic ranking, and the manual one. This justifies the use of SRμCT imaging to perform quantitative analysis of bone implant integration.

The state-of-practice technique of histological sectioning used today reveals information about only a small portion of the sample and the variance of that information is high depending on the cutting position. Furthermore, the outcome of the staining method may differ (as shown in Fig. 6) and the results depend on, e.g., the actual tissue (soft tissue or bone integration), the fixative used, the section thickness, the biomaterial itself (harder materials in general result more often in shadow effects). Such shortcomings, as well as other types of technical artifacts, make absolute quantifications and automatization very difficult.

SRμCT-devices require large-scale facilities and cannot be used routinely. The information is limited compared to histological sections, due to lower resolution and grayscale output only. However, the generated 3D volume gives a much broader overview and the problematic staining step is avoided. As shown in Sect. 3.1, the existing artifacts can be removed with satisfactory result and the acquired volumes are similar independent of the tissue type, allowing an absolute quantification.

6 Future Work

Future work involves developing methods for using the 3D data, e.g. estimating bone implant contacts and bone volumes around the whole implant. These measurements will much better represent the entire bone implant integration compared to 2D data. It is also of interest to further extract information from the image intensities, since density variations may indicate differences in the bone quality surrounding the implant.

Acknowledgment

Research technicians Petra Hammarström-Johansson and Ann Albrektsson are greatly acknowledged for skillful sample preparations. Also Dr. Ricardo Bernhardt and Dr. Felix Beckmann are greatly acknowledged. The authors would also like to acknowledge Professor Gunilla Borgefors and Dr. Nataša Sladoje. This work was supported by grants from The Swedish Research Council, 621-2005-3402 and was partly supported by the IA-SFS project RII3-CT-2004-506008 of the Framework Programme 6.

References

1. Donath, K.: Die trenn-dunnschliffe-technik zur herstellung histologischer präparate von nicht schneidbaren geweben und materialien. Der Präparator 34, 197–206 (1988)
2. Sarve, H., et al.: Quantification of Bone Remodeling in the Proximity of Implants. In: Kropatsch, W.G., Kampel, M., Hanbury, A. (eds.) CAIP 2007. LNCS, vol. 4673, pp. 253–260. Springer, Heidelberg (2007)
3. Sarve, H., et al.: Registration of 2D Histological Images of Bone Implants with 3D SRuCT Volumes. In: Bebis, G., et al. (eds.) ISVC 2008, Part I. LNCS, vol. 5358, pp. 1081–1090. Springer, Heidelberg (2008)
4. Siverigh, G.J., Elliot, P.J.: Interactive region and volume growing for segmenting volumes in mr and ct images. Med. Informatics 19, 71–80 (1994)
5. Elmoutaouakkil, A., et al.: Segmentation of Cancellous Bone From High-Resolution Computed Tomography Images: Influence on Trabecular Bone Measurements. IEEE Trans. on medical imaging 21 (2002)

6. Waarsing, J.H., Day, J.S., Weinans, H.: An improved segmentation method for in vivo uct imaging. Journal of Bone and Mineral Research 19 (2004)
7. Barrett, J.F., Keat, N.: Artifacts in CT: Recognition and avoidance. Radio Graphics 24, 1679–1691 (2004)
8. Van de Casteele, E., et al.: A model-based correction method for beam hardening in X-Ray microtomography. Journ. of X-Ray Science and Technology 12, 43–57 (2004)
9. Johansson, C., Morberg, P.: Cutting directions of bone with biomaterials in situ does influence the outcome of histomorphometrical quantification. Biomaterials 16, 1037–1039 (1995)
10. Smith, S., Brady, J.: SUSAN – a new approach to low level image processing. International Journal of Computer Vision 23, 45–78 (1997)
11. Otsu, N.: A threshold selection method from gray-level histograms. IEEE Transactions on Systems, Man, and Cybernetics 9, 62–66 (1979)
12. Borgefors, G.: Distance transformations in digital images. Computer Vision, Graphics, and Image Processing 34, 344–371 (1986)
13. Johnson, R.A., Wichern, D.W.: Applied Multivariate Statistical Analysis. Prentice-Hall, Englewood Cliffs (1998)
14. Johansson, C.: On tissue reactions to metal implants. PhD thesis, Department of Biomaterials / Handicap Research, Göteborg University, Sweden (1991)
15. Koplowitz, J., Bruckstein, A.M.: Design of perimeter estimators for digized planar shapes. Trans. on PAMI 11, 611–622 (1989)
16. Spearman, C.: The proof and measurement of association between two things. The American Journal of Psychology 100, 447–471 (1987)

Author Index

FSC
www.fsc.org
MIX
Papier aus verantwortungsvollen Quellen
Paper from responsible sources
FSC® C105338

If you have any concerns about our products,
you can contact us on
ProductSafety@springernature.com

In case Publisher is established outside the EU,
the EU authorized representative is:
Springer Nature Customer Service Center GmbH
Europaplatz 3, 69115 Heidelberg, Germany

Printed by Libri Plureos GmbH
in Hamburg, Germany